Advances in
MARINE BIOLOGY

VOLUME 33

Advances in
MARINE BIOLOGY

The Biology of Calanoid Copepods

by

J. MAUCHLINE

Dunstaffnage Marine Research Laboratory, Oban, Scotland

Series Editors

J.H.S. BLAXTER

Dunstaffnage Marine Research Laboratory, Oban, Scotland

A.J. SOUTHWARD

Marine Biological Association, The Laboratory, Citadel Hill, Plymouth, England

and

P.A. TYLER

Department of Oceanography, University of Southampton, England

ACADEMIC PRESS

San Diego London Boston
New York Sydney Tokyo Toronto

This book is printed on acid-free paper

Copyright © 1998 by ACADEMIC PRESS

All Rights Reserved
No part of this publication may be reproduced or transmitted in any form or by any means, electronic or mechanical, including photocopy, recording, or any information storage and retrieval system, without permission in writing from the publisher.

Academic Press
525 B Street, Suite 1900, San Diego, California 92101-4495, USA
http://www.apnet.com

Academic Press Limited
24–28 Oval Road, London NW1 7DX, UK
http://www.hbuk.co.uk/ap/

ISBN 0-12-105545-0 (paperback)
ISBN 0-12-026133-2 (hardback)

A catalogue record for this book is available from the British Library

Typeset by Keyset Composition, Colchester, Essex

Printed in Great Britain by WBC Book Manufacturers, Bridgend, Mid Glamorgan

98 99 00 01 02 03 WB 9 8 7 6 5 4 3 2 1

CONTENTS

PREFACE x

1. Introduction

1.1. The Scientific Literature 5
1.2. Environmental Sampling 8
1.3. Preservation of Samples 10
1.4. Methods of Study 11

2. External Morphology, Internal Anatomy

2.1. External Morphology of Adults 16
2.2. External Morphology of Young 22
2.3. Integument 28
2.4. Asymmetry 36
2.5. Abnormalities and Morphological Variation 36
2.6. Internal Anatomy 37

3. Taxonomy and Identification

3.1. Phylogeny of Calanoid Copepods 49
3.2. Identification of Genera and Species 57
3.3. Key to Genera of Platycopioid and Calanoid Copepods 65

4. World List of Platycopioid and Calanoid Copepods

4.1. Order Platycopioida Fosshagen, 1985 98
4.2. Order Calanoida Sars, 1902 99

5. Gut, Food and Feeding

5.1. Feeding Appendages 140
5.2. Food Capture 143
5.3. Food and Foraging in the Environment 149
5.4. Foraging Tactics 160
5.5. Feeding of Young Stages 166

5.6. Feeding Periodicity 167
5.7. Concluding Remarks 172

6. Physiology

6.1. Rates of Feeding and Ingestion 176
6.2. Respiration and Excretion 209
6.3. Metabolism 211
6.4. Osmotic Regulation, Salinity and Temperature 215
6.5. Responses to Environmental Variables 216
6.6. Concluding Remarks 218

7. Chemical Composition

7.1. Body Weight and Volume 221
7.2. Elements 236
7.3. Organic Components 245
7.4. Other Organic Components 249
7.5. Energy Content 250
7.6. Vitamins 250
7.7. Carotenoids 251
7.8. Enzymes 252

8. Reproduction

8.1. Seasonality in Breeding 253
8.2. Mating Behaviour 256
8.3. Spawning 266
8.4. Food Availability and Egg Production 289
8.5. Mortality of Eggs 292
8.6. Hatching of Eggs 293
8.7. Concluding Remarks 294

9. Growth

9.1. Laboratory Culture of Copepods 297
9.2. Development Time of Eggs 298
9.3. Growth 314
9.4. Growth Rates 338
9.5. Longevity 339
9.6. Concluding Remarks 344

10. Population Biology

10.1. Sampling 349
10.2. Demographic Analysis 352
10.3. Life History Patterns (Strategies) 381
10.4. Population Maintenance 392
10.5. Biomass of Populations 398

11. Behaviour

11.1. Swimming Activity 401
11.2. Spatial Distribution 423
11.3. Bioluminescence 434
11.4. Vertical Migration 439
11.5. Rhythms 452
11.6. Concluding Remarks 453

12. Distributional Ecology

12.1. Biomass of the Copepod Fauna 457
12.2. Associations of Copepods 460
12.3. Copepods of Pelagic Environments 470
12.4. Restricted Environments 487

13. Geographical Distribution

13.1. Introductions of Calanoids 505
13.2. Faunal Provinces and Large Marine Ecosystems 509
13.3. Identification of Species 516
13.4. Concluding Remarks 516

14. Copepods in Ecosystems

14.1. Ecosystems 519
14.2. Fate of Faecal Pellets 525
14.3. Perturbations within Ecosystems 526
14.4. The Biology of Copepods 527
References 531

Taxonomic Index 661
Subject Index 693
Cumulative Index of Titles 703
Cumulative Index of Authors 709

Dedicated to my wife Isobel

Preface

Calanoid copepods have been of intense interest to marine biologists for more than a century. Many scientific papers have been published over the years, but this is the first assessment of their biology, as a whole, that has been attempted.

Many colleagues throughout the world have encouraged and helped in the production of this work. My initial interest was stimulated by Dr J. H. Fraser and Dr S. M. Marshall many years ago. I wish to acknowledge, in particular, the helpful correspondence and/or discussions with G. A. Boxshall, J. M. Colebrook, F. D. Ferrari, A. Fosshagen, the late A. Fleminger, H. Grigg, L. R. Haury, C. C. E. Hopkins, K. Hülsemann, S. Kasahara, I. A. McLaren, S. Nishida, M. Omori, G. -A. Paffenhöfer, J. S. Park, T. Park, S. Razouls, H. S. J. Roe, K. Schulz, S. -I. Uye, J. C. Vaupel Klein, P. Ward, K. F. Wishner, and J. Yen. A special debt of gratitude is owed to Miss E. Walton, the Librarian of the Dunstaffnage Marine Laboratory, who has put up with my vagaries for years and obtained outside library loans of some very obscure publications; I especially thank her for her perseverance and patience. Finally, it is a pleasure to acknowledge the helpful comments, and aid in proof-reading of the manuscript, of the Editors Professor J. H. S. Blaxter, Professor A. J. Southward and Professor P. A. Tyler.

John Mauchline

1. Introduction

1.1. The Scientific Literature	5
1.2. Environmental Sampling	8
1.3. Preservation of Samples	10
1.4. Methods of Study	11

Copepods are probably the most numerous multicellular organisms on earth. They outnumber the insects although the insects are more diverse, having more species than copepods. They are aquatic animals, primarily marine, although they also occur in vast numbers in fresh water environments. Humes (1994) estimates that there are some 11,500 species, divided between about 200 families and 1650 genera, known at the end of 1993. He attempts to estimate the actual numbers of species on earth and suggests that as few as 15% of existing species are known at present.

The Copepoda form a subclass of the phylum Crustacea. The name copepod originates from the Greek words *kope*, an oar, and *podos*, a foot, and refers to the flat, laminar swimming legs of the animals. As Huys and Boxshall (1991) point out, there is no popular English name for them although the Norwegian *Hoppekrebs*, German *Ruderfusskrebs*, and the Dutch *Roeipootkreeft* reflect the derivation of the name Copepod. There are ten orders of copepods (Table 1) containing different numbers of families, genera and species:

The Platycopioida are marine, benthopelagic species, two living in anchialine caves in Bermuda.

The Calanoida are primarily pelagic, 75% are marine, 25% live in fresh water. Some marine species are benthopelagic or commensal.

The Misophrioida are primarily benthopelagic and inhabitants of anchialine caves – only two species are pelagic – and the Mormonilloida are pelagic marine species.

The Cyclopoida are divided between marine and fresh waters and can be pelagic, commensal or parasitic.

The Gelyelloida occur in karstic systems in France and Switzerland.

Table 1 Classification of copepods. The numbers of marine families (F), genera (G) and species (S) recognized in each order are indicated; these numbers are approximate because of the continuous addition of new taxa and modifications of older ones. After Huys and Boxshall (1991) and Humes (1994).

	F	G	S
Subclass Copepoda Milne-Edwards, 1840			
Infraclass Progymnoplea Lang, 1948			
Order Platycopioida Fosshagen, 1985	1	3	10
Infraclass Neocopepoda Huys & Boxshall, 1991			
Superorder Gymnoplea Giesbrecht, 1882			
Order Calanoida Sars, 1903	41^1	195^1	1800^1
Superorder Podoplea Giesbrecht, 1882			
Order Misophrioida Gurney, 1933	1	11^4	19^4
Order Cyclopoida Burmeister, 1834	12	80^2	450^2
Order Gelyelloida Huys, 1988	1	1	2
Order Mormonilloida Boxshall, 1979	1	1	2
Order Harpacticoida Sars, 1903	47	300^3	2500^3
Order Poecilostomatoida Thorell, 1859	46	>260	1570+
Order Siphonostomatoida Thorell, 1859	37	245	1430+
Order Monstrilloida Sars, 1903	1	4^4	74^4

[1] Excluding Diaptomidae and fresh water genera in other families.
[2] Marine and fresh water combined (Bowman and Abele, 1982).
[3] Approximate values derived from Bowman and Abele (1982).
[4] Approximate numbers derived from Razouls (1996).

The Harpacticoida are primarily marine species, 10% living in fresh waters. Most species are benthic, a few pelagic or commensal.

The Poecilostomatoida and Siphonostomatoida are marine, commensal or parasitic species.

The Monstrilloida are marine species that are pelagic as adults but parasitic when young.

The phylogenetic relationships of these orders are examined by Huys and Boxshall (1991) and reviewed by Ho (1990, 1994). There are several proposed cladograms illustrating possible linkages, one of which is given in Figure 1. An excellent summary of the development of current ideas on the evolutionary structure within the Copepoda is provided by Huys and Boxshall (1991). The Platycopioida superficially look like calanoid copepods because the division between the prosome and urosome is between the fifth pedigerous segment and the genital somite. This division is more anterior in all other copepods, being between the fourth and fifth pedigerous segments. The Platycopioida are nearest to the hypothetical ancestral stock of the Copepoda and the Calanoida are next. The gross morphology of the Calanoida is uniform (Figure 2) unlike that within other orders of the Copepoda (Figure 1 and Dudley, 1986; Huys and Boxshall,

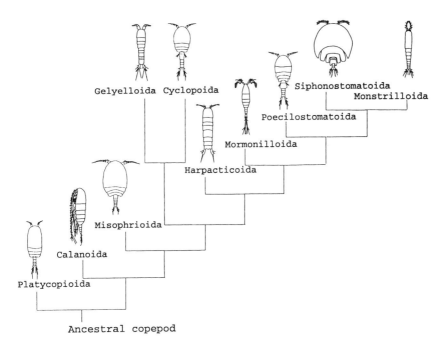

Figure 1 Phylogenetic relationships within the Copepoda. (After Huys and Boxshall, 1991; Ho, 1994.)

1991) where benthic, commensal and parasitic life styles have been adopted.

This volume examines the Platycopioida and Calanoida in detail but is primarily restricted to the marine and brackish water environments. There are, however, many fresh water species. They belong primarily to three families within the Calanoida: the Temoridae, Centropagidae and Diaptomidae. The genus *Senecella*, originally ascribed to the Pseudocalanidae but now to the Aetideidae, contains two species, one in north American fresh water lakes, the other in brackish waters of the Kara and Laptev Seas. The fresh water copepods are described in detail by Dussart and Defaye (1995) and reference to that work should be made for further information. Evolution within the Centropagidae is discussed by Maly (1996).

Calanoid copepods are of prime importance in marine ecosystems because many are herbivorous, feeding on the phytoplankton, and forming a direct link between it and fish such as the herring, sardine, and pilchard. Copepods are at the small end of the size spectrum of food of the baleen whales but sei, bowhead, right and fin whales consume large quantities of

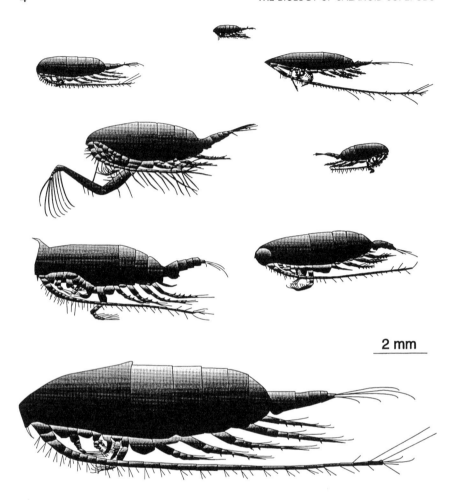

Figure 2 Body form of a variety of species of calanoid copepods. From top, and then left to right: *Acartia* sp., *Calanus finmarchicus*, *Rhincalanus nasutus*, *Pseudeuchaeta brevicauda*, *Aetideopsis multiserrata*, *Gaetanus latifrons*, *Cephalophanes refulgens*, and *Bathycalanus princeps*.

them in the north Atlantic, north Pacific and Antarctic Oceans (Gaskin, 1982). Copepods are also eaten by a vast variety of invertebrate species, both pelagic and epibenthic.

Pelagic copepods dominate the numbers of organisms caught in plankton samples from most sea areas, representing 55 to 95% of the numbers caught (Longhurst, 1985). They are most dominant in the Arctic and Antarctic Oceans and also over continental shelves in middle latitudes. Their numbers

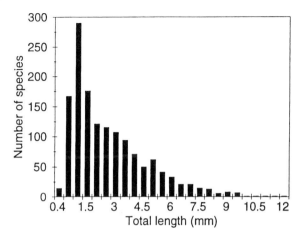

Figure 3 Size frequency distribution of calanoid copepod species.

relative to other organisms vary seasonally in middle and high latitudes. Their body size is small, most species having a body length of 0.5 to 2.0 mm (Figure 3). Consequently, their proportion in terms of biomass of plankton is lower, being in the range 25 to 80% depending on the region, season, and presence or absence of aggregations of other organisms such as siphonophores and euphausiids. The size record for a calanoid copepod is for a female *Bathycalanus sverdrupi* caught at 2000 to 2050 m depth in the Gulf of Guinea by Owre and Foyo (1967); it measured 18.0 mm in total length. The second largest copepod caught is a female *Bradycalanus pseudotypicus enormis* from 2893 m depth that measured 17.5 mm in total length (Björnberg, 1967a). One of the smallest species recorded is the cave-dwelling *Nanocopia minuta*; the female measured 0.27 mm and the male 0.25 mm in total length (Fosshagen and Iliffe, 1988).

1.1. THE SCIENTIFIC LITERATURE

The study of calanoid copepods is a complex and dynamic activity. The literature is huge and increasing all the time. There is no annotated bibliography to provide easy access although Vervoort (1986a,b, 1988) has listed, alphabetically by author, papers referring to any copepod whether calanoid or not. The list is very comprehensive but does not give direct access to literature on individual taxa. *Zoological Record* remains the principal source of such information in the pre-CD ROM eras.

The calanoid copepods, excluding the fresh water family Diaptomidae and the fresh water genera of the family Centropagidae (Table 9, pages 51–52), currently consist of some 41 families, 195 genera and 1800 species. Most species are extremely rare, some 1170 of them having 1–20 literature citations. A further 430 species have 20–50 citations and about 90 species have between 50 and 100 citations (Mauchline, unpublished). The remaining 72 most commonly cited species are listed in Table 2. The absolute number of citations is approximate because many references to the occurrence of copepods in the diets of other organisms have not been searched for. The list, however, indicates those species and genera that have received much attention. It reflects work in inshore coastal species and in the north Atlantic and north Pacific Oceans. It also shows which species are important in the economics of the oceans although there are one or two omissions; for example, the dominant species over large areas of the Antarctic Ocean, *Ctenocalanus citer*, *Metridia gerlachei* and *Drepanopus forcipatus*, are missing and reflect the relative lack of research there compared with other sea areas. Deep-sea species are also not prominent unless they are eurybathic and occur frequently in shallower depth horizons e.g. *Euchirella* spp., *Heterorhabdus* spp., *Undeuchaeta plumosa* (Table 2). Thus, knowledge about individual species varies considerably.

This volume is not a comprehensive review of all the literature but an attempt is made throughout to quote references that give easy access to secondary compilations on topics, whether they be on individual species or on ecological, physiological or other subjects. The reference lists of cited papers have been examined to evaluate coverage of their topic. Emphasis is placed on recent papers but older, significant contributions are also cited. References to descriptions of species are also treated in the same way in the world list in Chapter 4.

All aspects of the biology of calanoids are reviewed but some in more detail than others. The reactions of calanoids to concentrations of phytoplankton are described but no attempt is made to assess the information on estimates of the varying proportion of the phytoplankton stock grazed by copepods. This is a difficult field because of the dynamics of the phytoplankton and the selective capabilities of the copepods. Sautour *et al.* (1996), for example, found that the herbivorous copepods of the Gironde Estuary, dominated by *Paracalanus parvus* and *Temora longicornis*, grazed 17 to 21% of the total primary production and 9 to 14% of the phytoplankton stock. Some 70% of the phytoplankton stock, however, was less than 5 μm and too small to be available to the copepods. Consequently, they estimated that the copepods removed 35 to 68% daily of the size fraction of the stock available to them. Such estimates can only be approximate and can only be used to infer food excess or limitation for the copepods. The linking of diel vertical migration of the copepods to

Table 2 The most commonly cited species of calanoid copepods in the scientific literature. (Mauchline, unpublished).

100–200 citations
Acartia bifilosa
Acartia danae
Acartia discaudata
Acartia negligens
Calanoides acutus
Calanoides carinatus
Calanus glacialis
Calanus pacificus
Calanus propinquus
Calocalanus pavo
Calocalanus styliremis
Candacia aethiopica
Centropages furcatus
Cosmocalanus darwini
Ctenocalanus vanus
Eucalanus attenuatus
Eucalanus bungii
Eucalanus crassus
Eucalanus elongatus
Euchaeta acuta
Euchirella mesinensis
Euchirella rostrata
Eurytemora affinis

Eurytemora herdmani
Haloptilus longicornis
Heterorhabdus papilliger
Heterorhabdus spinifrons
Isias clavipes
Labidocera wollastoni
Lucicutia flavicornis
Mecynocera clausi
Mesocalanus tenuicornis
Neocalanus cristatus
Neocalanus plumchrus
Paracalanus aculeatus
Pleuromamma robusta
Pleuromamma xiphias
Pontellina plumata
Rhincalanus gigas
Rhincalanus nasutus
Scolecithricella minor
Scolecithrix danae
Tortanus discaudatus
Tortanus forcipatus
Undeuchaeta plumosa
Undinula vulgaris

200–300 citations
Acartia longiremis
Aetideus armatus
Anomalocera pattersoni
Calanus hyperboreus
Calanus (Nannocalanus) minor
Clausocalanus arcuicornis
Euchaeta marina
Metridia longa

Metridia lucens
Neocalanus gracilis
Pareuchaeta norvegica
Pleuromamma abdominalis
Pleuromamma gracilis
Pseudocalanus minutus
Temora turbinata

300–400 citations
Acartia tonsa
Centropages hamatus

Centropages typicus
Temora stylifera

400–500 citations
Calanus helgolandicus

Pseudocalanus elongatus

500–600 citations
Paracalanus parvus

Temora longicornis

>600 citations
Acartia clausi

Calanus finmarchicus

phytoplankton production and its consumption is complex because of variations in time and space. Longhurst et al. (1984) conclude that such a project in an area like the eastern Canadian archipelago is especially difficult because much of the phytoplankton sediments to the sea floor.

As mentioned above, copepods contribute to the diets of very many invertebrates, fish and whales. No attempt is made to list the species that prey on copepods although a few are mentioned when pertinent to the topic being discussed.

1.2. ENVIRONMENTAL SAMPLING

Beckmann (1984) concludes that a few oblique samples of deep-sea copepods in the Red Sea characterize large areas over extended time whereas many samples in space and time are required for the variable epipelagic, coastal and estuarine populations. Nobody would argue with this generalization. Copepods, like other planktonic organisms, are not randomly distributed in the sea but occur in patches both horizontally and vertically. This is discussed in some detail in the chapter on behaviour where their occurrence in patches, aggregations and swarms is described. This patchy distribution affects the sampling of a population as described by Wiebe (1971) and Wiebe and Holland (1968). The length of tow and the size of the net used are very important and can only be determined through pilot investigations and experience. At a fixed station on the Scotian Shelf, Sameoto (1978) found that the numbers of copepods caught, especially *Calanus* and *Pseudocalanus* species, were related to the tidal cycle. The period of observation was only over 26 h and he concluded that the tides carried a patch of these copepods past the sampling point and possibly returned them on an elliptical path past the sampling point on more than one tidal cycle. The effects, on sampling, of the transport of water by currents through a region are modelled and discussed by Power (1996). Broad considerations, therefore, of the region to be sampled for copepods must be examined along with the objectives of the sampling programme.

a. Are there marked gradients of temperature, salinity, depth, or tidal currents?
b. Will the copepods occur throughout the region to be sampled or are there species that are likely to have restricted distributions?
c. Is the sampling programme exploratory?
d. Are quantitative results in terms of biomass, numbers, or of horizontal or depth distributions of species required?

Answers to these questions will determine the sampling strategies to be adopted and the gear to be used. Good general introductions to sampling are given by Tranter and Fraser (1968) and Omori and Ikeda (1984). Nets for sampling pelagic, neustonic, and benthic copepods are discussed and illustrated as are the various methods of their deployment. Since then, Wiebe *et al.* (1985) describe new developments of the Multiple Opening/Closing Net and Environmental Sensing System (MOCNESS), an excellent and adaptable system for studying vertical and horizontal distributions of copepods quantitatively. Williams *et al.* (1983) use a double Longhurst/Hardy Plankton Recorder (LHPR) to resolve the vertical distributions of nauplii and copepodids of *Calanus helgolandicus*. The MOCNESS and LHPR are both for sampling offshore. An interesting continuous pump sampler that incorporates a plankton net as a collector and pumps the catch from the codend to the surface is described by Herman *et al.* (1984); profiles of the density of copepods in the surface 100 m of the ocean can thus be determined. Environmental probes can be mounted on the frame to provide simultaneous physical and chemical information. Herman (1992) adds an optical plankton counter to the codend; this counter is capable of sizing, and in some cases, identifying species or stages of copepods. A much simpler and less sophisticated sampler for quantitative investigation of shallow-water coastal copepods is described by Kršinić (1990). It is essentially a trap that can be opened and closed by messenger to sample the plankton in the volume of trapped water. A diver-operated device that can sample pelagic or benthopelagic copepods is described by Potts (1976); this idea could be modified in different ways, even to produce a very simple net that fits on a diver's arm (Kirkwood and Burton, 1987).

High-frequency acoustics, in the range of 100 kHz to 1 MHz, are capable of detecting individual zooplanktonic organisms as well as mapping patchiness in the pelagic realm. They have been used in studies of deep sound-scattering layers but the central problem is the identification of the species of copepod or plankton organism causing the scattering. Wiebe and Greene (1994) review current uses and the future potential of these methods.

Some copepods live in areas of the environment that are difficult to sample. Those associated with the surface film of the sea, neustonic species, are sampled by nets on floats at the surface (Omori and Ikeda, 1984). A sampler, not referred to by them, is the multiple net device of Schram *et al.* (1981) that samples contiguous subsurface layers. Under-ice samplers are described by Kirkwood and Burton (1987) and Nishiyama *et al.* (1987). A net-pump is used by Møhlenberg (1987) to sample copepods in the water column. Here, the water is pumped into the net which is deployed in the surface 25 m of the water column. This net could be used in a variety of shallow water environments and adapted for a diver.

The nature of the investigation, the characteristics of the sea area to be sampled, and the specifications of the boat or ship available will strongly affect the sampling methods and gear selected. Pilot investigations are strongly recommended. The same net will not collect adult and copepodid stages of copepods with the same efficiency, and one that samples adults will usually catch very few nauplii. Anderson and Warren (1991), for example, tested the catch rates of small and large Bongo nets for copepodids of *Calanus finmarchicus*. They discuss mesh sizes and mouth sizes of nets and their effects on catch rates and recommend that individual copepodid stages be targeted in sampling programmes. A mesh size of 75% of the body or prosome width of the nauplius or copepodid catches about 95% of those of that size in the water (Nichols and Thompson, 1991).

1.3. PRESERVATION OF SAMPLES

Steedman (1976) and Omori and Ikeda (1984) describe fixation and preservation procedures for plankton samples in detail. The best general fixative is formalin buffered with borax (sodium tetraborate); 30 g of borax to one litre of analytical reagent grade formalin, colloquially known as 40% formalin since that is the concentration of formalin in it. Plankton samples should be decanted from the bucket of the net into sample bottles of known volume. The settled volume of the plankton or copepods should not exceed 20% to 25% of the volume of the bottle. The sample plus associated sea water should fill less than 90% of the volume of the bottle. Buffered formalin is then added to fill the remaining 10% of the volume of the bottle, so resulting in a 4–5% solution of formalin in sea water. A clearly written label for the sample should be inserted, the bottle capped and then inverted gently several times to mix the formalin with the sample. The sample should remain in the formalin for at least 10 d. The formalin can then be drained off and the sample transferred to a preservative fluid. Formalin is detrimental to health and working with formalin-preserved samples is to be avoided.

The best preservative is a version of Steedman's fluid (Omori and Ikeda, 1984). The one used by the author for 20 years differs in that it has proportionately less formalin. This is because it is never used as a fixative but only as a preservative for copepods already fixed in 5% formalin. The formula for one litre of the fluid is:

40% buffered formalin	25 ml
Propylene phenoxetol	10 ml
Propylene glycol	100 ml
Filtered sea water	865 ml

The sample, which has been fixed in formalin, is transferred to the preservative fluid as follows. The fixed sample is gently decanted into a sieve and the formalin drained off. The sieve used depends on the size spectrum of the plankton sample. A simple method is to line a baker's sieve for flour with a sheet of the plankton gauze identical to that used in the original sampling net. The sample in the sieve should then be gently washed by passing filtered sea water through it several times. The sheet of plankton gauze plus its contained sample is then gently lifted from the sieve and the sample decanted into a container half-filled with the fixative. Once the entire sample has been transferred, the container is topped-up with fixative, the label (see next section), with the details of collection of the sample, inserted, and the container sealed.

The low formalin content of the preservative fluid makes the samples comfortable to work with. The copepods do not become brittle and so legs do not suffer damage. Internal tissues such as gonads preserve well and are in good condition even after 20 years in the fluid. Samples that have inadvertently been allowed to "dry out" are easily reconstituted by addition of further fluid. Stored samples, however, should be properly curated and the levels of preservative present in the containers inspected at intervals. The length of interval will depend on the environmental temperature that the samples are subjected to. The colours of the copepods and other organisms will survive preservation longer if the samples are stored in darkness.

1.4. METHODS OF STUDY

The stored samples must have labels in them. The amount of information on the labels will vary depending on the investigations being made and suggested formats of labels are given by Omori and Ikeda (1984). The labels are of good-quality paper; the best type of paper easily available is often the letter-headed notepaper of the institute or laboratory. It is good practice to insert two or more labels in the sample, one having as much detail as wished, the others simply having a sample identity number. This is done because some paper labels disintegrate during prolonged storage. Indelible inks or computer-printed labels should be used. The full details of each sample should be stored in a secure file.

There is a considerable advantage in separating the copepods from the other organisms in the samples if they are the ones of principal interest. The copepods can then be stored in vials that are placed in larger, reservoir containers. This allows easy curation during extended investigation of the taxonomy and distribution of species. This is when there is an advantage in having small labels with a sample number as opposed to large labels with full sampling details.

Working on quantitative samples often requires an additional label within the sample. This will give details of the number of specimens removed for further study. Identification to the species level sometimes requires detailed studies of sub-samples before the individuals in the original sample can be identified and counted. Quantitative investigations often require that representative sub-samples be used for analysis because it is not practicable to use the entire sample. Removal of such a sub-sample should be indicated on a label within the original sample.

Steedman (1976) and Omori and Ikeda (1984) describe in detail recommended procedures for examining copepods alive and in preserved samples. An appendix to Huys and Boxshall (1991) reviews a wide variety of such techniques. The use of stains is described in the above works. They can have quite specialized functions such as, for example, that of Nile Red used for detecting lipid storage within the bodies of copepods (Carman *et al.*, 1991).

A major requirement, especially in taxonomic studies of copepods, is the preparation of semi-permanent mounts of whole animals or dissected parts such as the appendages. There are a variety of media used and Koomen and Vaupel Klein (1995) and Stock and Vaupel Klein (1996) review their uses. Stock and Vaupel Klein (1996) recommend Reyne's fluid but it has a limited shelf-life and contains chloral hydrate which is poisonous. The author uses polyvinyl lactophenol, obtained commercially and with a longer shelf-life; it is tinted, before use, with the stain lignin pink and material to be mounted can be transferred directly to it from water. Its viscosity allows arrangement of appendages that is maintained when the cover slip is added. Such preparations are kept flat for several weeks and then stored on their sides in conventional slide cabinets. The edge of the cover slip may require flooding, on annual inspection, with additional mountant to counter evaporation. Conversely, a sealant can be applied around the edge of the cover slip at the time of preparation or when the slide enters storage.

The Scanning Electron Microscope is increasingly used to study such aspects as morphology of appendages, sensilla and even stomach contents of copepods. Felgenhauer (1987) describes the techniques involved in preparing copepods for examination by SEM while Toda *et al.* (1989) have developed a dry-fracturing technique for making observations on the internal anatomy and stomach contents.

Attempts have been made to automate counting and measuring of copepods, and even the identification of species. Image analysers have been investigated in this context with some success (Rolke and Lenz, 1984; Estep *et al.*, 1986; Noji *et al.*, 1991). Automation may help considerably in the future with processing of coastal samples with a low diversity of species, one or two of which are dominant. It is less promising for analysing oceanic samples of high diversity. Image analysers also allow biometrical studies, such as that of Jansá and Vives (1992) on the area presented by the dorsal aspect of species.

2. External Morphology, Internal Anatomy

2.1. External Morphology of Adults... 16
    2.1.1. Antennule (antenna 1)... 17
    2.1.2. Antenna (antenna 2)... 19
    2.1.3. Labrum and Labium... 19
    2.1.4. Mandible... 19
    2.1.5. Maxillule (maxilla 1)... 20
    2.1.6. Maxilla (maxilla 2)... 20
    2.1.7. Maxilliped... 20
    2.1.8. Swimming Legs.. 21
2.2. External Morphology of Young.. 22
    2.2.1. Egg... 22
    2.2.2. Nauplius... 23
    2.2.3. Copepodid.. 24
2.3. Integument.. 28
    2.3.1. The Setal System and Subintegumental Glands........................... 28
    2.3.2. Eyes and Frontal Organs.. 31
    2.3.3. Moulting.. 32
2.4. Asymmetry.. 36
2.5. Abnormalities and Morphological Variation.. 36
2.6. Internal Anatomy.. 37
    2.6.1. Endoskeleton.. 38
    2.6.2. Muscular System... 38
    2.6.3. Nervous System.. 38
    2.6.4. Circulatory System... 39
    2.6.5. Digestive System... 39
    2.6.6. Reproductive System... 41
    2.6.7. Excretory System... 46
    2.6.8. Oil Sac.. 47
    2.6.9. Chromosomes... 48

The overall body form of platycopioid and calanoid copepods is closely similar but different from those of other orders of copepods (Dudley, 1986). The former conform to the gymnoplean tagmosis in which a distinct division between the prosome and urosome is situated between the fifth pedigerous segment and the genital somite (Figure 4). All other copepods conform to the podoplean tagmosis in which an often less distinct separation of the

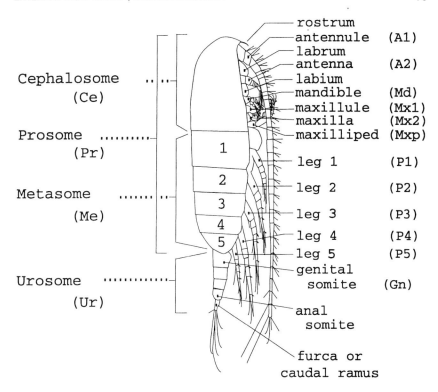

Figure 4 Diagrammatic illustration of the external morphology and appendages of a female calanoid copepod. The metasome has five clearly defined segments, numbered 1–5; this species has five pairs of swimming legs and so these five metasome segments are synonymous with pedigerous segments 1–5. Legs 1–5 are the swimming legs.

prosome and urosome is present, between the fourth and fifth pedigerous segments (Figure 1). Huys and Boxshall (1991) present a detailed comparative study of the morphology of all copepods and reference should be made to them for further information.

The lack of variety in the gross body form of platycopioid and calanoid copepods has required the full illustration of the animal and its appendages to be given within type descriptions of species. Thus, browsing of such taxonomic papers as Giesbrecht (1892) or Sars (1903, 1925) illustrates the amount of variety that does exist. Detailed descriptions of differences in external morphology are not reviewed here unless within a functional or broader context.

2.1. EXTERNAL MORPHOLOGY OF ADULTS

The body is divided into several regions, the cephalosome, metasome and urosome (Figure 4). Frequently, the first segment of the metasome is fused with the cephalosome, and/or the fourth and fifth segments of the metasome are fused. Thus, the metasome in some species may seem to have as few as three segments. The urosome consists of the genital somite and several segments posterior to it. The genital somite consists of fused segments that are separated in the corresponding males, and results in males apparently having an extra segment in the urosome (Figures 4, 5A). The cephalosome and metasome together are known as the prosome. This is a clearly defined part of the body and its length, from the anterior end of the cephalosome to the posterior lateral edge of metasome segment 5, is used as a direct measure of body length or size. This measurement is preferred to that of total body length because the urosome is often flexed, even damaged at times, causing larger errors when examining length/frequency distributions in statistical analyses of populations.

Copepods, like other crustaceans, have paired appendages that function in swimming, detection and obtaining food, and in mating. They are complex in form, and reference to Huys and Boxshall (1991) is required for the terminology applied to the constituent parts. Females and males are

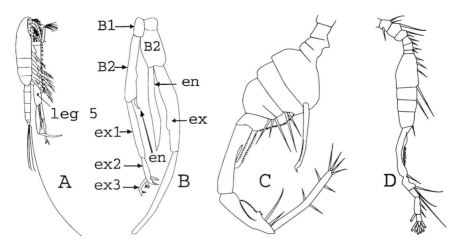

Figure 5 Diagrammatic illustration of the external morphology and some appendages of a male calanoid copepod. A, general lateral aspect of a male *Pareuchaeta norvegica*. B, pair of fifth legs of *P. norvegica* (key to components is in Figure 6, page 18). C, geniculate right antennule of a *Pontella* species. D, geniculate right antennule of a *Candacia* species.

distinguished by sexually dimorphic characters that usually develop during the later copepodid stages. Males are usually smaller in body size than females, have additional apparent segments in the urosome, often greatly modified fifth legs, and many have modified antennules (Gilbert and Williamson, 1983).

2.1.1. Antennule (antenna 1)

The antennule of females in the family Epacteriscidae can have as many as 27 segments, and in the Ridgewayiidae 26 segments, but in the typical female calanoid there are only 25 segments (Figure 6). Fusion of segments is present in many species (Huys and Boxshall, 1991), so that, for example, antennules of *Pontellopsis* species show as few as 16 segments. The antennules of the female and many males are bilaterally symmetrical. Males of families originally classed in the section Heterarthrandria, as opposed to Amphascandria or Isokerandria, by Giesbrecht (1892) and Sars (1903) have their antennules bilaterally asymmetrical. These classes within the Calanoida have now been abandoned (Huys and Boxshall, 1991). This asymmetry results from the right antennule being geniculate, that is knee-like with an articulation separating proximal and distal regions (Figure 5C and D); the left antennule is similar to those of the corresponding female. Males in families in the superfamily Arietelloidea, however, usually have the left antennule geniculate, the right being similar to those of the female. There are genera, within this superfamily, that have their left or right antennule geniculate and even species, *Pleuromamma* species for example, in which there is variation between individuals.

The antennule of the nauplius VI, like that of previous naupliar stages, has three segments but these are transformed to the 9 or 10 segments, depending on whether the distal two are fused, of the copepodid I (Hülsemann, 1991c). The proximal 6 segments of the antennule of the copepodid I generate all further segments of the adult antennule, the distal 6 or 7 remaining unaltered throughout the development of the sequential copepodids.

The boundaries between the antennular segments 2 to 25 have ring-like arthrodial membranes that allow limited flexure (Boxshall, 1985). The junction between segments 8 and 9 in many species, however, is modified, the distal part of the antennule breaking off easily (Bowman, 1978a). The relative lengths of the antennular segments vary little within a species. Sewell (1929, 1932) expresses the lengths of the various segments as parts per thousand of the whole length of the antennule, thus producing an antennular formula. This formula has been used by, for example, Vervoort (1963, 1965) and Boucher and Bovée (1970) but in more recent times by only

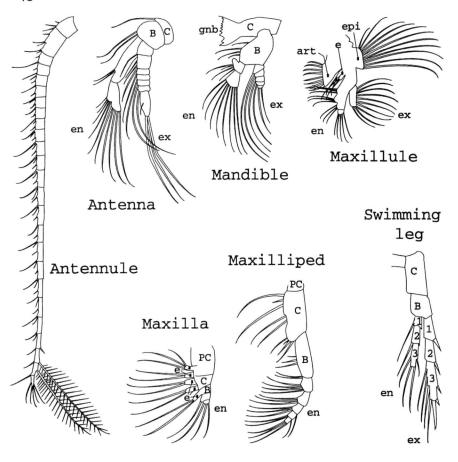

Figure 6 Diagrammatic representations of the appendages of a calanoid copepod. The swimming legs usually have developed endopods and exopods with up to three segments, numbered 1–3 here. Terminology after Huys and Boxshall (1991).
art, arthrite; B, basis; C, coxa; e, endite; en, endopod; epi, epipodite; ex, exopod; gnb, gnathobase; PC, praecoxa.

Soler *et al.* (1988). It is a cumbersome measurement to make but may still be useful.

The antennules are furnished with setae and sensilla or aesthetascs. The copepod often hangs vertically in the water column with the antennules held out laterally to slow down its sinking rate. The aesthetascs function to detect food, water disturbance and predators (Gill and Crisp, 1985a; Légier-Visser *et al.*, 1986; Jonsson and Tiselius, 1990; Kurbjeweit and Buchholz, 1991; Yen *et al.*, 1992; Bundy and Paffenhöfer, 1993; Lenz and Yen, 1993). The male antennules are used to grip the female during mating.

2.1.2. Antenna (antenna 2)

The antenna, unlike the antennule, is biramous, with an endopod and exopod (Figure 6). The endopod is usually of 3 segments, the second and third being partially fused. The exopod contains 8 or 9 segments and patterns of their fusion vary (Huys and Boxshall, 1991). The antennae, in conjunction with the other mouthparts, form an integral part of the food-gathering and handling mechanism of the copepod (Gill, 1987).

2.1.3. Labrum and Labium

The labrum and labium are not normally considered with the paired appendages but they form the margins of the mouth (Boxshall, 1985). The labrum is a muscular lobe, forming the anterior margin of the mouth, often ornamented with spines, and containing the eight labral glands, four on each side (Arnaud *et al.*, 1988a,b), opening on its posterior surface. These glands produce secretions that may bind the food together and initiate digestion in the buccal cavity. Nishida and Ohtsuka (1996) suggest that labral glands in species of the heterorhabdid genus *Heterorabdus* produce an anaesthetic or poison that is injected into prey through a hollow tooth (spine) in the mandible (see next section).

The paired lobes of the labium form the posterior and part of the lateral margins of the mouth. The paired lobes are derived by fusion of the paragnaths (Huys and Boxshall, 1991). The labium is also ornamented with rows of spines.

2.1.4. Mandible

The mandible is biramous (Figure 6), having an endopod of 2 segments and an exopod of 5 segments. The basal segment forms the gnathobase with its spined, distal edge for macerating the food. The development of these spines during the moult cycle has been investigated by Miller *et al.* (1990). The numbers and form of the spines (teeth), which have abrasive tips of silica (Sullivan *et al.*, 1975; Miller *et al.*, 1980), and the setulation of the endopod and exopod relate to the diet of individual species (Anraku and Omori, 1963; Ohtsuka *et al.*, 1996a) and Itoh (1970) has developed an "edge index" to quantify the differences. Schnack (1989) combines determination of the edge index of the gnathobase with the minimum intersetule distances on the maxillules and maxillae to draw conclusions about dietary potentials of species. The overall form of the gnathobase and the disposition of the spines is often such that it can be used to identify the species. The gnathobase resists digestion in the stomachs of predators and Karlson and Båmstedt (1994) have investigated their usefulness in estimating predation rates on populations of copepods.

The mandibles of the Heterorhabdidae have the ventral spine enlarged (Figure 25, key figs. 59, 60). Nishida and Ohtsuka (1996) state that this isolated spine is hollow with a subterminal pore and a basal opening. The basal opening is aligned with the cuticular pore of a large labral gland that is situated under the posterior face of the labrum. An anaesthetic or poison, secreted by the gland, is thought to be transferred, through its cuticular pore, into the basal opening of the mandibular spine. It then travels up the internal canal of the spine to be injected through the subterminal pore into the prey. Such a feeding technique has not previously been described in a copepod.

2.1.5. Maxillule (maxilla 1)

The maxillule (Figure 6) is a complex laminar appendage whose constituent parts are defined by Huys and Boxshall (1991). It is biramous with a 3-segmented endopod, some segments often fused, and a single segmented exopod. The setulation and overall form relate to the diet of the species (Anraku and Omori, 1963; Schnack, 1989; Ohtsuka *et al.*, 1996a).

2.1.6. Maxilla (maxilla 2)

The maxilla (Figure 6) is also a laminar appendage whose constituent parts are defined by Huys and Boxshall (1991). It is uniramous and 7-segmented. Its form also relates with the diet of the species, strong spines replacing setae in a few species (Landry and Fagerness, 1988; Ohtsuka *et al.*, 1996a).

2.1.7. Maxilliped

The maxilliped (Figure 6) is uniramous and 9-segmented according to Huys and Boxshall (1991), although most species have only 6 free segments in the endopod. It can be greatly developed as in *Pseudeuchaeta* species (Figure 2), and armed with setae or spines dependent upon the feeding strategy of the species.

2.1.8. Swimming Legs

The first four metasome segments of females and males always have paired, biramous swimming legs that are similar in both sexes. In some families, such as the Calanidae, a fifth pair of legs, similar to the first four pairs, is present. In other families, such as the Aetideidae and Euchaetidae, the fifth pair is usually absent in females but present, although considerably modified (Figure 5B), in males. The fifth pair of legs present in females can be considerably reduced in size and structure while that in males is normally enlarged as it functions to grasp the female during mating. The exopods and endopods of the five pairs of legs have a maximum of 3 segments each but their numbers may be reduced in one or more pairs of legs. The distribution of setae and spines on the legs also varies so that the morphology of the legs is very important in the identification of families, genera and species (see next chapter).

Sewell (1949) suggested a spine and setal formula to summarize the setation of appendages. It discriminates between spines, denoted by Roman numerals, and setae, given by Arabic numerals. Legs are examined in the order anterior to posterior, proximal before distal segments. Spines or setae on the outside of the segment are defined before those on the inside, those on the same segment being linked by a hyphen (Figure 7). Exopod and endopod segment 3 have a terminal armature that is interposed between the lateral ones so that the order is outer, terminal and inner armatures. This formula can be adapted for other appendages and may be useful in future, computerized, identification keys for species.

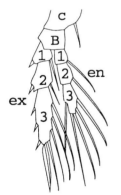

Coxa		0 - 1
Basis		0 - 0
exopod	1	I - 1
	2	I - 1
	3	III - I - 4
endopod	1	0 - 1
	2	0 - 1
	3	1 - 2 - 2

Figure 7 The spinal and setal formula of Sewell (1949). Spines are given by Roman, setae by Arabic numerals. See text for further explanation.

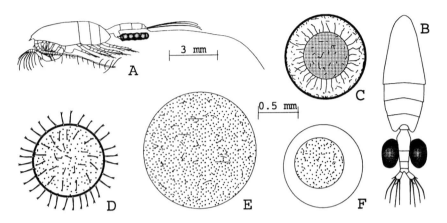

Figure 8 Eggs of calanoid copepods. A. *Pareuchaeta norvegica* with egg mass; B, *Valdiviella insignis* with eggs; C, egg of *Tortanus forcipatus* from sediment; D, egg of *Centropages abdominalis* from sediment; E, pelagic egg of *Calanus finmarchicus*; F, egg of *Acartia spinicauda*. A and B, 3 mm scale; C to F, 0.5 mm scale. (C and D, after Kasahara *et al.*, 1974; F, after Li Shaojing *et al.*, 1989.)

2.2. EXTERNAL MORPHOLOGY OF YOUNG

2.2.1. Egg

Eggs of copepods are either carried by the female attached in a mass to the genital opening or laid freely in the water column. The eggs carried by species such as *Euchaeta* and *Pareuchaeta* are enclosed in cuticular material that glues them together (Figure 8A). The egg mass is often referred to as an egg sac but there is little evidence that the eggs are carried in a bag. The most spectacular eggs are the two large ones carried by *Valdiviella insignis* (Figure 8B), a relatively common deep-sea species. Those eggs that are laid freely in the sea by many small coastal species can be of two types, subitaneous (non-resting) and resting eggs. The subitaneous eggs (Figure 8E, F) are usually relatively thin-walled and unadorned with any "spines". Resting eggs are thicker-walled and often sculptured or have surface "spines" (Figure 8C, D). They sediment to the sea bed where they remain for periods before hatching. Koga (1968) describes the eggs of 18 species, dividing them into those with floating devices (Figure 8D) and those without (Figure 8C, E, F). Further figures of eggs are given by Li Shaojing *et al.* (1989).

Hirose *et al.* (1992) describe the development of a multi-layered fertilization envelope in *Calanus sinicus*; it forms within the perivitelline space. Some eggs have a perivitelline space (Figure 8C, F) while others do not (Figure 8D, E). The number of membranes bounding the egg is not clear; Toda and Hirose (1991) figure sections in which they discern seven to eight layers. This seems excessive and three layers, a perivitelline, chitinous and cuticular, is a more reasonable interpretation, although there may be a degree of lamination in the outer two.

Identification of copepod eggs in the plankton to the species level is often relatively easy in coastal areas where dominance and size are often the key features. Resting eggs in sediments often have species-specific sculpturing of the membranes (Belmonte and Puce, 1994) and identification to species is again often possible. It is much more difficult to identify the eggs in offshore plankton samples.

A method of determining whether eggs are fertilized or not is given by Ianora et al. (1989) who use a fluorescent dye specific for cell nuclei. The unfertilized eggs have only the female nucleus whereas the recently fertilized egg has both female and male pronuclei.

2.2.2. Nauplius

Calanoid copepods have six naupliar stages, abbreviated to NI to NVI, except in some species of *Labidocera* and *Pseudodiaptomus* when the first is omitted and an NII emerges directly from the egg. The first 3 naupliar stages are true nauplii with 3 pairs of appendages, the antennules, antennae and mandibles. The later stages, however, are similar to the metanauplii of other Crustacea because they often have signs of "abdominal" segmentation and rudiments of more posterior appendages (e.g. Figure 9B). The successive naupliar stages within a species are identified by the progressive setation of the distal segment of the 3-segmented antennule (4-segmented when a basal segment is present), and by the progressive development of the armature of the posterior end of the body. The NI of all species has 3 setae on the distal segment of the antennule and 2 spines on the posterior end of the body (Figure 9A). By NVI, there are 9 to 17 setae and 10 or more spines respectively. A setal formula describes the setation of the distal segment of the antennule (Ogilvie, 1953) but it is often difficult to apply. It depends on the presence of a distal aesthetasc (Figure 9E) which is not always present or discernible. The dorsal and ventral setae are counted and the aesthetasc interposed; thus, the formula for the antennule in Figure 9E is 5a7. Difficulties arise when the aesthetasc is absent and/or when minute spines, as distinct from setae with setules, are present. Some authors count everything, others only the setae. These counts identify the naupliar stages within species and the formula has been used in an attempt to distinguish different species. It is most useful when the development of the other

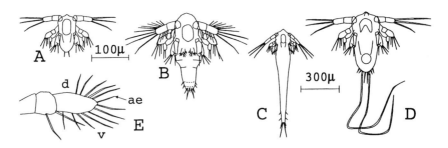

Figure 9 Nauplii of calanoid copepods. A, *Clausocalanus furcatus*, stage I (NI); B, *Paracalanus aculeatus*, NV; C, *Rhincalanus cornutus*, NIV; D, *Euchaeta marina*, NVI; E, antennule showing dorsal (d) and ventral (v) setae and terminal aesthetasc (ae). (After Björnberg, 1972, 1986b.)

appendages is included as well as the armature of the posterior part of the body (Faber, 1966; Björnberg, 1972; Sazhina, 1982). Body size alone often separates species in coastal regions with low diversity. Similarly, general body form can be used because some genera such as *Rhincalanus* (Figure 9C) have an elongated nauplius while that of *Euchaeta marina* has two conspicuous, long, thin setae posteriorly (Figure 9D).

The nauplii of some 83 species have been described (Table 3), and are those of only some 5% of known calanoids. Forty of these species, however, are also listed in the 72 most quoted species in Table 2 so that the development of many of the common species is known. Björnberg (1972, 1986a,b) argues that the form of the nauplius must be taken into account in formulating any taxonomic classification of the Calanoida. This is difficult at present because the nauplii of many families are completely unknown.

2.2.3. Copepodid

The NVI moults to the first of six copepodid stages, abbreviated CI to CVI, that resemble miniature adults. The CVI is the adult. The sequential stages are distinguished by the progressive development of the adult segmentation of the body, the increasing differentiation of the appendages, and successive increases in body size. Descriptions of the copepodids of species listed in Table 3 are given in many of the papers quoted there and by Ferrari (1988).

The terms cephalosome (head), prosome, metasome (thorax) and urosome (abdomen) used in Figure 4 do not correspond to the homologous segmentation of other crustacean orders. The cephalosome of copepods consists of the head fused with the first thoracic somite. Thus in the CI

EXTERNAL MORPHOLOGY, INTERNAL ANATOMY

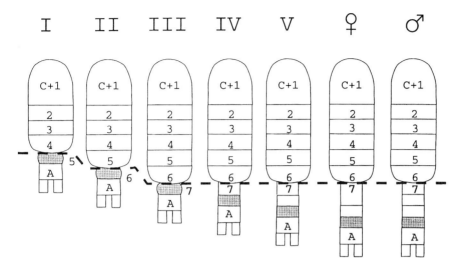

Figure 10 The sequential development of the apparent segmentation throughout the copepodid stages, CI to CV, the female and male. Various combinations of segments can be fused in different genera and species so that the apparent segmentation becomes reduced. The broken line indicates the division between the metasome and urosome. The shaded segment shows the new segment acquired at the previous moult. C + 1, cephalosome plus first thoracic segment; 2–7, thoracic segments; A, anal segment. (After Boxshall, 1985; Hülsemann, 1991a.)

(Figure 10), the body consists of the cephalosome plus four free segments and an anal segment (Figure 10A). Some species, e.g. *Pseudocalanus*, have only three free segments plus the anal segment (Table 4, and Corkett and McLaren, 1978). The numbering of the segments in Figure 10 acknowledges the incorporation of the first segment within the cephalosome (C + 1) (Boxshall, 1985; Hülsemann, 1991a). A new segment is added, immediately anterior to the anal segment, at each successive moult (Figure 10). The new segment in the CI and CII becomes incorporated in the metasome at the next moult (Figure 10) and the division between the metasome and urosome moves one segment posteriorly each time. The new segment in CIII is the 7th thoracic segment that remains in the urosome, the division between the metasome and urosome in calanoids being between the 6th and 7th thoracic segment. This is synonymous with the statement that this division is between the 5th pedigerous segment and the genital somite. Subsequent copepodid stages, CIV to CVI, add segments to the urosome (Figure 10). The numbers of apparent adult segments in both the metasome and the urosome (Table 4) can vary because various combinations of segments become fused. The female genital somite consists of at least two

Table 3 Sources of descriptions of nauplii of marine calanoid copepods.

Acartia bifilosa (Oberg, 1906); *A. californiensis* (Trujillo-Ortiz, 1986); *A. clausi* (Ogilvie, 1953; Klein-Breteler, 1982); *A. danae* (Björnberg, 1972); *A. grani* (Vilela, 1972); *A. lilljeborgi* (Björnberg, 1972); *A. longiremis* (Oberg, 1906); *A. negligens* (Björnberg, 1972); *A. tonsa* (Björnberg, 1972; Sazhina, 1982); *Aetideus armatus* (Matthews, 1964); *Calanoides carinatus* (Björnberg, 1972; Hirche, 1980); *Calanopia thompsoni* (Li Shaojing and Fang Jinchuan, 1984); *Calanus finmarchicus* (Björnberg, 1972); *C. helgolandicus* (Björnberg, 1972); *C. hyperboreus* (Sömme, 1934); *C. minor* (Björnberg, 1972); *Calocalanus pavo* (Björnberg, 1972); *C. styliremis* (Björnberg, 1972); *Candacia aethiopica* (Sazhina, 1982); *Candacia armata* (Bernard, 1964); *Centropages abdominalis* (Koga, 1960b); *C. chierchiae* (Sazhina, 1982); *C. furcatus* (Björnberg, 1972); *C. hamatus* (Oberg, 1906; Klein-Breteler, 1982); *C. kröyeri* (Sazhina, 1960); *C. typicus* (Lawson and Grice, 1970); *C. yamadai* (Koga, 1970); *Clausocalanus furcatus* (Björnberg, 1972; Sazhina, 1982); *Ctenocalanus vanus* (Björnberg, 1972); *Epilabidocera amphitrites* (Johnson, 1934b); *Eucalanus attenuatus* (Björnberg, 1972); *E. bungii* (Johnson, 1937); *E. crassus* (Björnberg, 1972); *E. elongatus* (Björnberg, 1972); *E. pileatus* (Björnberg, 1972); *Euchaeta marina* (Bernard, 1964); *Eurytemora affinis* (Katona, 1971); *E. americana* (Grice, 1971); *E. herdmani* (Grice, 1971a); *E. hirundo* (Björnberg, 1972); *E. hirundoides* (Björnberg, 1972); *E. pacifica* (Chiba, 1956); *E. velox* (Gurney, 1931); *Gladioferens pectinatus* (McKinnon and Arnott, 1985); *Labidocera acutifrons* (Sazhina, 1982); *L. aestiva* (Björnberg, 1972; Gibson and Grice, 1977); *L. bengalensis* (Ummerkutty, 1964); *L. brunescens* (Björnberg, 1972); *L. euchaeta* (Li Shaojing and Fang Jinchuan, 1983); *L. fluviatilis* (Björnberg, 1972); *L. jollae* (Johnson, 1935); *L. minuta* (Goswami, 1978b); *L. pavo* (Goswami, 1978b); *L. rotunda* (Onbé *et al.*, 1988); *L. trispinosa* (Johnson, 1935); *Limnocalanus grimaldi* (Lindquist, 1959); *Metridia lucens* (Ogilvie, 1953); *Microcalanus pusillus* (Ogilvie, 1953); *Neocalanus gracilis* (Sazhina, 1982); *N. tonsus* (Björnberg, 1972); *Paracalanus aculeatus* (Björnberg, 1972); *P. parvus* (Björnberg, 1972); *Pareuchaeta elongata* (Campbell, 1934; Lewis and Ramnarine, 1969); *P. norvegica* (Nicholls, 1934); *P. russelli* (Koga, 1960a); *Parvocalanus crassirostris* (Lawson and Grice, 1973); *Pleuromamma abdominalis* (Sazhina, 1982); *Pontella atlantica* (Sazhina, 1982); *P. meadi* (Gibson and Grice, 1976); *P. mediterranea* (Crisafi, 1965); *Pontellopsis brevis* (Björnberg, 1972); *P. occidentalis* (Johnson, 1965); *Pseudocalanus minutus* (Corkett and McLaren, 1978); *Pseudodiaptomus acutus* (Björnberg, 1972); *P. ardjuna* (Alvarez and Kewalramani, 1970); *P. aurivilli* (Ummerkutty, 1964); *P. binghami* (Goswami, 1978a); *P. coronatus* (Grice, 1969); *P. euryhalinus* (Johnson, 1948); *P. marinus* (Uye and Onbé, 1975); *Rhincalanus cornutus* (Björnberg, 1972); *R. gigas* (Björnberg, 1972); *R. nasutus* (Björnberg, 1972); *Ridgewayia klausruetzleri* (Ferrari, 1995); *Temora longicornis* (Corkett, 1967; Klein-Breteler, 1982); *T. stylifera* (Gaudy, 1961); *T. turbinata* (Koga, 1984); *Tortanus discaudatus* (Johnson, 1934a); *T. gracilis* (Björnberg, 1972); *Undinula vulgaris* (Björnberg, 1966, 1972); *Xanthocalanus fallax* (Matthews, 1964).

Table 4 Patterns of sequential development of the swimming legs and the segmentation of the urosome in species with different adult numbers of swimming legs and urosome segments.

Copepodid	Pairs of legs				Segments of the urosome					
I	2		2		2	1			1	
II	3		3		2	1			2	
III	4		4		2	1			2	
IV	4♀ 5♂		5♀ 5♂		3	2			3	
V	4♀ 5♂		5♀ 5♂		4	3		3♀ 4♂		
VI	4♀ 5♂		5♀ 5♂		5♀ 5♂	3♀ 4♂		2♀		5♂

fused segments. Thus the body segmentation can be used to define the copepodid stages.

The development of the pairs of swimming legs also identifies the individual copepodid stages (Table 4). Adults have 4 or 5 pairs of swimming legs, abbreviated as P1 to P5. Some adult females lack a P5, e.g. species of the Aetideidae and Euchaetidae. The legs do not suddenly appear fully developed to the adult condition in a single moult but can appear as a rudiment first and at successive moults become increasingly more complex (Ferrari, 1988). The CI may have two pairs of well-developed legs or the first pair may be more developed than the second. There is, however, always a progressive development of the legs in successive stages that can be used to identify them. In addition, sex can be determined at the CIV by the morphology of the legs in many species. The ontogenetic development, in the genus *Pontellina*, of the male P5, a complex grasping organ used during mating, is shown to contain phylogenetic information by Hülsemann and Fleminger (1975). Ferrari (1988, 1992, 1993a, 1995) reviews various aspects of the detailed development of the legs throughout the Calanoida and other copepods.

Another feature of the copepodids that is useful is the state of development of the segmentation and armature of the antennules but it is a more difficult character to use. The CI has an antennule of 8–13 segments but most species have 9 or 10. The CII has 11–18 segments, most having either 11–14 or 17 or 18. The CIII has 14–24, most having 19–24. The antennule of CIV can have as few as 16 segments although the majority of species have between 21 and 25. This character has a restricted value for identifying copepodid stages within a species but may be more useful for distinguishing the species of copepodids.

Males and females may be distinguished in CIV and CV in species that have the P5 modified in males as a grasping leg and in which the females have the P5 absent. Recognition of the sexes in other species, for example, in the Calanidae is much more difficult. One of the most detailed studies of sexual differentiation in the CV is that of Grigg and Bardwell (1982) and Grigg et al. (1981, 1985, 1987, 1989) in *Calanus finmarchicus*. They found that the CV has a bimodal distribution of prosome length, the larger CVs moulting to males while the smaller moult to females. This is surprising as adult males, in general, are smaller than the adult females. No clear evidence of bimodality in the CIV was present.

2.3. INTEGUMENT

The copepod integument consists of several layers (Bresciani, 1986). There is an outer, very thin epicuticle consisting of as many as four layers, sometimes covered with an outer, possibly, membranous layer. Inside the epicuticle, and sharply distinguished from it, is the procuticle comprising two thick and distinct layers, the exocuticle and the endocuticle, that extend to the underlying epidermis (Figure 12). Bresciani (1986) and Boxshall (1992) have adequately reviewed the structure of the integument and reference should be made to them for further information.

2.3.1. The Setal System and Subintegumental Glands

Structures occur on the surface of the integument and range from prominent head spines, posterior spinal extensions of the fifth metasomal segment, both present on some *Gaetanus* species (Figure 2), the aesthetascs of the antennules, integumental sensilla and openings of subintegumental glands, specialized and prominent structures on the maxillae of some genera, to minute groups of spinules and tufts of fine setae that are purely cuticular in nature. These features divide broadly into two categories. The first comprise those that are purely extensions of the cuticle itself and have no connection, neural or otherwise, with subintegumental tissues. The second are openings of subintegumental glands or seta-like sensilla that have ducts or neural connections through the integument. Not enough is known of the ultrastructure of many structures on the appendages and body to assign them confidently to one or other of these categories.

There have been several attempts to classify setae of crustaceans (Jacques, 1989; Watling, 1989) but their variety, even on an individual animal, is great. Campaner (1978a,b) figures setae of Aetideidae and Phaennidae and examination of the figures in any taxonomic work, such as

Sars (1925), adds further examples. Several distinctive types of setae or sensilla, especially on the maxillae and maxillipedes, are restricted to a few genera. Examples are the setae with button-like armature in the Augaptilidae that Krishnaswamy *et al.* (1967) think may act like cephalopod arms to hold prey, or the large, tufted structures of the maxillae of some genera in the families Phaennidae and Scolecitrichidae (Figure 28, Key-figure 172; Figure 29, Key-figure 202). Jacques (1989) points out that sensory functions in Crustacea mainly derive from the setal system and the current preliminary knowledge of the function of these structures in the Copepoda supports this statement. The current lack of knowledge results in some ambiguous terminology. Sensory setae are sensilla. The term sensilla, however, also includes pit organs (see Fleminger, 1973) which do not have a prominent seta extending from the surface of the integument. Consequently, sensillum is used here to describe any structure on the surface of the integument that has a neural, or suspected neural, connection with subintegumental regions; it may extend from the integument as a seta. Sensilla and gland openings are often referred to in the literature as integumental organs.

It is sensilla and subintegumental glands that are responsible for the pores or holes in an integument once it has been digested in hot aqueous potassium hydroxide. The digestion removes all the soft tissues such as the nerve connections and the walls of the ducts. These pores form a pattern over the digested integument that is species-specific and termed a pore signature. Sewell (1929, 1932), followed by Fleminger (1973), showed that these sensilla and gland openings are generally distributed in a bilaterally symmetrical pattern over the integument. Koomen and Vaupel Klein (1995) recommend immersion of copepods in 70% lactic acid at up to 100 °C for clearance of the integument and internal soft tissues. Digestion should be timed so that the soft tissues and sensilla are not destroyed. Vaupel Klein (1982a) examined their morphology in adult female *Euchirella messinensis* and distinguished some eight different structures. His list is not exhaustive as several other structures exist in other species and genera (Mauchline, 1977a; Guglielmo and Ianora, 1995). Vaupel Klein (1982a) in reviewing the literature, also discusses possible functions of the glandular pores and sensilla; these include mechanoreception, chemoreception, secretion of mucus, mucopolysaccharides, bioluminescent material, fatty or oily secretions to reduce body drag during swimming, and pheromone production.

Brunet *et al.* (1991) examined the fine structure of the segmentary integumental glands in *Hemidiaptomus ingens* using scanning and transmission electron microscopy combined with cytochemical tests. These authors review the earlier literature but their results highlight the inherent difficulties in deducing the function of the secretions, even from such a detailed structural examination. Several investigations describing the ultrastructure of sensilla and/or glands do not define their function conclusively (Elofsson, 1971; Gill, 1985, 1986; Koomen, 1991; Kurbjeweit and Buchholz, 1991; Bannister, 1993a; Bundy and Paffenhöfer, 1993; Weatherby et al., 1994). Yen et al. (1992), however, demonstrate, by electrophysiological techniques, that individual setae of the antennules can detect water velocities and even provide directional information. Rippingale (1994) shows how hair sensilla, distributed over the dorsal and lateral surfaces of the prosome of *Gladioferens imparipes*, act in concert through an unknown mechanism to attach the dorsal surface of the copepod to an underwater surface such as that of a seaweed.

Much more complex structures, purely mechanical in nature, exist in pontellid copepods. Ianora et al. (1992b) describe double horseshoe-shaped filamentous formations situated antero-dorsally on the cephalosome of copepodids and adults of *Anomalocera patersoni, Pontella mediterranea, P. lobiancoi, P. atlantica, Pontellopsis regalis, P. villosa, Pontellina plumata* and *Labidocera wollastoni*. They term it a surface attachment structure, its overall form being species-specific, that probably functions to attach these neustonic copepods to the surface film so conserving energy.

A few sexual differences in the integumental sensilla and gland openings have been found. The most obvious are between the antennules of males and females where one of the male antennules is geniculate. Nishida (1989) describes a cephalic dorsal hump that is peculiar to males and so far restricted to species in the families Calanidae, Megacalanidae, Mecynoceridae and Paracalanidae; he concludes that it plays an important role in mate recognition. One of the most curious and prominent integumental organs is the pigment knob of *Pleuromamma* species that occurs on either the left or right side of the first metasome segment. A detailed ultrastructural study by Blades-Eckelbarger and Youngbluth (1988) failed to discover its function.

Ontogenetic studies of the development of the pore patterns in copepodid stages have been made in *Eucalanus* species by Fleminger (1973), in *Pareuchaeta norvegica* and *Neocalanus* species by Mauchline and Nemoto (1977) and in *Pleuromamma* species by J.S. Park (1995a). The most detailed study is that of Park. She shows that the cephalosomal signature is completed in copepodid V but that the metasomal and urosomal signatures are not complete until the adult. Sexual differences in the signatures within species appear in copepodid IV but are primarily evident in the urosomes of the adults. Adult females have more species-specific components in the signatures than the corresponding males. Different species showed different rates of development of the adult signatures during the course of the copepodid stages, some features appearing earlier or later in one species than another.

2.3.2. Eyes and Frontal Organs

The nauplius eye is sited in the ventral median region of the anterior cephalosome (Figure 11). It consists of paired dorsolateral ocelli and a median ventral ocellus. Three bundles of axons connect them to the protocerebrum. The morphology of the eye varies between species, that in Figure 11 being relatively unspecialized and without lenses, as found in *Pareuchaeta norvegica* or *Calanus finmarchicus*. The ventral median ocellus of *Centropages* species has a lens, the dorsal ocelli being unmodified. The large eyes of *Cephalophanes refulgens* (Figure 2) are modified dorsolateral ocelli of the nauplius eye and similar modifications of these ocelli are responsible for the eyes of the pontellid copepods (Figure 24, Key-figures 3–5). Boxshall (1992) has reviewed the histology and function of these eyes but recourse to the original references on the individual species (Table 5) is required for the detailed descriptions.

Two other sensory structures occur in this region of the cephalosome and are connected by axons to the protocerebrum. The first are the paired Gicklhorn's organs (Figure 11) of unknown function. Boxshall (1992) raises

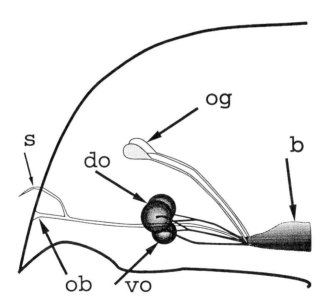

Figure 11 Nauplius eye and frontal organs of a copepod.
b, brain; do, paired dorsal ocelli of nauplius eye; ob, left organ of Bellonci, right not shown; og, organ of Gicklhorn; s, sensillum; vo, ventral ocellus of nauplius eye. (After Boxshall, 1992.)

Table 5 Species in which the nauplius eye and frontal organs have been described. References are given in parentheses.

Anomalocera patersoni	(4)	Epilabidocera amphitrites	(6)
Augaptilus sp.	(5)	Eucalanus elongatus	(1, 4)
Calanus finmarchicus	(5, 7, 9)	Euchaeta marina	(5)
Calanus glacialis	(9)	Labidocera acuta	(10)
Calanus hyperboreus	(5)	Labidocera wollastoni	(4)
Calanus marshallae	(9)	Neocalanus gracilis	(5)
Candacia aethiopica	(4)	Pareuchaeta norvegica	(5, 7)
Centropages furcatus	(3)	Pleuromamma abdominalis	(4)
Centropages typicus	(4)	Pontella mediterranea	(8)
Cephalophanes refulgens	(2)	Pontella spinipes	(10)
Chiridius armatus	(5, 7)	Pontellopsis regalis	(4)

(1) Esterly, 1908; (2) Steuer, 1928; (3) Krishnaswamy, 1948; (4) Vaissière, 1961; (5) Elofsson, 1966; (6) Park, 1966; (7) Elofsson, 1971; (8) Frasson-Boulay, 1973; (9) Frost, 1974; (10) Boxshall, 1992.

the question of whether they are homologous with the organs of Claus described by Esterly (1908) in *Eucalanus elongatus*. A more detailed discussion of this is given by Elofsson (1966, 1970) while homology is discussed by Frost (1974). The organs appear to have neurosecretory axons and photosensitive cells suggesting that any neurosecretions may be mediated by light. The second pair of organs, those of Bellonci (Figure 11), are described in most detail by Elofsson (1971) in *Pareuchaeta norvegica*, *Calanus finmarchicus* and *Chiridius armatus*. Their function is unknown although they may be chemosensory.

2.3.3. Moulting

Copepods, like all crustaceans, increase their body size by moulting (apolysis). This is a complex cyclical process described by Drach (1939) in decapod crustaceans but whose descriptions have been confirmed by J.S. Park (1995b) as applicable to calanoid copepods. The cycle in *Pleuromamma robusta* consists of postmoult, intermoult and premoult stages (Figure 12, Table 6). Postmoult copepods have soft integuments which harden during the intermoult stage. The soft postmoult stage is frequently recognizable in samples of copepods from the field (J.S. Park, 1995b). The integument thickens progressively throughout the cycle (Figure 12) and originates primarily from the development of the endocuticle (Table 7), especially in the adults where no development of a new integument takes

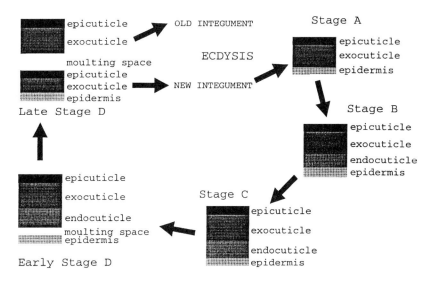

Figure 12 Diagrammatic illustrations of sections through the integument of *Pleuromamma robusta* at different stages in the moult cycle. Stages are defined in Table 6. (After J.S. Park, 1995b.)

place since the adults do not moult. During the cycle (Figure 12, early stage D) the old integument detaches from the epidermis and a moulting space (apolytic space) is formed. Enzymes are secreted into the space and digest up to 75% of the old integument (Roff et al., 1994). The products are recycled to form the chitin of the exocuticle (Figure 12, late stage D) of the new integument. Moulting then takes place and the exocuticle of the new integument continues to thicken (Figure 12, stage A) by *de novo* synthesis of chitin.

There are few direct studies of moulting in copepods. Hülsemann and Fleminger (1975) and Dexter (1981) confirm Currie's (1918) descriptions of the formation or replacement of setae at moulting. The new cuticle forms under the old one and the first sign of the new seta is the formation of a single-walled tube in the tissues below the existing seta (Figure 13A). This tube will become the proximal shaft of the new seta. A second tube then forms inside the first (Figure 13B) and this will become the distal shaft of the new seta. The new seta frees itself from the old seta by withdrawing into the underlying tissues of the body (Figure 13C). This withdrawal and the next phase of extrusion are enabled by the two eversion points (Figure 13, ep1 and ep2) that allow a turning-inside-out process to take place at these two positions. Extrusion of the new seta forces the old exuvium away

Table 6 The morphological characteristics of the different stages of the moult cycle in copepodids of *Pleuromamma robusta* are detailed. Illustrations of the stages are shown in Figure 12 and of setal development in Figure 13. (After Drach, 1939 and J.S. Park, 1995b.)

Stage of moult cycle	Characteristics of integument and setal development
Postmoult (Stages A and B)	1. Integument soft 2. Body appears transparent 3. Exocuticle hardens; endocuticle forms 4. Distal shafts of setae not fully emerged
Intermoult (Stage C)	1. Integument is firm to hard 2. Formation of endocuticle is complete 3. Distal shafts of setae fully emerged
Premoult (Stage D)	1. Integument is hard 2. Body appears opaque 3. Old endocuticle reabsorbed creating a moulting space between the old integument and the new epicuticle and exocuticle forming progressively 4. Formation of new setae and spines under old ones prior to next moult 5. Ecdysis follows

Table 7 The thickness (μm) of the integument during the moult cycles of female CV and adult female *Pleuromamma robusta*. (After J.S. Park, 1995b.)

Moult stage	Epicuticle		Endocuticle	
	CV	Female	CV	Female
Postmoult A	0.16	0.65	2.12	4.75
Postmoult B	0.17	0.79	2.84	6.40
Intermoult C	0.17	0.71	3.35	7.18
Premoult D	0.19	0.68	4.45	8.89

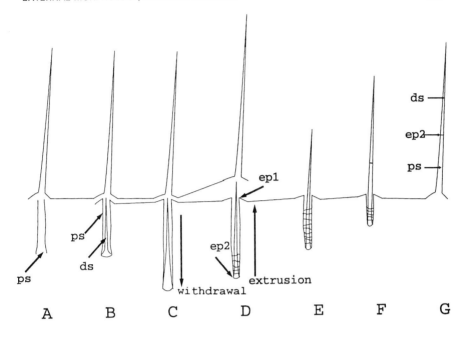

Figure 13 Diagrammatic representation of setogenesis in a copepod, showing successive stages A–G. Further explanation in the text. ds, distal shaft; ep1, ep2, eversion points along proximal shaft; ps, proximal shaft. (After Dexter, 1981.)

from the new integument (Figure 13D). The extrusion process builds up pressure on the old exuvium so that it finally splits along the dorsal region of the cephalosome and metasome and is shed completely from the copepod.

The dependence of copepods on the setal system for information on the surrounding environment suggests that they may suffer from sensory deprivation during at least part of the premoult period when the old setal system is disconnected and the new system is shielded by the old cuticle.

Moult cycle stages are defined by Miller and Nielsen (1988) in a study of the mandibles of *Neocalanus flemingeri* and *N. plumchrus*. They do not relate their identified phases of moulting to the stages defined in Table 6. The postmoult mandible is distinct because it has a long extension of the haemocoele. Intermoult mandibles do not have this haemocoele but have a uniform internal structure. The premoult mandible has the new mandible forming within it.

The presence of soft- and hard-bodied individuals in samples coupled with the histological state of the mandibles are useful tools in determining whether a copepod population is actively growing or not.

2.4. ASYMMETRY

Calanoid copepods are essentially bilaterally symmetrical but several characters in different genera and species are markedly asymmetrical. The urosome of males in genera such as *Pleuromamma* is twisted and the ornamentation is asymmetrical or individual segments are asymmetrical as in *Candacia* species. Garmew et al. (1994) describe asymmetrical rows of spines on the last metasomal segment of *Acartia tonsa*. Terminal setae of the caudal furcae can be asymmetrical as in the family Heterorhabdidae. The distribution of sensilla over the surface of the integument, generally symmetrical, has many examples of asymmetry (Mauchline, 1988b). The most detailed studies of asymmetry are those of Ferrari (1984, 1985) and Ferrari and Hayek (1990) on the genus *Pleuromamma* in which linked characters, present on one side of the body only, can change sides. The principal characters are the peculiar pigment knob (also called the black organ, button) present laterally in the first metasome segment, the male geniculate antennule and fifth pair of swimming legs, and the female genital somite.

2.5. ABNORMALITIES AND MORPHOLOGICAL VARIATION

Morphological variation exists within a species of copepod as it does in species of any other kind of animal (Riera, 1983). The most obvious character to vary is body size, populations showing seasonal, bathymetric and geographical changes, but these will be discussed later when examining growth and population dynamics of species. It is disconcerting to find that apparent intraspecific variation, on detailed study, results not infrequently from the presence of two closely related species. A most recent example of this is the separation of *Calanus jaschnovi* from *C. sinicus* by Hülsemann (1994). Confusion of this kind arises from the overall similarity of body form in the Calanoida, and the necessary separation of species on the basis of relatively detailed morphological features. Abnormal individuals do occur in samples, some of the abnormalities being just that. In large samples of *Pareuchaeta norvegica*, for example, a few stage V copepodids, usually markedly smaller than normal copepodids, have rudiments of the adult genital prominence, and look "unhealthy"; whether these survive to moult is unknown. There is often a very small proportion of individuals in samples of coastal species that are thin, small in size and atypical of the population as a whole. Such individuals are not rare and do not seem to be mentioned in the literature.

Individuals possessing both female and male characters, referred to as intersex or hermaphrodite, occur. Sewell (1929) records intersexes in *Paracalanus aculeatus, P. parvus, Parvocalanus crassirostris*, and *Acrocalanus inermis*, identified by the segmentation of the urosome and development of the fifth pair of legs; no examination of the gonads was made. Conover (1965) describes an intersex of *Calanus hyperboreus* in detail, including the development of a genital system which contains male and female elements. The antennules of the majority of copepods are sexually dimorphic but Fleminger (1985) has shown that females of species in the family Calanidae are dimorphic, most being the trithek morph, in which most antennular segments have two setae and one aesthetasc. Some, however, are the quadrithek morph in which odd-numbered segments 2b, 3, 5, 7 and 9 have two setae and two aethetascs, the normal male armature. There is no question of these being intersexes because they are thought to result from the sex-determination processes in the stage V copepodid stage.

The segmentation of appendages sometimes varies within a species as instanced by Drapun (1982) for *Metridia lucens* and Brylinski (1984a) for *Acartia* species in which the number of segments that can occur in the fifth legs is variable. In general, however, the segmentation and armature of the appendages are relatively constant although variation exists especially where rows of small spines are concerned. The male P5 of *Pareuchaeta norvegica*, for example, has a serrated lamella with 10 to 12 spines (teeth) present. Rows of spines on the last metasome segment and on the urosome of *Acartia clausi* in the Black Sea vary in number (Shadrin and Popova, 1994), individuals dividing between two general morphotypes. These spines also vary in number in *Acartia tonsa* where they are also asymmetrically distributed (Garmew *et al.*, 1994). The linked asymmetrical characters of *Pleuromamma xiphias* can be on either the right or left side of the body (Ferrari and Hayek, 1990). Obvious variation in numbers of articles is present in the cuticular ornamentation such as patches of spinules, rows of teeth and less variation is involved, although still present, in the setal system. Such variation shows itself in the pore signatures when the soft tissues have been digested from the integument (J.S. Park, 1996). Variation within species also exists in the silhouette of the cephalosome, in the genital prominence (e.g. that of *Pareuchaeta tonsa* described by Geinrikh, 1990), or in other regions of the body (Geinrikh, 1990).

2.6. INTERNAL ANATOMY

No detailed descriptions of differences in the internal anatomy of copepods are given here except where functional aspects are discussed later. The internal anatomy and histology of copepods have been reviewed by Brodsky et al. (1983), Blades-Eckelbarger (1986) and Boxshall (1992). The principal sources of the figures and descriptions in these papers are Lowe (1935) and Marshall and Orr (1955) on *Calanus finmarchicus* and Park (1966) on *Epilabidocera amphitrites*.

2.6.1. Endoskeleton

The endoskeleton provides the attachments for the muscles. It consists of two components, the endoskeleton proper, formed by invaginations of the exoskeleton into the body cavity, and the endoskeletal tendons (Park, 1966). The invaginations of the cuticle may remain hollow, apodemes, or be solid, apophyses. Park describes the endoskeleton of *Epilabidocera amphitrites* in some detail and Lowe (1935) that of *Calanus finmarchicus*. The two are similar but one pair of apodemes in *C. finmarchicus* and two pairs in *E. amphitrites* connect with each pair of swimming legs. The cells of the muscles are attached to the apodemes by epidermal tendinal cells, a detailed study of which has been made by Howse et al. (1992) in *Acartia tonsa*.

2.6.2. Muscular System

The main musculature of the prosome consists of the paired dorsal and ventral longitudinal muscles. The dorsal muscles consist of eight large fibres while the ventral ones consist of three fibres according to Park (1966) who reviews earlier literature. The urosome also has paired dorsal and ventral longitudinal muscles. Comparable descriptions of these muscles in *Euaugaptilus placitus* are given by Boxshall (1985). Further muscles, the extrinsic muscles of the appendages, are present within the body cavity and Park (1966) describes these in detail in *Epilabidocera amphitrites*, but gives no description of the intrinsic muscles of the appendages. Boxshall (1985), however, describes both the extrinsic and intrinsic muscles of the appendages of *Euaugaptilus placitus*. Less detailed descriptions of the main prosomal musculature, including the extrinsic muscles of the swimming legs, are given by Svetlichny (1988).

2.6.3. Nervous System

The nervous system consists of a brain, anterior to the oesophagus, which connects to the ventral nerve cord by two large circumoesophageal cords. The brain consists of protocerebral, deutocerebral and tritocerebral lobes.

The ventral nerve cord, posterior to the oesophagus, forms a fused trunk that extends to the posterior end of the prosome where it divides into a dorsal and ventral cord in the urosome, the dorsal cord dividing again into right and left halves that extend through the urosome to the caudal furcae. Lowe (1935) describes the nervous system of *Calanus finmarchicus* in detail, including the brain, sympathetic system, and the giant fibre system. Park's (1966) descriptions of the systems in *Epilabidocera amphitrites* are closely similar with small variations in detailed structure.

2.6.4. Circulatory System

The circulatory system of calanoid copepods is very simple, consisting of a muscular heart, an anterior aorta and a system of sinuses. The heart is located below the dorsal body wall of the second and third metasome segments. Lowe (1935) describes the heart, aorta and system of sinuses in *Calanus finmarchicus*. Park (1966) states that the heart and circulatory system of *Epilabidocera amphitrites* varies from that of *Calanus finmarchicus* in the number of ostia, the form of the pericardium and the arrangement of muscles in the wall of the heart. It has a single slit-like ostium as compared with the three venous ostia in the latter species. This results in a different pattern of blood flow from that described in *C. finmarchicus*. Descriptions of the ultrastructure of the hearts of *Anomalocera ornata* and *Pareuchaeta norvegica* have shown that they are similar in the two species (Howse *et al.*, 1975; Myklebust *et al.*, 1977).

2.6.5. Digestive System

The mouth is formed anteriorly by the labrum and posteriorly by the labium, both with toothed walls. The labrum contains the eight labral glands that discharge their contents through eight openings into the buccal cavity. Dorsal to the buccal cavity is the oesophagus, very short in many species, e.g. Figure 14B–D, and lined with chitin. Its wall has longitudinal folds, longitudinal muscles and strong circular muscles and can be dilated by muscles extending from it to various points on the body wall. The greater part of the alimentary canal of calanoid copepods is formed by the midgut which is divided into three regions (Figure 14, 1–3) as first described by Lowe (1935) and Park (1966). The midgut connects with the hindgut, lined with chitin, and its walls have longitudinal folds like those of the oesophagus.

The proportions of the three regions of the midgut vary, region 3 sometimes being considerably lengthened as in the mesopelagic species,

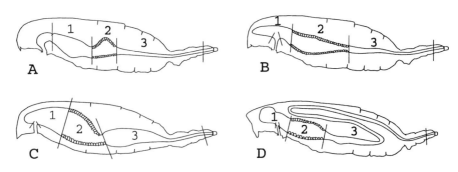

Figure 14 Diagrammatic structure of the alimentary canals of A, *Calanus minor*, B, *Calanus helgolandicus*, C, *Centropages typicus*, and D, *Lophothrix frontalis*. Mid-gut regions 1–3 are shown. The oesophagus is anterior, in A, or ventral, in B to D, to region 1. The hindgut is restricted to the posterior part of the urosome. (A to C, after Arnaud *et al.*, 1980; D, after Nishida *et al.*, 1991.)

Lophothrix frontalis (Figure 14D). Boxshall (1985) briefly described the alimentary canal of *Euaugaptilus placitus*; he divides the midgut into four regions but later revised (Boxshall, 1992) this division to three regions. It is difficult from his descriptions to decide whether the gut of *E. placitus* is peculiar or whether its apparent structure arises from rupturing of the vacuolar cells along a length of the wall. The midgut of calanoids is muscular with the circular muscles external to the longitudinal. Region 1 extends from the oesophagus to approximately the posterior end of the cephalosome, region 2 from the posterior region of the cephalosome to the first or second metasomal segment. Region 3 extends from there to join the hindgut in the posterior part of the urosome. Any extension of region 1 anterior to the oesophagus (Figure 14B–D) has been referred to as the midgut diverticulum. Ultrastructural studies by Arnaud *et al.* (1980), Nishida *et al.* (1991) and the review by Brunet *et al.* (1994) recognize several types of cells in the gut lining, R-, D-, F- and B-, and N-cells. The cellular structure of the three regions of the midgut is:

Region 1. Lined with a uniform layer of non-vacuolated, large microvillar cells with a smooth or rough endoplasmic reticulum (R- and F-cells).

Region 2. Glandular region lined with three types of cells. Large, columnar cells, each containing a large vacuole of secretion, the cytoplasm and nucleus being restricted to the base of the cell. These are B-cells and have narrow dense cells, D-cells, of unknown function between them. The third type of cell, F-cells, is columnar or cubical, fewer in number, and with a rough endoplasmic reticulum and phagosomes, a smaller oval nucleus occupying the centre of the cell.

Region 3. Cells are cuboidal and non-vacuolated like region 1 (R-cells) but undergo degeneration to become necrotic cells (N-cells).

Arnaud *et al.* (1980) conclude that R-cells are absorptive, that B-cells have both absorptive and excretory functions, while F-cells secrete digestive enzymes. Region 3 of the midgut of *Calanus helgolandicus*, according to Nott *et al.* (1985), has a valve in the fourth metasome segment. These authors described cyclical changes in the gut wall between feeding and non-feeding periods. Intracellular digestion of the gut contents takes place in the B-cells of region 2. When digestion is completed the cells burst into the lumen. The cuboidal cells of the posterior part of region 3, beyond the valve in metasome segment three, form a continuous lining when the copepod is not feeding. These cells start to disintegrate when the copepod feeds so that the membranes of the disintegrated cells become the peritrophic membrane of Gauld (1957a), binding the gut contents into a faecal pellet (Brunet *et al.*, 1994).

Baldacci *et al.* (1985b) examine the development of the gut wall in the copepodid stages of *Calanus helgolandicus*. There is seasonality in the development of the glandular areas of the midgut, it being much reduced in overwintering (resting) *Calanus* spp. according to Hallberg and Hirche (1980).

2.6.6. Reproductive System

2.6.6.1. Female

Females have a single median ovary located dorsal to the gut in the posterior part of the cephalosome and the first metasomal segment (Figure 15). Paired oviducts, often with large diverticula, originate antero-laterally and extend forwards, as diverticula, into the anterior cephalosome and backwards, on either side of the gut, to the urosome. The oviducts open into a single chamber, the genital antrum (or atrium) although in a few species such as *Scolecithrix danae* they open separately to the exterior (Blades-Eckelbarger, 1991a). The genital antrum has paired dorso-laterally directed pouches (Figure 15C), the spermathecae in which the spermatozoa are stored. According to Cuoc *et al.* (1989a,b), the fresh water diaptomids, *Hemidiaptomus ingens provinciae* and *Mixodiaptomus kupelwieseri*, do not have internal spermathecae but a temporary external seminal receptacle is formed during egg laying. The overall form of the female genital system is similar in all calanoid species (Hilton, 1931; Sömme, 1934; Lowe, 1935;

Marshall and Orr, 1955; Andrews, 1966; Park, 1966; Corkett and McLaren, 1978; Tande and Grønvik, 1983; Blades-Eckelbarger and Youngbluth, 1984; Batchelder, 1986; Razouls et al., 1986, 1987; Norrbin, 1994).

The posterior end of the ovary, the germinal site (Figure 15), contains oogonia undergoing mitotic division to become oocytes. The new oocytes occupy the region immediately anterior to the germinal site where their nuclei undergo changes (Hilton, 1931; Park, 1966). The oocytes then begin to grow in size and occupy the anterior region of the ovary and much of the oviducts except for the main tracts where the fully developed oocytes reside. As Park (1966) points out, there is no clear boundary between the ovary and the oviducts. Oogenesis has been studied ultrastructurally in *Labidocera aestiva* by Blades-Eckelbarger and Youngbluth (1984) who point out that the only previous studies are by light-microscope on *Eucalanus elong-*

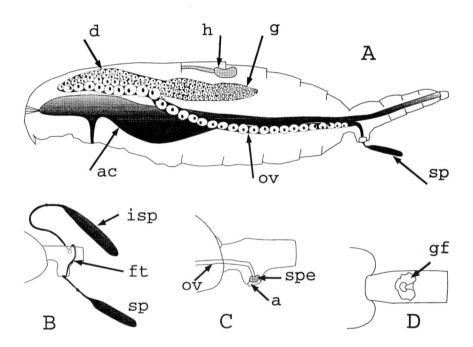

Figure 15 The female reproductive system of a calanoid copepod. A, Relative positions of the ovary and oviducts, alimentary canal and heart. B, The genital somite showing a directly and an indirectly placed spermatophore. C, Lateral view of genital somite. D, Ventral view of genital somite.
a, genital antrum; ac, alimentary canal; d, diverticula of oviducts; ft, fertilization tube; g, germinal site of ovary; gf, genital field; h, heart and aorta; isp, indirectly placed spermatophore; ov, left oviduct; sp, directly placed spermatophore; spe, left spermatheca.

atus by Heberer (1930), *Calanus finmarchicus* by Hilton (1931) and *Epilabidocera amphitrites* by Park (1966). There appear to be some differences in the mechanisms of vitellogenesis in the different species. Yolk formation is initiated by the accumulation of mitochondria in the perinuclear region of *Calanus finmarchicus* and *Epilabidocera amphitrites* and the yolk granules are formed in close association with the mitochondria. Blades-Eckelbarger and Youngbluth (1984), however, found no evidence of implication of mitochondria in yolk formation in *Labidocera aestiva*.

The ripe oocytes in the oviducts are covered by a very thin membrane, the perivitelline membrane. The process of fertilization is not clear (Park, 1966). The spermathecae open into the genital antrum and it is assumed that the eggs are fertilized as they pass through the antrum. The posterior regions of the oviducts, however, are glandular and are presumably responsible for investing the eggs with their outer membranes and the material that forms the "egg sac", in species that carry their eggs attached to the urosome. Thus, this material would be associated with the eggs before they pass into the antrum. Park (1966) describes large, paired glands in the posterior lateral regions of the metasome that have ducts that pass into the urosome and open into the oviducts before the genital antrum in *Epilabidocera amphitrites*.

The spermatozoa are transferred to the female in a spermatophore that usually discharges into the genital antrum and so into the spermathecae. Spermatophores are normally attached to the genital field (Figure 15D), and are known as direct placements. In some species, however, spermatophores can be indirectly placed on parts of the genital somite remote from the genital field. Most of these fail to discharge but some have a fertilization tube (Figure 15B) through which the spermatozoa discharge into the spermathecae. Direct and indirect placements will be discussed more fully later.

Marshall and Orr (1955) noted that a few CV of *Calanus finmarchicus* have a well-developed ovary with diverticula but, normally, the ovary is very small in this stage, with no diverticula and oviducts that are very thin and empty. Examining the same species in northern Norway, Tande and Hopkins (1981) found an undifferentiated gonad with paired, thin and empty gonoducts in the CIV; the right gonoduct of potential males had degenerated in the CV. Corkett and McLaren (1978) detected developing ovaries as early as the CIII in *Pseudocalanus* sp. Undifferentiated gonadal cells occur in the CI of *Calanus helgolandicus* according to Baldacci *et al.* (1985a). S. Razouls (1975) and Razouls *et al.* (1987), examining *Centropages typicus* and *Temora stylifera*, and Norrbin (1994), studying *Pseudocalanus acuspes* and *Acartia longiremis*, found first traces of the gonads in the CI but sexual differentiation did not occur until the CIV when the right gonoduct of potential males degenerates. It is probable that the early stages of development of the gonads takes place in most species in the early copepodid stages.

2.6.6.2. Male

There is a single testis in the male and only the left genital duct is developed (Figure 16). The rudimentary gonad in the early copepodids has paired gonoducts but the right one degenerates in the CIV in potential males (see above). The testis is located medially in the dorsal region of the first and second metasome segments, ventral to the heart in the same position as the ovary in Figure 15; Hopkins (1978) states that the testis of *Pareuchaeta norvegica* is in the anterior part of the cephalosome. Formation of the spermatophore within the male has been described in detail in a variety of species (Heberer, 1932; Park, 1966; Raymont *et al.*, 1974; Hopkins, 1978; Blades and Youngbluth, 1981). The posterior end of the testis contains the primary spermatocytes (Blades-Eckelbarger and Youngbluth, 1982) and spermatids are incorporated into large accessory cells in the anterior end of the testis before being released into the central lumen that merges with the *vas deferens*.

The genital duct comprises five identifiable parts (Marshall and Orr, 1955), but it can have two forms depending on whether a simple or complex spermatophore (Figure 16C and D) is produced. It emerges from the anterior region of the testis and proceeds posteriorly as the anterior *vas deferens* (Figure 16A, C) on the left side of the cephalosome. It is glandular, thick-walled, coiled or twisted and its narrow lumen contains a strand of spermatozoa and the seminal fluid, or core secretion, that is produced by its walls. The posterior part of the *vas deferens* also produces seminal fluid but, in addition, contributes material to begin formation of the wall of the developing spermatophore. The *vas deferens* is longer in, for example, pontellid species (Figure 16B) than in *Pareuchaeta* species (Figure 16A). The *vas deferens* leads into the thin-walled, less glandular, coiled seminal vesicle that contains a mass of spermatozoa and secretion and whose wall may secrete further material for the wall of the spermatophore; this part of the duct is a storage site (Blades and Youngbluth, 1981). A muscular sphincter separates the seminal vesicle from the highly glandular spermatophore sac. The proximal region of the spermatophore sac is called the former. The former is small in species with simple spermatophores and secretes the adhesive onto the neck of the spermatophore (Marshall and Orr, 1955; Hopkins, 1978). In species of the families Centropagidae and Pontellidae with complex coupling plates, the former is large (Figure 16B) and secretes seven or eight different secretions that are shaped by the template of the lumen into the coupling plates (Blades-Eckelbarger, 1991a). The newly formed spermatophore resides in the spermatophore sac until

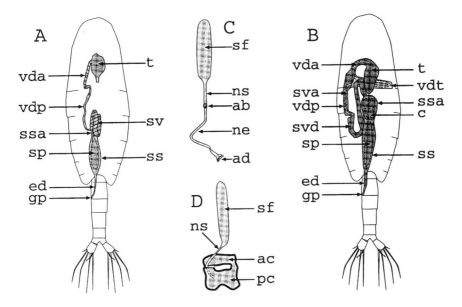

Figure 16 The reproductive system of a male copepod. A, male system producing a simple spermatophore. B, male system producing a complex spermatophore. C, simple spermatophore. D, complex spermatophore.
ab, adhesive body; ac, anterior coupling plate; ad, attachment disc; c, developing coupler; ed, *ductus ejaculatorius*; gp, gonopore; ne, extended neck; ns, short neck; pc, posterior coupling plate; sf, spermatophore flask; sp, developing spermatophore; ss, posterior part of spermatophore sac; ssa, anterior spermatophore sac or former; sv, seminal vesicle; sva, ascending seminal vesicle; svd, descending seminal vesicle; t, testis; vda, anterior *vas deferens*; vdp, posterior *vas deferens*; vdt, transverse *vas deferens*. (A and C after Hopkins, 1978; B and D, after Blades-Eckelbarger, 1991a.)

mating when it passes to the muscular *ductus ejaculatorius*, lined with chitin, immediately next to the genital opening.

The spermatophore contains the spermatozoa and secretions. The spermatozoa at the proximal end of the spermatophore are suspended in a secretion that allows them to resist swelling while those at the distal end are suspended in a different secretion and, once the spermatophore is transferred to the female, swell up forcing the proximal spermatozoa out into the female's spermathecae. The refractive index of these two secretions is different. The spermatozoa are not motile and have no flagella. The only available ultrastructural study of spermatogenesis in a calanoid copepod is that of Blades-Eckelbarger and Youngbluth (1982) on *Labidocera aestiva*. They discuss the energy source that allows the spermatozoa, first, to be stored in the male and, then, in the spermathecae of the female. The functions of the accessory cells of the anterior region of the testis are considered, especially their role as phagocytes in controlling the numbers of spermatids entering the *vas deferens*. Control is necessary because the developed spermatophore is stored by males in the spermatophore sac and there is no requirement for further spermatids until after mating.

Simple spermatophores (Figure 16C) consist of a flask-like container with a thin neck for attachment to the female. When this emerges from the male's gonoduct, it adheres to the male's left fifth leg by means of the adhesive body at the end of the short neck (Figure 16C). After attachment to the female, the extended neck (Figure 16C) is formed by extrusion of some of the contents of the spermatophore (Hopkins, 1978). The end of the extended neck forms the attachment disc (Figure 16C) which adheres to the genital somite of the female. A complex spermatophore, on the other hand, also has a flask-like container but the short neck has coupling plates, known as the coupler, and these are used to attach the spermatophore to the female. Most calanoids have simple spermatophores, complex ones occurring primarily in the Centropagidae and Pontellidae. The form of the coupler is unique to each species and fits the morphology of the female's genital somite like a "key and lock" (Fleminger, 1967; Lee, 1972). Detailed descriptions of the couplers in five species of *Centropages* are given by Lee (1972) and in many species of *Labidocera* by Fleminger (1967, 1975, 1979) and Fleminger and Moore (1977). Simple spermatophores of all species are closely similar in form although differences in the detailed shapes do exist (Miller, 1988).

The spermatophore of *Pseudocalanus* sp. contains about 350 spermatozoa (Corkett and McLaren, 1978). The spermatozoa are non-motile, oval in form and those of *Calanus hyperboreus* measure 6.5 μm in length (Conover, 1967) while those of *Pareuchaeta norvegica* are 16–20 μm long and 8–10 μm in diameter (Hopkins, 1978).

2.6.7. Excretory System

The excretory system of copepods consists of the maxillary glands which have been described by Lowe (1935) in *Calanus finmarchicus* and Park (1966) in *Epilabidocera amphitrites*. No antennary glands were found in the adults of either species although they were reported in a nauplius stage of *C. finmarchicus* by Grobben (1881). The maxillary glands are present in the lateral sinuses at the bases of the maxillae. They consist of an end-sac, a secretory tubule and an excretory duct that opens to the outside on the inner side of the basal segment of the maxilla. Lowe and Park describe the histology of the glands.

2.6.8. Oil Sac

Many calanoids store lipids in an oil sac that can extend throughout the dorsal half of the prosome in *Calanus finmarchicus*. Its position and shape vary between species (Ikeda, 1974). The sac can extend throughout the prosome as in *Calanus glacialis* or can be small and restricted to a single metasome segment as in *Eucalanus bungii* and *Paracalanus parvus*. The sac is different in *Euchaeta marina* where it lies between the midgut and ventral nerve cord in the metasome segments. The membrane of the sac is thin-walled and the size of the single sac varies seasonally (Sargent and

Table 8 The haploid number of chromosomes (n) found in different species of calanoid copepods. Authorities given in superscript.

Species	n	Species	n
Acartia centrura	6[2]	Labidocera euchaeta	10[6]
A. clausi	15[1]	L. kröyeri	10[3]
A. gravelyi	5[3]	L. laevidentata	10[6]
A. kerallensis	5[3]	L. madurae	10[6]
A. margalefi	15[1]	L. minuta	11[6]
A. negligens	6[2]	L. pavo	10[3]
A. plumosa	6[2]	L. pectinata	10[3]
A. southwelli	6[4]	L. pseudacuta	10[6]
A. spinicauda	6[2]	Labidocera sp.	10[6]
Calanopia aurivilli	11[9]	Neocalanus robustior	17[12]
C. elliptica	11[9]	Paracalanus aculeatus	6[7]
C. minor	11[9]	Pontella princeps	10[8]
Calanus finmarchicus	17[10]	Pontellina plumata	11[9]
C. helgolandicus	17 [10, 11]	Pontellopsis herdmani	10[8]
Centropages furcatus	3[3]	Pseudocalanus elongatus	16[14]
Eucalanus attenuatus	10[5]	P. minutus	16[14]
E. crassus	10[5]	Pseudodiaptomus aurivilli	6[4]
E. elongatus	10[5]	P. serricaudatus	11[4]
E. monachus	10[5]	Rhincalanus cornutus	10[5]
E. mucronatus	10[5]	R. nasutus	10[5]
E. subcrassus	10[5]	Temora discaudata	7[4]
Eurytemora affinis	10[13]	T. stylifera	7[4]
Labidocera acuta	10[3]	T. turbinata	3[3]
L. acutifrons	10[6]	Tortanus barbatus	6[3]
L. bataviae	10[6]	T. forcipatus	6[3]
L. bengalensis	10[6]	T. gracilis	6[3]
L. discaudata	10[4]		

[1]Alcaraz, 1976; Goswami and Goswami, [2]1973; [3]1974; [4]1978; [5]1979a; [6]1979b; [7]1982; [8]1984; [9]1985; [10]Harding, 1963; [11]Mullin, 1968; [12]Tsytsugina, 1974; [13]Vaas and Pesch, 1984; [14]Woods, 1969.

Henderson, 1986). The ultrastructure of the sac in *E. marina* and the storage sites of *Pleuromamma xiphias* are described by Blades-Eckelbarger (1991b) and Blades-Eckelbarger and Youngbluth (1991). Calanoids, such as *Pleuromamma xiphias*, have an additional storage site surrounding the anterior region of the midgut. The oil sac in *Calanus euxinus* represents 25 to 32% of the body volume (Vinogradov *et al.*, 1992), in *C. finmarchicus* about 20% (Plourde and Runge, 1993) and at its maximum in *Metridia pacifica* 20% (Hirakawa and Imamura, 1993).

2.6.9. Chromosomes

The haploid numbers of chromosomes in calanoid copepods range from 3 to 17 (Table 8). The number can vary within a genus. A few irregularities have been noted. Woods (1969) found two eggs of *Pseudocalanus* sp. with an apparent haploid number of 17 instead of 16 chromosomes. Harding and Marshall (1955) found large triploid nauplii of *Calanus finmarchicus* that hatched from larger-than-normal eggs. These larger eggs are relatively rare in the broods of *C. finmarchicus* and normally do not develop beyond the gastrula stage. Chromosomal aberrations were found in *Paracalanus aculeatus* coincidental with a solar eclipse (Goswami and Goswami, 1982). These authors suggested the aberrations may be connected with the high levels of ultraviolet and other electromagnetic rays present at that time.

3. Taxonomy and Identification

3.1. Phylogeny of Calanoid Copepods.	49
3.2. Identification of Genera and Species.	57
3.2.1. The Key to Genera.	57
3.2.2. Species Identification.	60
3.2.3. Pore Signatures.	61
3.2.4. Molecular Genetics.	63
3.3. Key to Genera of Platycopioid and Calanoid Copepods.	65

The first calanoid copepod to be named was *Monoculus finmarchicus* by Johan Ernst Gunnerus, Bishop of Trondheim, Norway nearly 250 years ago (Marshall and Orr, 1955). It is now known as *Calanus finmarchicus* (Gunnerus, 1765). The second species was named in 1792 as *Cyclops longicornis*, now referred to as *Temora longicornis* (O.F. Müller, 1792). Three more species were named in the 1830s: *Cyclops marina*, now *Euchaeta marina* (Prestandrea, 1833), *Anomalocera patersoni* (Templeton, 1837) and *Calanus hyperboreus* (Kröyer, 1838). By the end of the 1870s, 105 species, about 2.5 each year, had been named (Figure 17). The next 60 years, to the end of the 1930s, resulted in nearly 13 new species each year while from 1950 through to the end of the 1980s just over 20 new species were named each year. Will many more new calanoid species be discovered? Part of the decrease in numbers of species named in the 1980s may have resulted from the decrease in popularity of geographic and taxonomic studies and their replacement by studies of "rates and processes". The current decade may continue the decreasing trend since an upsurge would require exploration of new environments. There are certainly new species to be discovered in the pelagic and benthopelagic environments of the deep sea. Some coastal areas, such as Oceania, may yield more, certainly hyperbenthic species. It is probable, however, that the bulk of species of pelagic calanoid copepods are now known.

3.1. PHYLOGENY OF CALANOID COPEPODS

The platycopioid copepods consist of a single family, the Platycopiidae that contains 10 species divided between 3 genera (Table 9). The calanoid

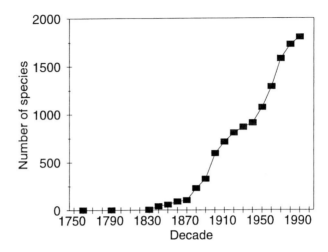

Figure 17 The numbers of new species of platycopioid and calanoid copepods named in each decade.

copepods, however, have 1800 marine species divided between 11 superfamilies. These superfamilies are very unequal in size (Table 9), the largest and most diverse being the Clausocalanoidea containing 40% of the species. The smallest is the monospecific Fosshagenioidea created by Suárez-Morales and Iliffe (1996) for the cave-dwelling species *Fosshagenia ferrarii*. This superfamily is closely related to the Centropagoidea and, according to the above authors, should be inserted between that superfamily and the Arietelloidea in Figure 18. The Calanoida are considered a monophyletic group (Park, 1986) and the superfamilies are distinguished from each other by morphological characters such as the numbers of spines, setae and segments on the swimming legs, structure of the antennules and antennae, and the form of the male's P5 (Figure 18).

Another approach to the phylogeny of copepods is through examination of the pore signatures of individual species. The integumental setal system and glands are distributed over the surface of the integument in species-specific patterns, the species pore signature. Corresponding generic, familial and superfamilial signatures were derived from the pore signatures of 249 species ascribed to 89 genera representing 28 families by Mauchline (1988b). This classification (Figure 19) corresponds to that derived from the other morphological characters (Figure 18). Björnberg (1972, 1986a,b) states that, in addition to the adult characters used to examine the phylogeny of copepods, the larval characters must also be taken into account. Not enough information on the nauplii and copepodids of a large enough variety of species is available to do this at present. Ferrari (1988) studied the

TAXONOMY AND IDENTIFICATION

Table 9 The composition of the Platycopioida and the superfamilies of the Calanoida. The number of marine species in each superfamily is shown in parenthesis, in each family and genus in superscript.

Superfamily	Constituent families and genera
Platycopioida (10)	Platycopiidae[10] *Antrisocopia*[1], *Nanocopia*[1], *Platycopia*[8]
Calanoida	
Epacterisciodea (4)	Epacteriscidae[4] *Enantiosus*[1], *Epacteriscus*[1], *Erobonectes*[2]
Pseudocyclopoidea (52)	Boholiniidae[2] *Boholina*[2] Pseudocyclopidae[30] *Pseudocyclops*[30] Ridgewayiidae[20] *Brattstromia*[1], *Exumella*[3], *Placocalanus*[5], *Ridgewayia*[11]
Arietelloidea (338)	Arietellidae[45] *Arietellus*[12], *Campaneria*[1], *Crassarietellus*[2], *Metacalanus*[4], *Paramisophria*[13], *Paraugaptiloides*[1], *Paraugaptilus*[7], *Pilarella*[1], *Rhapidophorus*[1], *Sarsarietellus*[2], *Scutogerulus*[1] Augaptilidae[131] *Augaptilina*[1], *Augaptilus*[7], *Centraugaptilus*[6], *Euaugaptilus*[73], *Haloptilus*[27], *Heteroptilus*[3], *Pachyptilus*[4], *Pontoptilus*[6], *Pseudaugaptilus*[3], *Pseudhaloptilus*[1] Discoidae[27] *Disco*[21], *Paradisco*[4], *Prodisco*[2] Heterorhabdidae[46] *Alrhabdus*[1], *Disseta*[5], *Hemirhabdus*[4], *Heterorhabdus*[30], *Heterostylites*[2], *Mesorhabdus*[3], *Microdisseta*[1] Hyperbionychidae[1] *Hyperbionyx*[1] Lucicutiidae[44] *Lucicutia*[44] Metridinidae[36] *Gaussia*[3], *Metridia*[23], *Pleuromamma*[10] Phyllopodidae[8] *Phyllopus*[8]
Fosshagenioidea (1)	Fosshageniidae *Fosshagenia*[1]
Centropagoidea (460)	Acartiidae[81] *Acartia*[78], *Paralabidocera*[3] Candaciidae[35] *Candacia*[31], *Paracandacia*[4]

Table 9 Continued

Superfamily	Constituent families and genera

Centropagoidea (continued)	Centropagidae[49]
	  Boeckella[FW], Calamoecia[FW], Centropages[32],
	  Gippslandia[1], Gladioferens[5], Hemiboeckella[FW],
	  Isias[3], Limnocalanus[3], Neoboeckella[FW],
	  Pseudoboeckella[FW], Pseudolovenula[FW],
	  Sinocalanus[5]
	Diaptomidae[FW]
	Parapontellidae[2]
	  Neopontella[1], Parapontella[1]
	Pontellidae[142]
	  Anomalocera[3], Calanopia[13], Epilabidocera[2],
	  Ivellopsis[1], Labidocera[47], Pontella[47],
	  Pontellina[4], Pontellopsis[25]
	Pseudodiaptomidae[80]
	  Archidiaptomus[1], Calanipeda[1],
	  Pseudodiaptomus[78]
	Sulcanidae[1]
	  Sulcanus[1]
	Temoridae[42]
	  Epischura[FW], Eurytemora[27], Ganchosia[1],
	  Heterocope[4], Lahmeyeria[1], Manaia[1], Temora[5],
	  Temoropia[3]
	Tortanidae[28]
	  Tortanus[28]

Megacalanoidea (135)	Calanidae[35]
	  Calanoides[6], Calanus[17], Canthocalanus[1],
	  Cosmocalanus[2], Mesocalanus[2], Neocalanus[6],
	  Undinula[1]
	Mecynoceridae[1]
	  Mecynocera[1]
	Megacalanidae[14]
	  Bathycalanus[8], Bradycalanus[5], Megacalanus[1]
	Paracalanidae[85]
	  Acrocalanus[6], Bestiolina[5], Calocalanus[48],
	  Delius[2], Ischnocalanus[4], Paracalanus[15],
	  Parvocalanus[5]

Bathypontioidea (22)	Bathypontiidae[22]
	  Alloiopodus[1], Bathypontia[15], Temorites[2],
	  Zenkevitchiella[4]

Eucalanoidea (24)	Eucalanidae[24] Eucalanus[20], Rhincalanus[4]
Ryocalanoidea (5)	Ryocalanidae[5] Ryocalanus[5]

Spinocalanoidea (47)	Spinocalanidae[47]
	  Damkaeria[1], Foxtonia[1], Isaacsicalanus[1],
	  Kunihulsea[1], Mimocalanus[10], Monacilla[4],

TAXONOMY AND IDENTIFICATION 53

Table 9 Continued

Superfamily	Constituent families and genera
Spinocalanoidea (continued)	Rhinomaxillaris[1], Sognocalanus[1], Spinocalanus[24], Teneriforma[3]
Clausocalanoidea (712)	Aetideidae[232] Aetideopsis[15], Aetideus[10], Azygokeras[1], Batheuchaeta[10], Bradyetes[4], Bradyidius[18], Chiridiella[18], Chiridius[11], Chirundina[3], Chirundinella[1], Comantenna[4], Crassantenna[2], Euchirella[26], Gaetanus[21], Gaidiopsis[1], Gaidius[12], Jaschnovia[2], Lutamator[2], Mesocomantenna[1], Paivella[2], Paracomantenna[3], Pseudeuchaeta[6], Pseudochirella[35], Pseudotharybis[6], Pterochirella[1], Senecella[2], Sursamucro[1], Undeuchaeta[6], Valdiviella[7], Wilsonidius[1] Clausocalanidae[37] Clausocalanus[15], Ctenocalanus[4], Drepanopus[4], Farrania[4], Microcalanus[2], Pseudocalanus[7], Spicipes[1] Diaixidae[12] Anawekia[3], Diaixis[9] Euchaetidae[105] Euchaeta[16], Pareuchaeta[89] Mesaiokeratidae[5] Mesaiokeras[5] Parkiidae[1] Parkius[1] Phaennidae[81] Brachycalanus[5], Cephalophanes[3], Cornucalanus[8], Onchocalanus[10], Phaenna[2], Talacalanus[2], Xantharus[1], Xanthocalanus[50] Pseudocyclopiidae[10] Paracyclopia[7], Pseudocyclopia[2], Stygocyclopia[1] Scolecitrichidae[167] Amallophora[3], Amallothrix[30], Archescolecithrix[1], Heteramalla[1], Landrumius[5], Lophothrix[7], Macandrewella[8], Mixtocalanus[3], Parascaphocalanus[1], Pseudophaenna[1], Puchinia[1], Racovitzanus[6], Scaphocalanus[34], Scolecithricella[38], Scolecithrix[4], Scolecocalanus[3], Scopalatum[5], Scottocalanus[15], Undinothrix[1] Stephidae[29] Miostephos[2], Parastephos[3], Stephos[24] Tharybidae[33] Neoscolecithrix[6], Parundinella[4], Rythabis[1], Tharybis[12], Undinella[10]

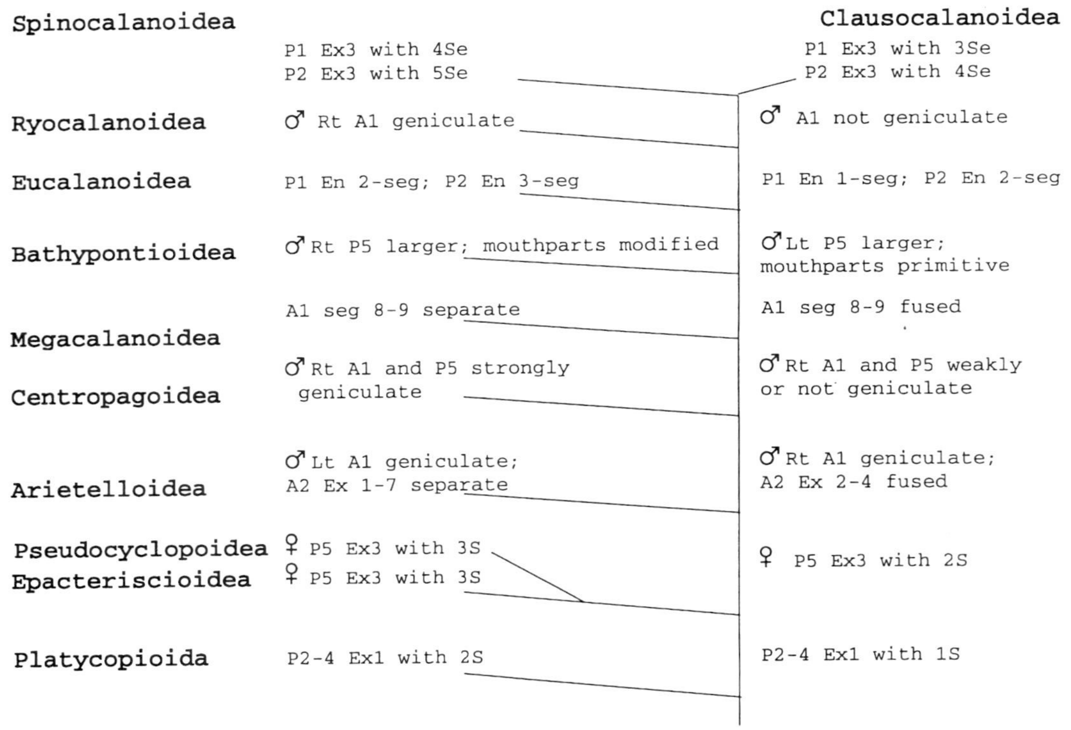

Spinocalanoidea		Clausocalanoidea
P1 Ex3 with 4Se		P1 Ex3 with 3Se
P2 Ex3 with 5Se		P2 Ex3 with 4Se

Ryocalanoidea   ♂ Rt A1 geniculate   ♂ A1 not geniculate

Eucalanoidea   P1 En 2-seg; P2 En 3-seg   P1 En 1-seg; P2 En 2-seg

Bathypontioidea   ♂ Rt P5 larger; mouthparts modified   ♂ Lt P5 larger; mouthparts primitive

   A1 seg 8-9 separate   A1 seg 8-9 fused

Megacalanoidea

Centropagoidea   ♂ Rt A1 and P5 strongly geniculate   ♂ Rt A1 and P5 weakly or not geniculate

Arietelloidea   ♂ Lt A1 geniculate; A2 Ex 1-7 separate   ♂ Rt A1 geniculate; A2 Ex 2-4 fused

Pseudocyclopoidea   ♀ P5 Ex3 with 3S

Epacterisciodea   ♀ P5 Ex3 with 3S   ♀ P5 Ex3 with 2S

Platycopioida   P2-4 Ex1 with 2S   P2-4 Ex1 with 1S

Figure 18 The phylogenetic relationships of the Platycopioida and the superfamilies of the calanoid copepods. The new monospecific superfamily Fosshagenioidea should probably be inserted between the Arietelloidea and Centropagoidea (Suárez-Morales and Iliffe, 1996).
A1, antennule; A2, antenna; En, endopod; Ex, exopod; Lt, left; P1–5, swimming legs 1–5; Rt, right; S, outer lateral spine; Se, seta; seg, segment. (After Andronov, 1974; Park, 1986.)

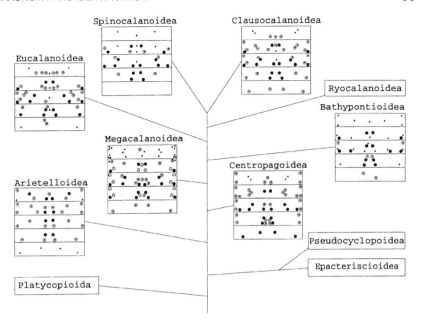

Figure 19 The pore signatures of the superfamilies of calanoid copepods arranged in the monophyletic format of Park (1986). No signatures are available for the Platycopioida, Epacteriscioidea, Pseudocyclopoidea and Ryocalanoidea. Small and large solid dots represent small and large pores respectively that occur in all species examined; ringed single dots are pores that occur in more than 50% of species examined; ringed groups of dots in Megacalanoidea are plates of pores. Further explanation in the text. (After Mauchline, 1988b.)

developmental patterns of segmentation of the swimming legs of calanoids relative to those of other orders of copepods. Within the calanoids, familial patterns of development exist but there are species and genera within families that do not conform. More information on legs and other appendages of copepodids is required, but in an agreed format, before the full significance of this approach to phylogeny can be assessed.

The phylogenetic relationships within individual genera have been examined in a variety of ways. Ferrari (1991) used patterns of development of the swimming legs and demonstrated two monophyletic groups within 10 species of the genus *Labidocera* and also among 14 genera of the Diaptomidae. The results of Ferrari's study on *Labidocera* differ from those of Fleminger (1967) and Fleminger *et al.* (1982) who used a combination of adult secondary sexual characters and geographic distribution. Matthews (1972) examined 35 characters in adults of the genera *Augaptilus*,

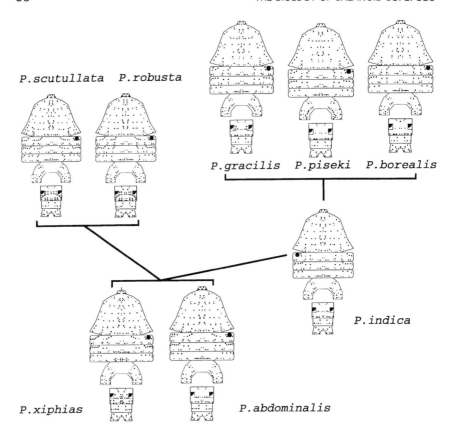

Figure 20 The pore signatures of eight *Pleuromamma* species divide into four groups, the most primitive consisting of *P. xiphias* and *P. abdominalis*. (After Park and Mauchline, 1994.)

Euaugaptilus, *Haloptilus* and *Pseudaugaptilus* and determined the similarity coefficients between species. This approach detects groupings of species but the data require a cladistic analysis to look at potential phylogenetic relationships within the genera. The species pore signatures can also be used to determine phylogenetic relationships as was done by Park and Mauchline (1994) for eight species of the genus *Pleuromamma* (Figure 20). A ninth species, *P. quadrungulata*, was not examined but, on the basis of other morphological characters, is probably associated with *P. scutullata* and *P. robusta*.

3.2. IDENTIFICATION OF GENERA AND SPECIES

The identification of individual species of calanoid copepods is relatively easy in many geographical regions where the diversity of species is low, such as estuarine and coastal waters. Regional keys are often available (Table 10) as a starting-point. Identification to the species level in offshore, and especially deep-sea, samples in which 200 or more species may be present is more difficult. The samples have to be sorted methodically as described in Chapter 1. Individuals belonging to many genera and species can be removed very quickly owing to their peculiar morphology but inevitably a residue of individuals remains that has to be processed through use of a key to genera, and finally, through reference to descriptions of species in the literature.

Zinntae Zo (1982) gives an example of a sequential key that uses alphanumeric words. It compresses the format used in the present key and can be adapted for computer programming. Automated systems for identification of species may be developed in the future. The image analyser may function in low-diversity samples or may be used to provide digitized information for generation of diffraction patterns of copepods (Zavala-Hamz *et al.*, 1996).

3.2.1. The Key to Genera

The 1810 marine species of platycopioid and calanoid copepods are ascribed to more than 198 genera. A full species list is given in the next chapter. This has resulted from the lists given by C. Razouls (1982, 1991, 1993, 1995) supplemented by the author's own files. Key papers on individual genera are referred to where available. C. Razouls provides information on modern synonymy within the Calanoida although some proposals are not accepted by all workers.

There are some geographical regional keys to species within a genus but very few keys that identify all species within any one genus. The species composition of several genera is not clear, as, for instance, that of the scolecitrichid genera *Amallothrix*, *Scaphocalanus*, *Scolecithricella* and *Scolecithrix* reviewed by Bradford *et al.* (1983) and detailed in the next chapter.

The generic key presented here is based on that of Bradford (1972) which recognizes 135 genera and, in turn, is based on that of Rose (1933) which identifies 100 genera. The present key recognizes some 200 genera but does not include genera that are restricted to fresh water environments. Detailed examination of the key demonstrates some problems within individual genera. Females and males of a genus usually key out at different points in

Table 10 Regional taxonomic works providing listings, descriptions, or sources of descriptions of species.

Region	References
Arctic Ocean	Brodsky, 1950; Brodsky *et al.*, 1983
Atlantic Ocean	
northeast	Giesbrecht, 1892; Sars, 1903, 1925; Rose 1933; Farran, 1948, 1951; Vervoort, 1952; Corral Estrada, 1972a,b
northwest	
Woods Hole	Wilson, 1932
Florida Current	Owre and Foyo, 1967
Mediterranean	Giesbrecht, 1892; Neunes, 1965; Razouls and Durand, 1991
east	
West Africa	Vervoort, 1963, 1965
South Africa	Carola, 1994
southwest	Björnberg, 1981; Guglielmo and Ianora, 1995
Antarctic Ocean	Vervoort (1951); papers by T. Park; C. Razouls, 1992, 1994
Pacific Ocean	
north	Brodsky, 1950; Gardner and Szabo, 1982; Brodsky *et al.*, 1983
northwest	
Japan	Mori, 1937; papers by O. Tanaka
China	Chen and Zhang, 1965; Zheng Zhong *et al.*, 1989
central	Wilson, 1950; Vervoort, 1964
Australia	Farran, 1936; Dakin and Colefax, 1940; papers of I.A.E. Bayly
New Zealand	Papers by J.M. Bradford; Bradford Grieve, 1994
East Indies	Vervoort, 1946
Malay Archipelago	Scott, 1909
Indian Ocean	Sewell, 1929, 1932, 1947

the key. Some 14 genera, however, have their females and/or males keying out at three or more points in the key. A flow diagram for the key (Figure 21) illustrates its structure. Distinctive genera and species are removed first. The number of segments present in the endopods of P3 and P4 divide the genera into three unequal groups. Endopods with one or different numbers

Figure 21 The general structure of the generic key showing morphological features used, in their approximate order of priority, starting at the top of the figure. Abbreviations are given in Table 11 (page 65).

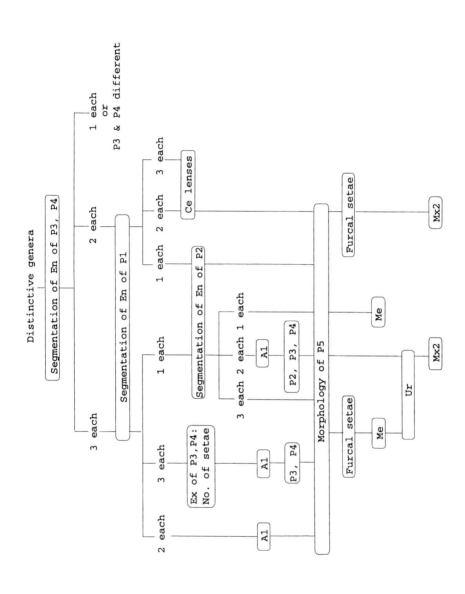

of segments on each side occur in 5 genera while those with two segments on each side occur in 24 genera. The remaining 170 or so genera have three-segmented endopods on the P3 and P4. Of these, 18 of them have a two-, 60 a three-, and over 80 a one-segmented endopod on the P1. The presence of a geniculate antennule identifies males, and whether it is on the left or right side separates the genera. The detailed form of the P5 in both females and males is one of the most distinctive characters of calanoid copepods used to separate the genera. The relative lengths of the furcal setae, numbers of segments in the metasome and urosome, and the structure of the maxilla are secondary characters identifying further genera.

3.2.2. Species Identification

Identification of species in samples from coastal regions is much easier than dealing with oceanic and deep sea samples where much greater diversity exists. General references for different geographical regions are given in Table 10. There has been considerable revision of genera and species. The genus *Calanus*, for example, one of the commonest genera in coastal and oceanic regions, has been revised several times, species being re-examined and resulting in the relatively newly recognized *C. agulhensis* De Decker *et al.* (1991), *C. euxinus* Hülsemann (1991) and *C. jaschnovi* Hülsemann (1994). Frost (1989) completely revised species in the genus *Pseudocalanus*. A re-examination of many other genera and species will probably result in splitting of some existing species and the recognition of new species. One genus in which this is very probable is *Acartia*. Garmew *et al.* (1994) describe geographical variation in the asymmetrically distributed spines on the prosome of *A. tonsa* in Chesapeake Bay and the coast of Peru; they did not create any new species. The problem is how much morphological difference is required before a new species is created. Some morphological characters carry more weight than others e.g. segmentation of the legs and armature of the mouthparts as opposed to variation in numbers of integumental spines. The coastal *A. clausi* from Woods Hole, Massachusetts is larger in size than the same species from the Pacific coast of Oregon but morphologically similar; Carillo *et al.* (1974) failed to cross-breed individuals from the two populations in the laboratory, suggesting that they should have specific status. Another coastal species, *Eurytemora affinis*, is very variable and Busch and Brenning (1992) studied its morphology in detail and concluded that *E. hirundoides* and *E. hirundo* should be regarded as synonyms.

The number of species within each genus is very variable (Figure 22). More than 120 genera have between 1 and 5 species and very few have more than 30 species; they are named in Figure 22. The largest genus is

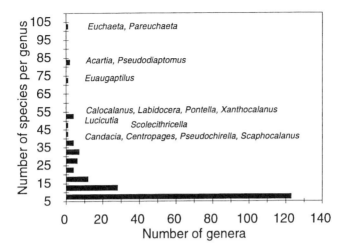

Figure 22   The number of genera with different numbers of species per genus. Those with over 30 species are named. *Euchaeta/Pareuchaeta* is treated as one genus.

Euchaeta/Pareuchaeta with 102 species but Park (1994a) has divided the species between these two genera and then grouped them within the genera. The genus *Acartia* also has groups of species within it, proposed as subgenera by Steuer (1923). Access to important taxonomic literature on each genus is given in the next chapter.

3.2.3. Pore Signatures

The distribution of integumental glands and sensilla over the integument of calanoid copepods can be used to identify species (Fleminger, 1973; Fleminger and Hülsemann, 1977, 1987; Campaner, 1984; Hülsemann and Fleminger, 1990; Hülsemann, 1991a, 1994; Park and Mauchline, 1994). The copepods are treated with hot potassium hydroxide, which digests the internal soft tissues of the body, including the ducts of glands and neural connections of sensilla that pass through the integument. Subsequent examination of the integument, stained with chlorazol black, allows mapping of the distributions of the pores or perforations left where the gland ducts and neural connections have been removed by the digestion. The pattern of distribution of these pores over the integument is unique to each species. This is illustrated in Figure 20 in which the signatures of *Pleuromamma* species separate them from each other and into phylogenetic groups.

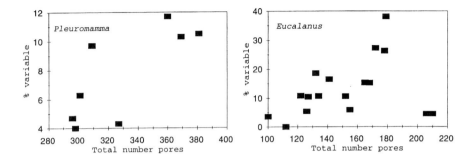

Figure 23 The percentages of the total numbers of pores that are variable in *Pleuromamma* and *Eucalanus* species. Data from J.S. Park (1995b) and Fleminger (1973).

The amount of phylogenetic information contained in the pore signature of a species is large. The signatures of the genus and family are contained within that of the species (Mauchline, 1988b). Defining the entire pore signature of a species destroys the body of the copepod and is exacting work. Hülsemann (1994) distinguishes adult females of 14 species of the genus *Calanus* on the basis of the pore signatures of the urosomal segments alone. The urosomal pore signatures of males and females of a species are different, as instanced by those of *Pleuromamma* species investigated by J.S. Park (1995a). No sexual differences were found in the cephalosomal signatures and only minor differences existed, in some species, in the metasomal signatures of the *Pleuromamma* species. Thus, signatures can be used to match males with females. Signatures of copepodids can also be used to identify them to species (J.S. Park, 1995a,b).

Fleminger (1973) examined the total numbers of pores in the integument of *Eucalanus subtenuis* at 60° intervals of longitude across the Atlantic, Indian and Pacific Oceans and demonstrated a progressive change. There is variation between individuals of a species in the total numbers of pores present. J.S. Park (1996) re-examined the variation found by Fleminger and compared it with the variation in the signatures that she found in the genus *Pleuromamma*. An unknown portion of this results from incomplete digestion of the soft tissues at the site of the pore so that the pore is not detected under the light microscope; this proportion is estimated as a few percent of the total pores present. The percentage variation in the total numbers of pores found in *Pleuromamma* and *Eucalanus* species is shown in Figure 23. The variation increases as the total number of pores increases, with exceptions. One exception among the *Pleuromamma* species is *P. abdominalis* with a pore number of 327. Two exceptions among the

Eucalanus species are *E. bungii* and *E. californicus*, both with 208 pores. Fleminger (1973) compared species within the species groups of the genus *Eucalanus* and concluded that the species within each group that had the most distinctive signature, relative to the others, also had the greatest geographic distribution. Re-working of his data shows that these species also have the greatest percentage of pores that are variable in their occurrence.

As mentioned above, the amount of information contained within the pore signatures is very large and is not as yet fully understood. There is no relationship between total pore number and body size within a genus (Fleminger, 1973). The pore signature consists of relatively non-subjective data relative to much morphological data that are used to compare species and subspecies. J.S. Park (1996) assesses the error involved in determining the pore signature of a species and finds that samples as small as five individuals will identify >95% of potential sites of pores. These data along with those derived from molecular genetics have obvious uses in solving many of the taxonomic difficulties at present existing within certain families of the Calanoida.

In some genera, identification of the females to species level is made principally on the shape of the genital somite and the morphology of the genital field e.g. *Euchaeta*, *Pareuchaeta* and *Valdiviella* as detailed by Zvereva (1975, 1976) and Tanaka and Omori (1968) while that of males derives from the morphology of the modified fifth legs. This contributes to a recurring problem throughout the Calanoida, the matching of the respective female and male of a species. Many species are only known from the female, a few from the male. Sex ratios are often very much in favour of females, the males often being of very rare occurrence in samples.

3.2.4. Molecular Genetics

Studies of protein and enzyme polymorphisms by electrophoretic techniques are very few in calanoid copepods. Manwell *et al.* (1967) were the first to demonstrate the value of these techniques by showing that they could clearly distinguish the three common northeastern Atlantic species *Calanus finmarchicus*, *C. helgolandicus* and *C. hyperboreus* from one another. The small body size of calanoids hindered further work but Sevigny and Odense (1985) used new techniques to detect 17 enzymes and resolve the isoenzyme systems of seven of these in the small copepods *Eurytemora herdmani* and *Temora longicornis*. A similar analytical procedure was used by Sevigny and McLaren (1988) to show that six *Calanus* species divided into three groups that corresponded with previous groupings based on morphological characters. Sevigny *et al.* (1989) extended these studies to *Pseudocalanus* species and demonstrated that the new and revised species in this genus (Frost, 1989) were clearly separated from each other by patterns of allozyme variation at the glucose phosphate isomerase locus; genome sizes were different (McLaren, 1989d). More recently, Bucklin and LaJeunesse (1994) and Bucklin *et al.* (1992, 1995) have studied six *Calanus* and three *Metridia* species, determining intra- and interspecific patterns in DNA sequences. Polymorphic populations of *Acartia* species were subjected to electrophoretic analysis and the presence of two species confirmed (McKinnon *et al.*, 1992). Cervelli *et al.* (1995) examined nine gene-enzyme systems in *Acartia* species in the Lagoon of Venice. They found clear genetic differentiation between *A. clausi* and *A. margalefi* but the latter had higher genetic variability. They suggest that a third species, similar morphologically to *A. margalefi*, may be present. Suzuki *et al.* (1995) extracted DNA from single individuals of 11 species of copepods and found that fragments of polymerase chain reaction (PCR) amplified 28S r DNA can be used to identify species.

The techniques can also be applied to study within-species variation. The variation in DNA sequence of *Calanus minor*, a common species extending from the Florida Straits, throughout the Gulf Stream and Sargasso Sea to the northeast Atlantic, distinguished a large and small form of this species which are genetically distinct and have different geographical distributions (Bucklin *et al.*, 1996).

In addition to the distinction of forms of a species, that may be at the subspecies level, considerable interest attends the quantification of the amount of gene flow between populations of a species over its geographical range. Bucklin and Marcus (1985) examine genetical variation within *Labidocera aestiva* across its geographical range in coastal waters of the eastern United States. Populations were examined at Woods Hole, Massachusetts, at Beaufort, North Carolina, and at Fort Pierce Inlet, Florida. Highly significant heterozygote deficiencies occurred between the populations indicating the possibility of restricted gene flow throughout the geographical range and/or differential selection within the populations at the three locations.

Finally, the techniques can be used to examine the degree of homogeneity within a population at a single geographical location to obtain a measure of the degree of isolation from or communication between populations in adjacent regions. Kann and Wishner (1996) examined the genetic structure of populations of *Calanus finmarchicus* in the Gulf of Maine and concluded that extensive gene flow takes place among the populations. Evidence of immigration to the Gulf was present in populations at its northern and southern limits.

Table 11 Abbreviations used in the generic key. The morphology of the adult and its appendages are shown in Figures 4 to 6.

A1	antennule
A2	antenna
B1	basipod segment 1 or coxa
B2	basipod segment 2 or basis
Ce	cephalosome
En	endopod
En1, 2, etc	endopod segments 1, 2, etc.
Ex	exopod
Ex1, 2, etc.	exopod segments 1, 2, etc.
Gn	genital somite of female
Md	mandible
Me	metasome
Me1, 2, etc.	metasome segments 1, 2, etc.
Mx1	maxillule
Mx2	maxilla
Mxp	maxilliped
P1, 2, etc.	paired swimming legs 1, 2, etc.
Pe	pedigerous segments
Pr	prosome
S	seta
Se1, 2, etc.	lateral spines on exopod of swimming leg
St	terminal spine on exopod of swimming leg
Ur	urosome

3.3. KEY TO GENERA OF PLATYCOPIOID AND CALANOID COPEPODS

This key is based on that of Bradford (1972) with a few modifications and many new genera added. The Figures 1 to 336 referred to within the key are those illustrated in the text Figures 24 to 32. The abbreviations used throughout the key are expanded in Table 11. The choices in the key are sequentially numbered, the number in parentheses referring to the number of the previous choice.

Key to Genera

(All figures referred to in the key below are "Key-figures" shown in Figures 24 to 32)

1.   Ce with lenses (Figs. 1–5)...2
       Ce without lenses..6

2(1).		Four lenses; two on each side of Ce; lenses round, sometimes difficult to see. Ce with lateral hooks (Fig. 3)................................ ....*Anomalocera* ♀ ♂ Two rounded lenses, sometimes difficult to see. Ce with or without lateral hooks (Fig. 4)...3
3(2).		En of P1 with 2 segments (some have 3 segments)..................... ....*Labidocera* ♀ ♂ En of P1 with 3 segments..4
4(3).		Rostrum massive (Fig. 5)..*Ivellopsis* ♀ ♂ Rostrum not particularly large..5
5(4).		Male P5 uniramous, 3 segmented, asymmetrical, right leg without chela...*Epilabidocera* ♂ Male right P5 with stout chela with sharp-pointed finger and thumb, left leg uniramous..*Pontella* ♂
6(1).		Mx2 and Mxp with long clusters of slender setae arranged like brushes (Fig. 6)...*Augaptilina* ♀ ♂ Mx2 and Mxp otherwise armed...7
7(6).		A1 short, equal to length of Ce, proximal segment expanded with large flattened process (Fig. 7).................*Placocalanus* ♀ ♂ A1 without these features...8
8(7).		Dorsal posterior of Ce with bilaterally symmetrical prominence (Fig. 8)...*Pterochirella* ♂ Dorsal Ce without this feature..9
9(8).		En of P3 and P4 with 3 segments...12 En of P3 and P4 with 2 segments...238 En of P3 and P4 with 1 segment...11 En of P3 with 2 segments, of P4 with 3 segments..*Foxtonia* ♀ ♂ En of P3 with 1 segment, of P4 with 2 segments.....................10
10(9).		P5 absent..*Disco longus* ♀ P5 present.......................................*Temorites discoveryae* ♀ ♂
11(9).		Ex of P1 has 3 segments.......................................*Heterocope* ♀ ♂ Ex of P1 has 2 segments..*Ganchosia* ♀ *Manaia* ♀ Ex of P1 has 1 segment. Mx2 armed with normal setae............. ....*Spicipes* ♀ Ex of P1 has 1 segment. Mx2 armed with strong curved setae (Figs. 13,14)..*Chiridiella* ♀ ♂
12(9).		En1 of P3 and P4 is very short (Fig. 9)..................*Sulcanus* ♀ ♂ En of P3 and P4 have segments of more or less equal size (Figs. 10–12)...13
13(12).		En of P1 with 3 segments..14 En of P1 with 2 segments..89 En of P1 with 1 segment...105

14(13).	Dark brown knob on right or left anterior lateral side of Me1 (Figs. 15,16)..*Pleuromamma* ♀ ♂ Knob absent..15
15(14).	En1 of P2 with an internal hook (Fig. 17)................................16 En1 of P2 without an internal hook...17
16(15).	Female Ur narrow (Fig. 18); anal segment of female and male without extensions (Fig. 18)....................................*Metridia* ♀ ♂ Female Ur expanded (Fig. 19); anal segment of female and male with backward directed process on each side (Fig. 19)...*Gaussia* ♀ ♂
17(15).	Ex3 of P3 and P4 with 2S, one of which is terminal, and an St (Fig. 10)...18 Ex3 of P3 and P4 with 3S and an St (Fig. 11).........................29 Ex3 of P3 and P4 terminates in a sharp point (Fig. 12)..*Archidiaptomus* ♀ ♂
18(17).	Left A1 of females and males has 27 segments........*Erebonectes* ♀ ♂ Left A1 of females and males has 25 or fewer segments........19
19(18).	St of Ex3 of P3 and P4 flattened externally with a smooth margin, not toothed (Fig. 10)..24 St of Ex3 of P3 and P4 strongly toothed or serrated externally (Fig. 20)...20
20(19). 21.	St of Ex3 of P3 and P4 strongly toothed (Fig. 20). St of Ex3 of P3 and P4 serrated (Fig. 21). Female P5 as in Fig. 22. Male P5 as in Fig. 23..*Gippslandia* ♀ ♂
21(20).	P5 biramous on both sides in male and female. En of male P5 have 3 segments..22 Female P5 uniramous on both sides, 3 or 4 segmented. Male P5 biramous on both sides, En rudimentary or absent, Ex of 2 or 3 segments...*Pseudodiaptomus* ♀ ♂
22(21).	Caudal rami with spinules..............................*Limnocalanus* ♀ ♂ Caudal rami without spinules..23
23(22).	Brackish and fresh water species primarily endemic to eastern Asia...*Sinocalanus* ♀ ♂ Brackish and fresh water species endemic to Australia and New Zealand...*Gladioferens* ♀ ♂
24(19).	B1 of P5, inner edge with small teeth (Fig. 24).......................25 B1 of P5, inner edge naked..26
25(24).	B2 of P2 and P3 with posterior spines on surface (Fig. 25)..*Cosmocalanus* ♀ Left P5 prehensile (Fig. 26)..............................*Cosmocalanus* ♂ B2 of P2 and P3 naked; P5 not prehensile..............*Calanus* ♀ ♂
26(24).	Ex2 of P2 in females and males with outer proximal edge

	evaginate (Fig. 27). Left P5 of male prehensile (Fig. 28)..*Undinula* ♀ ♂
	These characters absent..27
27(26).	B2 of P1 with distally directed seta on anterior surface modified into a proximally thickened spine (Fig. 29). *Canthocalanus* ♀ ♂
	This seta of normal plumose type (Fig. 30)..............................28
28(27).	En of right P5 with 8 setae (Fig. 31)..................*Neocalanus* ♀ ♂
	En of right P5 with 7 setae (Fig. 32)................*Mesocalanus* ♀ ♂
	En of P5 in female with 6 setae, of right P5 in male with no more than 6 setae..*Calanoides* ♀ ♂
29(17).	A1 only as long as the Ce..30
	A1, at least the left, at least as long as the second Pe............31
30(29).	Ex of female P5 of 3 segments; En of 1–3 segments (Fig. 33). P5 of male strongly chelate; En blade-shaped; left En has terminal spines (Fig. 34)..*Pseudocyclops* ♀ ♂
	In female, both Ex of P5 have 3 segments (Fig. 35); in male, left Ex has 2 segments (Fig. 36)...........................*Epacteriscus* ♀ ♂
	Ex of female P5 not, or not clearly, articulated (Fig. 37); En reduced, with one terminal seta. P5 of male with or without one-segmented En; Ex2 expanded (Figs. 38,39)...*Paramisophria* ♀ ♂
	Rhapidophorus wilsoni ♂
31(29).	Neither A1 modified as a prehensile organ.............................32
	One A1 modified as a prehensile organ, more or less obviously (Figs. 40,41)..65
32(31).	P5 has Ex and En on both sides..35
	P5 of different structure..33
33(32).	P5 biramous on at least one side...34
	P5 uniramous on both sides (Figs. 48–50)..............................60
34(33).	P5 bifurcate on each side as in Fig. 45...............*Sarsarietellus* ♀
	Ex of P5 with 2 segments, En with one segment or absent on one side and represented by a seta (Figs. 46,47).......*Scutogerulus* ♀
35(32).	Ex and En of P5 with 2 segments (Fig. 42)...........................36
	Ex and En of P5 with 3 segments (Figs. 33,43)......................37
	Ex of P5 with 3, En with 2 segments....................................46
	Ex of P5 with 3, En with 1 segment (Fig. 44,44a)..................49
36(35).	Body elongated...........................*Euaugaptilus hecticus* ♀
	Body robust..............................*Centraugaptilus pyramidalis* ♀
37(35).	Ex2 of P5 with a spine on the internal margin of form in Fig. 43..*Centropages* ♀
	This margin has a sabre-like spine or rudimentary seta (Figs. 51,52)..39
	Ex2 and Ex3 of P5 of form in Fig. 53...................................38

TAXONOMY AND IDENTIFICATION 69

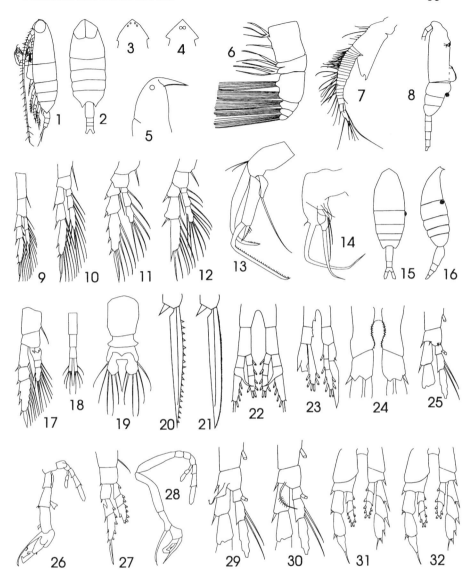

Figure 24 Key to genera of platycopioid and calanoid copepods, Key figs. 1–32.

38(37).	Ur 3-segmented..*Exumella* ♀	
	Ur 4-segmented...*Brattstromia* ♀	
39(37).	P5, internal margin of B2 with long seta; En1 and En2 with lobate sensory filaments (Fig. 54)............................*Alrhabdus* ♀	
	P5, these margins otherwise armed..40	
40(39).	On left furca one seta is much stronger and longer than the others...41	
	Furcal setae are symmetrical..52	
41(40).	Mx2 without strong hook setae (Fig. 55); one right furcal seta is much longer than the others..42	
	Mx2 with strong hook seta (Figs. 56,57); right furcal setae almost equal...43	
42(41).	Ur has 3 segments...*Enantiosus* ♀	
	Ur has 4 segments...*Disseta* ♀	
	Microdisseta ♀	
43(41).	Mx2 short and thickset, terminated by 2 very spiny, strong hook setae (Fig. 56). Mxp slender and elongate.........*Hemirhabdus* ♀	
	Hook setae of Mx2 not spiny. Mxp shorter (Figs. 57, 58).......44	
44(43).	Mx2 with more than 2 hook setae; its proximal lobes very reduced (Fig. 57); Md asymmetrical with one tooth well apart from the others, much larger and sickle-shaped (Figs. 59,60)..*Heterorhabdus* ♀	
	Mx2 with only 2 hook setae (Figs. 58,61); teeth on Md subequal..45	
45(44).	Proximal lobes of Mx2 and their setae little developed (Fig. 57)..*Heterostylites* ♀	
	These lobes and setae well developed (Fig. 61)......................... ....*Mesorhabdus* ♀	
46(35).	Furca four times as long as wide (Fig. 62)............... *Lucicutia* ♀	
	Furca at most two times as long as wide; often much shorter (Fig. 63)..47	
47(46).	Ur of 4 segments...48	
	Ur of 3 segments; body small (less than 1 mm total length), anchialine species..*Boholina* ♀	
48(47).	Rostrum absent. Total body length 4 mm or more.................... ....*Pontoptilus* ♀	
	Rostrum present. Total body length 1 mm or less..................... ....*Ridgewayia* ♀	
49(35).	Ex2 of P5 with one internal spine (Fig. 44)....................*Isias* ♀	
	Ex2 of P5 without internal spine..50	
50(49).	Rostrum present..51	
	Rostrum absent. Md very asymmetrical; one of them with a very strong lateral curved tooth (Figs. 65, 66). Body elongate............	

	....*Heteroptilus* ♀
51(50).	Rostrum bifurcate, distinct. Body very robust. Md greatly enlarged at the tip (Fig. 64).................................. *Pachyptilus* ♀
	Pseudhaloptilus longimanus ♀
	Rostrum of paired filaments. Body robust. Md symmetrical with 3 acute teeth (Fig. 66a)....................................*Crassarietellus* ♀
52(40).	Me4 and Me5 separated..53
	Me4 and Me5 fused...55
53(52).	Rostral filaments slender; Ex of P1 with 1, 1, 2 S (Fig. 67).....54
	Rostral filaments stout and sausage-like; Ex of P1 with 0, 0, 1 S (Fig. 57); second inner lobe of Mx2 without setae...................... ....*Bathycalanus* ♀ ♂
54(53).	B2 of P1 with hook (Fig. 69)............................*Megacalanus* ♀ ♂
	B2 of P1 without hook..*Bradycalanus* ♀ ♂
55(52).	Ur with 3 segments...56
	Ur with 4 segments...58
56(55).	Mx2 reduced to slim 3-segmented appendage (Fig. 70)...*Augaptilus* ♀
	Mx2 well developed...57
57(56).	Md with numerous teeth, often cut obliquely (Figs. 71,72). Rostrum most often small or absent, sometimes in form of two long thin filaments..*Euaugaptilus* ♀
	Md very elongated, with 2 thin curved teeth, and one smaller tooth between them (Fig. 73). Rostrum strong, bifurcated, projecting forward (Figs. 74,75). Setae on Mxp coiled, carrying special shields (Fig. 76)..*Centraugaptilus* ♀
58(55).	Ex2 of P5 with a long, unarmed, sabre-like seta on the internal margin. En3 of P5 with 5 setae (Fig. 51)...................*Lucicutia* ♀
	Ex2 of P5 without sabre-like seta on the internal margin.......59
59(58).	En3 of P5 with 4 setae (Fig. 52). Md a long rod, with fine teeth at its tip. Rostrum in the form of 2 long slender filaments..*Pseudaugaptilus* ♀
	En3 of P5 with 6 setae (Fig. 77)................................*Haloptilus* ♀
60(33).	P5 has B1 and B2 and a 3-segmented Ex (Fig. 78)..................... ....*Phyllopus* ♀
	P5 has no more than 3 segments (Figs. 49,50,79–82).............. 61
61(60).	P5 has one apical seta and one lateral seta (Fig. 49)................. ....*Paraugaptilus* ♀
	P5 with 2 or 3 segments (Figs. 50,79–82)................................62
62(61).	P5 terminal segment narrow and pointed (Fig. 50)..................... ....*Arietellus* ♀
	P5 terminal segment not narrow and pointed..........................63
63(62).	P5 with 2 or 3 segments (Figs. 79,80)..64

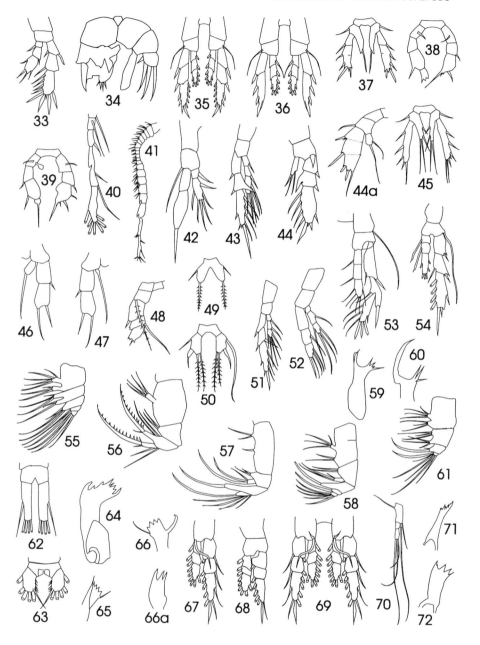

Figure 25 Key to genera of platycopioid and calanoid copepods, Key figs. 33–72.

	P5 of form in Fig. 81..*Hyperbionyx* ♀
	P5 of form in Fig. 82...*Paradisco* ♀
64(63).	A1 asymmetrical and reaching to end of prosome..................... ....*Metacalanus* ♀
	A1 asymmetrical and reaching to end of Ur............*Pilarella* ♀
65(31).	Right A1 prehensile..66
	Left A1 prehensile..70
66(65).	En of P5 with 3 segments on both sides (Figs. 83,84)..............68
	En of P5 is 3-segmented on right, one-segmented on left (Fig. 85)..*Brattstromia* ♂
	En of P5 rudimentary, one-segmented..67
	P5 very asymmetrical (Fig. 86)............................*Boholina* ♂
67(66).	P5 of form in Fig. 87..*Isias* ♂
	P5 of form in Fig. 88...*Ridgewayia* ♂
68(66).	Ex2 and Ex3 of right P5 unequal, forming a pincer (Fig. 83)...*Centropages* ♂
	Re3 of right P5 has spines forming a chela..........*Enantiosus* ♂
	These Ex segments of form in Fig. 89......................*Exumella* ♂
	These Ex segments nearly equal, without pincer....................69
69(68).	Ex2 of right P5 with internal spine (Fig. 84)......*Euaugaptilus* ♂
	Ex2 of right P5 without internal spine (Fig. 90)........ *Haloptilus chierchiae* ♂
70(65).	Ex and En of P5 with 3 segments on both sides (Figs. 91,92).... 71
	Ex of P5 with 3 segments, En rudimentary or absent (Figs. 93–95)..82
	Ex and En of left P5 with 3 segments; right P5 has Ex of 3 segments, En of 2 segments (Fig. 96).................. *Microdisseta* ♂
	Ex and En of left P5 with 3 segments; right P5 has Ex of 3 segments, En usually of 2 segments..........................*Lucicutia* ♂
71(70).	One seta on the left furca much longer and thicker than the others (Fig. 97)..72
	Furcal setae symmetrical, although one may be longer but not thicker than the others..76
72(71).	One seta on the right furca is much longer than the others (Fig. 97)..*Disseta* ♂
	Setae of right furca subequal...73
73(72).	B2 of right P5 with a long internal conical process, parallel to the En; Ex2 of right P5 very enlarged, both Ex3 segments terminated by a long point (Fig. 91)......................................*Heterostylites* ♂
	These features absent..74
74(73).	Ex3 of right P5 terminated by a long seta, straight and naked; Ex2 with a strong internal conical tooth (Fig. 92).......................

	....*Mesorhabdus* ♂ Ex3 of right P5 terminated by a little tooth or a claw (Figs. 98–100)..75
75(74).	Mx2 short and thick-set, with 2 strong hooked very spiny setae (Fig. 101). Ex of A2 clearly shorter than the En (Figs. 102)..*Hemirhabdus* ♂ Hooked setae on Mx2 not spiny. Ex of A2 hardly shorter than the En (Fig. 103)..*Heterorhabdus* ♂
76(71).	Terminal setae on the Mxp coiled, crook-shaped, with special shields (Fig. 76). Rostrum strong, bifurcated, and projecting forward (Figs. 74,75). Body thick-set...............*Centraugaptilus* ♂ These features absent..77
77(76).	Ex of P5 smooth, without spines or setae, except for a final point (Fig. 98)...*Pseudaugaptilus* ♂ Ex of P5 has external spines (Figs. 99,100)............................78
78(77).	En3 of P5 with 4 setae (Fig. 52); En2 without internal setae.79 En3 of P5 with 6 setae; En2 with one internal seta (Figs. 99,100)...80
79(78).	Md symmetrical, with numerous equal teeth (Fig. 104)..*Pontoptilus* ♂ Md asymmetrical, with one sickle-shaped tooth separated from the others (Fig. 66)..*Heteroptilus* ♂
80(78).	Right Ex2 of P5 without internal point (Fig. 99); never with shields on the Mxp setae...*Haloptilus* ♂ Right Ex2 of P5 with at least one internal point, sometimes small; numerous shields on the maxilliped setae (Fig. 100)...............81
81(80).	Mx2 reduced to a 3-segmented rod (Fig. 70).........*Augaptilus* ♂ Mx2 well developed...*Euaugaptilus* ♂
82(70).	En of P5 present on both sides (Figs. 94,95)...........................84 En of P5 on one side only (Figs. 105–107)..............................88 En of P5 absent on both sides (Figs. 93,108)...........................83
83(82).	P5 more or less symmetrical with 3 segments on each side (Fig. 93)..*Metacalanus* ♂ P5 asymmetrical, complex (Fig. 108)...................*Hyperbionyx* ♂
84(82).	En of left P5 with 1 segment (Figs. 94,95,110).........................85 En of left P5 with 2 segments (Fig. 109)..................................87
85(84).	En of left P5 bilobed (Fig. 94)................................*Arietellus* ♂ En of left P5 not bilobed..86
86(85).	En of right P5 with 1 segment (Fig. 95)............*Paraugaptilus* ♂ En of right P5 with 3 segments (Fig. 110)...............*Paradisco* ♂
87(84).	En of left P5 simple in form (Fig. 109)...............*Campaneria* ♂ En of left P5 distinctly 2-segmented, of form in Fig. 109a..*Paraugaptiloides* ♂

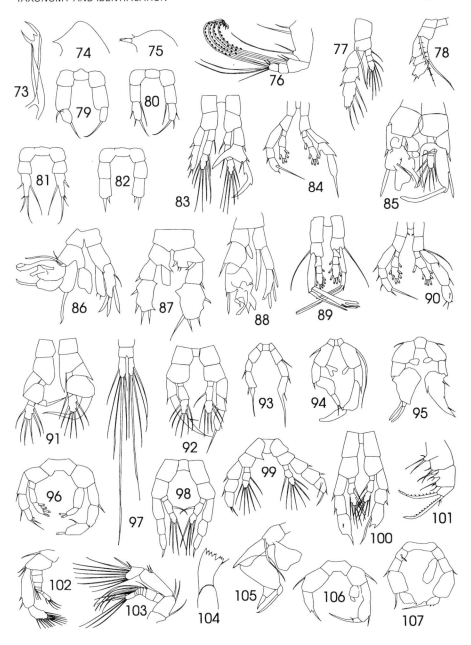

Figure 26 Key to genera of platycopioid and calanoid copepods, Key figs. 73–107.

88(82).	En of P5 large, in form of leaf; Ex segments greatly enlarged (Fig. 105)..*Phyllopus* ♂	
	En has 2 small segments; Ex segments little enlarged (Fig. 106)...*Sarsarietellus* ♂	
89(13).	A1 does not reach the hind margin of the Ce; first segment is very large (Fig. 111)...90	
	A1 longer than the Ce...91	
90(89).	Ur of 4 segments...*Platycopia* ♀ ♂	
	Ur of 5 segments...*Antrisocopia* ♀ ♂	
91(89).	P5 has En and Ex..92	
	P5 with one branch on each side, without plumose setae, or P5 entirely absent..96	
92(91).	Right En of P5 two-segmented, left three-segmented (Fig. 112)...*Lucicutia* ♂	
	Both En of P5 one- or two-segmented, reduced in size (Figs. 113,114), or absent...*Disco* ♂	
	Both En of P5 three-segmented...93	
	Both En of P5 one-segmented...95	
93(92).	Ur of 3 segments..*Euaugaptilus* ♀ ♂	
	Ur of more than 3 segments...94	
94(93).	Md very asymmetrical, right with large, curved tooth (Fig. 66)..*Heteroptilus* ♀ ♂	
	Md greatly enlarged at the tip, of form in Fig. 64... *Pachyptilus* ♀	
	Md symmetrical, similar to those in Figs. 71,72. .....*Lucicutia* ♀	
95(92).	Md greatly enlarged at the tip, of form in Fig. 64... *Pachyptilus* ♀	
	Md very asymmetrical, right with large, curved tooth (Fig. 66)..*Heteroptilus* ♀	
	Md normal, like Fig. 72; P5 as in Fig. 115................................... ....*Zenkevitchiella tridentae* ♀	
96(91).	Furca long and narrow, at least six times longer than wide (Fig. 116)..*Temora* ♀ ♂	
	Furca at most three times as long as wide.............................97	
97(96).	P3 and P4, surface of En naked...98	
	P3 and P4, surface of En with rows of small spines (Fig. 117)..101	
98(97).	P3 and P4, En2 with one seta, En3 with three setae 99	
	P3 and P4, En2 with one seta, En3 with 5 setae....................100	
	P3 and P4, En2 with two setae, En 3 with 7 setae.......*Temorites* ♀ ♂	
99(98).	Gn with posterior-lateral protrusions (Fig. 118)....................... .... *Prodisco* ♀ ♂	

	Gn without protrusions..*Disco* ♀
100(98).	Ex of P1 has 2 segments; Me segments have spines (Fig. 119). Female P5 present (Fig. 120); male P5 with left Ex (Fig. 121)..*Rhincalanus* ♀ ♂
	Ex of P1 has 3 segments; Me segments without spines. Female P5 absent; male P5 without En (Fig. 122)............*Eucalanus* ♀ ♂
	Ex of P1 has 3 segments; Me segments without spines. Female and male P5 uniramous, 3 segmented and modified as a grasping organ in males. Right antennule of female geniculate. Lives in caves..*Fosshagenia* ♀ ♂
101(97).	Ex3 of P2, P3 and P4 bordered externally with a row of small spines (Fig. 123)..102
	Ex3 of P2, P3 and P4 without these small spines..................103
102(101).	Female P5 reduced to knobs or lacking; only left P5 present in male..*Acrocalanus* ♀ ♂
	Female P5 has 2 or 3 segments (Fig. 124,125); P5 of male 2-segmented on right, 5-segmented on left (Fig. 126).................. ....*Paracalanus* ♀ ♂
103(101).	Gn swollen and projecting laterally (Fig. 127); male anal segment well-developed and swollen (Fig. 128)............. *Calocalanus* ♀ ♂
	Gn without lateral swellings but projecting ventrally (Fig. 129); anal segment of male as long as the two preceding segments (Fig. 130)..104
104(103).	P5 biramous in females and males (Figs. 131,132)...................... ....*Ischnocalanus* ♀ ♂
	Female has only left P5 present, of 2 segments (Fig. 133); male has only left P5 present, of 5 segments (Fig. 134).....*Delius* ♀ ♂
	Female P5 rudimentary (Fig. 135); male right P5 present (Fig. 136)..*Bestiolina* ♀ ♂
	P5 of female and male as in Figs. 137,138...........*Temoropia* ♀ ♂
105(13).	En of P2 with 3 segments..106
	En of P2 with 2 segments..108
	En of P2 with 1 segment. Mx2 with strong curved setae as in Figs. 13,14..*Chiridiella* ♀ ♂
	En of P2 with 1 segment. Mx2 without curved setae............ 218
106(105).	P5 biramous in females and males, although the one- segmented En may be absent on one side (Figs. 115,139)............................ ....*Zenkevitchiella* ♀ ♂
	P5 uniramous in females and males..107
107(106).	Female P5 has 2 segments as in Fig. 124. Male P5 as in Fig. 140..*Parvocalanus* ♀ ♂
	Female P5 has 5 segments (Fig. 141). Male P5 as in Fig. 142..*Mecynocera* ♀ ♂

	Female P5 has 4 segments (Fig. 131). Male P5 as in Fig. 132..*Ischnocalanus* ♀ ♂
108(105).	A1 short, does not reach the hind margin of the Ce; first segment very large (Fig. 111)..109
	A1 longer than the Ce..111
109(108).	P5 uniramous on each side (Figs. 143–146)..........................110
	P5 biramous on each side (Figs. 147,148).......... *Platycopia* ♀ ♂
110(109).	P5 as in Figs. 143 and 145...........................*Pseudocyclopia* ♀ ♂ *Stygocyclopia* ♀ ♂
	P5 as in Figs. 144 and 146.................................*Paracyclopia* ♀ ♂
111(108).	Mxp enormous, almost as long as the Pr (Fig. 149). Rostrum blunt and rounded...*Pseudeuchaeta* ♀ ♂
	Mxp not enormous..112
112(111).	Surfaces of the rami of P2,P3,P4, and especially the En of P3 and P4, have numbers of small spines (Fig. 150)..........................114
	Posterior surfaces of only P4 have small spines..................113
	Surfaces of these legs without small spines..................163
113(112).	Me5 has one or no posterior point (Fig. 151).....*Jaschnovia* ♀ ♂
	Me5 has two posterior points (Fig. 152), less prominent in males (Fig. 153)..*Neoscolecithrix* ♀ ♂
114(112).	Body wide, almost globular (Fig. 154)............*Phaenna* ♀ ♂
	Body elongated, elliptical..115
115(114).	P5 absent, occasionally rudimentary..116
	P5 present, sometimes small..126
116(115).	Ex3 of P2,P3,P4 with 5 internal setae................................117
	Ex3 of P2,P3,P4 with 4 internal setae................................118
117(116).	Rostrum present, bifurcated and thick (Figs. 155,156)................ ....*Monacilla* ♀
	Rostrum present, bifurcated and thin (Figs. 157,158)................. ....*Ryocalanus* ♀
	Rostrum absent (Fig. 159)................................*Spinocalanus* ♀
118(116).	Gn projecting ventrally (Fig. 160)......................................119
	Gn without such a projection..120
119(118).	Posterior corners of Me rounded......................*Scolecithrix* ♀
	Posterior corners of Me pointed, sometimes asymmetrical........ *Macandrewella* ♀
120(118).	A2 with Ex twice the length of the En (Figs. 291,292).........121
	A2 with Ex and En more or less equal in length (Fig. 260)....... 122
121(120).	A2 has Ex of 6 segments. Mxp has vermiform appendages on first segment...*Anawekia* ♀
	A2 has Ex of 7 segments. Mxp without vermiform appendages on first segment..*Diaixis* ♀

TAXONOMY AND IDENTIFICATION

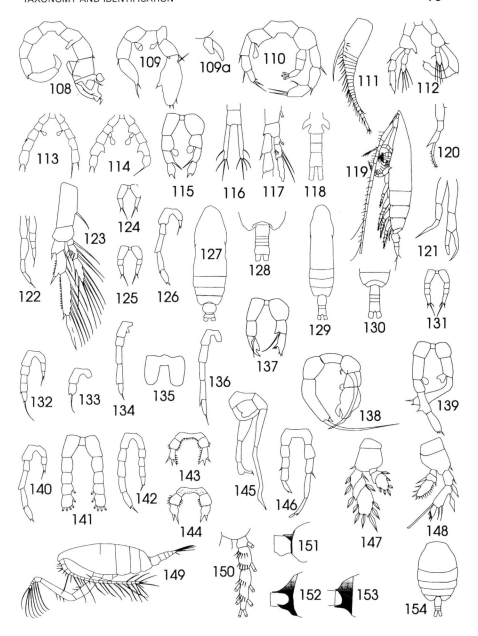

Figure 27 Key to genera of platycopioid and calanoid copepods, Key figs. 108–154.

122(120). Rostrum sausage-like or thick and wide, with or without 2 small points (Fig. 163). Mx2 with special terminal appendages (Fig. 162)...*Racovitzanus* ♀
Rostrum absent or small. Mx2 without special terminal appendages...123
123(122). Rostrum present...124
Rostrum absent...125
124(123). Rostrum small with 2 points, often divergent (Fig. 164)............
....*Bradyidius* ♀
Rostrum small with singly, bluntly rounded end........................
....*Sursamucro* ♀
125(123). Posterior borders of Ur segments fringed with fine spinules..*Azygokeras* ♀
These borders without fine spinules............... *Farrania orbus* ♀
126(115). Ur with 5 segments, often with very short anal segment. P5 asymmetrical with one or two branches on each side, often elongated...127
Ur with 4 segments. Leg 5 absent....................*Scaphocalanus* ♀
Ur with 4 segments. Only left P5 present, represented by a basal joint carrying a curved spine (Fig. 166)...........*Scolecocalanus* ♀
Ur of 4 segments, often with very small anal segment. P5 symmetrical with one branch on each side with 1–3 segments..146
127(126). Ex3 of P2,P3,P4 with 5 internal setae................................128
Ex3 of P2,P3,P4 with 4 internal setae................................131
128(127). P5 biramous on both sides...129
P5 uniramous on both sides, asymmetrical...........................130
129(128). Ex of P5 with 3 segments ending in points. Both En styliform (Fig. 167)...*Spinocalanus* ♂
Ex of P5 with 2 segments, not pointed. Left En lamellate (Fig. 168)..*Monacilla* ♂
130(128). P5 of form in Fig. 169..*Spinocalanus* ♂
P5 of form in Fig. 170 with vestigial En; or with no vestigial En and short right P5 of 5 segments, with a spine on outer edge of segment 3 and leg terminates in 2 unequal apical spines............
....*Ryocalanus* ♂
131(127). P5 biramous on one side only, En sometimes very small 132
P5 biramous on both sides (En sometimes very small).........134
P5 uniramous on both sides (may have vestigial endopodite on left leg) (Fig. 171)...*Bradyidius* ♂
Pseudotharybis dentatus ♂
P5 uniramous on both sides...136
132(131). Mx2 with worm-like appendages, without spiny hooked

	setae..133
	Mx2 with 2 large tufted appendages (Fig. 172)............................ ....*Mixtocalanus* ♂
133(132).	P5 long and thin, may or may not terminate on the left in a stylet (Fig. 174,175)...*Scolecithricella* ♂ P5 short and thick without stylet (Fig. 176)...........*Scolecithrix* ♂ P5 usually biramous on left; left leg with 3- segmented Ex shorter than the one-segmented En (Fig. 177)........................*Tharybis* ♂
134(131).	P5 of form in Fig. 178...*Xanthocalanus* ♂ P5 of form in Fig. 179.......................*Pseudotharybis robustus* ♂ P5 of other form...135
135(134).	P5 as in Fig. 180, with very reduced En................ *Talacalanus* ♂ P5 as in Fig. 181..*Xantharus* ♂ P5 with or without En, of form in Fig. 182..........*Drepanopus* ♂ P5 biramous on both sides, with more developed En (Figs. 183–189)..141
136(131).	Mx2 terminating in a strong claw, Mxp without claw or with 2 strong claws (Figs. 190–192)..137 Mx2 and Mxp without these claws, but may have a gently, curved long spine...138
137(136).	Mx2 ending in a strong claw, Mxp normal (Figs. 190, 192) ....*Onchocalanus* ♂ Mx2 ending in a strong claw and Mxp terminates in 2 claws (Figs. 190,191)..*Cornucalanus* ♂
138(136).	P5 robust with irregular segmentation and swellings (Fig. 193)..139 P5 otherwise formed and very asymmetrical (Figs. 194,195)...... 140
139(138).	Common basal segment of P5 with comb-like row of spines..... ....*Diaixis* ♂ Common basal segment of P5 without row of spines................ ....*Anawekia* ♂
140(138).	Right P5 short, of 3 segments (Fig. 194)........ *Scolecithricella* ♂ P5 of form in Figs. 194,195. Mx2 without claws but usually with sensory appendages...*Xanthocalanus* ♂
141(135).	P5 of form in Fig. 184...*Scopalatum* ♂ P5 of form in Fig. 196..*Scolecocalanus* ♂ P5 long and thin, right longer than left; both Ex 3-segmented, both En small and 1-segmented (Fig. 185)................*Farrania* ♂ P5 long and thin, left longer than right; right En with weak, terminal seta (Fig. 197)..*Azygokeras* ♂ P5 long and thin, its branches often terminating in points or stylets. Right En very short, left En of one segment, usually but

	not always shorter than the Ex (Fig. 183)............*Amallothrix* ♂
	Archescolecithrix ♂
	P5 otherwise formed...142
142(141).	En of left P5 shorter than Ex..143
	En of left P5 much longer than the Ex (Figs. 188,189).........145
143(142).	Rostrum present..144
	Rostrum absent..*Parascaphocalanus* ♂
144(143).	Right P5 has the Ex3 usually bifurcate and held at right angles to Re2 (Fig. 187)..*Macandrewella* ♂
	Right P5 does not have a bifurcate Ex3 (Fig. 186), although similar in configuration to that of the previous genus................. ....*Scottocalanus* ♂
145(142).	Left P5 En longer than Ex; right P5 En usually reaches Ex2, while Ex3 is long and blade-like (Figs. 188,189). Mx2 with 3 worm-like and 5 shorter brush-form appendages (Fig. 198)....... ....*Scaphocalanus* ♂
	Lophothrix ♂
	Left P5 En with 1 segment, right with one or two segments (Fig. 199)...*Racovitzanus* ♂
	Left P5 En one- or two-segmented, Ex 3-segmented with an inner process on segment (Fig. 200)..................... *Parundinella* ♂
146(126).	Mx2 terminates in sword-like setae and a strong toothed seta; there is a tuft of small worm-like setae associated with them (Fig. 201)..*Talacalanus* ♀
	Mx2 terminates in a strong curved claw (Fig. 190)................147
	Mx2 without this claw..148
147(146).	Mxp thin and long without spiny claws (Fig. 192)....................... ....*Onchocalanus* ♀
	Mxp with 2 strong claws more or less spiny (Fig. 191) ....*Cornucalanus* ♀
148(146).	Mx2 with two large appendages as tufts (Fig. 172)...............149
	Mx2 with one large appendage as a tuft (Fig. 202)...............150
	Mx2 otherwise armed..151
149(148).	Posterior corners of Me5 rounded......................*Heteramalla* ♀
	Posterior corners of Me5 extended into sharp points (Fig. 203)..*Puchinia* ♀
150(148).	P5 as in Fig. 204..*Scopalatum* ♀
	P5 as in Fig. 205...*Xanthocalanus* ♀
151(148).	Mx2 has only ordinary setae and some hooked setae (Fig. 165)..152
	Mx2 with special appendages, worm-like, brush-like, besides normal setae, and sometimes hooked or spiny setae (Fig. 162,198)...153

TAXONOMY AND IDENTIFICATION

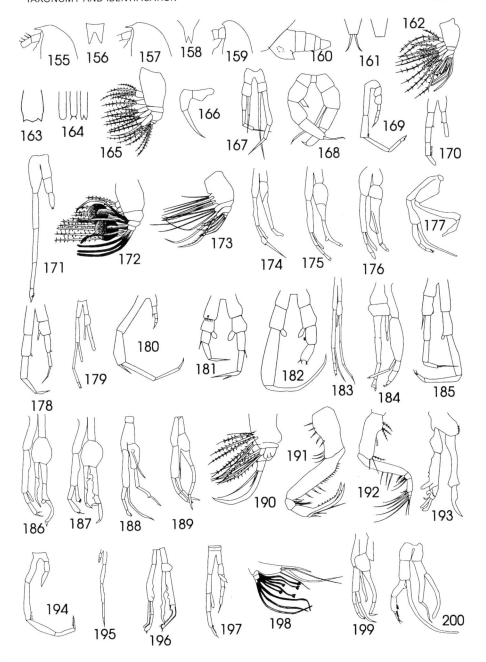

Figure 28 Key to genera of platycopioid and calanoid copepods, Key figs. 155–200.

152(151).	P5 as in Fig. 206...*Farrania* ♀	
	P5 as in Fig. 207, or with 3 terminal spines........ *Drepanopus* ♀	
	P5 as in Fig. 208...*Pseudotharybis* ♀	
153(151).	Mx2 has 3 worm-like appendages and 5 appendages terminated by a swollen knob...154	
	Mx2 with other kinds of appendages in the same or different numbers..155	
154(153).	P5 uniramous, 3-segmented although 2 or all segments may be fused; terminal segment with 2 to 4 spines, the innermost being longest (Fig. 209). Ex1 of P1 usually with an external spine...... ....*Amallothrix* ♀	
	....*Archescolecithrix* ♀	
	P5 uniramous, 3-segmented (Fig. 210). Ex1 of P1 without external spine...*Undinothrix* ♀	
155(153).	A1 very short, not extending past Me3 and very thick at the base (Fig. 211). P5 with 3 segments covered in spinules (Fig. 212) ....*Brachycalanus* ♀	
	A1 usually longer than the Pr. When shorter they are not swollen at the base. P5 with sparse spines...156	
156(155).	P5 absent or, if present, small and asymmetrical (Fig. 213) ....*Scolecithrix* ♀	
	P5 present and asymmetrical, 3-segmented of form in Fig. 214 ....*Mixtocalanus* ♀	
	P5 present and symmetrical, 3-segmented (Fig. 215)................. ....*Xantharus* ♀	
	P5 as in Fig. 216 with terminal segment longest; with 1 or 2 terminal spines as well as an articulated inner spine on the last segment...*Tharybis* ♀	
	Rythabis ♀	
	P5 minute and as in Fig. 217.........................*Parundinella* ♀	
	P5 present and symmetrical (Figs. 218–224)..........................157	
157(156).	P5 with one segment (Figs. 218)..158	
	P5 with 2 segments (Figs. 219)..160	
	P5 with 3 segments (Figs. 220–224)..161	
158(157).	Me4 and Me5 fused..159	
	Me4 and Me5 separate..*Macandrewella* ♀	
159(158).	P5 plate-like, wide and flat, with some short spines on each edge (Fig. 225)...*Scolecithricella* ♀	
	P5 elongated with a long seta on the internal margin (Fig. 218). Mx2 with 3 worm-like and 5 shorter brush-form appendages (Fig. 198)..*Scaphocalanus* ♀	
160(157).	P5 terminates in 3 or 4 short spines and often covered with spinules (Figs. 221, 226)...*Xanthocalanus* ♀	

P5 with a long internal seta, a large apical point, and usually one or two additional short spines (Figs. 218,219). Mx2 with 3 worm-like and 5 shorter brush-form appendages (Fig. 198) ....*Scaphocalanus* ♀
P5 imperfectly 3-segmented, with long sub-apical spine directed backwards and very small apical spine; without spinules (Fig. 227)..*Scottocalanus* ♀
P5 with one long internal seta and with or without one terminal spine (Fig. 228)..*Racovitzanus* ♀

161(157). Terminal segment of P5 much narrower, often shorter than the other segments, with 3 or 4 short spines. P5 often covered with spinules (Fig. 220)..*Xanthocalanus* ♀
Terminal segment of P5 as wide as or slightly narrower than the others, nearly always longer with 2 to 4 spines, of which at least the internal one is very long; P5 smooth, or at most with a few spinules (Figs. 222–224)..162

162(161). P5 3-segmented, terminal segment usually with 3 strong spines (Fig. 223). Rostrum strongly bifurcated (Fig. 229). Ex1 of P1 without outer spine..*Lophothrix* ♀
P5 3-segmented, distal segment elongate, usually with 4 spines (Fig. 230). Rostrum club- or plate-like, not strongly bifurcated (Fig. 231). Ex1 of P1 with outer spine..................*Landrumius* ♀
P5 usually 2-segmented, terminal segment with 3 or 4 spines (Fig. 224). Rostrum of 2 filaments, usually well developed (Fig. 232)..*Scaphocalanus* ♀
P5 3-segmented, terminal segment with 2 spines (Fig. 233). Rostrum a small knob, virtually absent....*Parascaphocalanus* ♀

163(112). B1, B2 and Ex of P2 and P3 broader than in P4; the B2 has a toothed edge enlarged like a calyx (Fig. 234)...........................
....*Clausocalanus* ♀ ♂
P2 and P3 without these features..164

164(163). External spines of Ex3 of P3 and P4 in the form of combs set in deep notches (Fig. 235)..................................*Ctenocalanus* ♀ ♂
These spines of the normal form..165

165(164). P5 symmetrical or absent..166
P5 asymmetrical..199

166(165). P5 present..167
P5 absent..173

167(166). Terminal segment of P5 claw-like or narrow and finger-like (Figs. 207,236–238)..168
P5 different, not as above..172

168(167). Terminal segment of P5 claw-like (Fig. 207)...... *Drepanopus* ♀
Terminal segment of P5 straight or slightly curved (Figs.

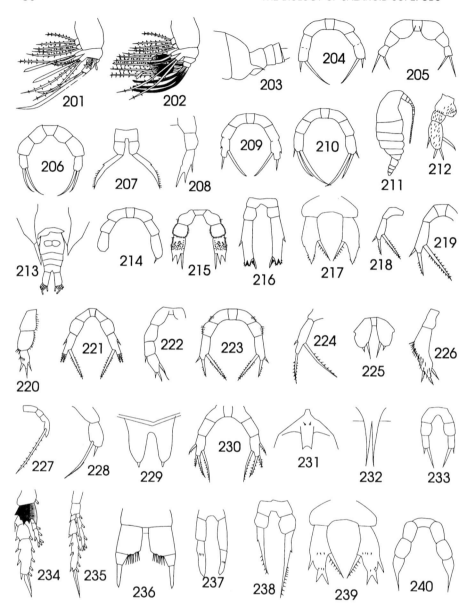

Figure 29 Key to genera of platycopioid and calanoid copepods, Key figs. 201–240.

	236–241)..169
169(168).	P5 wide at the base; terminal segment narrow and finger-like..*Pseudophaenna* ♀
	P5 minute, 2-segmented; terminal segment with one or more acuminate processes and one robust inner spine with ventral spinules (Figs. 217,239)......................................*Parundinella* ♀
	P5 otherwise formed (Figs. 216,237–241)................................170
170(169).	P5 as in Fig. 240. Ur of 3 segments......................*Miostephos* ♀
	P5 as in Figs. 237,238. Ur of 4 segments..................................171
171(170).	P5 usually symmetrical or at least having the two sides of nearly equal length; may be variously ornamented with spinules in rows or scattered (Fig. 237)..*Stephos* ♀
	P5 symmetrical as in Fig. 238.....................................*Parastephos* ♀
	P5 symmetrical as in Fig. 238 but with two terminal spine-like processes..*Mesaiokeras* ♀
172(167).	P5 as in Fig. 241 with 2 or 3 terminal spines and a lateral spine on each edge of the last segment.....................*Neoscolecithrix* ♀
	P5 as in Fig. 216 with terminal segment longest; with 1 or 2 terminal spines as well as an articulated inner spine on the last segment..*Tharybis* ♀
	P5 similar to that in Fig. 216 except that the 3 or 4 spines are terminal (Fig. 242)..*Undinella* ♀
	P5 minute and as in Fig. 217................................*Parundinella* ♀
173(166).	Ce with a median dorsal spine (Figs. 243–245)........*Gaetanus* ♀
	Head without spine..174
174(173).	Ex3 of P3 and P4 with 5 internal setae...................................175
	Ex3 of P3 and P4 with 4 internal setae...................................180
175(174).	Ex1-3 of P1 each with an external spine (Fig. 246)................176
	Ex1 of P1 without external spine..177
	Ex1 and Ex2 of P1 without external spines............................179
176(175).	Rostral plate with 2 terminal points (Fig. 247).*Sognocalanus* ♀
	Rostrum pointed, single and directed backwards.*Damkaeria* ♀
177(175).	Rostrum absent..*Mimocalanus* ♀
	Rostrum present...178
178(177).	Rostrum large (Fig. 248)..*Teneriforma* ♀
	Rostrum small (Fig. 249)..*Kunihulsea* ♀
179(175).	A1 reaching to anal segment; En of A2 longer than Ex; Mxp B1 with 5 setae..*Teneriforma naso* ♀
	A1 reaching to Me1; Ex of A2 longer than En; Mxp B1 with 2 setae..*Isaacsicalanus* ♀
180(174).	Internal margin of B1 of P4 naked or setose........................182
	Internal margin of B1 of P4 with spines (Figs. 250, 251).......181
181(180).	Ex1 and Ex2 of P1 incompletely fused........... *Pseudochirella* ♀

	Ex1 and Ex2 of P1 completely separate........... *Chirundinella* ♀
182(180).	Corners of Me5 rounded or with an obtuse point (Figs. 252,253)...183
	Corners of Me5 terminated by a sharp point........................188
183(182).	Gn asymmetrical with ventral and/or dorsal spine (Figs. 252,253)..*Undeuchaeta* ♀
	Gn symmetrical, without spines..184
184(183).	Terminal setae of A1 thick and annulated (Fig. 254).................. ....*Bradyetes* ♀
	These setae of ordinary type, not thickened and annulated....... 185
185(184).	Ex of P1 with 1 segment. Mx2 with strong, curved setae (Fig. 13)..*Chiridiella* ♀
	Ex1 of P1 with an external spine (Fig. 255)......................187
	Ex1 of P1 without external spine (Fig. 256).....................186
	Ex1 and Ex2 of P1 without external spines........*Mesaiokeras* ♀
186(185).	En of P1 with 4 setae.............................*Microcalanus* ♀
	En of P1 with 5 setae.....................................*Gaidius* ♀
187(185).	Mx2 with some setae developed as spines........ *Batheuchaeta* ♀
	Mx2 with ordinary setae....................*Pseudocalanus* ♀
188(182).	Rostrum with 2 points...189
	Rostrum with 1 point or absent...190
189(188).	Rostrum large, with 2 strong points (Fig. 257). Terminal seta of Ex3 of P2, P3 and P4 with numerous equal teeth (Figs. 258) ....*Aetideopsis* ♀
	Rostrum small, with 2 points. Setae of last 6 segments of A1 thick and annulated (Fig. 254). Mx2 without strong hooks........ ....*Bradyidius* ♀
190(188).	Ex of P1 with 3 S..191
	Ex of P1 with 2 S...*Gaidius* ♀
191(190).	Mx2 with worm-like appendages (Fig. 162).................*Diaixis* ♀
	Mx2 without these appendages...192
192(191).	Rostrum absent..193
	Rostrum present, sometimes very small..............................196
193(192).	A2 has Ex half the length of the En (Fig. 259)....................194
	A2 has Ex equal to or greater in length than the En (Fig. 260)..195
194(193).	Distal segment of Ex of A2 has one terminal seta..................... ....*Mesocomantenna* ♀
	Distal segments of Ex of A2 has 3 terminal setae...................... ....*Comantenna* ♀
195(193).	Posterior spines on Me5 stout and turned up dorsally (Fig. 261) ....*Paracomantenna* ♀

	Posterior spines on Me5 normal, projecting posteriorly...*Chiridius* ♀
196(192).	Me4 and Me5 separate..197
	Me4 and Me5 fused..*Lutamator* ♀
197(196).	Posterior spines of Me5 turned up dorsally (Fig. 261)..*Sursamucro* ♀
	Posterior spines of Me5 normal, projecting posteriorly........198
198(197).	Rostrum short, stout and rounded (Fig. 262).........*Gaidiopsis* ♀
	Rostrum very small as in Figs. 263–264..............*Crassantenna* ♀
199(165).	P5 of form in Fig. 237; may be ornamented with spinules in rows or scattered..*Stephos* ♀
	P5 of form in Fig. 238 but left leg much longer than right..*Parastephos* ♀
	P5 present on right side only, uniramous, 4-segmented and long (Fig. 265)...*Damkaeria* ♂
	P5 of another form..200
200(199).	P5 with or without reduced endopods, of form in Fig. 182 ....*Drepanopus* ♂
	P5 very large and complex; its segments irregular, enlarged with swellings, but without En (Fig. 266)..........................201
	P5 with its segments all narrow, often with one or two En present (Figs. 267–269)..204
201(200).	Ex1 of P1 with long, thin external spine (Fig. 246). P5 of form in Figs. 193,266..*Diaixis* ♂
	Ex1 of P1 without external spine..202
202(201).	Left P5 rudimentary (Fig. 270)...........................*Miostephos* ♂
	Both P5 well developed..203
203(202).	Penultimate segment of the left P5 cylindrical; last segment without appendages (Fig. 271). Right P5 terminates in a claw armed with spines................................... *Parastephos* ♂
	Penultimate segment of left P5 often swollen; last segment with 2 or many appendages (Fig. 272). Right P5 terminates in unarmed claw or mitten-like segment.........................*Stephos* ♂
204(200).	P5 without En, or with only one branch (Figs. 273, 274)......205
	P5 with at least one En, more or less rudimentary (Figs. 267–269) 210
205(204).	Mx2 with bush-like sensorial appendages (Fig. 275)...*Pseudophaenna* ♂
	Mx2 with only setae..206
206(205).	Ex of P1 with 2 S...207
	Ex of P1 with 3 S...208
207(206).	P5 present on both sides, left P5 is five or six segmented, right P5 three segmented (Fig. 273)............................*Microcalanus* ♂

	P5 present on both sides, left has 5, right 2 segments............... ....*Teneriforma* ♂ P5 present on both sides, variably asymmetrical with left longer than right; each branch essentially with two basal segments, 3 exopodal of which Ex1 and Ex2 may be partly fused............... ....*Mimocalanus* ♂ P5 has 5 segments on left but represented by a short stump on right...*Mesaiokeras* ♂
208(206).	Rostrum absent...*Chiridius* ♂ Rostrum present, variable in form..209
209(208).	Right P5 as long as left, needle like (Fig. 274)....................... ....*Pseudocalanus* ♂ Left P5 is 5 segmented and much longer than the 3 segmented right P5 (Fig. 276)...*Sognocalanus* ♂
210(204).	Both En present on P5..211 Right P5 without En. Left P5 with 3-segmented Ex that is half the length of the styliform En (Fig. 267)..................*Tharybis* ♂ Left P5 without En. Right leg with spinules along most of its length (Fig. 277)..*Neoscolecithrix* ♂
211(210).	En of P5 often reduced...213 En of P5 well developed. Mx2 may have special appendages...212
212(211).	Mx2 with one enlarged spine-like seta. P5 as in Fig. 278 ....*Parundinella* ♂ Mx2 with one enlarged spine-like seta. P5 as in Fig. 279........... ....*Undinella* ♂ Mx2 with 2 enlarged spine-like setae. P5 as in Fig. 280 ....*Batheuchaeta* ♂ Mx2 with 2 enlarged spine-like setae (Fig. 14). P5 as in Fig. 281 ....*Chiridiella* ♂ Mx2 greatly reduced. P5 as in Fig. 282................*Comantenna* ♂
213(211).	Ce with median dorsal spine (Figs. 243–245)...........*Gaetanus* ♂ Ce without median dorsal spine...214
214(213).	Rostrum with one point...215 Rostrum with 2 points...217 Rostrum absent. P5 as in Fig. 283...........................*Bradyetes* ♂
215(214).	Right and left P5 with Ex of 3 segments (Fig. 284)...*Gaidius* ♂ P5, left Ex of 3 segments, right Ex of 2 segments.................216
216(215).	Terminal segment of Ex of left P5 rounded distally (Fig. 285) ....*Pseudochirella* ♂ Terminal segment of Ex of left P5 is bilobed distally (Fig. 286) ....*Chirundinella* ♂
217(214).	One P5 has En tapering, the other rounded (Fig. 268)...............

TAXONOMY AND IDENTIFICATION

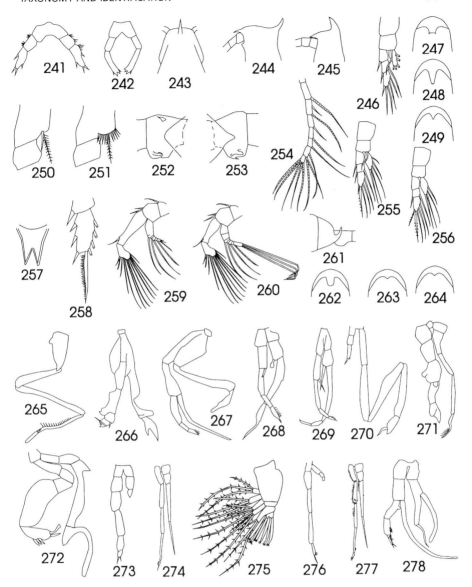

Figure 30 Key to genera of platycopioid and calanoid copepods, Key figs. 241–278.

	....*Aetideopsis* ♂
	Both En of P5 tapering (Fig. 269).........................*Bradyidius* ♂
218(105).	Ce with median dorsal spine (Figs. 243–245).......... *Gaetanus* ♀
	Ce without median dorsal spine..219
219(218).	P5 absent...220
	P5 present and asymmetrical...229
220(219).	Corners of Me5 extended in sharp points...............................221
	Corners of Me5 rounded or bluntly pointed..........................225
221(220).	B1 of P4 with one or many spines or spinules on the internal margin (Figs. 250,251). Pr divided into at least 5 segments..*Pseudochirella* ♀
	B1 of P4 with 2 transverse rows of small teeth (Fig. 287). Pr divided into 4 segments...*Paivella* ♀
	B1 of P4 without spines or spinules..222
222(221).	Rostrum with 2 points...223
	Rostrum absent..*Chiridius* ♀
	Rostrum with 1 point...*Gaidius* ♀
223(222).	Me4 and Me5 usually separate............................*Aetideopsis* ♀
	Me4 and Me5 fused..224
224(223).	Rostrum small (Fig. 288)..*Bradyidius* ♀
	Rostrum large, points fused or separated at the base (Figs. 289,290)...*Aetideus* ♀
225(220).	B1 of P4 with one or many spines or spinules on the internal margin (Figs. 250,251)...226
	B1 of P4 without spines or spinules..227
226(225).	En of A2 equal to or longer than half the Ex (Fig. 291). Ex of P1 with 3 segments..*Pseudochirella* ♀
	En of A2 at the most half the length of the Ex (Fig. 292). Ex of P1 with 2 segments...*Euchirella* ♀
227(225).	Ce with distinctive rostrum (Fig. 293). Two of the furcal setae always naked, usually longer than the others (Fig. 294,295)...... ....*Euchaeta* ♀
	Pareuchaeta ♀
	Ce without rostrum in Fig. 293. Two furcal setae not longer than the others...228
228(227).	Ce usually with high triangular crest...................*Chirundina* ♀
	Ce without crest...*Wilsonidius* ♀
229(219).	P5 without En (Figs. 296)...230
	P5 with En well developed (Figs. 297,298)...........................232
230(229).	P5 present only on left side (Fig. 296).......................*Aetidius* ♂
	P5 present on both sides...231
231(230).	P5 without trace of En; very asymmetrical, very short on right with a short terminal point (Fig. 171).................*Bradyidius* ♂

	P5 without trace of En; very asymmetrical, very elongated on left while right terminates in a long fine point (Fig. 299).................... ....*Paivella* ♂
232(229).	Ex of left P5 terminates in long, thin segment, often stylet shaped, and with a serrated lamella (Fig. 300,301). Rostrum as in Fig. 293...*Euchaeta* ♂ ....*Pareuchaeta* ♂ These features absent..233
233(232).	Ex of left P5 ends in a short pointed segment with tuft of hairs on the inside (Figs. 297,298)..234 Ex of left P5 rarely pointed; either without hairs, or when hairs are present never arranged in tufts..236
234(233).	The 2 En of P5 equally developed (Fig. 297)...... *Chirundina* ♂ En of left P5 less than half the length of the right En (Fig. 298)..235
235(234).	P5 complex, its basal segment swollen (Fig. 298)....................... .... *Undeuchaeta* ♂ P5 simple, little dilated at the base (Fig. 269)...... *Bradyidius* ♂
236(233).	Ex of P1 with 3 segments (Fig. 302)..........................*Gaidius* ♂ Ex of P1 with 2 segments, sometimes with traces of areas of fusion in the first (Fig. 303)...237
237(236).	En of A2 equal to or greater than half the length of the Ex (Fig. 291)..*Pseudochirella* ♂ En of A2 at most half the length of the Ex (Fig. 292)................. ....*Euchirella* ♂
238(9).	En of P1 with 1 segment...239 En of P1 with 2 segments...244 En of P1 with 3 segments...252
239(238).	En of P2 with 1 segment...241 En of P2 with 2 segments...240
240(239).	A1 shorter than the length of the Ce.................*Nanocopia* ♀ ♂ A1 much longer than the Ce............................*Eurytemora* ♀ ♂
241(239).	P5 absent..*Valdiviella* ♀ P5 present...242
242(241).	P5 with one 3-segmented branch on each side, the terminal segment with 3 or 4 spines on each side (Fig. 304).*Undinella* ♀ P5 small, one-segmented each side........................*Lahmeyeria* ♀ P5 large, of complex structure (Figs. 279,305).........................243
243(242).	P5 as in Fig. 279..*Undinella* ♂ P5 as in Fig. 305..*Valdiviella* ♂ P5 as in Fig. 306. Mx2 with strong curved setae as in Fig. 14...*Chiridiella* ♂
244(238).	Ce with 1 pair of lenses (Figs. 307,308)..............*Labidocera* ♀ ♂

94 THE BIOLOGY OF CALANOID COPEPODS

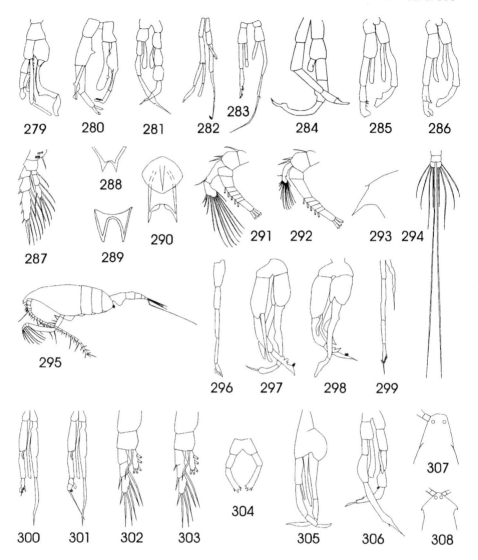

Figure 31 Key to genera of platycopioid and calanoid copepods, Key figs. 279–308.

	Ce without dorsal lenses..245
245(244).	P5 uniramous on both sides................................*Calanopia* ♀ ♂
	P5 biramous on both sides...246
246(245).	En of P5 three-segmented (Figs. 22,83)........... *Centropages* ♀ ♂
	En of P5 with only 1 segment or absent..............................247
247(246).	Furca with 3 plumose setae, often symmetrical. Mx2 squat with very long hooked setae (Fig. 309)....................*Bathypontia* ♀ ♂
	Furca with at least 4 symmetrical plumose setae..................248
248(247).	Mx2 shorter than Mxp. Furca long and narrow, at least six times longer than wide (Figs. 116,310).............................. *Temora* ♀ ♂
	Mx2 longer than Mxp (Fig. 311). Furca short (Fig. 312).......249
249(248).	Mx2 very large, with scythe-like setae (Fig. 311)..................250
	Mx2 not greatly enlarged, with thin, long, very spiny setae.251
250(249).	Female P5 end segment with terminal finger-like process which may be finely serrate on one or both margins (Fig. 314); 2 setae on the inner lateral margin of this segment. Male right A1 has no teeth in the geniculate region between segments 18 and 19 (Fig. 313); right P5 not chelate, ends in long feather-like seta..*Paracandacia* ♀ ♂
	Female P5 end segment not as above. Male right A1 toothed on segments 17 to 20 in geniculate region (Fig. 315); right P5 chelate...*Candacia* ♀ ♂
251(249).	Female P5 uniramous, 2 or 3 segmented, end segment bearing an internal spine and an external plumose seta (Fig. 316). Male P5 uniramous, 4-segmented on right, 5-segmented on left (Fig. 317) ....*Acartia* ♀ ♂
	Female P5 biramous, 1-segmented, En spine-like (Fig. 318). Male P5 uniramous, 4-segmented on right, 3-segmented on left (Fig. 319)..*Paralabidocera* ♀ ♂
252(238).	Ce with 2 or 4 lenses on the dorsal surface (Figs. 3,4,307,308, 320–322)...253
	Ce without lenses or lateral hooks...258
253(252).	Ce with lateral hooks and 4 lenses (Fig. 3,320).................... ....*Anomalocera* ♀ ♂
	Ce with 2 lenses (Figs. 321,322).............................254
254(253).	Ce without lateral hooks (Fig. 307)...................*Labidocera* ♀ ♂
	Ce with lateral hooks (Fig. 308).............................255
255(254).	Ur with 2 or 3 segments; neither A1 geniculate...................256
	Ur with 5 segments; right A1 geniculate..............................257
256(255).	Rostrum swollen at the base, with 1 rostral lens (Fig. 321) ....*Pontella* ♀
	Rostrum not swollen at the base, without lens (Fig. 322) ....*Epilabidocera* ♀

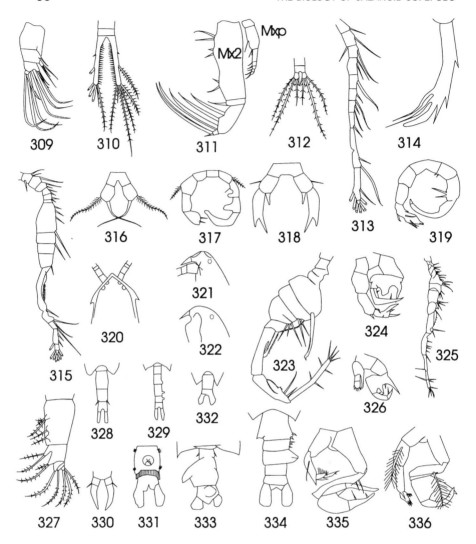

Figure 32 Key to genera of platycopioid and calanoid copepods, Key figs. 309–336.

257(255).	Terminal part of geniculate A1 with 2 segments (Fig. 323); no En on left P5 (Fig. 324)..*Pontella* ♂ Terminal part of geniculate A1 with 4 segments (Fig. 325); rudimentary En on left P5 (Fig. 326)....................*Labidocera* ♂	
258(252).	Mx2 powerful with strong curved setae (Fig. 327). Second segment female Ur with 2 dorsal spines (Fig. 328). Female P5 biramous. Male Ur 5-segmented, third and fourth segments with extensions on their right (Fig. 329)..................*Parapontella* ♀ ♂ Mx2 powerful with strong curved setae. Female P5 uniramous. Male Ur 4-segmented..*Neopontella* ♀ ♂ These combinations of features absent...................................259	
259(258).	Ur with 1, 2 or 3 segments..260 Ur with 5 segments..262	
260(259).	P5 uniramous on both sides, 2-segmented, terminating in a point or 1–3 prongs (Fig. 330)...*Tortanus* ♀ P5 with 2 branches on each side; En sometimes small 261	
261(260).	A1 with very plumose setae. Ur symmetrical. Furcae asymmetrical (Fig. 331). Gn with four groups of lateral spinules (Fig. 332) ....*Pontellina* ♀ A1 without very plumose setae. Ur not symmetrical. Furcae symmetrical. Gn without spinules (Fig. 333)........*Pontellopsis* ♀	
262(259).	Geniculate A1 little swollen, not globular in the middle............ ....*Tortanus* ♂ Geniculate A1 very swollen, globular in the middle (Fig. 323) 263	
263(262).	Third Ur segment has lump on the right margin (Fig. 334). External seta of B2 of P5 small (Fig. 335)............*Pontellopsis* ♂ Third Ur segment not lumpy. External seta of B2 of P5 long and very plumose (Fig. 336)..*Pontellina* ♂	

4. World List of Platycopioid and Calanoid Copepods

4.1. Order Platycopioida Fosshagen, 1985... 98
    4.1.1. Superfamily Platycopioidea... 99
4.2. Order Calanoida Sars, 1902... 99
    4.2.1. Superfamily Epacterisciodea... 99
    4.2.2. Superfamily Pseudocyclopoidea...................................... 100
    4.2.3. Superfamily Arietelloidea.. 101
    4.2.4. Superfamily Fosshagenioidea.. 109
    4.2.5. Superfamily Centropagoidea... 110
    4.2.6. Superfamily Megacalanoidea... 118
    4.2.7. Superfamily Bathypontioidea... 121
    4.2.8. Superfamily Eucalanoidea... 122
    4.2.9. Superfamily Ryocalanoidea.. 122
    4.2.10. Superfamily Spinocalanoidea.. 123
    4.2.11. Superfamily Clausocalanoidea...................................... 124

The following list of species of platycopioid and calanoid copepods, current in 1996, is based on that in C. Razouls (1982, 1991, 1993, 1995) and records from the author's own files. Razouls gives lists of synonyms, not reproduced here. Authorities for the taxonomic entities are given below but only references not quoted in Vervoort (1986a,b, 1988) are provided at the end of this book, primarily those published after 1985. Some references prior to 1985 are listed because they are cited for purposes other than taxonomy.

Notes are given for many genera to indicate sources of supplementary descriptions, reviews, and re-organization of them or taxonomic difficulty within them. The absence of such supplementary notes usually indicates that the information given within the type papers should be referred to.

4.1. ORDER PLATYCOPIOIDA FOSSHAGEN, 1985

This Order was established by Fosshagen in Fosshagen and Iliffe (1985).

4.1.1. Superfamily Platycopioidea

4.1.1.1. Family Platycopiidae *Sars, 1911*

This family is usually linked with the Stephidae and Tharybidae but there are difficulties in deciding its affinities and Lang (1948) suggested that it be placed in the sub-order Progymnoplea. This placement would recognize the primitive characters of the species (Razouls C., 1982). Further revision of its status is made by Fosshagen and Iliffe (1985, 1988).

Genus *Antrisocopia* Fosshagen, 1985
A. prehensilis Fosshagen, 1985.
Cave-dwelling in Bermuda. Described in Fosshagen and Iliffe (1985).

Genus *Nanocopia* Fosshagen, 1988
N. minuta Fosshagen, 1988.
Cave-dwelling in Bermuda. Described in Fosshagen and Iliffe (1988).

Genus *Platycopia* Sars, 1911
P. inornata Fosshagen, 1972; *P. orientalis* Ohtsuka and Boxshall, 1994; *P. perplexa* Sars, 1911; *P. pygmaea* Sars, 1919; *P. robusta* Andronov, 1985; *P. sarsi* Wilson, 1946, ♂; *P. tumida* (Wilson, 1935); *Platycopia* sp. Fosshagen, 1972.
Small, free-living hyperbenthic copepods.

4.2. ORDER CALANOIDA SARS, 1902

4.2.1. Superfamily Epacteriscioidea

4.2.1.1. Family Epacteriscidae *Fosshagen, 1973*

These are hyperbenthic copepods, some dwelling in caves.

Genus *Enantiosis* Barr, 1984
E. cavernicola Barr, 1984

Genus *Epacteriscus* Fosshagen, 1973
E. rapax Fosshagen, 1973

Genus *Erebonectes* Fosshagen, 1985
E. macrochaetus Fosshagen, 1994; *E. nesioticus* Fosshagen, 1985.
Described in Fosshagen and Iliffe (1985, 1994).

4.2.2. Superfamily Pseudocyclopoidea

4.2.2.1. Family Boholiniidae *Fosshagen and Iliffe, 1989*

Genus *Boholina* Fosshagen and Iliffe, 1989
B. crassicephala Fosshagen, 1989; *B. purgata* Fosshagen, 1989.

4.2.2.2. Family Pseudocyclopidae *Giesbrecht, 1893*

Genus *Pseudocyclops* Brady, 1872
P. arguinensis Andronov, 1986b; *P. australis* Nicholls, 1944; *P. bahamensis* Fosshagen, 1968a; *P. bilobatus* Dawson, 1977; *P. cokeri* Bowman and Gonzalez, 1961; *P. crassiremis* Brady, 1873; *P. gohari* Noodt, 1958; *P. kulai* Othman and Greenwood, 1989; *P. lakshmi* Haridas, Madhupratap and Ohtsuka, 1994; *P. latens* Gurney, 1947, ♀; *P. latisetosus* Sewell, 1932, ♂; *P. lepidotus* Barr and Ohtsuka, 1989; *P. lerneri* Fosshagen, 1968a; *P. magnus* Esterly, 1911, ♀; *P. mathewsoni* Fosshagen, 1968a; *P. minya* Othman and Greenwood, 1989, ♂; *P. mirus* Andronov, 1986b, ♂; *P. obtusatus* Brady and Robertson, 1873; *P. oliveri* Fosshagen, 1968a, ♀; *P. pacificus* Vervoort, 1964, ♂; *P. paulus* Bowman and Gonzalez, 1961; *P. pumilis* Andronov, 1986b; *P. reductus* Nicholls, 1944, ♀; *P. rostratus* Bowman and Gonzalez, 1961; *P. rubrocinctus* Bowman and Gonzalez, 1961; *P. simplex* Sewell, 1932; *P. spinulosus* Fosshagen, 1968a, ♀; *P. steinitzi* Por, 1968; *P. umbraticus* Giesbrecht, 1893; *P. xyphophorus* Wells, 1967.
Reviews and descriptions are given in: Vervoort, 1964; Tanaka, 1966; Fosshagen, 1968a; Andronov, 1986b; Barr and Ohtsuka, 1989; Othmann and Greenwood, 1989. These are benthopelagic species in coastal environments.

4.2.2.3. Family Ridgewayiidae *Wilson, 1958*

Shallow water, benthopelagic (?), some in caves (Fosshagen and Iliffe, 1991).

Genus *Brattstromia* Fosshagen, 1991
B. longicaudata Fosshagen, 1991.
Cave-dwelling. Description in Fosshagen and Iliffe (1991).

Genus **Exumella** Fosshagen, 1970
E. polyarthra Fosshagen, 1970; *E. mediterranea* Jaume and Boxshall, 1995a; *E. tuberculata* Grahame, 1979.

Genus **Placocalanus** Ohtsuka, Fosshagen and Soh, 1996
P. brevipes Ohtsuka, Fosshagen and Soh, 1996, ♂; *P. inermis* Ohtsuka, Fosshagen and Soh, 1996; *P. insularis* Fosshagen, 1970; *P. longicauda* Ohtsuka, Fosshagen and Soh, 1996; *P. nannus* Fosshagen, 1970. Ohtsuka *et al.* (1996b) revise the original definition of this genus by Fosshagen (1970).

Genus **Ridgewayia** Thompson and Scott, 1903
R. canalis (Gurney, 1927), ♂; *R. flemingeri* Othman and Greenwood, 1988b; *R. fosshageni* Humes and Smith, 1974; *R. gracilis* Wilson, 1958; *R. klausruetzleri* Ferrari, 1995; *R. krishnaswamyi* Ummerkutty, 1963; *R. marki* (Esterly, 1911); *R. marki minorcaensis* Razouls and Carola, 1996; *R. shoemakeri* Wilson, 1958; *R. typica* Thompson and Scott, 1903, ♀; *R. wilsoni* Fosshagen, 1970.

4.2.3. Superfamily Arietelloidea

4.2.3.1. Family Arietellidae *Sars, 1902*

Campaner (1977, 1984b) and Ohtsuka *et al.* (1994) review this family.

Genus **Arietellus** Giesbrecht, 1892
A. aculeatus (T. Scott, 1894); *A. armatus* Wolfenden, 1911; *A. giesbrechti* Sars, 1905; *A. minor* Wolfenden, 1911, ♀; *A. mohri* (Björnberg, 1975); *A. pacificus* Esterly, 1913, ♀; *A. pavoninus* Sars, 1905, ♀; *A. plumifer* Sars, 1905; *A. setosus* Giesbrecht, 1892; *A. simplex* Sars, 1905; *A. tripartitus* Wilson, 1950, ♀; *Arietellus* sp. Bradford, 1974.
Descriptions given by: Wolfenden, 1911; Sars, 1925; Brodsky, 1950; Wilson, 1950; Tanaka, 1964c; Owre and Foyo, 1967; Bradford, 1974b; Ohtsuka *et al.*, 1994.

Genus **Campaneria** Ohtsuka, Boxshall and Roe, 1994
C. latipes Ohtsuka, Boxshall and Roe, 1994, ♂.
This species was originally *Scutogerulus pelophilus*, ♂.

Genus **Crassarietellus** Ohtsuka, Boxshall and Roe, 1994
C. huysi Ohtsuka, Boxshall and Roe, 1994, ♀; *Crassarietellus* sp. Ohtsuka, Boxshall and Roe, 1994, ♂.

Genus *Metacalanus* Cleve, 1901
M. acutioperculum Ohtsuka, 1984; *M. aurivilli* Cleve, 1901; *M. curvirostris* Ohtsuka, 1985b; *M. inaequicornis* (Sars, 1903).
Descriptions given by: Thompson and Scott, 1903; Sars, 1903 (as *Scottula inaequicornis*); Ohtsuka, 1984, 1985b; Ohtsuka *et al.*, 1994.

Genus *Paramisophria* T. Scott, 1897
P. ammophila Fosshagen, 1968b; *P. cluthae* T. Scott, 1897; *P. fosshageni* Othman and Greenwood, 1992; *P. galapagensis* Ohtsuka, Fosshagen and Iliffe, 1993, ♂; *P. giselae* (Campaner, 1977), ♀; *P. itoi* Ohtsuka, 1985b; *P. japonica* Ohtsuka, Fosshagen and Go, 1991; *P. platysoma* Ohtsuka and Mitsuzumi, 1990; *P. ovata* Geinrikh, 1996; *P. reducta* Ohtsuka, Fosshagen and Iliffe, 1993; *P. rostrata* Geinrikh, 1996; *P. spooneri* Krishnaswamy, 1959; *P. variabilis* McKinnon and Kimmerer, 1985.
Ohtsuka *et al.* (1993a, 1994) review the genus and amend its definition. Descriptions in: Sars, 1903; Krishnaswami, 1959; Tanaka, 1966; Fosshagen, 1968b; Campaner, 1977 as *Parapseudocyclops giselae*; McKinnon and Kimmerer, 1985; Ohtsuka, 1985b; Ohtsuka and Mitsuzumi, 1990; Ohtsuka *et al.*, 1991, 1993a, 1994; Othman and Greenwood, 1992.

Genus *Paraugaptiloides* Otsuka, Boxshall and Roe, 1994
P. magnus (Ohtsuka, Boxshall and Roe, 1994), ♂.
This species was originally *Paraugaptilus magnus* Bradford, 1974b, ♂.

Genus *Paraugaptilus* Wolfenden, 1904
P. archimedi Gaudy, 1973, ♂; *P. bermudensis* Deevey, 1973; *P. buchani* Wolfenden, 1904; *P. indicus* Geinrikh, 1995; *P. meridionalis* Wolfenden, 1911, ♀; *P. mozambicus* Gaudy, 1965; *P. similis* A. Scott, 1909.
Descriptions given in: A. Scott, 1909; Wolfenden, 1911; Sars, 1925; Gaudy, 1965, 1973; Tanaka, 1964c; Deevey, 1973; Bradford, 1974b; Björnberg, 1975; Geinrikh, 1993; Ohtsuka *et al.*, 1994.

Genus *Pilarella* Alvarez, 1985
P. longicornis Alvarez, 1985, ♀.
Benthopelagic on shelf, 135 m depth. Descriptions in: Ohtsuka *et al.*, 1994.

Genus *Rhapidophorus* Edwards, 1891
R. wilsoni Edwards, 1891, possible ♂.
Closely related to *Paramisophria* species but Edwards' description is incomplete (see Fosshagen, 1968b and Ohtsuka *et al.*, 1994).

Genus *Sarsarietellus* Campaner, 1984
S. abyssalis Sars, 1905, ♀; *S. natalis* Geinrikh, 1993, ♀.
Descriptions in: Sars, 1925 as *Scottula abyssalis*, ♀; Ohtsuka *et al.*, 1994.

Genus *Scutogerulus* Bradford, 1969
S. pelophilus Bradford, 1969b, ♀.
S. pelophilus Bradford, 1969b, ♂ is now *Campaneria latipes* ♂. Descriptions in: Bradford, 1969b; Ohtsuka *et al.*, 1994.

4.2.3.2. **Family Augaptilidae** *Sars, 1905*

Genus *Augaptilina* Sars, 1920
A. scopifera Sars, 1920, ♀.
Description in: Sars, 1925.

Genus *Augaptilus* Giesbrecht, 1889
A. anceps Farran, 1908; *A. cornutus* Wolfenden, 1911, ♀; *A. glacialis* Sars, 1900; *A. lamellifer* Esterly, 1911, ♀; *A. longicaudatus* (Claus, 1863); *A. megalurus* Giesbrecht, 1889; *A. spinifrons* Sars, 1907.
Distinguished from *Euaugaptilus* by their reduced maxillule in females and males. Descriptions in: Sars, 1925; Brodsky, 1950; Tanaka, 1964b.

Genus *Centraugaptilus* Sars, 1920
C. cucullatus (Sars, 1905); *C. horridus* (Farran, 1908); *C. lucidus* (Esterly, 1911), ♂; *C. porcellus* Johnson, 1936; *C. pyramidalis* (Esterly, 1911), ♀; *C. rattrayi* (T. Scott, 1893).
Large, distinctive bathypelagic copepods. Leg 1 variable in female: En and Ex have 3 segments in *C. cucullatus*, *C. horridus* and *C. rattrayi*; both have 2 segments in *C. pyramidalis*; En has 2, Ex has 3 segments in *C. porcellus*. Descriptions in: Sars, 1925; Johnson, 1936; Brodsky, 1950.

Genus *Euaugaptilus* Sars, 1920
E. affinis Sars, 1920, ♀; *E. aliquantus* Park, 1993, ♀; *E. angustus* (Sars, 1905); *E. antarcticus* (Wolfenden, 1911); *E. atlanticus* Roe, 1975; *E. austrinus* Park, 1993; *E. brevirostratus* Park, 1993; *E. brodskyi* Hülsemann, 1967; *E. bullifer* (Giesbrecht, 1889); *E. clavatus* (Sars, 1907), ♀; *E. curtus* Grice and Hülsemann, 1967, ♂; *E. digitatus* Sars, 1920; *E. diminutus* Park, 1970, ♀; *E. distinctus* (Brodsky, 1950), ♀; *E. elongatus* (Sars, 1905); *E. facilis* (Farran, 1908); *E. fagettiae* Björnberg, 1975, ♀; *E. farrani* Sars, 1920, ♀; *E. fecundus* Tanaka and Omori, 1974, ♀; *E. filigerus* (Claus, 1863); *E. fundatus* Grice and Hülsemann, 1967, ♀; *E. gibbus* (Wolfenden, 1904); *E. gracilis* (Sars, 1905), ♀; *E. graciloides* Brodsky, 1950, ♀; *E. grandicornis* Sars, 1920, ♀; *E.*

hadrocephalus Park, 1993, ♀; *E. hecticus* (Giesbrecht, 1889); *E. hulsemannae* Matthews, 1972; *E. humilis* (Farran, 1926), ♀; *E. hyperboreus* Brodsky, 1950; *E. indicus* Sewell, 1932, ♀; *E. laticeps* (Sars, 1905); *E. latifrons* (Sars, 1907), ♀; *E. longicirrhus* (Sars, 1905); *E. longimanus* (Sars, 1905); *E. longiseta* Grice and Hülsemann, 1965; *E. luxus* Tanaka and Omori, 1974; ♀; *E. magnus* (Wolfenden, 1904); *E. malacus* Grice and Hülsemann, 1967, ♀; *E. marginatus* Tanaka, 1964b, ♀; *E. matsuei* Tanaka and Omori, 1967, ♀; *E. maxillaris* Sars, 1920; *E. mixtus* (Sars, 1907), ♀; *E. modestus* Brodsky, 1950, ♀; *E. niveus* Tanaka and Omori, 1974, ♀; *E. nodifrons* (Sars, 1905); *E. nudus* Tanaka, 1964b, ♀; *E. oblongus* (Sars, 1905); *E. pachychaeta* Matthews, 1972, ♀; *E. pacificus* Matthews, 1972; *E. palumbii* (Giesbrecht, 1889); *E. parabullifer* Brodsky, 1950, ♀; *E. paroblongus* Matthews, 1972, ♀; *E. penicillatus* Sars, 1920; *E. perasetosus* Park, 1993, ♀; *E. placitus* (A. Scott, 1909), ♀; *E. propinquus* (Sars, 1920), ♀; *E. pseudaffinis* Brodsky, 1950, ♀; *E. quaesitus* Grice and Hülsemann, 1967, ♀; *E. rectus* Grice and Hülsemann, 1967; *E. rigidus* (Sars, 1907); *E. roei* Matthews, 1972, ♀; *E. sarsi* Grice and Hülsemann, 1965, ♀; *E. similis* (Farran, 1908), ♀; *E. squamatus* (Giesbrecht, 1889); *E. sublongiseta* Park, 1970, ♀; *E. tenuicaudis* (Sars, 1905), ♀; *E. tenuispinus* Sars, 1920, ♀; *E. truncatus* (Sars, 1905), ♀; *E. unisetosus* Park, 1970, ♀; *E. validus* (A. Scott, 1909); *E. vescus* Park, 1970, ♀; *E. vicinus* Sars, 1920, ♀.

This genus has been reviewed by Matthews (1972), Tanaka and Omori (1974) and Park (1993). It is a difficult genus and no binary key to the species is available; the form of the mandibular gnathobase is important in species identification. The majority of species have a 3-segmented En and Ex in P1 to P5 in females and males. Nine species, including *E. distinctus* (formerly *Neoaugaptilus distinctus*, see Matthews, 1972), have a 2-segmented Ex in P1; *E. longicirrhus* has a 2 segmented Ex in P1 according to Sars (1925) but a 3-segmented P1 according to Roe (1975). Female *E. hecticus* are peculiar in having a 2-segmented En and Ex in P5. In males, the right A1 is geniculate in *E. hecticus*, *E. hyperboreus* and *E. squamatus*; all other known males have the left geniculate. Descriptions in: Farran, 1908; A. Scott, 1909; Sars, 1925; Brodsky, 1950; Tanaka, 1964b; Grice and Hülsemann, 1965, 1967; Hülsemann, 1967; Owre and Foyo, 1967; Tanaka and Omori, 1967, 1974; Park, 1970, 1993; Matthews, 1972; Björnberg, 1975; Roe, 1975.

Genus ***Haloptilus*** Giesbrecht, 1898

H. aculeatus (Brady, 1883), ♀; *H. acutifrons* (Giesbrecht, 1892); *H. angusticeps* Sars, 1907; *H. austini* Grice, 1959, ♀; *H. bulliceps* Farran, 1926, CIV, ♀ ♂; *H. caribbeanensis* Park, 1970, ♀; *H. chierchiae* (Giesbrecht, 1889); *H. fertilis* (Giesbrecht, 1892), ♂; *H. fons* Farran, 1908, ♀; *H. furcatus* Sars, 1920, ♀; *H. longicirrus* Brodsky, 1950, ♀; *H. longicornis* (Claus, 1863); *H. major* Wolfenden, 1911, ♀; *H. mucronatus* (Claus, 1863); *H. ocellatus*

Wolfenden, 1905; *H. orientalis* (Brady, 1883), CV; *H. ornatus* (Giesbrecht, 1892); *H. oxycephalus* (Giesbrecht, 1889); *H. pacificus* Chiba, 1956; *H. paralongicirrus* Park, 1970, ♀; *H. plumosus* (Claus, 1863), ♀; *H. pseudooxycephalus* Brodsky, 1950, ♀; *H. setuliger* Tanaka, 1964b, ♀; *H. spiniceps* (Giesbrecht, 1892); *H. tenuis* Farran, 1908; *H. validus* Sars, 1920, ♀; *Haloptilus* sp. Arcos, 1975, ♂.
Matthews (1972) and Park (1988) have reviewed this genus which has several doubtful species. There are no aberrant species with the exception of male *H. chierchiae* which have the right antennule geniculate (Sewell, 1947). Literature scattered and descriptions in: Brady, 1883; Wolfenden, 1911; Sars, 1925; Rose, 1933; Sewell, 1947; Brodsky, 1950; Vervoort, 1957; Grice, 1959, 1962a; Tanaka, 1960a, 1964b; Owre and Foyo, 1967; Park, 1968, 1970, 1988; Bradford, 1970b, 1971c; Tanaka and Omori, 1971; Stephen and Sarala Devi, 1973; Roe, 1975; Sarala Devi, 1977; Ali-Khan and Ali-Khan, 1984.

Genus *Heteroptilus* Sars, 1920
H. acutilobus (Sars, 1905); *H. attenuatus* (Sars, 1905), ♀; *Heteroptilus* sp. Vervoort, 1957, ♂.
Rare deep-sea genus. The male described by Sewell differs from the other males and females in having a 3-segmented endopod on P1. Descriptions in: Sars, 1925; Sewell, 1947.

Genus *Pachyptilus* Sars, 1920
P. abbreviatus (Sars, 1905), ♀; *P. eurygnathus* Sars, 1925, ♀; *P. lobatus* Sars, 1925, ♀; *P. pacificus* Johnson, 1936, ♀.
No review paper. Deep sea genus of distinctive species with peculiar mandibles. No males known. Descriptions in: Sars, 1925; Johnson, 1936; Ali-Khan and Ali-Khan, 1984.

Genus *Pontoptilus* Sars, 1905
P. lacertosus Grice and Hülsemann, 1967, ♀; *P. mucronatus* Sars, 1905; *P. muticus* Sars, 1905, ♀; *P. ovalis* Sars, 1907, ♀; *P. pertenuis* Sars, 1907, ♀; *P. robustus* Sars, 1905, ♂.
No review paper. Rare deep sea species. Descriptions in: Sars, 1925; Grice and Hülsemann, 1967.

Genus *Pseudaugaptilus* Sars, 1907
P. longiremis Sars, 1907; *P. orientalis* Tanaka, 1964b, ♀; *P. polaris* Brodsky, 1950, ♀.
Distinguished by their rod-like mandible. Descriptions in: Sars, 1925; Brodsky, 1950; Tanaka, 1964b.

Genus *Pseudhaloptilus* Wolfenden, 1911
P. longimanus (Wolfenden, 1906), ♀.
Keys out with the genus *Pachyptilus* and probably should be in that genus.
Description: Wolfenden, 1911.

4.2.3.3. Family Discoidae *Gordejeva, 1975*

Genus *Disco* Grice and Hülsemann, 1965
D. atlanticus Gordejeva, 1974a; *D. caribbeanensis* Gordejeva, 1974; *D. creatus* Gordejeva, 1975b, ♀; *D. curtirostris* Gordejeva, 1975b; *D. elephantus* Gordejeva, 1975b, ♀; *D. erythraeus* Gordejeva, 1974b, ♀; *D. fiordicus* Fosshagen, 1967; *D. hartmanni* Schulz, 1993, ♀; *D. inflatus* Grice and Hülsemann, 1965; *D. intermedius* Gordejeva, 1976; *D. longus* Grice and Hülsemann, 1965, ♀; *D. marinus* Gordejeva, 1974a, ♀; *D. minutus* Grice and Hülsemann, 1965, ♀; *D. oceanicus* Gordejeva, 1974a, ♀; *D. oviformis* Park, 1970, ♀; *D. peltatus* Gordejeva, 1974; *D. populosus* Gordejeva, 1974b, ♀; *D. robustipes* Gordejeva, 1974b, ♀; *D. tropicus* Gordejeva, 1974a, ♀; *D. vulgaris* Gordejeva, 1974b; *Disco* sp. Grice and Hülsemann, 1967, ♂.
Schulz (1993) reviews the genus. The species at present ascribed to this genus have different forms of legs. The numbers of En segments in P1 to P4 vary between species in both sexes. Descriptions in: Grice and Hülsemann, 1965, 1967; Fosshagen, 1967; Park, 1970; Gordejeva, 1974a,b, 1975b, 1976.

Genus *Paradisco* Gordejeva, 1975a
P. gracilis Gordejeva, 1975a; *P. grandis* Gordejeva, 1976; *P. mediterraneus* (Gordejeva, 1974b); *P. nudus* Schulz, 1993, ♀.

Genus *Prodisco* Gordejeva, 1975a
P. princeps Gordejeva, 1975a, ♀; *P. secundus* Gordejeva, 1975a, ♂.

4.2.3.4. Family Heterorhabdidae *Sars, 1902*

Genus *Alrhabdus* Grice, 1973
A. johrdeae Grice, 1973, ♀.
This species shares characters with the Heterorhabdidae and the Augaptilidae and its placement is questionable (Grice, 1973).

Genus *Disseta* Giesbrecht, 1889
D. coelebs Geptner, 1972, ♂; *D. grandis* Esterly, 1906; *D. magna* Bradford, 1971b, ♀; *D. palumboi* Giesbrecht, 1889; *D. scopularis* (Brady, 1883).
Geptner (1972b) reviews the Heterorhabdidae and suggests alternative classifications of the genera and species. His genus *Microdisseta* is adopted

here for *D. minuta* Grice and Hülsemann, 1965. Descriptions in: Esterly, 1906; Sars, 1925; Tanaka, 1964a; Bradford, 1971b; Geptner, 1972a,b.

Genus *Hemirhabdus* Wolfenden, 1911
H. falciformis Wolfenden, 1911, ♀; *H. grimaldii* (Richard, 1893); *H. latus* (Sars, 1905); *H. truncatus* (A. Scott, 1909), ♀.
Geptner (1972b) suggests retaining *H. grimaldii* in this genus but transferring the other three species to a new genus *Neorhabdus* Geptner, 1972b. Descriptions in: A. Scott, 1909; Wolfenden, 1911; Sars, 1925; Owre and Foyo, 1967; Geptner, 1972b.

Genus *Heterorhabdus* Giesbrecht, 1898
H. abyssalis (Giesbrecht, 1889); *H. atlanticus* Wolfenden, 1905; *H. austrinus* Giesbrecht, 1902; *H. brevicornis* (Dahl, 1894), ♀; *H. caribbeanensis* Park, 1970, ♀; *H. clausi* (Giesbrecht, 1889); *H. compactoides* Geptner, 1971; *H. compactus* (Sars, 1900); *H. egregius* Geptner, 1972a; *H. farrani* Brady, 1918; *H. fistulosus* Tanaka, 1964a; *H. lobatus* Bradford, 1971b; *H. longispinus* Davis, 1949, ♂; *H. medianus* Park, 1970; *H. nigrotinctus* Brady, 1918, ♂; *H. norvegicus* (Boeck, 1872); *H. pacificus* Brodsky, 1950; *H. papilliger* (Claus, 1863); *H. proximus* Davis, 1949; *H. pustulifera* Farran, 1929; *H. robustoides* Brodsky, 1950; *H. robustus* Farran, 1929; *H. spinifer* Park, 1970; *H. spinifrons* (Claus, 1863); *H. spinosus* Bradford, 1971b; *H. sub-spinifrons* Tanaka, 1964a; *H. tanneri* (Giesbrecht, 1895); *H. tenuis* Tanaka, 1964a, ♂; *H. tropicus* (Dahl, 1894); *H. vipera* (Giesbrecht, 1889).
Geptner (1972b) suggests re-organizing this genus to take account of the structure of the mouthparts. Descriptions in: Sars, 1903, 1925; Esterly, 1905; Brady, 1918; Davis, 1949; Brodsky, 1950; Vervoort, 1957; Tanaka, 1960a, 1964a; Owre and Foyo, 1967; Park, 1970; Bradford, 1971a,b; Geptner, 1971.

Genus *Heterostylites* Sars, 1920
H. longicornis (Giesbrecht, 1889); *H. major* (Dahl, 1894).
Descriptions in: Sars, 1925; Tanaka, 1964a.

Genus *Mesorhabdus* Sars, 1905
M. angustus Sars, 1907; *M. brevicaudatus* (Wolfenden, 1905); *M. gracilis* Sars, 1907.
Descriptions in: Sars, 1925; Sewell, 1947.

Genus *Microdisseta* Geptner, 1972
M. minuta (Grice and Hülsemann, 1965).

4.2.3.5. Family Hyperbionychidae *Ohtsuka, Roe and Boxshall, 1993*

Genus **Hyperbionyx** Ohtsuka, Roe and Boxshall, 1993
H. pluto Ohtsuka, Roe and Boxshall, 1993.

4.2.3.6. Family Lucicutiidae *Sars, 1902*

Genus **Lucicutia** Giesbrecht, 1898
L. anisofurcata Heptner, 1971; *L. anomala* Brodsky, 1950, ♀; *L. aurita* Cleve, 1904; *L. bella* Hülsemann, 1966; *L. bicornuta* Wolfenden, 1911; *L. biuncata* Heptner, 1971, ♀; *L. challengeri* Sewell, 1932; *L. cinerea* Heptner, 1971, ♀; *L. clausi* (Giesbrecht, 1889); *L. curta* Farran, 1905; *L. curvifurcata* Heptner, 1971, ♀; *L. flavicornis* (Claus, 1863); *L. formosa* Hülsemann, 1966, ♂; *L. gaussae* Grice, 1963; *L. gemina* Farran, 1926; *L. gigantissima* Heptner, 1971, ♀; *L. grandis* (Giesbrecht, 1895); *L. intermedia* Sars, 1905; *L. longicornis* (Giesbrecht, 1889); *L. longifurca* Brodsky, 1950; *L. longiserrata* (Giesbrecht, 1889); *L. longispina* Tanaka, 1963, ♀; *L. lucida* Farran, 1908; *L. macrocera* Sars, 1920; *L. magna* Wolfenden, 1903; *L. major* Wolfenden, 1911, ♀; *L. maxima* Steuer, 1904; *L. oblonga* Brodsky, 1950, ♂; *L. orientalis* Brodsky, 1950; *L. ovalis* (Giesbrecht, 1889); *L. pacifica* Brodsky, 1950; *L. pallida* Hülsemann, 1966, ♂; *L. paraclausi* Park, 1970; *L. parva* Grice and Hülsemann, 1965; *L. pellucida* Hülsemann, 1966, ♀; *L. pera* A. Scott, 1909; *L. polaris* Brodsky, 1950; *L. profunda* Brodsky, 1950; *L. pseudopolaris* Geptner, 1969; *L. rara* Hülsemann, 1966, ♂; *L. sarsi* Hülsemann, 1966, ♀; *L. sewelli* Tanaka, 1963; *L. uschakovi* Brodsky, 1950, ♂; *L. wolfendeni* Sewell, 1932.
Reviews by Hülsemann (1966, 1989) and Anonymous (1990). A key to species known before 1966 is given by Hülsemann (1966). Descriptions in: Wolfenden, 1911; Sars, 1925; Brodsky, 1950; Tanaka, 1963; Hülsemann, 1966; Owre and Foyo, 1967; Geptner, 1969, 1971, 1986; Park, 1970; Ali-Khan and Ali-Khan, 1982.

4.2.3.7. Family Metridinidae *Sars, 1902*

Genus **Gaussia** Wolfenden, 1905
G. asymmetrica Björnberg and Campaner, 1988, ♀; *G. princeps* (T. Scott, 1894); *G. sewelli* Saraswathy, 1973.
Large, distinctive oceanic species. Saraswathy (1973) reviews the nomenclature of this genus suggesting that *M. princeps* T. Scott, 1894 should be renamed *G. scotti* (Giesbrecht, 1897). Hülsemann (1988b) disagrees with this, *G. princeps* to be retained. Descriptions in: Saraswathy, 1973; Saraswathy and Bradford, 1980; Björnberg and Campaner, 1988, 1990.

Genus *Metridia* Boeck, 1864
M. alata Roe, 1975; *M. andraeana* Brady, 1918; *M. asymmetrica* Brodsky, 1950; *M. bicornuta* Davis, 1949, ♀; *M. boecki* Giesbrecht, 1889; *M. brevicauda* Giesbrecht, 1889; *M. calypsoi* Gaudy, 1963, ♂; *M. curticauda* Giesbrecht, 1889; *M. discreta* Farran, 1946, ♂; *M. effusa* Grice and Hülsemann, 1967; *M. gerlachei* Giesbrecht, 1902; *M. gurjanovae* Epstein, 1949; *M. ignota* Esterly, 1906, ♀; *M. longa* (Lubbock, 1854); *M. lucens* Boeck, 1864; *M. macrura* Sars, 1905; *M. okhotensis* Brodsky, 1950; *M. ornata* Brodsky, 1950; *M. pacifica* Brodsky, 1950; *M. princeps* Giesbrecht, 1889; *M. similis* Brodsky, 1950; *M. trispinosa* Brady, 1918, ♂; *M. venusta* Giesbrecht, 1889.
Mixture of coastal and oceanic species, some even bathypelagic. Two species, *M. andraeana* and *M. trispinosa*, have not been recorded since described by Brady. Descriptions in: Brady, 1918; Sars, 1903, 1925; Brodsky, 1950, 1976; Vervoort, 1957; Tanaka, 1960a, 1963; Grice and Hülsemann, 1965, 1967; Owre and Foyo, 1967; Roe, 1975; Ferrari, 1993b.

Genus *Pleuromamma* Giesbrecht, 1898
P. abdominalis (Lubbock, 1856); *P. borealis* (Dahl, 1893); *P. gracilis* (Claus, 1863); *P. indica* Wolfenden, 1905; *P. piseki* Farran, 1929; *P. quadrungulata* (Dahl, 1893); *P. robusta* (Dahl, 1893); *P. scutullata* Brodsky, 1950; *P. wolfendeni* Brady, 1918, ♀; *P. xiphias* (Giesbrecht, 1889).
Oceanic, epi- to mesopelagic species, some cosmopolitan (Saraswathy, 1986). One species, *P. wolfendeni*, has not been recorded since described by Brady. Descriptions in: Brady, 1918; Sars, 1903, 1925; Brodsky, 1950; Tanaka, 1963; Owre and Foyo, 1967; Bowman, 1971; Park and Mauchline, 1994.

4.2.3.8. **Family Phyllopodidae** *Brodsky, 1950*

Genus *Phyllopus* Brady, 1883
P. aequalis Sars, 1920; *P. bidentatus* Brady, 1883; *P. giesbrechti* A. Scott, 1909; *P. helgae* Farran, 1908; *P. impar* Farran, 1908; *P. integer* Esterly, 1911; *P. mutatus* Tanaka, 1964c, ♀; *P. muticus* Sars, 1907.
Descriptions in: Sars, 1925; Sewell, 1947; Brodsky, 1950; Wilson, 1950; Owre and Foyo, 1967.

4.2.4. **Superfamily Fosshagenioidea**

4.2.4.1. **Family Fosshageniidae** *Suárez-Morales and Iliffe, 1996*

Genus *Fosshagenia* Suárez-Morales and Iliffe, 1996
F. ferrarii Suárez-Morales and Iliffe, 1996

4.2.5. Superfamily Centropagoidea

4.2.5.1. Family Acartiidae Sars, 1900

Characteristic of estuarine and neritic environments throughout the world: only *Acartia danae* and *A. negligens* live in the open ocean (Tranter, 1977).

Genus *Acartia* Dana, 1846
A. adriatica Steuer, 1910; *A. africana* Steuer, 1915; *A. amboinensis* Carl, 1907; *A. australis* Farran, 1936; *A. bacorehuisensis* Zamora-Sánchez and Gomez-Aguirre, 1985; *A. baylyi* Greenwood, 1972; *A. bermudensis* Esterly, 1911; *A. bifilosa* (Giesbrecht, 1882); *A. bilobata* Abraham, 1970; *A. bispinosa* Carl, 1907, ♀; *A. bowmani* Abraham, 1976; *A. brevicornis* Brady, 1883; *A. californiensis* Trinast, 1976; *A. centrura* Giesbrecht, 1889; *A. chilkaensis* Sewell, 1919; *A. clausi* Giesbrecht, 1889; *A. danae* Giesbrecht, 1889; *A. denticornis* Brady, 1883; *A. discaudata* (Giesbrecht, 1881); *A. dubia* T. Scott, 1894; *A. dweepi* Haridas and Madhupratap, 1978; *A. ensifera* Brady, 1918; *A. enzoi* Crisafi, 1975; *A. erythraea* Giesbrecht, 1889; *A. fancetti* McKinnon, Kimmerer and Benzie, 1992; *A. floridana* Davis, 1948; *A. forcipata* Thompson and Scott, 1897; *A. fossae* Gurney, 1927; *A. giesbrechti* Dahl, 1894; *A. grani* Sars, 1904; *A. gravelyi* Sewell, 1919; *A. hamata* Mori, 1937, ♂; *A. hudsonica* Pinhey, 1926; *A. iseana* Ito, 1956; *A. italica* Steuer, 1910; *A. japonica* Mori, 1940; *A. jilletti* Bradford, 1976; *A. josephinae* Crisafi, 1975; *A. kempi* Sewell, 1914; *A. keralensis* Wellershaus, 1969; *A. latisetosa* (Krichagin, 1873); *A. laxa* Dana, 1853; *A. lefevreae* Bradford, 1976; *A. levequei* Grice, 1964; *A. lilljeborgi* Giesbrecht, 1889; *A. longipatella* Connell and Grindley, 1974; *A. longiremis* (Lilljeborg, 1853); *A. longisetosa* Brady, 1914; *A. macropus* Cleve, 1901; *A. major* Sewell, 1919; *A. margalefi* Alcarez, 1976; *A. minor* Sewell, 1919; *A. mossi* (Norman, 1878); *A. natalensis* Connell and Grindley, 1974; *A. negligens* Dana, 1849; *A. nicolae* Dussart, 1985; *A. omorii* Bradford, 1976; *A. pacifica* Steuer, 1915; *A. pietschmanni* Pesta, 1912, ♀; *A. plumosa* T. Scott, 1894; *A. ransoni* Rose, 1953; *A. remivagantis* Oliveira, 1945, ♂; *A. sarojus* Madhupratap and Harridas, 1994; *A. sewelli* Steuer, 1934; *A. simplex* Sars, 1905; *A. sinensis* Shen and Lee, 1963; *A. sinjiensis* Mori, 1940; *A. southwelli* Sewell, 1914; *A. spinata* Esterly, 1911; *A. spinicauda* Giesbrecht, 1889; *A. steueri* Smirnov, 1936; *A. teclae* Bradford, 1976; *A. tonsa* Dana, 1849; *A. tortaniformis* Sewell, 1912; *A. tranteri* Bradford, 1976; *A. tropica* Ueda and Hiromi, 1987; *A. tsuensis* Ito, 1956, ♀; *A. tumida* Willey, 1920.
Steuer (1923) reviews this genus. Madhupratap and Haridas (1994) question the status of the subgenera *Acanthacartia* Steuer, 1915; *Acartiella* Sewell, 1914; *Acartiura* Steuer, 1915; *Euacartia* Steuer, 1915; *Hypoacartia* Steuer,

1915; *Odontacartia* Steuer, 1915; *Paracartia* T. Scott, 1894 and *Planktacartia* Steuer, 1915. Redescriptions of *A. australis* and *A. japonica* are given by Ueda (1986a).

Genus ***Paralabidocera*** Wolfenden, 1908
P. antarctica (I.C. Thompson, 1898); *P. grandispina* Waghorn, 1979; *P. separabilis* Brodsky and Zvereva, 1976.

4.2.5.2. **Family Candaciidae** *Giesbrecht, 1892*

Genus ***Candacia*** Dana, 1846
C. armata (Boeck, 1873); *C. bipinnata* Giesbrecht, 1889; *C. bradyi* A. Scott, 1902; *C. caribbeanensis* Park, 1974; *C. catula* (Giesbrecht, 1889); *C. cheirura* Cleve, 1904; *C. columbiae* Campbell, 1929; *C. curta* (Dana, 1849); *C. discaudata* A. Scott, 1909; *C. elongata* (Boeck, 1873); *C. ethiopica* (Dana, 1849); *C. falcifera* Farran, 1929; *C. giesbrechti* Grice and Lawson, 1977; *C. grandis* Tanaka, 1964; *C. guggenheimi* Grice and Jones, 1960; *C. guinensis* Chahsavar-Archard and Razouls, 1983, ♀; *C. ketchumi* Grice, 1961; *C. longimana* (Claus, 1863); *C. magna* Sewell, 1932; *C. maxima* Vervoort, 1957; *C. nigrocincta* (Thompson, 1888); *C. norvegica* (Boeck, 1865); *C. pachydactyla* (Dana, 1849); *C. paenelongimana* Fleminger and Bowman, 1956; *C. parafalcifera* Brodsky, 1950; *C. pofi* Grice and Jones, 1960; *C. rotunda* Wolfenden, 1904; *C. samassae* Pesta, 1941; *C. tenuimana* (Giesbrecht, 1889); *C. tuberculata* Wolfenden, 1905; *C. varicans* (Giesbrecht, 1892).
Review by Grice (1963).

Genus ***Paracandacia*** Grice, 1963
P. bispinosa (Claus, 1863); *P. simplex* (Giesbrecht, 1889); *P. truncata* (Dana, 1849); *P. worthingtoni* Grice, 1981, ♀.

4.2.5.3. **Family Centropagidae** *Giesbrecht, 1892*

Evolution of the Australian genera of this family is discussed by Maly (1996).

Genus ***Boeckella*** de Guerne and Richard, 1889
Fresh water genus reviewed by Bayly (1979). It is not included in the key to genera in the Chapter 3.

Genus ***Calamoecia*** Brady, 1906
Fresh water genus reviewed by Bayly (1979). It is not included in the key to genera in the Chapter 3.

Genus *Centropages* Kröyer, 1849
C. abdominalis Sato, 1913; *C. acutus* McKinnon and Dixon, 1994; *C. alcocki* Sewell, 1912; *C. aucklandicus* Krämer, 1895; *C. australiensis* Fairbridge, 1944; *C. brachiatus* (Dana, 1849); *C. bradyi* Wheeler, 1900; *C. brevifurcus* Shen and Lee, 1963; *C. calaninus* (Dana, 1849); *C. caribbeanensis* Park, 1970; *C. chierchiae* Giesbrecht, 1889; *C. dorsispinatus* Thompson and Scott, 1903; *C. elegans* Giesbrecht, 1895; *C. elongatus* Giesbrecht, 1896; *C. furcatus* (Dana, 1852); *C. gracilis* (Dana, 1849); *C. halinus* McKinnon and Kimmerer, 1988; *C. hamatus* (Lilljeborg, 1853); *C. karachiensis* Haq and Fazal-ur-Rehman, 1973; *C. kroyeri* Giesbrecht, 1892; *C. longicornis* Mori, 1937; *C. mcmurrichi* Willey, 1920; *C. natalensis* Connell, 1981; *C. orsinii* Giesbrecht, 1889; *C. ponticus* Karawaev, 1895; *C. sinensis* Chen and Zhang, 1965; *C. tenuiremis* Thompson and Scott, 1903; *C. trispinosus* Sewell, 1914; *C. typicus* Kroyer, 1849; *C. velificatus* (de Oliveira, 1946); *C. violaceus* (Claus, 1863); *C. yamadai* Mori, 1937.
An essentially coastal genus. Descriptions in: Rose, 1933; Mori, 1937; Brodsky, 1950; Grice, 1962b; Tanaka, 1963; Vervoort, 1964; Owre and Foyo, 1967; Roe, 1975; Garcia-Rodriguez, 1985; Soler *et al.*, 1988.

Genus *Gippslandia* Bayly and Arnott, 1969
G. estuarina Bayly and Arnott, 1969.

Genus *Gladioferens* Henry, 1919
G. imparipes Thompson, 1946; *G. inermis* Nicholls, 1944; *G. pectinatus* (Brady, 1899); *G. spinosus* Henry, 1919; *G. symmetricus* Bayly, 1963.
Australian estuarine species associated with sediment and vegetation. Description in: Rippingale, 1994.

Genus *Hemiboeckella* Sars, 1912
Fresh water genus reviewed by Bayly (1979). It is not included in the key to genera in the Chapter 3.

Genus *Isias* Boeck, 1864
I. clavipes Boeck, 1864; *I. tropica* Sewell, 1924; *I. uncipes* Bayly, 1964.

Genus *Limnocalanus* Sars, 1863
L. grimaldii (de Guerne, 1886); *L. johanseni* Marsh, 1920; *L. macrurus* Sars, 1863.
Fresh and brackish environments. Descriptions in: Sars, 1903; Brodsky, 1950; Lindquist, 1961.

Genus *Pseudoboeckella* Mràzek, 1901
Fresh water genus reviewed by Bayly (1992), who treats it as synonym of

genus *Boeckella*, and Menu-Marque and Zúñiga (1994). It is not included in the key to genera in the Chapter 3.

Genus **Pseudolovenula** Marukawa, 1921
P. magna Marukawa, 1921
Not recorded since, although large at 5.1 mm total length. It is not included in the key to genera in the Chapter 3.

Genus **Sinocalanus** Burckhardt, 1913
S. doerrii (Brehm, 1909); *S. laevidactylus* Shen and Tai, 1964; *S. sinensis* (Poppe, 1889); *S. solstitialis* Shen and Lee, 1963; *S. tenellus* (Kikuchi, 1928)
Descriptions in: Hiromi and Ueda (1987), Zheng Zhong *et al.* (1989).

4.2.5.4. **Family Diaptomidae** *Sars, 1903*

Numerous genera, all essentially fresh water. Reviews in Reid (1987). The genera are not included in the key to genera in the Chapter 3.

4.2.5.5. **Family Parapontellidae** *Giesbrecht, 1892*

Genus **Neopontella** A. Scott, 1909
N. typica A. Scott, 1909.

Genus **Parapontella** Brady, 1878
P. brevicornis (Lubbock, 1857).

4.2.5.6. **Family Pontellidae** *Dana, 1853*

Genus **Anomalocera** Templeton, 1837
A. opalus Pennell, 1976; *A. ornata* Sutcliffe, 1949; *A. patersoni* Templeton, 1837.

Genus **Calanopia** Dana, 1853
C. americana F. Dahl, 1894; *C. aurivilli* Cleve, 1901; *C. australica* Bayly and Greenwood, 1966; *C. biloba* Bowman, 1957; *C. elliptica* (Dana, 1846); *C. herdmani* A. Scott, 1909; *C. media* Gurney, 1927; *C. minor* A. Scott, 1902; *C. parathompsoni* Gaudy, 1969; *C. sarsi* Wilson, 1950; *C. sewelli* Jones and Park, 1967; *C. seymouri* Pillai, 1969; *C. thompsoni* A. Scott, 1909.

Genus **Epilabidocera** Wilson, 1932
E. amphitrites (McMurrich, 1916); *E. longipedata* (Sato, 1913), ♂.

Genus *Ivellopsis* Claus, 1893
I. elephas (Brady, 1883).
Re-described by Wickstead and Krishnaswamy (1964).

Genus *Labidocera* Lubbock, 1853
L. acuta (Dana, 1849); *L. acutifrons* (Dana, 1849); *L. aestiva* Wheeler, 1901; *L. antiguae* Fleminger, 1979; *L. barbadiensis* Fleminger and Moore, 1977; *L. barbudae* Fleminger, 1979; *L. bataviae* A. Scott, 1909; *L. bengalensis* Krishnaswamy, 1952; *L. bipinnata* Tanaka, 1936; *L. brunescens* (Czerniavsky, 1868); *L. carpentariensis* Fleminger, Othman and Greenwood, 1982; *L. caudata* Nicholls, 1944; *L. cervi* Kramer, 1895; *L. dakini* Greenwood, 1978; *L. detruncata* (Dana, 1849); *L. diandra* Fleminger, 1967; *L. euchaeta* Giesbrecht, 1889; *L. farrani* Greenwood and Othman, 1979; *L. fluviatilis* F. Dahl, 1894; *L. insolita* Wilson, 1950; *L. jaafari* Othman, 1986; *L. japonica* Mori, 1935; *L. johnsoni* Fleminger, 1964; *L. jollae* Esterly, 1906; *L. kolpos* Fleminger, 1967; *L. kroyeri* (Brady, 1883); *L. laevidentata* (Brady, 1883); *L. lubbockii* Giesbrecht, 1889; *L. madurae* A. Scott, 1909; *L. minuta* Giesbrecht, 1889; *L. mirabilis* Fleminger, 1957; *L. moretoni* Greenwood, 1978; *L. nerii* (Kroyer, 1849); *L. orsinii* Giesbrecht, 1889, ♀; *L. panamae* Fleminger and Moore, 1977; *L. papuensis* Fleminger, Othman and Greenwood, 1982; *L. pavo* Giesbrecht, 1889; *L. pectinata* Thompson and Scott, 1903; *L. pseudacuta* Silas and Pillai, 1969; *L. rotunda* Mori, 1929; *L. scotti* Giesbrecht, 1897; *L. sinolobata* Shen and Lee, 1963; *L. tasmanica* Taw, 1974; *L. tenuicauda* Wilson, 1950; *L. trispinosa* Esterly, 1905; *L. wilsoni* Fleminger and Tan, 1966; *L. wollastoni* (Lubbock, 1857).
Reviews by Fleminger (1975, 1979).

Genus *Pontella* Dana, 1846
P. agassizi Giesbrecht, 1895; *P. alata* A. Scott, 1909, ♀; *P. andersoni* Sewell, 1912; *P. asymmetrica* Geinrikh, 1967; *P. atlantica* (Milne-Edwards, 1840); *P. cerami* A. Scott, 1909, ♂; *P. chierchiae* Giesbrecht, 1889; *P. cristata* Krämer, 1896, ♀; *P. danae* Giesbrecht, 1889, ♂; *P. denticauda* A. Scott, 1909; *P. diagonalis* Wilson, 1950; *P. elegans* (Claus, 1892); *P. fera* Dana, 1849; *P. forficula* A. Scott, 1909; *P. gaboonensis* T. Scott, 1894; *P. gracilis* Wilson, 1950, ♀; *P. hanloni* Greenwood, 1979, ♂; *P. indica* Chiba, 1956; *P. inermis* Brady, 1883, ♂; *P. investigatoris* Sewell, 1912; *P. karachiensis* Fazal-ur-Rehman, 1973; *P. kieferi* Pesta, 1933; *P. latifurca* Chen and Zhang, 1965; *P. lobiancoi* (Canu, 1888); *P. marplatensis* Ramirez, 1966; *P. meadii* Wheeler, 1900; *P. mediterranea* (Claus, 1863); *P. mimocerami* Fleminger, 1957; *P. natalis* Brady, 1915; *P. novae-zealandiae* Farran, 1929; *P. patagoniensis* (Lubbock, 1853); *P. pennata* Wilson, 1932; *P. polydactyla* Fleminger, 1957; *P. princeps* Dana, 1849; *P. pulvinata* Wilson, 1950; *P. rostraticauda* Ohtsuka, Fleminger and Onbe, 1987; *P. securifer* Brady, 1883; *P. sewelli* Geinrikh,

1987; *P. sinica* Chen and Zhang, 1965, ♀; *P. spinicauda* Mori, 1937; *P. spinipedata* Geinrikh, 1989; *P. spinipes* Giesbrecht, 1889; *P. surrecta* Wilson, 1950; *P. tenuiremis* Giesbrecht, 1889; *P. tridactyla* Shen and Lee, 1963; *P. valida* Dana, 1853; *P. whiteleggei* Kramer, 1896.

Genus **Pontellina** Dana, 1853
P. morii Fleminger and Hülsemann, 1974; *P. platychela* Fleminger and Hülsemann, 1974; *P. plumata* (Dana, 1849); *P. sobrina* Fleminger and Hülsemann, 1974.
Review by Fleminger and Hülsemann (1974).

Genus **Pontellopsis** Brady, 1883
P. albatrossi Wilson, 1950, ♀; *P. armata* (Giesbrecht, 1889); *P. bitumida* Wilson, 1950; *P. brevis* (Giesbrecht, 1889); *P. digitata* Wilson, 1950, ♀; *P. elongatus* Wilson, 1932; *P. globosa* Wilson, 1950, ♀; *P. herdmani* Thompson and Scott, 1903, ♀; *P. inflatodigitata* Chen and Shen, 1974; *P. krameri* (Giesbrecht, 1896), ♀; *P. laminata* Wilson, 1950, ♀; *P. lubbockii* (Giesbrecht, 1889); *P. macronyx* A. Scott, 1909; *P. occidentalis* Esterly, 1906; *P. pacifica* Chiba, 1953; *P. perspicax* Dana, 1849; *P. pexa* A. Scott, 1909, ♀; *P. regalis* Dana, 1849; *P. scotti* Sewell, 1932; *P. sinuata* Wilson, 1950; *P. strenua* (Dana, 1849); *P. tasmaniensis* Greenwood, 1978; *P. tenuicauda* (Giesbrecht, 1889); *P. villosa* Brady, 1883; *P. yamadai* Mori, 1937.

4.2.5.7. **Family Pseudodiaptomidae** *Sars, 1902*

Genus **Archidiaptomus** Madhupratap and Haridas, 1978
A. aroorus Madhupratap and Haridas, 1978.
Recorded in Cochin backwaters, India.

Genus **Calanipeda** Krichagin, 1873
C. aquae-dulcis Krichagin, 1873.
It is not included in the key to genera in the Chapter 3.

Genus **Pseudodiaptomus** Herrick, 1884
P. acutus (Dahl, 1894); *P. americanus* Wright, 1937, ♂; *P. andamanensis* Pillai, 1980; *P. annandalei* Sewell, 1919; *P. ardjuna* Brehm, 1953; *P. aurivilli* Cleve, 1901; *P. australiensis* Walter, 1987; *P. batillipes* Brehm, 1954; *P. baylyi* Walter, 1984; *P. binghami* Sewell, 1912; *P. bispinosus* Walter, 1984; *P. bowmani* Walter, 1984; *P. brehmi* Kiefer, 1938; *P. bulbiferus* (Rose, 1957); *P. bulbosus* (Shen and Tai, 1964); *P. burckhardti* Sewell, 1932; *P. caritus* Walter, 1986a; *P. charteri* Grindley, 1963; *P. clevei* A. Scott, 1909; *P. cokeri* Gonzalez and Bowman, 1965; *P. colefaxi* Bayly, 1966; *P. compactus* Walter,

1984; *P. cornutus* Nicholls, 1944; *P. coronatus* Williams, 1906; *P. cristobalensis* Marsh, 1913; *P. culebrensis* Marsh, 1913; *P. dauglishi* Sewell, 1932; *P. diadelus* Walter, 1986a; *P. dubius* Kiefer, 1936; *P. euryhalinus* Johnson, 1939; *P. forbesi* (Poppe and Richard, 1890); *P. galapagensis* Grice, 1964; *P. galleti* (Rose, 1957); *P. gracilis* (Dahl, 1894); *P. griggae* Walter, 1987; *P. hessei* (Mrázek, 1894); *P. heterothrix* Brehm, 1953; *P. hickmani* Sewell, 1912; *P. hypersalinus* Walter, 1987; *P. incisus* Shen and Lee, 1963; *P. inflatus* (Shen and Tai, 1964); *P. inflexus* Walter, 1987; *P. inopinus* Burckhardt, 1913; *P. ishigakiensis* Nishida, 1985; *P. jonesi* Pillai, 1970; *P. lobipes* Gurney, 1907; *P. longispinosus* Walter, 1989; *P. malayalus* Wellershaus, 1969; *P. marinus* Sato, 1913; *P. marshi* Wright, 1936; *P. masoni* Sewell, 1932; *P. mertoni* Früchtl, 1924; *P. mixtus* Walter, 1994; *P. nankauriensis* Roy, 1977; *P. nihonkaiensis* Hirakawa, 1983; *P. nostradamus* Brehm, 1933; *P. occidentalis* Walter, 1987; *P. ornatus* (Rose, 1957), ♀; *P. pacificus* Walter, 1986a; *P. panamensis* Walter, 1989; *P. pankajus* Madhupratap and Haridas, 1992; *P. pauliani* Brehm, 1951; *P. pelagicus* Herrick, 1884; *P. penicillus* Li Shaojing and Huang Jiaqi, 1984; *P. philippinensis* Walter, 1986a; *P. poplesia* (Shen, 1955); *P. poppei* Stingelin, 1900; *P. richardi* (Dahl, 1894); *P. salinus* (Giesbrecht, 1896); *P. serricaudatus* (T. Scott, 1894); *P. sewelli* Walter, 1984; *P. smithi* Wright, 1928; *P. spatulus* (Shen and Tai, 1964); *P. stuhlmanni* (Poppe and Mrázek, 1895); *P. tollingeri* Sewell, 1919; *P. trihamatus* Wright, 1937; *P. trispinosus* Walter, 1986a; *P. wrighti* Johnson, 1964.

Species inhabit fresh to hypersaline waters in most tropical and temperate coastal areas where they are primarily benthopelagic. Grindley (1984) and Walter (1986b) review aspects of this genus. Descriptions in: Pillai, 1980; Grindley, 1984; Walter, 1984, 1986a,b, 1987, 1989; Madhupratap and Haridas, 1992.

4.2.5.8. **Family Sulcanidae** *Nicholls, 1945*

Genus ***Sulcanus*** Nicholls, 1945
S. conflictus Nicholls, 1945.
Australian estuarine species.

4.2.5.9. **Family Temoridae** *Giesbrecht, 1892*

Genus ***Epischura*** Forbes, 1882
Fresh water genus. It is not included in the key to genera in the Chapter 3.

Genus *Eurytemora* Giesbrecht, 1881
E. affinis (Poppe, 1880); *E. americana* Williams, 1906; *E. anadyrensis* Borutzky, 1961; *E. arctica* Wilson and Tash, 1966; *E. asymmetrica* Smirnov, 1935; *E. bilobata* Akatova, 1949; *E. canadensis* Marsh, 1920; *E. composita* Sars, 1897; *E. foveola* Johnson, 1961; *E. gracilicauda* Attsatova, 1949; *E. gracilis* (Sars, 1898); *E. grimmi* Sars, 1897; *E. herdmani* Thompson and Scott, 1897; *E. hirundo* Giesbrecht, 1881; *E. hirundoides* (Nordquist, 1888); *E. inermis* (Boeck, 1864); *E. kieferi* Smirnov, 1931; *E. kurenkovi* Borutzky, 1961; *E. lacustris* (Poppe, 1887); *E. pacifica* Sato, 1913; *E. raboti* Richard, 1897; *E. richingsi* Heron and Damkaer, 1976; *E. thompsoni* Willey, 1923; *E. transversalis* Campbell, 1930; *E. velox* (Lilljeborg, 1853); *E. wolteckeri* Mann, 1940; *E. yukonensis* Wilson, 1953.
Busch and Brenning (1992) propose *E. hirundoides* as a synonym of *E. affinis*.

Genus *Ganchosia* Oliveira, 1946
G. littoralis Oliveira, 1946, ♀.

Genus *Heterocope* Sars, 1863
H. appendiculata Sars, 1862; *H. borealis* (Fischer, 1851); *H. saliens* (Lilljeborg, 1863); *H. septentrionalis* Juday and Muttowski, 1915.
Coastal, brackish and/or fresh water species. Descriptions in: Sars, 1903.

Genus *Lahmeyeria* Oliveira, 1946
L. turrisphari Oliveira, 1946, ♀.

Genus *Manaia* Oliveira, 1946
M. velificata Oliveira, 1946, ♀.

Genus *Temora* Baird, 1850
T. discaudata Giesbrecht, 1889; *T. kerguelensis* Wolfenden, 1911, ♂; *T. longicornis* (O.F. Müller, 1792); *T. stylifera* (Dana, 1849); *T. turbinata* (Dana, 1849).

Genus *Temoropia* T. Scott, 1894
T. mayumbaensis T. Scott, 1894; *T. minor* Deevey, 1972; *T. setosa* Schulz, 1986.
Review by Deevey (1972). Descriptions in: Wheeler, 1970.

4.2.5.10. **Family Tortanidae** *Sars, 1902*

Genus *Tortanus* Giesbrecht, 1898
T. barbatus (Brady, 1883), ♀; *T. bonjol* Othman, 1987; *T. bowmani* Othman,

1987; *T. brevipes* A. Scott, 1909, ♀; *T. capensis* Grindley, 1978; *T. compernis* Gonzalez and Bowman, 1965; *T. denticulatus* Shen and Lee, 1963; *T. derjugini* Smirnov, 1935; *T. dextrilobatus* Chen and Zhang, 1965; *T. digitalis* Ohtsuka and Kimoto, 1989; *T. discaudatus* (Thompson and Scott, 1897); *T. erabuensis* Ohtsuka, Fukuura and Go, 1987; *T. forcipatus* (Giesbrecht, 1889); *T. giesbrechti* Jones and Park, 1968; *T. gracilis* (Brady, 1883); *T. longipes* Brodsky, 1948; *T. lophus* Bowman, 1971; *T. murrayi* A. Scott, 1909; *T. recticauda* (Giesbrecht, 1889); *T. rubidus* Tanaka, 1965; *T. ryukyuensis* Ohtsuka and Kimoto, 1989; *T. scaphus* Bowman, 1971; *T. setacaudatus* Williams, 1906; *T. sheni* Hülsemann, 1988a; *T. sinensis* Chen, 1983; *T. spinicaudatus* Shen and Bai, 1956; *T. tropicus* Sewell, 1932; *T. vermiculus* Shen, 1955.

Ohtsuka and Kimoto (1989) review the subgenus *Atortus* Sewell, 1932. *Tortanus* species inhabit coastal waters of the world except the Antarctic, and eastern North Atlantic. Ohtsuka *et al.* (1995) re-describe *T. derjugini*.

4.2.6. Superfamily Megacalanoidea

4.2.6.1. **Family Calanidae** *Dana, 1849*

Bradford (1988) has reviewed this family and restructured the genera.

Genus *Calanoides* Brady, 1883
C. acutus (Giesbrecht, 1902); *C. carinatus* (Kröyer, 1849); *C. macrocarinatus* Brodsky, 1967; *C. natalis* Brady, 1914; *C. patagoniensis* Brady, 1883; *C. philippinensis* Kitou and Tanaka, 1969, ♀.

Genus *Calanus* Leach, 1819
C. agulhensis De Decker, Kaczmaruk and Marska, 1991; *C. australis* Brodsky, 1959; *C. chilensis* Brodsky, 1959; *C. euxinus* Hülsemann, 1991; *C. finmarchicus* (Gunnerus, 1765); *C. fonsecai* Oliveira, 1945, ♀; *C. glacialis* Jaschnov, 1955; *C. helgolandicus* (Claus, 1863); *C. hyperboreus* Kröyer, 1838; *C. jaschnovi* Hülsemann, 1994; *C. magellanicus* (Dana, 1853); *C. marshallae* Frost, 1974; *C. minor* (Claus, 1863); *C. pacificus* Brodsky, 1948; *C. propinquus* Brady, 1883; *C. simillimus* Giesbrecht, 1902; *C. sinicus* Brodsky, 1965.
Descriptions in: Sars, 1903; Brodsky, 1950; Rose, 1933; Park, 1968; Bradford, 1971c, 1972; Frost, 1974; Fleminger and Hülsemann, 1987; Hülsemann, 1991b; Bucklin *et al.*, 1995.

Genus *Canthocalanus* A. Scott, 1909
C. pauper (Giesbrecht, 1888).

Genus *Cosmocalanus* Bradford and Jillett, 1974
C. caroli (Giesbrecht, 1888); *C. darwini* (Lubbock, 1860).

Genus *Mesocalanus* Bradford and Jillett, 1974
M. lighti (Bowman, 1955); *M. tenuicornis* (Dana, 1849).
Descriptions in: Brodsky, 1950.

Genus *Neocalanus* Sars, 1925
N. cristatus (Kröyer, 1848); *N. flemingeri* Miller, 1988; *N. gracilis* (Dana, 1849); *N. plumchrus* (Marukawa, 1921); *N. robustior* (Giesbrecht, 1888); *N. tonsus* (Brady, 1883).

Genus *Undinula* A. Scott, 1909
U. vulgaris (Dana, 1849).

4.2.6.2. **Family Mecynoceridae** *Andronov, 1973*

Genus *Mecynocera* Thompson, 1888
M. clausi Thompson, 1888.
Description in: Corral Estrada, 1972a.

4.2.6.3. **Family Megacalanidae** *Sewell, 1947*

Genus *Bathycalanus* Sars, 1905
B. bradyi (Wolfenden, 1905), ♀; *B. eltaninae* Björnberg, 1967; *B. eximius* Brodsky, Vyshkvartseva, Kos and Markhaseva, 1983; *B. inflatus* Björnberg, 1967, ♀; *B. princeps* (Brady, 1883), ♀; *B. richardi* Sars, 1905; *B. sverdrupi* Johnson, 1958, ♀; *B. unicornis* Björnberg, 1967, ♀.
Michel (1994) has synonymized *B. bradyi* with *B. richardi*. Descriptions in: Sars, 1925; Sewell, 1947; Johnson, 1958; Björnberg, 1967a; Brodsky *et al.*, 1983.

Genus *Bradycalanus* A. Scott, 1909
B. gigas Sewell, 1947, ♀; *B. pseudotypicus* Björnberg, 1967, ♀; *B. p. enormis* Björnberg, 1967, ♀; *B. sarsi* (Farran, 1939); *B. typicus* A. Scott, 1909, ♀.
Descriptions in: A. Scott, 1909; Sewell, 1947; Björnberg, 1967a.

Genus *Megacalanus* Wolfenden, 1904
M. princeps Wolfenden, 1904.
Description in: Sars, 1925 as *M. longicornis*; Owre and Foyo, 1967.

4.2.6.4. Family Paracalanidae Giesbrecht, 1892

This family has been restructured by Andronov. *Parvocalanus* species have a single-segmented En on P1 and so key out along with *Mecynocera* species.

Genus ***Acrocalanus*** Giesbrecht, 1888
A. andersoni Bowman, 1958, ♀; *A. gibber* Giesbrecht, 1888; *A. gracilis* Giesbrecht, 1888; *A. indicus* Tanaka, 1960, ♂; *A. longicornis* Giesbrecht, 1888; *A. monachus* Giesbrecht, 1888.
Descriptions in: Sewell, 1929; Grice, 1962b.

Genus ***Bestiolina*** Andronov, 1991
B. amoyensis Li Shaojing and Huang Jiaqi, 1984; *B. inermis* (Sewell, 1912); *B. similis* (Sewell, 1914); *B. sinicus* (Shen and Lee, 1966); *B. zeylonica* (Andronov, 1972a).
This genus was formerly *Bestiola* Andronov, 1972a (see Andronov, 1991).

Genus ***Calocalanus*** Giesbrecht, 1888
C. aculeatus Shmeleva, 1987a; *C. adriaticus* Shmeleva, 1973; *C. africanus* Shmeleva, 1979, ♀; *C. alboranus* Shmeleva, 1979, ♀; *C. antarcticus* Shmeleva, 1978; *C. atlanticus* Shmeleva, 1975, ♀; *C. beklemishevi* Shmeleva, 1987b; *C. contractus* Farran, 1926; *C. curtus* Andronov, 1973; *C. dellacrocei* Shmeleva, 1987b; *C. elegans* Shmeleva, 1965; *C. elongatus* Shmeleva, 1968; *C. equilicauda* Bernard, 1958; *C. fiolentus* Shmeleva, 1978; *C. fusiformis* Shmeleva, 1978, ♀; *C. gracilis* Tanaka, 1956; *C. gresei* Shmeleva, 1973; *C. indicus* Shmeleva, 1974; *C. kristalli* Shmeleva, 1968, ♀; *C. latus* Shmeleva, 1968, ♀; *C. lomonosovi* Shmeleva, 1975; *C. longifurca* Shmeleva, 1975; *C. longisetosus* Shmeleva, 1965; *C. longispinus* Shmeleva, 1978; *C. minor* Shmeleva, 1975, ♀; *C. minutus* Andronov, 1973, ♀; *C. monospinus* Chen and Zhang, 1974; *C. namibiensis* Andronov, 1973, ♀; *C. nanus* Shmeleva, 1987a; *C. neptunus* Shmeleva, 1965; *C. omaniensis* Shmeleva, 1975; *C. ovalis* Shmeleva, 1967; *C. paracontractus* Shmeleva, 1974, ♀; *C. parelongatus* Shmeleva, 1979; *C. pavo* (Dana, 1849); *C. pavoninus* Farran, 1936; *C. plumatus* Shmeleva, 1967a, ♀; *C. pseudocontractus* Bernard, 1958, ♀; *C. pubes* Andronov, 1973, ♀; *C. pyriformis* Shmeleva, 1975, ♀; *C. regini* Shmeleva, 1987a; *C. sayademalja* Shmeleva, 1987b; *C. spinosus* Shmeleva, 1987b; *C. styliremis* Giesbrecht, 1888; *C. tenuiculus* Andronov, 1973, ♀; *C. vinogradovi* Shmeleva, 1987b; *C. vitjazi* Shmeleva, 1974, ♀; *C. vivesi* Shmeleva, 1979.
Descriptions in: Bernard, 1958; Corral Estrada, 1972a; Shmeleva, 1987a,b.

Genus *Delius* Andronov, 1972
D. nudus (Sewell, 1929); *D. sewelli* Björnberg, 1979.

Genus *Ischnocalanus* Bernard, 1963
I. equalicauda (Bernard, 1958); *I. gracilis* (Tanaka, 1956); *I. plumulosus* (Claus, 1863); *I. tenuis* (Farran, 1926), ♀.
Descriptions in: Tanaka, 1956b; Bernard, 1958; Corral Estrada, 1972a.

Genus *Paracalanus* Boeck, 1864
P. aculeatus Giesbrecht, 1888; *P. brevispinatus* Shen and Lee, 1966; *P. campaneri* Björnberg, 1980: *P. denudatus* Sewell, 1929, ♀; *P. gracilis* Chen and Zhang, 1965, ♀; *P. indicus* Wolfenden, 1905; *P. intermedius* Shen and Bai, 1956; *P. mariae* Brady, 1918; *P. nanus* Sars, 1907; *P. parvus* (Claus, 1863); *P. ponticus* (Krichagin, 1873); *P. quasimodo* Bowman, 1971; *P. serratipes* Sewell, 1912; *P. serrulus* Shen and Lee, 1963; *P. tropicus* Andronov, 1977, ♀.
Kang (1996) redescribes *P. parvus* and *P. indicus*.

Genus *Parvocalanus* Andronov, 1970
P. crassirostris (Dahl, 1894); *P. dubia* (Sewell, 1912); *P. elegans* Andronov, 1972; *P. latus* Andronov, 1972; *P. scotti* (Früchtl, 1923).
Descriptions in: Vervoort, 1963.

4.2.7. Superfamily Bathypontioidea

4.2.7.1. Family Bathypontiidae Brodsky, 1950

Genus *Alloiopodus* Bradford, 1969
A. pinguis Bradford, 1969, ♀.
Bradford's description is incomplete, only one of the five pairs of legs being undamaged in the two females examined; a more complete description is required before it can be entered in the key to genera.

Genus *Bathypontia* Sars, 1905
B. elegans Sars, 1920; *B. elongata* Sars, 1905; *B. intermedia* Deevey, 1973, ♀; *B. kanaevae* Björnberg, 1976, ♀; *B. longicornis* Tanaka, 1965, ♂; *B. longiseta* Brodsky, 1950, ♀; *B. major* (Wolfenden, 1911), ♀; *B. michelae* Deevey, 1979, ♀; *B. minor* (Wolfenden, 1911); *B. regalis* Grice and Hülsemann, 1967, ♂; *B. sarsi* Grice and Hülsemann, 1965; *B. similis* Tanaka, 1965; *B. spinifera* A. Scott, 1909; *B. unispina* Deevey, 1979, ♀; *Bathypontia* sp. Wheeler, 1970, ♀.

Genus ***Temorites*** Sars, 1900
T. brevis Sars, 1900; *T. discoveryae* Grice and Hülsemann, 1965.
Segmentation of P1 to P4 is different in these species.

Genus ***Zenkevitchiella*** Brodsky, 1955
Z. abyssalis Brodsky, 1955; *Z. atlantica* Grice and Hülsemann, 1965; *Z. crassa* Grice and Hülsemann, 1967, ♂; *Z. tridentae* Wheeler, 1970, ♀.
Differences between these species are discussed in the type papers. All have one-segmented En on P1 except *Z. tridentae* where they are two-segmented.

4.2.8. Superfamily Eucalanoidea

4.2.8.1. Family Eucalanidae *Giesbrecht, 1892*

Genus ***Eucalanus*** Dana, 1853
E. attenuatus (Dana, 1849); *E. bungii* Giesbrecht, 1892; *E. californicus* Johnson, 1938; *E. crassus* Giesbrecht, 1888; *E. dentatus* A. Scott, 1909; *E. elongatus* (Dana, 1849); *E. hyalinus* (Claus, 1866); *E. inermis* Giesbrecht, 1892; *E. langae* Fleminger, 1973; *E. longiceps* Matthews, 1925; *E. monachus* Giesbrecht, 1888; *E. mucronatus* Giesbrecht, 1888; *E. muticus* Wilson, 1950; *E. parki* Fleminger, 1973; *E. peruanus* Volkov, 1971; *E. pileatus* Giesbrecht, 1888; *E. quadrisetosus* Geletin, 1973, ♂; *E. sewelli* Fleminger, 1973; *E. subcrassus* Giesbrecht, 1888; *E. subtenuis* Giesbrecht, 1888.
Fleminger (1973) has reviewed this genus, while Geletin (1976) has proposed a new classification that divides the species between 3 genera: *Eucalanus* and 2 new genera, *Pareucalanus* Geletin and *Subeucalanus* Geletin. Arcos and Fleminger (1986) accept these genera as does Bradford-Grieve (1994). They have not been entered in the Key to genera here.

Genus ***Rhincalanus*** Dana, 1853
R. cornutus (Dana, 1849); *R. gigas* Brady, 1883; *R. nasutus* Giesbrecht, 1888; *R. rostrifrons* Dana, 1852.
Descriptions in: Rose, 1933; Tanaka, 1960a; Owre and Foyo, 1967.

4.2.9. Superfamily Ryocalanoidea

4.2.9.1. Family Ryocalanidae *Andronov, 1974*

Genus ***Ryocalanus*** Tanaka, 1956
R. admirabilis Andronov, 1992; *R. asymmetricus* Markhaseva and Ferrari,

1995, ♀; *R. bicornis* Markhaseva and Ferrari, 1995, ♀; *R. bowmani* Markhaseva and Ferrari, 1995, ♂; *R. infelix* Tanaka, 1956, ♂.

4.2.10. Superfamily Spinocalanoidea

4.2.10.1. Family Spinocalanidae *Vervoort, 1951*

Genus *Damkaeria* Fosshagen, 1983
D. falcifera Fosshagen, 1983.
This is a hyperbenthic copepod.

Genus *Foxtonia* Hülsemann and Grice, 1963
F. barbatula Hülsemann and Grice, 1963.

Genus *Isaacsicalanus* Fleminger, 1983
I. paucisetus Fleminger, 1983, ♀.
Description in: Schulz, 1987.

Genus *Kunihulsea* Schulz, 1992
K. arabica Schulz, 1992, ♀.

Genus *Mimocalanus* Farran, 1908
M. brodskii C. Razouls, 1974, ♀; *M. crassus* Park, 1970; *M. cultrifer* Farran, 1908; *M. damkaeri* Brodsky, Vyshkvartseva, Kos and Markhaseva, 1983; *M. heronae* Damkaer, 1975; *M. inflatus* Davis, 1949, ♀; *M. major* Sars, 1920, ♀; *M. nudus* Farran, 1908; *M. ovalis* (Grice and Hülsemann, 1965), ♀; *M. sulcifrons* Wheeler, 1970.
Review by Damkaer, 1975.

Genus *Monacilla* Sars, 1905
M. gracilis (Wolfenden, 1911), ♀; *M. tenera* Sars, 1907, ♀; *M. typica* Sars, 1905; *Monacilla* sp. Wheeler, 1970, ♂.
Damkaer (1975) has reviewed this genus. Descriptions in: Sars, 1925; Tanaka, 1956b; Schulz, 1987.

Genus *Rhinomaxillaris* Grice and Hülsemann, 1967
R. bathybia Grice and Hülsemann, 1967, ♀.
Grice and Hülsemann established this genus and species on the basis of a single damaged female; a more complete description is required before it can be entered in the key to genera.

Genus *Sognocalanus* Fosshagen, 1967
S. confertus Fosshagen, 1967.

Genus *Spinocalanus* Giesbrecht, 1888
S. abruptus Grice and Hülsemann, 1965, ♀; *S. abyssalis* Giesbrecht, 1888; *S. angusticeps* Sars, 1920; *S. antarcticus* Wolfenden, 1906; *S. aspinosus* Park, 1970, ♀; *S. brevicaudatus* Brodsky, 1950; *S. dispar* Schulz, 1987, ♀; *S. hirtus* Sars, 1907, ♀; *S. hoplites* Park, 1970, ♀; *S. horridus* Wolfenden, 1911; *S. longicornis* Sars, 1900; *S. macrocephalon* Brodsky, Vyshkvartseva, Kos and Markhaseva, 1983; *S. magnus* Wolfenden, 1904; *S. oligospinosus* Park, 1970, ♀; *S. polaris* Brodsky, 1950; *S. profundalis* Brodsky, 1955; *S. similis* Brodsky, 1950; *S. spinosus* Farran, 1908; *S. terranovae* Damkaer, 1975; *S. usitatus* Park, 1970, ♀; *S. validus* Sars, 1920; *Spinocalanus* sp. Grice and Hülsemann, 1967, ♀; *Spinocalanus* sp. A Roe, 1975, ♂; *Spinocalanus* sp. B Roe, 1975.
Descriptions in: Damkaer, 1975; Schulz, 1987.

Genus *Teneriforma* Grice and Hülsemann, 1967
T. meteorae Schulz, 1989, ♀; *T. naso* (Farran, 1936); *T. pentatrichodes* Schulz, 1989, ♀.
Descriptions in: Damkaer, 1975.

4.2.11. Superfamily Clausocalanoidea

4.2.11.1. Family Aetideidae *Giesbrecht, 1892*

Reviews by Brodsky (1950), Koeller and Littlepage (1976), Bradford and Jillett (1980). The most recent review by Markhaseva (1996) has not been obtained by the author and reference should be made to it in any work on the taxonomy of this family.

Genus *Aetideopsis* Sars, 1903
A. albatrossae Shih and Maclellan, 1981; *A. antarctica* (Wolfenden, 1908); *A. armata* (Boeck, 1872); *A. carinata* Bradford, 1969; *A. cristata* Tanaka, 1957; *A. divaricata* Esterly, 1911, ♀; *A. inflata* Park, 1978; *A. minor* (Wolfenden, 1911), ♀; *A. modesta* (With, 1915), ♀; *A. multiserrata* (Wolfenden, 1904); *A. retusa* Grice and Hülsemann, 1967, ♀; *A. rostrata* Sars, 1903; *A. trichechus* Vervoort, 1949, ♀; *A. tumorosa* Bradford, 1969; *Aetideopsis* sp. Bradford, 1972, ♀.
Review by Bradford and Jillett (1980).

Genus *Aetideus* Brady, 1883
A. acutus Farran, 1929; *A. arcuatus* (Vervoort, 1949); *A. armatus* (Boeck, 1872); *A. australis* (Vervoort, 1957); *A. bradyi* A. Scott, 1909; *A. divergens* Bradford, 1971; *A. giesbrechti* Cleve, 1904; *A. mexicanus* Park, 1974; *A. pseudarmatus* Bradford, 1971; *A. truncatus* Bradford, 1971.
Bradford (1971a) has completely revised this genus, and Bradford and Jillett

(1980) adopt Roe's (1975) inclusion of *Snelliaetideus* Vervoort, 1949 in this genus. Park (1978) does not mention Bradford's (1971a) revision.

Genus *Azygokeras* Koeller and Littlepage, 1976.
A. columbiae Koeller and Littlepage, 1976.
This species keys out along with *Farrania orbus* described by Tanaka (1956b) as *Drepanopsis orbus*.

Genus *Batheuchaeta* Brodsky, 1950
B. anomala Markhaseva, 1981, ♀; *B. antarctica* Markhaseva, 1986b, ♀; *B. brodskyi* Markhaseva, 1981, ♀; *B. enormis* Grice and Hülsemann, 1968, ♀; *B. gurjanovae* (Brodsky, 1955); *B. heptneri* Markhaseva, 1981, ♀; *B. lamellata* Brodsky, 1950; *B. peculiaris* Markhaseva, 1983; *B. pubescens* Markhaseva, 1986b; ♀; *B. tuberculata* Markhaseva, 1986b, ♀.
According to Vaupel Klein (1984), *B. enormis* does not belong to this genus.

Genus *Bradyetes* Farran, 1905
B. brevis Farran, 1905, ♀; *B. florens* Grice and Hülsemann, 1967, ♀; *B. inermis* Farran, 1905, ♀; *B. matthei* Johannessen, 1976.
Review by Bradford and Jillett (1980). Descriptions in: Grice, 1972. Male described by Johannessen (1976).

Genus *Bradyidius* Giesbrecht, 1897
B. angustus Tanaka, 1957, ♂; *B. armatus* Giesbrecht, 1897; *B. arnoldi* Fleminger, 1957; *B. bradyi* (Sars, 1903); *B. curtus* Markhaseva, 1993, ♀; *B. hirsutus* Bradford, 1976; *B. luluae* Grice, 1972; *B. pacificus* (Brodsky, 1950); *B. plinioi* Campaner, 1978; *B. rakuma* (Zvereva, 1976), ♀; *B. saanichi* Park, 1966; *B. similis* (Sars, 1903); *B. spinifer* Bradford, 1969; *B. styliformis* Othman and Greenwood, 1987; *B. subarmatus* Markhaseva, 1993; *B. tropicus* (Wolfenden, 1905); *Bradyidius* sp. Bradford, 1972, ♀; *Bradyidius* sp. Greenwood, 1977, ♀.
Reviewed by Markhaseva (1993).

Genus *Chiridiella* Sars, 1907
C. abyssalis Brodsky, 1950, ♀; *C. atlantica* Wolfenden, 1911, ♀; *C. bichela* Deevey, 1974, ♀; *C. bispinosa* Park, 1970, ♀; *C. brachydactyla* Sars, 1907; *C. brooksi* Deevey, 1974, ♀; *C. chainae* Grice, 1969, ♀; *C. gibba* Deevey, 1974, ♀; *C. kuniae* Deevey, 1974, ♀; *C. macrodactyla* Sars, 1907, ♀; *C. megadactyla* Bradford, 1971, ♀; *C. ovata* Deevey, 1974, ♀; *C. pacifica* Brodsky, 1950; *C. reducta* Brodsky, 1950; *C. sarsi* Markhaseva, 1983, ♀; *C. smoki* Markhaseva, 1983, ♀; *C. subaequalis* Grice and Hülsemann, 1965, ♀; *C. trihamata* Deevey, 1974, ♀.

The irregular reduction of segmentation in the legs causes difficulties in the key to genera. Bradford and Jillett (1980) review the genus.

Genus *Chiridius* Giesbrecht, 1892
C. armatus Boeck, 1872; *C. carnosus* Tanaka, 1957, ♂; *C. gracilis* Farran, 1908; *C. longispinus* Tanaka, 1957, ♂; *C. mexicanus* Park, 1975; *C. obtusifrons* Sars, 1903; *C. pacificus* Brodsky, 1950, ♀; *C. polaris* Wolfenden, 1911; *C. poppei* Giesbrecht, 1892; *C. subantarcticus* Park, 1978; *C. subgracilis* Park, 1975.
Review by Bradford and Jillett (1980).

Genus *Chirundina* Giesbrecht, 1895
C. antarctica Wolfenden, 1911, ♀; *C. indica* Sewell, 1929; *C. streetsi* Giesbrecht, 1895.
Review by Bradford and Jillett (1980).

Genus *Chirundinella* Tanaka, 1957
C. magna (Wolfenden, 1911).
Review by Bradford and Jillett (1980).

Genus *Comantenna* Wilson, 1924
C. brevicornis (Boeck, 1872); *C. crassa* Bradford, 1969, ♀; *C. curtisetosa* Alvarez, 1986, ♀; *C. recurvata* Grice and Hülsemann, 1970, ♀.
Descriptions in: Alvarez, 1986.

Genus *Crassantenna* Bradford, 1969
C. comosa Bradford, 1969, ♀; *C. mimorostrata* Bradford, 1969, ♀.

Genus *Euchirella* Giesbrecht, 1888
E. amoena Giesbrecht, 1888; *E. bella* Giesbrecht, 1888; *E. bitumida* With, 1915; *E. curticauda* Giesbrecht, 1888; *E. formosa* Vervoort, 1949; *E. galeata* Giesbrecht, 1888; *E. grandicornis* Wilson, 1950; *E. latirostris* Farran, 1929; *E. lisettae* Vaupel Klein, 1989, ♀; *E. maxima* Wolfenden, 1905; *E. messinensis* (Claus, 1863); *E. orientalis* Sewell, 1929; *E. paulinae* Vaupel Klein, 1980, ♀; *E. pseudopulchra* Park, 1976, ♀; *E. pseudotruncata* Park, 1976; *E. pulchra* (Lubbock, 1856); *E. rostrata* (Claus, 1866); *E. rostromagna* Wolfenden, 1911; *E. similis* Wolfenden, 1911, ♀; *E. speciosa* Grice and Hülsemann, 1968, ♀; *E. splendens* Vervoort, 1963; *E. tanseii* Omori, 1965, ♀; *E. truncata* Esterly, 1911; *E. unispina* Park, 1968; *E. venusta* Giesbrecht, 1888; *Euchirella* sp. Séret, 1979, ♀.
Review by Vaupel Klein (1984).

Genus *Gaetanus* Giesbrecht, 1888
G. antarcticus Wolfenden, 1905, ♀; *G. armiger* Giesbrecht, 1888; *G. brachyurus* Sars, 1907, ♀; *G. brevicaudatus* (Sars, 1907), ♀; *G. brevicornis* Esterly, 1906; *G. campbellae* Park, 1975; *G. curvicornis* Sars, 1905, ♀; *G. divergens* Wolfenden, 1911, ♀; *G. intermedius* Wolfenden, 1905; *G. kruppi* Giesbrecht, 1904; *G. latifrons* Sars, 1905; *G. microcanthus* Wilson, 1950, ♀; *G. miles* Giesbrecht, 1888; *G. minispinus* Tanaka, 1969, ♀; *G. minor* Farran, 1905; *G. paracurvicornis* Brodsky, 1950; *G. pileatus* Farran, 1904; *G. recticornis* Wolfenden, 1911, ♀; *G. tenuispinus* (Sars, 1900); *G. wolfendeni* Park, 1975, ♀; *Gaetanus* sp. Paiva, 1963, ♂.
Review by Bradford and Jillett (1980).

Genus *Gaidiopsis* A. Scott, 1909
G. crassirostris A. Scott, 1909, ♀.

Genus *Gaidius* Giesbrecht, 1895
G. affinis Sars, 1905; *G. brevirostris* Brodsky, 1950; *G. brevispinus* (Sars, 1900); *G. columbiae* Park, 1967; *G. inermis* (Sars, 1905); *G. intermedius* Wolfenden, 1905; *G. minutus* Sars, 1907, ♀; *G. pungens* Giesbrecht, 1895; *G. robustus* (Sars, 1905); *G. variabilis* Brodsky, 1950; *Gaidius* sp. Tanaka, 1969, ♀; *Gaidius* sp. Tanaka and Omori, 1970, ♂.
Review by Bradford and Jillett (1980).

Genus *Jaschnovia* Markhaseva, 1980
J. tolli (Linko, 1913); *J. johnsoni* Markhaseva, 1980.
Previously in the genus *Derjuginia* Jaschnov, 1947.

Genus *Lutamator* Bradford, 1969
L. elegans Alvarez, 1984, ♀; *L. hurleyi* Bradford, 1969, ♀.

Genus *Mesocomantenna* Alvarez, 1986
M. spinosa Alvarez, 1986, ♀.

Genus *Paivella* Vervoort, 1965
P. inaciae Vervoort, 1965; *P. naporai* Wheeler, 1970.

Genus *Paracomantenna* Campaner, 1978
P. gracilis Alvarez, 1986, ♀; *P. magalyae* Campaner, 1978, ♀; *P. minor* (Farran, 1905), ♀.
Descriptions in: Alvarez, 1986.

Genus *Pseudeuchaeta* Sars, 1905
P. arctica Markhaseva, 1986c; *P. brevicauda* Sars, 1905; *P. flexuosa* Bradford,

1969, ♀; *P. magna* Bradford, 1969, ♀; *P. major* (Wolfenden, 1911), ♀; *P. spinata* Markhaseva, 1986c, ♀.
Review by Bradford and Jillett (1980).

Genus *Pseudochirella* Sars, 1920
P. accepta Zvereva, 1976, ♀; *P. batillipa* Park, 1978; *P. bilobata* Vervoort, 1949, ♀; *P. bowmani* Markhaseva, 1986a, ♀; *P. calcarata* Sars, 1920, ♀; *P. cryptospina* Sars, 1905; *P. dentata* (A. Scott, 1909), ♀; *P. divaricata* (Sars, 1905), ♀; *P. dubia* (Sars, 1920); *P. elongata* (Wolfenden, 1911); *P. formosa* Markhaseva, 1989, ♀; *P. gibbera* Vervoort, 1949, ♀; *P. granulata* (A. Scott, 1909), ♀; *P. gurjanovae* Brodsky, 1955, ♂; *P. hirsuta* (Wolfenden, 1911); *P. limata* Grice and Hülsemann, 1968, ♀; *P. lobata* Sars, 1907, ♀; *P. mariana* Markhaseva, 1989, ♀; *P. mawsoni* Vervoort, 1957; *P. major* (Sars, 1907); *P. notacantha* (Sars, 1905); *P. obesa* Sars, 1920; *P. obtusa* (Sars, 1905); *P. pacifica* Brodsky, 1950; *P. palliata* (Sars, 1907), ♀; *P. polyspina* Brodsky, 1950; *P. pustulifera* (Sars, 1905); *P. scopularis* (Sars, 1905), ♀; *P. semispina* Vervoort, 1949, ♀; *P. spectabilis* (Sars, 1900); *P. spinosa* (Wolfenden, 1911), ♀; *P. squalida* Grice and Hülsemann, 1967, ♀; *P. tanakai* Markhaseva, 1989, ♂; *P. vervoorti* Tanaka and Omori, 1969, ♀; *Pseudochirella* sp. Bradford, 1972, ♀.
Review by Bradford and Jillett (1980). Markhaseva (1989) provides a key to the species. Vaupel Klein (1995) renames *P. fallax* Sars, 1920 as *P. major* (Sars, 1907). Vaupel Klein and Rijerkerk (1996) redescribe *P. obesa* and Vaupel Klein (1996) designates this species as the type species of the genus.

Genus *Pseudotharybis* T. Scott, 1909
P. brevispinus (Bradford, 1969), ♀; *P. dentatus* (Bradford, 1969), ♂; *P. magnus* (Grice and Hülsemann, 1970), ♂; *P. robustus* (Bradford, 1969); *P. spinibasis* (Bradford, 1969), ♂; *P. zetlandicus* T. Scott, 1909, ♀.
Review by Bradford and Jillett (1980).

Genus *Pterochirella* Schulz, 1990
P. tuerkayi Schulz, 1990, ♂.

Genus *Senecella* Juday, 1923
S. calanoides Juday, 1923; *S. siberica* Vyshkvartseva, 1994.
S. calanoides occurs in fresh water lakes of North America while *S. siberica* occurs in brackish waters of the Kara and Laptev Seas.

Genus *Sursamucro* Bradford, 1969
S. spinatus Bradford, 1969, ♀.

Genus *Undeuchaeta* Giesbrecht, 1888
U. bispinosa Esterly, 1911; *U. incisa* Esterly, 1911; *U. intermedia* A. Scott, 1909, ♀; *U. magna* Tanaka, 1957, ♀; *U. major* Giesbrecht, 1888; *U. plumosa* (Lubbock, 1856).
Review by Bradford and Jillett (1980).

Genus *Valdiviella* Steuer, 1904
V. brevicornis Sars, 1905, ♀; *V. brodskyi* Zvereva, 1975, ♀; *V. ignota* Sewell, 1929, ♂; *V. imperfecta* Brodsky, 1950, ♀; *V. insignis* Farran, 1908; *V. minor* Wolfenden, 1911; *V. oligarthra* Steuer, 1904.
Review by Bradford and Jillett (1980).

Genus *Wilsonidius* Tanaka, 1969
W. alaskaensis Tanaka, 1969, ♀.

4.2.11.2 **Family Clausocalanidae** *Giesbrecht, 1888*

Genus *Clausocalanus* Giesbrecht, 1888
C. arcuicornis (Dana, 1849); *C. brevipes* Frost and Fleminger, 1968; *C. dubius* Brodsky, 1950, ♂; *C. farrani* Sewell, 1929; *C. furcatus* (Brady, 1883); *C. ingens* Frost and Fleminger, 1968; *C. jobei* Frost and Fleminger, 1968; *C. laticeps* Farran, 1929; *C. latipes* Scott, 1894, ♂; *C. lividus* Frost and Fleminger, 1968; *C. mastigophorus* (Claus, 1863); *C. minor* Sewell, 1929; *C. parapergens* Frost and Fleminger, 1968; *C. paululus* Farran, 1926; *C. pergens* Farran, 1926.
Review by Frost and Fleminger (1968).

Genus *Ctenocalanus* Giesbrecht, 1888
C. campaneri Almeida Prado-Por, 1984; *C. citer* Heron and Bowman, 1971; *C. tageae* Almeida Prado-Por, 1984; *C. vanus* Giesbrecht, 1888.

Genus *Drepanopus* Brady, 1883
D. bispinosus Bayly, 1982; *D. bungei* Sars, 1898; *D. forcipatus* Giesbrecht, 1888; *D. pectinatus* Brady, 1883.
Reviews by Bayly (1982) and Hülsemann (1985a, 1991a). Spinules are present on Ex and En of P2 to P4 in some species but not in others. These are neritic species in the Antarctic Ocean.

Genus *Farrania* Sars, 1920
F. frigida (Wolfenden, 1911); *F. lyra* (Rose, 1937); *F. orbus* (Tanaka, 1956); *F. pacifica* (Brodsky, 1950), ♀.

Genus *Microcalanus* Sars, 1901
M. pusillus Sars, 1903; *M. pygmaeus* (Sars, 1900).

Genus *Pseudocalanus* Boeck, 1873
P. acuspes (Giesbrecht, 1881); *P. elongatus* (Boeck, 1864); *P. major* Sars, 1900; *P. mimus* Frost, 1989; *P. minutus* (Krøyer, 1845); *P. moultoni* Frost, 1989; *P. newmani* Frost, 1989.
This genus has been completely revised by Frost (1989).

Genus *Spicipes* Grice and Hülsemann, 1965
S. nanseni Grice and Hülsemann, 1965, ♀.

4.2.11.3. **Family Diaixidae** *Sars, 1903*

Othman and Greenwood (1994) redefine this family.

Genus *Anawekia* Othman and Greenwood, 1994
A. bilobata Othman and Greenwood, 1994; *A. robusta* Othman and Greenwood, 1994; *A. spinosa* Othman and Greenwood, 1994.

Genus *Diaixis* Sars, 1903
D. asymmetrica Grice and Hülsemann, 1970, ♀; *D. centrura* Connell, 1981; *D. durani* Corral Estrada, 1972; *D. gambiensis* Andronov, 1978; *D. helenae* Andronov, 1978; *D. hibernica* (A. Scott, 1896); *D. pygmaea* (T. Scott, 1899); *D. tridentata* Andronov, 1974, ♀; *D. trunovi* Andronov, 1978.
Descriptions in: Andronov, 1978.

4.2.11.4. **Family Euchaetidae** *Giesbrecht, 1892*

This family consists of distinctive and often large-sized species that have, on occasion, been ascribed to two genera *Euchaeta* and *Pareuchaeta*. Park (1994a) recounts the history of the 2 genera and re-examines their validity. He redefines the genera and ascribes 16 species to the genus *Euchaeta* and 82 to the genus *Pareuchaeta*. Most, but not all, species belong to 3 *Euchaeta* species groups and 6 *Pareuchaeta* species groups. Reference should be made to Park (1994a,b) for further information.

Genus *Euchaeta* Philippi, 1843
E. acuta Giesbrecht, 1892; *E. concinna* Dana, 1849; *E. indica* Wolfenden, 1905; *E. longicornis* Giesbrecht, 1888; *E. magniloba* Park, 1978, ♀; *E. marina* (Prestandrea, 1833); *E. marinella* Bradford, 1974; *E. media* Giesbrecht, 1888; *E. paraacuta* Tanaka, 1973; *E. paraconcinna* Fleminger, 1957; *E. plana* Mori, 1937; *E. pubera* Sars, 1907; *E. rimana* Bradford, 1974; *E. spinosa* Giesbrecht, 1892; *E. tenuis* Esterly, 1906; *E. wrighti* Park, 1968, ♀.

Genus *Pareuchaeta* A. Scott, 1909
P. abbreviata (Park, 1978); *P. abrikosovi* Geptner, 1971, ♀; *P. abyssalis* Brodsky, 1950; *P. abyssaloides* Geptner, 1987, ♂; *P. aequatorialis* Tanaka, 1958; *P. affinis* (Cleve, 1904), ♀; *P. alaminae* (Park, 1975), ♀; *P. altibulla* Park, 1994a; *P. anfracta* Park, 1994a; *P. antarctica* (Giesbrecht, 1902); *P. austrina* (Giesbrecht, 1902); *P. barbata* (Brady, 1883); *P. biloba* Farran, 1929; *P. birostrata* Brodsky, 1950; *P. bisinuata* (Sars, 1907); *P. bradyi* (With, 1915); *P. brevirostris* Brodsky, 1950; *P. bulbirostris* Heptner, 1987, CV; *P. californica* (Esterly, 1906); *P. calva* Tanaka, 1958; *P. comosa* Tanaka, 1958; *P. confusa* Tanaka, 1958; *P. copleyae* Park, 1994a; *P. dactylifera* (Park, 1978), ♀; *P. elongata* (Esterly, 1913); *P. eltaninae* (Park, 1978); *P. eminens* Tanaka and Omori, 1968, ♀; *P. erebi* Farran, 1929; *P. euryrhina* Park, 1994a; *P. exigua* (Wolfenden, 1911); *P. flava* (Giesbrecht, 1888); *P. glacialis* (Hansen, 1887); *P. gracilicauda* A. Scott, 1909; *P. gracilis* (Sars, 1905); *P. grandiremis* (Giesbrecht, 1888); *P. guttata* Heptner, 1971, ♀; *P. hanseni* (With, 1915); *P. hastata* Heptner, 1987, ♀; *P. hebes* (Giesbrecht, 1888); *P. implicata* Heptner, 1971, ♀; *P. incisa* (Sars, 1905); *P. investigatoris* Sewell, 1929; *P. kurilensis* Heptner, 1971, ♀; *P. longisetosa* Heptner, 1971, ♀; *P. malayensis* Sewell, 1929; *P. megaloba* Park, 1994a; *P. mexicana* Park, 1994a; *P. modesta* Brodsky, 1950, ♀; *P. norvegica* (Boeck, 1872); *P. oculata* Heptner, 1971, ♀; *P. orientalis* Brodsky, 1950, ♀; *P. papilliger* Park, 1994a; *P. parabbreviata* Park, 1994a; *P. paraprudens* Park, 1994a; *P. parvula* (Park, 1978); *P. pavlovskii* Brodsky, 1955, ♀; *P. perplexa* Heptner, 1987, ♂; *P. plaxiphora* Park, 1994a; *P. plicata* Heptner, 1971, ♀; *P. polaris* Brodsky, 1950, ♀; *P. prima* Heptner, 1971, ♀; *P. propinqua* (Esterly, 1906), ♀; *P. prudens* Tanaka and Omori, 1968, ♀; *P. pseudotonsa* (Fontaine, 1967); *P. rasa* Farran, 1929; *P. regalis* (Grice and Hülsemann, 1968); *P. robusta* (Wolfenden, 1911), ♀; *P. rotundirostris* Heptner, 1987, ♀; *P. rubicunda* (Farran, 1908), ♀; *P. rubra* Brodsky, 1950; *P. russelli* (Farran, 1936); *P. sarsi* (Farran, 1908); *E. scaphula* Fontaine, 1967; *P. scopaeorhina* Park, 1994a; *P. scotti* (Farran, 1908); *P. sesquipedalis* Park, 1994a; *P. sibogae* A. Scott, 1909, ♀; *P. similis* (Wolfenden, 1908); *P. simplex* Tanaka, 1958, ♀; *P. subtilirostris* Heptner, 1971, ♀; *P. tonsa* (Giesbrecht, 1895); *P. tridentata* Heptner, 1987, ♂; *P. triloba* Park, 1994a; *P. tuberculata* A. Scott, 1909; *P. tumidula* (Sars, 1905); *P. tycodesma* (Park, 1978); *P. vervoorti* (Park, 1978), ♀; *P. vorax* (Grice and Hülsemann, 1968); *P. weberi* A. Scott, 1909.

4.2.11.5. **Family Mesaiokeratidae** *Matthews, 1961*

Genus *Mesaiokeras* Matthews, 1961
M. heptneri Andronov, 1973; *M. kaufmanni* Fosshagen, 1978, ♂; *M. nanseni* Matthews, 1961; *M. semiplenus* Andronov, 1973; *M. tantillus* Andronov, 1973.

4.2.11.6. **Family Parkiidae** *Ferrari and Markhaseva, 1996*

Genus *Parkius* Ferrari and Markhaseva, 1996
P. karenwishnerae Ferrari and Markhaseva, 1996.

4.2.11.7. **Family Phaennidae** *Sars, 1902*

Genus *Brachycalanus* Farran, 1905
B. atlanticus (Wolfenden, 1904), ♀; *B. bjornbergae* Campaner, 1978, ♀; *B. minutus* Grice, 1972, ♀; *B. ordinarius* (Grice, 1973), ♀; *B. rothlisbergi* Othman and Greenwood, 1988a, ♀.
These are hyperbenthic copepods.

Genus *Cephalophanes* Sars, 1907
C. frigidus Wolfenden, 1911; *C. refulgens* Sars, 1907; *C. tectus* (Esterly, 1911), ♀.

Genus *Cornucalanus* Wolfenden, 1905
C. antarcticus Brodsky and Zvereva, 1976, ♀; *C. chelifer* (I.C. Thompson, 1903); *C. indicus* Sewell, 1929; *C. notabilis* Brodsky and Zvereva, 1976, ♀; *C. robustus* Vervoort, 1957, ♀; *C. sewelli* Vervoort, 1957, ♀; *C. simplex* Wolfenden, 1905, ♀; *Cornucalanus* sp. Séret, 1979.
Review by Bradford *et al.* (1983).

Genus *Onchocalanus* Sars, 1905
O. affinis With, 1915; *O. cristatus* (Wolfenden, 1904); *O. hirtipes* Sars, 1905; *O. latus* Esterly, 1911, ♀; *O. magnus* (Wolfenden, 1906); *O. paratrigoniceps* Park, 1983b; *O. scotti* Vervoort, 1950, ♀; *O. subcristatus* (Wolfenden, 1906), ♀; *O. trigoniceps* Sars, 1905; *O. wolfendeni* Vervoort, 1950, ♀.
Reviewed by Bradford *et al.* (1983).

Genus *Phaenna* Claus, 1863
P. spinifera Claus, 1863; *P. zetlandica* T. Scott, 1902, ♂.
Descriptions in: Bradford *et al.*, 1983.

Genus *Talacalanus* Wolfenden, 1911
T. greeni (Farran, 1905); *T. maximus* (Brodsky, 1950).
Bradford *et al.* (1983) reviews this genus. Descriptions in: Sars, 1925 as *Xanthocalanus greeni*; Tanaka and Omori, 1967 as *X. maximus*.

Genus *Xantharus* Andronov, 1981.
X. formosus Andronov, 1981.

Genus *Xanthocalanus* Giesbrecht, 1892
X. agilis Giesbrecht, 1892; *X. alvinae* Grice and Hülsemann, 1970, ♀; *X. amabilis* Tanaka, 1960, ♀; *X. antarcticus* Wolfenden, 1908, ♂; *X. borealis* Sars, 1900; *X. claviger* (T. Scott, 1909), ♂; *X. cornifer* (Tanaka, 1960), ♂; *X. crassirostris* (Tanaka, 1960), ♂; *X. difficilis* Grice and Hülsemann, 1965, ♀; *X. dilatus* Grice, 1962, ♀; *X. distinctus* Grice and Hülsemann, 1970, ♂; *X. echinatus* Sars, 1907; *X. elongatus* Grice and Hülsemann, 1970; *X. fallax* Sars, 1921; *X. giesbrechti* Thompson, 1903, ♀; *X. gracilis* Wolfenden, 1911, ♀; *X. groenlandicus* Tupitzky, 1982; *X. harpagatus* Bradford and Wells, 1983, ♀; *X. incertus* Sars, 1920, ♀; *X. irritans* (Tanaka, 1960), ♂; *X. kurilensis* Brodsky, 1950; *X. legatus* Tanaka, 1960, ♀; *X. macilenta* (Grice and Hülsemann, 1970), ♂; *X. macrocephalon* Grice and Hülsemann, 1970; *X. marlyae* Campaner, 1978, ♀; *X. medius* Tanaka, 1937, ♀; *X. minor* Giesbrecht, 1892; *X. mixtus* Sars, 1920, ♀; *X. multispinus* Chen and Zhang, 1965; *X. muticus* Sars, 1905, ♀; *X. obtusus* Farran, 1905, ♀; *X. oculata* (Tanaka, 1960), ♂; *X. paraincertus* Grice and Hülsemann, 1970; *X. pavlovskii* Brodsky, 1950, ♀; *X. pectinatus* Tanaka, 1960, ♂; *X. penicillatus* Tanaka, 1960, ♀; *X. pinguis* Farran, 1905; *X. polaris* Brodsky, 1950, ♀; *X. profundus* Sars, 1925, ♀; *X. propinquus* Sars, 1903, ♀; *X. pulcher* Esterly, 1911, ♀; *X. rotunda* (Grice and Hülsemann, 1970), ♂; *X. serrata* (Tanaka, 1960), ♂; *X. simplex* Aurivillius, 1898, ♂; *X. soaresmoreirai* Björnberg, 1975, ♂; *X. squamatus* Farran, 1936, ♀; *X. subagilis* Wolfenden, 1904; *X. tenuiremis* T. Scott, 1909, ♂; *X. tenuiserratus* Wolfenden, 1911, ♀; *X. typicus* (T. Scott, 1894), ♂.
Review by Bradford *et al.* (1983). Some of these species do not conform with the generic description e.g. *X. macroencephalon, X. paraincertus* which may belong to the Tharybidae (see Bradford *et al.*, 1983). Schulz and Beckmann (1995) suggest that *X. groenlandicus* should be transferred to the genus *Tharybis*.

4.2.11.8. **Family Pseudocyclopiidae** *Sars, 1903*

Genus *Pseudocyclopia* T. Scott, 1892
P. caudata T. Scott, 1894, ♀; *P. crassicornis* T. Scott, 1892; *P. giesbrechti* Wolfenden, 1902; *P. insignis* Andronov, 1986; *P. minor* T. Scott, 1892; *P. muranoi* Ohtsuka, 1992; *P. stephoides* I.C. Thompson, 1895.
Descriptions in: Andronov, 1986a.

Genus *Paracyclopia* Fosshagen, 1985
P. gitana Carola and Razouls, 1996; *P. naessi* Fosshagen, 1985.
Description in: Fosshagen and Iliffe (1985). A cave-dwelling species.

Genus *Stygocyclopia* Jaume and Boxshall, 1995b
S. balearica Jaume and Boxshall, 1995b.
A cave-dwelling species.

4.2.11.9. **Family Scolecitrichidae** Giesbrecht, 1892

Genus *Amallophora* T. Scott, 1894
A. elegans Wolfenden, 1911, ♀ and *A. impar* Wolfenden, 1911 may belong to the genus *Amallothrix*; *A. obtusifrons* Sars, 1905, ♀ may belong to the genus *Scaphocalanus*.

Genus *Amallothrix* Sars, 1925
A. arcuata (Sars, 1920); *A. dentipes* (Vervoort, 1951); *A. emarginata* (Farran, 1905); *A. falcifer* (Farran, 1926); *A. gracilis* (Sars, 1905); *A. hadrosoma* (Park, 1980); *A. indica* Sewell, 1929; *A. invenusta* Wilson, 1950, ♀; *A. longispina* Schulz, 1991; *A. parafalcifer* (Park, 1980); *A. pseudoarcuata* (Park, 1970), ♀; *A. pseudopropinqua* (Park, 1980); *A. robusta* (T. Scott, 1894), ♀.
Species probably in this genus: *Amallophora obtusifrons* Sars, 1905, ♀; *Amallothrix indica* Sewell, 1929; *A. invenusta* Wilson, 1950, ♀; *Scolecithricella curticauda* A. Scott, 1909, ♀; *S. denticulata* Tanaka, 1962, ♂; *S. incisa* Farran, 1929, ♀; *S. lanceolata* Tanaka, 1962, ♂; *S. lobata* Sars, 1920, ♀; *S. marquesae* Vervoort, 1965, ♀; *S. propinqua* Sars, 1920, ♀; *S. spinata* Tanaka, 1962, ♀; *S. timida* Tanaka, 1962, ♀; *Scolecithrix magnus* Wolfenden, 1911; *S. aculeata* Esterly, 1913, ♀; *S. elephas* Esterly, 1913, ♀; *S. medius* Wolfenden, 1911, ♀; *S. mollis* Esterly, 1913, ♀; *S. valens* Farran, 1926, ♀; *S. valida* Farran, 1908.
Review by Bradford *et al.* (1983).

Genus *Archescolecithrix* Vyshkvartseva, 1989b
A. auropecten (Giesbrecht, 1892).

Genus *Heteramalla* Sars, 1907
H. sarsi Roe, 1975, ♀.
Description in: Sars, 1925 as *H. dubia*.

Genus *Landrumius* Park, 1983a
L. antarcticus Park, 1983, ♀; *L. gigas* (A. Scott, 1909), ♀; *L. insignis* (Sars, 1920), ♀; *L. sarsi* (Wilson, 1950), ♀; *L. thorsoni* Björnberg, 1975, ♀.

Genus *Lophothrix* Giesbrecht, 1895
L. frontalis Giesbrecht, 1895; *L. humilifrons* Sars, 1905, ♀; *L. latipes* (T.

Scott, 1894); *L. quadrispinosa* Wolfenden, 1911, ♀; *L. similis* Wolfenden, 1911, ♀; *L. simplex* Wolfenden, 1911, ♀; *L. varicans* Wolfenden, 1911, ♀.
Review by Park (1983a). Males of *Lophothrix* and *Scaphocalanus* are closely similar.

Genus ***Macandrewella*** A. Scott, 1909
M. agassizi Wilson, 1950; *M. asymmetrica* Farran, 1936; *M. chelipes* (Giesbrecht, 1896); *M. cochinensis* Gopalakrishnan, 1973; *M. joanae* A. Scott, 1909; *M. mera* Farran, 1936, ♀; *M. scotti* Sewell, 1929; *M. sewelli* Farran, 1936.
Review by Gopalakrishnan (1973). Descriptions in: Campaner (1989).

Genus ***Mixtocalanus*** Brodsky, 1950
M. alter (Farran, 1929); *M. robustus* Brodsky, 1950; *M. vervoorti* (Park, 1980).
Genus revised by Vyshkvartseva (1989b).

Genus ***Parascaphocalanus*** Brodsky, 1955
P. zenkevitchi Brodsky, 1955.

Genus ***Pseudophaenna*** Sars, 1903
P. typica Sars, 1903.

Genus ***Puchinia*** Vyshkvartseva, 1989a
P. obtusa Vyshkvartseva, 1989a, ♀.

Genus ***Racovitzanus*** Giesbrecht, 1902
R. antarcticus Giesbrecht, 1902; *R. levis* Tanaka, 1961; *R. pacificus* (Esterly, 1905), ♀; *R. porrectus* (Giesbrecht, 1888), ♀; *Racovitzanus* sp. Grice and Hülsemann, 1967, ♂; *Racovitzanus* sp. Bradford, 1971, ♂.
Redefinition of genus by Park (1983a).

Genus ***Scaphocalanus*** Sars, 1900
Species definitely in this genus are: *S. acuminatus* Park, 1970, ♀; *S. acutocornis* Vyshkvartseva, 1987, ♀; *S. affinis* (Sars, 1905); *S. amplius* Park, 1970; *S. antarcticus* Park, 1982; *S. brevicornis* (Sars, 1900); *S. brevirostris* Park, 1970; *S. curtus* (Farran, 1926); *S. difficilis* Roe, 1975; *S. echinatus* (Farran, 1905); *S. farrani* Park, 1982; *S. invalidus* Hure and Scotto di Carlo, 1968; *S. longifurca* (Giesbrecht, 1888); *S. magnus* (T. Scott, 1894); *S. major* (T. Scott, 1894); *S. parantarcticus* Park, 1982; *S. paraustralis* Schulz, 1987, ♀; *S. pseudobrevirostris* Schulz, 1987, ♀; *S. similis* Hure and Scotto di Carlo, 1968; *S. subbrevicornis* (Wolfenden, 1911).
Species that may belong to this genus: *Amallophora elegans* Wolfenden,

1911, ♀; *A. impar* Wolfenden, 1911, ♀; *Amallothrix profunda* Brodsky, 1950, ♀; *Scaphocalanus angulifrons* Sars, 1920, ♀; *S. bogorovi* Brodsky, 1955; *S. californicus* Davis, 1949, ♀; *S. elongatus* A. Scott, 1909; *S. insignis* Brodsky, 1950, ♀; *S. insolitus* Wilson, 1950, ♀; *S. subelongatus* Brodsky, 1950, ♀. *Scolecithricella avia* Tanaka, 1962, ♂; *S. lobata* Sars, 1920, ♀; *S. obscura* Esterly, 1913, ♀; *S. polaris* Brodsky, 1950, ♀.
Bradford *et al.* (1983) have reviewed this genus. The majority of species have females with a fifth pair of legs present but the following do not: *S. amplius*, *S. curtus*, *S. similis* and *S. subcurtus*. The fifth legs of females usually have a basis and one segment but a few species have an additional segment. Males of *Scaphocalanus* and *Lophothrix* are closely similar.
Descriptions in: Tanaka (1961), Park (1970, 1982).

Genus *Scolecithricella* Sars, 1903
S. abyssalis (Giesbrecht, 1888); *S. cenotelis* Park, 1980; *S. dentata* (Giesbrecht, 1892); *S. minor* (Brady, 1883); *S. ovata* (Farran, 1905); *S. paramarginata* Schulz, 1991; *S. profunda* (Giesbrecht, 1892); *S. schizosoma* Park, 1980; *S. tropica* Grice, 1962, ♀; *S. vittata* (Giesbrecht, 1892).
The following species may belong in this genus: *Amallothrix farrani* Rose, 1942, ♀; *A. sarsi* Rose, 1942, ♀; *Scolecithricella globulosa* Brodsky, 1950; *S. lobophora* Park, 1970; *S. modica* Tanaka, 1962, ♀; *S. neptuni* Cleve, 1904; *S. obscura* Roe, 1975, ♀; *S. orientalis* Mori, 1937, ♀; *S. pacifica* Chiba, 1956, ♀; *S. pearsoni* Sewell, 1914; *S. unispinosa* Grice and Hülsemann, 1965, ♀; *S. vespertina* Tanaka, 1955, ♀; *Scolecithrix longipes* Giesbrecht, 1892, ♀; *S. longispinosa* Chen and Zhang, 1965; *S. marginata* Giesbrecht, 1888, ♀; *S. subdentata* Esterly, 1905, ♀; *S. subvittata* Rose, 1942, ♀; *S. tenuiserrata* Giesbrecht, 1892.
The following species have been placed in this genus, differ from the generic definition, and may have to be removed: *Scolecithricella aspinosa* Roe, 1975, ♀; *S. canariensis* Roe, 1975, ♀; *S. marquesae* Vervoort, 1965; *S. pseudoculata* Campaner, 1979; *S. spinacantha* Wilson, 1942; *Scolecithrix ctenopus* Giesbrecht, 1888; *S. fowleri* Farran, 1926; *S. grata* Grice and Hülsemann, 1967, ♀; *S. laminata* Farran, 1926; *S. maritima* Grice and Hülsemann, 1967; *S. tenuipes* T. Scott, 1894.
Genus reviewed by Bradford *et al.* (1983). The genus requires much further revision. Park (1980) gives a key to Antarctic species. Descriptions in: Ferrari and Steinberg (1993).

Genus *Scolecithrix* Brady, 1883
S. bradyi Giesbrecht, 1888; *S. danae* Lubbock, 1856. The following may also be in this genus: *S. birshteini* Brodsky, 1955; *S. nicobarica* Sewell, 1929. Reviewed by Bradford *et al.* (1983).

Genus *Scolecocalanus* Farran, 1936
S. galeatus Farran, 1936, ♀; *S. lobatus* Farran, 1936, ♀; *S. spinifer* Wilson, 1950.

Genus *Scopalatum* Roe, 1975
S. dubia (T. Scott, 1894), ♂; *S. farrani* Roe, 1975, ♀; *S. gibbera* Roe, 1975, ♀; *S. smithae* (Grice, 1962), ♀; *S. vorax* (Esterly, 1911), ♀.
Reviews by Roe (1975), Bradford *et al.* (1983) and Ferrari and Steinberg (1993). *Scopalatum vorax* has been found associated with larvacean houses (Steinberg *et al.*, 1994).

Genus *Scottocalanus* Sars, 1905
S. corystes Owre and Foyo, 1967; *S. dauglishi* Sewell, 1929; *S. farrani* A. Scott, 1909; *S. helenae* (Lubbock, 1856); *S. infrequens* Tanaka, 1969, ♀; *S. investigatoris* Sewell, 1929, ♂; *S. longispinus* A. Scott, 1909; *S. persecans* (Giesbrecht, 1892); *S. rotundatus* Tanaka, 1961; *S. securifrons* (T. Scott, 1894); *S. sedatus* Farran, 1936, ♀; *S. setosus* A. Scott, 1909, ♀; *S. terranovae* Farran, 1929; *S. thomasi* A. Scott, 1909; *S. thorii* With, 1915.
Review by Bradford *et al.* (1983) and Park (1983a).

Genus *Undinothrix* Tanaka, 1961
U. spinosa Tanaka, 1961, ♀.

4.2.11.10. **Family Stephidae** *Sars, 1903*

Genus *Miostephos* Bowman, 1976
M. cubrobex Bowman, 1976; *M. leamingtonensis* Yeatman, 1980.

Genus *Parastephos* Sars, 1903
P. esterlyi Fleminger, 1988; *P. occatum* Damkaer, 1971; *P. pallidus* Sars, 1903.

Genus *Stephos* T. Scott, 1892
S. antarcticus Wolfenden, 1908; *S. arcticus* Sars, 1909; *S. balearensis* Carola and Razouls, 1996; *S. canariensis* Boxshall, Stock and Sanchez, 1990; *S. deichmannae* Fleminger, 1957; *S. exumensis* Fosshagen, 1970, ♀; *S. fultoni* (T. and A. Scott, 1898); *S. gyrans* Giesbrecht, 1892, ♀; *S. kurilensis* Kos, 1972; *S. lamellatus* Sars, 1903; *S. longipes* Giesbrecht, 1902; *S. lucayensis* Fosshagen, 1970; *S. maculosus* Andronov, 1974, ♂; *S. margalefi* Riera, Vives and Gili, 1991; *S. minor* (T. Scott, 1892); *S. morii* Greenwood, 1978; *S. pacificus* Ohtsuka and Hiromi, 1987; *S. pentacanthos* Chen and Zhang, 1965, ♂; *S. robustus* Ohtsuka and Hiromi, 1987; *S. rustadi* Strömgren, 1969; *S.*

scotti Sars, 1903; *S. seclusum* Barr, 1984; *S. tropicus* Mori, 1942; *S. tsuyazakiensis* Tanaka, 1966.
Review by Riera *et al.* (1991).

4.2.11.11. **Family Tharybidae** *Sars, 1902*

Genus *Neoscolecithrix* Canu, 1896
N. antarctica Hülsemann, 1985b, ♀; *N. catenoi* Alvarez, 1985; *N. farrani* Smirnov, 1935; *N. koehleri* Canu, 1896; *N. magna* (Grice, 1972), ♀; *N. watersae* (Grice, 1972).
Review by Hülsemann (1985b). Some of these species have spinules on the surfaces of P4, others do not. Genus redefined by Alvarez (1985b).

Genus *Parundinella* Fleminger, 1957
P. dakini Bradford, 1973; *P. emarginata* Grice and Hülsemann, 1970; *P. manicula* Fleminger, 1957, ♀; *P. spinodenticula* Fleminger, 1957.
Review by Bradford *et al.* (1983).

Genus *Rythabis* Schulz, 1995
R. atlantica Schulz, 1995.
Described in Schulz and Beckman (1995).

Genus *Tharybis* Sars, 1903
T. altera (Grice and Hülsemann, 1970) ♂; *T. angularis* Schulz, 1995; *T. asymmetrica* Andronov, 1976, ♀; *T. compacta* (Grice and Hülsemann, 1970); *T. crenata* Schulz, 1995; *T. fultoni* Park, 1967; *T. macrophthalma* Sars, 1903; *T. magna* Bradford and Wells, 1983; *T. megalodactyla* Andronov, 1976; *T. minor* Schulz, 1981; *T. neptuni* (Cleve, 1904); *T. sagamiensis* Tanaka, 1960, ♀.
The following are possibly in this genus: *Xanthocalanus macrocephalon* Grice and Hülsemann, 1970, ♀; *X. paraincertus* Grice and Hülsemann, 1970, ♀. Schulz and Beckmann (1995) suggest that *X. groenlandicus* Tupitzky, 1982 should be transferred to *Tharybis*. Review by Schulz (1981) and Schulz and Beckmann (1995). Surface spinules on P2 to P4 may be few and very small in some species.

Genus *Undinella* Sars, 1900
U. acuta Vaupel Klein, 1970, ♀; *U. altera* Grice and Hülsemann, 1970, ♂; *U. brevipes* Farran, 1908; *U. compacta* Grice and Hülsemann, 1970; *U. frontalis* (Tanaka, 1937); *U. gricei* Wheeler, 1970, ♂; *U. hampsoni* Grice and Hülsemann, 1970; *U. oblonga* Sars, 1900; *U. spinifer* Tanaka, 1960, ♀; *U. stirni* Grice, 1971.
Review by Bradford *et al.* (1983). The segmentation of the legs varies between species and so the genus keys out at several places.

5. Gut, Food and Feeding

5.1. Feeding Appendages	140
5.2. Food Capture	143
5.2.1. Food Detection	145
5.2.2. Particle Feeding	146
5.2.3. Carnivory	148
5.3. Food and Foraging in the Environment	149
5.3.1. Natural Diets	149
5.3.2. Dissolved Organic Matter (DOM)	153
5.3.3. Detritus	153
5.3.4. Bacteria	154
5.3.5. Phytoplankton, Zooplankton	155
5.4. Foraging Tactics	160
5.4.1. Particulate Feeding	161
5.4.2. Predatory Feeding	162
5.4.3. Selectivity	164
5.5. Feeding of Young Stages	166
5.5.1. The Nauplii	166
5.5.2. Copepodids	167
5.6. Feeding Periodicity	167
5.6.1. Gut Fluorescence	167
5.6.2. Dietary Requirements	168
5.6.3. Short-term Periodicities	170
5.6.4. Diel Periodicities and Vertical Migration	170
5.6.5. Seasonal Periodicities	171
5.7. Concluding Remarks	172

Naupliar, copepodid and adult stages of calanoid copepods live in a wide range of environments: pelagic and benthopelagic, in coastal, shelf and oceanic regimes. The spectrum of potential food within these environments is subject to continuous change on scales measured in days and weeks. Changes in the population structure of calanoids, the incidence of naupliar, copepodid and adult stages, occur on scales of weeks and months. Increase in body size between egg and adult copepod, combined with ontogenetic development of the feeding appendages, results in different sources of food being utilized within those available at any one time.

Early studies of feeding mechanisms were somewhat simplistic in approach, as reviewed by Marshall (1973). Copepods with densely setose mouthparts were assumed to be herbivores whereas those with lightly setose mouthparts, often adorned with heavy spines, were assumed to be carnivores. The carnivores, for example most species of Euchaetidae, although often locally common, are not the dominant copepods in the sea. Consequently, emphasis naturally fell on studies of the common herbivores such as *Calanus* and *Pseudocalanus* species. In addition, production studies at the community or ecosystem level led to great interest in determining secondary production, especially that of copepods, and its dependence upon primary production. Estimation of grazing rates of calanoids on phytoplankton became of special interest. Many papers were published describing the filtering rates of species of copepods on named species of phytoplankton.

Assimilation coefficients were determined and the quality of different species of phytoplankton for growth of the copepods assessed. Much effort was expended in developing maintenance techniques for copepods, many species now being successfully grown through successive generations in the laboratory (see Table 47 on page 300).

Observation of feeding copepods using high-speed cinematography, however, has led to a questioning of the basic premises upon which many of the early experiments and observations were made (Price, 1988).

5.1. FEEDING APPENDAGES

The appendages used in feeding are the antennules, antennae, mandibles, maxillules and maxillae. They are most developed in CVs and adult females. Adult males of many species have reduced appendages and do not feed. Such appendages occur in the genera *Euchaeta*, *Euchirella*, *Gaetanus*, *Mesocalanus*, *Pareuchaeta*, *Pseudocalanus*, and *Rhincalanus* (Mazza, 1966; Mullin and Brooks, 1967; Lawson and Grice, 1973; Harding, 1974; Morioka, 1975; Schnack, 1978, 1982; Shuert and Hopkins, 1987; Marin, 1988a; Yen, 1988; Schnack-Schiel et al., 1991; Øresland and Ward, 1993). Both adult males and females of *Neocalanus cristatus*, in natural populations, have reduced mandibles and do not feed but, enigmatically, Hirakawa et al. (1995) found that they were developed in laboratory-reared individuals.

The antennules function primarily in predatory (raptorial) feeding, having sensilla that are used in the detection of prey; Landry (1980) removed the antennules of *Calanus pacificus* and found that the rate of feeding on phytoplankton was not affected but that predatory feeding rates were reduced, as they were in *Pareuchaeta norvegica* without antennules

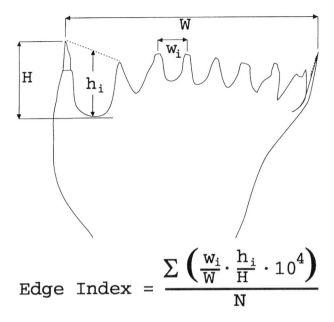

Figure 33 Calculation of the edge index of the mandible of a calanoid copepod. The space between each pair of teeth, w_i, is measured relative to the total width, W, of the mandible as is the height, h_i, of each relative to the total height, H, of the cutting edge. The sum of the products is divided by N, the number of teeth on the cutting edge. (After Itoh, 1970.)

(Yen, 1987). Schnack (1989), in reviewing the functional morphology of the mouthparts of calanoids, illustrates the three broad types of feeding appendages that occur. The first are those of filter or particle feeders, the second those of omnivores, and the third those of predatory feeders. The first two types of feeders have mouthparts that are: i) basically similar and ii) different from those of the third type, the predators. Mouthparts of the first two types are similar to those shown in Figure 6. Mouthparts of predators have a reduced segmentation and setation. Setae are often replaced by spines, especially on the maxillae (Figure 25, key-figs. 56–58). The form of the gnathobase of the mandibles, designed for cutting and grinding (Figure 6) in herbivores, changes to have an edge with prominent teeth (Figure 25, key-figs. 64–66), although they can still retain a grinding function (Turner, 1978; Zheng Xiaoyan and Zheng Zhong, 1989). Ohtsuka and Onbé (1991) describe such mandibles in the Pontellidae in a comparative study of the morphology of the mouthparts of that family.

Itoh (1970) introduced an Edge Index that describes the form of the cutting edges of the mandibles (Figure 33). The index ranges from about 100

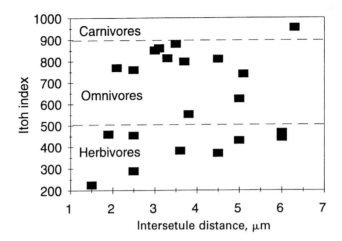

Figure 34 The edge index relative to the minimal intersetule spacing on setae of the maxillae of a variety of calanoid copepods. An edge index of <500 indicates a herbivorous, of 500 to 900 an omnivorous, and of >900 a carnivorous copepod. (Data from Schnack, 1982, 1989.)

to nearly 3000 in the 64 species that he investigated. He counted the teeth of the cutting edge (the mandible in Figure 33 having nine) and graphed the indices of the species against the numbers of teeth. The copepods clustered into three groups. From an examination of the reported diets of the species he concluded:

a. those with an index of <500 are herbivores;
b. those with an index between 500 and 900 are omnivores;
c. while those whose index is greater than 900 are carnivores.

The filtering setae of the maxillae have been examined in some detail as a means of estimating the sizes of particles that the mouthparts can handle (Marshall, 1973). These setae are armed with setules and a simple measurement is the minimal distance between setules. Schnack (1982, 1989) measured these distances in a variety of species (Figure 34) to show the correspondence between the Itoh index and the intersetule distances. Schnack (1989) found that *Arietellus setosus* had an index of 2830 and an intersetule distance of 7.5 μm; this species is not shown in Figure 34. There is no correlation between the index and the intersetule distance although addition of data on some 10 carnivores would produce a weak correlation at the 5% level. The weak correlation is not important. The position of a species relative to the two axes is interesting, as Schnack (1989) points out. Smaller intersetule distances are required to handle smaller particles while larger indices belong to more carnivorous species. Consequently, a herbivore with a large intersetule distance may have an omnivorous component to its diet, an omnivore with a small intersetule distance may be biased towards filtering and that with a large intersetule distance may be biased towards carnivory.

Intersetule distance of maxillary setae is indicative of the size of particle that can be removed from the water but minimum intersetule distance is not necessarily a measure of the minimum size of particle that can be retained (Bartram, 1980; Vanderploeg and Ondricek-Fallscheer, 1982). Seasonal variation in the intersetule distances occurs in adult female *Acartia longiremis, A. tonsa, A. bifilosa, Centropages hamatus, Pseudocalanus elongatus* and *Temora longicornis* but not necessarily in those of the adult males (Schnack, 1982). These changes are not correlated with seasonal changes in body size and therefore may relate to seasonal changes in particle sizes of available food.

The maxillary setae have been considered as a mechanical filtering screen (Marshall, 1973). Such filter feeding depended on the size and shape of particles for success. Visual observations by high-speed microcinematography (Alcaraz *et al.*, 1980; Rosenberg, 1980; Koehl and Strickler, 1981; Paffenhöfer *et al.*, 1982; Strickler, 1982; Cowles and Strickler, 1983; Price *et al.*, 1983; Price and Paffenhöfer, 1984, 1985) have demonstrated much more complexity within copepod feeding than simple filtering through a screen. Feeding is now recognized as a very active process involving flow fields (Yen and Fields, 1992; Bundy and Paffenhöfer, 1996).

5.2. FOOD CAPTURE

Poulet (1983) points out that "Marine copepods can potentially obtain food from any known stock of organic matter, in either dissolved or particulate form." Their feeding appendages and behaviour allow them to capture particles, phytoplankton or detritus, of a few microns in size or to attack living zooplankton such as chaetognaths, medusae or other copepods.

The copepods generate a feeding current that has been subject to much conjecture and discussion. Yen and Fields (1992) describe the flow field generated by *Temora longicornis* and in which nauplii of *Acartia hudsonica* become entrained. Flow velocities can be as great as 2 mm s^{-1}. The flow is laminar and the field funnel-shaped. Bundy and Paffenhöfer (1996) have studied the structure and generation of flow fields around free-swimming *Centropages velificatus*, an omnivore with a predilection for carnivory, and *Paracalanus aculeatus*, a more strictly herbivorous species. The copepod changes its swimming speed and orientation as it swims with the result that drag forces and flow fields are altered. The flow fields of the three *C.*

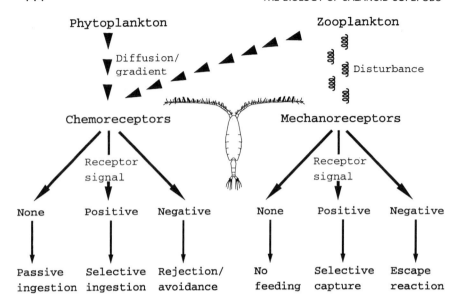

Figure 35 Flow diagram of a chemosensory model, modified from Poulet *et al.* (1986a), and incorporating a mechanosensory component, to illustrate possible feeding responses to different sensory stimuli in a copepod feeding on phytoplankton and zooplankton.

velificatus examined varied from each other and suggested a degree of plasticity that would allow the copepod to respond to micro-scale turbulence and different kinds of prey. This contrasted with the relatively constant flow field of the three *Paracalanus aculeatus* whose feeding current would entrain passive particles and where quick adaptation to capturing mobile food would not be required. The authors point out that more work is necessary to define intra- and interspecific differences in the generation and capabilities of flow fields.

Phytoplankton and zooplankton emit chemical signals that can be detected by the chemosensors of the copepod (Figure 35), through diffusion or along a diffusion gradient. Living zooplankton cause disturbances in the water that can be detected by the mechanoreceptors of the copepods. There is thus a bias for chemosensors to be used more in particle-feeding (herbivory, detritivory) and mechanoreceptors to be more functional in predatory feeding. No positive responses of receptors will still result in passive capture of particles in the feeding current, but there is unlikely to be an attack on active prey. Positive responses elicit selective capture of particles or motile prey, the option of rejection of the food being available to the copepod. Negative responses of chemoreceptors produce rejection or avoidance of the particles in the water while negative responses of mechanoreceptors elicit escape responses in the copepod. This is a very generalized concept of copepod feeding that may help in understanding the more detailed accounts that follow.

The size of food particle fed on by nauplii and copepodids varies with the size of the nauplii and copepodids. This is true for herbivorous, omnivorous and carnivorous modes of feeding. Hansen *et al.* (1994) find that the ratio of the body size of predatory copepod to that of the prey varies from about 10:1 to about 30:1, with a mean ratio of 18:1.

5.2.1. Food Detection

Copepods live in a chemical soup composed of sea water, soluble compounds, and living and inanimate particles, both organic and inorganic (Atema, 1985). Organic and inorganic particles can absorb and adsorb a variety of chemicals and materials. Living organisms, such as bacteria, nanoplankton and phytoplankton, emit chemicals, at least as excretory products. Chemosensors are known to be involved in detection and successful capture of food, often through detection of specific amino acids (Poulet and Marsot, 1978, 1980; Friedman, 1980; Poulet and Ouellet, 1982; Poulet *et al.*, 1986a, 1991; Gill and Harris, 1987; Gill and Poulet, 1988b; Tiselius, 1992). Buskey (1984) analysed the swimming patterns of *Pseudocalanus minutus* and showed that they altered in response to different particles and chemicals, inferring that both chemoreception and mechanoreception are involved in the recognition of potential food. Dead diatoms colonized by bacteria are selected in preference to sterile dead diatoms (Demott, 1988). Exudates and extracts of algae can have stimulatory or inhibitory effects on feeding (Alstyne, 1986; Huntley *et al.*, 1986). Further direct evidence of sensory feeding behaviour was found in *Calanus helgolandicus*, *Centropages typicus* and *Temora longicornis* by Gill and Poulet (1988a) and Poulet and Gill (1988); mechanoreception is used in handling individual particles and chemoreception instigates changes in the speed of the induced feeding current. Légier-Visser *et al.* (1986) modelled mechanoreception in copepods and showed that the pressure disturbance created by a diatom entrained in the feeding current may be detected by the copepod and provide information on its size and location; this, however, is questioned by Price (1988). Another model that examines the possible significance of chemical exudates from a diatom within the feeding current and their detection by the copepod is that of Andrews (1983).

Chemosensory faculties are less well developed in more strictly carnivorous copepods where the mechanosensory sensilla on the antennules are prominent. Yen *et al.* (1992) studied these receptors electrophysiologically in some 15 species belonging to different genera. Mechanical stimulation of the sensors elicited species-specific responses. Setae on the distal ends of the antennae may provide directional information. They found responses to stimuli up to the kHz range and hypothesize that response in this high frequency range may detect the rapid flicks of antennules of prey or potential predators. Such flicking of the antennules is used to renew the water at chemosensory sites (Atema, 1985). Yen *et al.* (1992) also found evidence of velocity detectors and the possibility that intensity and duration of a stimulus can be sensed. *Pareuchaeta norvegica* has setae on the third and thirteenth segments of the antennules that Yen and Nicoll (1990) consider may be designed to detect and identify the position of prey moving in three dimensions in the immediate environment of the copepod; this species (Yen, 1987) appeared to detect the tail-beats of cod larvae.

Jonnson and Tiselius (1990) describe *Acartia tonsa* feeding raptorially on ciliates. In feeding, this species alternates periods of sinking with short jumps (Tiselius and Jonsson, 1990). Swimming ciliates are detected at a range of 0.1 to 0.7 mm from the antennules of the sinking copepod. The copepod then turns towards the ciliate and captures it by movements of the antennae and mouthparts. Post-capture handling times of the prey are related to the size of the prey. Inert beads did not elicit a reaction from the copepod, inferring that it was responding to the movement of the ciliate, presumably through the antennular mechanoreceptors.

The small body size of calanoids hinders direct investigation of the functions of the different sensilla occurring on the appendages and integument of the body. Yen's electrophysiological studies of mechanoreceptors are more difficult to apply to the smaller chemoreceptors. Most information on chemoreceptors has been deduced from the whole copepod's reaction to introduced stimuli (Poulet *et al.*, 1986a). The actual receptors involved have not been identified although their location, i.e. on the antennules or on the feeding appendages, has sometimes been inferred. A discussion of the possible effects of turbulence in the water on mechanoreception is given by Yamazaki and Squires (1996).

5.2.2. Particle Feeding

Copepods generate a feeding current by movements of the antennae and maxillipeds which draw water towards the copepod (Paffenhöfer *et al.*, 1982; Cowles and Strickler, 1983; Price and Paffenhöfer, 1986a) and into a capture area (Koehl and Strickler, 1981). The function of such feeding currents in drawing material from outside the visual or chemosensory detection ranges of the copepods is discussed by Osborn (1996) who maintains that such currents increase the capabilities of the copepod for encountering food

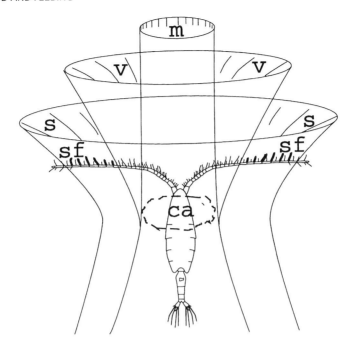

Figure 36   The model of the copepod feeding current conceived by Strickler (1985). It divides into three cores, the motion core (m), the viscous core (v) and the sensory core (s). Water in the sensory core passes over the sensory fields (sf) of the antennules. The capture area (ca) around the mouthparts is shown.

particles. The lateral movement of these mouthparts opens a cavity between them into which water plus food is drawn. The laminar flow of the feeding current acts like a conveyor belt dragging prey towards the mouth (Naganuma, 1996). Strickler (1985) has a conceptual model of what may take place (Figure 36). The central core of water passes directly to the capture area and to the chamber formed by the paired mouthparts. The maxillae act as "motile-particle deposition filters" (Rubenstein and Koehl, 1977) but just exactly how is as yet not understood.

Copepods live at low Reynolds numbers of about 1 which is between the higher numbers, up to 10^4, of the inertial environment and the lower numbers, down to 10^{-3}, of the viscous environment. The Reynolds number (Re) is a dimensionless ratio of inertial and viscous forces:

$$\text{Re} = \text{Inertial drag/viscous drag}$$

which can be simplified to (see Naganuma, 1996):

$$\text{Re} = \rho/\eta(LU)$$

where L in m is body length of an animal moving at a speed U in m s^{-1}. Fluid density, ρ in kg m^{-3}, of seawater at 20 °C is 1.025×10^3 while fluid viscosity, η in kg m^{-1} s^{-1}, at 20 °C is 1.1×10^{-3}. At a Reynolds number of 1 or less, the setae act as solid paddles (Koehl and Strickler, 1981) but Price and Paffenhöfer (1986a) and Cheer and Koehl (1987) show that water can flow through the setae depending on how the copepod moves them. The result is that small particles accumulate passively on, or in association with, the setae of the maxillae and are channelled to the mouth (Price and Paffenhöfer, 1984, 1986a). Larger particles, such as large diatoms, are individually detected by the copepod and actively seized, especially from the outer core, the sensory core (Figure 36). This is done by throwing the feeding appendages laterally, opening up the cavity between them. This "fling and clap" mechanism of Strickler (1985) has the effect of drawing the particle into the cavity.

Similar feeding mechanisms exist in euphausiids where the cavity is termed the food basket (Mauchline, 1989); it is constructed differently, the thoracic legs being involved. Once food is present in the basket, closing of the cavity by the appendages forces water through the setal sieves while retaining the food for handling by the antennal exopods and other mouthparts. Euphausiids enclose parcels of water in the food basket, express the water and retain the food.

The visual observations reported on copepods have a similar element of manipulation of parcels of water. Also, Price et al. (1983) describe the maxillae as periodically combing the feeding appendages, the setae acting as rakes to gather food particles; a similar activity in euphausiids is termed grooming and is a component of feeding to transfer particles from setular "sieves" to the mouth. The mouthparts of copepods are capable of capturing and handling or of rejection of single diatoms (see Paffenhöfer et al., 1982; Strickler, 1985). The copepod must be able to fix the position of the diatom relative to itself with considerable accuracy. The maxillipeds of *Eucalanus crassus* are used independently of each other, to catch and guide the diatom to the region of the mouth. The maxillae of other species can catch diatoms. Price and Paffenhöfer (1984) found that experienced *E. crassus* caught single diatoms more effectively than inexperienced, a learning process apparently being involved.

5.2.3. Carnivory

Feeding on suspended particles is a relatively passive process compared with predatory or raptorial feeding in which the copepod actively swims in a search pattern. Most studies of predation by copepods on living zooplankton have used species of Euchaetidae or Tortanidae. They do not create a feeding current but detect their potential prey by cruising in search of it (Greene and Landry, 1985; Greene, 1988; Yen, 1988; Uye and Kayano, 1994b). Females swim horizontally in a turn-and-search pattern. Predatory copepods such as the Euchaetidae have enlarged feeding appendages and mechanoreceptors on the antennules that may have a directional capability (Yen *et al.*, 1992). Landry (1980) suggests that the antennules of *Calanus pacificus* are important for locating prey in this normally herbivorous species.

Chemoreception is less well developed than in herbivorous and omnivorous copepods. Euchaetid species detect the hydrodynamic disturbances of moving prey (Yen, 1982, 1987, 1988), a prey species such as *Acartia fossae* moving in the vertical plane as opposed to the horizontal search plane of *Euchaeta rimana*. Large prey are held by the maxillae and maxillipeds and thrust to the mouth. Only parts of some large prey are consumed, the rest being discarded (Conover, 1966a,b; Uye and Kayano, 1994a,b). The use of Deborah numbers in considering encounter rates of predators and prey is suggested by Jenkinson and Wyatt (1992). They discuss and model aspects such as migration between patches of prey and smearing of potential scent trails by swimming.

Cahoon (1982) and Vaupel Klein and Koomen (1994) report on a mucus secretion used by *Euchirella* species to immobilize living prey; there are some 25 integumental organs in the roof of the oral cavity of *E. messinensis* that may be responsible for this secretion (Koomen, 1991). Vaupel Klein and Koomen have developed the idea of production of a mucus jet to entrap the prey outside the feeding area. This is difficult to envisage and Cahoon, observing living copepods, does not infer that a jet is produced. Further observations of living euchirellids are required to provide further information.

5.3. FOOD AND FORAGING IN THE ENVIRONMENT

5.3.1. Natural Diets

Examination of the food in the stomachs of copepods seems a relatively simple way of determining what they eat in the natural environment. There are, however, a number of difficulties that can arise. Harding (1974) reports on the stomach contents of more than 60 species collected from slope and deep waters of the Sargasso Sea. He lists the contents present in each species but points out that some contents are less liable to digestion than others and therefore may be more commonly recorded among the contents. Many of the species that he examined are carnivorous. Some carnivorous copepods suck out or bite parts of their prey, rejecting the rest. Thus, it is possible that there may be little recognizable evidence of the prey in the stomach. Feeding in the plankton nets, where there is a concentrated food supply, is also suspected but Harding rejects this as a source of error in his investigations. It is very difficult to detect net-feeding in copepods because of the maceration and grinding of the food by the mandibles. Examination of the food in the midgut and hindgut produces bias towards less digestible components.

Gowing and Wishner (1992) compare the merits of light and transmission electron microscopes for examining the gut contents and conclude that both instruments have their advantages. Models may be used to assess gross diets. Carlotti and Radach (1996), for example, modelled the annual cycle of *Calanus finmarchicus* in the North Sea and found that its growth characteristics could not be attained on an exclusive diet of phytoplankton; they had to consider an additional source of food, in this case pelagic detritus.

Consequently, the only way to gain an idea of natural food of most species is repetitive examination of stomach contents and faecal pellets of individuals taken in time series samples over the entire vertical range of the species and representative of the diel and seasonal cycles (Moloney and Gibbons, 1996). Light and electron microscopes are required (Turner, 1978, 1984a,b,c, 1985, 1987; Uchima, 1988; Voss, 1991; Gowing and Wishner, 1992; Urban *et al.*, 1992). The identification of particulate matter, among the identifiable components of the stomach contents, is a major problem. It is designated green remains or greenish or brownish debris in Table 12. Is the greater proportion of it obtained directly by particulate feeders from the suspended particulate matter in the water column? Tackx *et al.* (1995) suggest using a combination of the image analyser to quantify this material in the guts and a Coulter counter to quantify it in the water column.

The early studies of Lebour (1922) and Marshall (1924) on the stomach contents of copepods showed omnivory to be a feature. The diet varied seasonally, reflecting the food available. A summary of dietary components found by them in *Calanus* species (Table 12) indicates the variety of food eaten. Lebour also examined the stomachs of *Acartia clausi*, *Paracalanus parvus* and *Pseudocalanus elongatus* finding them to be herbivorous in contrast to the omnivorous *Calanus helgolandicus*, *Centropages typicus* and *Temora longicornis*. She reports *Anomalocera patersoni* and *Labidocera wollastoni* as carnivorous but also records diatoms in their stomachs. Ohtsuka (1985a), reviewing diets of pontellids, shows them to be predominantly carnivorous with a herbivorous element. Exclusively herbivorous copepods are probably quite rare: Davis (1977) records *Acartia longiremis* eating a small chaetognath in winter when phytoplankton is scarce and remarks on it as a vindication of Digby's (1954)

Table 12 Stomach contents of *Calanus helgolandicus* from Lebour (1922), who named it as *C. finmarchicus*, and of *C. finmarchicus* from Marshall (1924).

C. helgolandicus
Unidentified green remains not quantified
Flagellates: unidentified
Dinoflagellates: *Peridinium* sp.
Coccolithophores: *Pontosphaera huxleyi*
Diatoms: *Biddulphia, Chaetoceros, Coscinodiscus, Dytilium, Paralia, Phaeocystis, Rhizosolenia, Skeletonema, Thalassiosira*
Green algae: unidentified
Bits of copepods, some of *Paracalanus parvus*

C. finmarchicus
Unidentified greenish or brownish debris: <30–40% of contents
Silicoflagellates: *Distephanus, Dictyocha, Ebria*
Dinoflagellates: *Dinophysis, Gymnodinium, Peridinium, Phalacroma, Prorocentrum*
Coccolithophores: *Pontosphaera huxleyi*
Tintinnids: *Tintinnopsis ventricosa, Tintinnus subulatus*
Diatoms: *Biddulphia, Chaetoceros, Coscinodiscus, Dytilium, Fragilaria, Rhizosolenia, Skeletonema, Thalassiosira*
Radiolarians: *Acanthonia mülleri*
Bits of copepods

suggestion that supposed herbivorous species in higher latitudes must resort to a carnivorous diet in winter. Gaudy and Pagano (1989) propose a carnivorous diet in *A. tonsa* in winter in the Mediterranean.

Analyses of the stomach contents of a variety of species of copepods have been made (Table 13). There is frequently bias in lists of prey organisms from stomach contents, and more especially from faecal pellets. Centric diatoms are more persistent and more easily identified to species than ciliates. Whole prey is more easily named than fragments that may originate from more than one organism. Diversity within the diets is therefore difficult to quantify. Larger copepods can handle a broader range of sizes of food particles than smaller copepods and Hopkins (1985) found that larger copepods may have more diverse diets than smaller copepods, although quantifying diversity among the smaller components of the diet is often difficult and results in underestimation of them.

A completely different approach is that of Kleppel and Pieper (1984) and Kleppel *et al.* (1988a) who searched for biomarkers, among plant and animal pigments, of classes of food items. Carotenoids enabled them to class copepods as predominantly herbivorous or carnivorous. Carotenoids can also be used to distinguish different types of phytoplankton in the diets (Head and Harris, 1994). Long-chain and short-chain fatty acids and alcohols in copepods can act as marker lipids for herbivorous and

Table 13 Predominantly herbivorous, omnivorous and carnivorous genera and families of calanoid copepods as determined by examination of their stomach contents. Species are often detritivores when present in the benthopelagic environment (Gowing and Wishner, 1986, 1992; Steinberg, 1995).

Herbivorous:
Acartia	Lebour, 1922; Lillelund and Lasker, 1971; Davis, 1977; Turner, 1984c; Uchima, 1988
Calanoides	Hopkins, 1985, 1987
Calanus	Lebour, 1922; Marshall, 1924; Hopkins, 1985; Barthel, 1988; Ohman and Runge, 1994
Pseudocalanidae	Hopkins, 1985
Rhincalanus	Hopkins, 1985
Spinocalanidae	Hopkins, 1985; Gowing and Wishner, 1986, 1992

Omnivorous:
Acartia	Lonsdale *et al.*, 1979
Aetideopsis	Hopkins, 1985
Aetideus	Lillelund and Lasker, 1971; Robertson and Frost, 1977
Bathypontiidae	Gowing and Wishner, 1986, 1992
Bradyidius	Gowing and Wishner, 1992
Centropages	Lebour, 1922; Turner, 1987
Chiridius	Alvarez and Matthews, 1975; Gowing and Wishner, 1986
Eucalanus	Turner, 1984b; Ohtsuka *et al.*, 1993c
Euchirella	Hopkins, 1985; Gowing and Wishner, 1992
Gaetanus	Wickstead, 1962; Hopkins, 1985
Gaidius	Harding, 1974; Gowing and Wishner, 1992
Haloptilus	Hopkins, 1985
Limnocalanus	Warren, 1985
Lucicutiidae	Hopkins, 1985; Gowing and Wishner, 1992
Metridinidae	Wickstead, 1962; Hopkins, 1985; Gowing and Wishner, 1992
Paracalanus	Lebour, 1922; Turner, 1984b
Pseudochirella	Hopkins, 1985
Scolecitrichidae	Wickstead, 1962; Hopkins, 1985; Gowing and Wishner, 1986, 1992; Nishida *et al.*, 1991
Scopalatum vorax	Steinberg, 1995
Temora	Lebour, 1922; Turner, 1984a

Carnivorous:
Candaciidae	Wickstead, 1959; Lillelund and Lasker, 1971; Ohtsuka and Onbé, 1989; Ohtsuka and Kubo, 1991
Euaugaptilus	Harding, 1974
Euchaetidae	Wickstead, 1962; Hopkins, 1985, 1987; Shuert and Hopkins, 1987; Øresland, 1991; Øresland and Ward, 1993
Heterorhabdidae	Hopkins, 1985; Gowing and Wishner, 1992
Pachyptilus	Harding, 1974
Phaennidae	Hopkins, 1985; Gowing and Wishner, 1986, 1992
Pontellidae	Turner, 1978, 1984c, 1985; Ohtsuka, 1985a
Tharybidae	Gowing and Wishner, 1986, 1992
Tortanidae	Ohtsuka *et al.*, 1987a

omnivorous diets respectively (Hopkins et al., 1989; Graeve et al., 1994a,b). Hagen et al. (1995), however, find that the major lipid class stored by *Pareuchaeta antarctica* is wax esters while in *Euchirella rostromagna* it is phospholipids. They caution that in these two Antarctic species use of fatty acids and alcohols as trophic markers is not possible because of the different metabolism of the species. Lipids unique to bacteria and ciliates are transferred through a bacteria-ciliate-*Acartia tonsa* food chain and according to Ederington et al. (1995) may be useful markers for bacterial and ciliate consumption by the copepods. This is a promising line of research but currently at an early stage.

Most information on diets and feeding of copepods concerns coastal and epipelagic species. There is relatively little information on the diets of deep-sea species, Harding's (1974) work being an exception. One of the problems is that the copepods are in the nets for some time during hauling and, unless they are fixed and killed at depth, they may feed significantly on organisms in the nets. Gowing and Wishner (1992), while sampling benthopelagic deep-sea species, fixed them *in situ* with glutaraldehyde. They summarize the information available on the diets of deep-sea species which tend to be detritivorous although omnivorous and carnivorous diets are also common.

5.3.2. Dissolved Organic Matter (DOM)

As hard-bodied animals, calanoid copepods, like other crustaceans, have little or no ability to take up dissolved organic compounds from sea water through their epidermis (Poulet, 1983; Stephens, 1988). Early experiments were mostly carried out under non-sterile conditions and it is thought that bacteria on the cuticle were responsible for the apparent absorption reported by some authors. Autoradiographic studies with *Neocalanus plumchrus* showed some accumulation of glucose-derived tritium in the dermal glands and midgut, with transfer of small amounts to other tissue (Chapman, 1981). The initial uptake was in regions with thin or no cuticle, forming a small region of the body surface, and little nutritional advantage would be expected.

5.3.3. Detritus

Poulet (1983) has assessed the importance of detritus in the diets of copepods. He considers four types of detritus: (1) plant and animal detritus of planktonic origin; (2) detritus from coastal or estuarine surface water; (3)

structureless aggregates in deep water; and (4) detritus originating from coastal macrophytes. The first two types are consumed by copepods while the latter two are not. Faecal pellets are included in the first type and are readily consumed. Poulet makes the point that the type and state of ageing of the detritus are important to the copepod. Detritus is often suspected as an important source of nutrition for copepods (Roman, 1984; Sazhin, 1985; Finenko and Romanova, 1991) but its importance has not so far been quantified. This is equally true of faecal pellets (Poulet, 1983) whose production and degradation are discussed later. Ohtsuka and Kubo (1991) describe *Scolecithrix danae* as a saprophagous feeder and link this habit with the sensilla on the maxillae and maxillipeds; it feeds on discarded larvacean houses and carcasses of copepodids.

Analyses of stomach contents often note the presence of faecal pellets and/or detritus. The quantities available and the particle size distributions of the material make it readily available to the copepods and some species have been described as detritivores, especially in the benthopelagic environment. Its nutritional quality is undoubtedly very variable since it derives from such a variety of sources.

Mucous aggregates dominate the particulate matter of the water column above coral reefs. Gottfried and Roman (1983) found experimentally that *Acartia tonsa* ingested mucus at a rate equivalent to up to 81% body carbon $24\,h^{-1}$.

5.3.4. Bacteria

Poulet (1983) notes reports of bacteria in the diets of copepods. Free-living bacteria are probably too scarce in the environment, and too small, to be fed on directly by the copepods but they become accessible when they colonize detritus and their populations increase in density or when bacterio-aggregates are formed. A high load of suspended solids in the water column encourages larger populations of attached as opposed to free-living bacteria. Boak and Goulder (1983) found that *Eurytemora* sp. only exploited a small fraction of the total bacterial population in the Humber Estuary, satisfying only about 12% of the respiratory carbon demand. Lawrence *et al.* (1993) tested whether bacteria associated with particles could be utilized nutritionally by *Calanus pacificus* or whether they pass through the gut and contribute to the bacterial populations degrading the resulting faecal pellets. They found that some 30% of the ^{3}H label was absorbed by the copepod but that some bacteria did pass unharmed through the gut and aided in the degradation of the pellets. Bacteria occurring on the surface of the integument of copepods were present in the guts of the copepods and preferentially associated with the faecal pellets (Hansen and

Bech, 1996). The bacteria possibly passively colonize the gut during the course of filtration of the algae by the mouthparts.

Toxic blooms of cyanobacteria occur in certain regions. Sellner *et al.* (1994) found that they were not grazed on to any extent by *Acartia bifilosa* or *Eurytemora affinis* in the Gulf of Finland and that these copepods would not aid the disappearance of blooms of *Nodularia spumigena*.

5.3.5. Phytoplankton, Zooplankton

5.3.5.1. Diatoms

Kleppel (1993) points out that diatoms have been ascribed too much importance through the simplistic food chain "diatom → copepod → fish". It is now known that, although copepods do consume diatoms to a very great extent, they also eat many other kinds of organisms. Diatoms occur most prominently in the diets of copepods in highly productive systems, such as areas of upwelling, and are very much less prominent in oligotrophic waters such as the subarctic waters of the North Pacific. Thus, when diatoms are plentiful copepods eat them, sometimes preferentially, but will consume other organisms along with them. Some diatoms have deleterious effects on copepods, affecting egg production and hatching rates (Poulet *et al.*, 1994, 1995b; Laabir *et al.*, 1995a; Chaudron *et al.*, 1996; Ianora *et al.*, 1996; Richardson, 1997). The effects of exudates of diatoms on the grazing of copepods was studied by Malej and Harris (1993) but the role of such exudates in the natural environment has not yet been demonstrated. The harpacticoid copepod, *Tigriopus californicus*, is recommended by Shaw *et al.* (1994) as a bioassay organism for detecting exudates of phytoplankton that deter calanoids from feeding.

The colony-forming diatom *Thalassiosira partheneia*, has a cell size of about 9 μm but forms colonies of up to 5 cm in length. Schnack (1983) found that copepods were unable to feed on the entire colonies but consumed them once they had disintegrated.

The literature on experimental feeding of diatoms to copepods is very large and is reviewed later when examining estimates of grazing and ingestion rates of copepods.

5.3.5.2. Phaeocystis pouchetii

Phaeocystis pouchetii, world-wide in distribution, forms colonies and blooms that have been frequently reported as having toxic effects on other fauna. Copepods are now known to consume this alga quite readily but its nutritive value is as yet unclear. Bautista *et al.* (1992) conclude that *P. pouchetii* depresses the grazing rates of copepods so that copepods do not contribute to controlling the bloom. Further, they found an inverse relationship between the abundance of *P. pouchetii* and that of copepods but did not state whether the copepods died or were avoiding the bloom. Hansen and Boekel (1991) also found that the grazing of *Temora longicornis* was depressed in a *P. pouchetii* bloom. Conversely, Weisse (1983) found, experimentally, that maximum consumption rates of *P. pouchetii* were 48% of body carbon in *Acartia clausi* and 87% in *Temora longicornis*; the conclusion was that it is an important component of the diets of these copepods in spring and early summer. *Calanus hyperboreus* fed on colonies of 200 to 500 μm in size, ingesting daily rations of 8.1 to 12.4% body carbon (Huntley *et al.*, 1987b).

Other reports are cited in the papers referred to above and an as yet confusing picture obtains regarding toxicity and nutritional benefits of this alga.

5.3.5.3. *Protozoa*

Protozoa are an important component of the nanoplankton and microplankton in estuarine, coastal and oceanic waters (Stoecker and Capuzzo, 1990). They include dinoflagellates, nanoflagellates, ciliates, foraminiferans and radiolarians. The latter two, along with tintinnid ciliates, are commonly reported among stomach contents because parts of them are resistant to digestion. Most protozoans, however, are quickly digested and leave no recognizable remains among stomach contents. Gifford (1993) suggests that *Neocalanus plumchrus* obtains about 80% of its nutritional requirement from protozoans in the subarctic oceanic Pacific Ocean. Poulet (1983) reviews the food chains linking detrital bacteria predated by flagellates to copepods.

5.3.5.4. *Ciliates*

Ciliates do not persist in the stomachs of copepods for any length of time and population densities in natural waters are usually low. Tintinnids have been recorded among stomach contents (Table 12), and may be part of the diets of certain copepods (Robertson, 1983; Hopkins, 1987), and have been used in feeding experiments (Turner and Anderson, 1983). Tiselius (1989) and Stoecker and Capuzzo (1990) review the role of ciliates in the diets of copepods and suggest that they may be grazed in preference to diatoms; motility, size or chemical cues are invoked as possible reasons for the preference. The transient nature of blooms of ciliates in time and space contributes to the difficulties in assessing their nutritional importance. Montagnes *et al.* (1988) concluded that, at best, they might contribute 12% of the energy or food ration of copepods around the Isles of Shoals, Gulf of Maine. Burckhardt and Arndt (1987) suggest that their importance in landlocked coastal regions may be enhanced during periods of low phytoplankton concentration. In a study of the Dogger Bank, North Sea, Nielsen *et al.* (1993) showed that, during periods of marked stratification of the water column, the copepods were capable of consuming a substantial portion of the ciliate production.

The red-tide ciliate, *Mesodinium rubrum*, blooms in Southampton Water, England but Williams (1996) could detect no significant effect of it on resident copepods.

5.3.5.5. Dinoflagellates

Dinoflagellates, like diatoms, are a major component of the diets of calanoid copepods. Kleppel (1993) reviews the species that are ingested or rejected and further information is added in Table 14. Some species are of more nutritional value than others in terms of egg production of the copepod (Gill and Harris, 1987; Uye and Takamatsu, 1990; Razouls *et al.*, 1991).

Some flagellates named in Table 14 are responsible for phenomena such as red tides and yellow water. They produce toxins that may be accumulated in the body of a copepod. Turriff *et al.* (1995) found that although *Calanus finmarchicus* rejected *Alexandrium excavatum* as a food it still accumulated the toxins. There was no conclusive evidence, however, that the toxins were the only agent inhibiting feeding on *A. excavatum*. Similarly, Mallin *et al.* (1995) found that the estuarine dinoflagellate *Pfiesteria piscicida*, responsible for fish-kills, was consumed by *Acartia tonsa* but the only effect appeared to be the development of erratic swimming behaviour relative to that in the controls. Further, Carlsson *et al.* (1995) found that the okadaic acid of *Dinophysis acuminata* is potentially toxic to some but not all grazing copepods. Thus the results on red tide organisms are somewhat contradictory. This is emphasized by the work of Uye and Takamatsu (1990) who tested a range of flagellates, finding some to be rejected and others consumed by one or both of *Acartia omorii* and *Pseudodiaptomus marinus*.

Many dinoflagellate species are bioluminescent and the function of this is obscure. Esaias and Curl (1972) found that highly luminescent samples depressed copepod rates of grazing. The mechanism operating was not clear but they suggest that the light startles the copepod, allowing the dinoflagellate to escape. This is confirmed by Buskey and Swift (1983) who found that *Acartia hudsonica* reacted similarly to bioluminescence.

Table 14 Some dinoflagellate and red tide organisms that are ingested (I), rejected (R), regurgitated (RE) or avoided (A) by calanoid copepods. A few have a physiological reaction (P) such as decreased heart beat, feeding appendage movement or the development of erratic swimming behaviour. (Update of table in Kleppel, 1993.)

Species	Reaction	Authority
Alexandrium excavatum	R	Turriff *et al.*, 1995
	I	Santos, 1992
Ceratium dens	I	Kleppel, 1993
C. furca	I	Kleppel, 1993
Ceratium sp.	R	Kleppel, 1993
Chattonella antiqua	I	Uye, 1986
C. marina	I/R	Uye and Takamatsu, 1990
Chattonella sp.	I	Huntley, 1982
	I	Tsuda and Nemoto, 1984
Eutreptiella sp.	I/R	Uye and Takamatsu, 1990
Fibrocapsa japonica	I/R	Uye and Takamatsu, 1990
Gonyaulax acatenella	I	Kleppel, 1993
G. catenella	I	Kleppel, 1993
G. grindleyi	RE	Sykes and Huntley, 1987
G. polyhedra	I	Kleppel, 1993
	I	Jeong, 1994
G. sphaeroidea	I	Kleppel, 1993
G. spinifera	I/R	Uye and Takamatsu, 1990
G. tamarensis	I	Kleppel, 1993
	R	Kleppel, 1993
Gymnodinium flavum	R	Kleppel, 1993
	A	Huntley, 1982
G. nagasakiense	R	Uye and Takamatsu, 1990
G. sanguineum	I	Uye and Takamatsu, 1990
G. splendens	I	Kleppel, 1993
	A	Fiedler, 1982
Gyrodinium dorsum	I	Kleppel, 1993
G. resplendens	I	Kleppel, 1993
Heterocapsa triquetra	I	Uye and Takamatsu, 1990
Heterosigma akashiwo	I/R	Uye and Takamatsu, 1990
Olisthodiscus luteus	I/R	Uye and Takamatsu, 1990
Peridinium foliaceum	I	Kleppel, 1993
Pfiesteria piscicida	I	Mallin *et al.*, 1995
	P	Mallin *et al.*, 1995
Prorocentrum micans	I	Kleppel, 1993
	I	Uye and Takamatsu, 1990
P. minimum	I	Huntley, 1982
	I	Tsuda and Nemoto, 1984
P. triestinum	I	Uye and Takamatsu, 1990
Protoceratium reticulatum	I	Kleppel, 1993
	R	Kleppel, 1993
	I/R	Uye and Takamatsu, 1990

Table 14 Continued.

Species	Reaction	Authority
Pterosperma cristatum	I/R	Uye and Takamatsu, 1990
Ptychodiscus brevis	R	Kleppel, 1993
	P	Sykes and Huntley, 1987
Pyramimonas aff. amylifera	I/R	Uye and Takamatsu, 1990
Pyrophacus steinii	I/R	Uye and Takamatsu, 1990
Scrippsiella trochoidea	R	Kleppel, 1993
	P	Sykes and Huntley, 1987

5.3.5.6. *Coccolithophores, Foraminiferans and Radiolarians*

Coccolithophores are recorded among stomach contents of copepods (Table 12) because their calcareous scales resist digestion. Harris (1994), however, found that less than 50% of the calcite of coccolithophores occurred in the faecal pellets of the copepod, suggesting that it had been subjected to acid digestion.

Coccolithophores range in size from 2 to 20 μm (Ishimaru *et al.*, 1988). The commonest species available to copepods is *Emiliana huxleyi* which forms dense blooms detectable at the ocean surface by remote sensing techniques. Copepods actively graze *E. huxleyi* in the laboratory (Harris, 1994).

There are scattered records of foraminiferans and radiolarians in the diets but they are minor components.

5.3.5.7. *Metazoans*

Cannibalistic feeding of calanoid copepods has been reviewed and studied by Daan *et al.* (1988) and Hada and Uye (1991). It has been observed in *Acartia clausi, A. tonsa, Calanus pacificus, Centropages furcatus, Labidocera trispinosa, Rhincalanus nasutus, Sinocalanus tenellus, Temora longicornis, Tortanus discaudatus*. It is undoubtedly common within many populations of coastal copepods where their nauplii and copepodids dominate the plankton for short periods. Identification of species of copepods within the stomach contents of copepods is difficult because of the fragmentation by the mouthparts. Crustacean remains are often evident in stomachs but difficult to identify unless mandibular gnathobases or other distinctive parts of the prey species are found. Øresland (1991, 1995) and Øresland and Ward (1993) identified remains of *Metridia gerlachei, Calanoides acutus,*

Pareuchaeta spp., *Heterorhabdus* spp., *Microcalanus* spp. and *Drepanopus* sp. along with *Oncaea* spp. and *Oithona* spp. in the stomachs of four species of *Pareuchaeta* from the Antarctic.

Chaetognaths are probably attacked by calanoid copepods simply because they are common in the plankton but their importance in the diets of copepods is unknown. Wickstead (1959) shows *Candacia bradyi* attacking *Sagitta enflata* and records punctures in the body wall of the chaetognath caused by the spines on the maxillipeds of the copepod. This begs the question as to whether the copepods are capable of sucking out the body fluids of the chaetognath. Predatory copepods are known to bite other copepods and reject most of the body; simultaneous ingestion of body fluids and/or soft tissues is likely. Copepods of the family Candaciidae may be specialist carnivores feeding on chaetognaths, and Lawson (1977) reviews the evidence for this.

Larvaceans and their houses are preferentially selected by *Candacia* and *Paracandacia* species according to Ohtsuka and Onbé (1989) and Ohtsuka and Kubo (1991). *Scopalatum vorax* consumes houses of the giant larvacean, *Bathochordaeus* sp., at between 100 and 500 m depth off California (Steinberg *et al.*, 1994; Steinberg, 1995).

Fish eggs and larvae are potential prey of carnivorous copepods, and Bailey and Yen (1983), Turner *et al.* (1985) and Yen (1987) review reports of their predation by other zooplankton.

Most groups of planktonic organisms within the capture size of the copepods are potential prey. Hopkins (1987), examining the guts of copepods in the Antarctic, records the polychaete *Pelagobia longicirrata* and the mollusc *Limacina helicina* as being present.

5.4. FORAGING TACTICS

Primarily herbivorous species can be opportunistic carnivores (Turner, 1984a), omnivores can be opportunistic herbivores (Turner, 1984b) or carnivores, and carnivores can be opportunistic omnivores but rarely adopt a herbivorous diet. All can be opportunistic detritivores, especially in the benthopelagic environment (Gowing and Wishner, 1992). This results from the same species being able to use different techniques for feeding (Turner, 1987; Paffenhöfer, 1988; Turner and Roff, 1993). *Metridia pacifica* has fewer high-speed bursts and slower swimming speeds when feeding on phytoplankton while a cruising mode with frequent high-speed bursts is adopted in the absence of phytoplankton (Wong, 1988a). Greene (1988) summarizes the interaction of swimming and feeding (Figure 37) and the resulting dietary habit. The potential for the copepod to change from

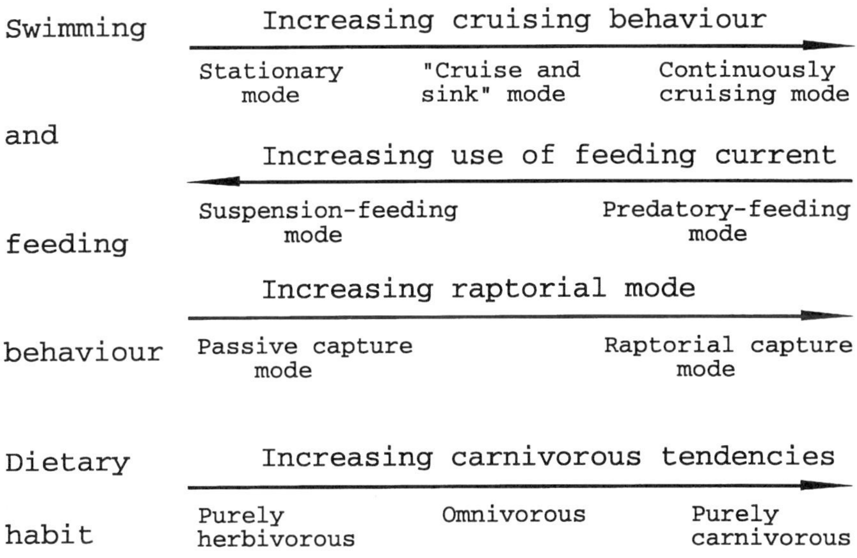

Swimming	Increasing cruising behaviour →		
	Stationary mode	"Cruise and sink" mode	Continuously cruising mode
and	Increasing use of feeding current →		
feeding	Suspension-feeding mode		Predatory-feeding mode
	Increasing raptorial mode →		
behaviour	Passive capture mode		Raptorial capture mode

Dietary	Increasing carnivorous tendencies →		
habit	Purely herbivorous	Omnivorous	Purely carnivorous

Figure 37 Foraging tactics of a copepod reflecting the coupled components of swimming and feeding behaviours and the resulting dietary habit. (After Greene, 1988.)

particle feeding to a predatory mode, or vice versa, is present (Kiørboe et al., 1996). A detailed analysis of swimming behaviour is made by Tiselius and Jonsson (1990): *Paracalanus parvus*, *Pseudocalanus elongatus* and *Temora longicornis* are slow-moving or stationary suspension feeders, *Centropages hamatus* and *C. typicus* are fast-swimming with periods of sinking, while *Acartia clausi* combine sinking with short jumps. The different swimming behaviours are an integral component of the feeding strategies and may have different hydrodynamic qualities. Static feeders may escape visual predators; swimming copepods may approach motile prey more stealthily than static feeders; sinking copepods are potential ambush predators, hydrodynamically quiet in approaching potential prey.

5.4.1. Particulate Feeding

Particulate- or suspension-feeding is possible in a stationary copepod or one that swims upwards and sinks, the cruise-and-sink mode of Figure 37. Copepods have a capability of searching the three-dimensional environment for potential food, a function of the cruise-and-sink mode. Paffenhöfer and Lewis (1990) contend, on experimental evidence, that there is increased sensitivity of the chemosensors of *Eucalanus pileatus* in response to a tenfold decrease in the cell concentration of phytoplankton; this results in a twofold increase in perceptive distance and a fourfold increase in perceptive volume. *Acartia tonsa*, under experimental conditions, locate thin layers of food by repetitive vertical jumps and are able to maintain themselves within the patch (Tiselius, 1992); feeding bout frequency is lower and jump frequency higher in the absence of food.

There is some evidence that copepods can conserve energy at low food concentrations by increasing periods of inactivity and decreasing the frequency of the fling-and-clap feeding process (Price and Paffenhöfer, 1985). The existence of a lower threshold concentration of food below which the copepods reduce their feeding activity or at which they commence feeding is present in some species (Kovaleva, 1989). Wlodarczyk *et al.* (1992) review its occurrence in a variety of species and conclude that there is no single threshold concentration common to all species. Sometimes there is a threshold for maximum clearance rate (Wlodarczyk *et al.*, 1992) and such may exist in *Acartia tonsa* studied by Støttrup and Jensen (1990).

5.4.2. Predatory Feeding

Calanoid copepods are non-visual, tactile predators that respond to the size and movement of potential prey (Yen, 1987), or create a feeding current, or flow field, that entrains small prey such as nauplii (Yen and Fields, 1992). Foraging tactics of predatory copepods are reviewed by Greene (1988) and Landry and Fagerness (1988) who conclude that selection of prey probably depends on an interaction between predator and prey size as concluded for *Tortanus forcipatus* by Uye and Kayano (1994b). This is illustrated in Figure 38 where the case shown on the left is probably more general than that on the right. Prey vulnerability is a function of encounter rate and the susceptibility of that prey. The escape capability of the prey (Mullin, 1979) is probably only a dominant factor when the prey is captured by the feeding current and much less dominant in the course of raptorial feeding. Body size and size of the feeding appendages of copepods are broadly related so that optimal prey size is related to the length of the maxilla (Landry and Fagerness, 1988); optimal prey length is approximately 80% of the length of the maxilla. Yen (1991) found that prosome length of copepods eaten by *Pareuchaeta antarctica* is about 65% of the length of the basis of its maxilliped (Figure 6). On the other hand, Hansen *et al.* (1994) found that the average ratio of the size of the predatory copepod to that of its prey was 18:1, ranging from 10:1 to 30:1, size being determined as the equivalent spherical diameter. Landry and Fagerness (1988) provide data on

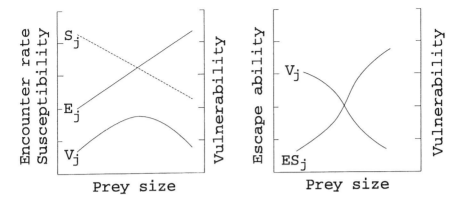

Figure 38 Prey capture by a copepod. Left, the general situation where encounter rate (E_j) of prey increases as the size of the prey increases, susceptibility (S_j) to capture decreases with increased prey size, with the result that a modal size of prey is most vulnerable (V_j) to predation by that predator. Right, a situation in which the prey's capability to escape (ES_j) governs its vulnerability (V_j) to capture by the predator. (After Greene, 1988.)

the preferred sizes of prey of *Calanus pacificus*, *Pareuchaeta elongata*, *Labidocera trispinosa* and *Tortanus discaudatus* on a body weight basis. The prey were all developmental stages of copepods. The following regression equation was calculated (r = 0.996, significant at the 1% level):

$$W_p = 0.0107 W_c + 0.0009$$

where W_p and W_c are the weights of the prey and predatory copepod respectively.

Predatory feeding involves cruising or a cruise-and-sink mode (Figure 37) until signals of the presence of potential prey are received by the mechanoreceptors of the copepod. *Euchaeta rimana* swims horizontally in a turn-and-search pattern (Yen, 1988) that maximizes encounter rates with potential prey. Once prey is detected, *E. rimana* can lunge at it (Yen, 1988) while it is actively moving; the lunge speed was as great as 60 body lengths s^{-1}. The handling of prey by *Pareuchaeta norvegica* differed between species (Båmstedt and Holt, 1978); the urosome of *Pseudocalanus elongatus* was eaten first while the anterior prosome of *Calanus finmarchicus* was consumed first. Uye and Kayano (1994b) state that *Tortanus forcipatus* and *T. gracilis* swim with a random, darting motion and pounce on prey, by extending the feeding appendages, when it enters detection range, about 0.5 mm. *Calanus propinquus* captures stationary cyclopoid copepods when its mouthparts make contact whereas *Metridia gerlachei* appeared to detect the presence of a stationary cyclopoid at a distance of about 1 mm. It then made several passes as if determining the precise location of the cyclopoid before catching it (Metz and Schnack-Schiel, 1995). *Acartia tonsa* detects swimming ciliates at a distance of 0.1 to 0.7 mm from the antennules (Jonsson and Tiselius, 1990) and immediately turns towards the ciliate before capturing it raptorially. See also Landry and Fagerness (1988) and Tiselius and Jonnson (1990).

Heterorhabdus species can inject prey with an anaesthetic or poison, produced by labral glands, through a hollow spine in the mandible (Nishida and Ohtsuka, 1996). This type of feeding is, as yet, unique to heterorhabdids.

5.4.3. Selectivity

Cinematographic studies have shown that copepods can reject individual particles but whether they can be selective grazers when suspension-feeding on phytoplankton is by no means clear. Marshall (1973), in reviewing the early literature, concludes that copepods presented with a mixture of sizes of food will tend to select the larger ones. Turner and Tester (1989) assess the evidence for and against selection. They conducted experiments with *Acartia tonsa*, *Centropages velificatus* and *Eucalanus pileatus* and concluded that the species of natural phytoplankton were eaten primarily in proportion to their abundance. They infer, from reviewing earlier literature, that other suspension-feeding copepods are also primarily non-selective in their diet. Moreover, their results, and those of Barthel (1988), do not support the contention that selection would be primarily evident when food is abundant, a tenet of optimal foraging theory.

Turner and Granéli (1992), however, found that *Acartia clausi* and *Centropages hamatus* selectively grazed flagellates that grazed the phytoplankton. *Acartia clausi* preferred the ciliate *Strombidium sulcatum* to the diatom *Thalassiosira weissflogii* in experiments done by Wiadnyana and Rassoulzadegan (1989). Huntley *et al.* (1983) show that *Calanus pacificus* has a clear preference for *Gyrodinium dorsum* rather than *Peridinium trochoidum*. Selection of the dominant modal size range within the spectrum of range of natural particles available occurs in *Acartia tonsa* according to Gaudy and Pagano (1989). Selection of modal sizes of particles in the environment, peak tracking, is also found in *Acartia* spp. and *Temora longicornis* by Tackx *et al.* (1989), in *Acartia omorii* by Tsuda and Nemoto (1988). Determining the effective size of phytoplanktonic organisms relative to feeding copepods is sometimes difficult because their length relative to their width varies and some have spines and some do not; Żurec and Bucka

(1994) compare the relevance of various measurements. The most selective feeders so far reported are species of *Candacia* and *Paracandacia* which predate larvaceans and their houses (Ohtsuka and Onbé, 1989; Ohtsuka and Kubo, 1991). Selection of food can be made by chemosensing, perceiving at a distance, or by rejection of unwanted particles at the mouth (Paffenhöfer *et al.*, 1995). Both mechanisms undoubtedly work in conjunction. Verity and Paffenhöfer (1996) review selectivity in particle feeders, showing that *Eucalanus pileatus* is selective. Heterotrophs were repeatedly preferred to autotrophs of similar cell-size. Houde and Roman (1987) and Kiørboe (1989) examined the nutritional quality of algae.

An extreme form of rejection of, or selection against, particles is the post-ingestion regurgitation of the dinoflagellate *Gonyaulax grindleyi* by *Calanus finmarchicus* found by Sykes and Huntley (1987) and of sterile beads by *Eurytemora affinis* found by Powell and Berry (1990). "Sloppy" feeding (Roy *et al.*, 1989) also occurs when the copepods fragment algal cells and debris escapes ingestion.

Selection of prey by predatory copepods is primarily based on the relationship between the size of potential prey and that of the predatory copepod as stated when discussing predatory feeding. Yen (1985), however, found that *Pareuchaeta elongata* selected *Pseudocalanus* spp. in preference to cyclopoid copepods, distinguishing them by differences in their activity patterns. Thus species-specific behavioural patterns, as well as body size, may influence dietary selection. Interestingly, she concluded that *Aetideus divergens* was not predated heavily probably because of its faster swimming speed and fast escape reaction. Nauplii of copepods often contribute to the diets of older stages of copepods; Lonsdale *et al.* (1979) found that *Acartia tonsa* predates its own nauplii much less than those of other copepods but suggests that their swimming ability may make them less vulnerable.

Evidence for selection of prey is relatively clear in predatory feeding, where there is a simple choice between acceptance or rejection of a single item. Selecting items while particulate-feeding is potentially more complex and there are many contradictory observations. Turner and Tester (1989) draw their conclusions that non-selective feeding is dominant from an environmental situation where a mix of phytoplankton is available to the copepod. The greater the number of species on offer, the less likely is selection to occur. The presence of a few disparate species that allow a degree of sorting on the basis of size, shape or taste may result in some selection. Situations in which many species are present, but few are dominant, may also result in some selection within the dominants; selection or non-selection among the other species will probably be masked. Characteristics of the microdistribution of the phytoplankton may influence the results. A copepod may have a temporary preference and/or rejection applying to one or two species of phytoplankton in its current ambit such that it will be attracted to one patch and avoid another. Detection by a copepod of individual phytoplankton cells increases with the size of the cells and can result in apparent selection or rejection of larger cells (Price and Paffenhöfer, 1985). On the other hand, Atkinson (1996) found that ciliates and dinoflagellates were cleared faster than centric diatoms of the same size suggesting that they were being fed on preferentially during periods of scarcity of diatoms. Evidence of selection of motile as opposed to sessile taxa of phytoplankton by *Calanus propinquus* and *Metridia gerlachei* is described by Atkinson (1995). The statement that particulate-feeding copepods are probably primarily non-selective feeders seems too broad and is not supported by the study of Verity and Paffenhöfer (1996).

5.5. FEEDING OF YOUNG STAGES

5.5.1. The Nauplii

Gauld (1959) and Marshall (1973) review earlier literature on the morphology and feeding of nauplii. According to Sekiguchi (1974), the nauplii of calanoid copepods divide into two types, those in which the gnathobase of the mandible is developed by nauplius IV and those in which it does not develop until the copepodid I. The nauplii of the latter type do not feed but survive on their oil sacs. The first nauplius stage to feed varies between species according to Sekiguchi (1974). The NI is the first feeding stage in *Pseudodiaptomus coronatus*; the NII in *Rhincalanus nasutus* and *Pontellopsis regalis*, in *Acartia* species, and in *Temora longicornis* (Landry, 1983; Klein Breteler *et al.*, 1994;); the NIII or NIV in many species; the NV in *Calanus hyperboreus*. Green *et al.* (1992) confirm that the nauplius III is the first feeding stage in *Calanus helgolandicus* and *Pseudocalanus elongatus* but considered that the nauplius II of the latter species might feed. According to the review of Sekiguchi (1974), nauplii of the following genera do not feed: *Aetideus*, *Bradyidius*, *Chiridius*, *Pareuchaeta*, *Candacia* and *Tortanus*. With the exception of the last two, these are primarily deep-water genera. Lewis and Ramnarine (1969), however, suggest that the NIII to NVI stages of *Pareuchaeta elongata* are feeding. High-speed cinematographic techniques have been used to examine the feeding methods of the nauplii (Paffenhöfer and Lewis, 1989). The antennae and mandibles of the nauplius are used for swimming and to create a weak feeding current. Phytoplankton cells are drawn towards the mouthparts and captured by them.

Nauplii graze smaller particles than the copepodids and adults. Estimates of naupliar grazing rates are reviewed by Uye and Kasahara (1983). Very little information is available on the diets of nauplii but they are recorded as feeding on phytoplankton and naupliar faecal pellets (Green *et al.*, 1992). Swadling and Marcus (1994) found that the NVI and the CI of *Acartia tonsa* fed selectively in 75% of experiments compared with the adults that fed selectively in 50% of experiments.

5.5.2. Copepodids

The stomach contents of copepodids II to VI of *Eucalanus bungii* contained the same food (Ohtsuka *et al.*, 1993c). Ontogenetic changes in the diets through the copepodid stages are related to biases derived from the increasing body size of successive stages or behavioural changes in the feeding patterns (Allan *et al.*, 1977; Paffenhöfer and Knowles, 1978; Paffenhöfer, 1984a). Both these mechanisms seem to be involved in Dexter's (1986) analyses of the feeding of copepodids of *Pseudocalanus* sp. and *Acartia clausi*.

Grazing and ingestion rates of copepodids are reviewed by Fernández (1979) and Uye and Kasahara (1983).

5.6. FEEDING PERIODICITY

Many copepods show a periodicity in their feeding activity. It can be irregular, a diel rhythm associated with a diel vertical migration to the surface waters rich in phytoplankton, or have seasonal components often associated with a dormant or resting phase in the life cycle. Such periodicities are detected by time series sampling of the environment at suitable frequencies and over relevant spans of time. Kleppel *et al.* (1988b) point out, however, that there is a high degree of asynchrony in feeding behaviour at the individual level and yet patterns do exist at the population level.

5.6.1. Gut Fluorescence

The presence or absence of food in the guts of the copepods in samples is usually determined by visual examination, a time-consuming, semi-quantitative method partially restricted to species with transparent integuments. Fluorometric methods measure chlorophyll and its derivatives within the gut contents of herbivorous euphausiids (Nemoto, 1968; Nemoto and Saijo, 1968). The method was adapted by Mackas and Bohrer (1976) for analysing the smaller guts of copepods collected in the field. Gut fluorescence is now widely used in studies of the periodicity of feeding of copepods.

There have, however, been criticisms of the measurements of fluorescence that primarily apply when using them to estimate ingestion rates of food. A proportion of the chlorophyll ingested is digested or reduced to non-fluorescing molecules (Lopez et al., 1988; Peterson et al., 1990b; Roy and Poulet, 1990; Head, 1992a; Head and Harris, 1992; Pasternak, 1994). The estimates of the quantities involved vary. Peterson et al. (1990a) concluded that none of the chlorophyll was destroyed or broken down into non-fluorescing components and review previous work advancing the same conclusion. They also evaluated earlier studies reporting significant losses. Head (1992a) examined pigment destruction and concluded that it is influenced by light and probably by species differences in physiological state and selectivity. Ingested chlorophyll a is subject to less destruction than chlorophyll c or diadinoxanthin (Head and Harris, 1992), all being partially destroyed during passage through the gut of *Calanus* spp. Some 10 to 80% of the chlorophyll a ingested was transformed into varying proportions of pyrophaeophytin a, pyrophaeophorbide a, and other products. Head and Harris (1994) suggest that the amount of chlorophyll a destroyed in the guts of copepods depends on the food concentration and the previous feeding history of the animal; the most extensive destruction occurs when food is scarce. Recently, Head and Harris (1996) conclude that destruction of the chlorophyll takes place at an early stage of feeding and that phaeopigments are not intermediates. They suggest that destruction is effected by two pools of enzymes, one within the copepods, that is primarily functional at low rates of ingestion, and one within the algae, that destroys more of the chlorophyll at higher rates of ingestion. Thus, determination of ingestion rates by measuring gut fluorescence without taking account of the destruction of the pigment is, according to Head and Harris, fundamentally flawed.

The problems associated with gut fluorescence analysis are pertinent to estimates of grazing and ingestion rates but apply much less to studies of diel and other periodicities in the feeding of copepods in the environment. The required accuracy for such observations is much less than required to determine grazing and ingestion rates of the copepods. Downs and Lorenzen (1985) derive values for recently ingested phytoplankton by determining carbon:phaeopigment ratios in faecal pellets.

5.6.2. Dietary Requirements

Kleppel (1993) examines some of the dietary requirements of copepods and points out that diversity in the diet is a nutritional requirement that changes ontogenetically through copepodids to adults, to provide for the production of lipid stores and reproductive products. For instance, Miralto et al. (1995)

found that diets of dinoflagellates, as opposed to diatoms, enhanced egg production and hatching success in *Centropages typicus*. Thus, there may be specific nutritional demands of the copepods at certain stages in their life histories. These various demands exist simultaneously in populations of copepods with short generation times (weeks) but will have seasonal components in species with longer generation times (months). Consequently, the simultaneous dietary demands on the trophic environment of communities of copepods in estuarine, coastal and shelf environments will be more variable, on average, than those of copepod communities in the deep sea or high latitude environments, with predominantly longer generation times. The dietary requirements of the copepods are satisfied through their feeding strategies which have been shown to be labile at the individual species level so that they exploit a wide variety of food resources.

The diversity of diet raises the question of just how accurate and, in the case of predominantly omnivorous and carnivorous species, how relevant measurement of gut fluorescence is in determining feeding periodicities in the environment. No studies seem to have been made of diel changes in the composition of the diets of copepods, especially in those performing a pronounced diel vertical migration. Do *Calanus* spp., for example, obtain copepod prey (Table 12) during the day at depth and exploit the phytoplankton at night when they have migrated upwards? Many studies consider only part of the copepod population. Studies of vertical migration are performed with pelagic nets that of necessity avoid the sea bed and studies of feeding migrations are often focused on the surface chlorophyll-rich layer. The proportion of the copepod population below the sampling range of the nets is often not considered. There may be a dynamic interchange within the vertical water column, individuals at the surface sinking or swimming downwards to be replaced by others migrating upwards, probably at a different stage in the short-term feeding cycle (Harris and Malej, 1986). Deeper elements of a population of a species may show markedly less food present in their stomachs than shallower living elements (Båmstedt, 1984) and interchange between the deep and shallow components has been shown to take place (Simard *et al.*, 1985). Mackas and Bohrer (1976) and Ishii (1990) found experimentally that *Acartia* spp., *Calanus* spp. and *Pseudocalanus minutus*, after starvation, filled their guts within one to two hours and then ceased or reduced feeding activity. The termination of nocturnal feeding was not correlated with the incident light regime in Arctic copepods studied by Head *et al.* (1985) and they suggest that satiation may have been the active factor.

Does such behaviour take place in the environment and would replete individuals sink in the water column to be replaced by others? Such short-term variations are considered next.

5.6.3. Short-term Periodicities

Short-term variability in feeding of calanoid copepods occurs on scales of minutes to a few hours (Dam et al., 1991; Paffenhöfer, 1994; Dagg, 1995). Variability as reviewed by these two authors is the variation exhibited by an individual that contributes to the variation within a population. There are short-term variations in the activity of the feeding appendages (on scales of milliseconds), in durations of feeding bouts (seconds and minutes), and in achieving full stomachs (1 to 2 h). Unknown variation arises from potential differences in the rates of food detection or recognition, especially in environments where it is patchy in distribution. Finally, there is the immediate feeding history of the individual spanning the previous one or two hours, the extremes of which will be whether it is starving or replete (Hassett and Landry, 1988). Copepods that have been starved exhibit enhanced feeding rates over those that are not starved. All of these factors contribute to the variation within the population but are obviously not great enough on many occasions to mask the detection of a feeding migration.

5.6.4. Diel Periodicities and Vertical Migration

Observations of diel feeding periodicities have been made recently by the following authors who cite the earlier literature (Hayward, 1980; Bautista et al., 1988; Daro, 1988; Ohman, 1988a; Tiselius, 1988; Dagg et al., 1989; Arinardi et al., 1990; Durbin et al., 1990; Ishii, 1990; Peterson et al., 1990b; Saiz and Alcaraz, 1990; Castro et al., 1991; Morales et al., 1991; Atkinson et al., 1992a,b; Perissinotto, 1992; Rodriguez and Durbin, 1992; Timonin et al., 1992; Wlodarczyk et al., 1992; Pagano et al., 1993; Drits et al., 1994; Landry et al., 1994a; Tang et al., 1994; Tsuda and Sugisaki, 1994; Hattori and Saito, 1995; Pasternak, 1995; Uye and Yamamoto, 1995; Atkinson et al., 1996a,b; Saito and Taguchi, 1996). The results are contradictory and it is difficult to generalize. Many diel vertical migrating species feed at night when they are in the phytoplankton-rich surface layers and feed to lesser degree at depth during the day. Populations of many species increase their feeding rate as they approach the surface at sunset and decrease it as they migrate downwards at dawn. Some have a decrease in feeding rate during the middle of the night resulting in a bimodal distribution in their feeding intensity (Simard et al., 1985; Pagano et al., 1993). In other investigations, i.e Dam (1986) on *Temora longicornis* and Drits et al. (1994) on *Calanoides acutus*, diel feeding rhythms were not correlated with a vertical migration. Scarcity of food also affects the rhythms (Boyd et al., 1980).

Diel vertical migrations are considered to be controlled by changing environmental light intensities, the animals migrating towards the surface as dusk approaches, remaining in the surface layers during darkness, and moving downwards at dawn. Stearns (1986) and Durbin *et al.* (1990), however, conclude that the migration of *Acartia tonsa* is controlled endogenously; conversely, Mobley (1987) found no endogenous rhythm of feeding in *Calanus pacificus* and Daro (1985) proposes an ontogenetic rhythm in *Pseudocalanus elongatus*. The classical explanation of such a migration is that animals feed in the surface layers at night and migrate to deeper regions at daylight to escape predation. Many species have been shown to conform to this pattern regularly or irregularly. Others apparently do not and a variety of explanations have been invoked. Atkinson *et al.* (1992a) found that active feeding began 8 h before dusk and continued until dawn in the Antarctic summer with a dark period of only 6 h. They suggest that daytime feeding may be necessary to satisfy food demand.

Harris and Malej (1986) found that a portion of the *Calanus helgolandicus* population resided in the upper 20 m of the water column during the day. There was a marked increase in numbers in this layer at night through vertical migration. They suggest, therefore, that these may be co-occurring migratory and non-migratory sub-populations with distinct metabolic characteristics. The portion living at the surface, however, could be an aberrant portion of the population subject to continuous interchange with the rest of the population. Nott *et al.* (1985) have shown that the digestive epithelium of the midgut lining disintegrates to form the peritrophic membrane of the faecal pellets. They suggest that the regeneration of the epithelium may impose a periodicity on the feeding bouts in a 24 h cycle. No evidence of this was found in *Acartia tonsa* by Hassett and Blades-Eckelbarger (1995); the B-cells responded to feeding with increased vacuole size but their life cycle was longer than 24 h.

Feeding cycles may be modified by the presence of predatory fish. Some copepods respond by decreasing their feeding activity, so becoming less mobile and more difficult for the fish to detect (Bollens and Stearns, 1992). The presence of predatory fish or of luminescing dinoflagellates can affect the timings of the diel vertical migration and so those of the feeding cycles (Buskey and Swift, 1983; Bollens *et al.*, 1994).

5.6.5. Seasonal Periodicities

Seasonal changes or periodicities in the feeding of calanoid copepods have been studied on few occasions. They are often inferred, especially in considering sources of nutrition during middle and high latitude winters, for species reputed to be primarily herbivorous. Small-sized herbivorous species actively migrated to shallower depths, where food was present, in winter in Saanich Inlet, Canada (Koeller et al., 1979); *Pseudocalanus minutus* was capable of feeding on winter flagellates.

Other overwintering populations of copepods are known to enter a resting or dormant stage, often as the copepodid V. The north Pacific *Neocalanus* spp. overwinter in the mesopelagic as copepodid V (Miller et al., 1984a; Miller and Clemons, 1988) and Tsuda and Sugisaki (1994) suggest that *Eucalanus bungii* in the same area feeds little and is dormant. Evidence of low digestive enzyme activities during the dormant phases of the life cycles is documented by Hassett and Landry (1990a). Coastal species such as *Calanus finmarchicus* have been found feeding in the benthopelagic environment in winter, particles of sediment occurring in the stomachs (Mauchline, unpublished). Hallberg and Hirche (1980) report that the digestive epithelium of the midgut in *Calanus* spp. is reduced in winter.

Poulet (1978) examined the feeding of small copepods in Bedford Basin, Nova Scotia over one year. All species tracked the commonest particle sizes which changed seasonally and ingestion was minimal in winter and maximal during the spring bloom. Ingestion rates change seasonally not only because of changes in the availability and quality of food but also because of seasonal changes in environmental temperatures and the body sizes of the copepods (Heerkloss and Ring, 1989; Dam and Peterson, 1991; Thompson et al., 1994).

5.7. CONCLUDING REMARKS

The literature on feeding of calanoid copepods is very large and the results often contradictory. The references quoted above are representative and their citation lists should be referred to for further amplification.

Huntley (1988) has produced two contrasting conceptions of feeding behaviour in copepods (Figures 39, 40). He describes them as paradigms of "feeding rate" which can be construed as "feeding behaviour" and incorporates feeding rate. He poses the question: "Are present or past environmental conditions more important in determining present feeding rate behaviour?". The dominant influences of the present (Figure 39) and the past (Figure 40) result in different conceptions of feeding behaviour. The formats of Figures 39 and 40 are the same but the heavy lines indicate where the greatest influences lie.

The environmental parameters shown are, for the most part, self-explanatory. Light, temperature and food abundance are those ambient around the copepod; food quality and size may or may not contain an element of choice or learning through the feed-back loop to "feeding history". Body weight influences the amount of food consumed and Huntley does not conceive it as having a feed-back loop to feeding history. Body

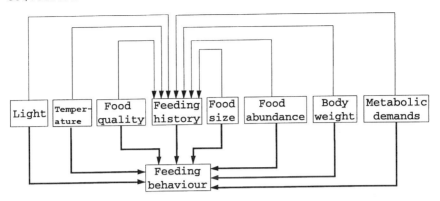

Figure 39 Concept of the current environmental factors that influence the feeding behaviour of a copepod more than its feeding history, which contains feedback from these factors in terms of conditioning and learning. (After Huntley, 1988.)

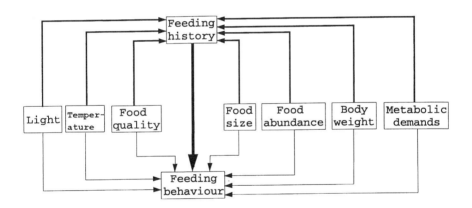

Figure 40 Concept in which the feeding history, in terms of conditioning and learning, influences the feeding behaviour more than the present environmental conditions. (After Huntley, 1988.)

weight, however, is a product of previous feeding history and the body has different dietary requirements at different stages in the life history. A requirement that is not satisfied because of feeding history will influence current feeding behaviour.

In Figure 39, the copepod is reacting instantaneously to its environment. Feed-back loops exist but are of little importance; the animal is an automaton responding to current cues such as temperature, light and, in estuarine situations, salinity. Such responses place the copepod in a certain part of the three-dimensional environment. Absence of food stimulates search patterns. Location of food through sensors stimulates feeding. Selection or non-selection for food quality and/or particle size comes into operation until satiation or ingestion of the available food, and the cycle begins again.

In Figure 40, the copepod is governed by its physiological state which is influenced by its feeding history. It may have an endogenous feeding rhythm reinforced by light cues. Food searching might be stimulated by hunger. Presence of predators may interfere with feeding. Ovarian maturation may create dietary demands while the laying of eggs at the surface or at depth may override the normal responses to environmental cues. What happens to feeding when mating is imminent? Many adult males have reduced mouthparts indicative at the very least of reduced feeding capabilities, and in the case of *Pareuchaeta* spp. of no feeding at all. Do such non-feeding males suggest that adult females can be sporadic feeders, responding to hunger rather than the immediate presence of food? As Huntley (1988) points out, the copepod in this paradigm is no longer a slave of its environment. Learning and memory in copepods have received virtually no attention but would feature significantly as a product of feeding history in this concept. Learning is known to be a feature in feeding behaviour. How long does it persist in terms of minutes, hours or days? Mayzaud *et al.* (1992), in studying digestion in *Drepanopus pectinatus*, suggest that feeding history extending back in time for 4 to 10 d may be operative in controlling acclimation responses of this species to changing trophic conditions. The challenge is to interrogate feeding in copepods against the background of this second paradigm.

Donaghay (1988) compares feeding of estuarine, shelf and oceanic species and suggests that their feeding behaviours are different. This arises because of differences in the morphology of the mouthparts, mechanisms and degrees of food selection or rejection, differences in the rates of adaptation to changing food availability, and the probable influence of their feeding history in defining the nutritional requirements of individuals. Such proposed differences along an onshore to offshore transect may have parallels in low to high latitude and in epipelagic to bathypelagic transects.

6. Physiology

6.1. Rates of Feeding and Ingestion 176
    6.1.1. Criticism of Laboratory Methods 176
    6.1.2. Rates of Feeding 178
    6.1.3. Concluding Remarks 207
6.2. Respiration and Excretion 209
6.3. Metabolism 211
6.4. Osmotic Regulation, Salinity and Temperature 215
6.5. Responses to Environmental Variables 216
    6.5.1. Temperature and Salinity 216
    6.5.2. Oxygen 217
    6.5.3. Light 217
    6.5.4. Pressure 218
6.6. Concluding Remarks 218

Various aspects of the physiology of copepods have been examined experimentally. Most attention has centred on feeding, attempts being made to determine rates of grazing, ingestion and assimilation of food. Much of this work is aimed at the estimation of grazing pressure of copepods on phytoplankton blooms. There is, however, an intrinsic interest in the feeding of copepods throughout the large literature on the subject. Chapter five reviewed how copepods locate, handle and collect their food. This chapter examines the rates at which they ingest and assimilate food as measured in laboratory experiments. It also examines gut passage times and the voidance of faecal pellets. Food demand is a function of the growth and metabolism of the copepod and rates of respiration and excretion are examined.

Marshall (1973) reviewed respiration and feeding of copepods, emphasizing the early experimental work on oxygen consumption, food selection, the quality of the food relative to nutritional requirements, volume of water filtered, and daily food ration. She made an overall assessment of the experimental conditions under which these observations have been made and compared the contradictory results from laboratory experiments and field observations. Her criticisms presage the more recent reservations expressed below.

6.1. RATES OF FEEDING AND INGESTION

6.1.1. Criticism of Laboratory Methods

Laboratory studies of feeding of copepods are large in number and variety but have produced contradictory results and the way forward is not clear (see Marshall, 1973). Strict adherence to a herbivorous, omnivorous, carnivorous, detrital or saprophagous diet by the vast majority of species is no longer tenable (Gifford and Dagg, 1988; Kleppel *et al*., 1988a). Such adherence is present over time scales of minutes and hours when a copepod is exploiting, for example, a phytoplankton bloom at night. Does it extend to days? How strict is such adherence? Does the copepod eat the occasional small copepod during the day at depth, the nutritional value of such food being greater than that of diatoms? This transition or switching between different feeding modes (Figure 37) was discussed in the chapter five.

The copepod in the natural environment will experience much greater rates of change, in space and time, of the food offered compared with laboratory experiments. These can not, at present, be replicated in the laboratory (Wirick, 1989a). Kleppel (1993), referring to Hitchcock (1982), states that, volume for volume, dinoflagellates contain 2 to 6 times more protein, 2.5 to 3.5 times more carbohydrate and 1.1 to 3.0 times more lipid than diatoms when axenically cultured. Such differences in nutritional quality must exist in the natural environment, not only between but also within taxonomic groups. Further, Kleppel advances cogent arguments that natural diets, of necessity, have to be diverse to satisfy nutritional demands which change markedly during the life history of a copepod. A much greater incorporation of the concepts of nutritional biochemistry with behavioural studies of feeding in copepods is required to gain a deeper understanding of the copepod's interaction with its trophic environment.

Many of the present criticisms of experimental estimates of grazing and ingestion rates of copepods are mentioned by Glasser (1984), Simard *et al*. (1985), Head *et al*. (1986), Paffenhöfer (1988), McClatchie (1992), Nejstgaard *et al*. (1995) and Moloney and Gibbons (1996). Doubts about the accuracy of estimates resulting from different experimental and computational methods have given rise to some direct comparisons (Hargis, 1977; Kiørboe *et al*., 1985a; Nöges, 1992).

Much of the problem from the copepodologist's point of view arises from the principal objective of much of the experimentation. It derives from the original simplistic *diatom → copepod → fish* food chain so that estimation of copepod grazing rates are frequently aimed at quantification of transfer of primary production into secondary production. This ecosystem approach is still dominant within the objectives of studying feeding of copepods (e.g. Dagg and Turner, 1982; Morales *et al*., 1993). The advent of the Coulter counter, gut fluorescence measurements, culture techniques and computers has encouraged a plethora of experiments with rather tenuous objectives and little conceptual thinking about the meaning of the results or how they are obtained (Glasser, 1984).

The employment of measurements of gut fluorescence (Head, 1988; Lopez et al., 1988; Morales et al., 1990) in environmental studies to measure feeding rates simply measures feeding rates on phytoplankton and ignores other sources of nutrition utilized by the copepods. The present criticisms of measurements of gut fluorescence that take no account of chlorophyll destruction within the gut are described in the chapter five. An alternative to gut fluorescence is the radioactive labelling of food with $Na^{14}C$ CO_3 for autotrophs and [methyl-^{3}H]methylamine hydrochloride for heterotrophs (White and Roman, 1991). The uptake of the latter was linear in both light and dark during the first hour of incubation and is suitable for measuring grazing at night and/or at depth; labelled bicarbonate is not transfixed by phytoplankton in the dark. According to Tackx and Daro (1993), the rates of ^{14}C uptake are related to the sizes of the phytoplankton and have to be corrected for when estimating grazing rates.

In vitro experiments to estimate grazing rates have bottle effects arising from a variety of sources. Roman and Rublee (1980) discuss some of these effects and specifically the impact of differential growth rates within mixed phytoplankton food organisms that give rise to difficulties in estimating grazing rates and degrees of particle size selection. The adaptation or acclimation of the copepods to the experimental conditions, the size of the bottles, the quantity and nutritional quality of the algae (Houde and Roman, 1987; Kiørboe, 1989; Roche-Mayzaud et al., 1991) all affect the results (Sautour, 1994). Phytoplankton colonies can be broken up and influence apparent filtration rates (Deason, 1980; Harbison and McAlister, 1980). Food concentration (Marin et al., 1986), the numbers of copepods in the container (Tackx and Polk, 1986), affect filtration rates. Time of day at which the experiments are carried out, ambient light intensity, and the history of exposure of the copepods to light all have an effect (Head, 1986).

One of the greatest differences between the laboratory container and the natural environment is that the container does not have the natural ranges of turbulence, diffusion, and substrate distributions that would have existed in the sea (Alcaraz et al., 1988; Wirick, 1989a,b; Costello et al., 1990; Marrasé et al., 1990; Saiz and Alcaraz, 1992a; Saiz et al., 1992a; Hwang et al., 1994; Kiørboe and Saiz, 1995; Saiz and Kiørboe, 1995; Strickler and Costello, 1996; Dower et al., 1997). Turbulence can increase food encounter rates of *Acartia tonsa* by a factor of 2.5 (Saiz, 1994). The observations are restricted to *Acartia* species and *Centropages hamatus*, species that can be expected to experience small-scale environmental turbulence. Turbulence may increase feeding rates but it can also cause the copepod to expend more energy in escape responses than is compensated for by the increased food intake. Hwang et al. (1994), in reviewing the topic, draw parallels with work on decapod crustaceans and suggest that feeding and escape responses are induced by different stimuli. They are incompatible because the copepod cannot feed during the escape response. Habituation of the escape response elicited by turbulence can result in a decline in its consequent frequency and so allow an increase in the amount of feeding. Capparoy and Carlotti (1996) model the effects of turbulence on *Acartia tonsa*.

The walls of the container also have an effect in providing a surface area that reacts with the copepod and the food organisms. Roman and Rublee (1980) review these effects and show that grazing rates in most such experiments are under-estimated because they decrease in the first few hours of incubation in experiments that have normally continued over 24 h. *Acartia tonsa*, in their experiments, filtered at a rate of about 1 l mg copepod dry weight^{-1} d^{-1} during the first 3 h but this had decreased to less than 0.4 l mg dry weight^{-1} day^{-1} after 12 h. Short incubation times are recommended. It has also been suggested that starved copepods be used as this would allow better estimation of maximum grazing rates (Dagg, 1983).

There are no continuous time series observations of the detailed species composition of the food eaten by a copepod over a time scale of one or two weeks. Garcia-Pamanes et al. (1991) show that the daily filtration rates of natural zooplankton populations fed ^{14}C-labelled phytoplankton can vary considerably. In *Calanus pacificus* the mean filtration rate was 23.5 ± 14.0 ml copepod^{-1} day^{-1} and in *Acartia tonsa* 25.3 ± 18.0 ml copepod^{-1} day^{-1}. Head (1986) considered that estimation of daily ingestion rate was probably best done by averaging values determined at intervals through a 24 h period which, at least, takes into account a diel feeding rhythm. New advances in technology, for example the automated *in vivo* fluorescence system described by McClatchie (1992), may provide continuous measurements of grazing rates over appropriate time scales.

The criticisms are many but primarily affect studies of the feeding rates and nutrition of the copepods themselves. They are not so pertinent to observations on the grazing rates of copepods on a phytoplankton bloom. The natural phytoplankton is often diverse and the differential growth rates of its components and the interactions between them characterize it. The point at issue is how fast the copepods graze this *milieu*, a conceptually simpler problem than determining the nutritional gain to the copepod.

6.1.2. Rates of Feeding

The factors involved in gathering and ingesting food are summarized in Figure 41 which places emphasis on slightly different aspects from those in

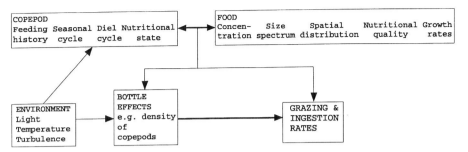

Figure 41 Interacting factors controlling grazing and ingestion rates of copepods in laboratory feeding experiments. The internal physiological status of the copepod determines the selection and rate of ingestion of food available. Environmental factors modify the behaviour of the copepod and are very important within experimental containers with which both the copepods and their food interact to influence the rates of grazing and ingestion.

Figure 40, also relevant, which examines overall feeding behaviour. In Figure 41, the copepod interacts with the food presented to it, whether *in vivo* or *in vitro*. Its feeding history, temporal position in its seasonal and diel cycles at the commencement of the observations (experiment), and its current nutritional state, whether starving or replete, or somewhere in between, will determine how it reacts to the food presented to it. Environmental factors such as light, temperature and turbulence affect the grazing rates of the copepods and obviously contribute to current feeding history of the copepod and its nutritional state. The food experimentally presented to or enclosed with the copepod also produces reactions from the copepod depending on its concentration, particle size distribution, spatial distribution and nutritional quality. Bottle or container effects originate from differences between its ambient environmental parameters and those of the natural environment of the copepod. They also originate within the experimental setup of the bottle in terms of the reaction of the copepod(s) and introduced food to the restricted environment and the interactions between them (e.g. differential growth rates) during the period of the observations.

Several measurements are used in the determination of feeding rates, that is how much the copepod consumes in unit time:

a. Filtering, clearance rate, or volume of water swept clear; this is the volume of water "filtered" by the copepod in unit time and is pertinent to estimating the potential grazing rates of copepods on a phytoplankton bloom.
b. Predation rate.

c. Ingestion rate: the amount of food passing through the stomach in unit time.
d. Gut filling time: the time it takes for a starved copepod to fill its gut.
e. Gut evacuation rate: the time it takes for a full gut to empty.
f. Egestion and the production of faecal pellets.
g. Digestion and assimilation of the food.

6.1.2.1. *Filtering or Clearance Rates*

The filtering or clearance rate is the volume of water (ml) filtered or swept clear by the copepod in unit time, usually per day. It is used in estimates of grazing on phytoplankton, but is subject to many criticisms, especially if the copepod also feeds by tactile encounter (Cushing, 1959, 1968) or raptorially. Clearance rate is relatively meaningless when considering the nutrition of the copepod (Marshall, 1973). There is, however, an interest in the relative volumes that species can filter in unit time as the act of filtering has a metabolic cost. Filtering rates are computed from the standard equations of Frost (1972) but Marin *et al.* (1986) developed a series of equations for use under different experimental conditions.

Marshall (1973) lists estimates of filtering rates of the different species made before that date. These are supplemented by more recent estimates on the same and different species in Figure 42. Some authors present a range of values for a species, others only an average rate. Both the maximum and average values are correlated with the total body length of the species.

The regression equations are:

$$\text{Maximum } FR = 68.306L - 11.804 \text{ r}^2, 0.096; n, 63$$

$$\text{Mean } FR = 50.555L - 24.152 \text{ r}^2, 0.174; n, 78$$

where FR is filtering rate in ml copepod^{-1} day^{-1}, and L is total body length in mm. Peters and Downing (1984) derived the following equation relating filtering rate to body dry weight:

$$\text{Log } FR = -1.245 + 0.534 \log W + 0.683 \log R - 0.067(\log R)^2$$

$$+ 0.0001C - 0.0002M$$

where FR is filtering rate in ml copepod^{-1} day^{-1}, W is dry weight of the copepod in μg, R is particle size of the food in μm^3, C is volume of experimental container in ml, and M is the duration of the experiments in minutes.

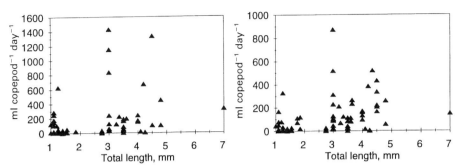

Figure 42 Filtering or clearance rates, measured both *in situ* and *in vitro*, related to the body length of copepods. Left, maximum recorded rates. Right, average recorded rates.
(Data from: Richman and Rogers, 1969; Esaias and Curl, 1972; Frost, 1972, 1974; Marshall, 1973; Hargis, 1977; Reeve and Walter, 1977; Richman *et al.*, 1977; Roman, 1977; Gamble, 1978; Schnack, 1979; Checkley, 1980a; Runge, 1980; Tomas and Deason, 1981; Dagg, 1983; Dagg and Wyman, 1983; Frost *et al.*, 1983; Hassett and Landry, 1983; Robertson, 1983; Turner and Anderson, 1983; Dam, 1986; Wang and Conover, 1986; Dagg and Walser, 1987; Stoecker and Egloff, 1987; Head *et al.*, 1988; Landry and Lehner-Fournier, 1988; Paffenhöfer and Stearns, 1988; Peterson *et al.*, 1988, 1990b; Wiadnyana and Rassoulzadegan, 1989; Stoecker and Capuzzo, 1990; Saiz *et al.*, 1992a; Landry *et al.*, 1994b.)

Peters and Downing (1984) found no evidence that filtering rates decline at both high and low concentrations of food as reported in earlier literature and, for example, by Kiørboe *et al.* (1985b) in *Acartia tonsa*. Marin *et al.* (1986) discuss the model of the relationship between filtering rate and food concentration and point out that since the rate of ingestion remains constant, the filtering rate will decrease above a certain critical concentration of food (Figure 43), this threshold probably being dependent on the species and developmental stage of the copepod and the size and species of food. A curve of this shape was obtained by Dagg and Walser (1987) for *Neocalanus plumchrus* feeding on *Thalassiosira weissflogii*. On the other hand, Frost *et al.* (1983) demonstrate a linear relationship in this copepod, and no relationship in *Neocalanus cristatus*, and ascribe the differences to the different patterns of setulation of the mouthparts of the two species. Abou Debs (1984) finds a continuous decrease in filtering rate with increasing food concentration in *Temora stylifera* as does Ohman (1987) in *Neocalanus tonsus* and Paffenhöfer (1988) in *Paracalanus* sp. Species such as *Acartia tonsa* and *Eucalanus hyalinus* have domed curves, indicated by the hatched extension in Figure 43; the range of food concentrations (modal range) at which filtering is maximal can be narrow or broad, dependent upon the species and food. These latter species with domed curves decrease

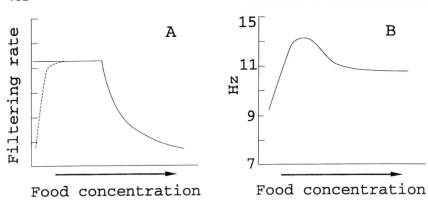

Figure 43 The relationship between filtering rate and food concentration. Left, the filtering rate declines at higher concentrations, the inflection point being referred to as the critical concentration. The hatched line indicates a filtering rate that also declines at low food concentrations. (After Marin *et al.*, 1986.) Right, the rate of fling-and-clapping of the appendages initially increases with food concentration and then declines at higher concentrations. (After Price and Paffenhöfer, 1986a,b.)

their filtering rates at low food concentrations. Paffenhöfer (1988) suggests that the different slopes and shapes of these curves are the result of the adaptation of the individual species to its trophic environment. Species exhibiting a continuous decrease in filtering rate, or a narrow modal range of maximal filtering rates, are adapted to low and high continuous supplies of food respectively. Those with a broad modal range are adapted to environments with a discontinuous food supply. Thus, some species will be able to survive starvation better than others.

The tolerance of species to starvation has been examined in a few cases (see Table 15). Survival time is inversely related to environmental temperature. The most detailed study is that of Tsuda (1994) who shows that tolerance increases throughout development in *Pseudocalanus newmani*. Species that enter a winter resting stage in their life cycles are considered to be able to resist starvation for several months, metabolizing stored fats.

Filtering rates of copepods, like other activities, increase with increasing temperature (Irigoien *et al.*, 1996). They can also increase with increasing particle size of the food. There are seasonal changes in clearance rates as there are in the concentrations and particle sizes of food available in the environment. Runge (1980) shows that the filtering rate of *Calanus pacificus* changes by a factor of about three between June and October. Prosome length also changes in this period but slightly out of phase so that the relationship between the two is not clear. Thus, do individuals of a species

Table 15 Starvation tolerances of calanoid copepods, adult females unless otherwise indicated, at different temperatures.

Species	Tolerance in days	Temp °C	Authorities
Acartia tonsa	6–10	15	Dagg, 1977
Calanoides carinatus	10–16	15	Borchers and Hutchings, 1986
Calanus finmarchicus CV	>21	15	Dagg, 1977
C. pacificus NIII	5–6	15	Fernández, 1979
NVI	4–6	15	Fernández, 1979
Centropages typicus ♀ ♂	3–6	15	Dagg, 1977
Eucalanus hyalinus	>28	20	Paffenhöfer, 1988
Paracalanus parvus	3–7	18	Checkley, 1980a
Pseudocalanus minutus	>16	15	Dagg, 1977
Pseudocalanus newmani			
Early nauplii	11.4 ± 1.0	5	Tsuda, 1994
Late nauplii	13.5 ± 2.5	5	Tsuda, 1994
CI	11.0 ± 3.2	5	Tsuda, 1994
CII	13.8 ± 3.4	5	Tsuda, 1994
CIII	19.5 ± 1.6	5	Tsuda, 1994
CIV	21.0 ± 4.0	5	Tsuda, 1994
CV	22.9 ± 4.2	5	Tsuda, 1994
CVI	23.5 ± 4.4	5	Tsuda, 1994
Early nauplii	8.4 ± 1.2	10	Tsuda, 1994
Late nauplii	9.7 ± 2.5	10	Tsuda, 1994
CI	8.2 ± 1.3	10	Tsuda, 1994
CII	10.0 ± 2.9	10	Tsuda, 1994
CIII	12.9 ± 2.0	10	Tsuda, 1994
CIV	15.6 ± 2.8	10	Tsuda, 1994
CV	16.6 ± 3.7	10	Tsuda, 1994
CVI	17.5 ± 5.0	10	Tsuda, 1994

have an innate rate of filtering that changes with season in response to the environment and, if so, how long does it persist in the laboratory experiments? Copepods, especially diel migrating species, have feeding rhythms. Respiration and excretion rates show diel rhythms corresponding to these feeding rhythms (Roman *et al.*, 1988; Checkley *et al.*, 1992; Cervetto *et al.*, 1993; Pavlova, 1994). Do these rhythms persist or are they lost in the laboratory situations?

One factor that overlies estimates of filtering rates is the pattern of feeding bouts of the copepods concerned. Price and Paffenhöfer (1986a) found that the proportion of time that *Eucalanus elongatus* spent flinging-and-clapping its feeding appendages did not change with food concentration but that the rate of movement of the appendages did change (Figure 43).

Table 16 Predation rates of calanoid copepods, in numbers of different species and stages, I to VI, of nauplii eaten day^{-1}. (After Daan *et al.*, 1988.)

Species	Naupliar prey		Number day^{-1}
Acartia clausi	*Acartia*	I	0.22
		II	0.16
		III	0.11
		IV	0.09
A. tonsa	*Acartia*	I–III	0.51
		IV–VI	0.25
	Scottolana	I–III	1.42
		IV–VI	1.03
	Oithona	I–III	0.82
		IV–VI	1.51
Centropages furcatus	*Pseudodiaptomus*	II–IV	10.5
	Temora	II–IV	11.9
	Centropages	II–IV	4.6
C. hamatus	*Acartia*	I–IV	5.3
Temora longicornis	*Temora*	I	4.3
		II	2.9
		III	1.3
T. stylifera	*Pseudodiaptomus*	II–IV	5.4
	Temora	II–IV	7.7
	Centropages	II–IV	3.5

The clearance rates of nauplii and copepodids are less than those of the respective adults simply because they are smaller in body size (Stoecker and Egloff, 1987; Berggreen *et al.*, 1988; Peterson *et al.*, 1990b). Characteristics of the appendages will also contribute to the rates.

6.1.2.2. Raptorial Feeding Rates

Relatively little information is available on the rates at which predatory copepods feed raptorially. Potential numbers of naupliar prey that can be eaten by several species of small copepods are reviewed by Daan *et al.* (1988) and shown in Table 16. Capture rates of starved copepods are higher than those that have fed recently (Yen, 1983). Attack rate increases with increased density of prey to an asymptotic level (Ambler and Frost, 1974; Robertson and Frost, 1977; Landry, 1978a; Yen, 1983, 1985, 1987, 1991; Daan *et al.*, 1988; Uye and Kayano, 1994a,b). Attack rate also increases with increasing environmental temperature (Uye and Kayano, 1994a,b) and with the developmental stage of the predator (Yen, 1985, 1991).

6.1.2.3. Ingestion Rates

Ingestion rates are often synonymous with filtering or clearance rates but expressed in different units. They are determined in the laboratory and environment by estimating the amount of food filtered from the water in unit time but the results are expressed in a variety of units referring to the food organisms in the water filtered:

a. Cells copepod^{-1} h^{-1};
b. 10^6 μm^3 copepod^{-1} day^{-1};
c. μg/mg copepod^{-1} day^{-1};
d. μg C/mg dry weight copepod^{-1} day^{-1};
e. μg C copepod^{-1} day^{-1};
f. μg N copepod^{-1} day^{-1};
g. % copepod body weight ingested/day^{-1};
h. % copepod C ingested/day^{-1};
i. % copepod N ingested/day^{-1};
j. % copepod protein ingested/day^{-1}.

These units are of varying practical value. (a) to (f) result in measures of ingestion rates of a copepod that can be translated to an environmental population in terms of its potential impact on a phytoplankton bloom. They are of little value in terms of the nutrition of the copepods. Dam and Peterson (1991) discuss the problems arising from the use of weight-specific measures of ingestion rates. (g) to (j) are the most valuable formats for deriving information on the feeding rates of the copepods. They constitute measures of the daily ration of a copepod. (g) expresses it in terms of the body weight of the copepod while (h) to (j) express it as a proportion of the copepod's carbon, nitrogen and protein contents.

Estimates of the daily ration of copepods measured as volume of food ingested per day (Table 17) range from 30 to 300×10^6 μm^3 copepod^{-1} day^{-1}. The volume of food ingested in unit time varies with the body size of the copepod. Heerkloss and Ring (1989) determined, under *in situ* conditions, that *Eurytemora affinis* consumed a seasonal maximum of 140 μg food/mg copepod^{-1} day^{-1}; their observations under laboratory conditions resulted in a seasonal maximum of about 2500 μg food/mg copepod^{-1} day^{-1}, the difference possibly being explained by pH and temperature regimes *in situ* and *in vitro*. There may also be a diel rhythm in feeding.

Ingestion rates in terms of μg C and μg N copepod^{-1} day^{-1} are shown in Table 18. Estimated daily rations of food are listed in Tables 19 for phytoplankton diets and in Table 20 for predatory diets as percentage body weight. The rations in Tables 21 and 22 are expressed as percentage of body carbon, body nitrogen and body protein ingested.

Table 17 Average daily ration of adult female calanoid copepods expressed as volume of food ingested day^{-1}.

Species	10^6 μm^3 ingested copepod^{-1} day^{-1}	Authority
Acartia clausi	30	Saiz *et al.*, 1992a
	125	O'Connors *et al.*, 1976
A. grani	70	Saiz *et al.*, 1992a
A. tonsa	80	Saiz *et al.*, 1992a
Centropages typicus	143	Dagg and Grill, 1980
Neocalanus plumchrus	300	Landry and Lehner-Fournier, 1988
Temora longicornis	72	O'Connors *et al.*, 1980

Table 18 Mean, minimum and maximum rates of ingestion by adult female calanoid copepods expressed as μg C and μg N ingested copepod^{-1} day^{-1}.

Species	μg C or N ingested copepod^{-1} day^{-1} mean	min	max	Authority
Carbon				
Acartia tonsa		0.3	1.81	Irigoien *et al.*, 1993
		0.4	2.06	Roman, 1977
A. bifilosa		0.13	0.55	Irigoien *et al.*, 1993
		1.3	1.65	Irigoien *et al.*, 1993
Calanoides acutus		4	5	Drits *et al.*, 1994
	9.4			Schnack, 1985
C. carinatus		14	28	Timonin *et al.*, 1992
		24	26	Peterson *et al.*, 1990a
Calanus finmarchicus		2.4	26.4	Tande and Båmstedt, 1985
	26			Gamble, 1978
C. glacialis		3	264	Tande and Båmstedt, 1985
C. pacificus	27			Frost, 1972
		24	72	Hassett and Landry, 1990b
Centropages hamatus	2.05		7.3	Conley and Turner, 1985
C. typicus	5.4	0.6	14.3	Saiz *et al.*, 1992a
	5.9			Dagg and Grill, 1980
Eurytemora affinis		0.16	1.45	Irigoien *et al.*, 1993
Pseudocalanus minutus		0.16	3.9	Poulet, 1974
Pseudodiaptomus marinus		0.8	5.5	Uye and Kasahara, 1983
Undinula vulgaris	70			Gerber and Gerber, 1979
Nitrogen				
Paracalanus parvus	1.1			Checkley, 1980a
Undinula vulgaris	8.7			Gerber and Gerber, 1979

Table 19 Daily ration of calanoid copepods, feeding on phytoplankton, expressed as % of copepod body weight ingested day^{-1}. Values are for adult females unless otherwise indicated.

Species		% body weight day^{-1}	Authority
Acartia clausi		33–45	Roman, 1977
		18.2	Båmstedt et al., 1990
	♂	14.4	Båmstedt et al., 1990
A. hudsonica	(10°C)	230	Paffenhöfer, 1988
	(15°C)	500	Paffenhöfer, 1988
A. tonsa		12–38	Roman, 1977
		6–81	Roman, 1977
		85	Roman, 1977
		123	Paffenhöfer, 1988
		360	Paffenhöfer, 1988
Calanus finmarchicus		17.8	Båmstedt et al., 1990
	CV	12.3	Båmstedt et al., 1990
C. helgolandicus		28–85	Roman, 1977
C. pacificus		16.8–18.4	Parsons et al., 1969
C. plumchrus	CIII–CIV	6–60	Parsons et al., 1969
	CV	14.8	Parsons et al., 1969
Metridia longa		3.1	Båmstedt et al., 1990
Paracalanus sp.		140	Paffenhöfer, 1988
Pseudocalanus elongatus		63–148	Roman, 1977
P. minutus		4.0	Parsons et al., 1969
P. minutus		2–55	Poulet, 1974
Pseudocalanus sp.		12.8	Båmstedt et al., 1990
	CV	11.3	Båmstedt et al., 1990

Daily ration as percentage body weight tends to decrease as body size increases as illustrated by the data for copepodids in Table 21 and references quoted by Paffenhöfer (1988). The decrease, under experimental conditions, is often not linear, for example see Klein Breteler's data on *Temora longicornis* in Table 21. Paffenhöfer (1988) discusses the possible significance of the irregularities in terms of learning phases and development of the appendages.

Ingestion rate increases with temperature (White and Roman, 1992b) to an asymptote or decreases at higher temperatures (Thébault, 1985). In some copepods ingestion rate increases with food concentration to an asymptotic level (Figure 44A) and consequently, there are seasonal changes in ingestion rate (Kleppel, 1992). The rate at which ingestion approached the asymptotic level increased with increasing protein and nitrogen content of the food (Libourel Houde and Roman, 1987). In other copepods, there is no evidence of an asymptotic level, at least at natural food concentrations (e.g. Tester and Turner, 1989).

Table 20 Daily ration of predatory calanoid copepods, feeding on different prey, expressed as % of copepod body weight ingested day^{-1}. Values are for adult females unless otherwise indicated.

Species	Prey	%	Authority
Aetideus divergens	*Artemia* nauplii	84	Robertson and Frost, 1977
Centropages typicus			
(15°C)	*Artemia* nauplii	48	Ambler and Frost, 1974
(80°C)	*Artemia* nauplii	22	Ambler and Frost, 1974
Pareuchaeta antarctica			
C VI	*Metridia gerlachei*	9.3	Yen, 1991
C V	*Metridia gerlachei*	9.2	Yen, 1991
C IV	*Metridia gerlachei*	21.5	Yen, 1991
C VI	*Microcalanus* spp.	0.4	Yen, 1991
C V	*Microcalanus* spp.	1.3	Yen, 1991
C IV	*Microcalanus* spp.	11.6	Yen, 1991
P. elongata	*Calanus pacificus* ♀	6.4	Yen, 1983
	Aetideus divergens	10	Yen, 1983
	Pseudocalanus ♀	17.1	Yen, 1983
	C. pacificus nauplii	1.4	Yen, 1983
P. norvegica	*Gadus morhua* larvae	10.5	Yen, 1987
Labidocera jollae	*Artemia* nauplii	27	Ambler and Frost, 1974
Metridia lucens	*Artemia* nauplii	24	Ambler and Frost, 1974
Tortanus discaudatus	*Artemia* nauplii	17	Ambler and Frost, 1974
	Artemia nauplii	24	Ambler and Frost, 1974

Copepod density, up to eight or nine individuals in 450 ml beakers containing 250 ml sea water, did not affect ingestion rates (Wong, 1988b). Experiments using combinations of two species, from *Calanus pacificus*, *Metridia pacifica* and *Pseudocalanus minutus*, provided no evidence of interference with the ingestion rates of individual species (Wong, 1988b). Wong, however, found that the ingestion rates of *P. minutus* were lower when the predator *Pareuchaeta elongata* was present in the beaker. This predator depressed the swimming behaviour of *Pseudocalanus minutus*, so reducing its feeding rate.

Ingestion rates can be estimated from observations of gut-fullness, with some qualifications (Penry and Frost, 1990). The rates can be determined if the gut clearance rate constant, K, is accurately known (Dam and Peterson, 1988). This constant is a reciprocal of the gut passage time as determined from starvation experiments (Figure 45). Ingestion rate I (amount of food animal^{-1} time^{-1}) is given by:

$$I = KG$$

where G is the gut contents (amount of food animal^{-1}) and K the gut

Table 21 Daily ration of calanoid copepods expressed as % of copepod body C ingested day^{-1}. Mean, minimum and maximum values are for adult females unless otherwise indicated.

Species	% of copepod C ingested/day			Authority
	mean	min	max	
Acartia clausi ♀	18			Båmstedt *et al.*, 1990
♂	14			Båmstedt *et al.*, 1990
A. longiremis	25			Kiørboe *et al.*, 1985a
A. tonsa	66	33	81	Roman, 1977
	150			Kiørboe *et al.*, 1985a
	22	1	152	Tester and Turner, 1988
	80			Durbin *et al.*, 1990
		70	302	Støttrup and Jensen, 1990
		3	96	Kleppel, 1992
		7	27	Irigoien *et al.*, 1993
Nauplii	280 ± 91			White and Roman, 1992b
CI–CIII	183 ± 64			White and Roman, 1992b
CIV–CV	96 ± 32			White and Roman, 1992b
CVI	58 ± 22			White and Roman, 1992b
		21	32	Houde and Roman, 1987
A. hudsonica		100	120	Deason, 1980
Nauplii	79 ± 38			White and Roman, 1992b
CI–CIII	10 ± 2			White and Roman, 1992b
CIV–CV	22 ± 17			White and Roman, 1992b
CVI	11 ± 2			White and Roman, 1992b
Aetideus divergens	21			Robertson and Frost, 1977
Calanoides acutus		2.2	2.7	Drits *et al.*, 1994
CV		5.6	27	Atkinson *et al.*, 1992b
	4.9	3.9	5.7	Froneman *et al.*, 1996
C. carinatus	59		126	Peterson *et al.*, 1990a
	40		58	Timonin *et al.*, 1992
	35		40	Gamble, 1978
Calanus finmarchicus				
CI, CII	148			Daro, 1980
CIII	115			Daro, 1980
CIV	83			Daro, 1980
CV	31			Daro, 1980
	5			Kiørboe *et al.*, 1985a
CI	40.3			Båmstedt *et al.*, 1991
CII	39.5			Båmstedt *et al.*, 1991
CIII	25.2			Båmstedt *et al.*, 1991
CIV	44.4			Båmstedt *et al.*, 1991
CV	10			Båmstedt *et al.*, 1991
CVI	17.6			Båmstedt *et al.*, 1991
	42–48			Ohman and Runge, 1994

Table 21   Continued

Species	% of copepod C ingested/day			Authority
	mean	min	max	
C. glacialis		9	54	Båmstedt et al., 1991
CI	75			Båmstedt et al., 1991
CII	27.9			Båmstedt et al., 1991
CIII	15.1			Båmstedt et al., 1991
CIV	14.5			Båmstedt et al., 1991
CV	16			Båmstedt et al., 1991
CVI	13.7			Båmstedt et al., 1991
C. hyperboreus		17	22	Båmstedt et al., 1991
CI	67.9			Båmstedt et al., 1991
CII	87.5			Båmstedt et al., 1991
CIII	119.8			Båmstedt et al., 1991
CIV	35			Båmstedt et al., 1991
CV	7.4			Båmstedt et al., 1991
CVI	10.6			Båmstedt et al., 1991
C. pacificus	85			Paffenhöfer, 1971
	40			Frost, 1972
	62			Robertson and Frost, 1977
	80			Hassett and Landry, 1990a
NIII		12	50	Fernández, 1979
NIV		45	85	Fernández, 1979
NV		70	130	Fernández, 1979
NVI		55	125	Fernández, 1979
CI		80	130	Fernández, 1979
C. propinquus	5.5	4.2	6.4	Froneman et al., 1996
C. simillimus				
CV	12			Atkinson et al., 1992b
Centropages hamatus		26	85	Kiørboe et al., 1982
	9			Kiørboe et al., 1985a
	14.4		51	Conley and Turner, 1985
NI–NVI		19	35	Tackx et al., 1990
CI–CIII		11	21	Tackx et al., 1990
CIV–CV		7	8	Tackx et al., 1990
CVI		23	29	Tackx et al., 1990
C. typicus	33.4			Dagg and Grill, 1980
		7	70	Dagg and Grill, 1980
		4	38	Dagg and Grill, 1980
	39	4	102	Saiz et al., 1992a
C. velificatus	2.9	0.1	31	Tester and Turner, 1988
Eucalanus pileatus	51.5			Paffenhöfer and Knowles, 1978
	1.3	0.1	2.8	Tester and Turner, 1988
Eurytemora affinis		50	250	Barthel, 1983
Labidocera aestiva	13.8		35	Conley and Turner, 1985

Table 21 Continued

Species	% of copepod C ingested/day			Authority
	mean	min	max	
Metridia gerlachei	25			Huntley and Escritor, 1992
	9.1	5.7	13.6	Froneman *et al.*, 1996
M. longa	3.1			Båmstedt *et al.*, 1990
	2.6	0.2	6.3	Båmstedt *et al.*, 1991
M. pacifica	2		22	Mackas and Burns, 1986
Neocalanus cristatus	5.8			Taguchi and Ishii, 1972
N. plumchrus	6.0			Taguchi and Ishii, 1972
N. tonsus		1.4	3.8	Ohman, 1987
Pseudocalanus sp.	113			Robertson and Frost, 1977
	7			Kiørboe *et al.*, 1985a
P. elongatus	140			Paffenhöfer and Harris, 1976
Rhincalanus gigas				
CV		1.5	2	Atkinson *et al.*, 1992b
	2	0.8	2.8	Froneman *et al.*, 1996
R. nasutus NI–CI		59	164	Corner (1972)
CI–CIV		27	69	Corner (1972)
CIV–CVI		24	45	Corner (1972)
Temora longicornis	80			O'Connors *et al.*, 1980
	18			Kiørboe *et al.*, 1985a
		0.7	13	Dam, 1986
NII	35			Klein Breteler *et al.*, 1990
NIII	35			Klein Breteler *et al.*, 1990
NIV	42			Klein Breteler *et al.*, 1990
NV	48			Klein Breteler *et al.*, 1990
NVI	58			Klein Breteler *et al.*, 1990
CI	85			Klein Breteler *et al.*, 1990
CII	128			Klein Breteler *et al.*, 1990
CIII	103			Klein Breteler *et al.*, 1990
CIV	138			Klein Breteler *et al.*, 1990
CV	170			Klein Breteler *et al.*, 1990
CVI ♀	103			Klein Breteler *et al.*, 1990
CVI ♂	65			Klein Breteler *et al.*, 1990
NI–NVI		43	129	Tackx *et al.*, 1990
CI–CIII		40	89	Tackx *et al.*, 1990
CIV–CV		43	44	Tackx *et al.*, 1990
CVI		57	58	Tackx *et al.*, 1990
T. stylifera		10	185	Abou Debs, 1984
CI–CV		34	38	Pagano *et al.*, 1993
females		63	66	Pagano *et al.*, 1993
males		27	29	Pagano *et al.*, 1993

Table 22 Daily ration of calanoid copepods expressed as a % of copepod body N, P or body protein ingested day^{-1}. Mean, minimum and maximum values are for adult females unless otherwise indicated.

Species	% of copepod N or protein ingested day-1			Authority
	mean	min	max	
Nitrogen				
Acartia tonsa	43			Durbin *et al.*, 1990
		14	26	Houde and Roman, 1987
Calanus finmarchicus				
CV–CVI spring	13.4			Corner, 1972
CV–CVI winter	3.1	4.5		Corner, 1972
C. helgolandicus	7.5			Corner *et al.*, 1972
C. pacificus	85			Hassett and Landry, 1990a
Centropages velificatus	24.5			Paffenhöfer and Knowles, 1980
N. tonsus		1.5	5.7	Ohman, 1987
Paracalanus sp.	48			Paffenhöfer, 1984b
CII	82			Paffenhöfer, 1984b
Phosphorus				
Calanus finmarchicus				
CV–CVI spring	17.6			Corner, 1972
CV–CVI winter	8	9.4		Corner, 1972
Protein				
Acartia tonsa		17	31	Houde and Roman, 1987

clearance rate constant (min^{-1}). This constant is referred to as the gut evacuation rate constant when gut passage times are estimated by methods other than starvation (Table 23). Ingestion rates derived from conventional grazing experiments are compared by Wang and Conover (1986) with those derived from evacuation rates; lower estimates resulted from the latter procedure.

Finally, Paffenhöfer *et al.* (1995) compared the ingestion rates of individual *Paracalanus aculeatus* and found significant variation between individuals, some ingesting twice as much as others. The problem of selection of particle sizes from mixed phytoplankton is examined by Ambler (1986b). *Paracalanus* spp., like many other particle-feeding copepods, select certain size classes of particles, the preferred sizes changing ontogenetically. She develops functions to quantify these potential changes in the effective food concentrations within the total amount of food available.

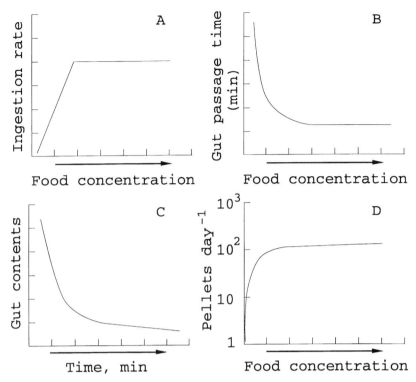

Figure 44 Ingestion and passage of food through the gut of a calanoid copepod.
A, ingestion rate related to food concentration: the rate increases to the critical food concentration after which it becomes asymptotic. B, gut passage time decreases as the food concentration increases. The equation for *Neocalanus plumchrus* is: Passage time (min) = 48.1 (Chl concentration in μg litre^{-1})$^{-0.44}$. (Dagg and Walser, 1987.) C, time course of evacuation of gut contents of a calanoid copepod transferred to filtered sea water. D, the number of faecal pellets voided per day by a copepod related to concentration of the food.

6.1.2.4. Gut Filling Time

Food is transferred from the mouth via the oesophagus to the anterior region of the midgut (Figure 14 on page 40). It accumulates there for 10 to 20 min (Arashkevich, 1977) before part of it passes posteriorly into the posterior region of the midgut where the faecal pellet is formed. The faecal pellet then passes out of the anus. Gut filling time is variable but usually between 20 and 60 min (Wang and Conover, 1986; Ohman, 1987).

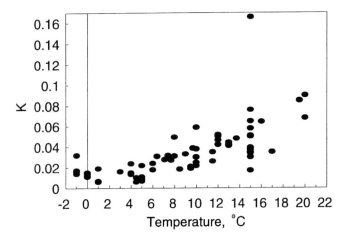

Figure 45 The relationship of the gut clearance rate constant (K) for calanoid copepods to environmental temperature. (The data from Dam and Peterson (1988) are supplemented by further data from: Simard *et al.*, 1985; Ellis and Small, 1989; Durbin *et al.*, 1990; Peterson *et al.*, 1990b; Saito *et al.*, 1991; Bautista and Harris, 1992; Saiz *et al.*, 1992a; Wlodarczyk *et al.*, 1992; Drits *et al.*, 1994.)

Table 23 Gut passage time. Food, labelled with ^{14}C, is fed for a few minutes and then the copepod is placed in a medium with unlabelled food. The time taken, minutes, for the labelled food to pass through the gut is determined. (After Arashkevich, 1977.)

Species	Passage time (min)	Temp °C
Acartia tonsa ?	35–60	17–19
Calanus minor	70–75	17–19
Centropages sp.	120–130	17–19
Clausocalanus mastigophorus	20–35	17–19
Eucalanus attenuatus	30–65	17–19
E. subtenuis	30–40	17–19
Neocalanus gracilis	30–40	17–19
Pleuromamma xiphias	25–30	17–19
Rhincalanus cornutus	40	17–19
R. nasutus	35–40	17–19
Scolecithrix danae	90–120	17–19

6.1.2.5. Gut Evacuation (Egestion) Rate

There are several methods of estimating the rate of egestion or evacuation of the gut contents of a copepod. The numbers of faecal pellets produced in unit time can be counted as described in the next section. Another method is to divide the mean gut pigment content by the amount of pigment occurring in the faecal pellets produced in 1 h (Dagg and Walser, 1987; Dagg *et al.*, 1989). A more commonly used method is to transfer copepods, that have been feeding, into filtered sea water and observe the decreasing amounts of food present at successive intervals thereafter. A curve of the form in Figure 44C results, and a gut evacuation rate constant (K) can be calculated. K is derived from the experimental data by calculating the regression of the natural logarithm of gut content relative to time from:

$$S_t = S_0 e^{-Rt}$$

where S_0 is the initial quantity of gut contents, S_t is the quantity of gut contents at time t, and R is the instantaneous evacuation rate with units of $1/t$ (Dagg and Wyman, 1983). Huntley *et al.* (1987a) found that a power model fitted their data on *Acartia* spp. better than the exponential model while Tiselius (1988) used a linear model.

This constant is temperature dependent, and Dam and Peterson (1988) correlated their own and previous measurements of gut evacuation rates to produce an equation that allows estimation of this constant when ambient temperature, T, is known and food is not limiting:

$$K = 0.0117 + 0.001794T \ (r^2 = 0.72, n = 44)$$

The data used by Dam and Peterson have been supplemented by additional data (Figure 45) and equations calculated with the outlier point at $K = 0.166$ included and excluded;

With outlier: $K = 0.00855 + 0.002853T \ (r^2 = 0.46, n = 70)$

Without outlier: $K = 0.00941 + 0.002575T \ (r^2 = 0.62, n = 69)$

The variation in this constant and at any one temperature is large, the currently exceptional value of 0.166 (Figure 45), found by Saiz *et al.* (1992a) in *Centropages typicus*, being as yet unexplained. Part of the variation in K derives from differences in the evacuation rates of different species under different food conditions. There is also considerable variation between individuals of a species in their gut evacuation times and the numbers of pellets produced (Paffenhöfer *et al.*, 1995a). Gut evacuation spans a period of 30 to 120 min and the time at which the curve (Figure 44C) enters the exponential phase varies. The constant, K, is usually only calculated from data representing the initial fast decrease period, the first 40 to 60 min of the observations (Atkinson et al., 1996b). Ellis and Small (1989) found no significant differences between values of K calculated from the initial 20 min and the total 90 min of gut clearance of *Calanus marshallae*. The use of these equations to approximate K is questionable in the case of an individual species of copepod because of the inherent variation evident in Figure 45.

Yet another method of estimating gut passage time is to use ^{14}C-labelled food, the copepod then being transferred to a medium with unlabelled food (Table 23). The values range from 20 min to just over 2 h. Dagg and Walser (1987) divided the gut content of pigment (ng pigment copepod^{-1}) by the rate of egestion (ng pigment hour^{-1}) and obtained a gut passage time of 23.4 min in *Neocalanus plumchrus*.

There is no seasonal change in gut evacuation rates nor does the concentration of food in which the animals were maintained prior to transfer to filtered sea water affect the rate (Tsuda and Nemoto, 1987; Ellis and Small, 1989). There is no diel cycle in the evacuation rate (Durbin et al., 1990) and Morales et al. (1990) found no convincing relationship between K and body size of the copepods.

6.1.2.6. Egestion and Faecal Pellets

Corner et al. (1986) tabulate data on the dimensions, volume, density and sinking rates of faecal pellets of many species and the numbers of pellets produced copepod^{-1} d^{-1}.

The number of faecal pellets produced in unit time increases more or less linearly with increased ingestion rate (Figure 46A); this linear relationship has been described by Reeve and Walter (1977), Gamble (1978), Ayukai and Nishizawa (1986), Ayukai (1990) and Tsuda and Nemoto (1990). The size of the pellets also increases, but to an asymptotic size, as food concentration increases (Figure 46B) as described by Dagg and Walser (1986) and Tsuda and Nemoto (1990). Estimates of the rates of faecal pellet production by a variety of species of copepods are shown in Table 24. The rate of production of faecal pellets can be used to estimate gut passage times. Assuming that two pellets occur simultaneously in the gut, then gut passage time is twice the interval between the production of a single pellet. Timonin et al. (1992) state that the amount of pigment in the gut of *Calanoides acutus* is equivalent to three pellets. The numbers of pellets produced per day range widely in Table 24. Estimates of between 100 and 150 are quite common. Such production rates represent 4 to 6 pellets h^{-1} during 24 hours of continuous feeding or about 8 to 12 h^{-1} if feeding is restricted to night time.

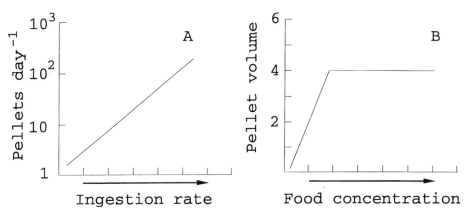

Figure 46 Faecal pellet production. A, the numbers of faecal pellets produced relative to ingestion rate. B, the volume of an individual faecal pellet relative to food concentration in the environment.

Respective gut passage times would then be about 20 to 30 min and 10 to 15 min. These gut passage times are very short (Table 23) in a normal situation when a diel rhythm of feeding is present, and even if 3, instead of 2, pellets are considered as the normal gut content. It again raises the importance of defining the frequency of feeding bouts of individuals over periods of a day and a few consecutive days.

Faecal pellets vary in shape and size according to the species of copepod, and size within a species can also be related to food concentration (Arashkevich and Cahoon, 1980). Photomicrographs of pellets of various species have been published: *Acartia clausi* and *A. tonsa* by Honjo and Roman (1978); *Calanus finmarchicus* by Honjo and Roman (1978); *C. helgolandicus* by Corner et al. (1986); *Centropages hamatus* by Lampitt et al. (1990) and Noji et al. (1991); *Neocalanus plumchrus* by Nagasawa (1992); *Pontella meadii* by Turner (1977); *Pseudocalanus newmani* by Nagasawa (1992). The size of pellet produced depends on the size of the nauplius, copepodid or adult producing it (Table 25). Uye and Kaname (1994) found the following relationship between faecal pellet size (PV) in μm^3 and prosome length (PL) in mm of the dominant copepods in the Inland Sea of Japan:

$$\text{Log } PV = 2.581 \log PL + 5.360$$

The equation derived from all the data in Table 25 is:

$$\text{Log } PV = 2.474 \log PL + 5.226 \quad r^2 = 0.902$$

Table 24 Faecal pellet production, number of pellets day^{-1}, and the temperatures at which the observations were made. Estimates are for adult females unless naupliar (N) or copepodid (C) stages are indicated. These data are supplementary to those of Corner *et al.* (1986).

Species	Number of pellets day^{-1}	Temp. °C	Authority
Acartia clausi	8–91		Roman and Rublee, 1981
	90.7 ±7.6	12	Honjo and Roman, 1978
A. tonsa	24.4 ±4.1	12	Honjo and Roman, 1978
	87	16	Butler and Dam, 1994
Calanoides carinatus	225	15	Peterson *et al.*, 1990a
	160–206	11	Timonin *et al.*, 1992
CV	144	11	Timonin *et al.*, 1992
Calanus helgolandicus	96	10	Corner *et al.*, 1972
NIII	48	15	Green *et al.*, 1992
NV/VI	58	15	Green *et al.*, 1992
C. pacificus	40	13	Ayukai, 1990
CI/II	100–165		Ayukai and Nishizawa, 1986
CIII	80–170		Ayukai and Nishizawa, 1986
CIV	50–160		Ayukai and Nishizawa, 1986
CV	40–150		Ayukai and Nishizawa, 1986
CVI	40–150		Ayukai and Nishizawa, 1986
Clausocalanus arcuicornis	35	20	Ayukai, 1990
Eucalanus pileatus	55–125	20	Paffenhöfer and Knowles, 1979
NIII	10–40	20	Paffenhöfer and Knowles, 1979
NIV/NV	95–120	20	Paffenhöfer and Knowles, 1979
Pseudocalanus elongatus			
NIII	52	15	Green *et al.*, 1992
NVI	116	15	Green *et al.*, 1992
P. newmani	82	4.5	Tsuda and Nemoto, 1990
Temora turbinata	10–169	20	Paffenhöfer and Knowles, 1979
CI/CII	ca 120	20	Paffenhöfer and Knowles, 1979

They also calculate the regression equation for the logarithmic relationship between faecal pellet volume (μm^3) and body weight of carbon (μg) in these species:

$$\text{Log } PV = 0.901 \log \mu g \text{ C} + 4.549$$

Using all the data in Table 25 gives the equation:

$$\text{Log } PV = 0.938 \log \mu g \text{ C} + 4.547 \quad r^2 = 0.803$$

Paffenhöfer and Knowles (1979) calculate comparable equations from the data in Table 25 for *Eucalanus pileatus* and *Temora longicornis*:

$$E.\ pileatus\ \text{Log}\ PV = 0.855 \log \mu g\ C + 4.900\ r^2 = 0.999$$

$$T.\ longicornis\ \text{Log}\ PV = 0.711 \log \mu g\ C + 4.780\ r^2 = 0.999$$

Faecal pellet volume within a species varies by a factor of 3 within any single size class of copepod. The numbers of faecal pellets produced in unit time also vary by a factor of 3, even when the copepods are restricted to a single diet. Thus, the use of faecal pellet production rates to estimate gut passage times is subject to the same amount of variation as the use of estimates of K, the gut clearance rate constant (Figure 45).

The production of faecal pellets by planktonic organisms in general and calanoid copepods in particular in many regions at particular seasons is so great that they constitute an identifiable component within the diets of many organisms, including the calanoid copepods themselves. Production of pellets by an individual copepod is often quantified as about 15% of the daily ration (Corner et al., 1986) but it varies considerably (Figure 47C), as it does when expressed as a percentage of body weight, carbon or nitrogen (described as % body content in Figure 47D). Their density is frequently greater than sea water (Figure 47B) and so they sink in the water column, transferring nutrients to greater depths. Their importance in the vertical flux of nutrients in the oceans is discussed in a later chapter.

Sinking rates of faecal pellets have been shown in laboratory studies to conform to Stokes Law:

$$v = (2/9)gr^2 \times ((\rho - \rho_0)/\eta)$$

where v = sinking velocity (cm s^{-1}), g = acceleration caused by gravity (980 cm s^{-1}), r = radius of pellet, ρ = assumed density of the pellet (1.20 g cm^{-3}), ρ_0 = density of sea water at 25 °C and 35‰ (~1.025 g cm^3), and η = the viscosity of the medium at 25 °C and 35‰ (0.010).

Paffenhöfer and Knowles (1979) calculate sinking rates (SR) relative to pellet volume ($10^4 \mu m^3$) from Stoke's Law and obtain the following equation:

$$\log SR = 0.698 \log \text{pellet volume}(\mu m^3) - 2.030$$

Density of pellets was assumed to be 1.20 g cm^{-3}.

Table 25 The volume of the faecal pellets of different species relative to the body size of the species expressed as prosome length (PL in mm) and weight of body carbon (μg C). These data are supplementary to those of Corner et al. (1986).

Species	Body size PL	μgC	Pellet $10^4 \mu m^3$	Authority
Acartia omorii	0.8	4.6	14	Uye and Kaname, 1994
A. tonsa	0.9	5.1	25	Honjo and Roman, 1978
	0.9	5.1	22	Butler and Dam, 1994
Calanus finmarchicus	3.5	115	230	Honjo and Roman, 1978
	3.5	140	93–226	Urban et al., 1993
NIII	0.4	1.4	1.0–2.1	Marshall and Orr, 1956
NIV	0.48	1.6	1.3–2.2	Marshall and Orr, 1956
NV	0.55	1.9	3.0–7.3	Marshall and Orr, 1956
NVI	0.61	2.2	6.1–7.0	Marshall and Orr, 1956
C. helgolandicus	3.2		110–150	Harris, 1994
NIII	0.4	1.4	0.56	Green et al., 1992
NV/VI	0.58	2.0	2.9	Green et al., 1992
Calanus sinicus	2.4	110	244	Uye and Kaname, 1994
Centropages abdominalis	1.2	13	36	Uye and Kaname, 1994
C. furcatus	1.4	25	69	Uye and Kaname, 1994
Eucalanus pileatus NIII		0.5	4.4	Paffenhöfer and Knowles, 1979
CII		5	31.5	Paffenhöfer and Knowles, 1979
		25	124.5	Paffenhöfer and Knowles, 1979
		25	75–185	Paffenhöfer and Knowles, 1979
Neocalanus plumchrus	2.2		176–390	Dagg and Walser, 1986
	4.0			Dagg and Walser, 1986
Paracalanus sp.		0.7	2.5	Paffenhöfer and Knowles, 1979
		1.6	5	Paffenhöfer and Knowles, 1979
		4.1	10.5	Paffenhöfer and Knowles, 1979
	0.67	2.5	8	Uye and Kaname, 1994

Table 25 Continued.

Species	Body size PL	Body size μgC	Pellet $10^4 \mu m^3$	Authority
Pareuchaeta norvegica				
CIV	3.0		110	Yen, 1987
CV	4.2		390	Yen, 1987
	5.7		940	Yen, 1987
Pseudocalanus elongatus	0.86	4.5	50	Harris, 1994
NIII	0.26		0.33	Green et al., 1992
NVI	0.44		0.46	Green et al., 1992
P. newmani	1.00		10	Tsuda and Nemoto, 1990
Pseudodiaptomus marinus	0.88	5.7	17	Uye and Kaname, 1994
Sinocalanus tenellus	0.93	6.5	19	Uye and Kaname, 1994
Temora longicornis				
NIV		0.1	1.2	Paffenhöfer and Knowles, 1979
CIII/IV		1	6	Paffenhöfer and Knowles, 1979
		2	9.9	Paffenhöfer and Knowles, 1979
T. stylifera	1.1	30	80	Paffenhöfer and Knowles, 1979
T. turbinata	1.0		13–44	Dagg and Walser, 1986
Undinula vulgaris	2.3	55	135	Paffenhöfer and Knowles, 1979

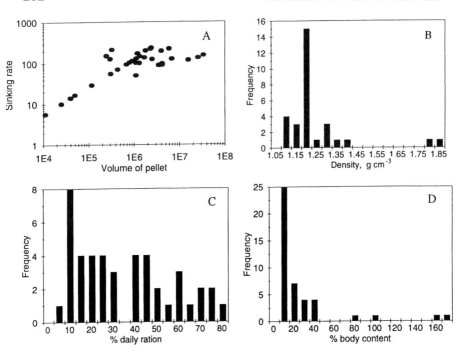

Figure 47 Faecal pellets. A, sinking rate in m day^{-1} relative to volume of the pellet in μm^3. B, frequency of estimates of different densities of faecal pellets. C, frequency of estimates of different daily weights of faecal pellets produced, expressed as percentages of the daily food ration. D, frequency of estimates of different daily weights of faecal pellets produced, expressed as percentages of body carbon or nitrogen content. (Data from Corner *et al.*, 1986; supplementary data in Figure A from Harris, 1994.)

Small *et al.* (1979) calculate equations for sinking rates measured in the laboratory at 14° C:

Small copepods, pellet volume range = 10 to 180 10^4μm^3:

$$\text{Log } SR = 0.374 \log \text{ pellet volume}(\mu\text{m}^3) - 0.416$$

Anomalocera patersoni, pellet volume range = 70 to 600 10^4μm^3:

$$\text{Log } SR = 0.461 \log \text{ pellet volume}(\mu\text{m}^3) - 1.045$$

Komar *et al.* (1981) have developed an equation from the data of Small *et al.* (1979):

$$w_s = ((1.21 \times 10^3)L^2)(L/D)^{-1.664}$$

where w_s = settling velocity in cm s^{-1}; L and D are length and diameter of the pellet in mm. Corner et al. (1986) review measurements of sinking rates and their tabulated data on those of calanoid copepods are graphed in Figure 47A to show the range of rates relative to the sizes of the pellets.

Sinking rates of faecal pellets can vary by threefold in pellets of the same size produced from the same diet (Turner, 1977). The size and density of faecal pellets vary seasonally (Urban et al., 1993), depending on the diet and its ability to be compacted within the pellet. Measurements of densities of faecal pellets are tabulated by Corner et al. (1986) and those for calanoids are illustrated in Figure 47B. The mean volumes of pellets of Calanus finmarchicus were 2.26 ± 1.37 in April, 1.45 ± 1.21 in June, and 0.93 ± 0.82 $10^6 \mu m^3$ in September while comparable mean densities were 1.11, 1.12, and 1.19 g cm^{-3}. The rates of sinking may, at present, be overestimated (Smetacek, 1980; Lampitt et al., 1990; Noji et al., 1991). Copepods, and other crustaceans, tend to fragment the pellets in the water column resulting in a decrease in their densities and of up to 50% in their sinking rates (Noji et al., 1991).

The density of faecal pellets and their nutritional quality are modified by bacteria. Bacteria rapidly colonize newly voided pellets (Jacobsen and Azam, 1984), bacterial degradation of pellets being faster at higher temperatures. Honjo and Roman (1978) found that the surface membrane of a pellet shows signs of degradation within 3 h at 20° to 25 °C and is almost completely degraded after 24 h. The membrane remains intact for 20 d at 5 °C. The types of bacteria, including indigenous (enteric) bacteria, associated with the pellets are reviewed by Delille and Razouls (1994) while Lee and Fisher (1992, 1994) examine the influence of bacteria on the release of trace elements from the pellets. Hansen et al. (1996) suggest that diatom-based faecal pellets are less suitable for the growth of bacteria than flagellate-based pellets. The latter degrade faster in surface waters in summer and autumn while the diatom-based pellets are liable to be degraded after sedimentation to the sea bed in spring. Nagasawa (1992) reviews the occurrence of enteric bacteria in the guts of copepods; they are lacking in *Pseudocalanus newmani* but are present in all other species examined, namely: *Acartia omorii, A. tonsa, Calanus pacificus, Centropages furcatus, Eucalanus bungii, Labidocera aestiva, Neocalanus plumchrus, Pleuromamma* sp., *Pontellopsis regalis.* Nott et al. (1985) suggest that bacteria occurring in the faecal pellets of *Calanus helgolandicus* are ingested with the food because no bacteria occurred within the empty guts of this species. Modification of the rates of degradation of the pellets may result from the fragmentation (coprorhexy) or partial fragmentation (coprochaly) of the pellets by copepods and other crustaceans (Lampitt et al., 1990; Noji et al., 1991).

6.1.2.7. Digestion and Assimilation

Honjo and Roman (1978) argue that the pH within the guts of copepods is close to that of sea water, about 8.2, because aragonite crystals and delicate coccolithophores show no evidence of dissolution on passage through the guts. This has recently been confirmed by Pond et al. (1995) using a pH-sensitive dye to determine the pH within different regions of the gut of *Calanus finmarchicus*. Average pH in starved animals was 6.86 to 7.19 while the lowest pH indicated was 6.11. The foregut was concluded to be the site of acid secretion but the pH in feeding animals was not acidic enough to digest significant quantities of the liths of the coccolithophore *Emiliana huxleyi*.

Copepods have a variety of digestive enzymes, listed and characterized by Mayzaud (1986a), the most important of which are amylase, laminarase, cellulase, carboxypeptidase and trypsin. The act of filling the stomach probably stimulates the F cells of the midgut to secrete the enzymes. Extracellular and intracellular (within the B cells) digestion takes place in the midgut, the products being absorbed by the R and D cells.

Does enzyme composition and secretion within the gut reflect the type and quantity of food? The presence of laminarase has been considered to indicate herbivory but Mayzaud (1986a) and Heerkloss and Ring (1989) point out that results are often contradictory. Mayzaud et al. (1992), and literature cited by them, reinforce the importance of the influence of feeding history on the response of the copepod to changing trophic conditions but also emphasize the governing role of the animal's internal metabolic requirements (Roche-Mayzaud et al., 1991). They suggest that its feeding history can form a physiological memory that influences acclimation responses. In other words, these represent additional controls, on the amounts of food ingested and digested, to the basic one of *body size: environmental temperature: respiration relationship*. Small coastal species with short generation times, low energy reserves and high metabolic demands will require to acclimate faster to changed trophic conditions than larger species, often deeper and more oceanic, with longer generation times, lipid reserves and lower metabolic demands.

There are diel and seasonal changes in enzyme activities but results are contradictory (Mayzaud et al., 1992). Oosterhuis and Baars (1985) found that short-term changes, measured in hours, of feeding activity of *Temora longicornis* are not reflected by comparable changes in the enzyme concentrations; they concluded that enzyme concentrations cannot be used as an index of feeding activity. Hassett and Landry (1990a) found that enzyme activities were high in spring in *Calanus pacificus* in Puget Sound. They and Hirche (1989a) document evidence of low enzyme activities in dormant or diapause stages in the life cycles of copepods.

Knowledge of digestion and assimilation can be gained from comparative studies of the chemical constitution of the food being ingested and the faecal pellets being voided (Cowie and Hedges, 1996). Degradation and mineralization of faecal pellets commence immediately they are voided into the environment. Roy and Poulet (1990) studied the chemistry of the ageing process within the pellets at 5° and 15 °C. Head (1992b) examined the chemical composition of faecal pellets relative to that of the particulate food and found that carbon assimilation efficiencies appeared to be correlated with concentrations of soluble carbohydrate in the diets. The molar composition of dissolved free amino acids in the faecal pellets depends on the type of food and on the hydrolysis of the proteins during digestion (Poulet et al., 1986b; Roy and Poulet, 1990). Dietary lipids are also modified by the digestive processes as shown by Prahl et al. (1984) in a study of the faecal lipids of *Calanus helgolandicus*.

The percentage weight of ingested food that is assimilated varies greatly and is difficult to determine according to Parsons et al. (1984) who review the subject. Assimilation efficiencies for herbivorous species have been reported as high as 60 to 95% and as low as 10 to 20%. In general, carnivorous species have higher assimilation efficiencies than herbivores and some values are given in Table 26. Assimilation efficiencies vary with the ash content of the food:

$$A = 87.8 - 0.73X$$

where A is the assimilation efficiency and X is the percentage ash per unit dry weight of the food (Conover, 1966b). Digestion can be selective at the molecular level e.g. for individual aldoses and amino acids (Cowie and Hedges, 1996).

Assimilation efficiencies for and bioaccumulation of metals by copepods are of interest in studies of marine pollution. Wang et al. (1996) examine the assimilation efficiencies of two neritic species, *Acartia tonsa* and *Temora longicornis*, for five metals – americium, cadmium, cobalt, selenium and zinc by radiolabelling of phytoplankton. Assimilation efficiencies were high, 70 to 95%, for Cd, Se and Zn but only about 40% for Co and 5% for Am. They briefly review studies of metal assimilation by copepods.

Net growth efficiencies, K_2, have been determined for a number of species and also for samples of mixed copepods (Table 27). The values vary considerably, partly because of the methods used and the element concerned. Parsons et al. (1984) review the perceived relationships between gross (K_1) and net (K_2) growth efficiencies:

$$K_1 = \Delta W/R\Delta t \qquad (K_2) = \Delta W/AR\Delta t$$

where $\Delta W/\Delta t$ represents growth per unit time, R is the food consumed, and

Table 26 Assimilation efficiencies for calanoid copepods.

Species	K_1	Authority
As % weight or volume assimilated		
Acartia tonsa	18–55	Roman, 1977
	10–45	Roman, 1977
Calanus finmarchicus	40–87	Conover, 1966a
C. helgolandicus	12–47	Gaudy, 1974
C. hyperboreus	44–70	Conover, 1966b
Centropages hamatus	18	Person Le-Ruyet et al., 1975
Centropages typicus	1–93	Gaudy, 1974
	34	Person Le-Ruyet et al., 1975
Chiridius armatus	91–98	Alvarez and Matthews, 1975
Euchirella rostrata	18	Conover, 1966a
Metridia gerlachei	80	Schnack, 1983
Paracalanus sp.	70	Paffenhöfer and Knowles, 1979
Pareuchaeta norvegica	91–94	Båmstedt and Holt, 1978
Pseudocalanus elongatus	10–30	Harris and Paffenhöfer, 1976a,b
Temora longicornis	16	Person Le-Ruyet et al., 1975
	10–40	Harris and Paffenhöfer, 1976a,b
T. stylifera	3–91	Gaudy, 1974
	28	Person Le-Ruyet et al., 1975
As % carbon assimilated		
Acartia tonsa	68	Gottfried and Roman, 1983
	44	Berggreen et al., 1988
Centropages typicus		
CI–CV	6–67	Razouls and Apostolopoulou, 1977
CVI	8.3	Razouls and Apostolopoulou, 1977
Eurytemora affinis	89	Barthel, 1983
Neocalanus cristatus	3.8	Taguchi and Ishii, 1972
N. plumchrus	5.11	Taguchi and Ishii, 1972
Temora stylifera CI–CV	1.5–31.3	Razouls and Apostolopoulou, 1977
CVI	6.9	Razouls and Apostolopoulou, 1977
Undinula vulgaris	85.8	Gerber and Gerber, 1979
	83.6	Le Borgne et al., 1989
As % nitrogen assimilated		
Acartia tonsa	36	Gottfried and Roman, 1983
Undinula vulgaris	89.2	Gerber and Gerber, 1979
	60.1	Le Borgne et al., 1989
As % phosphorus assimilated		
Undinula vulgaris	38.3	Le Borgne et al., 1989
As % iron assimilated		
Acartia tonsa	5–15	Hutchins et al., 1995

AR is the proportion of that food consumed that is available for growth after respiration and excretion have been accounted for. Efficiencies are affected by age, food concentration, temperature and other factors and so are difficult to determine with any accuracy. They can be estimated on a body weight basis or on a carbon or nitrogen basis, the method producing different results.

Head (1992b) found that light intensity did not affect the rate of carbon assimilation. Morales (1987) found a higher C:N ratio in faecal pellets of *Pseudocalanus* spp. and *Temora longicornis* than in their food, suggesting that nitrogen is assimilated more efficiently than carbon. Copepods have been said to maximize protein ingestion (Houde and Roman, 1987; Cowles *et al.*, 1988; Mayzaud *et al.*, 1992).

6.1.3. Concluding Remarks

Copepods eat to live, grow and reproduce. They are distributed across the spectrum of marine, estuarine and, indeed, fresh water environments. Consequently, they encounter a vast range of trophic conditions. They tend to be omnivorous rather than restricted feeders, resulting in individual species having the potential to recognize and exploit a wide variety of trophic resources. Ontogenetic changes in the diets within species result from increasing differentiation of the mouthparts and increased body size.

The rate of food ingestion increases to an asymptotic level as the concentration of food increases (Figure 44A). The rate of collection of food, filtering rate (Figure 43), is high at low food concentrations, or if initially low it increases to a high asymptotic level, but then with further increases in concentration of food to a critical concentration it decreases. This critical concentration of food, at the downward inflexion of the curve in Figure 43, is considered to correspond with the point at which the asymptotic level of ingestion rate is achieved in Figure 44A. The rate of procurement of food cannot exceed the rate at which the stomach empties as food passes posteriorly through the gut. The copepod must be able to procure more food in unit time than can be ingested into the gut and this is presumably one origin of the feeding bouts, viz. interruptions to the process of procuring food. The amount of food in the gut is a function of the rate of ingestion and the rate of gut evacuation as modelled by Dam *et al.* (1991). The rate of gut evacuation is linearly related to the ingestion rate (Figure 46A). In addition, the size of individual pellets produced increases with food concentration to an asymptotic level (Figure 46B). Also, faecal pellet size is very variable, as mentioned earlier. The rate of gut evacuation, therefore, probably does not restrict the rate of throughput of food within the gut. Consequently, the

Table 27 K_2, the net growth efficiency in terms of carbon (C), nitrogen (N) and phosphorus (P). Other values of K2 calculated from egg production and excretion are also given.

Species	$K_{2,C}$	$K_{2,N}$	$K_{2,P}$	K_2	Authority
Acartia spp.				42	Checkley et al., 1992
Calanoides acutus				23–72	Godlewska, 1989
Calanus finmarchicus	22.0			6–55	Conover and Huntley, 1991
		5.5			Corner et al., 1967
		43	22.4		Le Borgne, 1982
C. helgolandicus/C. finmarchicus		53.5	41		Butler et al., 1969
		38.6			Le Borgne, 1982
C. marshallae	0–58				Le Borgne, 1982
C. propinquus	48.7				Godlewska, 1989
Calanus sp.	26–53			25–68	Conover and Huntley, 1991
					Le Borgne, 1982
Centropages furcatus	8–59			33	Checkley et al., 1992
C. typicus CI–CV	11				Razouls and Apostolopoulou, 1977
CVI	24–64				Razouls and Apostolopoulou, 1977
Metridia gerlachei	57				Pagano et al., 1993
Rhincalanus gigas	11.9			28–62	Conover and Huntley, 1991
					Godlewska, 1989
					Conover and Huntley, 1991
Temora stylifera					
CI–CV	2–42				Razouls and Apostolopoulou, 1977
CVI	9				Razouls and Apostolopoulou, 1977
copepodids	27–40				Pagano et al., 1993
females	35–72				Pagano et al., 1993
males	31–33				Pagano et al., 1993
Undinula vulgaris	7.2	27.3	33.7		Le Borgne, 1982
		62.3			Le Borgne et al., 1989
Mixed copepods	46 (9–76)	61.6	26.4		Le Borgne, 1982
Small copepods	8.9	50			Le Borgne, 1982

controlling factors on the amount of food processed in unit time must be in the midgut where digestion takes place and the food is packaged into faecal pellets. The rate of digestion will vary dependent upon the quality of the food. The amount of digestion, whether partial or complete, may also vary dependent upon the amount of food the copepod can collect in unit time. At high ingestion rates, in for example a dense phytoplankton bloom, superfluous feeding may take place as argued by Turner and Ferrante (1979). Passage through this region allows more food to be collected and ingested.

The above concept does not take account of internal metabolic controls and feed-back loops affecting rates of collection, ingestion and digestion of food. The variability within and contradictions between the results of the many feeding experiments done on copepods have necessitated adoption of new approaches to such studies. The current metabolic requirements of individuals, shaped by their metabolic states, degree of starvation or satiation, with life history, seasonal and diel components, and governed overall by their recent feeding histories, are now being considered as major factors in the equations. Durbin et al. (1990) conclude that diel feeding rhythms are endogenous and controlled separately from diel vertical migration patterns, if they are present. The current concentrations of digestive enzymes within the gut reflect current feeding rates and behaviour but their low or high concentrations do not regulate feeding behaviour nor do the current concentrations predict potential for ingestion (Thompson et al., 1994).

Historical factors such as pattern of the environmental temperature regime, in situ food concentrations, copepod body size, and condition factor influence feeding of Acartia tonsa according to Thompson et al. (1994). Additional factors reviewed by them are type and size of food, size and developmental stage of the copepods, diel rhythms, presence of predators, and life history stage of the adults. Further, they conclude that ingestion rates should be modelled as varying by a factor of 2 or 3 and subject to modification by behavioural, physiological and environmental parameters. Their study considers many of the factors that have been identified in the last ten years as potentially important in interpreting the results of feeding experiments on copepods.

6.2. RESPIRATION AND EXCRETION

Copepods have no gills, and oxygen uptake and carbon dioxide release take place through the integument and in the hindgut. The site of excretion of the end products of catabolism is primarily the maxillary glands. Le Borgne

(1986) and others, such as Minkina and Pavlova (1992), review in detail methods of measurement, and factors affecting these measurements, of respiration and excretion rates. Factors discussed are the time after capture at which the measurement is made, how many animals are in the container (crowding effect), stress on the animals, their physiological state including effects of starvation, and container effects where the animals modify their external medium by contributing excretory products to it or by altering its oxygen partial pressure. Small-scale turbulence in an experimental environment enhances metabolic rates (Saiz and Alcaraz, 1992b) while the presence of dead ice-algae depressed it (Conover et al., 1988a).

The electron transport system (ETS) activity has been investigated as an indicator of rates of respiration on several occasions (Mayzaud, 1986b). Hernández-León and Gómez (1996) review the current results and discuss some of the problems inherent in this approach. Measurement of glutamate dehydrogenase (GDH) activity has been proposed as an index of excretion of ammonium. Mayzaud (1986b) discusses the variability inherent in this measure.

The respiration and excretion rates frequently show rhythms corresponding to diel feeding rhythms of the copepods (Harris and Malej, 1986; Checkley et al., 1992; Cervetto et al., 1993; Pavlova, 1994). They are temperature dependent and in many species vary seasonally, being high in the spring and summer and lower in the winter. This is especially true of species that have overwintering resting stages (Head and Harris, 1985; Båmstedt and Tande, 1988; Hirche, 1989a; Conover and Huntley, 1991). Latitudinal changes are present, metabolic activity being lower at higher latitudes (Ikeda and Mitchell, 1982; Hirche, 1987). Oxygen consumption and the quantities of excretory products, e.g. ammonia, vary with body size, larger animals consuming or producing more but the weight-specific oxygen consumption or excretion of ammonia decrease with increasing size of the animals, the smaller animals being more metabolically active (Marshall, 1973; Ikeda, 1974; Vidal, 1980c; Dagg et al., 1982; Ikeda and Mitchell, 1982; Paffenhöfer and Gardner, 1984; Båmstedt and Tande, 1985; Smith, 1988; Uye and Yashiro, 1988).

Ikeda (1970, 1974) gives regression equations relating oxygen consumption of planktonic species in general, including copepods, and distinguishes between boreal, temperate, subtropical and tropical species (Table 28). His data for calanoid copepods have been extracted and combined with data quoted by Marshall (1973) for a variety of copepods to provide comparable equations for calanoid copepods alone (Table 28, Figure 48). Regression equations for nitrogen excretion are given in Table 29. The slopes of the regression lines are between 0.6 and 0.92 with the exception of that for the line describing excretion rates in the Antarctic *Calanoides acutus* and *Metridia gerlachei* given by Huntley and Nordhausen (1995); this may be partly caused by the restricted size range of these two species. The intercepts of the regression lines for respiration and excretion are more variable than their slopes; Vidal and Whitledge (1982) discuss this aspect in some detail. Part of the variation in the intercepts can arise through body lipid being present or absent. Vidal and Whitledge (1982) show that intercepts have higher values when total dry body weight as opposed to lipid-free dry weight is used in calculations of metabolic rates. The former is then reflected by higher Q_{10} values than in the lipid-free situation. Consequently, lipid content must be considered in latitudinal comparisons of metabolic rates. The normal range of Q_{10} is 1 to 4 although values higher than 4 have been recorded in the literature (Hirche, 1987); values for copepods are usually between 2 and 4. A high Q_{10} in a species indicates temperature sensitivity and Hirche points out that there are seasonal changes in the Q_{10} related to the changing physiological condition of the organism.

Virtually no information is available on the physiology of meso- and bathypelagic species of copepods. It is known, however, that the Q_{10} of *Gaussia princeps* is greater at the lower temperatures and higher pressures prevalent at its daytime depths below 400 m (Childress, 1977). This means that its rate of oxygen consumption is more sensitive to changes of temperature at these depths. This is in contrast to the situation when it has migrated at night to depths of 200 to 300 m where pressure is the dominant factor governing rate of respiration.

6.3. METABOLISM

The substrate metabolized by copepods and other organisms consists of varying proportions of carbohydrates, lipids and proteins (Anderson, 1992). The ratios of the rates of oxygen consumption to ammonia excretion (O:N), ammonia excretion to phosphate excretion (N:P), and oxygen consumption to phosphate excretion (O:P) vary with the composition of the substrate used (Figure 49). Average atomic ratios were calculated by Ikeda (1977) but examination of the values of each ratio in the literature shows that they range widely (Figure 49). In general, high O:N and O:P atomic ratios are indicative of inclusion of a carbohydrate substrate, a low N:P ratio of inclusion of a lipid substrate, and a low O:N of inclusion of a protein substrate. Thus O:N atomic ratios of <20 result primarily from protein utilization while those of 50 or more result from metabolism of about equal proportions of protein and lipid. On the other hand, Vidal and Whitledge (1982) suggest that O:N atomic ratios >20 in tropical plankton, where lipid storage is minimal, result primarily from oxidizing substrates rich in carbohydrate. This contrasts with the situation in winter under food-limiting

Table 28 Regression equations describing the respiration rates of plankton in general, data from Ikeda (1974), and of calanoid copepods in particular, data from Ikeda (1970, 1974) and Marshall (1973). Respiration rates (R) are given as $\mu l\ O_2\ animal^{-1}\ h^{-1}$ and body dry weight (W) as mg. The sample size (n) and the regression coefficient (r) are given.

	Temperature range	n	Regression equation $\log R = b\log W + a$	r
Boreal	8.6°C(3.0–14.3)			
Plankton		78	$\log R = 0.783\log W + 0.057$	0.959**
Calanoid copepods		68	$\log R = 0.683\log W - 0.062$	0.953**
Temperate	15.0°C(11.7–17.5)			
Plankton		64	$\log R = 0.756\log W + 0.127$	0.955**
Calanoid copepods		31	$\log R = 0.787\log W + 0.245$	0.934**
Subtropical	20.2°C(17.3–22.5)			
Plankton		21	$\log R = 0.664\log W + 0.321$	0.915**
Calanoid copepods		5	$\log R = 0.852\log W + 0.541$	0.983**
Tropical	26.8°C(25.7–28.5)			
Plankton		98	$\log R = 0.595\log W + 0.481$	0.886**
Calanoid copepods		52	$\log R = 0.797\log W + 0.783$	0.926**

**Significant at 0.1% level.

Table 29 Regression equations describing the excretion rates of nitrogen for planktonic species in general and for calanoid copepods in particular. Excretion rate (E) is given as μg N animal^{-1} h^{-1} and body dry weight (W) as mg. From Vidal and Whitledge (1982). The sample size (n) and the correlation coefficient (r) are given.

	n	Regression equation logR = blogW + a	r
Boreal			
Plankton animals	43	LogE = 0.889logW + 1.906	0.953**
Copepods	28	LogE = 0.644logW − 1.395	0.894**[a]
	106	LogE = 0.234logW − 1.223	0.607**[b]
	91	LogE = 1.02 logW + 0.23	0.889**[c]
Subtropical			
Plankton animals	77	LogE = 0.836logW + 1.773	0.974**
	72	LogE = 0.926logW + 1.535	0.741**[d]
Undinula vulgaris	25	LogE = 0.830logW + 1.694	0.994**
Centropages typicus	18	LogE = 0.779logW + 1.428	0.848**
Tropical			
Copepods	27	LogE = 0.844logW − 0.385	0.829**[a]

**Significant at 0.1% level; [a]Ikeda, 1974; [b]Huntley and Nordhausen, 1995; [c]Smith, 1988; [d]Bishop and Greenwood, 1994.

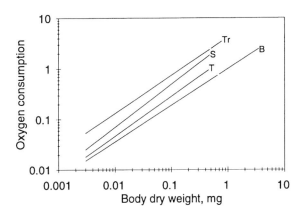

Figure 48 Respiration rates of calanoid copepods, μl O$_2$ animal^{-1} h^{-1}, relative to their body dry weight in mg. Regression lines are given for boreal (B), temperate (T), subtropical (S) and tropical (Tr) species. The relevant equations are given in Table 28. (After Ikeda, 1974.)

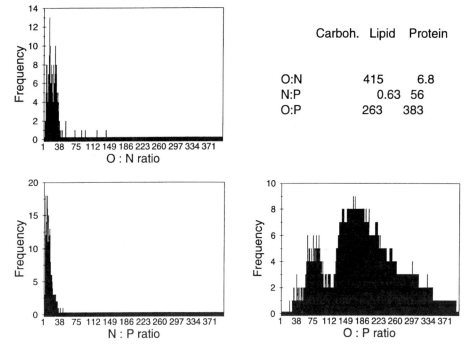

Figure 49 The ranges of the O:N, N:P and O:P atomic ratios calculated from the rates of respiration, ammonia excretion and phosphate excretion in calanoid copepods. Average atomic ratios were calculated by Ikeda (1977) and are tabulated top right. The data for the graphs are the single or range of values given for individual species by: Ikeda and Mitchell, 1982; Gaudy and Boucher, 1983, 1989; Båmstedt and Tande, 1985, 1988; Pagano and Gaudy, 1986; Mayzaud and Conover, 1988; Ikeda and Skjoldal, 1989; Le Borgne *et al.*, 1989; Conover and Huntley, 1991.

conditions in high latitudes where such O:N atomic ratios result from metabolism of stored lipids.

These metabolic ratios or quotients are at present relatively crude indicators of the quality of food being catabolized by copepods (Mayzaud and Conover, 1988). Measurement of the quantities of digestive enzyme activity present provides supplementary information, amylase catabolizing the hydrolysis of starch and trypsin that of protein. The ratios can be further amplified by determining the turnover rates, percentage loss per day, of body carbon, nitrogen and phosphorus (Båmstedt and Tande, 1985; Ikeda and Skjoldal, 1989; Huntley and Nordhausen, 1995). Losses range from about 0.5 to 9.0% d^{-1}, varying according to the substrate being used and also the metabolic activity of the copepod. Turnover rates are usually less in larger than smaller copepods, in overwintering and intermittently feeding species than in actively feeding and growing copepods.

Further work on these metabolic ratios, coupled with parallel observations on digestive enzymes and turnover rates of elements, is required to amplify the usefulness of this approach. Flint *et al.* (1991) suggest that there may be physiological classes of copepods. Some species, such as in the genus *Eucalanus*, have depressed rates of metabolism relative to other copepods; they have used the term "jelly-bodied copepods" for these species. There are certainly physiological classes within a species on many occasions, and Båmstedt and Tande (1985) suggest that studies of metabolism would benefit from observations at the single-individual level.

Energy budgets have been estimated for several species of copepods. Båmstedt *et al.* (1990), for example, calculated them for *Acartia clausi*, *Calanus finmarchicus* and *Pseudocalanus* sp. in Kosterfjorden, western Sweden, and concluded that copepods have a potential for growth of up to 8.5% d^{-1}. Abou Debs (1984) describes energy budgets for *Temora stylifera* in terms of carbon and nitrogen. Pagano and Saint-Jean (1994) give a metabolic budget for *Acartia clausi* in a tropical lagoon.

6.4. OSMOTIC REGULATION, SALINITY AND TEMPERATURE

Relatively few studies of the osmoregulation of copepods have been made (Brand and Bayly, 1971; Farmer, 1980; Roddie *et al.*, 1984). Endemic estuarine species exhibit hyper-osmoregulation at low salinities and hypo-osmoregulation at high salinities while oceanic and coastal marine euryhaline species are osmoconformers. These latter species have blood that is isotonic with the external medium. Osmoregulatory functions, and possibly also ionic regulation, in the euryhaline species are probably performed at the cellular level. Bayly (1969) reports that the ionic composition of the blood is modified relative to that of the external medium, magnesium being maintained below the external concentration while sodium is maintained above it.

Osmotic stress caused the patterns of proteins synthesized by *Eurytemora affinis* to change (Gonzalez and Bradley, 1994). Additional heat stress resulted in synthesis of another set of proteins. The proteins were characterized by their molecular weight. Proteins of 60–63 and 75–85 kD (kilodalton) were common in animals stressed at 30 °C and referred to as heat shock proteins. Proteins of 30 kD occurred at hyperosmotic salinities at 15 °C but not under heat stress at 30 °C; they appear to be unique to osmotically stressed *E. affinis*.

6.5. RESPONSES TO ENVIRONMENTAL VARIABLES

6.5.1. Temperature and Salinity

Salinity and salinity-temperature interactions of copepods control their distributions in coastal and estuarine situations. Temperature, salinity and T/S distributions characterize the geographical distributions of many species. Earlier studies on reactions of copepods to changing temperature and salinity are referred to by Rippingale and Hodgkin (1977), Roddie *et al.* (1984) and Ough and Bayly (1989). Copepods are more tolerant of changes in salinity when fed than when starved, inferring that they require energy to cope with the stressful circumstances, energy that may be required for the osmoregulatory processes.

Pagano and Gaudy (1986) found that salinity variations affected respiration but not excretion rates of the brackish water species *Eurytemora velox*. Lower temperatures result in higher survival rates over a broader range of salinity changes than higher temperatures. Lance (1964) found evidence that adult male *Acartia discaudata*, *A. clausi* and *Centropages hamatus* are less tolerant of changing salinity than the adult females. Slow changes of salinity and temperature allow acclimatization (Bhattacharya, 1986) which can result in an increase in the ranges of tolerance.

Upper and lower limits of thermal tolerance of copepods usually demonstrate that individual species are fit to live in the environment that they inhabit. This is well illustrated by the temperature tolerances determined for *Calanus hyperboreus*, *C. glacialis*, *Metridia longa* and *C. finmarchicus* by Hirche (1987); the range between the minimal and maximal temperatures tolerated by each species was about the same, 10 °C, but there was a shift in the order that the species are listed from a colder to a warmer range of temperature. Species living in variable coastal environments, especially estuaries, have wider thermal and salinity tolerances than oceanic species. Temperature tolerance, at least, may be extended in a species such as *Eurytemora affinis* through acclimation. B.P. Bradley and co-workers have made a detailed study of the thermal tolerance and acclimation of this estuarine species whose populations are subject to an annual temperature range of 0 to 30 °C (Tepper and Bradley, 1989). They developed a non-destructive, short-term assay for temperature tolerance (Bradley, 1976, 1986a; Bradley and Davis, 1991). It consists of measuring the time elapsed between administration of a temperature shock to the animal and its becoming comatose. This assay has shown that survival and temperature tolerance are related both physiologically and genetically (Bradley *et al.*, 1988).

The direct effects of temperature on feeding, respiration, development and growth are discussed elsewhere.

6.5.2. Oxygen

Copepods normally live in well-oxygenated environments. Many small coastal species, however, produce resting eggs that sink to bottom muds that may be anoxic. Effects of oxygen concentrations on hatching of eggs are discussed later. A few oceanic species appear adapted to deep oxygen minimum layers. Herman (1984), reviewing earlier information and his own, shows that oxyclines curtail the downward extent of diel vertical migrations of some species but not others.

6.5.3. Light

Exposure to strong natural light is harmful to most calanoid copepods. Marshall et al. (1935) found that respiration rate doubled during exposure of *Calanus finmarchicus* to light. Threshold light intensities detected by *Acartia tonsa* are 2.8×10^{11} photons $m^{-2} s^{-1}$ (Stearns and Forward, 1984a). Buskey et al. (1989) measured thresholds of 9.5×10^{12} and 7.7×10^{10} photons $m^{-2} s^{-1}$ in *Pleuromamma gracilis* and *P. xiphias* respectively. Maximal spectral sensitivity of the mesopelagic *Pleuromamma* species was in the blue-green around 480 nm (Buskey et al., 1989). Maximal spectral sensitivity of the coastal species *Acartia tonsa* was over a broader range, 453 to 620 nm (Stearns and Forward, 1984a).

The *Pleuromamma* species are bioluminescent and Buskey et al. (1987) show that they are sensitive to dim flashes of blue light.

Several species have been shown to be sensitive to the plane of polarization of the downwelling daylight in the sea. Umminger (1968) found that *Centropages hamatus*, *Labidocera aestiva* and *Pontella meadii* oriented in response to the plane of polarization but that *Acartia tonsa*, *Eucalanus monachus* and *Pseudodiaptomus coronatus* did not. There was no correlation between the presence of complex, as opposed to simple, eyes in the response.

Biologically harmful mid-ultraviolet irradiation, UV-B 290 to 320 nm, can penetrate to depths of at least 20 m in the sea. Karanas et al. (1979) demonstrated experimentally that the reproductive capability of *Acartia clausi* was impaired by exposure to UV-B radiation but that the harmful effects were not transmitted to their surviving offspring (Karanas et al., 1981). Difficulties in extrapolating to the natural environment are discussed by them and they conclude that enough incident UV-B radiation will be present at 1 m depth to cause significant mortality of early and late naupliar stages of this calanoid. Bollens and Frost (1990) could not demonstrate a direct connection between UV-B irradiation and an avoidance of the surface layer during diel vertical migration of *A. hudsonica*. A natural defence against UV-B radiation is conferred by UV-absorbing mycosporine-like amino acids which have been shown to occur in many marine organisms and in the Antarctic calanoid *Calanus propinquus* by Karentz et al. (1991).

6.5.4. Pressure

Some calanoid copepods are sensitive to small changes in hydrostatic pressure (Knight-Jones and Qasim, 1967; Lincoln, 1971). Macdonald et al. (1972) found that the neustonic species, *Anomalocera patersoni*, was more sensitive to pressure changes than several bathypelagic species tested.

6.6. CONCLUDING REMARKS

The literature on different aspects of the physiology of calanoid copepods is very large. A summary of the results is presented above citing modern literature that, in turn, cites the earlier work. The number of different species, aspects of whose physiology have been described, is small principally because of the difficulties encountered in maintaining them in the laboratory. Culture of individual species has been achieved relatively recently (see Table 47, p. 300) and has allowed experimental observation. The most frequently cultured species belong to the coastal genera *Acartia*, *Centropages*, *Eurytemora*, *Pseudocalanus* and *Pseudodiaptomus*. These are all small and have relatively short generation times. Larger species (Table 47), such as *Calanus*, have been maintained in the laboratory with some success. There are, however, potential problems and criticisms involved with such cultures.

Major differences exist between the conditions of culture and those pertaining in natural environments. Many are known and can be included in the interpretations of the experimental results. Others are only suspected. One of the principal handicaps is the restricted diets available for the cultured copepods. These are derived from food organisms that themselves are in cultures. Many organisms that are included in the natural diets of copepods have not been cultured and little is known about the selection that a copepod makes in the natural environment. Cultures of copepods sustained for several generations can suddenly terminate themselves. An axenic culture of the harpacticoid copepod, *Tigriopus fulvus*, was successfully bred through 11 generations by Mauchline and Droop (unpublished) and suddenly died. This raised questions of possible vitamin deficiency or whether the maintenance under a constant temperature and light regime was harmful. Might a seasonal component be required within such long-term cultures? Were periods of low and high temperature required or were marked changes necessary in the food composition parallel to those of the natural environment?

Much experimental work using a wider range of species remains to be done. Nutritional requirements of copepods relative to stages in their life histories are suspected but not defined. The significance of social or behavioural interactions between individuals and species, and the role of integumentary organs and sensilla in mediating them, within laboratory situations is not known. More experiments, such as those of Wong (1988b), examining interactions could be very interesting.

7. Chemical Composition

7.1. Body Weight and Volume 221
   7.1.1. Wet Weight 221
   7.1.2. Water Content 226
   7.1.3. Dry Weight 227
   7.1.4. Ash-free Dry Weight 233
   7.1.5. Ash Weight 233
   7.1.6. Volume 235
7.2. Elements 236
   7.2.1. Carbon 236
   7.2.2. Nitrogen 240
   7.2.3. Carbon:Nitrogen Ratio 242
   7.2.4. Hydrogen 242
   7.2.5. Phosphorus 242
   7.2.6. Atomic Ratios 243
   7.2.7. Mineral Composition 243
7.3. Organic Components 245
   7.3.1. Protein 245
   7.3.2. Lipids 245
   7.3.3. Carbohydrate 248
   7.3.4. Chitin 248
7.4. Other Organic Components 249
   7.4.1. Free Amino Acids 249
   7.4.2. Nucleic Acids 249
   7.4.3. ATP (Adenosine triphosphate) 249
7.5. Energy Content 250
7.6. Vitamins 250
7.7. Carotenoids 251
7.8. Enzymes 252

The subject matter of this chapter is supplementary to "*The Biological Chemistry of Marine Copepods*" (Corner and O'Hara, eds, 1986). The integral chapters of this work are repeatedly referred to here as they review and assess knowledge of most aspects of the subject prior to 1985. The emphases within Corner and O'Hara (1986) reflect the interests of the individual authors and so some additional topics are discussed here along with a review of relevant information published after 1985.

The chemical composition of many species, especially in medium and high latitudes, changes seasonally and often in a very marked manner. Bottrell and Robins (1984) found that seasonal changes in prosome length were smaller than those of body dry weight in *Calanus helgolandicus*. Dry weight was logarithmically related to prosome length but significantly different equations were generated for each of eight cruises; those for February through to August are given in Table 33 (p. 228). Carbon and nitrogen contents reflected the seasonal changes in body dry weight. Dry body weight also changes ontogenetically, especially in terms of energy stores. Båmstedt (1986, 1988b) discusses these changes in some detail, including the amount of variation in constituents found at the individual level within a population. He reviews seasonal changes in water content, body wet and dry weights, carbon and nitrogen contents and the carbon:nitrogen ratio. He also discusses comparable changes in ash, protein, lipid, carbohydrate and chitin contents. Energy contents of copepods also vary seasonally, as reviewed by Båmstedt, being dependent primarily on the supply of food. The result of all this variation is that only the water and total organic content of copepods vary little between species, all other components showing much greater variation.

7.1. BODY WEIGHT AND VOLUME

7.1.1. Wet Weight

Estimates of wet or fresh body weight of calanoid copepods, like those of other organisms, have an experimental error associated with the varying amounts of water retained on the surfaces of the body. The normal method for determining wet weight is to rinse the animals in fresh or de-ionized water to remove adhering salts. They are then placed on absorbent blotting paper and gently agitated on it, with the aid of a fine hair brush, to remove interstitial and surface water from the body in as constant a manner as possible to reduce error. The animal, or animals, are then weighed.

Body weight can be derived from measurements of body volume by applying a factor for specific gravity, 1.025 by Chojnacki (1983). Geometrical formulae for calculating body volume are given later. Kuz'micheva (1985) gives a formula for calculating body wet weight (W in mg) from measurements of body length (L in mm), where L is the total length from the anterior of the head to the posterior end of the caudal furcae:

$$W = aL^3$$

where $\qquad a = (0.295 - 0.194L/l + 0.398c/l)^2$

where L is as defined above, l is the length of the prosome and c its width in mm. The value of a changes between different copepodid stages and between seasons. Kuz'micheva shows that the value of a decreases regularly with advancing copepodid stage in some species and not in others and suggests that further study of its properties might be useful. Svetlichny (1983) uses the formula:

$$W = Kld^2$$

where W is the wet weight, l and d are the length and width of the prosome, and K is a coefficient that varies; K was 0.534 ± 0.057 for the 20 species studied, 0.56 ± 0.03 and 0.58 ± 0.01 for females and males respectively of 68 species in the literature. The application of this coefficient estimated wet weight with an accuracy of $\pm 15\%$.

Wet weight varies approximately as the cubic function of body length. Divergence of the exponents of body length from 3 (Table 30) arises from three principal sources. The first two are experimental error and individual variation in the measurements and the third is caused by allometric growth where body width increases or decreases at the expense of length. Most estimates of wet weight are made on formalin-preserved animals because of the difficulties of sorting and identifying species when they are still fresh. Bird and Prairie (1985) discuss some of the problems involved in the statistical descriptions of length-weight relationships while Cohen and Lough (1981) standardize the equations, in the papers reviewed by them, to the exponential form. The equations in Table 30 are on a $\log_{10}$ basis because the double logarithmic (power curve) relationship usually gives the best fit. The exponents of both total and prosome length vary from about 2.5 to 3.3 (Table 30) and Gruzov and Alekseyeva (1970) suggest that this results from the shape of the copepod. They ascribe species into 4 groups (Table 31): Group 1 species have a heavy prosome and short urosome e.g. *Scolecithrix danae*; Group 2 contains the majority of species e.g. Calanidae; Group 3 species have an elongated prosome e.g. Metridinidae and Euchaetidae; Group 4 species with the prosome elongated or flattened dorso-ventrally, examples occurring in many genera. The use of the four equations in Table 31 allows estimation of wet weight of species with an accuracy of ± 15 to $\pm 25\%$ (Gruzov and Alekseyeva, 1970).

Pearre (1980) argues that the prosome width rather than length is a better correlate of body wet weight. A large number of species were examined (Table 32) and the correlation coefficient is best when prosome width is used. No such advantage existed in tropical copepods studied by Chisholm and Roff (1990a), correlation coefficients for prosome length being equal to or better than those for prosome width in size/weight regressions.

Table 30 Relationship of wet or fresh weight (WW in µg) to prosome length (PL in µm) or to total length (TL in mm) in different species.

Species	Equation	Authority
Aetideus armatus	Log WW = 3.049logPL − 7.267	Shmeleva, 1965
Calanoides acutus	Log WW = 4.117logTL − 11.539	Mizdalski, 1988
C. carinatus	Log WW = 2.923logTL − 7.280	Gruzov and Alekseyeva, 1970
Calanus helgolandicus	Log WW = 3.561logTL − 9.525	Williams and Robins, 1982
Calanus minor	Log WW = 3.103logPL − 7.350	Shmeleva, 1965
	Log WW = 3.121logTL − 7.796	Gruzov and Alekseyeva, 1970
Calanus propinquus	Log WW = 3.208logTL − 8.094	Mizdalski, 1988
Calanus tenuicornis	Log WW = 2.757logPL − 6.519	Shmeleva, 1965
Calocalanus pavoninus	Log WW = 2.772logPL − 6.488	Shmeleva, 1965
Candacia curticauda	Log WW = 3.458logTL − 8.759	Gruzov and Alekseyeva, 1970
C. pachydactyla	Log WW = 3.012logTL − 7.506	Gruzov and Alekseyeva, 1970
Centropages typicus	Log WW = 3.213logPL − 7.589	Shmeleva, 1965
C. violaceus	Log WW = 2.484logPL − 5.599	Shmeleva, 1965
Clausocalanus arcuicornis	Log WW = 2.826logPL − 6.548	Shmeleva, 1965
C. furcatus	Log WW = 2.489logPL − 5.638	Shmeleva, 1965
C. paululus	Log WW = 2.800logPL − 6.356	Shmeleva, 1965
Ctenocalanus vanus	Log WW = 2.654logPL − 6.108	Shmeleva, 1965
Eucalanus attenuatus	Log WW = 2.789logTL − 6.903	Gruzov and Alekseyeva, 1970
E. crassus	Log WW = 3.187logTL − 8.097	Gruzov and Alekseyeva, 1970
E. pseudoattenuatus	Log WW = 2.741logTL − 6.763	Gruzov and Alekseyeva, 1970
E. subtenuis	Log WW = 2.827logTL − 7.088	Gruzov and Alekseyeva, 1970
Euchaeta marina	Log WW = 3.555logTL − 9.490	Gruzov and Alekseyeva, 1970
E. paraconcinna	Log WW = 3.742logTL − 10.187	Gruzov and Alekseyeva, 1970
Euchirella curticauda	Log WW = 2.710logTL − 6.338	Gruzov and Alekseyeva, 1970
E. pulchra	Log WW = 2.020logTL − 3.826	Gruzov and Alekseyeva, 1970
E. rostrata	Log WW = 2.035logTL − 3.907	Gruzov and Alekseyeva, 1970

Table 30 Continued.

Species	Equation	Authority
E. splendens	Log WW = 2.967logTL − 7.192	Gruzov and Alekseyeva, 1970
Gaidius tenuispinus	Log WW = 3.223logTL − 8.096	Mizdalski, 1988
Heterorhabdus farrani	Log WW = 3.550logTL − 9.274	Mizdalski, 1988
Labidocera acutifrons	Log WW = 3.162logTL − 7.981	Gruzov and Alekseyeva, 1970
L. trispinosa	Log WW = 2.747logPL − 6.324	Vlymen, 1970
Lucicutia flavicornis	Log WW = 3.327logPL − 8.007	Shmeleva, 1965
Mecynocera clausi	Log WW = 2.599logPL − 6.146	Shmeleva, 1965
Metridia gerlachei	Log WW = 3.643logTL − 9.756	Mizdalski, 1988
Neocalanus gracilis	Log WW = 0.502logPL + 0.994	Shmeleva, 1965
	Log WW = 3.055logTL − 7.631	Gruzov and Alekseyeva, 1970
N. robustior	Log WW = 3.086logTL − 7.730	Gruzov and Alekseyeva, 1970
Paracalanus aculeatus	Log WW = 4.536logPL − 11.213	Shmeleva, 1965
P. parvus	Log WW = 2.681logPL − 6.159	Shmeleva, 1965
Pareuchaeta antarctica	Log WW = 3.177logTL − 8.282	Mizdalski, 1988
P. elongata	Log WW = 3.061logPL − 7.427	Morioka, 1975
P. gracilis	Log WW = 3.214logTL − 8.469	Gruzov and Alekseyeva, 1970
P. hebes	Log WW = 2.901logPL − 6.847	Shmeleva, 1965
	Log WW = 4.183logTL − 11.641	Gruzov and Alekseyeva, 1970
Pleuromamma abdominalis	Log WW = 2.977logTL − 7.438	Gruzov and Alekseyeva, 1970
P. gracilis	Log WW = 2.505logPL − 5.547	Shmeleva, 1965
P. robusta	Log WW = 3.133logTL − 7.882	Gruzov and Alekseyeva, 1970
Rhincalanus cornutus	Log WW = 2.635logTL − 6.460	Gruzov and Alekseyeva, 1970
R. gigas	Log WW = 4.042logTL − 11.557	Mizdalski, 1988
Scolecithrix danae	Log WW = 2.460logTL − 5.352	Gruzov and Alekseyeva, 1970
NI-NVI	Log WW = 2.950logTL − 0.830	Ostrovskaya et al., 1982
CI-CVI	Log WW = 2.488logTL − 0.921	Ostrovskaya et al., 1982

Temora stylifera		Log WW = 2.057logPL − 4.042	Shmeleva, 1965
Undinula vulgaris		Log WW = 3.177logTL − 7.974	Gruzov and Alekseyeva, 1970
18 species	CI	Log WW = 2.874logPL − 6.707	Shmeleva, 1965
21 species	CII	Log WW = 2.831logPL − 6.614	Shmeleva, 1965
30 species	CIII	Log WW = 2.714logPL − 6.316	Shmeleva, 1965
34 species	CIV	Log WW = 2.738logPL − 6.313	Shmeleva, 1965
41 species	CV	Log WW = 2.649logPL − 6.049	Shmeleva, 1965
37 species	CVI ♀	Log WW = 2.719logPL − 6.220	Shmeleva, 1965
51 species	CVI ♂	Log WW = 2.695logPL − 6.121	Shmeleva, 1965
Total stages of 51 species		Log WW = 2.810logPL − 6.523	Shmeleva, 1965
17 species	CII-CVI	Log WW = 3.204logTL − 8.269	Mizdalski, 1988

Table 31 Relationship of total length (TL) to body wet weight (WW) in species of copepods grouped according to the ratio of length:width of the prosome ($L_{PR}:D_{PR}$) and the ratio of the length of the prosome:urosome ($L_{PR}:L_U$). After Gruzov and Alekseyeva (1970).

Group	$L_{PR}:D_{PR}$	$L_{PR}:L_U$	Equation
1	<2.2:1	3.7–5.2:1	Log WW = 2.721logTL − 6.223
2	2.0–2.8:1	3.0–4.0:1	Log WW = 2.919logTL − 7.105
3	2.6–3.1:1	1.9–2.6:1	Log WW = 3.008logTL − 7.576
4	3.0–4.0:1	Variable	Log WW = 3.068logTL − 7.974

Table 32 Regression equations of Pearre (1980) comparing the use of total length (TL), prosome length (PL) and prosome width (D) as a measure of the body wet weight of adult female copepods. The correlation coefficient, r, is given.

Number of species	Equation	r
89	Log WW = 2.9878logTL − 1.4568	0.9711
84	Log WW = 2.8514logPL − 1.0050	0.9839
89	Log WW = 2.8782logD + 0.1510	0.9889

7.1.2. Water Content

The water content of copepods has been reviewed by Båmstedt (1986) who collated earlier information. Water content of an individual varies because it replaces lost organic material such as ovarian eggs and lipid stores. Båmstedt shows that it is most variable in high, and least variable in low, latitude species but, nevertheless, water content is probably not related to latitude. He plotted a frequency distribution of published values and found that they approached a normal distribution. The values range from about 67 to 92% of body wet weight with a modal range of 82 to 84%. Water content of deep-living species is high but it may also be high in surface-living species. Flint et al. (1991) find that *Eucalanus hyalinus* and *E. inermis* have high water contents and consequent low organic contents; they term them jelly copepods. Morris and Hopkins (1983) record 92.2% of wet weight as water in the former species.

More recently, Ikeda (1988) measured 81.4 ± 0.9% of wet weight as water in the Antarctic mesopelagic *Calanus propinquus*. Ikeda and Skjoldal (1989) found 68.0 to 79.7% wet weight as water in four species from the Barents Sea. A more comprehensive analysis of Antarctic copepods has been made by Mizdalski (1988). Some 20 species were examined including

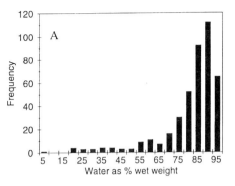
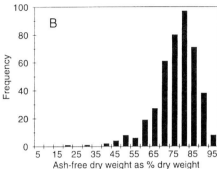

Figure 50 Frequency distributions of the determinations of A, water content and B, ash-free dry weight of the 20 species of adults and copepodids analysed by Mizdalski (1988).

determinations of the water content of copepodid stages. Frequency distributions of the 419 determinations made on 17 of the species show (Figure 50, A) a wider range of values than obtained by Båmstedt (1986). Modal values are between 85 and 95%, higher than Båmstedt's data. The overall mean water content is 78.7 ± 15.2% dry body weight.

7.1.3. Dry Weight

The same methods are used to determine dry weight of copepods as for determination of the wet weight except that the animals are removed from the blotting paper and placed on aluminium planchets in an oven at 60 °C and dried until a constant weight is obtained. The animals used should be freshly caught from the sea because fixed or frozen copepods have reduced dry weights relative to those of freshly caught ones (Williams and Robins, 1982; Böttger and Schnack, 1986; Giguere *et al.*, 1989). Weight losses caused by fixation are frequently as great as 30 to 40%, the equations relating dry weight to prosome length of fresh and fixed *Eurytemora affinis* obtained by Böttger and Schnack (1986) being given in Table 33. In addition, Omori (1978a) suggests that rinsing in fresh or de-ionized water causes reduction in the dry and ash-free dry weights; filtered sea water should be used. Böttger and Schnack (1986) review the literature and discuss many aspects of fixation and the determination of body dry weight.

The body dry weight relative to body length has been determined in many species (Table 33). The correlations are all significant at the 0.1% level.

Dry body weight varies seasonally (e.g. Ohman, 1988a; Castel and

Table 33 Relationship of dry weight (DW in μg) to prosome length (PL in μm), total length (TL in μm), or metasome length (ML in μm) in different species.

Species		Equation	Authority
Acartia clausi			
	CV–CVI	Log DW = 2.864logPL − 7.472	Robertson, 1968
	NI–NVI	Log DW = 2.848logTL − 7.265	Durbin and Durbin, 1978
	CI–CVI	Log DW = 3.095logPL − 8.195	Durbin and Durbin, 1978
	CVI	Log DW = 3.625logPL − 9.784	Durbin and Durbin, 1978
	CIII–CVI	Log DW = 2.72logPL − 7.096	Landry, 1978b
	NI–NVI	Log DW = 3.213logTL − 7.681	Klein Breteler *et al.*, 1982
	CI–CVI	Log DW = 2.967logPL − 7.719	Klein Breteler *et al.*, 1982
	NI–NVI	Log DW = 2.640logTL − 6.77	Uye, 1982a
	CI–CVI	Log DW = 3.36logPL − 9.09	Uye, 1982a
	CI–CVI	Log DW = 3.06logPL − 8.12	Uye, 1982b
	NI–NVI	Log DW = 2.147logTL − 5.510	Hay *et al.*, 1988
	CI–CVI	Log DW = 3.252logPL − 8.785	Hay *et al.*, 1988
	NI–NVI	Log DW = 3.208logTL − 7.644	Hay *et al.*, 1991
	CI–CVI	Log DW = 2.965logPL − 7.713	Hay *et al.*, 1991
	♀	Log DW = 4.548logPL − 12.562	Christou and Verriopoulos, 1993a
	♂	Log DW = 4.088logPL − 11.174	Christou and Verriopoulos, 1993a
	CI–CVI	Log DW = 3.181logPL − 8.590	Christou and Verriopoulos, 1993a
		Log DW = 2.659logPL − 7.114	Cataletto and Fonda Umani, 1994
A. hudsonica	CI–CVI	Log DW = 2.720logPL − 7.000	Middlebrook and Roff, 1986

A. tonsa	♀	Log DW = 1.296logPL − 3.082	Heinle, 1966
	♀	Log DW = 6.215logPL − 17.256	Durbin et al., 1983
	NI–NVI	Log DW = 3.314logTL − 8.508	Berggreen et al., 1988
	CI–CVI	Log DW = 2.921logPL − 7.958	Berggreen et al., 1988
	♀	Log DW = 2.659logPL − 7.114	Cataletto and Fonda Umani, 1994
	♀	Log DW = 2.995logPL − 1.158	Thompson et al., 1994
A. tranteri	♀	Log DW = 3.658logPL − 9.707	Kimmerer and McKinnon, 1987a
A. tsuensis	CI–CVI	Log DW = 3.27logPL − 8.88	Uye, 1982b
Calanoides acutus			
	CIII–CVI	Log DW = 3.310logTL − 9.552	Mizdalski, 1988
	CI–CVI	Log DW = 2.965logTL − 8.355	Godlewska, 1989
C. carinatus	CI–CVI	Log DW = 3.031logPL − 7.989	Verheye, 1991
Calanus finmarchicus			
	CV–CVI	Log DW = 3.264logPL − 8.714	Robertson, 1968
		Log DW = 3.147logPL − 7.274	Fransz and Van Arkel, 1980
		Log DW = 3.600logPL − 8.039	Fransz and Van Arkel, 1980
		Log DW = 3.581logTL − 9.890	Diel and Klein Breteler, 1986
	CI–CVI	Log DW = 3.460logPL − 9.553	Hay et al., 1991
	CI–CVI	Log DW = 2.134logPL − 4.822	Karlson and Bŏmstedt, 1994
	CI–CVI		
C. helgolandicus			
	CIII–CVI	Log DW = 2.790logTL − 7.370	Williams and Robins, 1982
February	CIV–CVI	Log DW = 3.103logPL − 8.719	Bottrell and Robins, 1984
March	CII–CVI	Log DW = 3.134logPL − 8.833	Bottrell and Robins, 1984
May	CII–CVI	Log DW = 3.318logPL − 9.218	Bottrell and Robins, 1984
June	CII–CVI	Log DW = 2.864logPL − 7.454	Bottrell and Robins, 1984
August	CIII–CVI	Log DW = 2.500logPL − 6.621	Bottrell and Robins, 1984
October	CIII–CVI	Log DW = 1.762logPL − 3.875	Bottrell and Robins, 1984
January	CIV–CVI	Log DW = 2.470logPL − 6.524	Bottrell and Robins, 1984
	CI–CVI	Log DW = 2.691logPL − 6.883	Hay et al., 1991
C. hyperboreus		Log DW = 3.389logPL − 9.719	Hirche and Mumm, 1992

Table 33 Continued.

Table 33 Continued.

Species		Equation	Authority
C. marshallae	CIV–CVI	Log DW = 3.942logTL − 11.526	Peterson, 1986
C. propinquus	CIII–CVI	Log DW = 3.314logTL − 9.237	Mizdalski, 1988
	CIII–CVI	Log DW = 2.990logTL − 8.196	Godlewska, 1989
C. sinicus		Log DW = 2.66 logPL − 6.68	Uye, 1982b
Centropages abdominalis			
	CI–CVI	Log DW = 3.00logPL − 7.89	Uye, 1982b
C. hamatus			
	NI–NVI	Log DW = 2.236logTL − 5.546	Klein Breteler *et al.*, 1982
	CI–CVI	Log DW = 2.449logPL − 6.098	Klein Breteler *et al.*, 1982
C. typicus	NI–CVI	Log DW = 2.243logTL − 5.568	Hay *et al.*, 1991
	CI–CVI	Log DW = 2.451logPL − 6.103	Hay *et al.*, 1991
Drepanopus pectinatus			
	CI–CVI	Log DW = 3.064logPL − 8.169	S. Razouls, 1985
Euchaeta plana and *E. concinna*			
		Log DW = 2.62 logPL − 6.47	Uye, 1982b
Eurytemora affinis			
♀		Log DW = 3.407logPL − 9.039	Heinle and Flemer, 1975
	CI–CVI	Log DW = 2.567logPL − 6.631	Burkill and Kendall, 1982
Fresh	CI–CV	Log DW = 3.24logPL − 8.51	Böttger and Schnack, 1986
Fixed	CI–CV	Log DW = 2.95logPL − 7.90	Böttger and Schnack, 1986
Fresh	CVI ♀	Log DW = 1.78logPL − 4.17	Böttger and Schnack, 1986
Fixed	CVI ♀	Log DW = 2.11logPL − 5.29	Böttger and Schnack, 1986
Fresh	CVI ♂	Log DW = 2.66logPL − 6.84	Böttger and Schnack, 1986
Fixed	CVI ♂	Log DW = 2.28logPL − 5.94	Böttger and Schnack, 1986
	CI–CVI	Log DW = 2.441logPL − 6.095	Escaravage and Soetaert, 1993
E. herdmani	CI–CVI	Log DW = 2.96logPL − 7.604	Middlebrook and Roff, 1986

E. velox	CI–CVI	Log DW = 3.03logPL − 4.846	Pagano, 1981b
	♀	Log DW = 3.00logPL − 4.818	Pagano, 1981b
	♂	Log DW = 3.00logPL − 4.689	Pagano, 1981b
Gaidius tenuispinus			
	CIII–CVI	Log DW = 3.143logTL − 8.466	Mizdalski, 1988
Heterorhabdus austrini			
	CV–CVI	Log DW = 2.685logTL − 7.139	Mizdalski, 1988
H. farrani	CIV–CVI	Log DW = 2.560logTL − 6.730	Mizdalski, 1988
Metridia gerlachei			
	CIII–CVI	Log DW = 2.803logTL − 7.684	Mizdalski, 1988
M. longa		Log DW = 3.017logPL − 7.968	Hirche and Mumm, 1992
M. lucens	NI–NVI	Log DW = 2.155logTL − 5.737	Hay *et al.*, 1991
	CI–CVI	Log DW = 3.062logPL − 8.073	Hay *et al.*, 1991
Paracalanus parvus			
	NI–NVI	Log DW = 2.285logTL − 5.965	Hay *et al.*, 1991
	CI–CVI	Log DW = 2.738logPL − 6.934	Hay *et al.*, 1991
Pareuchaeta antarctica			
	CI–CVI	Log DW = 2.929logTL − 7.979	Mizdalski, 1988
E. elongata	NIII–CVI	Log DW = 2.629logPL − 6.873	Greene and Landry, 1985
Pleuromamma abdominalis		Log DW = 0.560logML − 0.790	Bennett and Hopkins, 1989
P. gracilis and *P. piseki*		Log DW = 0.910logML − 2.422	Bennett and Hopkins, 1989
P. xiphias		Log DW = 1.140logML − 3.281	Bennett and Hopkins, 1989
Pseudocalanus elongatus			
	NI–NVI	Log DW = 0.989logTL − 2.712	Hay *et al.*, 1988
	CI–CVI	Log DW = 3.346logPL − 8.899	Hay *et al.*, 1988
	NI–NVI	Log DW = 2.231logTL − 5.483	Hay *et al.*, 1991
	CI–CVI	Log DW = 2.732logPL − 6.916	Hay *et al.*, 1991
Pseudocalanus minutus		Log DW = 3.649logPL − 9.87	McLaren, 1969

Table 33 Continued.

Table 33 Continued.

Species	Equation	Authority
Pseudocalanus sp.		
CII–CVI	Log DW = 3.640logPL − 9.846	Corkett and McLaren, 1978
NI–NVI	Log DW = 2.269logTL − 5.570	Klein Breteler *et al.*, 1982
CI–CVI	Log DW = 2.730logPL − 6.912	Klein Breteler *et al.*, 1982
CI–CVI	Log DW = 4.612logTL − 14.660	Davis, 1984a
♀	Log DW = 2.402logPL − 5.984	Ohman, 1985
Pseudodiaptomus hessei		
Nauplii	Log DW = 1.389logTL − 3.66	Jerling and Wooldridge, 1991
Copepodids	Log DW = 2.392logTL − 6.54	Jerling and Wooldridge, 1991
Rhincalanus gigas		
CII–CVI	Log DW = 3.347logTL − 9.911	Mizdalski, 1988
CIII–CVI	Log DW = 2.967logTL − 7.816	Godlewska, 1989
R. nasutus CI–CVI	Log DW = 3.600logTL − 10.661	Mullin and Brooks, 1967
Sinocalanus tenellus	Log DW = 2.73logTL − 7.42	Uye, 1982b
Temora longicornis		
CV–CVI	Log DW = 1.792logPL − 4.101	Robertson, 1968
NI–NVI	Log DW = 2.167logTL − 5.534	Klein Breteler *et al.*, 1982
CI–CVI	Log DW = 3.064logPL − 7.696	Klein Breteler *et al.*, 1982
CI–CVI	Log DW = 3.085logPL − 7.168	Chojnacki, 1986
NI–NVI	Log DW = 2.045logTL − 5.239	Hay *et al.*, 1988
CI–CVI	Log DW = 2.815logPL − 7.181	Hay *et al.*, 1988
NI–NVI	Log DW = 2.179logTL − 5.567	Hay *et al.*, 1991
CI–CVI	Log DW = 3.059logPL − 7.682	Hay *et al.*, 1991
Tortanus forcipatus	Log DW = 2.50logPL − 6.11	Uye, 1982b
9 species	Log DW = 2.761logTL − 7.448	Hirota, 1981
26 species	Log DW = 2.620logTL − 6.519	Gaudy and Boucher, 1983
20 species	Log DW = 2.546logTL − 6.697	Mizdalski, 1988
10 species	Log DW = 2.891logPL − 7.467	Uye and Matsuda, 1988
Not stated	Log DW = 2.225logTL − 5.489	White and Roman, 1992b

Feurtet, 1989; Conover and Huntley, 1991). Change can often take place on short time scales, especially so in short-lived species, and results in wide scatter in regressions of dry weight on body length such that no useful relationship exists (Razouls and Razouls, 1976). Greene *et al.* (1993), in a detailed study of the seasonal changes in body length, weight and carbon content, show that the body weights of the CI of *Calanus helgolandicus* and *Pseudocalanus elongatus* decreased by 37.6% and 20.5% respectively between May and August but body lengths decreased only by 4.1 and 2% respectively over the same period. The dry weights, like body lengths, show geographical variations, and Conover and Huntley (1991) tabulate weights of copepodids and adults of many species showing such variation.

7.1.4. Ash-free Dry Weight

The ash-free dry weight of a copepod is the weight of total organic matter present in it. It is obtained by subtracting the ash weight from the dry weight. Båmstedt (1986) reviews earlier work and shows that organic matter ranges from 70 to more than 98% of dry weight. He plots a frequency distribution of the published values and finds that, unlike values for water content, the data are skewed to the higher end of the range. The maximum modal range is 92 to 94% of dry weight. Surface-living species in high latitudes have a different organic matter content from other copepods; there is much more variability. Latitudinal differences are less obvious than in the case of water content.

Hirche and Mumm (1992) plotted ash-free dry weight against prosome length in the Arctic copepods, whereas Mizdalski (1988) examined Antarctic species (Table 34). The latter author has made a comprehensive analysis of the ash-free dry weight content of some 20 species of Antarctic copepods, copepodid stages included. The frequency distribution of the 420 determinations is shown in Figure 50, B. The modal values are between 75 and 85% dry weight, lower than those indicated by Båmstedt (1986). The overall mean is 73.6 ± 11.6% of dry weight.

7.1.5. Ash Weight

Båmstedt's (1986) review of the weight of total organic matter in copepods, the ash-free dry weight, infers that the modal ash content must be 6 to 8% of body dry weight. Hirota (1981) found that ash represented an average of only 5.8% of dry weight in nine species from the Sea of Japan. Morris and Hopkins (1983), on the other hand, examining 27 species of calanoids from the Gulf of Mexico found the average ash content to be slightly higher, 9.1%

Table 34 Relationship of ash-free dry weight (AFDW in μg) to prosome length (PL in μm), total length (TL in μm), or metasome length (ML in μm) in different species.

Species		Equation	Authority
Calanoides acutus	CIII–CVI	Log AFDW = 3.641logTL − 10.870	Mizdalski, 1988
Calanus finmarchicus	CI–CVI	Log AFDW = 3.590logPL + 0.852	Fransz and Diel, 1985
Calanus finmarchicus and C. glacialis			
C. hyperboreus		Log AFDW = 3.442logPL − 9.395	Hirche and Mumm, 1992
		Log AFDW = 3.358logPL − 9.659	Hirche and Mumm, 1992
C. propinquus	CIII–CVI	Log AFDW = 3.467logTL − 9.879	Mizdalski, 1988
Gaidius tenuispinus	CIII–CVI	Log AFDW = 3.313logTL − 9.171	Mizdalski, 1988
Heterorhabdus farrani	CIV–CVI	Log AFDW = 2.477logTL − 6.620	Mizdalski, 1988
Metridia gerlachei	CIII–CVI	Log AFDW = 2.927logTL − 8.268	Mizdalski, 1988
M. longa		Log AFDW = 3.099logPL − 8.292	Hirche and Mumm, 1992
Pareuchaeta antarctica	CI–CVI	Log AFDW = 3.048logTL − 8.522	Mizdalski, 1988
Rhincalanus gigas	CII–CVI	Log AFDW = 3.572logTL − 10.891	Mizdalski, 1988
20 species	CII–CVI	Log AFDW = 2.650logTL − 7.202	Mizdalski, 1988

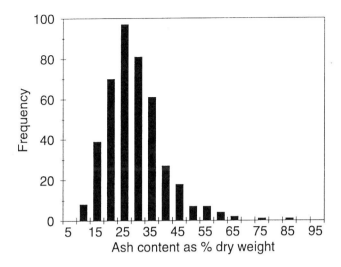

Figure 51 The frequency distribution of the determinations of ash content of the 20 species of adults and copepodids analysed by Mizdalski (1988).

of dry weight. Adult females of four species from the Barents Sea averaged 8.5% of dry weight as ash (Ikeda and Skjoldal, 1989).

More than 400 determinations of the ash content of some 20 species of adult and copepodid stages of Antarctic copepods have been made by Mizdalski (1988). The overall mean ash content of the species is 26.4 ± 10.5% of body dry weight and the modal values are between 20 and 35% (Figure 51), much higher than inferred by the data reviewed by Båmstedt (1986). Masuzawa *et al.* (1988) records the ash content of *Neocalanus plumchrus* as 5.8% dry weight.

7.1.6. Volume

The relative constancy of the overall body form of calanoid copepods allows approximation of their body volume by geometrical means. The prosome is an ellipsoid while the urosome represents a cylinder so that body volume, V, is given by:

$$V = 1/6 \; \pi L W^2 + \tfrac{1}{4}\pi L w^2$$

where L is the total length of the prosome or urosome, W is the maximum width of the prosome, and w is the mean width of the urosome. This formula is used to measure biovolumes of *Centropages typicus* and *Temora stylifera* by Razouls and Razouls (1976). An alternative formula is given by Chojnacki (1983):

$$V = \pi(1/6 L_{PL} h^2 + \tfrac{1}{4} L_U R^2 + 1/6 L_{an} r^2)$$

where V is volume, h is prosome height, L_{PL} is prosome length, L_U is urosome length, L_{an} is length of antennule, R is diameter of urosome, r is diameter of antennule.

Shmeleva (1965) found that the geometric determination of volume, coupled with the assumption that the specific gravity of calanoids is 1, underestimated body weight in most species. It did not underestimate weight in all species suggesting that this method is not suitable for between-species comparisons.

Mauchline (unpublished) measured the volumes (V) of 20 species of copepods in the genera *Acartia* (1 species), *Calanus* (2 species), *Euaugaptilus* (1 species), *Euchaeta* and *Pareuchaeta* (10 species), *Euchirella* (2 species), *Pseudocalanus* (1 species), *Pseudochirella* (1 species), *Undeuchaeta* (1 species), *Valdiviella* (1 species) using the Berardi apparatus (Berardi, 1953). A total of 38 determinations were made. The regression equation is:

$$\text{Log } V \text{ (mm}^3) = 3.164 \log PL(\mu m) - 10.690 \quad r = 0.972$$

where PL is the prosome length.

7.2. ELEMENTS

As Båmstedt (1986) points out, most attention has centred on the determination of carbon in copepods with less attention paid to hydrogen, nitrogen and phosphorus. Carbon content can be used to estimate biomass and energy content of species and systems while nitrogen and phosphorus are of interest in studies of nutrient regeneration.

7.2.1. Carbon

Båmstedt (1986) plots a frequency distribution of published values of carbon content and finds that they approximate to a normal distribution with modal values of 40 to 46% dry weight; the total range is from 28 to 68% dry weight. The mean carbon content, on the basis of these data, is 44.7% of body dry weight. Surface-living species in both low and medium latitudes have lower carbon contents while high latitude deep-living species tend to have the most carbon present. Conover and Huntley (1991) review carbon contents of Arctic and Antarctic species and conclude that they are generally within the range 40 to 55% of body dry weight; the mean value of their tabulated data is 47.3 ± 6.0. Mean values of carbon in seven calanoids in the northwestern Mediterranean were 36.6 to 46.5% dry weight (Gorsky et al., 1988) while 9 species in the Inland Sea of Japan had values of 45.9 to 49.3% (Hirota, 1981) and *Neocalanus plumchrus* from the Sea of Japan had 39.1 (Masuzawa et al., 1988). Pagano and Saint-Jean (1993) record 43 ± 3.4% dry weight as carbon in *Acartia clausi* in a tropical lagoon while Ohman and Runge (1994) find 54.2% as carbon in *Calanus finmarchicus* from the Gulf of St Lawrence. Supplementary values, within the published ranges, are also given by Ikeda and Skjoldal (1989) for *Calanus* species and *Metridia longa* in the Barents Sea; by Miller (1993b) for *Neocalanus flemingeri* of 41.9 to 46.8% in the Gulf of Alaska; by Ikeda (1988) for *Calanus propinquus* and *Metridia gerlachei* of 43.6 and 45.3 respectively; and by Huntley and Nordhausen (1995), for four Antarctic species, of 36 to 54%.

As Kankaala and Johansson (1986) point out, length–mass regressions are frequently used in studies of biomass because of their convenience. There is, however, great variation within these relationships in time and space. In addition, the accuracy with which estimations of length, dry weight and carbon can be made are different. Kankaala and Johansson (1986) find that length reflects carbon content best when the animals are actively growing through the naupliar and copepodid stages; the correspondence is less exact in the CV and CVI where gonads and lipid stores are prominent. They also demonstrate that the best relationship varies from a linear relationship, log–linear to a log–log relationship, the logarithm being $\log_{10}$ or $\log_e$. Thus, there are inherent inaccuracies in using regressions such as those listed in Table 35. They are, however, useful as starting-points or as rough estimates, to be confirmed by measurements made from the actual populations being studied.

The seasonal range in the carbon content of *Acartia tonsa* is 32.8 to 48.8% (Ambler, 1985) while that of *A. clausi* is 41.2 to 45.3 (Kerambrun, 1987) and 40.8 to 49.8% body dry weight (Cataletto and Fonda Umani, 1994). Tanskanen (1994), examining the carbon content of *A. bifilosa* monthly (Table 35), found that 75 to 88% of the variation can be explained by body length changes. Ohman et al. (1989) show that seasonal carbon content reflects seasonal changes in body dry weight of *Neocalanus tonsus*. Green et al. (1993) present a detailed study of the seasonal changes in the carbon content of *Calanus helgolandicus* and *Pseudocalanus elongatus* in the English Channel.

Stable isotope ratios, $^{13}C/^{12}C$, usually reflect the ratios in the food and so are of interest in studies of food webs. The values of $\delta^{13}C$ range from

Table 35 Relationship of carbon weight (µg) to prosome length (µm) or total length (µm) in different species.

Species		Equation	Authority
Acartia bifilosa			
February	NI–VI	Log C = 2.372logTL − 6.402	Tanskanen, 1994
April	NI–NVI	Log C = 2.300logTL − 6.262	Tanskanen, 1994
November	CI–CVI	Log C = 3.018logPL − 8.322	Tanskanen, 1994
December	CI–CVI	Log C = 2.982logPL − 8.289	Tanskanen, 1994
February	CI–CVI	Log C = 2.708logPL − 7.554	Tanskanen, 1994
April	CI–CVI	Log C = 2.181logPL − 6.141	Tanskanen, 1994
May	CI–CVI	Log C = 2.918logPL − 7.985	Tanskanen, 1994
June	CI–CVI	Log C = 2.508logPL − 6.879	Tanskanen, 1994
July	CI–CVI	Log C = 2.910logPL − 7.972	Tanskanen, 1994
August	CI–CVI	Log C = 2.642logPL − 7.346	Tanskanen, 1994
September	CI–CVI	Log C = 2.639logPL − 7.455	Tanskanen, 1994
October	CI–CVI	Log C = 2.643logPL − 7.353	Tanskanen, 1994
Acartia clausi	NII–NVI	Log C = 2.83logTL − 7.48	Landry, 1978b
	CI–CVI	Log C = 2.71logPL − 7.50	Landry, 1978b
	CI–CVI	Log C = 3.08logPL − 8.51	Uye, 1982b
		Log C = 3.032logPL − 8.556	Cataletto and Fonda Umani, 1994
	CVI♀	Log C = 3.055logPL − 8.444	Ayukai, 1987
A. omorii	NI–NVI	Log C = 2.64logTL − 7.12	Liang and Uye, 1996
	CI–CVI	Log C = 3.08logPL − 8.51	Liang and Uye, 1996
A. tonsa	NI–NVI	Log C = 3.319logTL − 8.519	Berggreen *et al.*, 1988
	CI–CVI	Log C = 2.919logPL − 7.953	Berggreen *et al.*, 1988
	CVI♀	Log C = 2.476logPL − 0.698	Thompson *et al.*, 1994
A. tsuensis	CI–CVI	Log C = 3.03logPL − 8.52	Uye, 1982b
Calanus finmarchicus	Egg–CVI	Log C = 2.540logTL − 6.811	Davis, 1984a

Species	Stage	Equation	Reference
C. helgolandicus	CI	$C = 0.321 \pm 0.002 DW$	Paffenhöfer, 1976
	CIII	$C = 0.350 \pm 0.001 DW$	Paffenhöfer, 1976
	CV	$C = 0.369 \pm 0.001 DW$	Paffenhöfer, 1976
	CVI	$C = 0.371 \pm 0.003 DW$	Paffenhöfer, 1976
	CI–CVI	$C = 0.372 DW - 0.248$	Paffenhöfer, 1976
Calanus sinicus		$\log C = 2.64 \log PL - 7.00$	Uye, 1982b
	NI–NVI	$\log C = 2.588 \log TL - 6.827$	Uye, 1988
	CI–CVI	$\log C = 3.378 \log PL - 9.416$	Uye, 1988
Centropages abdominalis	NI–NVI	$\log C = 2.10 \log TL - 6.02$	Liang et al., 1996
	CI–CVI	$\log C = 2.97 \log PL - 8.19$	Uye, 1982b
Euchaeta plana and E. concinna		$\log C = 2.45 \log PL - 6.25$	Uye, 1982b
Eurytemora affinis			
Bothnian Bay	CI–CVI	$\log C = 2.83 \log PL - 7.694$	Kankaala and Johansson, 1986
Baltic	CI–CVI	$\log C = 2.13 \log PL - 5.87$	Kankaala and Johansson, 1986
Limnocalanus macrurus		$\log C = 1.65 \log PL - 4.452$	Kankaala and Johansson, 1986
Paracalanus parvus	Egg–CVI	$\log C = 1.966 \log TL - 5.404$	Davis, 1984a
Paracalanus sp.	NI–NVI	$\log C = 1.333 \log CL - 4.132$	Uye, 1991
	CI–CVI	$\log C = 3.128 \log PL - 8.451$	Uye, 1991
Pseudocalanus minutus	CIII–CVI	$\log C = 3.640 \log PL - 10.155$	McLaren, 1969
Pseudocalanus sp.	Egg–CVI	$\log C = 2.260 \log TL - 6.123$	Davis, 1984a
Pseudodiaptomus marinus			
	NI–NVI	$\log C = 2.000 \log TL - 5.67$	Uye et al., 1983
	CI–CVI	$\log C = 3.171 \log PL - 8.63$	Uye et al., 1983
Rhincalanus nasutus	CI–CVI	$\log C = 4.30 \log TL - 13.368$	Mullin and Brooks, 1967
Sinocalanus tenellus		$\log C = 2.71 \log PL - 7.67$	Uye, 1982b
		$\log C = 2.33 \log TL - 6.62$	Kimoto et al., 1986a
		$\log C = 3.12 \log PL - 8.74$	Kimoto et al., 1986a
Tortanus forcipatus	NI–NVI	$\log C = 2.61 \log PL - 6.80$	Uye, 1982b
	CI–CVI	$\log C = 2.04 \log PL - 5.10$	Uye and Kayano, 1994b

−20.4 to −33.7 (Saupe et al., 1989; Rau et al., 1991; Hobson and Welch, 1992).

7.2.2. Nitrogen

Båmstedt (1986) shows that published values of nitrogen content range from 5.2 to 15.8% of dry body weight, with most in the range 7 to 13%. There are several potential modes in the frequency distribution of the measurements that lead Båmstedt to suggest a separation into three distributions with central values at 7, 10 and 13% nitrogen, an effect that may be latitudinal with more nitrogen at high latitudes. A few correlations between nitrogen content and body length have been made (Table 36).

Nitrogen content was studied seasonally in CII to adults of *Calanus helgolandicus* by Bottrell and Robins (1984) who found that they follow the same seasonal pattern as the changes in body dry weight. A similar situation was found by Ohman et al. (1989) in *Neocalanus tonsus* and Cataletto and Fonda Umani (1994) in *Acartia clausi*. Seasonal values of nitrogen in *Acartia tonsa* range from 8.7 to 16.4% of body dry weight (Ambler, 1985) and from 11.3 to 12.3 in *A. clausi* (Kerambrun, 1987).

Conover and Huntley (1991) review nitrogen content of Arctic and Antarctic species and conclude that Antarctic species generally have more, 8.5 to 12.5%, than Arctic species, 5.2 to 11.2%; the overall mean content was 8.4 ± 2.2% dry weight. Mean values of nitrogen in seven calanoids in the northwestern Mediterranean ranged from 9.0 to 11.9% dry weight (Gorsky et al., 1988) while nine species had values ranging from 10.9 to 13.1% in the Inland Sea of Japan (Hirota, 1981) and *Neocalanus plumchrus* had 9.6% in the Sea of Japan (Masuzawa et al., 1988). Pagano and Saint-Jean (1993) record 10.8 ± 0.7% dry weight as nitrogen in *Acartia clausi* in a tropical lagoon and Ohman and Runge (1994) find 54.2% as carbon in *Calanus finmarchicus* from the Gulf of St Lawrence. Miller (1993b) records 6.59 to 7.44% for *Neocalanus flemingeri* in the Gulf of Alaska. *Calanus* species and *Metridia longa* in the Barents Sea have values ranging from 6.7 to 11.2% (Ikeda and Skjoldal, 1989). *Calanus propinquus* and *Metridia gerlachei* have 12.5 and 11.4% dry weight as nitrogen (Ikeda, 1988) while a further four Antarctic species have 5.4 to 8.9% dry weight as nitrogen (Huntley and Nordhausen, 1995).

Stable isotope ratios of $^{15}N/^{14}N$ in copepods reflect the ratios in their food and so are of interest on studies of food webs. The values of $\delta^{15}N$ range from 1.7 to 9.2‰ (Rau et al., 1991; Hobson and Welch, 1992). Values of $\delta^{15}N$ in excreted ammonia and $\delta^{15}N$ in bodies of zooplankton, including copepods, are linearly related (Checkley and Miller, 1989) by the equation:

$$\delta^{15}N \text{ excreted ammonia} = 0.96\ \delta^{15}N \text{ body} - 2.7$$

Table 36 Relationship of body content of nitrogen (N) and phosphorus (P) in µg to prosome length (µm) in different species.

Species	Equation	Authority
Nitrogen		
Acartia clausi CI-CVI	Log N = 3.07logPL − 9.06	Uye, 1982b
	Log N = 0.643logPL − 2.052	Cataletto and Fonda Umani, 1994
A. tonsa ♀	Log N = 3.067logPL + 0.164	Thompson et al., 1994
A. tsuensis CI-CVI	Log N = 2.92logPL − 8.85	Uye, 1982b
Calanus sinicus	Log N = 2.50logPL − 7.10	Uye, 1982b
Centropages abdominalis CI-CVI	Log N = 3.02logPL − 8.88	Uye, 1982b
Euchaeta plana and *E. concinna*	Log N = 2.82logPL − 8.08	Uye, 1982b
Sinocalanus tenellus	Log N = 2.82logPL − 8.55	Uye, 1982b
Tortanus forcipatus	Log N = 2.82logPL − 8.02	Uye, 1982b
Phosphorus		
10 species	Log P = 2.90logPL − 9.34	Uye and Matsuda, 1988

7.2.3. Carbon:Nitrogen Ratio

Båmstedt (1986) finds that modal values of the carbon:nitrogen ratio (C:N ratio), as determined from published data, deviate considerably from those inferred by the distributions of carbon and nitrogen weights as defined above. The frequency distribution of published values of the C:N ratio range from about 3 to 13 but are strongly skewed towards the lower end of the distribution, many being between 3 and 5. The C:N ratio shows an increase from low to high latitudes. Conover and Huntley (1991) review C:N ratios in Arctic and Antarctic copepods and find means of 6.8 ± 2.9. Values for *Acartia tonsa* are between 3.04 and 4.88 (Ambler, 1985). Mean values of C:N in seven calanoids in the northwestern Mediterranean ranged from 3.5 to 5.2 (Gorsky *et al.*, 1988). Pagano and Saint-Jean (1993) find a value of 4.6 ± 0.2 in *Acartia clausi* in a tropical lagoon and Ohman and Runge (1994) find 5.16 ± 0.28 in *Calanus finmarchicus* from the Gulf of St Lawrence. Miller (1993b) records ratios of 6.0 to 6.4 for *Neocalanus flemingeri* in the Gulf of Alaska. Ratios in four Antarctic species range from 4.4 to 8.6 (Huntley and Nordhausen, 1995).

A seasonal study of *Acartia clausi* in the Mediterranean by Kerambrun (1987) and Cataletto and Fonda Umani (1994) show that C:N ratios range between 3.55 and 3.75 and between 1.73 and 4.54 respectively. The seasonal changes in the C:N ratio of *Neocalanus tonsus* reflect the seasonal changes in body dry weight, carbon and nitrogen contents (Ohman *et al.*, 1989).

7.2.4. Hydrogen

Båmstedt (1986) tabulates published values for the hydrogen content of marine copepods. They range from 3.2 to 10.3% of body dry weight. He points out that hydrogen is associated with lipid rather than protein and shows that lipid-storing species had an average of 8.5% while species that do not have significant lipid store have an average of 6.4% of body dry weight as hydrogen.

The hydrogen content of nine species in the Inland Sea of Japan ranged from 6.7 to 7.2% (Hirota, 1981). Miller (1993b) found 6.25 to 7.33% dry weight *Neocalanus flemingeri* in the Gulf of Alaska as hydrogen and Ikeda (1988) 6.7 to 6.9% in *Calanus propinquus* and *Metridia gerlachei* in the Antarctic. Hydrogen content varied seasonally between 4.98 and 6.15% dry weight in *Acartia clausi* (Kerambrun, 1987).

7.2.5. Phosphorus

There are few data on the concentrations of phosphorus in marine copepods (Båmstedt, 1986). Those that are available suggest that it rarely exceeds 1%

of body dry weight with a mean value of 0.76% (number of observations is 9). Masuzawa et al. (1988) find 0.33% dry weight as phosphorus in the Sea of Japan. Uye and Matsuda (1988), however, record 0.7 to 1.76% dry weight as phosphorus among ten species of copepods in the Inland Sea of Japan and relate these higher values to others in the earlier literature. Båmstedt (1986) briefly discusses phosphorus pools in copepods while Uye et al. (1990a) examine phosphorus regeneration rates relative to body sizes of the copepods in the Inland Sea of Japan. A regression equation is given in Table 36 for the phosphorus content of copepods relative to their prosome length.

Ikeda (1988) found 0.6 to 0.7% dry weight as phosphorus in *Calanus propinquus* and *Metridia gerlachei*. The "jelly" copepods *Eucalanus hyalinus* and *E. inermis*, already noted, have a high water content and consequent low phosphorus content (Flint et al., 1991).

7.2.6. Atomic Ratios

Atomic ratios of O:N and O:P, the ratio of oxygen consumed to nitrogen or phosphorus excreted, measure metabolic rates as reviewed by Mayzaud and Conover (1988).

Gaudy and Boucher (1983) measured atomic ratios in 26 species of copepods from the Indian Ocean. O:N ranged from 2.71 to 28.64, O:P from 19.62 to 410.04, and N:P from 2.27 to 37.04. Ikeda and Skjoldal (1989), reviewing earlier literature, find O:N ratios in *Calanus* species and *Metridia longa* to range from 18.3 to 92. Seasonal ranges in the C:H ratio were 7.06 to 8.61 (Kerambrun, 1987).

7.2.7. Mineral Composition

Båmstedt (1986) reviews published values of trace elements in copepods and mixed zooplankton. The data are few and some are questionable, the overall ranges for copepods alone being given in Table 37. Masuzawa et al. (1988) determine trace elements in *Neocalanus plumchrus* (Table 37). There is some variation between the two sets of values and further work is required.

Assimilation efficiencies of Fe, Zn and Mn by *Acartia tonsa* fed on radiolabelled diatom and flagellate cultures were 7 to 18% for Fe, 32 to 37% for Zn and 3 to 10% for Mn (Hutchins and Bruland, 1994). Selenium was assimilated with a 97.1 ± 1.5% efficiency (Fisher and Reinfelder, 1991). Retention efficiencies by *Anomalocera patersoni* of various metals were estimated by Fisher et al. (1991); they were 4.5% for Am, 30% for Cd, 21%

Table 37 Trace element contents of copepods, in $\mu g\,g^{-1}$. The values from Båmstedt are for copepods only and those from Masuzawa *et al.* are for *Neocalanus plumchrus*.

Trace element	Båmstedt (1986)	Masuzawa *et al.* (1988)
Ag	0.1–3	0.14
Al	20–3000	
As		3.6
Ba	5–200	
Be	0.1–0.3	
Br		200
Ca	1000–12 900	5500
Cd	1.6–7.5	
Cl		29 500
Co	1–1.5	0.042
Cr	4–7	0.29
Cs		0.023
Cu	9–200	
Fe	55–4000	33
Ga	1–2	
Hg	0.06–0.16	
I		27
K	0.007–0.015	2600
Li	3–15	
Mg	500–10 900	4200
Mn	3.3–50	2.3
Na	500–94 200	19 700
Mo	5	
Ni	0.5–20	
Pb	6–30	
Rb		0.9
Sb		0.032
Sc		0.009
Se		0.8
Si	500–10 000	
Sn	3–7	
Sr	60–290	146
Ti	4–40	
V	2–7	
Zn	62–1500	47
Zr	1–7	

for Hg, 0.8% for Po, and 48% for Zn. Lee and Fisher (1994) show that copepods are more efficient at making Ag, Cd, Co, Pb, and Po available from diatoms for remineralization than micro-organisms.

7.3. ORGANIC COMPONENTS

The major organic components of copepods, as in other animals, are proteins, with their constituent amino acids, lipids and carbohydrates. Båmstedt (1988b) reviews the variability of levels of protein and lipid and the protein:lipid ratio between individuals within species of Barents Sea copepods. The levels of these components may vary by a factor of five between comparable individuals.

7.3.1. Protein

Båmstedt (1986) plots a frequency distribution of published values of protein contents of copepods. They range from 24 to 82% of body dry weight and show high frequencies over a range of about 30 to 70% dry weight. A minimum mode at 32 to 36% appears to originate from deep-living species in low and middle latitudes but the origins of a high mode at 52 to 56% dry weight are not clear.

Willason et al. (1986) records 42.75 to 54.5% dry weight as protein in CV and adult *Calanus pacificus*. Timonin et al. (1992) finds 3.1 to 12.9% wet weight (ca. 15 to 65% dry weight) as protein in CV and adult *Calanoides carinatus*. Drits et al. (1994) for *C. acutus* and Kosobokova (1994) for *Calanus propinquus* give protein contents of individuals but not the sizes of the individuals; consequently, their data cannot be compared with those of others. Flint et al. (1991) find normal levels of protein in *Calanus* species, 32.5 to 53% dry weight, but very low levels, 4 to 5.5%, in the jelly copepods (those with a high water content) of the genus *Eucalanus*.

Proteins synthesized during stress (stress proteins) are discussed by Bradley et al. (1992) and Gonzalez and Bradley (1994). They found different responses to salinity or osmotic stress as opposed to stress caused by change of environmental temperature.

Båmstedt (1986) briefly discusses the amino acid composition of copepods from the few data available.

7.3.2. Lipids

Frozen samples, frequently stored for periods of months, of copepods from the field tend to be used for lipid analysis because the species of interest has to be sorted from mixed zooplankton. Ohman (1996) found that freezing caused little damage to the lipids but prolonged storage at −15 °C did. He recommends rapid freezing in liquid nitrogen followed by storage at temperatures below 70 °C. A hydrophobic fluorophore, Nile red, was used by Carman et al. (1991) to provide quantitative measurements of lipids in benthic copepods of 1 to 10 μg dry weight. They suggest that it may be an alternative analytical method for lipids in a variety of organisms including copepods.

Båmstedt (1986) reviews the published information on lipid content of copepods. He plots a frequency distribution of the lipid contents and shows that they range from 2 to 61% in low and medium latitude species and from 8 to 73% dry body weight in high latitude species. Small-sized low and medium latitude species have low lipid contents, probably corresponding to a mode at 8 to 12% in Båmstedt's frequency distribution. Lipid content varies seasonally in high latitudes and values within a species can range widely. Båmstedt discusses the ranges within genera and species and its possible relationship with the depth of occurrence of the species.

The chemistry of the lipids is reviewed by Sargent and Henderson (1986) and Sargent and Falk-Petersen (1988). Nauplii often have residual lipid stores from the egg but these are soon used and the later nauplii and early copepodids normally have less lipid, probably predominantly structural phospholipids, present in their bodies. Some species do show a distinct increase in the content of lipid during the developmental sequence of copepodids (Kattner and Krause, 1987; Hagen, 1988). The CIV, and especially the CV, develop lipid stores that are later used for a variety of purposes. Sargent and Henderson tabulate data on 28 species, the lipid content of which ranges from 3 to 74% dry body weight, the constituent wax esters comprising 9 to 92% and the triacylglycerols 0 to 30% of the lipid weight. Also present are the structural phospholipids, that can on occasion dominate the lipids (Ohman, 1988a), and the free fatty acids and sterols. The composition of the wax esters, comprising fatty acids and fatty alcohols, have been analysed in a number of calanoids and they review these in detail. They also describe possible pathways of biosynthesis of wax esters and their constituent fatty alcohols by copepods fed on diets lacking these compounds. Lipid-rich species tend to live in higher latitudes than species with less lipid (Shchepkina et al., 1991). Wax ester storage is associated with a herbivorous diet, detritivorous and carnivorous copepods tending to store less lipids. Conover and Huntley (1991) review wax ester and triacylglycerol content of Arctic and Antarctic copepods and state that more investigation of Antarctic species is required. Some Antarctic species have surprisingly low levels of lipids present and they are dominated by triacylglycerols in *Calanus propinquus* (Hagen, 1988; Schnack-Schiel et al., 1991; Hagen et al., 1993; Kattner et al., 1994) and *Euchirella rostromagna* (Hagen, 1988; Hagen et al., 1995).

Lipid stores are a seasonal feature of the CV and adults of many species, the stores showing marked seasonal changes in size (e.g. Kattner et al., 1994; Hagen and Schnack-Schiel, 1996). They also vary with food concentrations (Håkanson, 1987) and the individual fatty acids stored are influenced by the diet (Kattner et al., 1989; Graeve et al., 1994a,b; Fahl, 1995; Kattner and Hagen, 1995). The classes of stored lipids can also change seasonally. Norrbin et al. (1990) show that stores of *Pseudocalanus acuspes* are largest in the autumn and always predominated by wax esters. Stores of *Acartia longiremis*, however, although also being greatest in the autumn, are dominated by triacylglycerols in the autumn but by wax esters in the spring and summer.

The fatty acid and alcohol compositions of the wax esters are reviewed by Sargent and Henderson (1986) and Sargent and Falk-Petersen (1988) and recourse to these papers should be made for further information. Calanoid wax esters are characterized by the 20:1(n-9) and 22:1(n-11) fatty alcohols and being rich in 18:4(n-3), 20:5(n-3) and 22:6(n-3) polyunsaturated fatty acids. The fatty acids and alcohols change progressively through the copepodids of *Calanus finmarchicus*; earlier stages have greater concentrations of 16:0 acid and alcohol whereas adults have more 20:1 and 22:1 acids and alcohols (Kattner and Krause, 1987). The composition of the major fatty acids and alcohols of *C. glacialis*, *C. finmarchicus* and *C. hyperboreus* conform to this description (Clarke et al., 1987; Kattner et al., 1989; Kattner and Graeve, 1991; Hirche and Kattner, 1993). Recent analyses of the composition of the fatty acids and fatty alcohols of *Pseudocalanus acuspes* and *Acartia longiremis* are given by Norrbin et al. (1990). The lipids of these small copepods are dominated by short chain (16 and 18) fatty acids and alcohols and the 20 and 22 chains are relatively unimportant although there are seasonal changes in the occurrence of many individual acids and alcohols. Fraser et al. (1989) examine the fatty acid and alcohol composition of the lipids of *Calanus finmarchicus*, *Pseudocalanus* sp. and *Temora longicornis*, the last species storing predominantly triacylglycerols. Lipid composition of *Neocalanus tonsus* is defined by Ohman et al. (1989). Hagen et al. (1993) describe the fatty acids and alcohols of *Calanoides acutus* and *Calanus propinquus* and Hagen et al. (1995) analyse the composition of the lipids of *Pareuchaeta antarctica* and *Euchirella rostromagna* and find marked differences between the species. Kattner et al. (1994) and Fahl (1995) provide additional studies of the composition of the lipids of *Calanoides acutus*, *Calanus propinquus*, *Pareuchaeta antarctica*, *Metridia gerlachei* and *Rhincalanus gigas* and discuss the composition against the background of development, diet and ecology. Recently, Ward et al. (1996a,b) have shown that the lipid stores in the Antarctic *Rhincalanus gigas* are predominantly wax esters while those of *Calanus simillimus* are triacylglycerols.

Pathways of biosynthesis of wax esters by copepods are described by

Sargent and Henderson (1986) and Sargent and Falk-Petersen (1988) and amplified by Kattner and Hagen (1995).

The function of the lipid stores is reviewed by Sargent and Henderson (1986) who discuss aspects such as food reserves, transference to the eggs, and buoyancy implications. They are mobilized during periods of food scarcity (Kovaleva and Shadrin, 1987; Attwood and Peterson, 1989; Pavlova et al., 1989). The triacylglycerols are mobilized faster than the wax esters and could perform a different metabolic function (Sargent and Henderson, 1986). The wax ester content of a copepod may indicate its long-term feeding history while the triacylglycerol content may reflect its short-term feeding history. Triacylglycerols may be obtained directly in the diet from phytoplankton but this has not been confirmed. These lipid stores are prominent in the CV and CVI and are used by females to form reproductive tissues and by the males in physical activity. They also function at times of starvation or fluctuating food supply and may be involved in buoyancy regulation.

Drits et al. (1994) for *Calanoides acutus* and Kosobokova (1994) for *Calanus propinquus* give lipid contents of individuals but not the sizes of the individuals with the result that their data cannot be compared with those of others. Jelly copepods, those with a high water content, such as *Eucalanus hyalinus* and *E. inermis* have lower lipid contents than normal copepods (Flint et al., 1991).

7.3.3. Carbohydrate

Båmstedt (1986) tabulates the information on the carbohydrate content of copepods but treats chitin separately as is done here. Carbohydrate contents range from 0.2 to 5.1 (mean 2.0%) of body dry weight. Low latitude species may have higher concentrations than higher latitude species.

7.3.4. Chitin

Chitin consists of a single monomer, N-acetylglucosamine, although a portion may be deacetylated to glucosamine. Roff et al. (1994) describe the pathway of chitin synthesis from glucose and present a radiochemical method to determine its rate of synthesis. The method was tested on *Daphnia magna* which was fed ^{14}C-labelled algae and incorporated [^{14}C]N-acetylglucosamine.

Båmstedt (1986) tabulates published values of the chitin contents of copepods and shows that they range from 2.1 to 9.3% of body dry weight, mean 4.6%.

CHEMICAL COMPOSITION 249

7.4. OTHER ORGANIC COMPONENTS

7.4.1. Free Amino Acids

Båmstedt (1986) reviews knowledge of the occurrence and composition of the free amino acid pool in copepods. They are primarily used in osmotic regulation of the animals. The constituent amino acids in the pool are different between estuarine and truly marine species, the total pool increasing with environmental salinity. In general, glycine, alanine, arginine, lysine, proline and taurine are the most important acids.

7.4.2. Nucleic Acids

Båmstedt (1986) lists RNA and DNA contents of copepods. The range of RNA content is 0.09 to 68 and of DNA is 0.6 to 39.2 μg mg dry weight^{-1}. Average contents, in terms of percentage dry body weight, are 2% for RNA and 1.8% for DNA. Båmstedt suggests that seasonal changes in the RNA concentration, increasing values from winter towards summer, indicate better conditions for growth.

RNA and DNA contents of copepods have been investigated as indicators of rates of growth. Ota and Landry (1984), examining *Calanus pacificus*, found that the results were too contradictory to predict the growth rates of this species and conclude that RNA concentrations are not useful in predicting growth in field populations. This followed the earlier work of Båmstedt and Skjoldal (1980) on a variety of copepods as well as other planktonic organisms. They provide regression equations describing the RNA content relative to body dry weight in five species of copepods at different times of the year. RNA content as μg RNA mg body dry weight^{-1} ranged from about 2 to 50, depending on season, and was inversely correlated with body dry weight. Nakata *et al.* (1994) demonstrated a positive correlation between the RNA:DNA ratio in adult female *Paracalanus* sp. and rates of egg production.

McLaren and Marcogliese (1983) counted nuclei in a variety of copepods and found that the NI of all species has about 2000 nuclei while the CI has between 9600 and 13 000 nuclei. The larger species have much larger nuclei but size of nucleus, equivalent to the DNA content, is not necessarily correlated with body size.

7.4.3. ATP (Adenosine triphosphate)

Båmstedt (1986) tabulates concentrations of ATP in copepods, ranging from 2.8 to 18.4 μg mg dry weight^{-1}, with a mean value in 13 different species of

7.6 µg. The average carbon-to-ATP ratio is then 58.5, assuming a mean carbon content of copepods of 44.7% body dry weight.

7.5. ENERGY CONTENT

Morris and Hopkins (1983) calculated the caloric density, in kcal g^{-1}, by using the conversion factors 5.7 kcal g^{-1} for protein, 9.3 kcal g^{-1} for lipid and 4.0 kcal g^{-1} for carbohydrate. They tabulate caloric density for 20 species of calanoids and found it to range from 0.238 to 1.23 kcal g^{-1} wet weight and from 3.06 to 7.10 kcal g^{-1} dry weight. The jelly copepod, *Eucalanus hyalinus*, had the lowest caloric density. *Acartia clausi* has a seasonal range of caloric density from 4.28 to 5.00 kcal g^{-1} dry weight (Kerambrun, 1987).

Båmstedt (1986) reviews the published information on the energy content of copepods. He expresses energy in terms of J mg^{-1}, equivalent to 0.239 cal mg^{-1} (Salonen *et al.*, 1976), the measure formerly used. The energy content of a copepod depends on the proportion and composition of the organic matter present. The average energy content of animal protein is 23.63, for lipid 39.35 and for carbohydrate 17.16 J mg^{-1}. Båmstedt discusses the limitations of these factors especially in viewing the energy in the copepod as transferable to a predator.

A frequency distribution of the published values for energy content of copepods (Båmstedt, 1986) shows that it ranges from 9 to 31 J mg^{-1}. The distribution has modes at 12 to 14 related to species living in low and medium latitudes, at 18 to 22 for surface-living species in high latitudes, and at 26 to 28 J mg^{-1} for deep-living species at high latitudes. There is a distinct trend of increasing energy content with increasing latitude. There is much variation, especially in species living in medium latitudes where the total range is 9 to 31 J mg dry weight^{-1}.

7.6. VITAMINS

Few observations have been made on the occurrence of vitamins in calanoid copepods. Fisher (1960) discusses the possible transfer of vitamin D to cod from a diet of copepods but, with the exception of vitamin A, no information on other vitamins in copepods had been recorded. No vitamin A was found by Fisher (1964) in *Calanus finmarchicus* though he records 4.4 µg g^{-1} wet weight in *Euchirella curticauda*, 20 µg g^{-1} in *Gaetanus kruppi* and possible positive records in *G. pileatus*.

Poulet et al. (1989) measured the following amounts of vitamin C in μg g^{-1} wet weight: *Anomalocera patersoni*, 11.6; *Acartia clausi*, 30 to 260; *Calanus helgolandicus*, 80 to 538; *Temora longicornis*, 125 to 368. Vitamin C content varies with body weight of *Calanus helgolandicus*. It was also seasonally correlated with the occurrence of phytoplankton. Further, the carnivorous *Anomalocera patersoni* had lower concentrations than the other species. Hapette and Poulet (1990) extended investigations of the occurrence of vitamin C to other planktonic organisms and show that calanoid copepods are among the most important carriers of the vitamin. They report 53 to 110 μg g^{-1} in *Centropages typicus* and 9.29 to 13.83 μg g^{-1} in *Drepanopus pectinatus*. Nauplii of *Acartia clausi* and *Temora longicornis* had 201 to 235 μg g^{-1}. These are the first studies of vitamin C in plankton and much more information is required to define its significance to the copepods and their role in the marine food chains.

7.7. CAROTENOIDS

Many copepods, especially meso- and bathypelagic species, are red to orange in colour. Ohman et al. (1989), however, found that orange-red *Neocalanus tonsus* had carotenoid pigments only as a minor part of the total lipids. They detected astaxanthin, the normal body carotenoid in zooplankton whose chief precursor is the plant carotenoid β-carotene. Kleppel et al. (1985) show that concentrations change dielly and are related to feeding activity. The most detailed study and review of carotenoids in copepods is that of Bandaranayake and Gentien (1982) who record the following among four species:

Centropages furcatus Astaxanthin, the esters being most common.
Temora turbinata   Astaxanthin, the esters being most common.
Undinula vulgaris   Astaxanthin, the esters being most common.
Pareuchaeta russelli Phoenicoxanthin, β-doradexanthin, astaxanthin, the esters most common.
Canthaxanthin, crustaxanthin, β,β-carotene-3,3',4-triol, and 2'-norastaxanthin ester were less common.

This work supports the much earlier work of Fisher (1964) who found carotenoid concentrations ranging from traces to 1133 μg g^{-1} wet weight among 80 species of calanoids. He concluded that the predominant and probably the only carotenoid present in most of the species was astaxanthin or its esters. Zagalsky et al. (1983) examined an astaxanthin-protein with a single positive band at 660 nm in *Anomalocera patersoni*.

7.8. ENZYMES

Enzymes, digestive and other, in copepods have been reviewed by Mayzaud (1986a,b) and recourse to these papers should be made for further information. The correlation, or its lack, between the digestive enzyme activity and feeding activity has been discussed here when considering feeding of the copepods.

8. Reproduction

8.1. Seasonality in Breeding 253
8.2. Mating Behaviour 256
    8.2.1. Attraction of the Sexes 256
    8.2.2. Mating Position 258
    8.2.3. Attachment of the Spermatophore 261
    8.2.4. Spermatophore Production 264
8.3. Spawning 266
    8.3.1. Egg Laying 266
    8.3.2. Resting Eggs in Sediments 268
    8.3.3. Egg and Brood Sizes 272
    8.3.4. Fecundity 282
8.4. Food Availability and Egg Production 289
    8.4.1. Quality and Quantity 290
    8.4.2. Stored Lipids 292
    8.4.3. Rates of Egg Production 292
8.5. Mortality of Eggs 292
8.6. Hatching of Eggs 293
8.7. Concluding Remarks 294

The internal anatomy, histology and development of the female and male reproductive systems are described in Chapter 2. This chapter discusses the mating behaviour and the production of eggs.

8.1. SEASONALITY IN BREEDING

Middle to higher latitude species of copepods usually show seasonality in breeding. This may result from the seasonal occurrence of small coastal species originating from resting eggs in the sediment. There can be several generations within a single breeding season, each generation comprising overlapping cohorts resulting from successive broods of eggs produced by individual females within the generations. Generation times are usually of the order of two weeks to two months. In species that have generation times spanning several months to one or two years, adult males and females can occur throughout the year and seasonality is reflected in the presence of sexually active males and females within restricted time-windows. In addition, there are often seasonal differences in the numbers of eggs produced per brood or clutch (Gaudy, 1971; Tourangeau and Runge, 1991; Diel and Tande, 1992). Seasonality in the deposition of diapause eggs to the sediments is usually pronounced (Uye, 1983; Marcus and Fuller, 1989; Næss, 1991a; Tiselius *et al.*, 1991).

The periods of breeding can be defined by several methods. The presence of eggs attached to females or in the water column can be monitored in time-series samples that define time-limits on the breeding season. Quantitative sampling can often estimate the number of cohorts produced within a generation and the number of sequential generations resulting. In many species, overlapping of cohorts and generations confuses the results, even within species where a single generation is thought to be involved. Further resolution of these components of the population can often be obtained by examining the degrees of maturity of the gonads of both the males and females, and the presence of attached spermatophores or full spermathecae on or in the females.

The spermatophores can often be seen quite clearly within the male gonoduct, their presence indicating potential for mating. Tande and Hopkins (1981), Tande and Grønvik (1983) and Norrbin (1991, 1994) show that copepodid V males have to be examined, as well as the adult, in high latitude species such as *Calanus finmarchicus*, *Metridia longa*, *Microcalanus pusillus*, *M. pygmaeus*, and *Pseudocalanus acuspes* where the male reproductive system can develop considerably in this stage. The percentage of CVs of *Metridia longa* with presumptive spermatophores within the *ductus ejaculatorius* increased from zero in August to 75% in November; they did not occur at any other time of year. The incidence of fully formed spermatophores in the *ductus ejaculatorius* of adult males can be determined, if necessary, by using a clearing agent such as lactic acid. The detection of males that had transferred their spermatophores to the females is described by Tande and Grønvik (1983). The spermatophore sac in such males is almost transparent, containing some undefined material, in contrast to the dark spermatophore resident in this region of males about to mate. Thus, the presence of presumptive spermatophores in CVs, that of ripe ones in adult males and the detection of spent males is a progressive situation that allows close definition of the breeding season.

Comparable examinations of the ovaries result in definition of the season of egg production. As in the males, CV females should also be examined because the ovaries can develop in this stage (Tande and Hopkins, 1981; Razouls *et al.*, 1987; Smith, 1990; Norrbin, 1991). Analyses of females can

Table 38 Sources of illustrations of stages of ovarian development in calanoid copepods.

Stages	Species	Authority
I. Dormant	*Calanus finmarchicus*	Marshall and Orr, 1955
	Neocalanus spp.	Miller and Clemons, 1988
	Acartia longiremis	Norrbin, 1994
II. Developing	*Calanus finmarchicus*	Marshall and Orr, 1955
	Calanoides carinatus	Romano, 1993
	Acartia tonsa	Romano, 1993
	Paracalnus parvus	Romano, 1993
	Neocalanus spp.	Miller and Clemons, 1988
	Rhincalanus gigas	Ommanney, 1936
	Acartia longiremis	Norrbin, 1994
	Calanus pacificus	Razouls *et al.*, 1991
	Metridia longa	Tande and Grønvik, 1983
	Sinocalanus tenellus	Kimoto *et al.*, 1986b
III. Ripe	*Calanus finmarchicus*	Marshall and Orr, 1955
	Neocalanus spp.	Miller and Clemons, 1988
	Rhincalanus gigas	Ommanney, 1936
	Temora stylifera	Razouls *et al.*, 1986
	Sinocalanus tenellus	Kimoto *et al.*, 1986b
IV. Spent	*Calanus finmarchicus*	Marshall and Orr, 1955
	Neocalanus spp.	Miller and Clemons, 1988
	Rhincalanus gigas	Ommanney, 1936
	Calanus pacificus	Razouls *et al.*, 1991
	Temora stylifera	Razouls *et al.*, 1986
Overwintering	*Acartia longiremis*	Norrbin, 1994

define the breeding season more accurately than data from males which often have spermatophores ready for transference one or two weeks before the females are ready to lay eggs, and for a similar period after the females have ceased to lay eggs. Examination of the gross morphology of the ovaries of a species from a time series of samples results in a subjective division of them into several stages of maturity. Staining techniques with chlorohydric carmine and borax carmine are used by Ward and Shreeve (1995) and Romano (1993). Species in which different ovarian stages have been determined and illustrated are given in Table 38. Estimation of the relative incidences of the different ovarian stages of maturation within the time series of samples results in a progressive set of data that allows definition of the breeding season. Plourde and Runge (1993) demonstrate a direct correlation between the proportion of females in the population with ovaries in an advanced stage of maturation and the numbers of eggs laid female^{-1} d^{-1}; this they term the reproductive index. Webber and Roff (1995a) use an even simpler reproductive index, the proportion of females with egg masses and/or spermatophores attached relative to those with neither.

A different approach is the use the RNA/DNA ratio in tissues or whole body extracts of females. The amount of DNA in nuclei of body cells is constant within a species but concentrations of RNA fluctuate because it is involved in synthesis of proteins. Nakata *et al.* (1994) review earlier work and, from their own studies on *Paracalanus* sp. in a frontal system of the Kuroshio, demonstrate a linear relationship between daily egg production per female and the RNA/DNA ratio in body tissues. They suggest that the temporal changes that they found in the RNA/DNA ratio may indicate changes in rates of egg production in response to phytoplankton abundance on small temporal and spatial scales.

8.2. MATING BEHAVIOUR

Mating behaviour consists of a sequence of events that are more or less followed by all species (Blades-Eckelbarger, 1991a). They are: 1, the attraction of the male to the female; 2, the initial capture of the female by the male; 3, the adoption of the mating position and; 4, the transfer and attachment of the spermatophore. The spermatophore discharges its contents into the female's spermathecae or, in species that do not have spermathecae, into a mass or plug formed by secretions from the spermatophore itself. The male usually releases the female immediately after transfer of the spermatophore but clasping may continue in a few species for some time after mating.

8.2.1. Attraction of the Sexes

There is circumstantial evidence that the female produces sex-attractant pheromones that are detected by chemosensory means by the male (Uchima and Murano, 1988). Strickler and Costello (1996), quoting S.P. Colin, state that males accelerate over distances of 10 cm to overtake and mate with females whose scent trails they have detected. The pheromones may be glycoproteins according to Snell and Carmona (1994). The male responds by changing its pattern of swimming to an erratic movement referred to as searching or mate-seeking behaviour (Blades-Eckelbarger, 1991a). The male frequently approaches the female from behind, that is downstream of the water currents produced by her feeding and swimming. Mechanosensory detection by the males of water disturbance by the female has also been suggested (Fleminger, 1967). There are records of males being attracted to females of other species. Nishida (1989) and Blades-Eckelbarger (1991a) conclude that initial chemo- and/or mechanosensory recognition of females is not always effective; secondary chemosensory recognition, as the male gains the mating position, together with the mismatch of the male's fifth legs and the morphology of the female's genital field, may all combine to prevent hybridization. Species in the families Centropagidae and Pontellidae have spermatophores with complex coupling plates that match the morphology of the female's genital field and prevent hybridization (Fleminger, 1967).

Most attention has centred on the male being attracted to exudates, or pheromones, of the female but van Duren and Videler (1996) found that the female of *Temora longicornis* probably reacts to exudates of the male. The female increased the numbers of small hops that it made when placed in water that had been inhabited by males. These small hops could act as strong hydromechanical signals from receptive females to the males.

Males, of course, must be sexually mature with at least one ripe spermatophore in the genital duct ready for transference to the female. As stated in Chapter 2, males have a single testis and only the left genital duct is developed. The state of sexual maturity of the female at mating varies between species. Most have gravid ovaries. Males of the cyclopoid *Oithona davisae* preferentially mate with gravid females (Uchima and Murano, 1988) and so do *Pseudodiaptomus* species according to Jacoby and Youngbluth (1983) and *Diaptomus* species according to Watras (1983). This situation may be quite general among copepods. Species in the oceanic metridinid genus *Pleuromamma*, however, demonstrate a different situation. Adult females and males have different depth distributions in the water column (Hayward, 1981; J.S. Park, personal communication). Park (1995b) found that the vertical distribution of mature males corresponded to that of the CV females. Mating takes place immediately after the CV female moults to the adult with a soft integument and long before the ovaries are gravid. Griffiths and Frost (1976) found that male *Calanus pacificus* and *Pseudocalanus* species react to newly moulted females by a change in their swimming behaviour. Consequently, newly moulted females may be the target in other species.

Capture of the female has been observed in a variety of genera (Jacoby and Youngbluth, 1983): *Acartia, Centropages, Eurytemora, Labidocera, Limnocalanus, Pseudodiaptomus, Temora*. These genera all have a geniculate antenna (Figure 5) with which the male grasps the female's urosome, caudal furcae or caudal setae. This is the normal initial act of capture. The initial act of capture in genera without a geniculate antenna has not been described.

Interspecific mating has been observed in the laboratory between conspecific species of *Pseudodiaptomus* by Jacoby and Youngbluth (1983) and has been inferred in environmental samples of *Pareuchaeta* species. Ueda (1986b), however, was unsuccessful in inducing the co-occurring *Acartia omorii* and *A. hudsonica* to transfer spermatophores between them.

8.2.2. Mating Position

The male adjusts his position relative to that of the female after the initial capture. Such adjustments often consist of erratic and fast movements such that the actual instant of transfer of the spermatophore is difficult to observe. The male is generally above the female so that the modified fifth legs (Figure 52) grasp the female's urosome. The orientation of the male,

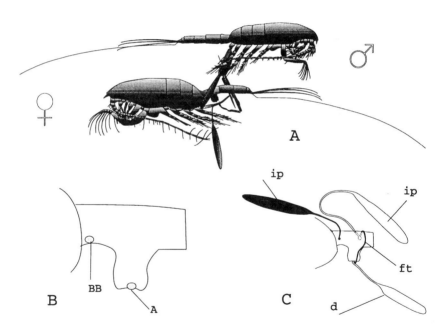

Figure 52 Spermatophore transference by the male to the female copepod. A, probable mating position in *Pareuchaeta norvegica* and many other species of calanoid copepods. B, genital somite of female showing position of direct (A) and indirect (BB) placement of spermatophores that are successful in fertilizing the eggs. C, direct (d) and indirect (ip) placements of spermatophores; one of the indirectly placed spermatophores remains full but the other empties through a fertilization tube (ft). (Partially after Hopkins *et al.*, 1978.)

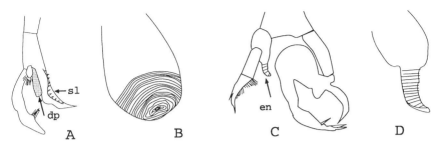

Figure 53 The fifth legs of male calanoids. A, terminal part of left P5 of a male *Pareuchaeta norvegica* showing the row of spines on the serrated lamella (sl) and the digitiform process (dp). B, enlargement of the digitiform process of *P. norvegica* showing ridging. C, male P5 of *Labidocera aestiva* showing the endopod (en) of the left leg. D, enlargement of the endopod of *L. aestiva* showing ridging. (B, after Hopkins *et al.*, 1978; C and D after Blades and Youngbluth, 1979.)

relative to that of the female, varies. In *Centropages typicus* and *Pseudodiaptomus* species the male is on his side, is located above the female and faces in the opposite direction to the female, so that only their urosomes overlap (Gauld, 1957b; Blades, 1977; Jacoby and Youngbluth, 1983). He grasps her urosome with the chela of his fifth leg and releases the hold of his antennule on her. *Labidocera aestiva* behaves in virtually the same manner except that the male is not lying on his side but is dorsal side up (Blades-Eckelbarger, 1991a). Blades and Youngbluth (1979) describe pre-copulatory stroking of the area of the female's urosome that has pit-pores. The stroking is done with the ridged left endopod (Figure 53C, D). The function of this behaviour, which has not been described in any other species, remains unclear. According to Blades and Youngbluth, these pores may secrete a solvent to dissolve the adhesive that attaches the spermatophore. The stroking by the male may clean the surface in cases where a previous mating and production of a brood had taken place. The male left fifth leg of *Pareuchaeta norvegica* has a digitiform process that has a ridged "thumbprint" (Hopkins *et al.*, 1978). This is on the third exopodal segment (Figure 53A, B) and may also be used to clean the integument of the female prior to adhesion of the spermatophore.

The mating of species in families without a geniculate antenna is not so well known. Griffiths and Frost (1976) found that male *Calanus pacificus* and *Pseudocalanus* sp. reacted with changed swimming patterns to newly moulted females but, although spermatophores were attached by the latter species, no description of mating is given. *Pareuchaeta norvegica* was examined by Hopkins *et al.* (1978) and several potential mating positions determined. Mating has not been observed in this species. Mating is usually over in a short time, a few seconds to a few hours in *Eurytemora velox*, *Centropages typicus* and *Acartia tonsa* but may last for several hours or even days in *Pseudodiaptomus euryhalinus* (Gauld, 1957b; Fleminger, 1967; Blades, 1977). It lasts from 30 s to over 1 h in *Eurytemora affinis* according to Katona (1975) and in some *Pseudodiaptomus* species (Jacoby and Youngbluth, 1983).

Some males, e.g. those of *Pareuchaeta norvegica*, have a spermatophore held on the third exopodal segment of the left, fifth leg ready for attachment to the female before mating takes place (Hopkins *et al.*, 1978). Males of *Eurytemora affinis* behave similarly (Katona, 1975) but many species retain the spermatophore internally in the *ductus ejaculatorius* until mating. Blades (1977) describes the spermatophore of *Centropages typicus* being extruded from the genital pore of the male by muscular contractions of the genital duct. It is then held by the fifth legs.

Species that carry their eggs attached to the urosome in an egg mass (sac) usually produce more than one brood of eggs. Those that extrude their eggs freely into the environment often produce successive batches, a batch once a day or irregularly. It is not clear how many times a female requires to be mated. Spermatozoa are known to remain viable for some time. For example, Katona (1975) found that female *Eurytemora affinis* could produce viable nauplii up to 57 d after transference of the spermatophore. A single female *Calanus finmarchicus*, that mated immediately after moulting from the CV, can produce batches of eggs over a period of 60 to 80 d (Marshall and Orr, 1955). Female *C. hyperboreus* and *C. pacificus* mate only once (Conover, 1967; Runge, 1984). Corkett and Zillioux (1975) state that a single insemination of female *Pseudocalanus minutus* and *P. elongatus* suffices for the production of up to 10 successive egg sacs. Uye (1981) found that a single mating is probably sufficient to fertilize all the eggs produced by a single female *Acartia clausi*; remating in this species is infrequent (Ianora *et al.*, 1996). Other species producing successive broods have been shown to require remating before each brood is produced. These are: *A. tonsa* observed by Wilson and Parrish (1971) and Parrish and Wilson (1978); fresh water diaptomid species according to Watras (1983); *Eurytemora affinis* (Heinle, 1970; Katona, 1975); *Sinocalanus tenellus* (Kimoto *et al.*, 1986b) and *Temora stylifera* (Ianora *et al.*, 1989). It may be that females that are mated immediately after moulting from the CV mate only once whereas females that are mated when the ovaries are gravid may require remating for each brood of eggs.

Males are usually smaller than females. There are seasonal variations in the body sizes of the males and females, especially in short-lived coastal and epipelagic species. Body size also varies within single populations and is most pronounced when there is more than one cohort or generation represented. The question then arises as to whether production of fertilized, viable eggs is more successful when mating pairs have a size relationship close to the modal or normal value for that species. This has not been tested in marine species but has been examined in diaptomids (DeFrenza et al., 1986; Grad and Maly, 1988, 1992). Males select females on the basis of body size in some species and not in others. Hart and McLaren (1978) demonstrate that small males mate with small females, and large with large, in *Pseudocalanus* sp.

8.2.3. Attachment of the Spermatophore

The functional morphology of spermatophores and their transfer to the female have been reviewed by Blades-Eckelbarger (1991a). Spermatophores of species in the families Centropagidae and Pontellidae have complex coupling plates by which they are attached by the males to the females (Heberer, 1937; Fleminger, 1957, 1967, 1979; Lee, 1972; Blades, 1977; Blades and Youngbluth, 1979, 1980; Blades- Eckelbarger, 1991a). The morphology of these plates varies between species and corresponds to the morphology of the genital complex of the respective females of the species. Coupling plates, but relatively simple in form, are described in *Gaussia asymmetrica* (family Metridinidae) by Björnberg and Campaner (1988) who also point out that, as Lee (1972) found in many of the species in the genus *Centropages*, there are no spermathecae. Lee suggests that a ventral sac, an integral part of the complex coupler in *Centropages* species, is distended by material expelled from the neck of the spermatophore; the spermatozoa are then extruded from the spermatophore in a thin-walled tube that coils through this material. The detailed mechanism of storage of the spermatozoa in *G. asymmetrica* is unknown.

The spermatophores of the vast majority of species in the other families of calanoid copepods are simple tube-like flasks without complex coupling devices (Heberer, 1937; Marshall and Orr, 1955; Vaupel Klein, 1982b, 1989; Reddy and Devi, 1985, 1989, 1990; Miller, 1988; Devi and Reddy, 1989a,b; Blades-Eckelbarger, 1991a). These are attached to the females by adhesive secretions extruded from the spermatophore itself during its transfer to the female by the male (Blades-Eckelbarger, 1991a).

8.2.3.1. Number Attached

In calanoid copepods, the majority of females receive only one spermatophore during the production of one brood of eggs. There are, however, records of females receiving more than one. Gibbons (1933, 1936) found female *Calanus finmarchicus* with as many as 15 spermatophores.

Diaptomids are recorded with as many as nine spermatophores (Reddy and Devi, 1985, 1989, 1990; Devi and Reddy, 1989a,b). *Clausocalanus paululus* can have two (Frost and Fleminger, 1968) while *Labidocera diandra*, described by Fleminger (1967) can have four. A female *Cosmocalanus darwini* is figured with three spermatophores by Chiba (1953) and Ferrari (1980) states that *Pseudochirella squalida* can also have three spermatophores. Multiple placements were observed in *Limnocalanus macrurus* by Roff (1972). Katona (1975) found up to 31 spermatophores on a single female *Eurytemora affinis*, confirming the earlier observation of Lucks (1937) of multiple placement in this species. Hammer (1978) found multiple placement in *Acartia tonsa* and Jacoby and Youngbluth (1983) in *Pseudodiaptomus* species. It is, however, in the Euchaetidae that multiple placement of spermatophores has been most commonly reported (Sewell, 1947; Zvereva, 1976; Hopkins and Machin, 1977; Ferrari, 1978; Bradford, 1981; Ferrari and Dojiri, 1987; Ward and Robins, 1987; Mauchline, 1994b).

8.2.3.2. *Placement Position*

The male normally attaches the spermatophore over the genital pore within the genital field of the first somite of the urosome. This placement is frequently referred to as the direct placement (Figures 52, B; 54, site A) (Hopkins and Machin, 1977) or correct placement (Ferrari and Dojiri, 1987). There are, however, indirect or incorrect placements recorded in species from a variety of families (Gibbons, 1936; Lucks, 1937; Giron, 1963; Fleminger, 1967; Katona, 1975; Zvereva, 1976; Blades, 1977; Hopkins and Machin, 1977; Ferrari, 1978, 1980; Schweder, 1979; Bradford, 1981; Vaupel Klein, 1982b; Ferrari and Dojiri, 1987; Blades-Eckelbarger, 1991a). The sites in Figure 54 are labelled in order of precedence. The direct placement site A on the genital field and the indirect site BB are commonest with the indirect placement sites B through to E sub-dominant. Indirect sites F through to S are rare. Placements on the right-hand side of the genital somite are rare with the exception of site C. Bradford (1981), however, records a common placement site in *Pareuchaeta erebi* and *P. similis* as G but could not confirm that it was the dominant site in these species. The two terms "indirect" and "incorrect" arise from the possibility, or lack of possibility, of the spermatozoa from spermatophores in such placements being able to fertilize the eggs. Most indirectly placed spermatophores remain full (Figure 52C) but some have a fertilization tube (Figure 52C) connecting them to the genital opening. These spermatophores empty through the tube and the spermatozoa are successful in fertilizing the eggs. Such fertilization tubes are described in *Pareuchaeta* species by Zvereva

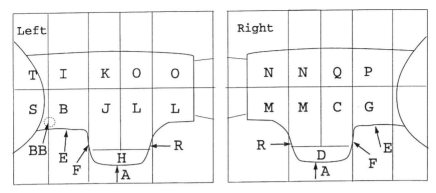

Figure 54 Diagrammatic illustration of positions of placement of spermatophores on the left and right side of the genital somite of female calanoid copepods. The alphabetical series labels areas in order of decreasing importance in *Pareuchaeta* species, site A being the dominant, direct placement site over the genital opening. (Based on Hopkins and Machin, 1977.)

(1976), Hopkins and Machin (1977), Bradford (1981) and Mauchline (1994b), in *Pseudochirella* species by Ferrari (1980), in *Euchirella* species by Vaupel Klein (1982b, 1989), and in *Undinula vulgaris* by Blades-Eckelbarger (1991a).

Spermatophores directly placed at site A in *Pareuchaeta* species tend to have shorter flasks than indirectly placed spermatophores (Hopkins, 1978; Bradford, 1981; Ferrari and Dojiri, 1987; Mauchline, 1994b). The significance of this polymorphism of the spermatophores of this genus is discussed fully but conjecturally by Ferrari and Dojiri (1987).

Direct placement of spermatophores takes place in most species but not all. Direct attachment of spermatophores at site A (Figure 54) is the dominant placement found in populations of *Pareuchaeta norvegica* in Loch Etive, western Scotland and the alternative indirect placement at site BB (Figure 54) is the second commonest placement. Placement in deepwater populations of this species in the Rockall Trough, however, is predominantly at site BB. Ferrari (1978) found the dominant sites in populations off Delaware, eastern United States to be B, E and G (Figure 54). These results infer variation in mating behaviour, and possibly in the positions of the male and female during mating, within a species. Other species where the dominant site of placement is indirect are *Pareuchaeta antarctica*, *P. erebi*, *P. gracilis*, *P. similis*, *Euchirella messinensis* and *Undinula vulgaris* (Mauchline, 1994b). All of these species develop fertilization tubes.

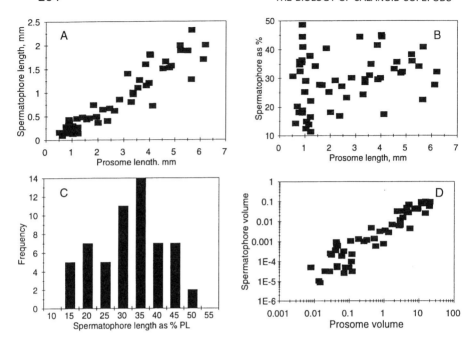

Figure 55 Spermatophores of calanoid copepods. A, the length of the spermatophore relative to prosome length in different species; the relationship is spermatophore length (mm) = 0.345. Prosome length (mm) − 0.087, n = 58, r = 0.942. B, the length of the spermatophore expressed as a percentage of the length of the prosome in different species. C, the frequency occurrence of species with spermatophores representing different percentages of prosome length. D, the volume (mm^3) of the spermatophore relative to the body volume (mm^3) in different species; the relationship is log spermatophore volume (mm^3) = 1.098 log body volume (mm^3) − 2.426, n = 58, r = 0.931.
Data from: Marshall and Orr, 1955; Chiba, 1956; Comita and Tommerdahl, 1960; Frost and Fleminger, 1968; Katona, 1975; Bradford, 1981; Reddy and Devi, 1985, 1989, 1990; Devi and Reddy, 1989a,b; Frost, 1989; Belmonte *et al.*, 1994; Mauchline, 1994b, unpublished; Rayner, 1994.

8.2.4. Spermatophore Production

The length of the spermatophores is correlated with the prosome length of the males in between-species comparisons within the Euchaetidae (Mauchline, 1995) and in calanoid copepods in general (Figure 55A). Spermatophore length of the 58 species examined (Figure 55B) represented 29 ± 9% (range 11–48%) of prosome length. Small-sized species showed greater variability in spermatophore size than the larger-sized species (Figure 55B). A frequency diagram of spermatophore size relative to prosome length suggests a bimodal distribution (Figure 55C), one mode being centred at 25%, the other at 35% of prosome length.

Hopkins (1978) states that a spermatophore of *Pareuchaeta norvegica* represents 0.2% of wet and 0.9% of dry body weight. Mauchline (1995), examining 12 species of this genus, estimated that a spermatophore represented 0.753 ± 0.474% of the body volume and so, approximately, of the body wet weight. The 58 calanoid species analysed in Figure 55D have spermatophores representing 0.545 ± 0.462% of body volume while those of the 30 Euchaetidae represented 0.513 ± 0.318% of body volume. There is a lower limit to the body size of calanoid copepods but these small-sized species show the greatest variability in the sizes of their spermatophores (Figure 55B, D).

Mauchline (1995) considered that a male might invest the same amount of energy in producing spermatophores as a female invests in at least one brood of eggs, although this is contradicted by Gaudy *et al.* (1996) working on *Acartia tonsa*. Female Euchaetidae produce a brood of eggs equivalent to 12.114 ± 2.213% of body volume. The number of spermatophores representing 10% of the male's body volume in the 12 north Atlantic Euchaetidae is 18.9 ± 12.3 (range 5.6 to 44.8) (Mauchline, 1995). The 30 species of Euchaetidae examined here have a mean of 30 ± 26 spermatophores (range 7 to 128) equivalent to 10% of body volume. The numbers of spermatophores, equivalent to 10% of body volume, vary greatly between species and genera. For example two *Pseudocalanus* species have 63 and 68 (data from Frost, 1989); 13 *Clausocalanus* species have 119 ± 115 (range 11 to 382) (data from Frost and Fleminger, 1968); nine species of diaptomids have 14 ± 9 (range 6 to 34) (data from Reddy and Devi, 1985, 1989, 1990; Devi and Reddy 1989a, b); *Calanus finmarchicus* has 130 (Marshall and Orr, 1955); *Rhincalanus nasutus* has 7 (Chiba, 1956); and *Acartia tonsa* has 127 (Belmonte *et al.*, 1994). Thus, there is a range in the potential energetic costs to the male of producing spermatophores.

The numbers of successive spermatophores produced by a male in unit time has been estimated experimentally on a few occasions. Ianora and Poulet (1993) isolated pairs consisting of a male and female *Temora stylifera* and found that the male can produce between one and three spermatophores per day. They illustrate a typical male as producing 95 spermatophores in a period of 50 d. This, however, was recorded in the laboratory and males of this species in the natural environment may not normally survive long enough to produce this number. The longevity of individual males, relative to that of females, is very difficult to assess but males are frequently assumed to be shorter-lived. Parrish and Wilson (1978) state that male *Acartia tonsa* live one third to one half as long as females in culture.

As mentioned earlier, multiple placement of spermatophores occurs in some species in the genera *Acartia, Calanus, Clausocalanus, Eurytemora, Labidocera, Limnocalanus, Pareuchaeta, Pseudochirella* and *Pseudodiaptomus*. An examination of multiple placement within the genera *Euchaeta* and *Pareuchaeta* by Mauchline (1994b) suggests that it occurs in epipelagic and mesopelagic species that make diel vertical migrations and not in the non-diel migrating mesopelagic or in the bathypelagic species. Thus, multiple placement in this genus is associated with diel vertical movements within the populations and a strong seasonality in the population parameters. Species in the genera listed above are also all coastal, epipelagic or mesopelagic diel migrators subject to marked environmental variability. Most indirectly placed spermatophores do not have fertilization tubes and so are not involved in fertilization; they must, therefore, be surplus to requirements. This excess production of spermatophores has not been recorded in a bathypelagic species for which reproductive efficiency is undoubtedly highly selected for. Sex ratios, males to females, within populations of ten species of the Euchaetidae that were partitioned bathymetrically in a 2000 m water column did not change significantly with modal depth of occurrence (Mauchline, 1995). Consequently, high sex ratios are not a prerequisite for multiple placement.

8.3. SPAWNING

Most studies of spawning have been made on temperate and high latitude species. One area of interest is the possibility that laying of the eggs is timed to take advantage of phytoplankton increases or blooms. The association of spawning events with concentrations of phytoplankton are not clearly demonstrated. Maturing females require a good supply of food for maturation of the ovaries and the horizontal and vertical distributions of herbivorous species would be expected to show some relationship with phytoplankton concentrations. There would then be a coincident relationship between spawning events and phytoplankton distribution. A few species, such as *Calanoides acutus*, show no correlation between environmental chlorophyll concentrations and maturation of the ovaries (Lopez *et al.*, 1993); this may be a function of adequate alternative and preferred food at depth and of lipid stores in an overwintering CV.

8.3.1. Egg Laying

Most species of pelagic calanoid copepods broadcast their eggs freely into the sea but others carry them attached to the ventral side of the genital somite until nauplii hatch out. Some benthopelagic species may broadcast their eggs freely into the sea in close association with the sediment surface to which they adhere individually. Webb and Weaver (1988) point out that broadcasting of the eggs as opposed to attachment of them in an egg mass to the female has been evolved within marine calanoids but not within the harpacticoid and cyclopoid copepods. They argue that this indicates relatively low mortality through predation of free eggs in the water column.

Eggs of all species extruding from the genital pore are elongated or pear-shaped and often initially stick together. Marshall and Orr (1955) show the eggs of *Calanus finmarchicus* emerging from the genital pore in a string and remaining attached until they round off. Eggs of *C. hyperboreus* are squeezed from the genital opening in pairs forming a ribbon from which single eggs separate within a few minutes; the eggs were laid at a rate of 30 to 35 min^{-1} (Conover, 1967). Pear-shaped eggs laid by *Acartia clausi*, separate and round-off after a few minutes (Koga, 1973). Eggs of *Neocalanus plumchrus* are extruded as two ribbons and after a few minutes become rounded and separated from the ribbons (Fulton, 1973). In *Sinocalanus tenellus*, the clutch of about 20 eggs form an egg mass attached to the ventral side of the genital somite but after a few minutes separate freely into the water (Kimoto *et al.*, 1986b). Yang (1977) states that the extruded eggs of *Parvocalanus crassirostris* separate from the ribbons as groups of 2 to 8, most commonly 4, the eggs sticking to each other. Species in genera such as *Euchaeta, Euchirella, Eurytemora, Pareuchaeta, Paraugaptilus, Pseudocalanus, Pseudochirella, Pseudodiaptomus* and *Valdiviella* carry the eggs in an egg mass attached to the genital somite until the nauplii hatch out. There seems to be some irregularity about the formation of egg masses. *Euaugaptilus magnus* has an egg mass in the Rockall Trough (Mauchline, 1988a) but no other species in this genus has been observed to have one. *Chiridius gracilis* forms an egg mass (MacLellan and Shih, 1974) but *C. armatus* does not (Matthews, 1964).

The egg mass is often called an egg sac, but in calanoids there is no evidence that the eggs are contained in a membrane. Rather, the secretions that form the outer membrane of eggs of free spawners seem to be more copious in egg mass carriers so that the eggs as they are laid do not separate but form into a mass attached to the genital somite. The majority of calanoids produce a single, laminar egg mass with the eggs more or less arranged in a single layer (Figure 8, A). Some species in the genus *Pseudodiaptomus* produce a pair of asymmetrical egg masses, the right-hand one being absent or containing only a few eggs while the left one is larger with up to 17 eggs (Jacobs, 1961; Jacoby and Youngbluth, 1983).

Matthews (1964) describes egg laying in the benthopelagic *Chiridius armatus* and states that, in the broadcast eggs, the outer layer of the egg is adhesive and the egg will stick to any surface that it comes in contact with.

Table 39 Spawning times of calanoid copepods.

Species	Time	Authority
Acartia spp.	night	Stearns *et al.*, 1989
		Checkley *et al.*, 1992
		Cervetto *et al.*, 1993
Calanoides carinatus	night	Armstrong *et al.*, 1991
Calanus finmarchicus	1200–1600	Marshall and Orr, 1955
	dawn	Runge, 1987
C. helgolandicus	midnight, midday	Laabir *et al.*, 1995b
C. pacificus	night, dawn	Mullin, 1968; Runge, 1985
C. propinquus	night	Kosobokova, 1994
C. sinicus	2200–0600	Uye, 1988
Centropages furcatus	night	Checkley *et al.*, 1992
Labidocera aestiva	night	Marcus, 1985a
Metridia pacifica	night, dawn	Mullin, 1968; Runge, 1985
Paracalanus sp.	dawn	Uye and Shibuno, 1992

Eggs of *Aetideus armatus* are also adhesive. In contrast, he found that the eggs of another benthic species, *Bradyidius bradyi*, are attached in groups to the substratum and not singly as in the former species.

There may be a diel rhythm of egg laying (Table 39), several species being known to spawn at night. A few species, however, have no rhythm e.g. *Acartia grani* which produces eggs continuously a few at a time and not in distinct clutches (Rodríguez *et al.*, 1995).

8.3.2. Resting Eggs in Sediments

Diapause eggs are produced by a wide variety of crustaceans, especially in fresh waters (Dahms, 1995; Alekseev and Fryer, 1996). Most investigations of resting eggs in marine calanoid copepods have been made in northern temperate waters, a few in subtropical waters and none in tropical regions (Marcus, 1996).

Some 44 species in the families Acartiidae, Centropagidae, Pontellidae and Temoridae have been shown to produce two types of eggs, the normal, or subitaneous, eggs and resting, or diapause, eggs (Table 40). The two types may or may not be distinguishable morphologically from each other (Kasahara *et al.*, 1974). The resting eggs of *Acartia* species and *Pontella mediterranea*, for example, differ from the subitaneous eggs in having long spines and being slightly larger in size (Grice and Gibson, 1981; Belmonte and Puce, 1994), while those laid by species such as *Labidocera aestiva*, *Anomalocera patersoni* and *Tortanus forcipatus* are indistinguishable in

Table 40 List of species of marine calanoid copepods in which resting eggs have been found. (Updated from Uye, 1985.)

Family/species	Locality	Reference
Acartiidae		
Acartia bifilosa	Northern Baltic	Viitasalo, 1994
	Northern Baltic	Katajisto, 1996
	Jiaozhou Bay, China	Zhong and Xiao, 1992
A. californiensis	Yaquina Bay, Oregon	Johnson, 1980
	Mission Bay, California	Uye and Fleminger, 1976
A. clausi	Inland Sea, Japan	Kasahara et al., 1975a
	Onagawa Bay, Japan	Uye, 1980a
	Jakle's Lagoon, Washington	Landry, 1975a
	Mission Bay, California	Uye and Fleminger, 1976
	Northern California	Marcus, 1990
	Buzzards Bay, Mass.	Marcus, 1984a
	Western Norway	Naess, 1991b, 1996
A. erythraea	Inland Sea, Japan	Kasahara et al., 1975a
A. hudsonica	Rhode Island, USA	Sullivan and McManus, 1986
	Buzzards Bay, Mass.	Marcus, 1984a
A. josephinae	Mediterranean	Belmonte and Puce, 1994
A. latisetosa	Mediterranean	Belmonte, 1992
A. longiremis	N California	Marcus, 1990
A. pacifica	Inland Sea, Japan	Uye, 1985
	Jiaozhou Bay, China	Zhong and Xiao, 1992
	Xiamen, China	Marcus, 1996
A. sinjiensis	Inland Sea, Japan	Uye, 1985
A. spinicauda	Xiamen, China	Marcus, 1996
A. steueri	Onagawa Bay, Japan	Uye, 1980a
A. teclae	Western Norway	Naess, 1996
A. tonsa	N California	Marcus, 1990
	La Jolla, California	Uye and Fleminger, 1976
	Northeast US	Zillioux and Gonzalez, 1972
	Rhode Island, USA	Sullivan and McManus, 1986
	Buzzards Bay, Mass.	Marcus, 1984a
	Florida, USA	Marcus, 1989
	Northern Gulf of Mexico	Marcus, 1991
	Argentina	Sabatini, 1989
	S Baltic	Arndt and Schnese, 1986
	SW Baltic	Madhupratap et al., 1996
	Mediterranean	Gaudy, 1992
A. tsuensis	Inland Sea, Japan	Uye, 1985
Acartia spp.	Italian coast	Belmonte et al., 1995
Centropagidae		
Centropages abdominalis	Inland Sea, Japan	Kasahara et al., 1975a
C. furcatus	Florida, USA	Marcus, 1989

Table 40 Continued.

Family/species	Locality	Reference
C. hamatus	Buzzards Bay, Mass.	Marcus, 1984a
	Georges Bank	Marcus, 1996
	Florida, USA	Marcus, 1989
	White Sea	Pertzova, 1974
	North Sea	Lindley, 1990
	Western Norway	Naess, 1991b, 1996
	SW. Baltic	Madhupratap et al., 1996
	English Channel	Lindley, 1990
C. ponticus	Black Sea	Sazhina, 1968
C. tenuiremis	Xiamen, China	Marcus, 1996
C. typicus	North Sea	Lindley, 1990
C. yamadai	Inland Sea, Japan	Kasahara et al., 1975a
Centropages sp.	Italian coast	Belmonte et al., 1995
Sinocalanus tenellus	Inland Sea, Japan	Hada et al., 1986
Pontellidae		
Anomalocera ornata	Northern Gulf of Mexico	Marcus, 1996
A. patersoni	Mediterranean	Ianora and Santella, 1991
Calanopia americana	Northern Gulf of Mexico	Marcus, 1996
C. thompsoni	Inland Sea, Japan	Kasahara et al., 1975a
Epilabidocera amphitrites	Northern California	Marcus, 1990
E. longipedata	Yaquina, Oregon	Johnson, 1980
Labidocera aestiva	Woods Hole, Mass.	Grice and Lawson, 1976
	Vineyard Sound, Mass.	Marcus, 1979
	Buzzards Bay, Mass.	Marcus, 1984a
	Florida	Marcus, 1989
	Northern Gulf of Mexico	Marcus, 1996
L. bipinnata	Inland Sea, Japan	Uye et al., 1979
L. scotti	Florida, USA	Marcus, 1989
L. trispinosa	La Jolla, California	Uye, 1985
L. wollastoni	Irish Sea	Lindley, 1990
	English Channel	Lindley, 1986
	Mediterranean	Grice and Gibson, 1982
Labidocera sp.	Italian coast	Belmonte et al., 1995
Pontella meadi	Woods Hole, Mass.	Grice and Gibson, 1977
	Northern Gulf of Mexico	Marcus, 1996
P. mediterranea	Black Sea	Sazhina, 1968
	Mediterranean	Grice and Gibson, 1981
	Mediterranean	Santella and Ianora, 1990

Table 40 Continued

Family/species	Locality	Reference
Temoridae		
Eurytemora affinis	Yaquina Bay, Oregon	Johnson, 1980
	Rhode Island, USA	Marcus *et al.*, 1994
	Western Norway	Naess, 1991b, 1996
	Northern Baltic	Viitasalo, 1994
	Northern Baltic	Katajisto, 1996
	SW Baltic	Madhupratap *et al.*, 1996
	Lake Ohnuma, Japan	Ban and Minoda, 1991
E. americana	Buzzards Bay, Mass.	Marcus, 1984a
	Rhode Island, USA	Marcus *et al.*, 1994
E. pacifica	NW Pacific	Solokhina, 1992
	Onagawa Bay, Japan	Uye, 1985
E. velox	*Mediterranean	Gaudy and Pagano, 1987
Temora longicornis	Georges Bank	Marcus, 1996
	Irish Sea	Lindley, 1990
	English Channel	Lindley, 1990
	Western Norway	Naess, 1996
	North Sea	Lindley, 1990
	*Northern Baltic	Viitasalo and Katajisto, 1994
Tortanidae		
Tortanus derjugini	Xiamen, China	Chen and Li, 1991
T. discaudatus	Northern California	Marcus, 1990
	Rhode Island, USA	Marcus *et al.*, 1994
T. forcipatus	Inland Sea, Japan	Kasahara *et al.*, 1975a
	Xiamen, China	Chen and Li, 1991

*Probable records.

gross form and size although there are differences at the histological level (Kasahara and Uye, 1979; Marcus and Fuller, 1986; Santella and Ianora, 1990; Ianora and Santella, 1991). One major histological difference between the two types of eggs is often the presence of a thicker chorion in the diapause eggs (Ban and Minoda, 1991; Ianora and Santella, 1991; Lindley, 1992). The diameter of subitaneous eggs of *Eurytemora affinis* is 79.9 ± 2.7 μm compared with 86.1 ± 2.2 μm for diapause eggs (Ban and Minoda, 1991). Resting or diapause eggs do not develop immediately but sink to the sea bed. Copepods can be induced to lay diapause eggs by being subjected to high population densities (Kasahara and Uye, 1979; Ban and Minoda, 1994). A combination of shortening day-length and lowering temperatures is also stimulatory (Marcus, 1982a,b; Ban, 1992; Hairston and Kearns, 1995). Ban (1992) suggests that conditions under which the nauplii develop determine whether the resultant adults produce diapause eggs.

Diapause eggs can become buried in the sediments but can remain viable for as long as 40 years (Marcus et al., 1994). Hairston and Brunt (1994) suggest that diapause eggs of the fresh water *Leptodiaptomus minutus* remain viable for two or more decades. Diapause eggs of *Eurytemora affinis* were considered viable after 7 to 8 years and possibly for as long as 18 to 19 years by Katajisto (1996). In general, however, diapause eggs are seasonal in occurrence and the interval between deposition in the sediments and hatching is measured in months. Duration of the diapause has been shown to vary according to the season at which the eggs are laid (Marcus, 1987; Ban and Minoda, 1991) but this was not found in those of *Acartia* species by Uye (1980a). Frequently, they require a period of chilling before they will hatch (Marcus and Fuller, 1986; Sullivan and McManus, 1986; Marcus, 1989; Ban and Minoda, 1991; Marcus, 1995). Burying in the sediments can delay hatching (Kasahara et al., 1975b; Ban and Minoda, 1992). Organisms such as polychaete worms can cause bioturbation of the sediments, consuming but not digesting eggs at depth and depositing them on the sediment surface where they may hatch (Marcus and Schmidt-Gengenbach, 1986). The vertical distribution of the eggs in the sediment following a bioturbation event may depend on the grain size of the sediments (Marcus and Taulbee, 1992).

The sinking rates of the eggs are not simply a function of Stoke's Law but affected by turbulence within the water column and the density of the eggs, which may change through osmoregulatory processes. Miller and Marcus (1994) found that eggs of *Acartia tonsa* laid in water of salinity 15‰ had a density of $1.066\,\text{g cm}^{-3}$ while those in water of 31‰ had a density of $1.086\,\text{g cm}^{-3}$, water temperature in both cases being 20 °C. Measurements of density and sinking rates of eggs are given in Table 41. Subitaneous eggs were less dense than diapause eggs of *Labidocera aestiva*. According to Conover et al. (1988b), the eggs of the high latitude *Calanus hyperboreus* are laid at depth but are buoyant and rise to the water/ice interphase where they complete their development.

8.3.3. Egg and Brood Sizes

8.3.3.1. *Estimation of Brood Sizes*

Brood size and fecundity are two different measurements. Brood size is the number of eggs forming a single egg mass in species that carry their eggs attached to the urosome. Such species often produce successive egg masses and fecundity of these species is the sum of the numbers of eggs in all the successive egg masses, that is the total number of eggs produced by the

Table 41 Sinking velocities of subitaneous (S) and diapause (D) eggs.

Species		Density g cm^{-3}	Velocity cm s^{-1}	Authority
Acartia clausi	?		0.0373	Uye, 1980a
A. steueri	?		0.0703	Uye, 1980a
A. tonsa	S	1.066–1.086	0.0150–0.0278	Miller and Marcus, 1994
Calanus finmarchicus	S	1.074 ± 0.0002	0.0393	Salzen, 1956
	S		0.0289–0.0324	Melle and Skjoldal, 1989
Labidocera aestiva	S	1.081–1.133	0.0463	Marcus and Fuller, 1986
	D	1.101–1.181	0.0610	Marcus and Fuller, 1986

female in her lifetime. The determination of brood size and fecundity in species that lay their eggs freely into the sea is much more difficult. Many authors, e.g. Bautista *et al.* (1994), Liang *et al.* (1994), Kiørboe and Sabatini (1995), Poulet *et al.* (1995a), term the numbers of eggs laid during one laying event as the fecundity but this is a single clutch or brood possibly equivalent to a single egg mass of females that carry egg masses. A free-spawning species lays successive clutches of eggs often at intervals of roughly 24 h, as described later when discussing rates of production of eggs.

Runge (1987) distinguishes different stages in the cycle of oogenesis in the ovaries of preserved *Calanus finmarchicus*. This technique has been applied in other crustaceans, e.g. euphausiids (Mauchline and Fisher, 1969), with variable results. The principal problem is discovering the rates of development of oocytes in the ovary and their rates of maturation and subsequent laying. Some oocytes, usually the smaller ones, will not mature to be laid within the current egg mass or clutch with the result that it is often easy to overestimate brood and clutch sizes from counting oocytes.

Daily rates of egg production in free spawners are usually determined by incubating mature females and counting the eggs produced every 24 h. An experimental shipboard egg production chamber in which eggs, to avoid cannibalism, sediment through a mesh into a counting chamber is described by White and Roman (1992a). Rates of egg production are sometimes determined in the natural environment by the so-called egg ratio method. Counts of adult females and eggs are made from quantitative samples collected with plankton nets. The numbers of eggs are then divided by the numbers of adult females to give the numbers of eggs per female. The egg ratio method is subject to some potential errors not least of which is the sampling efficiencies of the nets used to collect two very differently sized components (Siefert, 1994). Clutch sizes determined from incubation are normally greater than from the egg ratio, as instanced by Beckman and Peterson (1986), Peterson and Kimmerer (1994) and Liang et al. (1994). It is tempting to ascribe the differences to predation of eggs in the natural environment or to sinking of diapause eggs out of the depth range of the samplers. This is acceptable only when sampling errors for eggs and adults have been accounted for. Eggs, which are passive, and adults, which are mobile, are patchy in distribution on the sampling scales used and their degrees of patchiness are, to a certain extent, independent of each other.

8.3.3.2. *Size of Eggs*

Free-spawning copepods produce smaller eggs relative to their body volume than those that carry their eggs in egg masses (Mauchline, 1988a). Information on the egg sizes of Euchaetidae, along with sources of data on other species, is given in Table 42.

In the following analysis, the copepods are grouped into the Euchaetidae which form egg masses, other species that form egg masses, all species carrying egg masses combined, and free spawners. The diameter of eggs, relative to the prosome length of the females, is different between those carried in egg masses and those freely spawned (Figure 56). The Euchaetidae and other species with egg masses are identified in Figure 56 although the regression line is given for both combined. The relevant regression data are given in Table 43. Likewise, free-spawned eggs have a smaller volume relative to the volume of the female than those carried in egg masses (Figure 57). Egg diameter and egg volume can be expressed as a percentage of prosome length and body volume respectively and the mean values and their standard deviations are given in Table 44. The ranges of these values for eggs in egg masses and those freely spawned are illustrated in Figure 58.

There is more scatter in the results obtained from freely spawned eggs than in those from eggs that occur in egg masses although all the correlations are significant at the 0.1% level. The large, egg-carrying species *Valdiviella insignis* carries two large eggs (Figure 8B), the sizes of which are identified in Figures 56 and 57. Their diameter represents 15% of prosome length but only 4% of body volume. Eggs in sacs have an average diameter of 11.6% of prosome length compared with the 7.2% for eggs freely spawned (Table 44). Eggs in sacs have a volume equivalent to 1.1 to 1.4% of the body volume of the female compared with a value of 0.4% for freely spawned eggs. Kiørboe and Sabatini (1994, 1995) determined egg and body size on a

Table 42 Prosome length (mm), egg diameter (mm) and average numbers of eggs per egg mass of *Euchaeta* and *Pareuchaeta* spp. (Mauchline, 1992a and unpublished.) Some species were examined from different geographical regions. Comparable measurements for other species can be found in the papers cited in the footnote to this table.

Species	Prosome length	Egg diameter	Egg number
Genus *Euchaeta*			
E. acuta	2.760	0.360	16
	2.948	0.337	17.5
E. concinna	2.235	0.2	20
E. indica	1.81	0.264	9
	1.88	0.25	10
	2.03	0.295	5
E. longicornis	2.008	0.218	15
	2.034	0.248	15
E. marina	2.518	0.283	22
	2.52	0.27	12
	2.764	0.275	11
	2.764	0.276	21
	2.764	0.325	8
E. media	2.635	0.3	19
	2.814	0.297	12.5
E. paraconcinna	1.91	0.272	4
E. pubera	2.981	0.365	9.5
	3.03	0.382	6
E. spinosa	4.67	0.523	9.5
	4.735	0.518	11
Genus *Pareuchaeta*			
P. abbreviata	4.413	0.686	6.5
P. barbata	5.349	0.64	6.33
	5.375	0.64	7.65
	5.424	0.63	7.8
	5.647	0.66	9
P. birostrata	5.81	0.68	10
P. bisinuata	3.726	0.43	5.5
P. confusa	5.212	0.609	8
P. elongata	4.658	0.479	17
	5.3	0.5	25
P. barbata f. *farrani*	7.6	0.77	9
P. glacialis	7.25	0.564	57
P. gracilis	4.451	0.458	16
	4.513	0.45	11.5
P. hanseni	6.337	0.87	7
P. hebes	2.306	0.298	15
P. kurilensis	4.171	0.63	6

Table 42 Continued.

Species	Prosome length	Egg diameter	Egg number
P. norvegica	4.997	0.421	32.5
P. rasa	5.5	0.673	9
P. rubra	5.482	0.668	6
	5.576	0.7	10
P. russelli	2.81	0.29	19.9
P. sarsi	6.564	0.66	18
P. scotti	3.874	0.596	5.5
P. simplex	2.55	0.24	8

Sources: Scott, 1909; Johnson, 1934a, 1961; Sewell, 1947; Marshall and Orr, 1954; Østvedt, 1955; Chiba, 1956; Koga, 1960a,b, 1968; Jacobs, 1961; Grindley, 1963; Matthews, 1964; Bernard, 1965; Gaudy, 1965, 1992; McLaren, 1966; Sazhina, 1968, 1985; Corkett and McLaren, 1969, 1970, 1978; Grice, 1969; Lawson and Grice, 1970, 1973; Valentin, 1972; Zillioux and Gonzalez, 1972; Fulton, 1973; Kasahara et al., 1974; Corkett and Zillioux, 1975; Morioka, 1975; Gibson and Grice, 1976; Grice and Lawson, 1976; Uye and Fleminger, 1976; Dagg, 1978; Goswami, 1978a; Landry, 1978b; Uye et al., 1979, 1982; Checkley, 1980a,b; Uye, 1981, 1983; S. Razouls, 1982, 1985; Castel et al., 1983; Jacoby and Youngbluth, 1983; Abou Debs, 1984; Kimmerer, 1984; Runge, 1984; Lawrence and Sastry, 1985; Reddy and Devi, 1985, 1989, 1990; Arnott et al., 1986; Crawford and Daborn, 1986; Kimoto et al., 1986b; Trujillo-Ortiz, 1986; Vidal and Smith, 1986; Gaudy and Pagano, 1987; Hirche and Bohrer, 1987; Mobley, 1987; Ohman, 1987; Smith and Lane, 1987; Vuorinen, 1987; Ianora and Scotto di Carlo, 1988; McLaren et al., 1988; Peterson, 1988; Devi and Reddy, 1989a, b; Fransz et al., 1989; Frost, 1989; Hirche, 1989b, 1990, 1991; Jonasdottir, 1989; Rijswijk et al., 1989; Marcus, 1990; Paul et al., 1990; Santella and Ianora, 1990; Smith, 1990; Ban and Minoda, 1991; Huntley and Escritor, 1991; Ianora and Santella, 1991; Tiselius et al., 1991; Castel and Fuertet, 1992; Checkley et al., 1992; Diel and Tande, 1992; Ianora et al., 1992a; Uye and Shibuno, 1992; McKinnon and Thorrold, 1993; Park and Landry, 1993; Plourde and Runge, 1993; Hirakawa and Imamura, 1993; Ianora and Poulet, 1993; Kosobokova, 1993, 1994; Lopez et al., 1993; Nielsen et al., 1993; Belmonte and Puce, 1994; Dam et al., 1994; Conway et al., 1994; Kiørboe and Sabatini, 1994; Rayner, 1994; Viitasalo and Katajisto, 1994; Ward and Shreeve, 1995; Madhupratap et al., 1996.

carbon basis and found that eggs in egg masses are larger than those freely spawned (Table 43).

8.3.3.3. *Size of Broods*

Brood size is the number of eggs laid in a single spawning event, i.e. the production of an egg mass by a Euchaetid or a single spawning of a free-spawner. It is assumed that in most free-spawners the eggs ripen in the

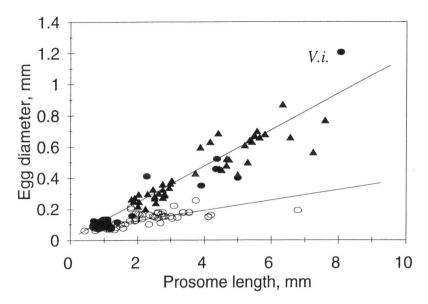

Figure 56 The relationship of egg diameter to prosome length in calanoid copepods of the genera *Euchaeta* and *Pareuchaeta* (shaded triangles), other species carrying egg masses (shaded circles) and in free-spawners (open circles). The regression lines are for all egg mass carriers and for free-spawners. *V.i.*, *Valdiviella insignis*.

oviducts and all such ripe eggs are laid at a single spawning. The volume of a brood is the product of the volume of a single egg and the number spawned. It is easily determined in species that carry egg masses. It is less easily determined in free-spawners because the majority of such determinations must of necessity be made under laboratory conditions. Numbers of eggs in egg masses vary between species. They range from about 5 to 60 in the Euchaetidae (Table 42) and other egg-carrying genera. The number of eggs spawned by free-spawning females in a single spawning event also varies between species, usually between 3 and 50. These spawning events often take place once in each 24 h period and are synonymous with the rates of egg production measured by many authors as eggs female^{-1} d^{-1}.

The relationship between brood volume and body volume in species that produce egg masses and in free-spawners is examined in Figure 59 and Tables 43 and 44. There is a significant correlation between brood volume and body volume in species that produce egg masses but none in the free-spawners. There are, however, a lack of data on large-sized free-spawners such as species of the Megacalanidae, the largest species examined here being *Calanus hyperboreus*. Brood volume averages about 12% body

Table 43 Relationships between egg size and body size in species of Euchaetidae and other species with egg masses, separately and combined, and in free spawning species. Values for the x coefficient (m), constant (c), and the correlation coefficient (r) are given along with the number of observations and the corresponding number of species.

Relationship	m	c	r	n	Number of species
Egg diameter (mm) on Prosome length (mm)					
Euchaetidae	0.103	0.048	0.906	44	29
Others with egg masses	0.132	−0.021	0.959	43	29
Euchaetidae and Others	0.115	0.002	0.951	87	58
Free spawners	0.032	0.060	0.768	86	57
Log Egg volume (mm^3) on log Body volume (mm^3)					
Euchaetidae	0.866	−1.914	0.928	44	29
Others with egg masses	0.912	−2.014	0.940	43	29
Euchaetidae and Others	0.931	−1.974	0.972	87	58
Free spawners	0.541	−2.841	0.866	86	57
Brood volume (mm^3) on Body volume (mm^3)					
Euchaetidae	0.098	0.047	0.891	45	29
Others with egg masses	0.039	0.019	0.992	38	28
Euchaetidae and Others	0.080	0.073	0.856	83	57
Free spawners	0.005	0.045	0.311	48	27
Log Egg size (μg C) on log Body size (μg C)*					
Species with egg masses	0.930	−1.841	0.933	21	
Free spawners	0.621	−1.859	0.866	41	
Log Weight specific fecundity (that is the number of eggs as μg C female^{-1}day^{-1}) related to log body weight of female (as μg C).*					
Species with egg masses	−0.260	−0.850	0.921	10	
Free spawners	−0.262	−0.474	0.752	35	

*Kiørboe and Sabatini (1995)

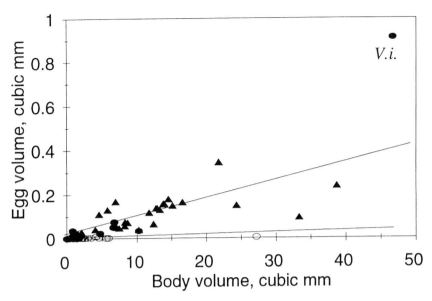

Figure 57 The relationship between the volume of the egg and body volume of the female in calanoid copepods of the genus *Euchaeta* and *Pareuchaeta* (shaded triangles), other species carrying egg masses (shaded circles) and in free-spawners (open circles). The regression lines are for all egg mass carriers and for free-spawners. *V.i.*, *Valdiviella insignis*.

volume in Euchaetidae but is highly variable in other egg-carrying genera (Table 44). Much of the variation is caused by small species in the genera *Eurytemora* and *Pseudodiaptomus* where egg masses often represent about 40% of body volume. Estimated brood size in free-spawners is about 10.5% suggesting that the single spawning event of such a calanoid is equivalent to an egg mass of, for example, a euchaetid species. There is an inverse relationship with body weight when the brood or clutch size is transformed to a weight-specific value as illustrated by Kiørboe and Sabatini (1995), whose equations are given in Table 43.

The above correlations are made between species but egg size and number relative to body size can also vary within a species (Pond *et al.*, 1996). The diameter of the eggs of crustaceans such as mysids varies seasonally, being smaller in summer than in winter and spring (Mauchline, 1980). Lawrence and Sastry (1985) found such seasonal variation in the eggs of *Tortanus discaudatus* and Crawford and Daborn (1986) in the eggs of *Eurytemora herdmani* but no such relationship was demonstrated in *Pareuchaeta norvegica* by Nemoto *et al.* (1976) nor in *Centropages abdominalis* by Liang *et al.* (1994). Egg diameter is related to prosome

Table 44 The diameter of the egg relative to prosome length (PL) of the female, the volume of the egg and of the egg mass or clutch relative to the volume (BV) of the female, expressed as a percentage; standard deviations of the means are given.

	Egg diameter as % of PL	Egg volume as % of BV	Brood volume as % of BV
Euchaetidae	11.632 ±1.827	1.141 ±0.569	12.219 ±5.001
Others with egg masses	11.593 ±2.387	1.427 ±0.901	22.061 ±17.661
Euchaetidae and Others	11.613 ±2.110	1.282 ±0.761	16.725 ±13.360
Free spawners	7.219 ±2.221	0.396 ±0.407	10.438 ±11.699

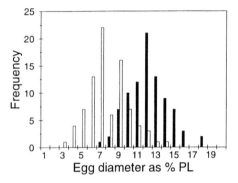
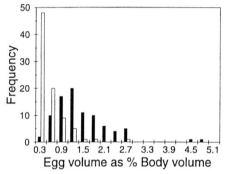

Figure 58 The frequency distributions of, left, egg diameter as a percentage of prosome length, and, right, egg volume as a percentage of body volume of the female in species carrying egg masses (shaded histograms) and free spawners (open histograms).

length of *Paracalanus* sp. according to Uye and Shibuno (1992) and Hart and McLaren (1978) demonstrated such a relationship in *Pseudocalanus* sp. but there is a marked seasonal component in their data.

The numbers of eggs in a brood or egg mass and in a clutch or single spawning event increases with increasing prosome length or weight of the female within a number of species. A positive correlation has been found in the following species that carry egg masses: *Pareuchaeta norvegica* by Nemoto et al. (1976) and C.C.E. Hopkins (1977); *Eurytemora affinis* by Crawford and Daborn (1986) and Ban (1994): *Gladioferens imparipes* by Rippingale and Hodgkin (1974); *G. pectinatus* by Arnott et al. (1986); *Pseudocalanus* spp. by Corkett and McLaren (1969), Corkett and Zillioux (1975) and Ohman (1985); *Pseudodiaptomus marinus* by Uye et al. (1982).

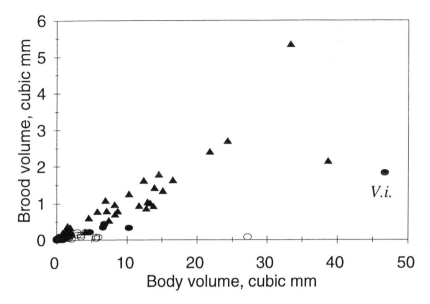

Figure 59   The relationship between brood or clutch volume and volume of the female in calanoid copepods of the genus *Euchaeta* and *Pareuchaeta* (shaded triangles), other species carrying egg masses (shaded circles) and in free spawners (open circles). *V.i.*, *Valdiviella insignis*.

Likewise, a positive correlation has been demonstrated in the clutch sizes of the free-spawning species: *Acartia clausi* by Landry (1978b); *A. tonsa* by Durbin *et al.* (1983); *Calanus glacialis* by Hirche (1989b) and Smith (1990), *C. pacificus* by Runge (1984) and Mobley (1987); and *Neocalanus plumchrus* by Fulton (1973). No correlation, however, was found in *Calanus marshallae* by Peterson (1988), *C. pacificus* by Mullin (1991a), nor in *Centropages typicus* by Dagg (1978). These observations have to be made within restricted seasonal time-windows, if not within single samples, because there is seasonal variation in brood and clutch sizes in some species resulting from food-limiting and non-limiting conditions (e.g. Frost, 1985). There are also seasonal changes in body size in many species. Thus, seasonal components have to be eliminated from the observations.

8.3.3.4. *Seasonal Variation*

The numbers of eggs in an egg mass or clutch vary seasonally and partially correspond to seasonal changes in the body sizes of adult females. This has been quantified in the following egg mass carriers: *Pareuchaeta elongata* by Lewis and Ramnarine (1969); *Eurytemora affinis* by Bradley (1986b) and

Castel and Feurtet (1992); *E. herdmani* by Crawford and Daborn (1986); *Pseudocalanus minutus* by Koeller *et al.* (1979); *Pseudodiaptomus marinus* by Uye *et al.* (1982). Clutch size in free-spawners also varies seasonally as found in *Acartia clausi* by Ianora and Scotto di Carlo (1988), Ianora and Buttino (1990), Tiselius *et al.* (1991) and Kiørboe and Nielsen (1994), *A. hudsonica* by Sullivan and McManus (1986), *A. grani* by Rodríguez *et al.* (1995), *A. tonsa* by Beckman and Peterson (1986), Sullivan and McManus (1986) and Gaudy (1992), *Calanoides carinatus* by Armstrong *et al.* (1991), *Calanus finmarchicus* by Diel and Tande (1992) and Kiørboe and Nielsen (1994), *C. helgolandicus* by Pond *et al.* (1996), *C. marshallae* by Peterson (1988), *Centropages hamatus* by Tiselius *et al.* (1991) and Kiørboe and Nielsen (1994), *C. typicus* by Smith and Lane (1987), Ianora *et al.* (1992a) and Kiørboe and Nielsen (1994), *Paracalanus parvus* by Kiørboe and Nielsen (1994), and *Temora longicornis* by Fransz *et al.* (1989), Peterson and Kimmerer (1994) and Kiørboe and Nielsen (1994). This seasonality in egg mass and clutch size usually reflects a build-up to an optimal time for egg laying and a tailing-off from this optimum.

There may also be variations in the brood and clutch sizes of a species when examined in different geographical regions. The within-sample variation in the estimates of clutch size at any single location, taking seasonal and trophic resources into account, makes comparisons very difficult. This is reinforced by Tiselius *et al.* (1995) who found that *Acartia tonsa* in culture, in a field population in Öresund, Denmark and in a field population in the eastern United States behaved differently; egg production in the United States population fluctuated in accordance with a strong aggregation in patches of food. Regional differences in diel and aggregational behaviour will contribute to variation in daily egg production between sites.

8.3.4. Fecundity

The interval between successive egg masses is longer than that for successive clutches of free-spawning species because the eggs of the former are larger than free-spawned eggs and a second egg mass cannot be produced until the nauplii have hatched out from the previous one. Egg production is measured in terms of numbers of eggs female^{-1} d^{-1} in both species with egg masses and free spawners (Table 45). The rates are about 7.5 times higher in free spawners than in species with egg masses (Kiørboe and Sabatini, 1995). The weight-specific fecundity, that is the weight of eggs day^{-1} unit female body weight^{-1}, decreases with female body weight, the regression equations being given in Table 43. Clutches of eggs of *Calanus finmarchicus* are spawned once every 24 h, approximately, in British waters but only once every 2 or 3 d in high latitudes (Diel and Tande, 1992). Tourangeau and Runge (1991) found that *C. glacialis* produced clutches at 3 to 12 d intervals in May but at 3 or 4 d intervals in June under the ice in Hudson Bay while Hirche (1989b) found intervals of 1 to 3 d at 0 °C in the laboratory. Hirche and Niehoff (1996) found that a single *C. hyperboreus* spawned seven times in two months at 0 °C, producing more than 1000 eggs. The small coastal species *Centropages typicus* produced a clutch of 53 eggs and *Temora stylifera* a clutch of 33 eggs once every 4 to 6 d according to S. Razouls (1982). Such variation in the rates of clutch production must reflect rates of development of ripe eggs from the germinal sites of the ovary. Such rates may be temperature- or food-dependent.

Some estimates of fecundity of different species of calanoids are given in Table 46. They are, of necessity, laboratory estimates and are therefore approximate. Species with egg masses in Table 46 have produced 9 to 10 successive broods in the laboratory over periods of about 3 months but whether they normally survive this length of time in the natural environment is often questionable. Likewise, some of the maximal estimated fecundities of free-spawners (Table 46) may also be unrealistic. Fecundities of species carrying egg masses is undoubtedly less, on average, than that of free-spawners. Probably the most extreme case is that of the bathypelagic *Valdiviella insignis* living at low temperature and producing only two eggs per brood. Development time of the embryo and sequential broods is liable to be depressed by the low temperature prevalent at bathypelagic depths. Park and Landry (1993), however, caution that simple temperature adjustments of development times of temperate species may not accurately estimate development times of tropical species; they may also produce biased projections for deep-sea species. This results from the different nutritional environments present and the metabolic adaptations of the species.

Species with egg masses probably have fecundities within the range 10 to 200 eggs, assuming an average of about four successive broods, while free spawners probably produce 30 to 700 eggs. Conover (1988) concludes that fecundity of the different species in the genera *Calanus* and *Neocalanus* is about the same, about 450 to 700 eggs. These are relatively large-sized species living in middle to high latitudes and their fecundities would be expected to be at the upper end of the range of free-spawners. The effects of crowding on the fecundity of *Centropages typicus* were examined by Miralto et al. (1996). Population densities of the order of $10^6 \, m^{-3}$ caused a reduction in the numbers of eggs produced. Such high densities are really only equivalent to those found in swarms (Table 66, pp 426–427), a situation in which eggs are unlikely to be laid. Swarms are usually present during daylight and probably function as protection against visual predators. They tend to disperse at night when spawning is probably more common (Table 39).

Table 45 Production of eggs by egg-carrying and free-spawning calanoids expressed as numbers of eggs per day and as an estimate of production by expressing the dry weight of eggs d^{-1} as a proportion of the female's body dry or carbon weight. The temperature (T°C) at which the measurements were made is also given.

Species	T°C	Eggs day^{-1}	Daily production	Authority
Egg-carrying species				
Eurytemora affinis	14		0.10	Escaravage and Soetaert, 1993
E. herdmani	9		0.45	McLaren and Corkett, 1981
	9.5		0.33	McLaren and Corkett, 1981
	13		0.68	McLaren and Corkett, 1981
E. velox	10–20	2–11	0.04–0.16	Gaudy and Pagano, 1987
Pareuchaeta antarctica			0.154	Ward and Robins, 1987
P. norvegica	15	2.77	0.023	C.C.E. Hopkins, 1977
Pseudocalanus elongatus	15	4.75	0.078	Corkett and Zillioux, 1975
	15	5.53	0.09	Paffenhöfer and Harris, 1976
	15	2.7	0.11	Frost, 1985
P. minutus	15	4	0.10	Dagg, 1977
P. moultoni	15	7.5	0.08	Jonasdottir, 1989
P. newmani	15	5.6	0.06	Jonasdottir, 1989
Pseudocalanus sp.	8–15		0.10–0.16	Corkett and McLaren, 1978
	0.5–6.0	7.5	0.048–0.054	Vidal and Smith, 1986
Pseudodiaptomus marinus	10	3–17	0.04–0.26	Uye *et al.*, 1982
	20	12	0.183	Uye *et al.*, 1983

Species	Temp	Value	Reference
Free-spawning species			
Acartia californiensis	15	20	0.15	Trujillo-Ortiz, 1990
A. clausi	15	10.6	0.14	Iwasaki et al., 1977
	15	40	0.35	Landry, 1978b
	15	23	0.38	Uye, 1981
	17.5	40–50	0.61 ± 0.05	Saiz et al., 1992a
	17.5	40–50	0.70 ± 0.05	Saiz et al., 1992a
	17.5	40–50	0.36 ± 0.04	Saiz et al., 1992a
A. grani	15	20	0.32	Sekiguchi et al., 1980
A. hudsonica	summer		0.09–0.283	Sullivan and McManus, 1986
	winter		0.09–0.236	Sullivan and McManus, 1986
A. omorii	10	30	0.19	Uye, 1981
	13		0.40	Ayukai, 1987
A. steueri	15	17–37	0.29	Uye, 1981
A. tonsa	15	18–26	0.19	Corkett and Zillioux, 1975
	15	50	0.29	Dagg, 1977
	15	34	0.47	Kiørboe et al., 1985a
	15	22	0.28	Ambler, 1986
	summer		0.145–0.718	Sullivan and McManus, 1986
	winter		0.054–0.095	Sullivan and McManus, 1986
			0.2–0.7	White and Roman, 1992a
			0.68	Dam et al., 1994
A. tranteri	20	40–50	0.37 ± 0.03	Saiz et al., 1992a
Acartia spp.	20	12	0.34	Kimmerer and McKinnon, 1987a
Acrocalanus gibber	5		0.20	Landry, 1978b
A. gracilis	30		0.13–0.41	McKinnon and Thorold, 1993
	20		0.03–0.21	McKinnon and Thorold, 1993
Calanoides acutus	3.5	26	0.04–0.05	Ward and Shreeve, 1995
C. carinatus	15	20	0.32	Borchers and Hutchings, 1986
		41.6	0.66	Armstrong et al., 1991

Table 45 Continued.

Table 45 Continued.

Species	T °C	Eggs day^{-1}	Daily production	Authority
Calanus australis	15	40	0.13	Attwood and Peterson, 1989
C. finmarchicus		42	0.045	Hirche, 1990
		40–70	0.055	Hirche, 1989b, 1990
		11–45	0.05–0.08	Ohman and Runge, 1994
C. glacialis		14.6	0.033	Hirche and Bohrer, 1987
		50–80	0.049	Hirche, 1989b
	−2.3		0.013	Tourangeau and Runge, 1991
		68	0.077	Smith, 1990
		67	0.08	Hirche *et al.*, 1994
C. helgolandicus		<33	0.16	Pond *et al.*, 1996
C. hyperboreus	4–6	17	0.01	Conover, 1967
	−2.3	57	0.042	Smith, 1990
	0	51–150		Hirche and Niehoff, 1996
C. marshallae	0.5–6.0	75	0.06	Vidal and Smith, 1986
C. pacificus	15	42	0.12	Peterson, 1988
		33–46	0.10–0.14	Runge, 1984
	15	28	0.09	Razouls *et al.*, 1991
C. propinquus	0	20–30	0.024	Kosobokova, 1994
C. simillimus	3.5	25	0.04–0.05	Ward and Shreeve, 1995
Centropages abdominalis	9–20	39–142	0.13–0.48	Liang *et al.*, 1994
	9–20		0.19–0.70	Liang *et al.*, 1994
	10		0.30	Liang *et al.*, 1994
C. hamatus	15		0.31	Klein Breteler *et al.*, 1982
	17	109	0.248	Fryd *et al.*, 1991

C. typicus	15	5–230	0.01–0.16	Dagg, 1978
	8–20	28–55	0.03–0.06	Smith and Lane, 1985
	9–18.5	23–76	0.01–0.06	Smith and Lane, 1987
	15	60	0.16	Davis and Alatalo, 1992
	17		0.339	Fryd et al., 1991
Eucalanus bungii	0.5–6.0	129	0.09	Vidal and Smith, 1986
Neocalanus plumchrus		56	0.04	Fulton, 1973
Paracalanus parvus	18	18–65	0.37	Checkley, 1980a
Paracalanus sp.	10	6	0.08	Uye and Shibuno, 1992
Rhincalanus gigas	3.5	33	0.04–0.05	Ward and Shreeve, 1995
Sinocalanus tenellus	10	19	0.12	Kimoto et al., 1986a
	20	60	0.37	Kimoto et al., 1986b
Temora longicornis	15	16.5	0.044	Corkett and Zillioux, 1975
	15	32	0.069	Harris and Paffenhöfer, 1976a
			0.77	Peterson and Dam, 1996
T. stylifera	15	35	0.17	Abou Debs and Nival, 1983
Undinula vulgaris		20–67	0.021–0.051	Park and Landry, 1993

Table 46 Estimates of fecundity of calanoid copepods. Extreme values for free spawners, usually derived from culturing for more than one month under conditions of plentiful food, are given in parentheses.

Species	Number of eggs	Authority
Species with egg masses		
Eurytemora affinis	117	Katona, 1975
E. hirundoides	129	Poli and Castel, 1983
E. velox, winter	311	Gaudy and Pagano, 1987
spring	109	Gaudy and Pagano, 1989
Pseudocalanus elongatus	106	Corkett and McLaren, 1969
	567	Sazhina, 1971
P. minutus	180	Corkett and McLaren, 1969
Pseudodiaptomus coronatus	113	Jacobs, 1961
Free spawners		
Acartia clausi	260 (1281)	Sazhina, 1971
A. tonsa	435	Zillioux and Gonzalez, 1972
	718	Parrish and Wilson, 1978
Anomalocera patersoni	1050	Sazhina, 1971
	25–143	Ianora and Santella, 1991
C. finmarchicus, polar	743	Hirche, 1990
Atlantic	586 (3101)	Hirche, 1990
St Lawrence	913	Plourde and Runge, 1993
N Norway	600	Diel and Tande, 1992
C. glacialis	340 (1274)	Hirche, 1989b
C. helgolandicus	420	Sazhina, 1971
	691	Gatten *et al.*, 1980
C. hyperboreus	450 (3800)	Conover, 1967
	300 (>1000)	Hirche and Niehoff, 1996
C. pacificus	613–691	Mullin and Brooks, 1967
	2267	Paffenhöfer, 1970
Centropages ponticus	325	Sazhina, 1971
C. typicus	(>1800)	Davis and Alatalo, 1992
Labidocera brunescens	440	Sazhina, 1971
Neocalanus plumchrus	535	Fulton, 1973
	123–719	Kosobokova, 1994
N. tonsus	285–453	Ohman, 1987
Paracalanus parvus	390	Sazhina, 1971
	560	Checkley, 1980a
Pontella mediterranea	880	Sazhina, 1971
Rhincalanus nasutus	103–355	Mullin and Brooks, 1967
Sinocalanus tenellus	2531	Kimoto *et al.*, 1986b

8.4. FOOD AVAILABILITY AND EGG PRODUCTION

Rates of egg production by a female are influenced by a number of factors, the principal one being the availability of food. Egg production, measured as eggs female^{-1} d^{-1}, increases as food available increases to achieve an asymptotic level (e.g. Durbin et al., 1983; Dam et al., 1994). Quality as well as quantity of food available is important, high quality encouraging production of successive egg masses and clutches (Cahoon, 1981; S. Razouls, 1981, 1982; Abou Debs and Nival, 1983; Durbin et al., 1983; Donaghay, 1985; Kiørboe et al., 1985b; Runge, 1985; Ambler, 1986a; Beckman and Peterson, 1986; Kimoto et al., 1986b; Razouls et al., 1986, 1991; Bellantoni and Peterson, 1987; Hirche and Bohrer, 1987; Peterson and Bellantoni, 1987; Peterson et al., 1988; Attwood and Peterson, 1989; Rijswijk et al., 1989; White and Dagg, 1989; Nival et al., 1990; Sciandra et al., 1990; Støttrup and Jensen, 1990; Armstrong et al., 1991; Tiselius et al., 1991; Walker and Peterson, 1991; Davis and Alatalo, 1992; Diel and Tande, 1992; Kleppel, 1992; Uye and Shibuno, 1992; Ban, 1994; Bautista et al., 1994; Dam et al., 1994; Jónasdóttir, 1994; Nakata et al., 1994; Laabir et al., 1995a; Ianora et al., 1995; Jónasdóttir et al., 1995; Kleppel and Burkart, 1995; Miralto et al., 1995; Rodríguez et al., 1995; Pond et al., 1996).

Herbivorous copepods are often associated with phytoplankton patches and blooms. Correlation, however, between the co-occurrence of phytoplankton and egg laying in copepods is difficult to prove in practice, though shown by some observers (see Williams and Lindley, 1980a,b; Bird, 1983; Richardson, 1985; Kiørboe et al., 1990; Bautista et al., 1994; Nielsen and Hansen, 1995) and not others (Melle and Skjoldal, 1989; Diaz Zaballa and Gaudy, 1996; Pond et al., 1996). The lack of correlation may be because chlorophyll *a* is an imperfect index of the availability of phytoplankton or, more likely, that the copepods are also feeding on organisms other than phytoplankton. Conversely, egg production may be affected by food availability in some species and not in others. Frost (1985), for example, found that bloom conditions of phytoplankton accelerated egg production in *Calanus pacificus* but not in the co-occurring *Pseudocalanus* sp. in Dabob Bay, Washington. The co-occurring *Calanus finmarchicus* and *Pseudocalanus elongatus* in Loch Striven, western Scotland, responded in the same way.

A few observations have shown that environmental temperature rather than phytoplankton abundance controls egg production (Uye, 1981; Abou Debs and Nival, 1983; Runge, 1985; Kiørboe et al., 1988; Hirakawa, 1991; White and Roman, 1992a). Daily rates of egg production increase with temperature to a maximum but then decrease with further increases in temperature. Thus, there is an optimal range of temperature for egg production of a species which, in practice, is less limiting to egg production than the ambient food supplies.

8.4.1. Quality and Quantity

In describing food and feeding of copepods earlier, it was noted that defining the diets of copepods, even so-called herbivorous species, was difficult because species do not appear to be exclusively herbivorous, omnivorous, carnivorous or detritivorous but any one species can only be defined as predominantly one or the other. A predominantly herbivorous species can feed on phytoplankton and/or heterotrophic microplankton, including ciliates, dinoflagellates and rotifers (Stoecker and Egloff, 1987; Ohman and Runge, 1994). Dam et al. (1994) conclude that heterotrophic feeding contributes some 40% of the energy required for egg production in *Acartia tonsa* and there is little reason to assume that this species is peculiar. A modelling investigation of egg production in the Irish Sea (Prestidge et al., 1995) indicated that energy was being obtained from food other than phytoplankton.

A diet of *Skeletonema costatum* resulted in sterility or death of *Temora stylifera*; Ianora et al. (1995), Poulet et al. (1994, 1995b) and Laabir et al. (1995a) suggest that a diatom diet may produce compounds that inhibit embryogenesis and question whether diatoms as a food really regulate the production of copepods. Eggs of *Acartia clausi* exposed to high concentrations of extracts of diatoms had a low hatching success (Ianora et al., 1996) while those of *A. tonsa* were not affected (Jónasdóttir and Kiørboe, 1996). *Calanus helgolandicus* feeding on dense cultures of *Phaeodactylum cornutum* in the laboratory produced eggs with an increasingly low hatching success (Poulet et al., 1995b). Deformities were also present in nauplii of those eggs that hatched. High diatom concentrations increased fecundity but inhibited egg hatching while the reverse was true at low concentrations (Chaudron et al., 1996). Diatoms as food or extracts of diatoms reduced egg viability in *Calanus pacificus* (Uye, 1996). Laabir et al. (1995a), working in coastal waters off Roscoff and in the English Channel, found highest occurrence of abnormal eggs and nauplii of *Calanus helgolandicus* during spring and mid-summer, periods when diatoms dominated the diet. Jónasdóttir (1994) found that age of the diatom culture influenced ability of the eggs of *Acartia* species to hatch.

Donaghay (1985) emphasizes that feeding history immediately prior to egg production, coupled with ambient food concentrations, influences the numbers of eggs produced. Egg production is not instantaneous and it is logical that modal feeding conditions may range from optimal to suboptimal throughout the course of time required for ovarian development. Successful oogenesis coupled with an optimal nutritional state of the female, a result of feeding history, must influence the response of the copepod to ambient conditions of food quantity and quality by affecting the timing and number of eggs laid. Attwood and Peterson (1989) suggest that the intermittent food supply in the upwelling region of the Benguela Current may depress egg production, and so population development, of *Calanus australis*. The successful coastal *Acartia tonsa* lives in an environment with a patchy distribution, in time and space, of food. Fluctuating food supply affects its rates of egg production. Calbet and Alcaraz (1996) found that increased periods of starvation, between 3 and 5 d, resulted in increased periods of recovery before normal rates of egg production were resumed. Alternating periods of low and high concentrations of food on 12 and 24 h cycles did not repress egg production but a 48 h cycle did. Thus, this species has very limited buffering against low food availability.

Båmstedt (1988b) demonstrated considerable variation in characteristics such as body weight, reproductive state and metabolic activity between individuals within various species of copepods in Norwegian fjords and the Barents Sea. He concludes that part of this variability may be genetical but some, maybe the greater part, may result from the nutritional history of the individuals. Cultured *Centropages typicus* under conditions of plentiful food supply, maintained their egg production rate at 60 eggs female^{-1} d^{-1} throughout a reproductive period of 30 days at which time the experiment was terminated by Davis and Alatalo (1992). This resulted in a fecundity of 1800 eggs and it could have been larger had the experiment been extended, there being no age-related effect apparent.

The chemical composition of the food has not been investigated relative to egg production to any degree (Pond *et al.*, 1996). Jónasdóttir *et al.* (1995) and Jónasdóttir and Kiørboe (1996) found that fatty acids with high w3:w6 ratios and low 20:5 to 22:6 ratios are required by *Acartia hudsonica*, *A. tonsa* and *Temora longicornis* for production of eggs.

There is a lag time between food being ingested and being converted into production of eggs. This probably varies between species (Kiørboe *et al.*, 1985b; Hirche and Bohrer, 1987; Kiørboe and Nielsen, 1990; Tester and Turner, 1990) and individuals so giving rise to some of the variation found between the numbers of eggs produced per unit time. Tester and Turner (1990) estimated the lag time to be 16 to 17 h in *Centropages velificatus* and *Labidocera aestiva* but 67 to 91 h in *Anomalocera ornata* and *Centropages typicus*. Marcus (1988) found diel changes in gonad size, egg production and food availability; this probably occurs in most migrating copepods and can be linked with the fact that many copepods spawn at night, some towards dawn (Table 39). The results of incubation experiments to determine how many eggs a female produces per day are influenced by the reproductive status of the females at the time of the experiment; some females will be gravid while others may not be (Ianora, 1990).

Food may be a limiting factor on egg production as Tourangeau and Runge (1991) suggest is the case in populations of *Calanus glacialis* living under the ice in Hudson Bay and dependent upon blooms of ice microalgae. Jónasdóttir (1989) found that *Pseudocalanus newmani* has its maximal rate of egg production at much lower food concentrations than *P. moultoni*. This infers that food-limiting conditions will differ between species and some will be more tolerant to low concentrations than others.

8.4.2. Stored Lipids

In some high latitude species, egg production is more or less independent of ambient food concentrations. These species overwinter, often as copepodid V, and mobilize lipid stores for egg production. Huntley and Escritor (1991) suggest that this happens in the Antarctic *Calanoides acutus* and Smith (1990) finds it in *Calanus glacialis* and *C. hyperboreus* in the Fram Strait, Greenland Sea. Hirche and Kattner (1993) suggest that there may be two components in the population of female *C. glacialis* in the western Barents Sea; the first comprises overwintering two-year-old females dependent on lipid stores for egg production in the spring and, the second, one-year-old females, derived from overwintering CVs that depend on ambient food supplies for egg production in the summer. Ohman (1987) draws similar conclusions in a study of *Neocalanus tonsus* in which mesopelagic dwelling winter females utilized lipids for egg production while epipelagic dwelling spring females were dependent on ambient food supply.

8.4.3. Rates of Egg Production

Daily rates of egg production for a variety of species are given in Table 45 for species that carry their eggs and for free-spawners. The rates for those that carry their eggs are lower, ranging from about 0.05 to 0.68, with an average of about 0.17. Free-spawners have a higher average rate of production of about 0.22. Peterson and Dam (1996) state that this value averages 0.37 in terms of nitrogen in herbivorous species, assuming that all assimilated nitrogen is channelled into egg production in adult females.

8.5. MORTALITY OF EGGS

Mortality of calanoid copepod eggs in the natural environment is usually considered to be heavy through predation and other causes. Ianora *et al.*

(1992a) demonstrate seasonal variation in the viability of hatching eggs of *Centropages typicus*. There were also differences between years; they determined a mean annual value for percentage of viable eggs in 1989 as 73 ± 28% and for 1990 as 84 ± 22%. The lack of remating of the female after production of the initial clutch or clutches of eggs is often blamed for the appearance of non-viable eggs. Ianora *et al.* (1995) and Miralto *et al.* (1995), however, suggest that an alternative reason may be the presence of inhibitory compounds derived from a diet of diatoms that inhibit embryogenesis.

Marcus (1984) and Marcus and Schmidt-Gegenbach (1986) found that subitaneous and diapause eggs of *Labidocera aestiva* passed through the guts of the predatory polychaetes *Capitella* sp. and *Streblospio benedicti* with no effect on their viability. These authors suggest that deposition of these eggs on the sea-bottom in faecal pellets of such benthic predators would enhance their potential for hatching. The survival of subitaneous eggs of *Eurytemora herdmani* following ingestion and defaecation by the Atlantic silversides *Menidia menidia* exceeded 90% (Redden and Daborn, 1991). More recently, Flinkman *et al.* (1994) fed female *Eurytemora affinis*, with egg masses attached to their genital somites, to Baltic herring and found that nauplii hatched from the faeces of the herring. Hatching success was estimated to be at about the 60% level. After passage through the gut of larval turbot, 74% of eggs of *Eurytemora affinis*, 64% of eggs of *E. velox*, and 20% of eggs of *Pseudocalanus elongatus* were viable (Conway *et al.*, 1994); gut passage time through the larval fish varied between 2 and 6 h.

These results may suggest that the greatest sources of mortality of eggs will arise from predators such as crustaceans that macerate their food.

8.6. HATCHING OF EGGS

Hatching of the eggs of a variety of calanoid copepods is described and figured by Marshall and Orr (1954). Davis (1968) reviews later descriptions but little new information has been added. The nauplius is visible within the egg membranes and its limbs begin to twitch occasionally in the period immediately prior to hatching. Individual egg membranes are not discernible at this stage. At hatching, the outer membrane, which is usually thicker than the inner, splits and the nauplius in the inner membrane bulges outwards through the split. The inner membrane containing the nauplius swells so that it is much larger than the original egg. The split outer membrane separates from the swollen inner one which is now spherical. The nauplius within it twitches actively and ruptures this now diaphanous membrane and swims away. The swelling of the inner membrane is considered to be effected by osmotic forces that cause the outer membrane to split. Yang (1977) found that the inner membrane of hatching nauplii of *Parvocalanus crassirostris* tightly enclosed the nauplius when the outer membrane was split and sloughed off. Only then did this inner membrane swell.

Marshall and Orr (1954) point out that the eggs contained in egg masses often change colour as they develop. This is most obvious in a *Pareuchaeta* species such as *P. norvegica*. Eggs that are about to be extruded from the female gonoducts are dark blue in colour and are this colour once they are laid and attached to the female. Just prior to hatching, however, they turn orange in colour. Such orange egg masses immediately hatch if removed from the female and placed in a watch glass of warm sea water.

The factors controlling development and hatching of eggs of calanoid copepods are slightly different depending upon whether the eggs are subitaneous or diapause and are laid freely into the sea or are retained in egg masses attached to the urosome of the female.

Temperature is the primary controlling factor of dormancy of diapause eggs (Grice and Marcus, 1981; Ban and Minoda, 1991). Diapause eggs tend to be produced during periods of falling environmental temperatures at the end of the growing season. Those of *Acartia hudsonica* will not hatch at the ambient temperatures at which they are laid but only after subjection to periods of low temperature (Sullivan and McManus, 1986). Johnson (1967) showed that cooling of the eggs of *Tortanus discaudatus* delayed hatching but that immediately they were warmed hatching started. Grice and Marcus (1981) review the effects of temperature on the eggs of 16 of the species that produce diapause eggs. They also point out that these eggs are tolerant of salinity changes but are sensitive to the low oxygen concentrations that they often meet within sediments. Low oxygen concentrations can delay embryological development (Lutz *et al.*, 1992, 1994; Marcus and Lutz, 1994).

8.7. CONCLUDING REMARKS

The response of increased egg production to increased food supply has a lag time. This is usually measured experimentally in incubation vessels where concentrations of food are decreased and increased and the resulting numbers of eggs laid determined by counting. The radioactive labelling of food and its time course through the metabolic system has rarely been used. ^{14}C-labelled phytoplankton was used by Smith and Hall (1980) who found that the tracer was present in ovarian tissue after a delay of 24 h.

Gravid females of many species readily lay their eggs in the laboratory.

A recent development has been the routine incubation on board ship of freshly caught adult females. The numbers and rate of production of the eggs is determined and equated to production of the population. This is based on the assumption that adult females do not grow in body size but the equivalent of that energy is directed into production of the eggs. It also assumes that this growth rate is equal to the growth rates of the actively growing earlier copepodid stages. Estimates of daily egg production in terms of female body dry weight or carbon content are given in Table 45. These will be discussed further when examining the growth rates of the copepods in Chapter 9.

9. Growth

9.1. Laboratory Culture of Copepods 297
9.2. Development Time of Eggs 298
9.3. Growth 314
    9.3.1. Growth Process 314
    9.3.2. Body Size Parameters 327
9.4. Growth Rates 338
9.5. Longevity 339
9.6. Concluding Remarks 344

Calanoid copepods, like other crustaceans, increase their body size by moulting. In the vast majority, a first nauplius (NI) hatches from the egg and develops through five moults to the sixth nauplius (NVI). Species in the genus *Pseudodiaptomus* omit the NI, a NII hatching directly from the egg so that these species have only five successive nauplii. The NVI moults to the first copepodid (CI) which passes through five successive moults to become the CVI or adult stage.

Studies of growth and development of copepods can be made by several methods:

a. time series sampling of field populations and identification of the nauplii, copepodids and sequential cohorts and/or generations;
b. laboratory rearing of species;
c. new techniques, not yet developed, using specific enzyme activities and molecular biological techniques, examples of such techniques being the measurement of the enzyme aspartate transcarbamylase (Hernández-León *et al.*, 1995) and DNA polymerase activities.

Food availability in conjunction with environmental temperature influence body size of copepods. McLaren (1969) argues that only temperature directly affects size, food being only relevant when it is inadequate. This is not supported by Ban (1994) who found that food availability for *Eurytemora affinis* in Lake Ohnuma, Japan is more important than temperature in influencing growth.

9.1. LABORATORY CULTURE OF COPEPODS

Historical aspects of the culture of marine calanoid copepods are reviewed by Paffenhöfer and Harris (1979). Culturing through multiple generations finally succeeded in the 1960s but the species were neritic in habit, or in the case of *Pseudodiaptomus coronatus* (Jacobs, 1961), not strictly planktonic. The first truly pelagic species to be cultured were *Rhincalanus nasutus* by Mullin and Brooks (1967) and *Calanus helgolandicus*, now recognized as *C. pacificus*, by Paffenhöfer (1970).

Descriptions of apparatus for culturing and observing copepods are given by Paffenhöfer and Harris (1979), Yassen (1981), Omori and Ikeda (1984), Arndt *et al.* (1985), Jerling and Wooldridge (1991), and Davis and Alatalo (1992). Some are complex with through-flow systems (e.g. Klein Breteler and Laan, 1993) with control and monitoring of the quality and quantity of suspended food. Others are relatively simple and maintain the copepods in a healthy and food-rich state to determine times of development from one stage to the next, and so generation time. A large variety of organisms can be used as food, singly or in mixtures, the most common being laboratory cultures of diatoms and dinoflagellates; nauplii, including the ubiquitous *Artemia* nauplii, have been used in attempts to culture predatory species such as *Euchaeta marina*. Paffenhöfer and Harris (1979) tabulate the species of copepods and the food organisms used in cultures. They also review the conditions of culture and discuss many aspects such as quality and quantity of food, the potential benefits of re-circulating culture vessels, and the relevance of the results gained from cultures to the biology of the species within the natural environment. Huntley *et al.* (1987c) found that the composition of the diet of nauplii of *Calanus pacificus* altered their development and survival times. Klein Breteler (1980) recommends the inclusion of heterotrophic dinoflagellates among the food organisms to help control fouling of the culture medium and tanks. Many of the problems have already been discussed in Chapter 6 because of the difficulties of supplying a natural diet to copepods in the laboratory. Paffenhöfer and Harris (1979) review these problems in the context of trying to obtain realistic estimates of feeding, respiration and growth rates and of generation times. Tiselius *et al.* (1995) compare results from cultured and field populations of *Acartia tonsa*. Differences in rates of egg production are thought to be caused by the cultured copepods having lost the diel feeding rhythm of the animals in the field. Further, continuously cultured copepods are subject to different selective factors compared with those in the natural environment, the most obvious being the desirability for a short generation time and high reproductive rate. They conclude that the cultured animals are satisfactory for obtaining relative results but that deriving absolute values for field parameters is more difficult.

The primary reason for culturing copepods is to allow experimental examination of aspects of their physiology such as development through the successive naupliar stages, growth rates, duration of generations, reproduction and feeding. Mass culture of copepods is also relevant to mariculture (Ohno and Okamura, 1988). Fish larvae are carnivorous and so small copepods such as *Pseudodiaptomus marinus* are potential food (Iwasaki and Kamiya, 1977).

Species that have been successfully cultured are listed in Table 47. The majority are still neritic species although considerable success has been achieved with *Calanus* species and *Calanoides carinatus*. Greatest success has been with the species in the genera *Acartia*, *Centropages*, *Eurytemora*, *Paracalanus*, *Pseudocalanus*, *Pseudodiaptomus* and with *Temora longicornis*.

9.2. DEVELOPMENT TIME OF EGGS

Observations have been made in the laboratory on the duration of the embryological development of the eggs, that is the time elapsing between the egg being laid and its hatching to the NI. There has been some criticism of the results gained in the laboratory. Hart and McLaren (1978) found that eggs of *Pseudocalanus* sp. caught in the sea in the spring and reared in the laboratory take longer to hatch than those obtained in the summer at warmer environmental temperatures. They also found that larger females, at all times of the year, produced larger and slower developing eggs than smaller females. Rearing of females at reduced temperatures, however, resulted in the production of larger eggs whose development time was not extended, so inferring the possibility of a genetic component. Landry (1975b) states that the development time of the eggs of *Acartia clausi* is affected by the temperature that the parent female has experienced. He suggests that winter-acclimated animals require more than one generation to adapt to summer conditions. Tester (1985, 1986), however, found that acclimation of egg hatching time in *A. tonsa* is only about 24 to 48 h at 20°C.

McLaren *et al.* (1968) show that the effects of salinity on the development rate of eggs of *Pseudocalanus minutus* are negligible relative to those of temperature. The same is true of *Acartia* species examined by Uye (1980a) although specific differences exist suggesting that eggs of *A. steueri* may be less tolerant of salinity alteration.

Present evidence suggests that laboratory measurements of development rates of eggs are representative of those that will occur in the environment. Temperature controls the development time within any one species, the relationship between development time (D in days) and temperature (T in °C) being adequately described by Bělehrádek's (1935) empirical equation relating physiological rates and temperature:

$$D = a(T - \alpha)^b$$

where a, α and b are fitted constants. Equations describing the duration of development of the eggs of a variety of species are given in Table 48. Further discussion of this equation is given in the next section when considering rates of development between hatching and adulthood.

The average development time of eggs is different for those carried in egg masses as opposed to those freely spawned into the sea. Kiørboe and Sabatini (1994) review development times of eggs in egg masses of *Pseudocalanus* and *Eurytemora* species and of free-spawned eggs of *Paracalanus*, *Calanoides* and *Calanus* species and found that the former take about twice as long as the latter, 2 d as opposed to 24 h. Egg development times within these two categories were independent of egg size (Kiørboe and Sabatini, 1995). Eggs in egg masses ranged from about 80 to 130 μm in diameter while freely spawned eggs ranged from about 70 to 190 μm between the species examined by them. This is unrepresentative of the size range of eggs in egg masses, which is about 70 to 700 μm (Figure 56, p. 277), but representative of that of freely-spawned eggs. Eggs in egg masses are larger than freely spawned eggs and a relationship between size of egg and development time might be expected. McLaren *et al.* (1989c) found that the development times of the eggs of four *Pseudocalanus* species is related to their size and states that this relationship also exists within a species between different geographical locations. Development times of eggs of three *Calanus* species is also size-dependent (McLaren *et al.*, 1969) and later McLaren *et al.* (1988) show that there is a good correlation between the value of a and egg and body sizes expressed as the DNA content of adult females in 6 species of *Calanus* (Figure 60). The relationship may exist within a species, genus and between closely related genera but is not a general relationship within the calanoids (Corkett and McLaren, 1970). This is unexpected because of the general inverse relationship between size and metabolic rate; Steele and Steele (1975) demonstrate it specifically for the duration of embryonic development between orders of crustaceans ranging from Copepoda to Decapoda. The selection of a constant value of b in Bělehrádek's equation, -2.05 by McLaren *et al.* (1969), means that development time, D, can be easily compared with egg size by plotting values of a against egg diameter or weight (Figure 60).

The eggs of calanoids do not develop synchronously and consequently

Table 47 Development times of species of copepods cultured in the laboratory at designated temperatures. A few data are derived by other means from field samples (**) at approximate or indeterminate temperatures. Others are for a series of stages and not the complete generation time.

Species	T°C	Notes	Generation time (d)	Authority
Acartia californiensis	15		20.7	Landry, 1983
A. clausi	15		22	Klein Breteler et al., 1982
	9	NI–CVI	25.9	Hay et al., 1988
	15		19.2	Klein Breteler, 1980
	10		37	Klein Breteler and Schogt, 1994
	15		23	Klein Breteler and Schogt, 1994
	20		15	Klein Breteler and Schogt, 1994
	25		11	Klein Breteler and Schogt, 1994
	18		30	Nassogne, 1970
	20		12.5	Landry, 1975c
	20		20	Person-Le Ruyet, 1975
	15		20	Iwasaki et al., 1977
	20		19	Iwasaki et al., 1977
	13.1		27	Uye, 1980b
	16.4		21	Uye, 1980b
	20.3		15	Uye, 1980b
	17–23		30	Rippingale and Hodgkin, 1974
	13–25		28.6	Christou and Verriopoulos, 1993a
Tisbury Great Pond			56–70**	Deevey, 1948
Raritan Bay			47–64**	Jeffries, 1976
Wadden Sea			35–38**	Martens, 1981
Central North Sea			66–202**	Fransz et al., 1984
Loch Striven	8.5–13.5		28–42**	McLaren, 1978
English Channel			35–42**	Digby, 1950

A. biflosa	Mundaka Estuary, Spain		25**	Villate et al., 1993
A. discaudata	Wadden Sea		35-38**	Martens, 1981
A. grani		17-21	26	Vilela, 1972
A. sinjiensis		28-30	5-6	Doi et al., 1994
A. steueri		16.8	31.1	Uye, 1980b
		20.3	20.7	Uye, 1980b
		23	16.2	Uye, 1980b
A. tonsa		15.5	13	Heinle, 1966
		22.4	9	Heinle, 1966
		25.5	7	Heinle, 1966
		13-16	9-11	Heinle, 1969
		17	25	Zillioux and Wilson, 1966
		5-28	9-12	Ogle, 1979
		15	20.3	Landry, 1983
		16-18	13	Berggreen et al., 1988
		20	ca. 9	Paffenhöfer, 1991
	Tisbury Great Pond		42-49**	Deevey, 1948
	Delaware Bay		30**	Deevey, 1960a
	Raritan Bay		26-47**	Jeffries, 1976
A. tsuensis		27.7	8	Ohno et al., 1990
Calanoides carinatus		15	21-22	Hirche, 1980
		7	75	Borchers and Hutchings, 1986
		13	23	Borchers and Hutchings, 1986
		18	75	Borchers and Hutchings, 1986
		15.5	18.3	Peterson and Painting, 1990
		19.5	12	Peterson and Painting, 1990

Table 47 Continued.

Table 47 Continued.

Species	T°C	Notes	Generation time (d)	Authority
Ghanaian waters			14-18**	Mensah, 1974
Calanus australis	15.5		20.3	Peterson and Painting, 1990
	19.5		16	Peterson and Painting, 1990
C. finmarchicus	10	NI-CV	39	Diel and Klein Breteler, 1986
	6	to CV	63	Tande, 1988a
	14		28	Marcus and Alatalo, 1989
	19		21	Marcus and Alatalo, 1989
	11		45.5**	McLaren, 1978
Flemish Cap			43-51**	Anderson, 1990
Scotian Shelf	2		97**	Sameoto and Herman, 1990
Scotian Shelf	6		55**	Sameoto and Herman, 1990
Scotian Shelf	10		35**	Sameoto and Herman, 1990
Scotian Shelf	12		29**	Sameoto and Herman, 1990
C. finmarchicus/helgolandicus	10		39	Thompson, 1982
	15		25	Thompson, 1982
C. hyperboreus	4-6		110-120	Conover, 1967
C. marshallae	10		64	Peterson, 1986
	11		62	Peterson, 1986
	15		36	Peterson, 1986
C. minor Ivory Coast			19-23**	Binet, 1977
C. pacificus	12		36	Mullin and Brooks, 1967
	15		23	Mullin and Brooks, 1970
	15		18-36	Paffenhöfer, 1970
	15		19.8	Landry, 1983
	15		22-30	Hirakawa, 1979

Candacia armata		16.5	to CI	Bernard, 1965
Centropages abdominalis	Inland Sea of Japan			
Generation 1		15.5		20	Liang et al., 1996
Generation 2		13.7		28	Liang et al., 1996
Generation 3		10.6		36	Liang et al., 1996
Generation 4		9.8		40	Liang et al., 1996
Generation 5		12.0		32	Liang et al., 1996
C. chierchiae Ivory Coast				18-19**	Binet, 1977
C. furcatus Ivory Coast				17-18**	Binet, 1977
C. hamatus		20		21	Person-Le Ruyet, 1975
		15		19.7	Klein Breteler, 1980
		17	NII-CV	16.1	Fryd et al., 1991
Central North Sea				95-135	Fransz et al., 1984
German Bight		20		38	Martens, 1980
Loch Striven				20-25**	McLaren, 1978
C. typicus		10		34-51	Smith and Lane, 1985
		10		49	Smith and Lane, 1987
		15		33	Smith and Lane, 1987
		17	NII-CV	13.3-14.6	Fryd et al., 1991
		15		20-25	Davis and Alatalo, 1992
		18		50	Nassogne, 1970
		18-19		19-23	Lawson and Grice, 1970
New York Bight		20		25	Person-Le Ruyet, 1975
				50-60**	Smith and Lane, 1987
Banyuls-sur-Mer		10		ca. 100	Razouls, 1974
Banyuls-sur-Mer		15		ca. 50	Razouls, 1974

Table 47 Continued.

Table 47 Continued.

Species	T°C	Notes	Generation time (d)	Authority
Banyuls-sur-Mer	20		ca. 30	Razouls, 1974
C. velificatus	28		19.5	Chisholm and Roff, 1990b
Ctenocalanus citer Weddell Sea			54**	Fransz, 1988
C. vanus	18		35	Nassogne, 1970
Eucalanus hyalinus	20		28-35	Paffenhöfer, 1991
E. pileatus	20		18-21	Paffenhöfer, 1991
Euchaeta marina		Ivory Coast	21-27**	Binet, 1977
E. paraconcinna		Jamaica	20-24***	Webber and Roff, 1995b
Eurytemora affinis		Ivory Coast	16-20**	Binet, 1977
	10	Woods Hole	33	Katona, 1970
	15		20	Katona, 1970
	20		15	Katona, 1970
	25		12	Katona, 1970
	10		29.2	Vijverberg, 1980
	15		20.3	Vijverberg, 1980
Patuxent River	10		34.1	Heinle and Flemer, 1975
	15		21.9	Heinle and Flemer, 1975
	20		16.5	Heinle and Flemer, 1975
	25		11	Heinle and Flemer, 1975
Southampton, UK	10		28	Katona, 1970
	15		17	Katona, 1970
	20		12	Katona, 1970
Finland	10		27	Vuorinen, 1982
	15		18.5	Vuorinen, 1982
	20		15.5	Vuorinen, 1982

Location	Size		Value	Reference
	14		28-31	Vuorinen, 1987
	6-14		36.8	Vuorinen, 1987
Finland	5		63.6	Heerkloss et al., 1990
	10		33.9	Heerkloss et al., 1990
	15		22.2	Heerkloss et al., 1990
	20		16.8	Heerkloss et al., 1990
Schlie	10		29.8	Hirche, 1974
	15		23.1	Hirche, 1974
	20		19.4	Hirche, 1974
	25		16.9	Hirche, 1974
Schelde Estuary	8		57.6	Escaravage and Soetaert, 1993
	10		26.8	Escaravage and Soetaert, 1993
	14		17.4	Escaravage and Soetaert, 1993
	17		14.2	Escaravage and Soetaert, 1993
	20		13.5	Escaravage and Soetaert, 1993
Gironde, France	10		43.3	Poli and Castel, 1983
	15		27.7	Poli and Castel, 1983
	20		20.9	Poli and Castel, 1983
	25		16.1	Poli and Castel, 1983
	♀ 10	Lake Ohnuma	22.8	Ban, 1994
	15		11.4	Ban, 1994
	20		9.3	Ban, 1994
	♂ 10		20.6	Ban, 1994
	15		10.1	Ban, 1994
	20		8.0	Ban, 1994
	15	Lake Ohnuma	13-14	Ban and Minoda, 1994

Table 47 Continued.

Table 47 Continued.

Species		T°C	Notes	Generation time (d)	Authority
E. americana		4		40	Grice, 1971
E. herdmani		10		36	Katona, 1970
		15		19	Katona, 1970
		20		19	Katona, 1970
		4		39	Grice, 1971a
		7		34-36	Grice, 1971a
E. velox		10		55	Gaudy and Pagano, 1987
		15		42	Gaudy and Pagano, 1987
		20		31	Gaudy and Pagano, 1987
		10		28	Nagaraj, 1988
		15		18	Nagaraj, 1988
		20		12	Nagaraj, 1988
Gladioferens imparipes		15		28	Rippingale and Hodgkin, 1974
		25		12	Rippingale and Hodgkin, 1974
G. pectinatus		18		19.6-43.2	Arnott et al., 1986
Labidocera trispinosa		15		34	Landry, 1983
L. wollastoni		18-21		28-32	Grice and Gibson, 1982
Limnocalanus macrurus	Char Lake			180-210**	Roff and Carter, 1972
Metridia pacifica	Gulf of Alaska			90-120**	Batchelder, 1985
Microcalanus pygmaeus	Loch Striven			ca 77**	McLaren, 1978
Neocalanus plumchrus	Gulf of Alaska		CI-CV	100-120**	Miller and Nielsen, 1988
Paracalanus aculeatus		28		19.5	Chisholm and Roff, 1990b
		15	NI-CVI	18.6	Landry, 1983
P. parvus		12		25-30	Davis, 1984a,b
		18		ca. 18	Uye, 1991
		20		12-15	Paffenhöfer, 1991

Species	Location		Value	Reference
	Georges Bank		25-30**	Davis, 1984b
Paracalanus sp.		15	19.8	Uye, 1991
		17.5	17.6	Uye, 1991
Pareuchaeta elongata		0.5	355	Ikeda and Hirakawa, 1996
Pontella meadi		20	18-25	Gibson and Grice, 1976
Pseudocalanus acuspes	Bedford Basin, Nova Scotia			
		1.3	95**	McLaren et al., 1989b
		4	64-93**	McLaren et al., 1989b
P. elongatus		15	37	Katona and Moodie, 1969
		12.5	28	Paffenhöfer and Harris, 1976
		5	62	Thompson, 1982
		10	29	Thompson, 1982
		15	22	Thompson, 1982
		10	32.5	Hay et al., 1988
		15	15	Klein Breteler et al., 1990
		5	59	Klein Breteler et al., 1995
		10	31	Klein Breteler et al., 1995
		15	19	Klein Breteler et al., 1995
		20	24	Klein Breteler et al., 1995
		8-12.5	21-34.5** NI-CVI	McLaren, 1978
	Central North Sea		35-69	Fransz et al., 1984
P. minutus		11.9	38.4	Corkett, 1970
Pseudocalanus sp.		5	60-70	Davis, 1984b
Pseudodiaptomus acutus		♀ 24-26	15.0 ± 1.4	Jacoby and Youngbluth, 1983
		♂ 24-26	12.1 ± 0.7	Jacoby and Youngbluth, 1983

Table 47 Continued.

Table 47 Continued.

Species	T°C	Notes	Generation time (d)	Authority
P. cokeri	♀ 24-26		12.9 ± 0.8	Jacoby and Youngbluth, 1983
	♂ 24-26		9.8 ± 0.5	Jacoby and Youngbluth, 1983
P. coronatus	20		25	Jacobs, 1961
	♀ 24-26		12.4 ± 0.4	Jacoby and Youngbluth, 1983
	♂ 24-26		11.2 ± 0.4	Jacoby and Youngbluth, 1983
P. hessei	♀ 16		27.9	Jerling and Wooldridge, 1991
	♂ 16		27.3	Jerling and Wooldridge, 1991
	♀ 20		21.8	Jerling and Wooldridge, 1991
	♂ 20		21.1	Jerling and Wooldridge, 1991
	♀ 23		18.0	Jerling and Wooldridge, 1991
	♂ 23		17.6	Jerling and Wooldridge, 1991
	♀ 26		–	Jerling and Wooldridge, 1991
	♂ 26		14.3	Jerling and Wooldridge, 1991
P. marinus	20		24	Uye et al., 1983
	20		23	Uye and Onbé, 1975
Rhincalanus nasutus	12		28-49	Mullin and Brooks, 1967
	10-15		22-53	Mullin and Brooks, 1970
	15		25.4	Landry, 1983
Sinocalanus tenellus	6.2		80.2	Kimoto et al., 1986a
	9.9		38	Kimoto et al., 1986a
	14.9		21.2	Kimoto et al., 1986a
	20.1		11.9	Kimoto et al., 1986a
	22.6		9.3	Kimoto et al., 1986a
	27.1		7.5	Kimoto et al., 1986a
Sulcanus conflictus	♀ 18		27.8	Ough and Bayly, 1989
	♂ 18		25.5	Ough and Bayly, 1989

Species	Location		Value	Reference
Temora longicornis		♀ 25	22.5	Ough and Bayly, 1989
		♂ 25	18.5	Ough and Bayly, 1989
		12.5	24-33	Harris and Paffenhöfer, 1976a
		15	20.6	Klein Breteler, 1980
		10	23	Klein Breteler and Gonzalez, 1986
		15	16	Klein Breteler and Gonzalez, 1986
		20	14	Klein Breteler and Gonzalez, 1986
		9-10	31.4	Hay et al., 1988
		15	13	Klein Breteler et al., 1990
		20	21	Person-Le Ruyet, 1975
	Loch Striven	8.5-13.5	NI-CVI	McLaren, 1978
	English Channel		35-39**	Digby, 1950
	Southern Bight, North Sea		45-55**	Daro and Gijsegem, 1984
	Central North Sea		14-18**	Fransz et al., 1984
	Southern North Sea	5-10	46-105**	Fransz et al., 1989
	Southern North Sea	7-12	45**	Fransz et al., 1989
	Southern North Sea	12-18	35**	Fransz et al., 1989
	German Bight	20	50**	Martens, 1980
	Long Island Sound		21**	Peterson, 1985
	Long Island Sound		32-59**	Peterson and Kimmerer, 1994
T. stylifera		16-20.5	30-62**	Yassen, 1981
	Banyuls-sur-Mer	15	15-20	Razouls, 1974
	Banyuls-sur-Mer	20	ca. 60	Razouls, 1974
	Ivory Coast		ca. 30	Binet, 1977
T. turbinata		28	17-20**	Chisholm and Roff, 1990b
	Ivory Coast		19.5	Binet, 1977
Undinula vulgaris	Ivory Coast		17-18**	Binet, 1977
	Jamaica		20-21**	Webber and Roff, 1995b
			23-33**	

**Field estimates.

Table 48 Development time relative to environmental temperature of eggs to hatching (embryonic development), and the time from hatching to the CI. The duration of generations relative to temperature is also given. The equation used is that of Bělehrádek (1935) relating physiological rates and temperature: $D = a(T - \alpha)^b$ where D is the duration in days, T is the temperature in °C, a, b and α are fitted constants.

Species	Location	Equation	Authority
Development time of eggs			
Acartia clausi	Nova Scotia	$D = 1163(T + 8.2)^{-2.05}$	McLaren et al., 1969
	L. Striven, Scotland	$D = 1442(T + 10.49)^{-2.05}$	McLaren, 1978
	Onagawa Bay, Japan	$D = 650(T + 5.8)^{-2.05}$	Uye, 1980a
A. grani	S.E. Spain	$D = 28\,902(T + 2.99)^{-2.14}$	Guerrero et al., 1994
A. longiremis	Trømso, Norway	$D = 1008(T + 8.70)^{-2.05}$	Norrbin, 1996
A. steueri	Onagawa Bay, Japan	$D = 747(T + 3.2)^{-2.05}$	Uye, 1980a
A. tonsa	Narragansett Bay	$D = 489(T - 1.8)^{-2.05}$	McLaren et al., 1969
Calanus finmarchicus	Trømso	$D = 1122(T + 14.1)^{-2.05}$	Corkett et al., 1986
	Nova Scotia	$D = 691(T + 10.60)^{-2.05}$	Corkett et al., 1986
C. glacialis	Frobisher	$D = 1491(T + 14.5)^{-2.05}$	Corkett et al., 1986
	Nova Scotia	$D = 975(T + 13.04)^{-2.05}$	Corkett et al., 1986
	Nova Scotia	$D = 1067(T + 12.97)^{-2.05}$	McLaren et al., 1988
C. helgolandicus	S North Sea	$D = 1014(T + 10.94)^{-2.05}$	Corkett et al., 1986
C. hyperboreus	Nova Scotia	$D = 1575(T + 14.40)^{-2.05}$	Corkett et al., 1986
C. marshallae	Seattle	$D = 831(T + 11.01)^{-2.05}$	McLaren et al., 1988
C. pacificus	Seattle	$D = 608(T + 7.39)^{-2.05}$	McLaren et al., 1988
C. sinicus	Inland Sea of Japan	$D = 545(T + 5.7)^{-2.05}$	Uye, 1988
Centropages abdominalis	Inland Sea of Japan	$D = 159(T + 3.18)^{-1.58}$	Liang et al., 1996
C. furcatus	Jamaica	$D = 422(T - 3.7)^{-2.05}$	McLaren et al., 1969
C. typicus	Woods Hole	$D = 1068(T + 9.37)^{-2.05}$	McLaren et al., 1989a
Eurytemora affinis	Lake Ohnuma, Japan	$D = 268.8(Y + 3.4)^{-1.68}$	Ban and Minoda, 1991
E. herdmani	Nova Scotia	$D = 1640(T + 10.40)^{-2.05}$	Corkett and McLaren, 1970
Metridia longa	Nova Scotia	$D = 1099(T + 15.1)^{-2.05}$	McLaren et al., 1969

Paracalanus parvus	California	$D = 432(T + 2.97)^{-2.25}$	Checkley, 1980b
Paracalanus sp.	Inland Sea of Japan	$D = 140(T + 2.2)^{-1.85}$	Uye, 1991
Pseudocalanus acuspes	Nova Scotia	$D = 1949(T + 12.59)^{-2.05}$	McLaren et al., 1989c
P. elongatus	L. Striven, Scotland	$D = 536(T + 9.99)^{-1.68}$	McLaren, 1978
P. minutus	Nova Scotia	$D = 2144(T + 13.40)^{-2.05}$	Corkett and McLaren, 1970
	Nova Scotia	$D = 2338(T + 13.90)^{-2.05}$	McLaren et al., 1989c
Large form	Ogac Lake	$D = 82(T - 5.00)^{-0.95}$	McLaren, 1966
P. moultoni	Nova Scotia	$D = 1889(T + 12.03)^{-2.05}$	McLaren et al., 1989c
P. newmani	Nova Scotia	$D = 1572(T + 11.30)^{-2.05}$	McLaren et al., 1989c
Pseudodiaptomus hessei	South Africa	$D = 403(T - 0.4)^{-1.64}$	Jerling and Wooldridge, 1991
P. marinus	Inland Sea of Japan	$D = 448(T - 1.0)^{-1.80}$	Uye et al., 1982
Sinocalanus tenellus	Inland Sea of Japan	$D = 499(T + 3.2)^{-1.76}$	Hada et al., 1986
Sulcanus conflictus	Lakes near Melbourne	$D = 5240(T - 0.2)^{-1.58}$	Ough and Bayly, 1989
Temora longicornis	Nova Scotia	$D = 1346(T + 10.40)^{-2.05}$	Corkett and McLaren, 1970
T. stylifera			
Autumn generation	Villefranche-sur-mer	$D = 3.53(T - 12)^{-0.50}$	Abou Debs and Nival, 1983
Spring generation	Villefranche-sur-mer	$D = 45.51(T + 0.5)^{-1.14}$	Abou Debs and Nival, 1983
	Nova Scotia	$D = 1785(T + 9.0)^{-2.05}$	McLaren et al., 1969
Tortanus discaudatus	Nova Scotia	$D = 2307(T - 9.40)^{2.12}$	McLaren, 1966

Development time (D) from hatching to CI

Acartia clausi	L. Striven, Scotland	$D = 6866(T + 10.49)^{-2.05}$	McLaren, 1978
Calanus finmarchicus	Nova Scotia	$D = 8882(T + 13.04)^{-2.05}$	Corkett et al., 1986
	Nova Scotia	$D = 6419(T + 10.60)^{-2.05}$	McLaren et al., 1988
C. glacialis	Nova Scotia	$D = 8825(T + 12.97)^{-2.05}$	McLaren et al., 1988
C. helgolandicus	S North Sea	$D = 7042(T + 10.94)^{-2.05}$	McLaren et al., 1988

Table 48 Continued.

Table 48 Continued.

Species	Location	Equation	Authority
C. hyperboreus	Nova Scotia	$D = 13\,532(T + 14.4)^{-2.05}$	Corkett et al., 1986
	Nova Scotia	$D = 13\,532(T + 14.4)^{-2.05}$	McLaren et al., 1988
C. marshallae	Seattle	$D = 9353(T + 11.01)^{-2.05}$	McLaren et al., 1988
C. pacificus	Seattle	$D = 3830(T + 7.39)^{-2.05}$	McLaren et al., 1988
C. sinicus	Inland Sea of Japan	$D = 582(T + 0.7)^{-1.44}$	Uye, 1988
Eurytemora herdmani	Nova Scotia	$D = 5227(T + 10.4)^{-2.05}$	Corkett and McLaren, 1970
Pseudocalanus minutus	Nova Scotia	$D = 9224(T + 13.4)^{-2.05}$	Corkett and McLaren, 1970
Pseudodiaptomus marinus	Inland Sea of Japan	$D = 1756(T - 1.0)^{-1.80}$	Uye et al., 1983
Temora longicornis	Nova Scotia	$D = 8313(T + 10.4)^{-2.05}$	Corkett and McLaren, 1970
T. stylifera			
Autumn generation	Villefranche-sur-mer	$D = 24.00(T - 12.0)^{-0.50}$	Abou Debs and Nival, 1983
Spring generation	Villefranche-sur-mer	$D = 293.16(T + 0.5)^{-1.14}$	Abou Debs and Nival, 1983
Duration of generation			
Acartia clausi	L. Striven, Scotland	$D = 14\,748(T + 10.49)^{-2.05}$	McLaren, 1978
	Onagawa Bay, Japan	$D = 11\,170(T + 5.8)^{-2.05}$	Uye, 1980b
		$D = 1695(T - 2.33)^{-1.52}$	Klein Breteler and Schogt, 1994
A. steueri	Onagawa Bay, Japan	$D = 14\,450(T + 3.2)^{-2.05}$	Uye, 1980b
Calanus finmarchicus	L. Striven, Scotland	$D = 31\,630(T + 14.10)^{-2.05}$	McLaren, 1978
	L. Striven, Scotland	$D = 6779(T + 9.43)^{-1.68}$	McLaren, 1978
C. sinicus	Inland Sea of Japan	$D = 1258(T + 0.7)^{-1.44}$	Uye, 1988
Centropages abdominalis	Inland Sea of Japan	$D = 2123(T + 3.18)^{-1.58}$	Liang et al., 1996
Eurytemora herdmani ♀	Nova Scotia	$D = 11\,465(T + 10.40)^{-2.05}$	McLaren and Corkett, 1981
♂	Nova Scotia	$D = 10\,562(T + 10.40)^{-2.05}$	McLaren and Corkett, 1981

Paracalanus parvus	Scotian Shelf	$D = 12\,430(T + 2.97)^{-2.25}$	McLaren et al., 1989a
Paracalanus sp.	Inland Sea of Japan	$D = 4210(T + 2.2)^{-1.85}$	Uye, 1991
Pseudocalanus acuspes	Nova Scotia	$D = 22\,591(T + 12.59)^{-2.05}$	McLaren et al., 1989c
P. elongatus	L. Striven, Scotland	$D = 5174(T + 9.99)^{-1.68}$	McLaren, 1978
	North Sea	$D = 9398(T - 8)^{-1.98}$	Klein Breteler et al., 1995
P. minutus	Nova Scotia	$D = 22\,331(T + 11.45)^{-2.05}$	McLaren et al., 1989c
	Nova Scotia	$D = 19\,350(T + 13.40)^{-2.05}$	McLaren, 1974
	Nova Scotia	$D = 21\,494(T + 13.40)^{-2.05}$	McLaren, 1978
P. moultoni	Nova Scotia	$D = 28\,668(T + 13.90)^{-2.05}$	McLaren et al., 1989c
P. newmani	Nova Scotia	$D = 23\,559(T + 12.03)^{-2.05}$	McLaren et al., 1989c
Pseudodiaptomus marinus	Inland Sea of Japan	$D = 16\,358(T + 11.30)^{-2.05}$	McLaren et al., 1989c
Temora longicornis	L. Striven, Scotland	$D = 3638(T - 1.0)^{-1.80}$	Uye et al., 1983
	Wadden Sea,	$D = 16\,988(T + 10.40)^{-2.05}$	McLaren, 1978
	Netherlands	$D = 98(T + 2.9)^{-0.62}$	Klein Breteler and Gonzalez, 1986

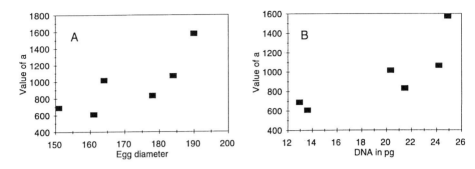

Figure 60 Development times of eggs. Relationship of the value of *a* in Bělehrádek's equation to A, egg diameter in μm and to B, DNA content of adult females. (After McLaren *et al.*, 1988.)

there is a range in the embryonic duration for each species. There is also evidence, obtained in *Calanus marshallae* by Peterson (1986), that slow or fast development of an egg is reflected in slow or fast development of the subsequent nauplii and copepodids. This may be a genetic component, deriving from heterozygosity within the eggs. This fast or slow development occurs not only between eggs of different clutches but also within a clutch.

Each calanoid species has a range of temperature within which successful embryological development takes place. The optimal range varies between species and results in the majority of the eggs hatching. Uye (1991) shows that hatching success of eggs of *Paracalanus* sp. is better than 60% between 7.5 and 21.3°C but less than 30% at temperatures higher than 22.6°C. Oceanic species can be expected to be much less eurythermic than *Paracalanus* sp. and may have a restricted temperature window for successful hatching.

9.3. GROWTH

9.3.1. Growth Process

Crustaceans grow in a stepwise manner because of the rigidity of the integument that has to be shed. The new soft integument accommodates the growth in body size that has taken place during the previous stage and after moulting. Consequently, studies of the growth rates of crustaceans involve determination of the intermoult periods and the associated growth factors for each instar.

Table 49 Comparison of the mean and its standard deviation and of the ranges of prosome lengths (mm) of copepodids and adult female *Pleuromamma robusta* with soft (postmoult) and hard (premoult) bodies. (After J.S. Park, 1995b.)

Copepodid	Soft bodied		Hard bodied	
	mean	range	mean	range
CII	0.61 ± 0.02	0.56–0.63	0.63 ± 0.11	0.61–0.66
CIII	0.83 ± 0.03	0.76–0.87	0.85 ± 0.02	0.80–0.92
CIV	1.08 ± 0.04	0.97–1.14	1.10 ± 0.32	0.83–1.17
CV ♀	1.47 ± 0.08	1.39–1.54	1.52 ± 0.08	1.37–1.62
CVI ♀	1.97 ± 0.07	1.88–2.11	2.10 ± 0.09	1.71–2.24

9.3.1.1. Moulting

The moult cycle and its effects on the structure of the integument are shown in Figure 12 (p. 33) and Table 6 (p. 34). Growth in body size of the animal during the intermoult period has been shown to occur in mysids by stretching of the integument (Mauchline, 1973). Potential intermoult growth of copepods was examined in *Pleuromamma robusta* by J.S. Park (1995b) who measured prosome lengths of soft (postmoult) and hard (intermoult and premoult) bodied copepodids and adult females (Table 49). The mean prosome length of hard-bodied individuals within a copepodid stage always exceeds that of the soft-bodied individuals although the differences are very small.

The changes in the thickness of the integument (Figure 12, p. 33; Table 7, p. 34) are reflected by changes in the chitin content of the copepod. Båmstedt and Matthews (1975) found that chitin content ranges from 0.11 to 0.19 mg female^{-1}, 0.04 to 0.10 mg male^{-1} and 0.06 to 0.08 mg copepodid V^{-1} of *Pareuchaeta norvegica*. This variation was not related to body weight or external dimensions of the copepod and presumably reflects the stages of the intermoult cycle.

The percentage of nauplii of *Calanus pacificus* within any one naupliar stage that moulted sucessfully decreased with the age of the nauplii in that stage (Lopez, 1991). Carlotti and Nival (1991, 1992a,b) examine when moulting takes place within individual copepodids of *Temora stylifera* and *Centropages typicus*. Individuals of the same brood were found to have varying durations within any one copepodid. The distribution of durations is asymmetrical such that the modal duration was significantly less than the mean, ranging from about 80 to 90% of the mean. Some individuals had stage durations twice to four times as long as the modal ones. No information is available on the size distributions within the stages relative to the stage durations. Carlotti and Nival suggest that there is a critical

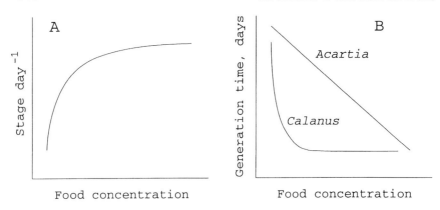

Figure 61 Effect of food concentration on A, the number of developmental stages passed through by *Centropages typicus* per day and B, on the generation time of *Acartia clausi* and *Calanus pacificus*. (After Landry, 1978b; Vidal, 1980b; Davis and Alatalo, 1992)

weight that initiates moulting and that the bulk of the individuals within a copepodid stage attain this at a similar, or modal, age. This concept is incorporated in the model of development and growth of an individual copepod proposed by Bosch and Gabriel (1994). Development rate, in terms of stages d^{-1}, is known to increase to an asymptotic level with increasing concentration of food (Davis and Alatalo, 1992). Alternatively, generation time decreases linearly in *Acartia clausi* and to an asymptotic level in *Calanus pacificus* with increasing food concentration (Landry, 1978b; Vidal, 1980b). The shape of the curves are shown in Figure 61.

Natural physiological variation in growth rates among the copepodids results in some being slower in attaining moulting status. The presence of a distinct modal age for moulting indicates the homogeneous nature of these individuals. The minority, with a range of slower growth rates, gains new members from the fast modal group at each successive moult. Carlotti and Nival (1992b) infer that this results in an increase in the standard deviation. This is not true if the standard deviation is expressed as a percentage of the mean. The largest percentage value is associated with the CIII, 60% in *Centropages hamatus* (Table 50) and 70% in *Temora stylifera* (see Calotti and Nival, 1991). The CIII of *Calanus marshallae* is also the most variable (Peterson, 1986). This variation within the CIII is interesting but can not be explained at present. Ovarian development was noticed as early as CI in these species by Razouls *et al.* (1987) and the sexual differentiation known to take place in CIV may be linked to the extended duration of CIII.

Experimental observation of moulting frequency may be difficult with some species. Miller and Nielsen (1988) found that *Neocalanus flemingeri*

Table 50 Mean durations and their standard deviations of copepodid stages of *Centropages hamatus* maintained at 15 °C. The standard deviation (SD) is expressed as a percentage (%) of the mean duration. The number (n) of individuals examined is given.

Stage	Duration (days)	SD as %	n
CI	2.73 ± 0.58	21.2	112
CII	2.21 ± 0.91	41.2	90
CIII	2.48 ± 1.47	59.3	76
CIV	2.93 ± 1.17	39.9	64
CV	3.47 ± 1.32	38.0	52
Adult female	14.80 ± 4.59	31.0	28
Adult male	16.27 ± 3.80	23.4	24

and *N. plumchrus* lost the plumose caudal setae, and most also lost the rest of the tail fan as well, during the process of capturing them in nets. These setal injuries prevented the copepods from moulting because adhesion took place between the new and old integuments. Miller *et al.* (1984b) discuss some of the problems associated with determining moulting frequency and stage duration experimentally soon after capture of the copepods at sea. They found some evidence of bursts of moulting activity at night.

9.3.1.2. *Intermoult Duration*

Intermoult or stage duration is the period measured in hours or days between successive moults in the developmental sequence of the copepod through the naupliar and copepodid stages. More attention has been paid to stage duration than to growth factors between one stage and the next. Stage duration is relatively easily observed in the laboratory in species that can be successfully cultured (Table 47, p. 300). The observations can be made on individual animals or on batches when the mean duration of the stage is taken as the time required for 50% of the individuals to moult to the next stage (e.g. Fryd *et al.*, 1991) or for the first appearance of the next stage e.g. Klein Breteler *et al.* (1982). The stage durations under conditions of food saturation have been examined at several constant temperatures in a variety of species in the laboratory. Predictive models have been proposed to describe the various patterns found. Alternatively, Runge *et al.* (1985) describe how duration of stages can be derived from determinations of the rates of moulting within a species.

(a). *Models of development* The following conceptual models describing the development of copepods have been proposed.

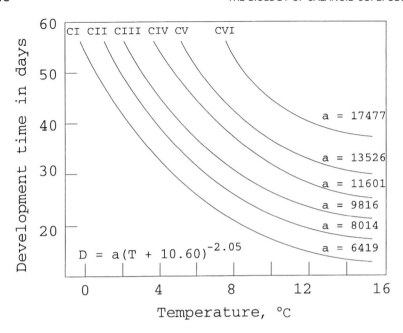

Figure 62 Bĕlehrádek's equations used to draw curves for development times from egg laying to the CI to the CVI (the adult) of *Calanus finmarchicus*. The value of b in Bĕlehàrdek's equation is kept constant at -2.05 resulting in the increasing values of a as shown for each copepodid. (After Corkett *et al.*, 1986.)

(i). Equiproportional development

Corkett (1984) introduced the term equiproportional growth to describe the development, relative to time, of copepods at any given temperature. Each developmental stage occupies the same proportionate amount of time relative to the egg development time at that temperature. Thus, given that the development times of the eggs are determined at three or more temperatures the appropriate Bĕlehrádek's equation can be derived (Table 48, p. 310). Use of this equation in conjunction with experimentally determined development times of the older stages at a single temperature can predict development times of the older stages at any selected temperature. A family of curves can be generated (Figure 62) predicting development times of individual stages at different temperatures. It must be emphasized that this approach is empirical and the experimental results often deviate somewhat from the generated curve. Thompson (1982), instead of using Bĕlehrádek's equation, used:

$$\mathrm{Log}_e D = a - bT$$

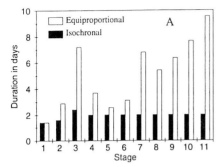 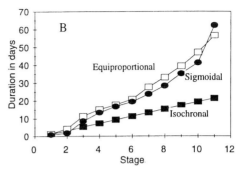

Figure 63 Models of development. A, the durations of individual stages (1–6, naupliar; 7–12, copepodid) during equiproportional development where they are different and in isochronal development where they are more or less equal. B, cumulative curves of equiproportional and sigmoidal development and the linear pattern of isochronal development. Data from Miller *et al.* (1977), Thompson (1982) and Peterson (1986).

which gives equally useful results. The former equation, however, is in more general use. The assumption of equiproportional development and the adoption of a common value of b, -2.05, allows easy comparisons of developmental times of different species.

A typical form of equiproportional development is illustrated in Figure 63, based on the development of *Calanus helgolandicus* at 7.55°C described by Thompson (1982). The data are shown in histogram form (Figure 63A) to contrast the unequal durations of the stages and in cumulative form to show the shape of the developmental curve. Species that have or nearly have equiproportional development are shown in Table 51.

(ii). Isochronal development
Miller *et al.* (1977) coined the term isochronal development to describe the development of *Acartia* species. All stages have virtually the same duration. Development progresses linearly with time (Figure 63B) although the earlier non-feeding naupliar stages may diverge from the linearity. The developmental rate at different temperatures is predicted in the same way as for equiproportional species by the use of Bělehrádek's equation. Species that have been shown to have isochronal or nearly isochronal development are listed in Table 51.

(iii). Sigmoidal development
The sigmoidal pattern of development described by Peterson (1986) results when the rates of development of the early non-feeding naupliar stages are markedly shorter than those of later nauplii and when the later copepodid stages have longer development times than earlier copepodids. The growth

Table 51 Growth patterns of calanoid copepods. The species listed are those that approximate more or less to the pattern of development.

Species	Authority
Equiproportional development	
Calanoides carinatus	Peterson and Painting, 1990
Calanus finmarchicus	Corkett *et al.*, 1986
C. helgolandicus	Corkett *et al.*, 1986
C. pacificus	Corkett *et al.*, 1986
C. sinicus	Uye, 1988
Centropages typicus	Fryd *et al.*, 1991
Isochronal development	
Acartia californiensis	Trujillo-Ortiz, 1990
A. clausi	Landry, 1975c
	Miller *et al.*, 1977
	Uye, 1980b
	Klein Breteler *et al.*, 1982
	Klein Breteler and Schogt, 1994
A. steueri	Uye, 1980b
A. tonsa	Heinle, 1969
	Berggreen *et al.*, 1988
A. tsuensis	Ohno *et al.*, 1990
Centropages hamatus	Klein Breteler *et al.*, 1982
C. typicus	Davis and Alatalo, 1992
Eurytemora affinis	Katona, 1971
E. herdmani	Corkett and McLaren, 1970
Paracalanus aculeatus	Chisholm and Roff, 1990b
Pseudocalanus elongatus	Klein Breteler *et al.*, 1982
	Klein Breteler *et al.*, 1994
	Klein Breteler *et al.*, 1995
Temora longicornis	Klein Breteler *et al.*, 1982
T. turbinata	Chisholm and Roff, 1990b
Sigmoidal development	
Calanus marshallae	Peterson, 1986
Non-conformist development	
Acartia clausi	Christou and Verriopoulos, 1993b
Calanus australis	Peterson and Painting, 1990
C. finmarchicus	Tande, 1988a
C. helgolandicus	Thompson, 1982
Centropages hamatus	Fryd *et al.*, 1991
Paracalanus sp.	Uye, 1991
Pseudocalanus elongatus	Thompson, 1982
Pseudodiaptomus hessei	Jerling and Wooldridge, 1991
P. marinus	Uye *et al.*, 1983
Temora longicornis	Klein Breteler and Gonzalez, 1986

curve resulting from this pattern of development is shown in Figure 63B. The only example of this pattern is that of *Calanus marshallae* (Table 51).

(iv). Non-conformist development

The development of some species does not conform with the equiproportional rule (Peterson and Painting, 1990). The stage durations vary somewhat irregularly. Re-examination of species already allotted to other developmental patterns may be found to vary from them when methods of determining the stage durations have become more refined. Species whose developmental patterns are not within any of the above conceptual patterns are listed as non-conformist in Table 51.

(b). *Comparison of models* There have been other models tested to describe the development of copepods (Hart, 1990; Guerrero *et al.*, 1994; Blanco *et al.*, 1995; McLaren, 1995). They have not, however, been applied to any extent and the ones in common use above serve to define life history parameters empirically. Understanding the developmental patterns is another matter and new models may be developed for this purpose.

The models have been derived from laboratory observations and, although extremely useful for empirically determining stage and generation durations, smooth out irregularities in the durations of the stages. These irregularities, some accounted for in the sigmoidal pattern of development, seem to be a true feature of the development. The non-feeding nauplii have shorter durations than feeding nauplii while the first feeding stage has a prolonged duration (Landry, 1983). This immediately introduces variations in the pattern of development between species. Some species, probably relatively few, begin feeding in the NI (e.g. *Pseudodiaptomus coronatus*) while others begin in the NII (e.g. *Acartia* species, *Rhincalanus nasutus*). Probably the majority of species commence feeding in the NIII or NIV but a few, such as *Calanus hyperboreus*, may not commence until the NV. Sekiguchi (1974) relates commencement of feeding of nauplii to the development of the gnathobase of the mandible. According to Sekiguchi, nauplii of *Acartia* species do not feed but Landry (1983) states that the NII and later nauplii feed; thus the concept of isochronal development from at least the NII in these species seems reasonable (Table 51). The possible commencement of feeding in the NI of *Pseudodiaptomus* species may result in an apparently isochronal development. There are reports of extended durations of the NVI, which is still a non-feeding stage in the Euchaetidae (Sekiguchi, 1974). In *Pareuchaeta norvegica*, early nauplii have durations of about 24 h while the N VI persists for about 5 days (Nicholls, 1934). Nauplii of *P. russelli* have durations of about 12 h but the NVI persists for 24 to 30 h (Koga, 1960a). The duration of the NVI in *Calanus helgolandicus*, however, was markedly shorter than other naupliar stages except that of NI or NII, at four of the five culture temperatures while the same was true of the NVI of *Pseudocalanus elongatus* at the ten culture temperatures used by Thompson (1982).

Development of the copepodids also shows potential for variation. Discussion of the development of the ovary in Chapter 2 mentions observations of undifferentiated gonadal cells as early as CI. The ovaries may commonly start appreciable development as early as CIII and certainly in CIV when secondary sexual characteristics develop in many species. Marked ovarian and other development in the CV is considered responsible for the extended duration of this stage. Klein Breteler *et al.* (1994) found consistent differences in between-stage durations when the results from different cultures of three species of copepods were examined suggesting that they were not experimental errors.

Landry (1983) summarizes the overall development of copepods as follows:

a. the non-feeding naupliar stages, number variable between species, have shorter durations than later stages;
b. the first feeding nauplius stage, NIII, has an extended duration but the situation in species where the NI or NII are the first feeding stage is unknown;
c. the feeding naupliar stages and early copepodid stages have about the same rates of development;
d. the CV has an extended duration.

The individual variation in stage durations (Table 50) within a culture present problems in determining the true duration. Peterson and Painting (1990) and Klein Breteler *et al.* (1994) review the different methodologies used in determining stage duration and make a plea for standardization to the "median development time (MDS)". This is calculated for each stage from regression analysis of data relating stage frequency to time. This is important in understanding the observed development patterns and relating them to the physiology and growth of the animal. In addition, Carlotti and Nival (1991) advise determination of the standard deviation of each stage duration.

Mean times for the development of stages and of generations of a copepod can be estimated from cohort analysis of copepods in cultures or in the field. Trujillo-Ortiz (1995) reviews Landry's (1978b) iterative method and develops a quadratic method that he states is more accurate.

Landry (1983) suggests that strict isochronal development is probably very rare but Klein Breteler *et al.* (1994) suggest that it is quite common, experimental errors confusing the issue. Strict conformation to equiproportional development may also be rare. Nevertheless, these concepts allow estimates of stage and generation times that are extremely useful although their extrapolation to determine mortality and production rates within the environment can only be made with caution. Laboratory experiments are normally done under conditions of constant temperature while the stages will experience changing temperature in the environment. Pedersen and Tande (1992) found that, at low temperatures, cultures of nauplii, of *Calanus finmarchicus*, subjected to rises of temperature of 0.1 and $0.2\,°C\,d^{-1}$ developed better than the culture of nauplii kept at constant temperature. They review evidence for alteration of rates by variable environmental conditions.

Determination of the mean development time (MDS) for each stage and its standard deviation will allow much more realistic conjecture of the factors causing one stage to be longer or shorter than another. Combining such data with comparable data on the increment at each successive moult, described in the next section, is crucial to an understanding of the processes involved in the growth and development of the copepod. Peterson (1986) found that eggs of *Calanus marshallae* within a single clutch had a range of development times. A fast-developing egg resulted in fast-developing later stages while an egg that had a longer duration of development resulted in slower growing later stages. He also compared median development rates of different clutches and found that there were fast and slow developing clutches of eggs. These, however, did not reflect fast and slow development times respectively to adulthood as was found for eggs within the same clutch. No comparable data are available on the variation of body size-at-stage for these experiments.

9.3.1.3. *Moult Increment*

The increment of body size at moulting has received less attention than the durations of the intermoult period. The increment can be expressed in two ways, as an increment or as a growth factor:

(i) Increment = (postmoult size − premoult size)/premoult size
(ii) Growth factor = postmoult size/premoult size

The moult increment has values of 0.1 to 0.5 and the growth factor values of 1.1 to 1.5 for body length; both have the same values on a weight basis, namely from less than zero, a negative value, to about 2.0 (Figure 64).

Corkett and McLaren (1978) state that the average increment of length is 0.17 to 0.28 with a mean value of 0.22. This is equivalent to a 1.8 increase in body weight. A review of data available in the literature (Figure 64) shows that this value is approximately correct. Length measurements of nauplii and copepodids are not directly comparable because of the development of the urosome within the copepodids; this contributes a greater component to body length than to body weight. The increment of the NVI when it moults to the CI is 0.391 ± 0.201 ($n = 26$) based on total length of CI, and 0.269 ± 0.270 ($n = 15$), based on prosome length of CI (Table 52).

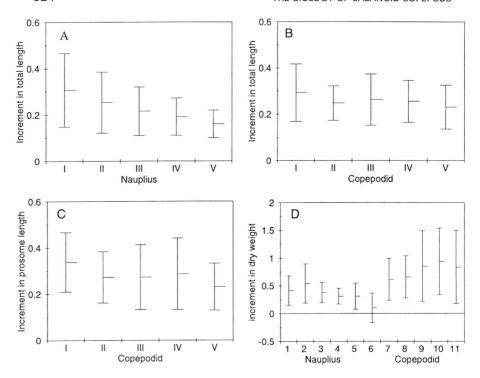

Figure 64 Growth of copepods at moulting. A, increment on premoult total body length of nauplii I to V and, B, of copepodids I to V. C, increment on premoult prosome length of copepodids I to V. D, increment on premoult body dry weight of nauplii I to VI (1–6) and copepodids I to V (7–11). The means and standard deviations are given in Table 52.

Total and prosome length data from: Johnson, 1934a, 1935, 1937, 1948, 1965; Nicholls, 1934; Lindquist, 1959; Koga, 1960a,b, 1970, 1984; Gaudy, 1961; Mazza, 1964, 1965; Ummerkutty, 1964; Andrews, 1966; Björnberg, 1966, 1967b; Heinle, 1966; Matthews, 1966; Grice, 1969, 1971; Alvarez and Kewalrami, 1970; Heron and Bowman, 1971; Katona, 1971; Pillai, 1971, 1975a,b; Vilela, 1972; MacLellan and Shih, 1974; Makarova, 1974; Morioka, 1975; Uye and Onbé, 1975; Gibson and Grice, 1976, 1977; Bakke, 1977; Goswami, 1977, 1978a,b; Bakke and Valderhaug, 1978; Durbin and Durbin, 1978; Hirche, 1980; Burkill and Kendall, 1982; Grice and Gibson, 1982; Kimmerer, 1983; Uye *et al.*, 1983; Razouls, 1985; Trujillo-Ortiz, 1986; Hay *et al.*, 1988, 1991; Ohman, 1988a; Razouls and Razouls, 1988; Frost, 1989; Sabatini, 1990; Hülsemann, 1991a; Verheye, 1991; Longhurst and Williams, 1992; Hirche *et al.*, 1994; Ferrari, 1995.

Weight data from: Heinle, 1966; Dagg and Littlepage, 1972; Durbin and Durbin, 1978; Vidal, 1980a; Williams and Lindley, 1980b; Burkill and Kendall, 1982; Klein Breteler *et al.*, 1982; Uye *et al.*, 1983; Davis, 1984b; McLaren, 1986; Berggreen *et al.*, 1988; Hay *et al.*, 1988, 1991; Ohman, 1988a; Jerling and Wooldridge, 1991; Verheye, 1991; Longhurst and Williams, 1992; Christou and Verriopoulos, 1993a; Escaravage and Soetaert, 1993.

Table 52 Increments of total and prosome lengths and of body weights of nauplii and copepodids. Sources of data given in Fig 64.

Stage	Mean	S.D.	n	Mean	S.D.	n	Mean	S.D.	n
	Total length			Prosome length			Body weight		
Nauplii									
NI	0.307	0.158	39				0.418	0.265	12
NII	0.253	0.132	52				0.543	0.355	13
NIII	0.215	0.105	52				0.380	0.182	13
NIV	0.191	0.080	52				0.312	0.143	14
NV	0.160	0.059	52				0.313	0.235	14
NVI	*			**			0.103	0.266	17
Females									
CI	0.293	0.124	37	0.338	0.128	80	0.617	0.380	29
CII	0.247	0.074	35	0.272	0.111	90	0.660	0.381	30
CIII	0.263	0.110	35	0.273	0.140	94	0.855	0.638	34
CIV	0.255	0.090	35	0.287	0.155	90	0.940	0.601	35
CV	0.230	0.095	28	0.231	0.101	86	0.837	0.659	35
Males									
CIII	0.218	0.149	23	0.241	0.152	44			
CIV	0.224	0.091	24	0.235	0.111	43			
CV	0.145	0.120	20	0.040	0.399	42			

Regression equations:
Nauplii: Increment = -0.036 Stage number $+0.332$ r, 0.9768***
Log increment = -0.069 stage number -0.452 r, 0.9947***

* on CI total length 0.391 ± 0.201 n = 26.
** on CI prosome length 0.269 ± 0.270 n = 15.
***Significant at the 1% level.

The increments in body length of nauplii of 46 species of calanoid copepods are shown in Figure 64; there are no values for the NI to NII moult in several species and several values exist for later moults of a few species. The average increments decrease at successive moults. The increments are correlated with stage number better on a log-linear than a linear scale (Table 52). This is the pattern of decrease in sizes of increments found in shrimp-like decapods, mysids and euphausiids (Mauchline, 1977b, 1980). The increments in the copepodids, on a total or prosome length basis (Figure 64B, C), are not correlated with stage number (Table 52). Likewise, there is no correlation between the weight increments and stage number (Figure 64D). What is of interest is the change in the magnitude of the weight increments between the nauplii and copepodids (Figure 64D), the latter having notably larger increments. Also, the NVI, when it moults to the CI, has an exceptionally small increment, often negative. There is, of course, a loss of body weight at moulting through casting of the old integument. The weight of the old integument in *Calanus pacificus* represents between 2.8 and 5.1% of body carbon of the premoult stage (Vidal, 1980b). This weight is close to the range of experimental errors for the determinations of body weights of the copepods and, maximally, represents only 20% of the moult increment (Figure 64D). Consequently, its inclusion or exclusion should not affect the determinations of rates of growth in body weight of the copepods significantly. It should, however, be included in considerations of total carbon budgets of developing copepods.

The adult males are usually smaller in body size than the adult females. The increments of the CIII to the male CIV, the male CIV to CV, and the CV to CVI are given in Table 52 but the number of determinations are relatively few. The increment between the CV and CVI male is smaller than those of the corresponding females, whether based on total or prosome length. The variation in the increment of prosome length, as indicated by the standard deviation, is large and the increment may be negative.

Thus, there is no general model describing the growth increments at moulting of copepodids of calanoid copepods. This presumably devolves from the wide range of body form in terms of nearly spherical to almost tubular prosomes coupled with abbreviated to extremely elongated urosomes. Allometric growth between copepodid stages is therefore quite pronounced in some species. The nauplii are much more uniform in body shape and the increments tend to decrease logarithmically at successive moults as they have been shown to do in a variety of other crustaceans.

9.3.1.4. *General Concept of Growth of Calanoids*

The various models proposed to describe the intermoult periods of successive stages of development of calanoid copepods only describe the development accurately in a very few species. Other species approximate to one or other models but most seem to deviate either in the non-feeding nauplii, the NVI or the late copepodids. Likewise, there are no successful models describing the successive increments of the stages at moulting. Shrimp-like crustaceans, those with a caridoid facies (Mauchline, 1977b, 1980), have intermoult durations that increase logarithmically and moult increments that decrease logarithmically at each successive stage in the life cycle. Generally, a single equation describes the intermoult periods but two, one for larvae and a second for juveniles and adults, describe the moult increments. This reflects two succesive phases of development. This concept is illustrated in Figure 65 for calanoids. The intermoult period of *Calanus helgolandicus* tends to increase logarithmically in the CII to CV; the duration of the NI also lies on this line but the later nauplii deviate markedly. The moult increments of the NII to NV and those of the CI and CIII to CV

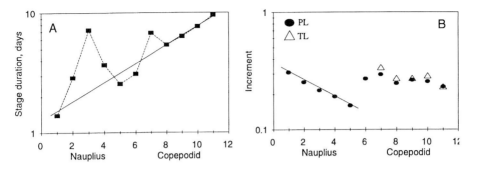

Figure 65 Equiproportional development of *Calanus helgolandicus*. A, stage durations and, B, the premoult increment of nauplii and copepodids, derived from the data in Figure 64A and B, plotted on log-linear scales. The solid lines connect stages fitting regressions, the hatched lines those that deviate from the regressions. Data from Thompson (1982).

decrease logarithmically but on different intercepts suggesting two phases of development. Those of the NI and CII deviate from their respective lines while the NVI is situated between the two phases of development.

Ivanova (1973) also recognized two phases. The body length of each stage and its duration are expressed as percentages of the length of the adult female and of the total development time between egg and adult respectively. She used overall mean values derived from 12 species of calanoids and 8 species of cyclopoids. The data for nauplii are plotted as log percentage length on percentage duration while those of the copepodids are on log-log scales. The growth of the nauplii is fitted by an exponential function while that of the copepodids is best fitted by a power function. The fit of her data is not perfect even though this model contains inclusive values that would tend to smooth irregularities.

Bosch and Gabriel (1994) develop a model that predicts isochronal and equiproportional development. It assumes that the rate of development of the new integument controls the time at which moulting takes place, that is there is a threshold weight for the new integument that initiates moulting to the next stage.

9.3.2. Body Size Parameters

9.3.2.1. *Body Length*

Total body length, measured from the anterior end of the prosome to the posterior end of the caudal furcae, and prosome length, measured from the anterior end of the prosome to the posterior lateral end of the prosome (Figure 4, p. 15), are the two commonest measures of body size. Other measurements, such as the distance between the anterior end of the prosome to the base of the caudal furcae (Deevey, 1960b), have occasionally been used. Length measurements are more convenient to make than those of body wet or dry weights. They are satisfactory measures of body size within a species but some difficulties arise when comparative studies between different species are being made. The copepods illustrated in Figure 2 (p. 4) have a variety of body forms. The last segment of the metasome often has a spinous extension, e.g. *Gaetanus latifrons*, or may be extended posteriorly to a point or the front of the head may be extended. Determining comparable measurements between species is sometimes difficult.

Body length graphed against successive stage number (Figure 66A) demonstrates phases in the growth of females. In *Acartia* species the nauplii I to VI form the first phase with copepodids I to IV and V to VI forming a second and third phase respectively. Growth of *Calanus marshallae* is more complex, nauplii I and II, III to VI and copepodids I and II and III to VI forming four phases of growth. Other species show similar phases of development with some minor variation. Graphing of these data on a log-linear basis (Figure 66C) results in a highly significant relationship that can be used to predict approximate lengths of the stages from measurements of a few stages. Phases are still obvious, the nauplii being distinguished from the copepodids by the discontinuity between them. Nauplii I and II tend to differ from later nauplii and, in *C. marshallae*, copepodids I to IV differ from V and VI.

Relating total length to age of developing females (Figure 66B) also demonstrates phases in the growth but they vary from those found relative to stage of development. Transformation of these data to a log-linear basis (Figure 66D) provides no useful predictive capability. The major problem is the discontinuity between the nauplii and copepodids because of the development of the urosome in the latter.

Body length of the adults is a product of the increments at successive moults during development. The length achieved during an intermoult period, however, bears some relation to the duration of the period and the duration of the period is strongly influenced by environmental temperature (Figure 62, p. 318; Table 48, p. 310). Sabatini (1989) shows that temperature is more important than available food in influencing body length of *Acartia tonsa*. Adult female *Pseudocalanus minutus* reared from C III isolated from the field are larger when reared at lower than higher temperatures; the moult increments of body length increase from about 43% at 12°C to about 60% at 6°C (McLaren, 1974). *Eurytemora herdmani* grow larger at lower temperatures in the laboratory (McLaren and Corkett, 1981). There are

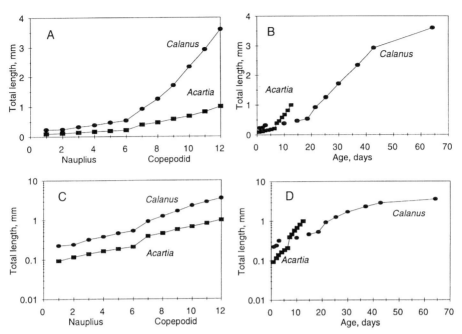

Figure 66 Increase in total length of developmental stages of females relative to A, stage of development and B, age on linear scales and, C and D, on log-linear scales. *Acartia* species exhibit isochronal and *Calanus marshallae* sigmoidal development (see Figure 63). Regression equations for the relationships are:
A. tonsa Log TL = 0.099 Stage number -1.158 r, 0.988
C. marshallae Log TL = 0.118 Stage number -0.858 r, 0.991
(Based on data of: Heinle, 1966; Peterson, 1986; Berggreen *et al.*, 1988.)

many instances cited in the literature of increased body size at lower temperature, reflected by the occurrence of larger individuals of a species at higher latitudes. The relationship of length to environmental temperature is linear in females (Table 53) but Lock and McLaren (1970) found no effect of temperature on body size of male *Pseudocalanus minutus*.

Body length, total or as prosome length, varies seasonally and there are numerous studies describing this (see for example: Uye *et al.*, 1982; Brylinski, 1984b; Crawford and Daborn, 1986; Hada *et al.*, 1986; Smith and Lane, 1987; Chisholm and Roff, 1990a; Mauchline, 1992a, 1994a). In the equations for species examined by Deevey (1960b) in Table 53, she related prosome length in field samples of copepods to the mean environmental temperature one month previous to sampling; the other equations in Table 53 are derived from the temperatures prevalent at the time of sampling the copepods. Liang and Uye (1996) and Liang *et al.* (1996) found that

Bělehrádek functions describe the relationship of mean body size of each naupliar and copepodid of *Acartia omorii* and *Centropages abdominalis* when related to the temperature at the time of sampling. These temperature relationships explain much of the seasonal and geographic variation in the body size of a species.

A dimensionless measure of variation in body length is the standard deviation of the mean expressed as a percentage of the mean. This can be used to examine or compare variation in body size in cultured copepods or in field samples of species. For instance, annual mean prosome lengths and their corresponding standard deviations are derived from time series samples of different species, the samples being representative of all seasons of the year. The standard deviations are then expressed as percentages of the annual means and in Figure 67, are related to the modal depth of occurrence of the species. The standard deviations are largest in coastal and least in bathypelagic populations, contrasting the highly variable coastal environments with the much more conservative deep sea.

(a). *Influence of other factors on body length* Sometimes little correlation is found between body size and environmental temperature at the time of sampling, as instanced by the data of Hada *et al.* (1986) on populations of *Sinocalanus tenellus* in a brackish water pond in Japan. Lack of correlation probably arises from two sources; the temperature changes markedly within the generation time of the copepod and/or there are major fluctuations in the availability of suitable food. For example, Durbin *et al.* (1983) found that there was no correlation between body dry weight and prosome length in *Acartia tonsa* in Narragansett Bay during the summer. Egg production fluctuated widely and they conclude that the growth was food-limited. This resulted in an almost constant prosome length throughout the summer even although environmental temperatures changed. The variation in the environmental conditions, especially of food availability, results in individuals having a variable condition factor as instanced by the constancy of prosome length but the variability of body dry weight. Ingested energy is directed to storage reserves or directly to egg production rather than to increase in body length.

An adequate supply of suitable food is a prerequisite for development and growth. Klein Breteler and Gonzalez (1982) question the dominance of temperature as a determinant of body size; they found that food concentration influenced body size in several species and that it explained 80% of the variation in size of *Centropages hamatus* in culture. Diel and Klein Breteler (1986), studying populations of *Calanus* species both in the field and experimentally, conclude that development and growth can be arrested by changes in the quality of food available. Development time and generation time are influenced by food (Figure 61, p. 316) and they in turn reflect growth in length. Evans (1981) found that some 83% of the variance

Table 53 Relationship of prosome length, PL in μm, to environmental temperature, T in °C.

Species	Stage	Equation	Authority
Acartia clausi	CI	PL = 392 − 3.1T	Durbin and Durbin, 1978
	CII	PL = 491 − 4.6T	Durbin and Durbin, 1978
	CIII	PL = 567 − 4.8T	Durbin and Durbin, 1978
	CIV ♀	PL = 683 − 6.5T	Durbin and Durbin, 1978
	CIV ♂	PL = 658 − 6.3T	Durbin and Durbin, 1978
	CV ♀	PL = 807 − 6.9T	Durbin and Durbin, 1978
	CV ♂	PL = 745 − 5.1T	Durbin and Durbin, 1978
	CVI ♀	PL = 955 − 7.5T	Durbin and Durbin, 1978
	CVI ♂	PL = 851 − 7.2T	Durbin and Durbin, 1978
Acartia tonsa	CVI ♀	PL = 984 − 15.9T	Uye, 1982a
	CVI ♀	PL = 1045 − 11.6T	Heinle, 1969
	CVI ♀	PL = 1014 − 12.0T	Ambler, 1985
	CVI ♀	PL = 1069 − 14.4T	Cataletto and Fonda Umani, 1994
Acrocalanus gibber	CVI	PL = 1338 − 19.4T	McKinnon and Thorrold, 1993
Calanoides carinatus	CVI ♀	PL = 2480 − 18.0T	Binet and Suisse de Saint Claire, 1975
	CVI ♀	PL = 2707 − 28.9T	Petit and Courties, 1976
Calanus minor	CVI ♀	PL = 1880 − 11.0T	Ashjian and Wishner, 1993a
C. sinicus	CIV	PL = 1564 − 8.0T	Huang et al., 1993
	CV	PL = 2443 − 30.7T	Huang et al., 1993
	CVI ♀	PL = 2807 − 33.6T	Huang et al., 1993
	CVI ♂	PL = 2582 − 28.8T	Huang et al., 1993
Centropages typicus	CVI	PL = 2455 − 20.3T	Uye, 1988
	CVI ♀	PL = 1242 − 10.5T	Deevey, 1960b
	CVI ♀	PL = 1290 − 13.1T	Deevey, 1960b
Drepanopus pectinatus	CVI ♀	PL = 972 + 73.7T	Razouls and Razouls, 1988
Gladioferens imparipes	CVI ♀	PL = 1022 − 9.1T	Rippingale and Hodgkin, 1974
Paracalanus sp.	CVI ♀	PL = 850 − 9.9T	Uye, 1991
Pseudocalanus minutus	CVI ♀	PL = 1212 − 28.3T	Deevey, 1960b
Temora longicornis	CVI ♀	PL = 1309 − 45.5T	Deevey, 1960b

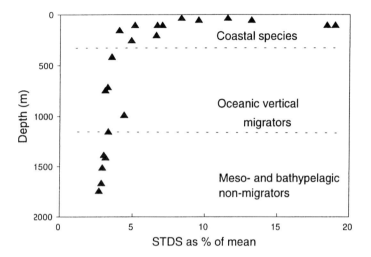

Figure 67 The standard deviation of the mean annual prosome length of different species of copepods expressed as a percentage of the mean annual prosome lengths. Mean annual prosome length was derived from time-series samples of each species published by various authors. The species of meso- and bathypelagic non-migrators are all *Pareuchaeta* species. The mesopelagic migrators are *Euchaeta* and *Pareuchaeta* species and *Calanus hyperboreus*. Coastal species belong to genera such as *Acartia*, *Centropages*, *Microcalanus*, *Pseudocalanus* and *Temora*. (After Mauchline, 1992a.)

of body size on *Temora longicornis* in the North Sea was explained by the abundance of the diatom *Thalassiosira* and only 7% by the temperature. There may, however, be a delayed effect of changes in food availability on the consequent changes in body size, as much as two weeks in *Temora longicornis* according to Sander and Moore (1983).

Salinity has an effect on growth and development in estuarine species. Gaudy *et al.* (1988) and Pagano and Saint-Jean (1989) found that body length of female *Acartia clausi* increased with salinity to an asymptotic length at between 10 and 15‰. There is likely to be an optimal window of salinity for successful development and growth of an estuarine species. Salinity and temperature tolerances interact and Nagaraj (1988) found that nauplii of *Eurytemora velox* tolerated low salinities better at higher temperature and high salinities better at lower temperatures.

Ohno *et al.* (1990) also found that the prosome length of *Acartia tsuensis* decreased at increasing density in their cultures. This may have reflected decreasing food rations per individual thus actually reflecting a degree of food limitation.

Studies of heritable characteristics in copepods have been made by

McLaren (1976) and Corkett and McLaren (1978). The genetic component of the determination of body length within a species is still not defined. Klein Breteler *et al.* (1990) found that body size of species cultured in the laboratory over many generations increased or decreased depending upon the species. Field-collected males crossed into the cultures tended to restore the body size of *Temora longicornis*, suggesting a possibility of a genetic component. The culture conditions themselves, without a genetic component, but through nutritional deficiencies or lack of temperature changes, may result in a cumulative change in body size.

9.3.2.2. *Body Weight*

The discontinuity between morphology of nauplii and copepodids is not obvious when growth is examined in terms of dry body weight (Figure 68). Dry weight increases in a nearly exponential manner with sequential stages of development. The copepods illustrated in Figures 68 and 69 are an *Acartia* species, based on data on the isochronal *A. tonsa*, and *Calanus marshallae* which has a sigmoidal pattern of development. The data are shown on a semi-logarithmic plot in Figure 68C and D where the deviation from a truly exponential increase in weight can be seen in *C. marshallae*. It is very small, especially between NV and CVI, the nauplii deviating most. The data for *Acartia* species deviate less, as first shown by Miller *et al.* (1977) for *A. tonsa* and by McLaren and Corkett (1981) for *Eurytemora herdmani*. Plotting the same data on a linear and semi- logarithmic basis against age in days (Figure 69) results in similar deviation from exponential growth. The data for the *Acartia* species are closer to exponential than those of *C. marshallae*. Near exponential growth with time (Figure 69C) was also found for *A. tonsa* by Miller *et al.* (1977) and Berggreen *et al.* (1988) and for *Centropages typicus* and *C. hamatus* by Fryd *et al.* (1991). Deviating curves, similar to that for *Calanus marshallae* in Figure 69D, were found for *Temora longicornis* by Harris and Paffenhöfer (1976a), for *Calanus sinicus* by Uye (1988) and for *Paracalanus* sp. by Uye (1991). The body of a copepod can be conceived as consisting of two compartments, the structural and the storage compartments (Harris, 1983). The storage compartment is usually considered as the stored lipids or oil sac. McLaren (1986) suggests that species that do not have significant quantities of stored lipids, e.g. *Acartia* and *Eurytemora* species, exhibit exponential growth in weight, while those that store marked quantities of lipids show close to exponential growth if the lipid stores are ignored. Carlotti *et al.* (1993) discuss the existence of exponential growth further but show that after the CIV there is deviation that at present is ascribed to storage compartments and not to structural growth.

Long-lived species have discontinuous growth curves as illustrated by

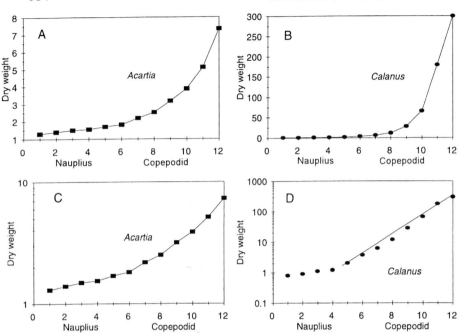

Figure 68 Increase in body dry weight (μg) of A, *Acartia tonsa* and B, *Calanus marshallae* relative stage of development (stage number) on a linear basis. C and D are the same data plotted on a log-linear basis. Species are the same as in Figure 66. The regression equation for the relationships in C and D are:
A. tonsa
NV to CVI Log DW = 0.090 Stage number -0.275 r, 0.984
NI to CVI Log DW = 0.064 Stage number -0.040 r, 0.957
C. marshallae
NV to CVI Log DW = 0.322 Stage number -1.391 r, 0.995
(Based on data of: Heinle, 1966; Berggreen *et al.*, 1988; Peterson, 1988.)

that of *Calanus glacialis* (Figure 70). High latitude species frequently overwinter in a resting copepodid stage, most frequently but not always the CV. The durations of these stages, relative to the others in the developmental sequence, is extended but even then, as Carlotti *et al.* (1993) point out, the structural growth of the stages, adjusted to take account of resting stages, approaches the exponential.

As mentioned previously, the loss of weight through casting of the old integument at each moult represents some 2 to 5% of body weight (see Vidal, 1980b), a small proportion of the increment in body weight that takes place at each moult. Its effect, therefore, on the computed curves of growth in weight will be minimal.

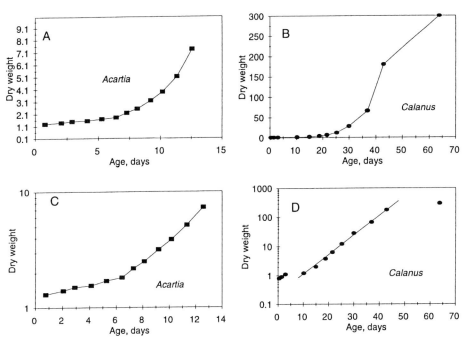

Figure 69 Increase in body dry weight (μg) of A, *Acartia tonsa* and B, *Calanus marshallae* relative to age on a linear basis. C and D are the same data plotted on a log-linear basis. Species are the same as in Figure 66. The regression equations for the relationships in C and D are:
A. tonsa
NI to CVI Log DW = 0.061 Age − 0.033 r, 0.954
Calanus marshallae
NIV to CV Log DW = 0.068 Age − 0.669 r, 0.998
(Based on data of: Heinle, 1966; Berggreen *et al.*, 1988; Peterson, 1988)

Body dry weight, like body length, fluctuates seasonally and is related to environmental temperature (see examples in Table 54). Castel and Feurtet (1989) show that the dry weights of CI to CVI each range by an annual factor of two in *Eurytemora affinis* in the Gironde Estuary; least change occurs in the CI but the amplitude increases progressively to the adult females and males. The seasonal change in dry weight usually, but not always, reflects corresponding seasonal changes in body length. Body dry weights in *Centropages typicus* and *Temora stylifera* from the Gulf of Lions (Banyuls-sur-Mer) varied strongly over short time periods such that derived length to dry weight relationships were of little value (Razouls and Razouls, 1976). No correlation between body length and dry weight of adult *Temora longicornis* in the Southern Bight of the North Sea was found by Daro

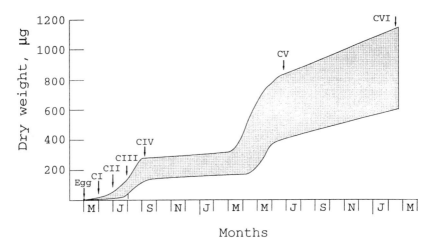

Figure 70 Probable growth curve of a biennial species, *Calanus glacialis*, showing the times of development of the copepodids. (After Slagstad and Tande, 1990.)

and Gijsegem (1984) although a significant correlation was present in copepodids I to III. Durbin *et al.* (1983) found no correlation between body dry weight and prosome length in *Acartia tonsa* in a food-limited population in Naragansett Bay.

These changes in body weight, seasonal and otherwise, relative to body length have initiated development of the concept of a condition factor. The commonest condition factor (CF), first used by Durbin and Durbin (1978), is:

$$CF = (\text{Dry weight } (\mu g) \times 10^{-1})/(\text{Prosome length (mm)})^3$$

This gives a factor for *Acartia clausi* of between 0.75 to 1.25 (Durbin and Durbin, 1978), of 0.44 to 0.57 (Ayukai, 1987), and of 0.7 to 1.0 (Christou and Verriopoulos, 1993a). Comparable factors for *A. tonsa* are between 1.4 and 2.8 and Durbin *et al.* (1983) show that the factor is unrelated to prosome length but increases with increasing food availability to an asymptotic level in this species.

An alternative condition factor is used by Oh *et al.* (1991):

$$CF = (\text{Wet weight (mg)} \times 10^2)/(\text{Prosome length (mm)})^3$$

This formula yielded condition factors of 5.06 and 5.64 for *Neocalanus cristatus* in Sagami Bay and the northern North Pacific respectively.

Table 54 Relationship of body dry weight, W in μg, to environmental temperature, T in °C.

Species	Stage	Equation	Authority
Acartia clausi	CI	W = 0.677 − 0.00349T	Durbin and Durbin, 1978
	CII	W = 1.044 − 0.00469T	Durbin and Durbin, 1978
	CIII	W = 1.630 − 0.00498T	Durbin and Durbin, 1978
	CIV ♀	W = 2.815 − 0.00673T	Durbin and Durbin, 1978
	CIV ♂	W = 2.534 − 0.00666T	Durbin and Durbin, 1978
	CV ♀	W = 5.436 − 0.00759T	Durbin and Durbin, 1978
	CV ♂	W = 4.124 − 0.00590T	Durbin and Durbin, 1978
	CVI ♀	W = 11.217 − 0.00877T	Durbin and Durbin, 1978
	CVI ♂	W = 6.612 − 0.00801T	Durbin and Durbin, 1978
A. tonsa	CVI ♀	W = 8.67 − 0.25T	Cataletto and Fonda Umani, 1994
A. tranteri	CVI ♀	W = 7.30 − 0.18T	Kimmerer and McKinnon, 1987a
Temora longicornis	CVI ♀	W = 26.31 − 0.49T	Daro and Gijsegem, 1984
	CVI ♂	W = 27.86 − 0.78T	Daro and Gijsegem, 1984

9.3.2.3. Allometric Growth

The general proportions of the body of copepods vary from species to species. This is amply illustrated in Frost's (1989) study of species in the genus *Pseudocalanus*. He presents various biometric measurements, including the relationship of urosome length to prosome length which is different between *P. acuspes* and *P. minutus*. Prosome length is approximately 0.75 of total length in many species. Some species, however, have elongated or abbreviated urosomes and in them the ratio deviates significantly from 0.75. Interspecies comparisons of lengths or widths of component parts of the body, for instance the width of the mandibles (Karlson and Båmstedt, 1994), do not result in useful correlations with body length because of the variation in morphology between species. Such differences, however, can be used to separate closely related species (Fleminger, 1967a; Frost, 1974; Grigg *et al.*, 1987) or to examine sexual differentiation within a species (Grigg *et al.*, 1981, 1985).

Within-species comparisons are much more frequently justified although there is allometric growth of the prosome and urosome during development. Karlson and Bådmstedt (1994) found direct correlation between the width of the mandible and prosome length of the CI to CVI females of *Calanus finmarchicus*; the mandible of the male tended to be smaller relative to prosome length of the male. A detailed study of variation and allometry, involving 21 separate measurements of parts of *Temora stylifera* from different regions, is made by Riera (1983); correlations with environmental temperature and season are examined.

9.4. GROWTH RATES

Rates of growth of copepods are temperature dependent (Huntley and Lopez, 1992). Rates of growth in terms of developmental stages are calculated as stages d^{-1} or the proportion of body length d^{-1}. The most meaningful expression of growth, however, is as weight-specific growth rate d^{-1}, the increase in body weight d^{-1} given as a proportion of the body weight of the female or stage of development being considered. Kiørboe and Sabatini (1995) show that the weight-specific growth rates of nauplii, CI to CVI and CIV or CV to CVI are independent of the body size. A similar result was obtained by Huntley and Lopez (1992) when specific growth rates are related to the body weight of adult females (Figure 71A). Their data are reduced here to mean values where multiple results are quoted by them for a single species. Examples of weight-specific growth rates of species are given in Table 55. Conversely, Peterson and Hutchings (1995) find that the maximum specific growth rates of *Calanus agulhensis* and *C. pacificus* decrease as body weight increases through the successive developmental stages (Figure 72).

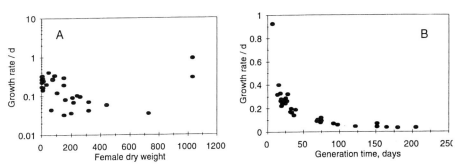

Figure 71 The specific growth rate per day relative to A, female dry body weight in μg and B, generation time in days. (After Huntley and Lopez, 1992.)

This is approximately true of many species although considerable variation exists. The weight-specific growth rate is related to generation time (Figure 71B) so that longer-lived copepods grow at slower rates.

Adult females do not grow appreciably in body size and eggs are the only form of production. Sekiguchi *et al.* (1980), examining *Acartia clausi hudsonica*, suggested that rates of egg production are predictable from growth rates. The converse relationship, the prediction of growth rates from rates of egg production, is of practical value in the field for the quick estimation of growth rates (McLaren and Leonard, 1995; Hay, 1995; Poulet *et al.*, 1995a). Gravid females are incubated, often on board ship, and the rate of egg production determined. McLaren and Leonard (1995) caution that the correspondence between the rate of egg production and growth rate has not been fully validated and that some further work is advisable. They discuss the current problems of the method in considerable detail. Further, this relationship (Figure 72) means that growth rates of adult females will not be representative of those of earlier stages. Fransz and Diel (1985) found highest values in NII and NIII and again in CI to CIII of *C. finmarchicus*; thus a continuous decrease in growth rates is by no means the rule. Examples of these rates are given for egg production in Table 45 (p. 284) and for growth rates in Table 55. Growth rates can also be estimated from the difference between the weights of the egg and adult female and the generation time (Huntley and Lopez, 1992; Kiørboe and Sabatini, 1995).

9.5. LONGEVITY

Longevity of adult copepods is difficult to estimate in the field. Laboratory estimates are often made under conditions of excess food and constant

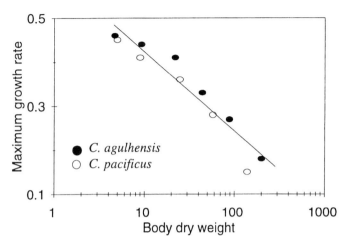

Figure 72 Maximum specific growth rates of CI to CVI of *Calanus agulhensis* and CI to CV of *C. pacificus* relative to body dry weight (μg). The equation for the line is:

$$G_{max} = 0.602 - 0.182\log W \quad r = 0.958$$

(After Peterson and Hutchings, 1995.)

temperature. Uye (1982a) calculates that adult female *Acartia clausi* have longevities ranging seasonally from about 10 days to 1 to 2 d. Longevities determined in laboratory cultures at identical ranges of temperature are 68 days and about 27 d respectively. Uye considers that the abbreviated longevities in the wild result from predation. Longevity of female and male *A. tranteri* appeared similar, having median values of 26 d below 17°C and 4 d above 20°C (Kimmerer and McKinnon, 1987a). Peterson (1985) considers that female *Temora longicornis* in Long Island Sound lives for 3.1 d on average but its longevity varies seasonally; about 10 d in March, less than one day in May, but then increases to about 10 d by late June after which it decreases to zero in late July, when water temperatures exceed the lethal limit (20°C) for this species. These field estimates are much shorter than the 45 d at 12°C and 50 d at 7°C found by Harris and Paffenhöfer (1976a) and Peterson (1985) in the laboratory. Female *Calanoides carinatus* in Ghanaian waters lives for some 32 to 33 d (Mensah, 1974). Laboratory estimates of longevity of some species are given in Table 56.

Parrish and Wilson (1978) state that male *Acartia tonsa* only live one third to one half as long as females. Adult males of calanoid copepods are usually smaller in body size than adult females. Gilbert and Williamson (1983) review the female:male ratios of body length in 24 families of calanoid copepods and find that they range from 1.08 in the Scolecitrichidae to 1.41 in the Eucalanidae; the mean ratio is 1.13. They also plot the ratios

Table 55 Weight-specific growth rates d^{-1} (growth d^{-1}) of calanoid copepods expressed as a proportion of the female's body dry or carbon weight. Temperature (T °C) and relevant developmental stages are given.

Species	Stage	T °C	Growth d^1	Authority
Acartia clausi		20	0.34–0.36	Landry, 1978b
		20	0.34–0.36	Uye, 1982a
A. tonsa		20	0.53	Miller et al., 1977
		25.5	0.75	Miller et al., 1977
	NI–CVI	16–18	0.44	Berggreen et al., 1988
A. tranteri			0.11	Kimmerer and McKinnon, 1987a
Acartia sp.		8–10	0.13	Landry, 1978b
Eucalanus bungii	CI–CIII	0.5–6.0	0.11	Vidal and Smith, 1986
Calanus finmarchicus	CIV		0.048	Miller and Nielsen, 1988
C. helgolandicus	NI–CI	15	0.06–0.27	Green et al., 1991
C. marshallae	NI–NIV	10	0.05	Peterson, 1986
	NV–CV	10	0.176	Peterson, 1986
	CV–CVI	10	0.024	Peterson, 1986
	CI–CIII	0.5–6.0	0.14–0.15	Vidal and Smith, 1986
C. pacificus	NI–CI	15	0.30	Mullin and Brooks, 1970
	NI–CI	15	0.29	Paffenhöfer and Harris, 1976
	NI–CI	15	0.41	Paffenhöfer and Harris, 1976
	NII–CI	15	0.17	Fernandez, 1979
	CII	8	0.188	Vidal, 1980a
	CIII	8	0.171	Vidal, 1980a
	CIV	8	0.164	Vidal, 1980a
	CV	8	0.118	Vidal, 1980a
	CII	12	0.334	Vidal, 1980a
	CIII	12	0.292	Vidal, 1980a
	CIV	12	0.219	Vidal, 1980a
	CV	12	0.131	Vidal, 1980a
	CII	15.5	0.413	Vidal, 1980a

Table 55 Continued.

Species	Stage	T°C	Growth d[1]	Authority
Centropages hamatus	CIII	15.5	0.355	Vidal, 1980a
	CIV	15.5	0.277	Vidal, 1980a
	CV	15.5	0.149	Vidal, 1980a
	NI–NVI	17	0.264 ± 0.01	Fryd et al., 1991
	CI–CV	17	0.288 ± 0.08	Fryd et al., 1991
	Egg–CV	17	0.267 ± 0.04	Fryd et al., 1991
C. typicus	NI–NVI	17	0.340 ± 0.01	Fryd et al., 1991
	CI–CV	17	0.376	Fryd et al., 1991
	NI–CV	17	0.360	Fryd et al., 1991
C. velificatus	CI–CVI	28	0.63	Chisholm and Roff, 1990b
Eucalanus bungii	CI–CIII		0.11	Vidal and Smith, 1986
Euchaeta marina	NI–CVI		0.07–0.38	Webber and Roff, 1995b
	NII–CV	14	0.15	Escaravage and Soetaert, 1993
E. herdmani	NI–CVI	9	0.41	McLaren and Corkett, 1981
	NI–CVI	9.5	0.43	McLaren and Corkett, 1981
	NI–CVI	13	0.62	McLaren and Corkett, 1981
Metridia pacifica	CI–CV	0.5–6.0	0.13–0.15	Vidal and Smith, 1986
Neocalanus cristatus	CI	0.5–6.0	0.072–0.079	Vidal and Smith, 1986
	CII	0.5–6.0	0.067–0.078	Vidal and Smith, 1986
	CIII	0.5–6.0	0.059–0.077	Vidal and Smith, 1986
	CIV	0.5–6.0	0.040–0.065	Vidal and Smith, 1986
	CV	0.5–6.0	0.031–0.047	Vidal and Smith, 1986
N. flemingeri	CV		0.10	Miller and Nielsen, 1988
N. plumchrus			0.035–0.140	Fulton, 1973
	CI	0.5–6.0	0.143	Vidal and Smith, 1986
	CII	0.5–6.0	0.141	Vidal and Smith, 1986
	CIII	0.5–6.0	0.133	Vidal and Smith, 1986

Species	Stage	Value	Reference
Paracalanus aculeatus	CIV	0.5–6.0	Vidal and Smith, 1986
	CV	0.5–6.0	Vidal and Smith, 1986
	CV		Miller and Nielsen, 1988
P. indicus	CI–CVI	28	Chisholm and Roff, 1990b
		0.63	Kimmerer and McKinnon, 1989
P. parvus		0.22	Checkley, 1980a
Pareuchaeta elongata		0.38	
	CIII	18	Dagg and Littlepage, 1972
	CIV	6-10	Dagg and Littlepage, 1972
	CV	6-10	Dagg and Littlepage, 1972
	CVI	6-10	Dagg and Littlepage, 1972
		6-10	
		0.15	
		0.128	
		0.019	
		0.01	
Pseudocalanus elongatus	NI–CI	15	Green et al., 1991
	NI–CVI	12.5	Fransz et al., 1991
	CII–CV	0.5–6.0	Vidal and Smith, 1986
		0.08–0.24	
		0.17–0.26	
		0.13	
Pseudocalanus sp.	NII	20	Uye et al., 1983
Pseudodiaptomus marinus	NIII	20	Uye et al., 1983
	NIV	20	Uye et al., 1983
	NV	20	Uye et al., 1983
	NVI	20	Uye et al., 1983
	CI	20	Uye et al., 1983
	CII	20	Uye et al., 1983
	CIII	20	Uye et al., 1983
	CIV	20	Uye et al., 1983
	CV	20	Uye et al., 1983
		0.283	
		0.309	
		0.246	
		0.213	
		0.105	
		0.568	
		0.567	
		0.219	
		0.277	
		0.342	
Rhincalanus nasutus	NI–CI	15	Mullin and Brooks, 1970
Sinocalanus tenellus	NI–NVI	20	Kimoto et al., 1986a
	CI–CVI	20	Kimoto et al., 1986a
		0.64	
		0.53	
		0.74	
Temora longicornis	NI–CVI	12.5	Fransz et al., 1991
T. turbinata	CI–CVI	28	Chisholm and Roff, 1990b
Undinula vulgaris	CVI	30	Gerber and Gerber, 1979
	NI–CVI		Webber and Roff, 1995b
		0.13–0.18	
		0.48	
		0.048	
		0.11–0.49	

for individual species on a frequency distribution and show that modal values are 1.0 to 1.2. Males tend to develop faster from the egg, that is they have shorter generation times than females. Drits *et al.* (1994) considers that female *Calanoides acutus* in the Antarctic may achieve adulthood at more than one year old while males will complete development in less than a year. Adult male *Centropages typicus* usually developed two to four days before the females (Smith and Lane, 1985).

Maximum longevity is related to temperature as instanced by:

$$\text{Longevity (days)} = 1334(T + 0.7)^{-0.86}$$

for *Calanus sinicus* where T is temperature in °C (Uye, 1988).

9.6. CONCLUDING REMARKS

Growth rates of copepods, like those of other crustaceans, have to be estimated by time series observations of cohorts within populations and/or by laboratory rearing of the species. Roff *et al.* (1994), however, have developed a radiochemical method for determining the rate of chitin synthesis that can be applied to crustaceans as individuals or as populations. They fed ^{14}C-labelled algae to *Daphnia magna* and measured the rate of its incorporation into the integument as [^{14}C]N-acetylglucosamine. Further development of this method may allow estimates of production in marine copepods comparable to those of primary production.

Copepods are small animals, the largest being about 12 mm in total length (Figure 3, p. 5). There are presumably constraints on body size, a topic examined by Myers and Runge (1986) and Runge and Myers (1986). They suggest that the decrease in body size with temperature is caused by increasing rates of mortality. This devolves from the fact that growth rates increase with increasing temperature yet resulting body sizes decrease. They quote Miller *et al.* (1977) who conclude that "If an animal is forced through its fixed quota of molts quickly by high temperature, it simply has no chance to grow large". This is not a problem peculiar to copepods but of many other organisms as well. In copepods, body size is a result of the intermoult duration and the growth increment at moulting. These are both inversely related to temperature. They are modified by the quantity and quality of food available. Food, however, is most likely to be limiting in higher latitudes at lower temperature where selection for greater body size exists. There may also be genetic constraints (McLaren and Corkett, 1978). A diapause or resting stage in the life cycle (Figure 70), usually the CV but sometimes the CIII and CIV are involved, also tends to occur at higher

Table 56 Longevity of adult calanoid copepods in days, estimated in the laboratory.

Species	Female	Male	Authority
Acartia clausi	30.2 ± 12.4		Ianora *et al.*, 1996
A. tonsa	26		Paffenhöfer, 1991
Centropages typicus	14.8	16.3	Carlotti and Nival, 1992b
Eucalanus hyalinus	>60		Paffenhöfer, 1991
E. pileatus	30		Paffenhöfer, 1991
Eurytemora affinis	20		Vuorinen, 1987
Paracalanus parvus	11		Paffenhöfer, 1991
Pseudodiaptomus acutus	15.0 ± 1.4	12.1 ± 0.7	Jacoby and Youngbluth, 1983
P. cokeri	12.9 ± 0.8	9.8 ± 0.5	Jacoby and Youngbluth, 1983
P. coronatus	12.4 ± 0.4	11.2 ± 0.4	Jacoby and Youngbluth, 1983

latitudes. This represents an interruption to growth and development and yet, once it is over, the previous pattern of moulting and growth is resumed (Figure 70). Runge and Myers (1986) suggest that there may be a general coupling between temperature, rates of growth and mortality. Vidal (1980d) suggests that smaller-sized species of copepods optimize growth and use food more efficiently at higher temperatures, even under oligotrophic conditions, while larger-sized species optimize growth and use food more efficiently at lower temperatures, thus contributing an explanation for geographical and vertical patterns in the body sizes of copepods.

The conception of rates of growth should possibly incorporate terms describing rates of approach to sexual maturity. Evidence of gonad formation is present as early as the CI in some species and in the CIV in most. What controls the formation and maturation of the gonads and how are the processes linked to growth and temperature? Is there a quantifiable proportion of mortality directly linked to successful reproduction, the end-point of growth and development?

Perhaps the most difficult area to quantify is the genetically-transmitted components involved in growth and development. Early work by McLaren (1976) suggests that inherited traits influencing, for example, rates of mortality in *Eurytemora herdmani*, vary between progeny of different females. Tepper and Bradley (1989) discuss the maintenance of genetic variability within a population of *E. affinis* and its interaction with the physiological adaptation of the individual.

At a lower level of organization, McLaren and Marcogliese (1983) found that the numbers of cells in nauplii of different species in different genera approximate to 2000 while the corresponding numbers in CI of a similar range of species are 9600 to 13,000. The sizes of nuclei are not closely related to body size of the different species across the genera. Consequently, increased body size up to at least the CI must reflect the differences in the quantities of cytoplasm associated with the nuclei. McLaren *et al.* (1989c) showed that DNA per nucleus in six *Pseudocalanus* species and seven *Calanus* species constitute a highly-significant non-random series not attributable to polyploidy. They suggest that it is a quantum series with a unit of about 4.19 pg DNA. The sizes of nuclei within the six *Pseudocalanus* species is related to body size of the species. Consequently, there may be nucleotypic control of body size and other life history parameters within a genus.

10. Population Biology

10.1. Sampling	349
10.2. Demographic Analysis	352
10.2.1. Seasonal Changes in Stage Structure	353
10.2.2. Size Frequency	354
10.2.3. Breeding Seasons	354
10.2.4. Generation Time	363
10.2.5. Sex Ratio	363
10.2.6. Mortality	366
10.2.7. Control of Population Size	370
10.2.8. Production	373
10.3. Life History Patterns (Strategies)	381
10.3.1. Tropical and Subtropical Patterns	382
10.3.2. Temperate Patterns	383
10.3.3. High Latitude Patterns	384
10.3.4. Deep-sea Patterns	390
10.4. Population Maintenance	392
10.4.1. Annual Fluctuations	393
10.4.2. Inter-annual Fluctuations	394
10.5. Biomass of Populations	398

The natural life cycles of copepods have been studied for many years with varying success. Most attention has been given to coastal genera such as *Acartia*, *Centropages*, *Pseudocalanus*, and *Temora*. Some offshore groups such as the Calanidae and Metridinidae have received attention because of their importance in the economics of the oceans. Others, such as *Pareuchaeta norvegica*, normally oceanic or inhabitants of the slope, have been studied because they also form isolated populations in inshore areas such as fjords. The advent of laboratory cultivation of many coastal species in the last 20 years or so has accelerated our understanding of aspects of the life histories of a variety of species through influencing the interpretations of results gained from the field samples.

Field studies of the biology of copepods are not easy for a variety of reasons, some of which are discussed in the next section. Extrapolation of results gained from laboratory culture to such populations is often difficult because of the stable and food-rich conditions of the culture regimes. For instance, Corkett and McLaren (1978) point out that an adult female *Pseudocalanus* sp. can produce up to 10 or so clutches of eggs under culture conditions but that this is probably unlikely in the field because of natural mortality. Likewise, longevity of breeding adults in cultures is frequently measured in weeks whereas it may be as short as a few days in many wild populations. The data from cultures do, however, provide estimates of the effects of changing environmental temperature on development times and so on lengths of generations.

Population dynamics are studied by observing the changes in the mean values for the population of different parameters, one of the commonest of which is body size, relative to time. There is considerable variation in the mean size that can be expressed as the standard deviation of the mean. This measure does not define the source of the variation which may arise from heterozygosity within the population and from varying composition of the population. A population may be relatively isolated and contain individuals primarily endemic to that region. Other populations, in coastal as well as oceanic regions, recruit individuals from upstream regions and export individuals downstream. Laboratory cultures can define expected variation of a parameter between individuals.

There are various ways of approaching or planning field investigations depending on the aspects of the biology of the copepod being investigated. Analysis of the generation time in a few species and environments is a simple matter of determining the seasonal occurrence of gravid females or females with egg masses attached. A high level of synchrony in the timing of egg laying by the females in the population can sometimes extend to the second and subsequent broods of that generation, making determination of the life history relatively easy. Such a situation, however, tends to be rare and, in practice, there are overlapping generations and overlapping broods within a generation and a much more complex analysis is necessary. Corkett and McLaren (1978) have described variations in the life history of *Pseudocalanus* sp. between different environments. This species has a life history of 1 to 2 years in high latitudes and analysis of its life history is simply defined by counting the incidence of developmental stages and adults against time. In lower latitudes, with overlapping broods and generations, counting of individual stages is combined with measurements of the body lengths of the stages. The body lengths of the stages change seasonally and often contiguous generations can be separated on the basis of their body sizes. Such data, coupled with the frequency of stages against time, are often successful in identifying successive broods and generations. There are, however, some inherent problems in most investigations, some of which are discussed in the next section.

10.1 SAMPLING

Studies of the population biology of a species of copepod in the field depend on obtaining representative samples of the population. Such samples are very difficult to obtain, for a variety of reasons. The specific difficulties encountered are dependent upon the objectives of the sampling programme. A programme that is designed to examine the life history of a coastal species is often located within a relatively enclosed bay, fjord or in a delimited region of an estuary. Gagnon and Lacroix (1981) discuss the variability between time-series samples taken in such an environment and provide an interpretative model. The hope of obtaining representative samples is increased if the environment within which the population lives is restricted topographically. The population is more or less isolated from other populations of the same species such that immigration and emigration are not a major feature and can often be ignored. Such a population can be subjected to time-series sampling with some confidence that the same population is being sampled on each visit. Equal confidence is usually lacking in studies sited in more open coastal or oceanic situations. For example, Atkinson (1989a,b) discusses problems of interpreting data from samples of the six major species around South Georgia where his winter sampling grid lay within the polar front.

There are, however, still difficulties in the sampling of enclosed and isolated populations. The environments of such populations are often looked upon as large, natural experimental tanks. The distribution of copepods within experimental tanks is often uneven, there being "wall effects", as discussed in Chapter 6. Such effects can also be present within fjords since they often have deep and shallow areas with slopes and shelves. A species is not usually evenly distributed throughout the entire volume of the fjord but has centres of population associated with certain features such as deep basins, shelves or low salinity regions. Consequently, between-sample variation in numbers of, for example, adult females caught, can be significant. The numbers of adult males caught will also vary, and sometimes almost independently of those of the females, because of behavioural differences. Determination of sex ratios can then have large errors caused by unrepresentative sampling.

Mauchline (1994a) examines the percentage variation in several parameters of populations of *Pareuchaeta norvegica* in the fjordic environment of Loch Etive, western Scotland, and in the oceanic environment of the Rockall Trough, northeastern Atlantic Ocean (Table 57). Multiple placement of spermatophores takes place in this species, the females frequently having several, 8 to 9, attached simultaneously to the urosome; a 2.5 or 4.5% error in the mean numbers attached is negligible (Table 57). Estimating the

Table 57 Comparison of the variation between contiguous samples of *Pareuchaeta norvegica* in the fjordic environment of Loch Etive and in the oceanic environment of the Rockall Trough. The standard deviations of each contiguous pair of samples in Loch Etive was expressed as a percentage of the mean of that pair for each population parameter; the overall mean and its standard deviation for the percentage values of the 46 contiguous pairs of samples are shown for each parameter. Similar calculations were made for the samples from the Rockall Trough which were not contiguous throughout but comprised groups of replicate samples taken on successive cruises. Standard deviations were calculated as percentages of the mean for each group of replicate samples for each population parameter. Overall means and standard deviations were obtained from these percentage values. The comparison between the fjordic and oceanic population of this species is, therefore, approximate. (After Mauchline, 1994a.)

	Loch Etive	Rockall Trough
No. of spermatophores female^{-1}	2.65 ± 2.27	4.40 ± 2.40
% of females with spermatophores	12.79 ± 15.98	23.54 ± 18.45
Seasonal numbers of females	12.96 ± 12.35	30.88 ± 15.85
% of females with eggs	16.20 ± 16.59	43.66 ± 21.42
Sex ratio	17.31 ± 15.51	37.06 ± 18.03

mean number of females with one or more spermatophores attached is subject to a greater error, about 13 or 24%. A surprising result was the relative restriction of the error in estimating the mean number of adult females present in the population in Loch Etive, the error in the Rockall Trough being much greater. The largest sampling errors, however, are associated with the estimates of the mean number of females carrying eggs attached to the urosome and determination of the sex ratio (Table 57). The larger error associated with the egg masses partially derives from the ease with which they become detached during the course of sampling. Retrieval of such egg masses from the residues of the samples from Loch Etive was more common than from those from the Rockall Trough. Predation on egg masses within the nets during hauling from depth in the Rockall Trough is considered a possible contributory factor to the greater error in the samples from there. The large error associated with the determination of sex ratio probably arises from differences in the spatial distribution of the males relative to females in both environments. Thus, there are errors associated with the estimation of all parameters but they are greater for the oceanic population than for the fjordic.

Another major difficulty is the comparative quantitative sampling of nauplii and copepodids of a species. Frequently, the net used adequately to sample later copepodids and adults is of too coarse a mesh to sample early copepodids and nauplii. Even if the one net catches all stages, the efficiency with which it catches early nauplii and late copepodids will be different. So far, there has been no solution to this problem and so estimates of the mortality of the different stages cannot be directly determined. Frequently, mortality is assumed to take place at a constant rate throughout development, a situation that can be argued as rather unlikely.

Lastly, the frequency at which sampling of wild populations is done in the course of time-series investigations is critical and is directly linked to the rate at which events in the life cycle are taking place. Weekly or two-weekly sampling of a large species with a generation time measured in months is satisfactory but for species with generation times of 2 or 3 weeks more frequent sampling is required. The work-load involved in the analyses of the samples then increases dramatically, especially if both counts and measurements of the individual stages are required. In addition, the location of the samples in time and space within the environment has to be determined through exploratory investigations. For instance, the most representative samples may be obtained during hours of darkness when the population has migrated vertically into the surface layers. This is especially true if a component, often CV or adults, is thought to be associated with the sediment surface, the benthopelagic or hyperbenthic environment, during the day and not available to conventional zooplankton sampling nets. Further, some measure of the horizontal, as well as vertical, patchiness of the different stages is required in designing the sampling programme. Uye (1982a), in studying the populations of *Acartia tonsa* in Onagawa Bay, Japan, surveyed the entire Bay four times a year but located his standard station, sampled at two-weekly intervals, in the restricted embayment at the head of the Bay.

Some environments have special features that affect the sampling processes. Fjords, for example, are elongated environments often subject to strong tidal inflows and outflows. Loch Etive in western Scotland is such an environment. *Pareuchaeta norvegica, Calanus finmarchicus* and *Acartia clausi* can all occur in the same samples, but representativeness of those samples for each individual species varies because of the different efficiency of the nets in catching the different species and the different centres of distribution of the species within the loch. Further, the tidal flow, whether it is coming in or going out, affects the sampling. A net towed towards the head of the loch catches very different numbers of copepods from a net towed towards the mouth of the loch unless there is instrumentation to ensure that the net is fishing the same depth horizons and the same amount of water in both directions. This difference between samples is present even in oblique hauls taken contiguously but in different directions relative to the axis of the loch. The solution in Loch Etive was to take a north- and south-going sample as a pair and combine them for analysis. Such peculiarities will exist in other sampling regions and should be investigated and incorporated in the design of the principal sampling programme before it begins.

10.2. DEMOGRAPHIC ANALYSIS

An introduction to the demographic analysis of copepods is provided by McLaren (1978). He re-analyses the counts and measurements of the developmental and adult stages of the species of copepods occurring in Loch Striven, western Scotland published by Marshall et al. (1934) and Marshall (1949).

Analysis of raw data from time-series observations presents difficulties, especially when overlapping cohorts and generations are present. Schematic examples of contrasting life histories of two copepods are shown in Figure 73. One has a simple cycle in which a single brood of eggs is produced each breeding season and results in a new generation of adults. The second (Figure 73B) is the much more common and complex situation where individual females produce more than one brood of eggs, these successive broods contributing cohorts of adults within the resulting summer generation. Some females, within the population, will produce more successive broods than others. The production of broods is not normally synchronous and so production of first broods by females in the population will extend over a period of days if not weeks. There is variation in the development times of the eggs within a brood, and even more between broods. This variation is enhanced for the individual times required for the developmental sequence between egg and adult. Consequently, the adulthood of individuals will be achieved at different ages and any apparent cohorts of eggs will result in less apparent cohorts of adults.

Interpretation of stage and length frequency data obtained from time-series samples of populations of copepods is often difficult as mentioned above. Several models have been developed to aid interpretation of such data. Hay et al. (1988) use a model to describe the birth, growth and mortality rates of small copepods, reared in enclosures; the data contained noise and required smoothing. Batchelder and Miller (1989) and Batchelder and Williams (1995) model the population dynamics of *Metridia pacifica* and *M. lucens* respectively and compare observed seasonal frequencies of copepodid stages with those generated by the model. Aksnes and Høisæter (1987), Hairston et al. (1987) and Saunders and Lewis (1987) discuss the obtaining of data for life tables from cohort analysis of populations of copepods.

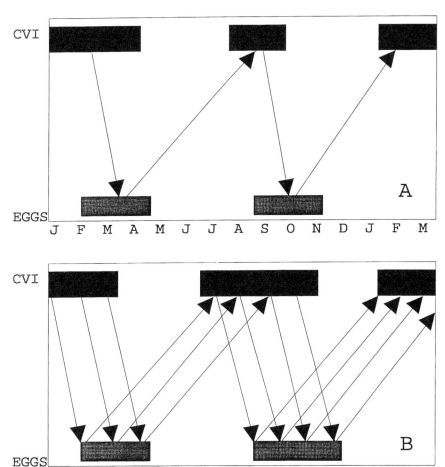

Figure 73 Schematic life histories of copepods. A, a species that produces a single brood of eggs in spring and summer resulting in discrete generations of adults. B, a species that produces up to three broods of eggs per female in spring and up to four broods of eggs in summer so that successive and overlapping cohorts of adults result; accurate estimates of generation times in this species are difficult.

10.2.1. Seasonal Changes in Stage Structure

The seasonal changes in numbers of each developmental stage, the nauplii bulked together, of many species have been made e.g. Marshall (1949), Uye (1982a). Such studies are often necessary to distinguish the number of generations occurring in a year and have been made by many of the authors listed in Table 58. They are not always successful, however, as instanced by that of Binet (1977) of eight calanoid species off the Ivory Coast. Numbers of CI to CVI were counted twice weekly for 14 months and more than 20 cohorts identified in a 12-month period in four of the species, the other four having between 11 and 18 cohorts. These cohorts could not be grouped into generations even with this frequency of sampling. These populations are even more complex than the schematic one shown in Figure 73B.

Analysis of seasonal changes in the incidence of developmental stages is a very labour-intensive process and is really only successful in species with a very simple life history. It is applicable to species with up to three or four generations per year if they are separated in time as Corkett and McLaren (1978) illustrate.

10.2.2. Size Frequency

Size-frequency distributions of adults and CVs often provide further information that can be used to help identify generations. The size distribution of adult females is sometimes bimodal, indicating that females with different origins are present. This is especially true when there are females from an overwintering population persisting in the spring among adult progeny from that winter population. The bimodality of size reflects the different temperature regimes controlling the growth of the females in the two modes. Successive broods of eggs produced in the spring by an overwintering population of adults can result in a polymodal distribution of progeny if the environmental temperature progressively increases throughout the period of development of the successive broods. Body size tends to decrease with increasing environmental temperature and results in adult progeny of such successive broods with progressively smaller modal body sizes. They also tend to have successively shorter development times. Both these factors contribute to overlapping in the timing of adulthood as well as of the body sizes of the cohorts.

10.2.3. Breeding Seasons

Information on the breeding seasons of species carrying eggs can be obtained by examining the seasonal frequency of females with an egg mass and with spermatophores attached to the urosome (Mauchline, 1994a). Comparable information can be obtained from genera that lay their eggs freely in the sea by examining developmental stages of the ovary as well as the occurrence of spermatophores in the males and attached to the females. Such data are really only helpful in determining generation times in annual

Table 58 Number of years generation^{-1} or generations year^{-1} of calanoid copepods in the natural environment.

Species	Location	Generations Years	Generations Year^{-1}	Authority
Acartia clausi	Jackles Lagoon		5	Landry, 1978b
	Tisbury Great Pond		2	Deevey, 1948
	Long Island Sound		4	Conover, 1956
	Raritan Bay		3	Jeffries, 1976
	Loch Striven, W. Scotland		4	McLaren, 1978
	Western North Sea		7	Evans, 1977
	English Channel		6	Digby, 1950
	Gulf of Marseille		6	Gaudy, 1972
	Gulf of Marseille		6	Benon, 1977
	Saronikos Gulf, Greece		4–5	Christou and Verriopoulos, 1993b
	Black Sea		7	Greze and Baldina, 1964
	Onagawa Bay, Japan		6	Uye, 1982a
	Maizuru Bay, Japan		4	Ueda, 1978
	Western North Sea		4	Evans, 1977
A. longiremis	Adriatic Sea		3	Shmeleva and Kovalev, 1974
A. negligens	Japan Inland Sea		9	Liang and Uye, 1996
A. omorii	Biscayne Bay, Florida		ca. 11	Woodmansee, 1958
A. tonsa	Delaware Bay		4	Deevey, 1960a
	Long Island Sound		4	Conover, 1956
	Raritan Bay		6	Jeffries, 1976
	Bahía Blanca, Argentina		7	Sabatini, 1989
	Baltic		8–9	Arndt and Schnese, 1986
	Gulf of Marseille		6	Gaudy, 1962
	Mediterranean		6–8	Gaudy, 1992
Calanoides acutus	Antarctic		1	Andrews, 1966

Table 58 Continued.

Table 58 Continued.

Species	Location	Generations		Authority
		Years	Year^{-1}	
	Antarctic		1	Marin, 1988a
	Antarctic		1	Huntley and Escritor, 1991
	Scotia Sea		1	Atkinson, 1991
	Antarctic		1	Drits et al., 1994
	Croker Passage		1	Żmijewska, 1993
	E Weddell Sea		1	Schnack-Schiel and Hagen, 1994
C. carinatus	Ghana, 5°N		>3	Mensah, 1974
	Ivory Coast		4–6	Binet and Suisse de Sainte Claire, 1975
	Congo Shelf		6	Petit and Courties, 1976
	S Benguela		6–7	Verheye et al., 1991
Calanus chilensis	Bay of San Jorge, Chile		6	Escribano and Rodriguez, 1994
C. finmarchicus	Igloolik, Canadian Arctic		1	Grainger, 1959
	W Greenland		1	Maclellan, 1967
	Conception Bay, Newfoundland		3 (4)*	Davis, 1982
	Gulf of Maine		3	Fish, 1936
	Scotian Shelf		1 (2)*	McLaren and Corkett, 1986
	Scotian Shelf		2	McLaren et al., 1989a
	Scotian Shelf		4	Sameoto and Herman, 1990
	Fram Strait		1	Diel, 1991
	NW Barents Sea		1	Tande et al., 1985
	Balsfjorden, Norway		1	Tande, 1982
	Korsfjorden, Norway		2 (3)*	Matthews et al., 1978
	Lindåspollene, W. Norway		1	Aksnes and Magnesen, 1983
	Loch Striven, W. Scotland		3	McLaren, 1978
	Southwest of British Isles		1	Williams and Conway, 1988

Species	Location			Reference
C. glacialis	W. Greenland		1	Maclellan, 1967
	Sea of Okhotsk		1	Safronov, 1989
	Resolute Bay, Arctic	2–3		Conover et al., 1991
	Nova Scotia		1	Runge et al., 1985
	Scotian Shelf, Nova Scotia		1	McLaren et al., 1989a
	East Greenland	2		Diel, 1991
	Spitsbergen	2		Diel, 1991
	NW Barents Sea	2		Tande et al., 1985
	Barents Sea	1–2		Tande, 1991
C. helgolandicus	English Channel		3?	Green et al., 1993
	Gulf of Marseille		3?	Gaudy, 1972
	Adriatic		5	Vučetic, 1966
C. hyperboreus	Resolute Bay, Arctic	2–3(4)		Conover et al., 1991
	Fram Strait	2		Diel, 1991
	Korsfjorden, Norway		1	Matthews et al., 1978
C. marshallae	SE Bering Sea		2	Osgood and Frost, 1994b
	Dabob Bay, Washington		1	Osgood and Frost, 1994b
	Oregon coastal upwelling		3–4	Osgood and Frost, 1994b
C. minor	Gulf of Marseille		2	Gaudy, 1962
	Gulf of Marseille		5	Gaudy, 1972
	Adriatic Sea		4	Shmeleva and Kovalev, 1974
C. pacificus	Hokkaido, Japan		2	Hirakawa, 1979
	Dabob Bay, Washington		3	Osgood and Frost, 1994b
C. propinquus	Antarctic		1	Marin, 1988a
	Croker Passage		1	Żmijewska, 1993
	S Weddell Sea		2	Drits et al., 1993; Kosobokova, 1994
	E Weddell Sea			Schnack-Schiel and Hagen, 1994
C. simillimus	Scotia Sea		1–2	Atkinson, 1991
C. sinicus	Yellow, East China Sea		3	Hülsemann, 1994
	South China Sea		1–2	Hülsemann, 1994

Table 58 Continued.

Table 58 Continued.

Species	Location	Generations Years	Generations Year^{-1}	Authority
Calocalanus elegans	Adriatic Sea		4	Shmeleva and Kovalev, 1974
C. pavo	Adriatic Sea		3	Shmeleva and Kovalev, 1974
Centropages abdominalis	Japan Inland Sea		6	Liang et al., 1996
C. hamatus	Loch Striven, W. Scotland		8	McLaren, 1978
	Roscoff		5	Person-Le Ruyet et al., 1975
C. typicus	Banyuls-sur-Mer		7	Razouls, 1974
	Banyuls-sur-Mer		7	Razouls and Razouls, 1976
	Gulf of Marseille		5	Gaudy, 1962
	Gulf of Marseille		5	Gaudy, 1972
	Gulf of Marseille		6	Gaudy, 1984
	Adriatic Sea		5	Shmeleva and Kovalev, 1974
	Adriatic Sea		4	Shmeleva and Kovalev, 1974
Chiridius armatus	Korsfjorden, Norway		2	Bakke and Valderhaug, 1978
	Korsfjorden, Norway		1	Båmstedt, 1988a
Clausocalanus arcuicornis	Gulf of Marseille		5	Gaudy, 1972
	Adriatic Sea		3–5	Shmeleva and Kovalev, 1974
C. furcatus	Gulf of Marseille		5	Gaudy, 1972
C. paululus	Adriatic Sea		5	Shmeleva and Kovalev, 1974
Ctenocalanus citer	East Weddell Sea		1	Schnack-Schiel and Mizdalski, 1994
C. vanus	Adriatic Sea		4–5	Shmeleva and Kovalev, 1974
Drepanopus bispinosus	Burton Lake, Antarctica		1	Wang, 1992
D. pectinatus	Kerguelen Archipelago		4	Razouls and Razouls, 1988
Epilabidocera amphitrites	Kamchatka		2	Safronov, 1991
	Bering Sea		1	Heinrich, 1982
Eucalanus bungii	Gulf of Alaska	2(1–3)		Miller et al., 1984

Species	Location		Reference
Eurytemora affinis	Elbe Estuary	3	Peitsch, 1993
E. velox	Camargue, France	3	Pagano, 1981a
Haloptilus longicornis	Adriatic Sea	3	Shmeleva and Kovalev, 1974
Ischnocalanus plumulosus	Adriatic Sea	4	Shmeleva and Kovalev, 1974
Limnocalanus macrurus	Char Lake, NW Territories	1	Roff and Carter, 1972
Lucicutia flavicornis	Adriatic Sea	4	Shmeleva and Kovalev, 1974
Mecynocera clausi	Adriatic Sea	4	Shmeleva and Kovalev, 1974
Mesocalanus tenuicornis	Adriatic Sea	3	Shmeleva and Kovalev, 1974
Metridia gerlachei	Bransfield Strait	1?	Huntley and Escritor, 1992
	Croker Passage	2	Żmijewska, 1993
Metridia longa	West of Spitsbergen	2?	
	Spitsbergen	1	Diel, 1991
	Balsfjorden, N Norway	1	Diel, 1991
	Kosterfjorden	2	Grønvik and Hopkins, 1984
Metridia lucens	Scotian Shelf	4	Båmstedt, 1988a
	Dabob Bay, Washington	3	McLaren *et al*, 1989a
Metridia pacifica	N Bering Sea	1	Osgood and Frost, 1994b
	W Bering Sea	4	Heinrich, 1962b
	NE Pacific	3	Heinrich, 1962b
	NE Pacific	3	Batchelder, 1985
	SW Sakhalin	4	Batchelder and Miller, 1989
	Sea of Japan	1	Fedotova, 1975
	Sea of Japan	1	Hirakawa and Imamura, 1993
Microcalanus pygmaeus	Ellesmere Island	1	Sunami and Hirakawa, 1996
	Loch Striven, W Scotland	3	Cairns, 1967
	East Weddell Sea	2?	McLaren, 1978
Neocalanus cristatus	Bering Sea	1	Schnack-Schiel and Mizdalski, 1994
	Gulf of Alaska	1	Heinrich, 1962b
			Miller *et al.*, 1984

Table 58 Continued.

Table 58 Continued.

Species	Location	Generations		Authority
		Years	Year^{-1}	
N. flemingeri	Bering Sea		2	Heinrich, 1962b
	Gulf of Alaska		1	Miller and Clemons, 1988
	Sea of Japan		1–2	Miller and Terazaki, 1989
	Adriatic Sea		3	Shmeleva and Kovalev, 1974
N. gracilis	Gulf of Alaska		1	Miller *et al.*, 1984
N. plumchrus	Gulf of Alaska		1	Miller and Clemons, 1988
	Strait of Georgia		1	Fulton, 1973
	Sea of Japan		1	Miller and Terazaki, 1989
Paracalanus nanus	Adriatic Sea		4	Shmeleva and Kovalev, 1974
P. parvus	English Channel		6	Digby, 1950
	Gulf of Marseille		6	Gaudy, 1972
	Gulf of Marseille		6	Benon, 1977
	Adriatic Sea		5–6	Shmeleva and Kovalev, 1974
Pareuchaeta antarctica	South Georgia		2?	Ward and Robins, 1987
	Croker Passage		1	Żmijewska, 1993
P. gracilis	Rockall Trough		2	Mauchline, 1994a
P. hebes	Adriatic Sea		4	Shmeleva and Kovalev, 1974
P. norvegica	W Scotland		2	Mauchline, 1994a
	Rockall Trough		1	Mauchline, 1994a
	Korsfjorden, Norway		1	Båmstedt and Matthews, 1975
	Korsfjorden, Norway		2	Bakke, 1977
	W Scotland		2	C.C.E. Hopkins, 1982
P. pseudotonsa	Rockall Trough		1	Mauchline, 1994a

Species	Location	Number	Reference
Paralabidocera antarctica	Syowa Station, Antarctica	1	Tanimura *et al.*, 1984, 1996
Pleuromamma gracilis	Adriatic Sea	3–4	Shmeleva and Kovalev, 1974
Pseudocalanus acuspes	Resolute Bay, Arctic	1	Conover *et al.*, 1991
	Bedford Basin, Nova Scotia	3	McLaren *et al.*, 1989b
P. elongatus	White Sea	2	Pertzova, 1981
	Norwegian Sea	1	Corkett and McLaren, 1978
	Norway	4–5	Corkett and McLaren, 1978
	Loch Striven, W Scotland	6	Corkett and McLaren, 1978
	Western North Sea	5	Evans, 1977
	English Channel	9	Corkett and McLaren, 1978
		6?	Green *et al.*, 1993
P. minutus	Ogac Lake, Baffin Island	1	McLaren, 1969
	N Labrador	2	Corkett and McLaren, 1978
	Bedford Basin, Nova Scotia	1	McLaren *et al.*, 1989b
P. moultoni	Browns Bank, Scotian Shelf	2	McLaren *et al.*, 1989b
P. newmani	Browns Bank, Scotian Shelf	3	McLaren *et al.*, 1989b
	Emerald Bank, Scotian Shelf	5	McLaren *et al.*, 1989a
Pseudocalanus sp.	Ellesmere Island	2	Cairns, 1967
	Foxe Basin, N Canada	1–2	Corkett and McLaren, 1978
	Dabob Bay, Washington		Ohman, 1985
Rhincalanus gigas	Antarctic	8	Ommanney, 1936
	Antarctic	2	Marin, 1988a
	Scotia Sea	1 or 2	Atkinson, 1991
	Croker Passage	1 or 2	Żmijewska, 1993
	E Weddell Sea	1 or 2	Schnack-Schiel and Hagen, 1994
Stephos longipes	Antarctic	1	Kurbjeweit *et al.*, 1993
	Antarctic	1	Schnack-Schiel *et al.*, 1995

Table 58 Continued.

Table 58 Continued.

Species	Location	Years	Generations Year^{-1}	Authority
Temora longicornis	Long Island Sound		3	Peterson, 1985
	Long Island Sound		2 (3?)*	Peterson and Kimmerer, 1994
	Loch Striven, W Scotland		5	McLaren, 1978
	Western North Sea	1972	4	Evans, 1977
		1973	5	Evans, 1977
		1971	6	Evans, 1977
	English Channel		5	Digby, 1950
	Roscoff		5	Person-Le Ruyet *et al.*, 1975
T. stylifera	Banyuls-sur-Mer		6–7	Razouls, 1974
	Banyuls-sur-Mer		7	Razouls and Razouls, 1976
	Gulf of Marseille		5	Gaudy, 1962
	Gulf of Marseille		5	Gaudy, 1972
	Adriatic Sea		5	Shmeleva and Kovalev, 1974

*Possible alternative.

or biannual breeders with relatively narrow seasonal windows of breeding. The numbers of successive broods or clutches by individual females can rarely be determined from these observations unless the breeding season extends over several months and has peak periods within it.

10.2.4. Generation Time

The generation times of many species have been investigated in wild populations and in cultures in the laboratory (Table 47, p. 300). McLaren (1978) and Uye (1982a), using the Bĕlehrádek's equations relating generation time to environmental temperature (Table 48, p. 310), predict the generation times of species in Loch Striven, western Scotland and of *Acartia clausi* in Onagawa Bay and compare them with those observed in the natural populations. There is good agreement between observed and predicted times.

Such experimental determination of generation times can be used in the interpretation of time series of stage frequency and body sizes gained from field populations.

10.2.5. Sex Ratio

The determination of the sex ratio within a population can have a relatively large error for a variety of reasons. Mauchline (1994a) found that estimates of the mean ratio varied on average by 17% in samples from the fjordic population of *Pareuchaeta norvegica* in Loch Etive while the error was 37% in samples of the same species from the oceanic environment of the Rockall Trough. Ferrari and Hayek (1990) took six series of four replicate samples of *Pleuromamma xiphias* seasonally in the oceanic environment southwest of Hawaii and found that estimates of the error ranged between 20 and 33%. A portion of this error derives from behavioural differences between males and females. They often occupy different depth horizons as Bennett and Hopkins (1989) found for species of the genus *Pleuromamma*. Seasonal changes in the sex ratio occur, especially in species with longer generation times of months rather than weeks. The longevity of males is probably shorter than that of the females (Table 56, p. 345). There may also be differences in the aggregational behaviour of males and females but no information on this aspect seems to be available. Consequently, samples representative of season and total environment of the population are required to examine the sex ratio. The shorter life span of males is at least partly responsible for bias in sex ratios in favour of females.

There is still a lack of knowledge on the factors determining the sex of an individual. The classical study of Takeda (1950) on the harpacticoid copepod *Tigriopus japonicus* has not been repeated on calanoid copepods. He concluded that the physiological state of the NVI influenced the later differentiation into a male or female. Higher temperature causing an increased rate of development of the NVI results in an increased frequency of males in the CI; lower rates of development increase the frequency of females. First traces of the gonads have been found in the CI in a variety of species of calanoids, namely *Acartia longiremis, Calanus helgolandicus, Centropages typicus, Pseudocalanus acuspes,* and *Temora stylifera* (Baldacci *et al.,* 1985a; Razouls *et al.,* 1987; Norrbin, 1994) but sex could not be identified until the CIV when the right gonoduct of potential males degenerated. Thus, Takeda's results cannot be directly applied to calanoids in general.

Sex ratios vary seasonally and Ferrari and Hayek (1990) list references to studies on marine, brackish and fresh water species. Consequently, statements such as that of Tschislenko (1964) that common species have ratios close to 1:1 and that rare species are dominated by females are difficult to confirm because there may be critical seasonal windows when the ratios alter dramatically. The same applies to the findings of Moraitou-Apostolopoulou (1969), who suggests that the ratio of males to females is high in species that prefer warmer, and very low in species that prefer colder waters. The ratios vary relative to the densities of the populations (Alcaraz and Wagensberg, 1978; Saraswathy and Santhkumari, 1982; Kouwenberg, 1993). In many species, as in Figure 74, the male-to-female ratio is highest during the periods of lowest female abundance or when the population is less dense. Kouwenberg (1993) suggests that herbivorous species are dominated by females, that there is less dominance in omnivorous species and that carnivorous species have ratios close to 1:1. This seems a very broad generalization and requires confirmation. The seasonal changes in the sex ratios in the CIV, CV and adult CVI *Pareuchaeta norvegica* were examined in Loch Etive, western Scotland, by C.C.E. Hopkins (1982). Males dominated the CIV in all months except February to May and the CV except in March to June (Figure 74). The adults were dominated by females but approached a 1:1 ratio in February to April. An almost identical pattern of change, in magnitude and timing, in the seasonal sex ratios, occurred in *Pseudocalanus acuspes* in Balsfjorden, northern Norway but the few data available on the co-occurring *Acartia longiremis* suggest that it may have a different pattern (Norrbin, 1994). A different pattern is also suggested for *Metridia longa* by the partial series of data given by Tande and Grønvik (1983). Ferrari and Hayek (1990) examined the sex ratios in the CIV to CVI of *Pleuromamma xiphias* but their samples are not frequent enough to define seasonal changes with any accuracy. This is an interesting approach to sex determination and further work is required, including better

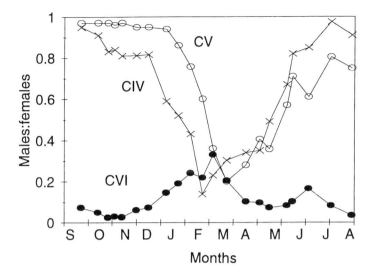

Figure 74 Seasonal variation in the sex ratio, percentage of males, of CIV, CV and adult CVI of *Pareuchaeta norvegica* in Loch Etive, western Scotland. (After C.C.E. Hopkins, 1982.)

estimations of the comparative durations of female and male CIV and CV. These are all species in which secondary sexual characteristics develop in the CIV. Identification of the sexes in, for example, the Calanidae, is frequently not possible until the adult. Grigg *et al.* (1981, 1985, 1987, 1989) has made a detailed study of the CV of *Calanus finmarchicus* and recognizes developing males and females.

The sex ratio of populations of copepods raised in the laboratory can be variable. Peterson (1986) found only one male among 137 adult *Calanus marshallae* grown from eggs in the laboratory. Peterson reviews the earlier literature, pointing out that no males were raised in cultures of *Calanoides carinatus*, *Calanus helgolandicus*, or *C. pacificus* but were present in other cultures of *C. pacificus* (0.25 to 0.40) and *Rhincalanus nasutus* (ratio, 0.44 to 0.50). Conversely, cultures with normal sex ratios do occur as Arnott *et al.* (1986), for example, found in the case of *Gladioferens pectinatus*.

Hada *et al.* (1986) mentions previous studies, including that of Katona (1970) on *Eurytemora* species, that concluded that temperature was important in determining the sex ratio. Heinle (1969) found that sex ratio and population density are directly correlated in *Acartia tonsa*. Hada *et al.* (1986), however, found no clear relationship between sex ratio and these variables in *Sinocalanus tenellus*. An interesting observation by Heinle (1970) is that the sex ratio increased, in cultures of *Acartia tonsa*, in favour of females at the highest rates of predation.

Moore and Sander (1983), in reviewing work on sex ratios, conclude that single explanations of variations in sex ratios within and between species are unrealistic and that many factors are involved ranging from genetical, environmental and the responses of the species to the environment. The result is that sex ratios are often flexible and, being so, confer options on the populations. Sex reversal is discussed by Moore and Sander (1983) but the only example was an apparent reversal in *Pseudocalanus elongatus* caused by parasitism. Fleminger (1985) reviews the subject more extensively. He presents evidence, on the basis of finding dimorphic females in the populations, that *Calanus* species may switch sex. This hypothesis requires confirmation.

10.2.6. Mortality

Some of the problems involved in obtaining quantitative and representative samples of a population of *Eurytemora affinis* for estimation of rates of mortality in an estuarine situation are described by Peitsch (1993). The sampling strategy is of paramount importance as stated earlier in this Chapter. Ohman (1986), in a detailed analysis of the rates and sources of mortality in populations of *Pseudocalanus* sp. in Dabob Bay, Washington emphasizes sampling problems during discussion of his results.

The importance of placing more emphasis on estimations of mortality when studying the dynamics of natural populations is expanded by Ohman and Wood (1995, 1996). They discuss the various factors determining or influencing rates of mortality and suggest that mortality rather than fecundity can operate to control size of the population, as they found in *Pseudocalanus newmani*. This contrasts with the study of Mullin (1991b) who finds that rates of egg production rather than mortality control population numbers of *Calanus pacificus* and *Rhincalanus nasutus*. These species, however, spawn their eggs freely into the sea and have a greater range in the daily rate of egg production and a higher rate of mortality of eggs (Mullin, 1993; Kiørboe and Sabatini, 1994) than species such as *Pseudocalanus newmani* that carries an egg mass.

Perhaps the most difficult rate of mortality to determine is that of the eggs of a species in the natural environment and yet it is often suspected as being very significant, especially through predation by young fish (Mullin, 1995) and copepods. Peterson and Kimmerer (1994) concluded that cannibalism of eggs of *Temora longicornis* in Long Island Sound accounted for a mortality of 99% of the eggs per day. Other studies have noted heavy mortality, 80 to 99%, during periods of intense breeding (e.g. Ianora and Buttino (1990) in *Centropages typicus* and *Acartia clausi*) but the causes could not be determined. The evidence, presented earlier when discussing

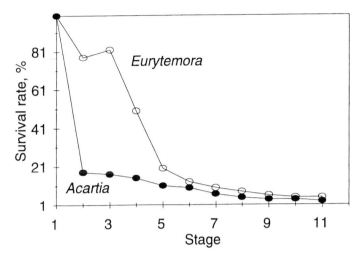

Figure 75 Survival curves for *Acartia clausi* in Onagawa Bay and *Eurytemora affinis* in the Elbe Estuary. (After Uye, 1982a; Peitsch, 1993.)

the mortality of eggs in Chapter 8, of eggs passing unharmed through the digestive tracts of fish and other organisms has also to be assessed. How widespread is such survival?

Likewise, there are major difficulties in determining the rate of mortality of older stages, also often caused by planktivorous fish. Slagstad and Tande (1990) briefly discuss potential mortality of *Calanus glacialis* caused by feeding migrations of capelin in the Arctic.

Life tables and survivorship curves are used by Gehrs and Robertson (1975) to describe the demography of populations of the fresh water copepod *Diaptomus clavipes*. An evaluation of their use in toxicity studies on *Eurytemora affinis* by Allan and Daniels (1982) emphasizes the need to obtain data over the entire life cycle. Such tables were used by Uye (1982a) in a study of populations of *Acartia clausi*. Survivorship curves are a useful graphic summary allowing visual comparisons of patterns of mortality over the life cycle. Two curves are illustrated in Figure 75 and show different patterns of mortality over the course of development. A schematic survival curve for a cohort of copepods (Figure 76) is derived by Poulet *et al.* (1995a) from a review of data in recent publications, including those of Uye (1982a) on *A. clausi* shown in Figure 75. The development is divided into four phases that have decreasing rates of mortality. The highest rates of mortality are in the egg phase, with phases II and III having less mortality. The least mortality is associated with the CII to CVI phase in which it is usually considered as constant. This curve does not take predation of the cohort into

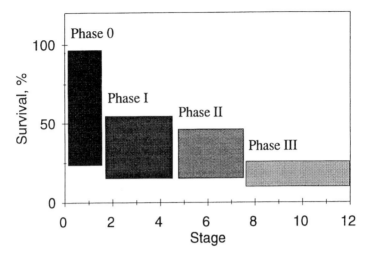

Figure 76 Rates of survival associated with different phases of development of a cohort of copepods. Phase 0 is the egg from laying to hatching; Phase I is the NI to NIII, the NI and NII being non-feeding stages; Phase II is the NIII to CI; Phase III is the CI to adult. This curve does not take predation into account. (After Poulet *et al.*, 1995a.)

account but illustrates potential natural mortality from primarily physiological reasons. The pattern of mortality in natural populations of *Pseudocalanus newmani* over two years is described by Ohman and Wood (1995, 1996). There was a close similarity between the two years, mortality increasing from the egg to a maximum at CI, followed by low mortality in CII to CIV and a further maximum in CV. This species carries an egg mass and so mortality of eggs will reflect that of the adult females. Rates of mortality on the CI and CV are very variable suggesting that these stages in this species are sensitive.

In between-species comparisons, rates of mortality are usually considered to be inversely related to body size but Kiørboe and Sabatini (1995) suggest that larger copepods do not suffer lower mortality than smaller ones. This suggestion is, however, questioned by Aksnes (1996). Many species of copepods have an ontogenetic vertical distribution, discussed elsewhere, larger stages within a species living deeper, and probably liable to reduced mortality at depth relative to the stages higher in the water column. Vertical migration, especially diel, will affect rates of mortality from surface-living visual predators. There is considerable variation within and between species in the timings and extents of diel migrations. There is also the variation in the occurrence and behaviour of the predators to be considered. Consequently, there is liable to be considerable variation in rates of mortality linked with the behaviour of individuals and species. The arguments for mortality having a degree of size dependence are strong but the behaviour of individuals and species of copepods coupled with that of their predators is probably the primary forcing mechanism on rates of mortality and may frequently obscure or cancel any perceived correlations.

Castel and Feurtet (1992) found mean mortality rates in nauplii, copepodids and adults of *Eurytemora affinis* in the Gironde Estuary of 0.184, 0.535 and 0.327 d^{-1} respectively. They conclude that the lower mortality of nauplii is associated with their shorter development time relative to that of the copepodids. Gaudy (1992), on the other hand, found higher rates of mortality in nauplii than copepodids of *Acartia tonsa*. Mortality rates in an Australian population of *A. tranteri* were low, especially in the late nauplii and the copepodids, mainly through a lack of significant predation (Kimmerer and McKinnon, 1987a).

Population mortality rates have been estimated in a number of species: Matthews *et al.* (1978) find a daily rate of about 0.1 for *Calanus finmarchicus* and a low rate of about 0.005 in *C. hyperboreus* in Korsfjorden, western Norway.

Mortality rates in laboratory culture have been estimated for the following species: *Acartia clausi*, *Centropages hamatus* and *Temora longicornis* by Klein Breteler (1980).

Mortality will increase as a cohort ages (Carlotti and Nival, 1992a,b). It is also influenced by sharp temperature changes (Tande, 1988a). Kiørboe and Sabatini (1995), in reviewing growth and fecundity, suggest that mortality rates are independent of the body size of the copepod. One of their premises is that fecundity is independent of body size, as discussed in Chapter 8, where it is suggested that the range of body size examined by them is unrepresentative of calanoids. Aksnes (1996) also criticizes their conclusions on rates of mortality and points out that the rates of mortality of a species are influenced by a wide range of factors, some dependent on, some independent of, body size.

There are seasonal changes in the rates of mortality in many populations of copepods. Myers and Runge (1983) developed a model, using life-history theory, that describes the seasonal trends in mortality of a population of *Acartia clausi*. Mortality rates in copepod populations are consistently high in the late summer and autumn in temperate coastal areas (Kiørboe and Nielsen, 1994). This is the period when temperatures are decreasing, phytoplankton is becoming less available, breeding activity is becoming less intense and resting or diapause eggs are produced. There can also be marked predation pressure on the populations at this time.

Mass mortality of copepod populations has been suspected on at least two occasions (Lee and Nevenzel, 1979; Volkman *et al.*, 1980). Waxy deposits found on the shores of Bute Inlet, British Columbia were suspected to originate from populations of *Neocalanus plumchrus* while milky water in the North Sea had a lipid composition closely resembling that of *C. finmarchicus*. Both situations would be explained by mass mortalities of the *Calanus* species. The likely causes of such mortality are conjectural; the formation of low salinity gyres in coastal and fjordic situations, after heavy fresh water run-off, is possible. It is much more difficult, having determined that man-made pollution was not responsible, to account for the mortality in the open North Sea.

Copepod carcasses are often recorded in plankton samples. According to Harding (1973), bacterial decomposition of dead copepods takes place within 11 d at 4 °C in Scotian Shelf waters, and within 3 d at 22 °C in Sargasso Sea waters. He suggests that carcasses at the surface last one day in subtropical waters and six days in temperate regions. Comparable rates of microbial decomposition of carcasses of *Anomalocera patersoni* were determined by Reinfelder *et al.* (1993) who also examined rates of release of trace metals from the carcasses.

A density of carcasses, possibly representing mass mortality, occurred in the upwelling region off Cape Blanc, northwest Africa (Weikert, 1977). The species involved were primarily *Temora stylifera* and *T. turbinata*. The mortality was ascribed to violent and irregular mesoscale upwelling at the shelf edge. Large numbers of carcasses of *Neocalanus cristatus* occurred below the thermocline in the Sea of Japan and Terazaki and Wada (1988) ruled out disease or predators as the cause. Sudden changes in temperature, possibly followed by starvation, may have been responsible. The possibility exists, however, that these carcasses had accumulated over a period of at least a year as a consequence of natural rates of mortality and slow rates of degradation. Haury *et al.* (1995) found copepod carcasses, that they illustrate, associated with banks, ridges and seamounts. Some are eviscerated, some with parts of the cephalosome missing, while others are empty husks, all consistent with damage inflicted by predators. Haury *et al.* conclude that these remains arise from concentrations of predators associated with the flanks of the shallow regions, migrating vertically at night to feed. High incidences of carcasses were also associated with coral reefs off Eilat, Israel and attributed to predation (Genin *et al.*, 1995).

10.2.7. Control of Population Size

Sizes of populations of copepods are controlled by a variety of factors including genetic constitution, fecundity, rates of growth and development, food availabilty, predation and parasites. Other factors such as marked changes in the physical and chemical environments and immigration and emigration to and from populations are also active. The direct effects of all these factors are very difficult to quantify in natural populations and frequently involve time lags in their effects. For example, the effects of changing food regimes on egg production is not instantaneous but initially affects the metabolism of the ovary and is later reflected in the rate of laying of eggs.

10.2.7.1. Genetic Aspects

Length, weight and lipid composition vary seasonally, influenced primarily by changes in environmental temperature and food availability. Durations of embryonic development and of naupliar and copepodid stages also vary seasonally, reflecting seasonal temperature changes. These variations result in variations in generation length, brood and clutch sizes and so fecundity. In addition, there are variations in physiological tolerance to factors such as temperature and salinity. Such variation is discussed elsewhere and primarily reflects changing seasonal conditions of the environment of the populations.

There is, however, a genetic component to variation. Individuals within a population of a species at any one time have a range of body size, stage durations, brood or clutch sizes etc. Siblings show a range of values for each parameter but Marcus (1985b) shows that the ranges can be different when two sets of siblings are compared, thus inferring a genetic contribution from the respective sets of parents. Distinct differences in rates of egg production and the degree of hatching success occurred after only six generations. Tepper and Bradley (1989) present two models describing the maintenance of genetic variance within a population of *Eurytemora affinis*.

10.2.7.2. Seasonal Changes in Brood Sizes and Growth Rates

Brood and clutch sizes change seasonally in middle and high latitudes, as discussed earlier. These changes affect the potential rates of recruitment. Fransz *et al.* (1989), in examining fecundity as a controlling factor of population size, integrate it with postulated rates of mortality. Mullin (1991b) examines the balance between rates of egg production and mortality of developmental stages and suggests that rates of egg production are more important than mortality in controlling population numbers in *Calanus pacificus* and *Rhincalanus nasutus* off southern California.

Seasonal variations in growth rates contribute significantly to rates of recruitment and are as important as variation in rates of mortality in influencing population size (Davis *et al.*, 1991).

10.2.7.3. *Food*

Scarcity of food can affect a population of a copepod in various ways. It can affect growth in body size and so clutch size and fecundity (Walker and Peterson, 1991). In general, oceanic populations of copepods may be food-limited at times but coastal populations are unlikely to be (Huntley and Boyd, 1984). Davis and Alatalo (1992) and Huntley and Lopez (1992), in reviewing rates of production of copepods, suggest that food may not limit growth in the field. This is further reinforced by the three-year study of the dominant coastal copepods in an enclosed ecosystem by Sullivan and Ritacco (1985) who found no immediate response to changes in feeding regimes. The quality of food, rather than its quantity, can affect growth and Anderson and Hessen (1995), examining potential carbon and/or nitrogen limitation to growth, conclude that more complex components of the diet, such as essential amino acids, might control production at times.

Ohman (1985) found no evidence that *Pseudocalanus* sp. in Dabob Bay, Washington was food-limited. Peterson (1985) found that egg production of *Temora longicornis*, in Long Island Sound, was food-limited and suggests that phytoplankton production controls growth. *Acartia tonsa* in Naragansett Bay, Rhode Island, was food-limited (Durbin *et al.*, 1983). A comparative study of the oceanic species *Neocalanus plumchrus* in the Bering Sea and at Ocean Station P in the Gulf of Alaska by Dagg (1991) shows that lack of food retarded development in the latter.

10.2.7.4. *Predation*

Predation is often considered to control numbers of adult copepods but it is extremely difficult to obtain quantitative measurements. Purcell *et al.* (1994) suspected that the gelatinous zooplankton common in Chesapeake Bay would limit the abundance of *Acartia tonsa* but found this to be untrue, other factors presumably controlling the populations. Hutchings *et al.* (1991) present evidence in the Benguela ecosystem that swarming euphausiids can remove 60 to 90% of the copepod populations each day. This would represent predation rates in limited parts of the copepod populations. *Pseudocalanus* sp. in Dabob Bay, Washington, were predator-limited as deduced by Ohman (1986) from studies of the diets of *Pareuchaeta elongata*, *Euphausia pacifica* and *Sagitta elegans* coupled with measurements of the mortality rates of *Pseudocalanus* sp. This is a detailed study, integrating quantitative environmental and experimental data, and should be examined prior to planning any similar study on the same or another copepod. Steele and Henderson (1995) model predation control of a population discussing match-mismatch characteristics of prey and predator frequencies of occurrence as referred to by Bollens *et al.* (1992a). Studies of rates of mortality caused by predation involve quantitative studies of the population dynamics of the copepod and of the various predators in time and space. Further, they have to include behavioural characteristics of the copepod and of the predators in time and space. Rates of predation by, for example, planktivorous fish are usually conjectural even on populations of copepods in restricted environments such as fjords. Heavy mortality of the eggs can effectively control the population size of some species in some sea areas as found by Peterson and Kimmerer (1994) in *Temora longicornis* in Long Island Sound. Only some 10% of eggs laid reached the first nauplius, the mortality being thought to be caused by cannibalism. The role of benthic faunas as predators of *Acartia hudsonica* was tested by Sullivan and Banzon (1990) and found to be operative, demonstrating that the predation by the benthos may have to be accounted for in coastal situations.

10.2.7.5. *Parasites*

Copepods are parasitized by a large variety of organisms but little is known about the actual rates of mortality caused by them. Kimmerer and McKinnon (1990) suggest that mortality can be significant. They examined the incidence of a parasitic dinoflagellate, *Atelodinium* sp., that infects *Paracalanus indicus* in Port Phillip Bay, Australia, and suggest that rates of mortality of adult females caused by it average 7% d^{-1} but can be as high as 41% d^{-1}.

10.2.8. Production

The most comprehensive introduction to the measurement of production of populations of aquatic animals is that of Winberg (1971). Two later papers, Banse and Mosher (1980) and Tremblay and Roff (1983b), examined annual Production/Biomass (P/B) ratios generally and in copepods in particular. They evoke a discussion and some criticism in a subsequent series of papers (Banse, 1984; Roff and Tremblay, 1984; McLaren and Corkett, 1984). Banse and Mosher (1980) show that annual P/B ratios are related to body mass, over a range of 10^5, when the ratios in a variety of invertebrates and vertebrates are compared; they also demonstrate temperature dependence of the ratios. Variation within the P/B ratio to body mass relationship can accommodate a lack of correlation within a restricted taxonomic group such as calanoid copepods, with a range of body mass of 10^2. In calanoids, P/B ratio is strongly influenced by environmental temperature. Thus, there are several methods of determining P/B ratios. Middlebrook and Roff (1986) estimate annual P/B ratios of populations of *Acartia hudsonica* and *Eurytemora herdmani* by several of the methods and conclude that McLaren and Corkett's (1981) provide the best estimates for calanoid copepods. They are based on Bělehrádek's equations (Table 48, p. 310), when food is not limiting and development times are accurately known. It is also assumed that the copepods grow exponentially and that the rate of production of eggs is the same as the rate of body growth of the earlier copepodids. The temperature dependence of the ratios means that two similar-sized species, one dominant in spring and the other in autumn, have different environmental temperature regimes and so different P/B ratios, as Conover and Poulet (1986) showed for copepods in the Bedford Basin. The length of the active growing season, the occurrence of non-breeding individuals or resting stages in the population, both affect the values of the annual P/B ratios. Consequently, it is now normal to calculate daily P/B ratios and such ratios are shown for a variety of species in Table 59. The geographical location of the populations is indicated as is the range of environmental temperature where this is available. The ratios range widely. Small, coastal temperate species have ratios between about 0.02 and 0.35 compared with ratios of about 0.01–0.15 for offshore species. The highest ratios occur in tropical species or a population subject to a seasonal maximum in temperature, such as the ratio of 1.25 for *Acartia tonsa* in Narragansett Bay (Table 59).

Recently, Huntley and Lopez (1992) have estimated production of copepods by using data on generation time and weights of eggs and adults to calculate growth rates. This then provides a measure of the rate of production from the equation:

$$P = Bg$$

where P is the production as mass area^{-1} d^{-1} or mass volume^{-1} d^{-1}, B is the biomass (mass area^{-1} or mass volume^{-1}) and g is the weight-specific growth rate (mass mass^{-1} d^{-1}). Generation time is inversely related to temperature (Figure 77) and, because body size is also inversely related to temperature, body size differences are also contained in this relationship. Species in the genera *Calanus* and *Neocalanus*, which are larger than most of the others investigated, are primarily at the low end of the temperature range in Figure 77. The weight-specific growth rate calculated from egg weight, adult weight and generation time is correlated with temperature (Figure 78). Further, the specific growth rates of the species are inversely related to generation time (Figure 79). Consequently, this means that the growth rate is explained by temperature alone and there is no requirement to take account of body size. Thus, having calculated P and measured biomass, the P/B ratio can be calculated for the population of a species or extended to communities of mixed species.

Table 59 Daily P:B ratios of calanoid copepods.

Species	Location	T°C	P:B ratio	Authority
Acartia bifilosa	Gdansk Bay	7–18	0.03–0.12	Ciszewski and Witek, 1977
A. clausi	Jackles Lagoon	8–20	0.12–0.23	Landry, 1978b
	Chesapeake Bay		0.5	Heinle, 1966
	North Sea		0.02–0.13	Fransz *et al.*, 1991
	Kattegat		0.011	Kiørboe and Nielsen, 1990
	W Mediterranean		0.15	Pagano and Saint-Jean, 1989
	Black Sea		0.04	Zaika, 1968
	Black Sea		0.13	Porumb, 1974
	Black Sea		0.04–0.23	Uye, 1982a
	Sea of Azov		0.063	Uye, 1982a
	Gulf of Guinea	ca.15–18	0.01–0.68	Pagano and Saint-Jean, 1989
	Onagawa Bay	5–22	0.17–0.23	Uye, 1984
A. hudsonica	Nova Scotia	15	0.05–0.33	Uye, 1982a
	Narragansett Bay	3–22	0.34	Sekiguchi *et al.*, 1980
	Skagerrak		0.10–0.67	Durbin and Durbin, 1981
A. longiremis	Chesapeake Bay	ca.26	0.12	Peterson *et al.*, 1991
A. tonsa	Narragansett Bay	17–23	0.47	Heinle, 1966
	Southern Baltic	10–15	0.79–1.25	Durbin and Durbin, 1981
	Etang de Berre, France		0.27	Arndt and Schnese, 1986
	Marseilles lagoon		0.28	Pagano and Saint-Jean, 1989
	Westernport Bay		0.13–0.44	Gaudy, 1989
A. tranteri	Hawaii	22–27	0.11	Kimmerer and McKinnon, 1987a
Acrocalanus inermis			0.07–0.36	Kimmerer, 1983
Calanipeda aquae-dulcis	Black Sea		0.09	Zaika, 1968
Calanoides carinatus	S Benguela		0.07–0.23	Verheye, 1991
	S Benguela		0.19	Walker and Peterson, 1991
Calanus agulhensis	Agulhas Bank		0.05–0.23	Peterson and Hutchings, 1995

Table 59 Continued.

Species	Location	T°C	P:B ratio	Authority
C. finmarchicus ♀	Spitsbergen		0.01	Diel, 1991
	Scotian Shelf		0.018	Tremblay and Roff, 1983b
	Lindåspollene, Norway		0.07–0.37	Aksnes and Magnesen, 1988
	Kattegat		0.058	Kiørboe and Nielsen, 1990
	Skagerrak		0.10	Peterson et al., 1991
C. glacialis	Scotian Shelf		0.01	Tremblay and Roff, 1983b
	Scotian Shelf		0.02	McLaren et al., 1989a
	Barents Sea		0.03–0.05	Slagstad and Tande, 1990
♀	Spitsbergen		0.03	Diel, 1991
♀	East Greenland		0.02–0.03	Diel, 1991
C. helgolandicus	Black Sea	ca.15–18	0.15	Uye, 1984
C. hyperboreus	Scotian Shelf		0.008	Tremblay and Roff, 1983b
	Scotian Shelf		0.02	McLaren et al., 1989a
C. pacificus	La Jolla, Ca.	ca.12–15	0.03–0.11	Mullin and Brooks, 1970
	Toyama Bay, Japan		0.04	Morioka, 1981
C. sinicus	Inland Sea, Japan	10	0.08	Huang et al., 1993
		15	0.17	Huang et al., 1993
		20	0.26	Huang et al., 1993
Candacia armata	Coastal Pacific	8–26	0.03–0.5	Huang et al., 1993
	Scotian Shelf		0.02	Tremblay and Roff, 1983b
	Scotian Shelf		0.09	McLaren et al., 1989a
Centropages brachiatus	S Benguela		0.09–0.32	Walker and Peterson, 1991
C. hamatus	North Sea		0.04–0.05	Fransz et al., 1991
	Kattegat		0.07–0.1	Kiørboe and Nielsen, 1990
	Lindåspollene, Norway		0.03–0.35	Aksnes and Magnesen, 1988
C. kröyeri	Black Sea		0.05	Zaika, 1968

Species	Location		Value	Reference
C. typicus	Scotian Shelf		0.036	Tremblay and Roff, 1983b
	Scotian Shelf		0.09	McLaren et al., 1989a
	Skagerrak		0.33	Peterson et al., 1991
	Banyuls-sur-Mer		0.061	Razouls, 1974
Clausocalanus sp.	Scotian Shelf		0.058	Tremblay and Roff, 1983b
	Scotian Shelf		0.18	McLaren et al., 1989a
Cosmocalanus darwini	Indian Ocean		0.08–0.16	Sazhina, 1986
Ctenocalanus vanus	S Benguela		0.04–0.09	Walker and Peterson, 1991
Eucalanus pseudoattenuatus	Indian Ocean		0.10–0.13	Sazhina, 1986
Euchaeta marina	Jamaica		0.10–0.47	Webber and Roff, 1995b
	Indian Ocean		0.04–0.20	Sazhina, 1986
Eurytemora affinis	Chesapeake Bay		0.08	Allan et al., 1976
	Bristol Channel			
	NVI		0.09	Burkill and Kendall, 1982
	CI		0.06	Burkill and Kendall, 1982
	CII		0.05	Burkill and Kendall, 1982
	CIII		0.14	Burkill and Kendall, 1982
	CIV		0.02	Burkill and Kendall, 1982
	CV		0.01	Burkill and Kendall, 1982
	Population	6–14	0.03–0.13	Burkill and Kendall, 1982
	Southern Baltic		0.20	Arndt, 1989
	Elbe Estuary		0.11–0.30	Peitsch, 1995
	Gironde Estuary		0.09	Castel and Feurtet, 1989
	Schelde Estuary		0.09	Escaravage and Soetaert, 1993
E. herdmani	Halifax	8–18	0.16–0.17	McLaren and Corkett, 1981
Metridia lucens	Toyama Bay, Japan		0.06	Morioka, 1981
	Scotian Shelf		0.025	Tremblay and Roff, 1983b
	Scotian Shelf		0.09	McLaren et al., 1989a
Neocalanus plumchrus	Sea of Japan		0.181	Shushkina et al., 1974
	Toyama Bay, Japan		0.03	Morioka, 1981

Table 59 Continued.

Table 59 Continued.

Species	Location	T °C	P:B ratio	Authority
Paracalanus parvus	Toyama Bay, Japan		0.07	Morioka, 1981
	Scotian Shelf		0.06	Tremblay and Roff, 1983b
	Scotian Shelf		0.20	McLaren et al., 1989a
	Lindåspollene, Norway		0.28–0.43	Aksnes and Magnesen, 1988
	Kattegat		0.09–0.11	Kiørboe and Nielsen, 1990
	Skagerrak		0.24	Peterson et al., 1991
	Black Sea		0.07	Greze et al., 1968
	Black Sea		0.09	Porumb, 1974
	S Benguela		0.12–0.17	Walker and Peterson, 1991
Pseudocalanus elongatus	North Sea		0.07–0.14	Fransz et al., 1991
	Kattegat		0.058	Kiørboe and Nielsen, 1990
	Lindåspollene, Norway		0.01–0.31	Aksnes and Magnesen, 1988
	Gdansk Bay	3–7	0.01–0.05	Ciszewski and Witek, 1977
	Black Sea		0.16	Greze et al., 1968
	Black Sea		0.09	Porumb, 1974
P. minutus	Toyama Bay, Japan		0.06	Morioka, 1981
	Scotian Shelf		0.052	Tremblay and Roff, 1983b
P. newmani	Scotian Shelf		0.15	McLaren et al., 1989a
Pseudocalanus spp.	Sea of Japan		0.268	Shushkina et al., 1974
	Skagerrak		0.18	Peterson et al., 1991

Pseudodiaptomus hessei	South Africa	15	0.15	Jerling and Wooldridge, 1991
		20	0.23	Jerling and Wooldridge, 1991
		25	0.30	Jerling and Wooldridge, 1991
P. marinus	Inland Sea, Japan	10	0.03	Uye et al., 1983
		25	0.24	Uye et al., 1983
Rhincalanus nasutus	S Benguela		0.03–0.15	Walker and Peterson, 1991
Temora longicornis	North Sea		0.05–0.11	Fransz et al., 1991
	Kattegat		0.05–0.09	Kiørboe and Nielsen, 1990
	Lindåspollene, Norway		0.02–0.33	Aksnes and Magnesen, 1988
	Skagerrak		0.05	Peterson et al., 1991
Temora stylifera	Banyuls-sur-Mer		0.053	Razouls, 1974
Undinula vulgaris	Pacific atoll	29	0.09	Gerber and Gerber, 1979
	Tikehau Atoll		0.34	Le Borgne et al., 1989
	Jamaica		0.16–0.63	Webber and Roff, 1995b.

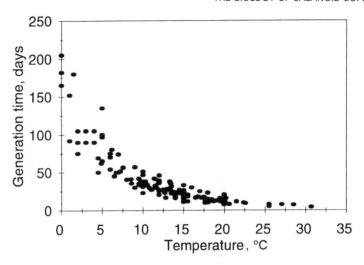

Figure 77 The generation time in days (D) of different species of copepods related to environmental temperature (T °C). The relationship is described by the equation:

$$D = 128.8\, e^{-0.120T}$$

(After Huntley and Lopez, 1992.)

Pagano and Saint-Jean (1989) found that daily P/B ratios in populations of *Acartia clausi* over the year in a coastal lagoon decreased exponentially as the mean body weight increased, a reflection of Banse and Mosher's (1980) relationship for animals in general. Uye (1982a) shows that the daily P/B ratios in *A. clausi* are related linearly to environmental temperature. Daily P/B ratios are positively correlated with environmental temperature as Uye (1982a) shows in *A. clausi*, Uye *et al.* (1983) in *Pseudodiaptomus marinus*, and Jerling and Wooldridge (1991) in *P. hessei*.

Production of zooplankton in general, and copepods in particular, in temperate seas with annual ranges of temperature greater than 10 °C used to be considered as food-limited. Current evidence suggests that this may be so in restricted areas and populations but that, more generally, production is controlled by environmental temperature (Davis, 1987). A simple model predicting biomass when its increase is a function of temperature, and mortality is virtually absent, is described by Samain *et al.* (1989). The population dynamics of *Calanus finmarchicus* are modelled by Miller and Tande (1993); the model assumes that development is controlled by temperature, and that food-limitation is unimportant, and so uses the Bělehrádek equations in Table 48 (p. 310). Peterson and Kimmerer (1994), however, found that egg production of *Temora longicornis* in Long Island Sound was directly linked to phytoplankton blooms; production of eggs at

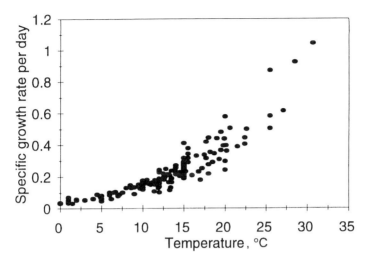

Figure 78 The specific growth rate d^{-1} (g) of species of copepods calculated from the weight of egg and the adult and the generation time of each species and related to environmental temperature (T °C). The equation for the relationship is:

$$g = 0.0445\ e^{0.111T}$$

(After Huntley and Lopez, 1992.)

other times was very nearly zero. This will result in a fluctuating, non-breeding component of the population and fluctuating P:B ratios. Such fluctuating P:B ratios can be inferred for copepods in the North Sea (Hay, 1995).

10.3. LIFE HISTORY PATTERNS (STRATEGIES)

The use of "Life history strategies" as a title for this section is controversial (Rothlisberg, 1985). The etymological derivation of the word "strategy" is from the Greek words *stratos* army and *agō* lead. Its appropriation from the battlefield by ecologists to describe the variety of means by which a species or population survives, invades or retreats in the natural environment does have an attraction.

The environmental conditions encountered by copepods in tropical, temperate and high latitudes are very different and are reflected in different life history patterns. There are, however, a great variety of environments ranging from brackish to fully saline, from shallow to deep, from sheltered to exposed within any one band of latitude. Environmental temperature

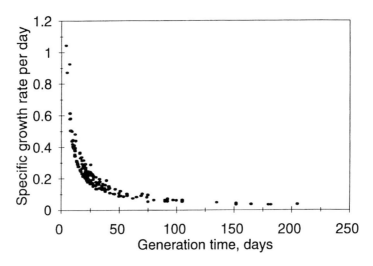

Figure 79 The specific growth rate d^{-1} (g) of species of copepods related to their generation time in days (D). The relationship is described by the equations:

$$g = 0.025 + (4.327/D) \qquad r^2 = 0.91$$
$$\text{Log } g = 0.551 - 0.895 \log D \qquad r^2 = 0.959$$

(After Huntley and Lopez, 1992.)

tends to decrease between tropical and high latitudes on a more-or-less regular basis and is the principal factor governing rates of metabolic processes such as growth and generation times. Environmental temperature also tends to decrease with depth and many species use this feature during the course of ontogenetic and/or diel migrations.

Calanoid copepods occur in all regions of the oceans and have exploited the pelagic, shallow hyperbenthic merging to the deep benthopelagic, fully saline extending to brackish, and even fresh water, environments. Species and populations have developed a range of life history patterns enabling them to survive in broad and/or restricted geographical regions, and some in quite restricted niches such as caves.

10.3.1. Tropical and Subtropical Patterns

The principal feature of tropical species of copepods is that there is no seasonality in their breeding patterns. Many species breed continuously but others are somewhat irregular and, according to Moore and Sander (1977), can breed randomly throughout the year. Commenting on Heinrich's (1962b) classification of breeding patterns of copepods, they conclude that the species they examined divide into three categories: continuous breeders, discontinuous but protracted breeders, and discontinuous, sporadic breeders.

Tropical species in these three categories all tend to produce successive broods and generations and it is very difficult in practice to identify the numbers of broods and generations by conventional analysis of the natural populations. In addition, Chisholm and Roff (1990a) found no seasonal variation in the prosome lengths of copepodid stages of *Centropages velificatus*, *Paracalanus aculeatus* or *Temora turbinata* although prosome lengths of adult females of the last species did vary significantly. Consequently, body size is of little potential value in identifying successive generations. Chisholm and Roff (1990b) suggest that the generation times of tropical species are comparable to those of temperate species and do not reflect the higher temperatures that they experience. This may result from adaptation to higher temperatures or possibly a reflection of food limitation.

Calanoides carinatus, in the upwelling region of the Ivory Coast, adopts a diapause CV at the end of the cold season that sinks to deep water until the next cold season (Binet and Suisse de Sainte Claire, 1975). This adoption of a diapause CV is a feature of this species farther south in the Benguela Current but is the only record of a tropical to temperate species with such a life cycle. No investigation of the possible production of diapause eggs by tropical calanoids has been made (Marcus, 1996) although their occurrence in a few subtropical regions is known (Table 40, p. 269).

10.3.2. Temperate Patterns

Species living in temperate as opposed to tropical and subtropical regions are subject to lower temperature regimes. There is a tendency for them to be larger in body size and to have longer generation times but many of the smaller species have generation times comparable to those of tropical species. In addition, there is marked seasonality in the environment such that breeding is seasonal although it may be for a very protracted period, extending for 10 or 11 months with contained periods of higher and lower activity. Successive broods and generations, as in tropical species, are often a feature, especially among small coastal species. The majority of smaller Mediterranean species have 4 to 6 generations per year in the field (Table 58, p. 355), although some may have up to 12 or so. Their life histories are then difficult to determine from conventional analysis, although body sizes of copepodids may change significantly between generations, especially where seasonal fluctuations in environmental temperature are larger.

The adoption of a diapause CV coupled with an ontogenetic migration is a feature of high latitude species (Table 60). *Calanoides carinatus*, however, in the upwelling environment of the Southern (Verheye et al., 1991) and Northern (Arashkevich et al., 1996) Benguela Current has a true diapause CV and resting female to allow its population to be maintained in this environment.

Temperate latitudes grade into high latitudes and species producing diapause eggs (Table 40, p. 269), resting copepodids and resting adults (Table 60) are common. The occurrence of these resting stages and their importance within the life histories of the copepods are discussed in the next section. Diapause is a common phenomenon throughout the Crustacea and is reviewed by Alekseev and Fryer (1996).

10.3.3. High Latitude Patterns

The pronounced seasonality of environmental variables in high latitudes results in an equally pronounced seasonality in the biology of the resident copepod species. Growth rates are slowed, development times extended and breeding periods restricted seasonally. Conventional population analysis can often describe the biology of populations successfully although there are still difficulties, especially when individual females produce successive broods. Few species have simple life cycles such as that described by Roff and Carter (1972) for *Limnocalanus macrurus* in Arctic fresh water lakes; the adults overwinter to produce eggs and nauplii in January to April while the copepodids appear successively in March to May, the first becoming adults in June to form the overwintering population. More normally, the pronounced winter conditions have resulted in the development of so-called overwintering strategies to ensure survival of populations. These take two principal forms: the adoption of a resting or diapause stage within the developmental sequence of copepodids coupled with an ontogenetic, seasonal, vertical migration or the production of a resting or diapause egg that sinks through the water column to reside in the mud of the sea bed. Diapause eggs are restricted to coastal and shelf species (Table 40, p. 269). In offshore species, the production of deep-living, resting copepodid stages, coupled with an ontogenetic migration, is not adopted by all species. Huntley and Escritor (1992), for example, show that no ontogenetic migration takes place in populations of the common Antarctic species *Metridia gerlachei* and it may not enter diapause. Grønvik and Hopkins (1984) also suggest that the north Atlantic *M. longa* overwinters in an active state. Some species of *Acartia* have populations that do not appear to produce resting eggs (Table 60) but pass the winter in a resting adult stage with stored spermatozoa (Norrbin, 1994).

10.3.3.1. Overwintering Copepodids

The occurrence of resting or diapause copepodids has been described in several species (Table 60). The commonest diapause stage is the CV although other stages, which may only be subject to arrested development because of low temperature regimes, occur in winter (Table 60). Diapausing copepodids are subject to arrested development, have depressed respiration and excretion rates, do not usually feed and often have reduced epithelium in the gut associated with reduced digestive enzyme activity (Hallberg and Hirche, 1980; Hirche, 1983, 1989a; Alldredge et al., 1984; Båmstedt et al., 1985; Hassett and Landry, 1985, 1990a; Mayzaud, 1986a; Smith, 1988; Bathmann et al., 1990, 1993; Drits et al., 1994; Arashkevich et al., 1996). In addition, large stores of lipids are often present, especially in herbivorous species (e.g. Clarke and Peck, 1991; Schnack-Schiel et al., 1991; Kattner et al., 1994; Dahms, 1995; Arashkevich et al., 1996). These stores are used for survival through the winter and subsequent egg laying (e.g. Ohman, 1987). Egg laying in advance of the spring phytoplankton bloom can take place in a number of species, such as *Calanus glacialis* and *C. hyperboreus* in the Greenland Sea, through use of these lipid stores for gonad maturation (Smith, 1990). Båmstedt and Ervik (1984), in a comparative study, found that *Calanus finmarchicus* enters diapause in the winter and depends on stored lipids while the co-occurring *Metridia longa* does not, but feeds continuously.

The two factors suggested as controlling entry and emergence from diapause are photoperiod and temperature. Watson (1986) suggests that these factors act independently, especially in fresh water environments.

Induction of diapause in multi-voltine species must take place in the generation prior to that which diapauses (Norrbin, 1996). An early copepodid of this penultimate generation is assumed to be the sensitive stage at which "switching" to the diapause mode takes place. This may be in response to decreasing day-length. Increasing day-length may cue termination of diapause in shallow, marine, coastal areas but no such changes in the seasonal light regime are present at 500 to 2000 m depth in oceanic regions where, for example, diapause CVs of *Calanus plumchrus* live. Some other cue or cues are necessary but what they may be is conjectural (Ianora and Santella, 1991; Miller and Grigg, 1991; Miller et al., 1991). Arashkevich et al. (1996) review the possibility of endogenous mechanisms of diapause control in *Calanoides carinatus* in the Benguela ecosystem.

Huntley et al. (1994) found that the late copepodids of *Calanoides acutus* in Antarctica emerge from diapause within a relatively short period of time before the spring bloom of phytoplankton, so inferring a degree of synchrony. Miller et al. (1990) and Pedersen et al. (1995) show that the

Table 60 Deep-living, overwintering stages of calanoid copepods associated with an ontogenetic migration. The resting stages for species with life cycles of more than a year are indicated for each successive winter.

Species	Location	Stage	Authority
Acartia clausi	Japan Inland Sea	Nauplius?	Uye, 1980a
A. longiremis	Balsfjorden, N Norway	CVI	Norrbin, 1994, 1996
A. steueri	Japan Inland Sea	Nauplius?	Uye, 1980a
Calanoides acutus	Antarctic	CIV to adult	Marin, 1988a,b
	Antarctic	CIV to adult	Atkinson, 1991
	Antarctic	CIV, CV	Huntley and Escritor, 1991
	Antarctic	CIV to adult	Schnack-Schiel and Hagen, 1995
C. carinatus	Southern Benguela	CV	Verheye *et al.*, 1991
	Northern Benguela	CV	Timonin *et al.*, 1992
	Northern Benguela	CV	Arashkevich *et al.*, 1996
	Somalia	CV	Smith, 1995
Calanus finmarchicus	Arctic	CIII to CV	Tande, 1982
	Greenland Sea	CV	Smith, 1988; Hirche, 1989a
	W Barents Sea	CIV, CV	Pedersen *et al.*, 1995b
	SW of United Kingdom	CV?	Williams and Conway, 1988
	Gulf of Maine	CIV, CV	Miller and Grigg, 1991
C. glacialis	Barents Sea		
	1st winter	CIII, CIV	Tande *et al.*, 1985
	2nd winter	CV	Tande *et al.*, 1985
C. helgolandicus	SW of United Kingdom	CV?	Williams and Conway, 1988
C. hyperboreus	Greenland Sea	CV	Hirche, 1989a
C. marshallae	Dabob Bay, Washington	CV	Osgood and Frost, 1994b
C. pacificus	Dabob Bay, Washington	CV	Osgood and Frost, 1994b
C. propinquus	Antarctic	CIV, CV	Marin, 1988a,b
	Antarctic	CIII to adult	Schnack-Schiel and Hagen, 1995

Species	Location	Stage	Reference
C. simillimus	Antarctic	CV	Atkinson, 1991
Drepanopus bispinosus	Antarctic freshwater	nauplii (?)	Wang, 1992
Eucalanus bungii	W Bering Sea	CIV, CV	Geinrikh, 1962a
	NE Pacific		
	1st winter	CIII, CIV	Miller et al., 1984a
	2nd winter	CV	Miller et al., 1984a
	3rd winter	female	Miller et al., 1984a
E. monachus	Somalia	CV	Smith, 1995
Metridia pacifica	W. Bering Sea	CIV, CV	Geinrikh, 1962a
	Sea of Japan	CV	Hirakawa and Imamura, 1993
Neocalanus cristatus	W Bering Sea	CIII, CIV	Geinrikh, 1962a
	NE Pacific	CV	Miller et al., 1984a
N. flemingeri	NE Pacific	female	Miller and Clemons, 1988
	Sea of Japan	CIV	Miller and Terazaki, 1989
N. plumchrus	NE Pacific	CV	Miller et al., 1984a
Paralabidocera antarctica	Syowa, Antarctica	NIV, NV	Tanimura et al., 1996
Pseudocalanus acuspes	Resolute Bay, Arctic	CIII, CIV	Conover and Huntley, 1991
	Resolute Passage	CIV, CV	Fortier et al., 1995
	Balsfjorden, N Norway	CIV, CV	Norrbin, 1994, 1996
	White Sea	CIII to CV	Pertzova, 1981
P. elongatus	Saanich Inlet, Canada	CIV, CV	Koeller et al., 1979
P. minutus	Balsfjorden, N Norway	CV	Norrbin, 1994
Rhincalanus gigas	Antarctic		
	1st winter	CIII, CIV	Marin, 1988a, b
	2nd winter	CV, adult	Marin, 1988a, b
	Antarctic	CIII to CV	Atkinson, 1991
Stephos longipes	Antarctic	CIV, CV	Kurbjeweit et al., 1993

morphology and histology of the mandibles can be used to detect the approaching emergence from diapause. The teeth of the new mandible form, within the mandible, under the old ones before there is any sign elsewhere in the body of the copepod that it is about to moult. Pedersen *et al.* note that individuals are not fully synchronous in this regard. This links with a statement made by Watson (1986) that "termination of diapause has frequently been recognized as the primary factor synchronizing actively growing stages with favourable growing conditions". Much further work is required on factors controlling entry to and emergence from diapause as well as on those controlling gonad maturation and egg laying.

10.3.3.2. *Eggs in Sea Bottom Muds*

Many of the species of copepods listed in Table 40 (p. 269) can be seasonally absent from regions where they are extremely common at other times, as Marshall (1949) found in populations of *Centropages hamatus* and *Temora longicornis* in Loch Striven, western Scotland. This is because resting eggs, described in Chapter 8, are produced under adverse environmental conditions, frequently on a marked seasonal basis. Such conditions are not only the onset of winter, and its associated low temperature regime, but also the high summer temperatures that can prevent the eggs from hatching and they then sink to the sediments (Sullivan and McManus, 1986). Temperature, rather than food availability, is the controlling factor in diapause egg production although Ban and Minoda (1994) experimentally show that crowded populations of *Eurytemora affinis* are induced by the presence of their own metabolic products to produce diapause eggs.

Lindley and Hunt (1989) and Lindley (1990) examine the distributions of *Labidocera wollastoni* and *Centropages hamatus* in the North Sea and around the British Isles relative to the distributions of open-sea sediments at depths between 20 and 50 m and deeper sediments. Resting eggs in the sediments were more important to the survival and resultant population distribution of *L. wollastoni* than of *C. hamatus*. The distribution of the latter species owed much to the persistence of overwintering populations in areas where phytoplankton survives at low temperatures as to reservoirs of diapause eggs in the sediments. Areas of deposition of diapause eggs in or on the sediments might be expected to be correlated with certain size fractions of sedimented inorganic particles having a sinking velocity equal to that of the eggs. Areas of deposition of eggs and particles would then be the same and determined by the physical hydrography of the region. This correlation is assumed rather than proven (Marcus and Fuller, 1989), even within a sea area such as Buzzards Bay.

Undoubtedly diapause eggs in sediments can represent a sanctuary allowing recolonization of a pelagic regime that, for one reason or another, has failed to maintain the adult population. Eggs in sediments can remain viable for more than 40 years (Marcus et al., 1994) given that they do not dry out or freeze (Næss, 1991b). Marcus and Fuller (1989) find that diapause eggs of *Labidocera aestiva* in Buzzards Bay sink to the bottom and become buried in the sediments progressively during the autumn and winter. Those buried deepest are then dependent on bioturbation of the sediments to raise them back to the surface where their development may be stimulated. The duration of diapause must be dependent, to a considerable degree, on such extrinsic factors.

Lindley (1992) goes as far as to suggest that the success of the Centropagoidea in fresh water environments results from their adoption of the diapause egg as a means of maintaining populations in adverse environments. Some shallow coastal regions heat up in summer to temperatures lethal to the adult copepods. Resting eggs are produced, sediment to the bottom, and replace the population once conditions become favourable. The more common situation is the production of resting eggs in the autumn, the virtual absence of a pelagic population of the species during the winter, and a reappearance of the species in the spring when the resting eggs hatch.

10.3.3.3. *Resultant Life Histories*

Complex life histories with short generation times and overlapping broods and generations exist among small species in genera such as *Acartia*, *Centropages*, *Pseudocalanus* and *Temora*. Populations of these species may have up to eight generations per breeding season. Some produce diapause eggs that overwinter in the sediments giving rise to, or augmenting existing low-density, pelagic populations in the spring and summer. Life histories tend to be simpler at higher latitudes because of the decreasing seasonal window within which they can actively grow. Conover (1988) compares the life histories in northern high latitudes and concludes that they are primarily of three types: smaller species with multigenerations in relatively rich boundary regions that have winter temperatures well above 0°C; annual species with a single generation in oceanic gyres and northern boundary areas where ice is not a major problem; and species that take more than a year to reach sexual maturity and breed, probably predominantly confined to Arctic waters proper. Arctic and sub-Arctic, along with Antarctic and sub-Antarctic, species are often large in body size and tend to produce a single generation per year. A diapause CV may or may not be present in winter, but, whether it is or not, the life cycle is relatively simple. It is more difficult to determine whether a species has a 2-, 3- or 4-year life cycle because all stages are present at some time of year and many simultaneously so.

Bathymetric time-series samples are required to elucidate ontogenetic, vertical migrations and these often have to be coupled with examination of the state of maturation of the gonads in CVs as well as adults (e.g. Miller *et al.*, 1984a; Tande *et al.*, 1985; Miller and Clemons, 1988; Smith, 1990).

A simple life history is shown in Figure 80A involving diapause and an ontogenetic vertical migration. The copepod has one generation per year, in the spring, and the CV enters diapause in the autumn to mature the following spring. A species having this life history is *Calanoides acutus* in the Antarctic. Other species such as *Calanus simillimus* and *Rhincalanus gigas* have a spring and summer generation, the CVs resulting from the summer breeding entering diapause in the autumn (Figure 80B). Some species with this general pattern can have more than two generations per year, the CVs produced from the last breeding period entering diapause. As described above, diapausing CVs are characterized by reduced levels of metabolic activity and do not feed to any extent. Consequently, without physiological measurements, it is difficult to distinguish between true diapausing winter stages and those in which development is arrested through low temperature. Pertzova (1981), for instance, states that *Pseudocalanus elongatus* has two generations per year (Table 58, p. 355) and that the second generation overwinters in the CIII to CV (Table 60, p. 386). Other records of overwintering or resting in earlier stages than the CV are also given in Table 60, even for *Acartia* species and the fresh water *Drepanopus bispinosus* in the naupliar stages, but it is questionable as to whether these are true diapause stages.

The pattern of ontogenetic vertical migration in Figure 80A can be extended for two or more years to describe life histories of species with longer generation times (Figure 81).

These are simplified gross patterns, and considerable variation exists in time and space. The same species often has a different pattern in different geographical regions. The detailed bathymetric distribution of the copepodids often varies within wide limits and the stage of development reached in unit time also varies. Hirche (1991) raises the question as to how overwintering copepods in high latitudes, with the absence of downwelling light during several winter months, locate and maintain their overwintering depths. Bathymetric temperature distributions are not consistent enough. Nothing is known of the use copepods make of bathymetric pressure changes. Other potential factors are conjectural or unknown.

10.3.4. Deep-sea Patterns

The general conception of deep-sea species are those that live at depths greater than about 200 m. Here, however, deep-sea species refers only to

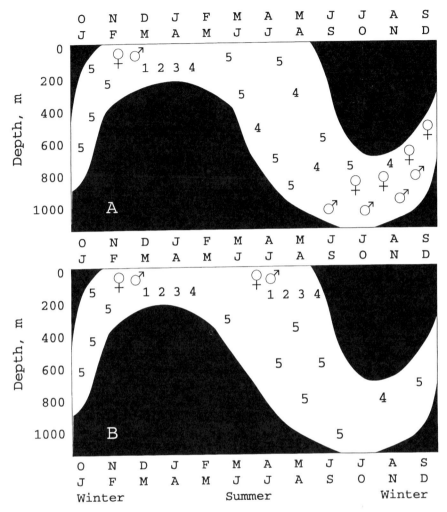

Figure 80 Schematic representations of the ontogenetic seasonal migration, breeding and development of copepodids 1 to 5 of high-latitude species. A, a species with a single generation per year. B, a species with two generations per year, one in the spring, the other in the summer. The corresponding spring, summer and winter months are shown for north and south latitudes. (After Atkinson, 1991.)

those species that live in the bathypelagic environment, below depths of about 500 m and do not perform diel vertical migrations. Included with them are mesopelagic non-diel migrating species. These latter species, like the truly bathypelagic species, do not normally perform the type of ontogenetic migration described in Figures 80 and 81 but tend to live at depths between

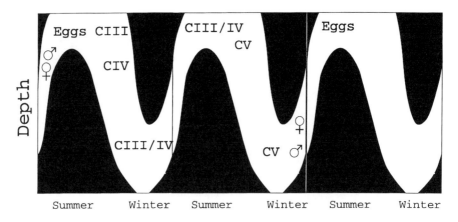

Figure 81 Schematic representation of the ontogenetic seasonal migrations, breeding and development of copepodids III to V of a high-latitude species with a two-year life cycle.

about 100 and about 600 to 800 m depth. The mesopelagic and bathypelagic species are subject to decreasing seasonality as the modal depth of their populations increases (Mauchline, 1992a). Little is known about their life histories except that they breed continuously throughout the year at relatively low environmental temperatures that are assumed to lengthen their development and, so, generation times. Some, like *Valdiviella insignis*, which produces only two eggs per brood (Figure 8, p. 22), have very small brood sizes. Some may have life spans of several years, some may produce a single brood, while others may produce successive broods at irregular intervals (Mauchline, 1992b, 1994a,b, 1995).

The amplitude of seasonal change in environmental temperatures decreases with depth, being less than 1°C at depths greater than 500 m in the Rockall Trough (Mauchline, 1991). There are no seasonal changes in the breeding capacity within the populations so that there is always a proportion of the individuals sexually mature and breeding at any one time. Variation in the body lengths of individuals within the populations is much reduced relative to that in epipelagic or coastal populations. Further, variation in the population sizes with time is likewise reduced because there are no peak periods of recruitment nor major fluctuations in predation pressure.

10.4. POPULATION MAINTENANCE

The principal factors enabling maintenance of a population of pelagic copepods vary between species and habitat. Deposition of diapause eggs in the sediment allows a species to survive mass mortality in the pelagic environment, the new pelagic population being regenerated at a later date from these eggs when a more favourable environment is present (e.g. Marcus et al., 1994). Diapausing copepodids allow synchronization of active development with favourable environmental conditions, including availability of suitable and adequate supplies of food. High fecundity, through large brood size, production of successive broods by individual females, and short generation times counteract environmental factors such as heavy predation and/or emigration from the local environment. Behaviour of individuals is also involved and will be discussed more fully in the next chapter.

Estuarine, coastal and brackish water species are small in body size with opportunistic life styles that allow them to survive unfavourable conditions and exploit favourable ones. The longer-lived offshore epipelagic and mesopelagic species tend to be larger in body size and less resilient in the face of unfavourable environmental conditions that they meet less often than the previous species. The longest-lived species are those in the high latitude and the bathypelagic environments. Those in high latitudes have developed resting copepodid stages and/or lipid stores that aid survival in adverse winter environments; their fecundity is often high, allowing them to exploit favourable conditions in the spring and summer. The deep-sea species, in an environment with relatively stable environmental and biological characteristics, but with parsimonious resources, have developed strategies of energy conservation; among these are a reduction in fecundity, higher water content in the body contributing efficiently to larger body size and to a lower metabolic rate.

10.4.1. Annual Fluctuations

Little change takes place in the biomass of populations where breeding is aseasonal as in tropical and bathypelagic environments. This, however, does not apply to tropical upwelling environments where a seasonality is imposed on the populations by the periods of active upwelling. Seasonally breeding populations show wide fluctuations in biomass and numbers of individuals. This is especially true of fecund species with short generation times. In general, the longer the generation time of a species the less pronounced are the fluctuations in population biomass and numbers. Superimposed on these fluctuations are variations caused by the seasonal or selective activity of predators.

Many of the studies listed in Table 58 (p. 355) provide examples of the extent of these annual changes in sizes of populations, the extents varying from species to species, between environments and also, as referred to in the next section, between successive years.

10.4.2. Inter-annual Fluctuations

According to Castel (1993), time-series must be >10 years in extent before any valid results can be obtained. This conclusion, however, is somewhat optimistic as can be seen from the attempts to draw conclusions from the 35-year observations of the Continuous Plankton Records (e.g. Colebrook, 1985a,b; Broekhuizen and McKenzie, 1995).

There are inter-annual variations in the numbers and biomass of populations of individual species of copepods, especially in coastal and estuarine regions. A selection of studies of such variations is given in Table 61. A major difficulty is trying to ascribe causes for such variation. The zooplankton of the northeast Atlantic, including the North Sea, declined in numbers during the period 1948 to 1980 and Colebrook (1985b) found no clear trend in surface temperatures in the North Sea during these years but a possible relationship with the occurrence of westerly weather over the British Isles. Colebrook et al. (1984) and Roff et al. (1988) conclude that the downward trend in numbers might be a product of some unidentified factor or factors affecting its overwintering stocks, and possibly derived from the incidence of westerly weather. Dickson et al. (1988, 1992), on the other hand, invoke north and east winds, the former being linked with delayed spring phytoplankton blooms and consequent shortening of the growing season for zooplankton, the latter with westward transport of low-salinity water from Skagerrak and subsequent reinforced conditions of vertical stratification of the water column. These are complex concepts as reviewed by Fransz et al. (1991). In contrast, Jossi and Goulet (1993) show that there is an increasing trend in the numbers of *Calanus finmarchicus* in the northeast continental shelf of the United States over the last 30 years, although no such increase is evident in the other species of copepods that they examined. In another study, the meanderings of the Gulf Stream between 1966 and 1992 were correlated with the inter-annual changes in numbers of copepods in a region to the north of Scotland except in the last four years, 1989 to 1992 (Hays et al., 1993).

The inter-annual fluctuations in numbers of herbivorous copepods were not related to comparable fluctuations in the quantities of phytoplankton (Gieskes and Kraay, 1977). Predators such as planktivorous fish can cause notable mortality of copepods as mentioned earlier. The inter-annual fluctuations in size of the population of a migrating predatory fish may be determined by fluctuations in the parameters of an environment distant from the environment in which the subsequent mortality of the copepods takes place. Copepods are not the only planktonic organisms showing inter-annual fluctuations, Lindley et al. (1995) having shown parallel changes in the numbers of echinoderm larvae present in the plankton.

One of the difficulties in considering the copepods of a region as a

Table 61 Inter–annual studies on calanoid copepods.

Species	Location	Years of observation	Authority
Acartia bifilosa	N Baltic Sea	12–18	Viitasalo, 1994; Viitasalo *et al.*, 1995b
	SW France	14	Castel, 1993
A. californiensis	San Francisco Bay	4	Ambler *et al.*, 1985
	Yaquina Bay, Oregon	7	Frolander *et al.*, 1973
A. clausi	San Francisco Bay	4	Ambler *et al.*, 1985
	NE Atlantic	32	Colebrook, 1982
	North Sea	32	Colebrook, 1982
	Gulf of Naples	7	Mazzocchi and Ribera d'Alcalà, 1995
	Adriatic Sea	10	Regner and Vučetic, 1980
		15	Regner, 1991
	Gulf of Trieste	11	Cataletto *et al.*, 1995
Acartia spp.	SE North Sea	28	Gieskes and Kraay, 1977
	W North Sea	15	Roff *et al.*, 1988
Calanus finmarchicus	Scotian shelf	29	Jossi and Goulet, 1993
	US northeast shelf	70	Sherman *et al.*, 1983
	US northeast shelf	29	Jossi and Goulet, 1993
	W Gulf of Maine	8	Meise-Munns *et al.*, 1990
	Georges Bank	8	Meise-Munns *et al.*, 1990
	E North Atlantic	30	Planque and Fromentin, 1996
	E North Atlantic	30	Planque and Fromentin, 1996
	North Sea	30	Brander, 1992
	European Shelf	31	Williams *et al.*, 1994
	North Sea	34	Broekhuizen and McKenzie, 1995

Table 61 Continued

Species	Location	Years of observation	Authority
Calanus helgolandicus	European Shelf	31	Williams *et al.*, 1994
	E North Atlantic	30	Fromentin and Planque, 1996
	E North Atlantic	30	Planque and Fromentin, 1996
	Gulf of Trieste	11	Cataletto *et al.*, 1995
Centropages hamatus	N Baltic Sea	12–18	Viitasalo, 1994; Viitasalo *et al.*, 1995a
C. kröyeri	Adriatic Sea	10	Regner and Vučetić, 1980
	Gulf of Trieste	11	Cataletto *et al.*, 1995
C. typicus	Scotian shelf	29	Jossi and Goulet, 1993
	US northeast shelf	70	Sherman *et al.*, 1983
	US northeast shelf	29	Jossi and Goulet, 1993
	W Gulf of Maine	8	Meise-Munns *et al.*, 1990
	Georges Bank	8	Meise-Munns *et al.*, 1990
	Gulf of Naples	7	Mazzocchi and Ribera d'Alcalà, 1995
	Adriatic Sea	10	Regner and Vuvetić, 1980
	Gulf of Trieste	11	Cataletto *et al.*, 1995
C. violaceus	Gulf of Trieste	11	Cataletto *et al.*, 1995
Ctenocalanus vanus	Gulf of Trieste	11	Cataletto *et al.*, 1995
Eurytemora affinis	N Baltic Sea	12–18	Viitasalo, 1994; Viitasalo et al., 1995b
	S Baltic Sea	7	Heerkloss *et al.*, 1993
	SW France	14	Castel, 1993
E. americana	Yaquina Bay, Oregon	7	Frolander *et al.*, 1973
Limnocalanus macrurus	N Baltic Sea	12–18	Viitasalo, 1994; Viitasalo *et al.*, 1995b
Mesocalanus tenuicornis	Gulf of Trieste	11	Cataletto *et al.*, 1995
Metridia lucens	Scotian shelf	29	Jossi and Goulet, 1993
	US northeast shelf	29	Jossi and Goulet, 1993

Paracalanus parvus	Yaquina Bay, Oregon	7	Frolander et al., 1973
	Gulf of Naples	7	Mazzocchi and Ribera d'Alcalà, 1995
	Adriatic Sea	10	Regner and Vuvetić, 1980
	Gulf of Trieste	11	Cataletto et al., 1995
Pseudocalanus elongatus	N Baltic Sea	12–18	Viitasalo, 1994; Viitasalo et al., 1995b
	NE Atlantic	32	Colebrook, 1982
	NE Atlantic, North Sea	35	Colebrook et al., 1984
	North Sea	32	Colebrook, 1982
	W North Sea	15	Roff et al., 1988
	North Sea	32	Dickson et al., 1992
	Gulf of Trieste	11	Cataletto et al., 1995
P. minutus	Scotian shelf	29	Jossi and Goulet, 1993
	US northeast shelf	70	Sherman et al., 1983
	US northeast shelf	29	Jossi and Goulet, 1993
Pseudocalanus sp.	Yaquina Bay, Oregon	7	Frolander et al., 1973
Temora longicornis	N Baltic Sea	12–18	Viitasalo, 1994; Viitasalo et al., 1995b
	Wadden Sea	19*	Fransz et al., 1992
	W North Sea	15	Roff et al., 1988
	Gulf of Trieste	11	Cataletto et al., 1995
T. stylifera	Gulf of Naples	7	Mazzocchi and Ribera d'Alcalà, 1995
	Adriatic Sea	10	Regner and Vučetić, 1980
	Gulf of Trieste	11	Cataletto et al., 1995
Temora sp.	SE North Sea	28	Gieskes and Kraay, 1977

*Discontinuous sampling.

homogeneous entity is that they are rarely such. The fauna of copepods usually comprises an association of species that exhibit a wide variety of environmental preferences. Consequently, different species will react in different ways to changes in the environmental factors. Viitasalo *et al.* (1994) studied the reactions of the two dominant species, *Acartia bifilosa* and *Eurytemora affinis*, in the northern Baltic Sea to changing environmental conditions and factors such as temperature, salinity and depth and found that there were not only differences between the two species but also between the reactions of different stages and generations of the same species. Different patterns of inter-annual fluctuations had been previously documented by Colebrook (1978, 1982) in *Acartia clausi* and *Pseudocalanus elongatus*. An analysis of a 34-year block of data by Broekhuizen and McKenzie (1995) for the North Sea shows no effect of the dynamics of up-current populations on their down-current counterparts. Hydrographically similar, though geographically separate, regions of the North Sea have similar patterns of seasonal dynamics.

The different responses of species is further shown by populations of *Calanus finmarchicus* and *C. helgolandicus* in the eastern North Atlantic. The numbers of the former species have decreased over the period 1962 to 1992 whereas those of the latter have increased (Planque and Fromentin, 1996). Coherence within the data is greater for *C. finmarchicus* whereas the upward trend in numbers of *C. helgolandicus* is more irregular, groups of a few years deviating from the general trend. Fromentin and Planque (1996) relate the changes to changes in the North Atlantic Oscillation.

Thus, although some of these time-series extend for more than thirty years, it is still difficult to recognize causes and effects. This may partly arise through not asking the correct questions. There is always the possibility that the correct parameters of the organisms and the environment are not being measured. Much further work on even longer time-series is required.

10.5. BIOMASS OF POPULATIONS

Biomass is a measure of the size of a population and is used in studies of rates of production, and in assessing population demands or impact on the environment. It is difficult to measure because it depends upon representative and quantitative sampling. Individuals in a population of a copepod are normally patchily distributed both in the horizontal and vertical planes. Diel vertical migration must be taken into account. Densities of individuals in a population, especially in temperate and high latitudes, vary seasonally with consequent changes in the biomass. There may also be a seasonal ontogenetic vertical migration. Degrees of aggregation of individuals can also vary seasonally, contributing to errors in the estimations. Consequently, most estimates of biomass are approximate. It is often calculated by transforming counts of copepods to wet, dry or carbon weight through use of body length to weight regression equations. It can be expressed as the weight under 1 m² of sea surface or as per m³.

Tranter (1973), in examining seasonal and geographical variation in biomass, uses a dimensionless coefficient of variation for comparing values. The standard deviations of the monthly or quarterly mean estimates of biomass are expressed as a percentage of the overall annual mean biomass. Thus, he shows that the variation in the seasonal biomass of the population of *Calanus hyperboreus* in the Norwegian Sea decreases progressively from the surface to about 1500 m depth while that of *Rhincalanus gigas* in the Antarctic shows a peak of variation at 200 m, a decrease at 400 m and a progressive increase to depths of 900 m. This coefficient can also be used to compare the sizes of between-sample variation at the same location at different times or at different locations at the same time. It will also give a measure of the degree of patchiness, a major contributor to sampling error, between populations.

11. Behaviour

11.1. Swimming Activity	401
11.1.1. Swimming Pattern	402
11.1.2. Swimming Speed	407
11.1.3. Body Density and Sinking Rate	409
11.1.4. Escape Reaction	414
11.1.5. Response to Predators	419
11.1.6. Rhythmic Activity	420
11.1.7. Energetic Cost	421
11.2. Spatial Distribution	423
11.2.1. Aggregations	424
11.2.2. Swarms	428
11.2.3. Schools	429
11.2.4. Multispecies Aggregations and Swarms	429
11.2.5. Mechano- and Chemoreception Distances	432
11.3. Bioluminescence	434
11.3.1. Luminescent Glands	434
11.3.2. The Luminous Secretions	436
11.3.3. Luminescent Behaviour	437
11.4. Vertical Migration	439
11.4.1. Behaviour of Different Species	442
11.4.2. Thermoclines, Haloclines and Oxyclines	443
11.4.3. Abrupt Bottom Topography	444
11.4.4. Ontogenetic Migrations	445
11.4.5. Cues for Diel Migration	446
11.5. Rhythms	452
11.6. Concluding Remarks	453

This chapter examines what is known about the behaviour of calanoid copepods. Some aspects have already been described. Feeding behaviour is discussed in Chapters 5 and 6 while mating behaviour is described in Chapter 8.

Behaviour here is defined as activities of the individual and population that result in a degree of independence from the physical and chemical environment and also allow successful reproduction and guard the population from predation or extinction. It is the entire behaviour of the animal in its natural habitat over time (Hamner, 1985). The escape reaction translates the copepod from the viscous world of low Reynolds numbers to the inertial world of higher Reynolds numbers. Exploitation of eddies and other physical features of the hydrography and bottom topography removes the copepod from the concept of passively drifting plankton and gives it a measure of control over its movement and, in the context of its population, dispersal. Communication between individuals of a species and between species, and social behaviour, maintains the individuals in aggregations, swarms and schools. Omori and Hamner (1982) review some of the difficulties, including sampling strategies, of describing and assessing such behavioural traits.

Individuals within populations of copepods are not randomly distributed in the three-dimensional ocean. The Longhurst-Hardy plankton recorder and pump sampling have been employed to examine their spatial distributions. The sizes and/or densities of patches can change along transects showing that various scales of patchiness are operable simultaneously (Wiebe, 1970; Fasham *et al.*, 1974; Tsuda *et al.*, 1993). Patchiness is not restricted to the horizontal plane but occurs on scales of 10 m in the vertical plane (Haury, 1976a). It is influenced by the physical and chemical characteristics of the water masses but, as Hamner (1988) points out, patchiness in the open, as opposed to coastal, pelagic environment may result from behavioural aggregation. Coastal environments are generally shallow and populations have intimate contact with the sea-bed, at least during daylight hours. Here, the physical hydrographical environment has a strong influence as does the topography of the sea-bed but the behaviour of individual organisms is still a major influence.

The behaviour of the individual is neglected in all studies of diel vertical migration of copepods. These studies sample the numbers of species inhabiting different depth horizons of the water column at different times of the day and night. The data are then analysed to show what proportion of the populations migrate towards the surface and over what time-scale and depth-range. These data do not show the distance migrated by an individual, whether the same individuals migrate night after night or only sporadically, whether the day-time population, which often inhabits a wide bathymetric zone, migrates upwards as an integrated unit or whether selected individuals from different depth horizons migrate to the surface. Hamner (1988) makes a plea for more studies of the behaviour of individuals rather than of populations.

11.1. SWIMMING ACTIVITY

Most copepods swim by rapid beating of the antennae, the mandibular palps, the maxillules and the maxillae. The antennules are active in producing a jumping motion. The musculature involved in swimming is described in detail by Boxshall (1985). Gill and Crisp (1985b) and Gill (1987) describe rates of beating of the antennae in swimming and escape modes of several species while Svetlichny (1991) estimates beat cycles, angular velocities and tractive forces of the appendages of *Calanus helgolandicus* by using filming techniques. Articulations within the setae of the appendages and also in the caudal setae are thought by Vaupel Klein (1986) to allow the setae to be rigid during the power stroke and flexible during the recovery stroke. The advent of video recording has allowed objective investigation of swimming patterns but tethering of the copepods can affect the results (Hwang *et al.*, 1993). Micro-impedance units can also be used to determine activity of appendages (Gill and Crisp, 1985a; Gill and Poulet, 1986, 1988a). Swimming takes a variety of forms but all species have periods of inactivity, when they sink passively.

A variety of postures are adopted by different species of calanoids during swimming. *Centropages* species swim with the urosome flexed dorsally (Gauld, 1966) and occasional beats of the swimming legs are used to change direction. Most copepods swim with the ventral side downwards, but some species, such as *Acartia* and *Pseudocalanus*, have no particular orientation with respect to gravity, while others, such as *Temora*, roll spirally (Gauld, 1966). *Calanus* species often hang in the water column with the head upwards or downwards (Bainbridge, 1952) and the antennules stretched out laterally. Benthopelagic species of *Paramisophria* and *Stephos* swim with their left-hand sides ventrally (Ohtsuka and Hiromi, 1987; Ohtsuka and Mitsuzumi, 1990; Ohtsuka, 1992). The left lateral side of the metasome of *Paramisophria platysoma* is compressed and flat and so very suitable for lying on the sediment surface in areas of active water currents; the left antennule is longer than the right and is held out anteriorly from the body during swimming over the sediment while the shorter right one is orientated slightly posteriorly. Another benthopelagic species, *Pseudocyclopia muranoi* swims normally and continuously with its ventral side downwards by beating its antennae, mandibular palps and, probably, the outer lobes and the exo- and endopods of the maxillules; the antennules are stretched laterally and swimming direction is changed, not by flexing the swimming legs, but by bending the urosome in the new direction (Ohtsuka, 1992). Yen (1988) found that *Euchaeta rimana* also changes the direction of swimming by altering the aspect of the urosome.

11.1.1. Swimming Pattern

11.1.1.1. Nauplius

The nauplii have three pairs of appendages used in swimming, the antennules, the antennae and the mandibles (Figure 82). The antennules,

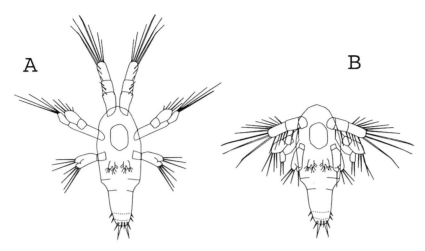

Figure 82   Ventral view of the nauplius, showing the three pairs of appendages used in swimming, the antennules, antennae and mandibles. A, the nauplius with the appendages in the resting position. B, the nauplius then moves the appendages in posteriorly directed arcs to swim.

unlike the other two pairs of appendages, do not have plumose setae. Three forms of swimming are present (Gauld, 1959; Paffenhöfer *et al.*, 1996). The first is a slow, gliding movement performed with the antennules in the resting position (Figure 82) while the antennae and mandibles sweep backwards and forwards (Gauld, 1959). The second consists of rapid, darting movements in which all three pairs of appendages sweep backwards and forwards rapidly in what Gauld considered to be a metachronal rhythm. A third form is described by Van Duren and Videler (1995) in nauplii of *Temora longicornis* and Paffenhöfer *et al.* (1996) in *Centropages velificatus* as a cruise-and-pause form of swimming. Nauplii can escape predation by using an escape reaction which is probably an extension of the rapid darting movement mentioned above.

11.1.1.2.   *Copepodid*

Metridia longa, M. pacifica, Centropages hamatus, Temora longicornis, T. turbinata swim almost continuously (Gill, 1987; Hirche, 1987; Wong, 1988a; Hwang and Turner, 1995), gliding, looping and swimming in circles both upwards and downwards, usually in a smooth, gliding motion (Figure 83A). This is often termed slow swimming (Gauld, 1966). It is primarily effected by the antennae which move in a rotary fashion. The mandibular palps have

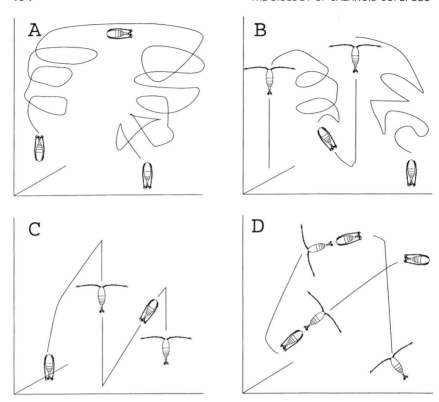

Figure 83 Swimming patterns of copepods. Copepods swim in three dimensions, usually with the ventral side downwards. The patterns are shown diagrammatically to display the attitude of the antennules.
A, cruising: the copepod swims in an irregular path, including upward and downward spirals.
B, cruise and sink: this is probably the most economic pattern of swimming for a negatively buoyant copepod. Bouts of swimming are interrupted by periods of passive sinking.
C, hop and sink: this is a much more jerky form of swimming than the cruise-and-sink pattern. It is referred to as jerky swimming when performed more or less in the horizontal plane.
D, jumping: this pattern consists of fast jumps powered by the antennules interposed by periods of rest.

a rotary movement of lesser amplitude and it is the forwards and backwards beat of the palp that will primarily produce propulsion; the maxillules and maxillae are also involved. These are the feeding appendages and feeding and swimming can take place simultaneously, the appendages generating feeding currents as well as propulsion. Cruising speed can also be varied in many species but just how the appendages change gear is not understood in detail (Greene, 1988). In addition to the species named above, this type of swimming has been observed in the genera *Calanus, Euchaeta, Pareuchaeta, Eurytemora, Isias, Limnocalanus, Neocalanus, Paracalanus* and *Pseudocalanus* (Gauld, 1966; Wong and Sprules, 1986; Greene, 1988; Yen, 1988; Tiselius and Jonsson, 1990). These genera have antennae with the exopod and endopod about equal in length, with the exopod pointing upwards, often curving over the back of the animal, and the endopod pointing outwards and downwards.

Calanus finmarchicus and *C. glacialis* move in a loop or spiral downwards or upwards thus accentuating part of the normal cruising pattern described above. This pattern has been described as a cycloid path by Kittredge *et al.* (1974) who suggest that it may function to enable the copepodid to follow a pheromone excretion to its source, i.e. a male *Labidocera jollae* locating a female. Katona (1973) terms this type of swimming as mate-seeking behaviour.

Cruise-and-sink swimming activity is the same as cruising but with periods of sinking interposed, usually with the antennules outstretched (Figure 83B). It is often referred to as hop-and-sink behaviour (Haury and Weihs, 1976). It has been proposed as the most energetically conservative form of swimming for a negatively buoyant copepod. It is a common pattern in *Calanus* and other species.

Hop-and-sink swimming is often performed in the laboratory in association with the bottom of the aquarium. *Calanus finmarchicus* sinks with the body pointing vertically, and the antennules outstretched, and hops upwards when the caudal setae touch the bottom (Figure 83C). This can take place repeatedly in this situation but also occurs in the water column (Hirche, 1987), where *Calanus* species can hang with the antennules outstretched. It is, however, not a common mode of swimming, being more or less peculiar to *Calanus* species (Gauld, 1966) although the motionless sinking combined with short jumps described in *Acartia clausi* by Tiselius and Jonsson (1990) seems closely similar. Laboratory observations of swimming patterns have been shown to be influenced by the amount of freedom and space that the experimental animal has for swimming (Pavlova *et al.*, 1982). Container-effects are bound to be present in most experimental situations. The hop or jump rate of copepods is used as a measure of activity but, as would be expected, is a function of temperature and, to some extent, the amount of food present (Shadrin *et al.*, 1983).

Gill (1987) states that *Acartia clausi* and *Anomalocera patersoni* swim in a jerky fashion with periods of rest in between. Such movements are probably the result of forwards and backwards beating of the antennae with little rotational movement. Gauld (1966) states that the backward strokes of the appendages are made with the appendages held stiffly and the setae spread while in the forward, or recovery, strokes the terminal segments of the appendages are flexed and the setae closed together. The resting position is with the appendages at the posterior end of the strokes. Gauld found that, in addition to the two species above, *Labidocera wollastoni* and *Parapontella brevicornis* are jerky swimmers. In these genera, the exopod of the antenna is shorter than the endopod.

Jumping is performed faster than other activities and always involves strokes of the antennules (Figure 83D). It is assumed that, after the power stroke of the antennules, they are then held close to the body to reduce drag. Hirche (1987) states that, in many jumps of *Calanus hyperboreus*, the antennules appear to be wrapped around the body and have a flagellum-like beat producing propulsion of the animal. Jumping, or darting upwards, of *Acartia tonsa* results from a thrust of the antennules and, sometimes, the swimming legs (Jonsson and Tiselius, 1990).

The escape reaction is a fast jump away from the stimulus causing it. It results from the combined actions of the antennules and swimming legs. It is so fast that it has not as yet been analysed properly by the video equipment available. The urosome of some species may also be involved in the reaction. It transports the copepod several body lengths from the stimulus.

Swimming and feeding are inseparable as the former enables the latter, whether the copepod is feeding on particles or engaged in carnivory. The generation of feeding currents by copepods is discussed in Chapter 5. Marrase *et al.* (1990) briefly review the topic while Bundy and Paffenhöfer (1996) investigate flow fields generated by the swimming appendages of a herbivorous and a carnivorous species. The latter show that copepods can change their body attitude as they swim so changing the properties of the flow fields around them. Swimming patterns alter depending on the availability and type of food (Buskey, 1984; Wong and Sprules, 1986; Wong, 1988a; B. Hansen *et al.*, 1991; Tiselius, 1992; Van Duren and Videler, 1995). *Metridia pacifica* shows more high-speed bursts of swimming when food is absent while *Limnocalanus macrurus* increases the amount of circuitous swimming. This probably represents an efficient pattern for moving towards and searching for other regions where food is available, especially when the slower cruising mode of swimming dominates. Absence of food results in other species, such as *Acartia tonsa*, *Calanus pacificus* and *Centropages typicus*, reducing swimming activity, so conserving energy (Frost, 1975; Reeve and Walter, 1977; Cowles and Strickler, 1983). The presence of bioluminescent dinoflagellates causes *Acartia hudsonica* to increase its swimming speed and decrease the time spent cruising so tending to move the copepods away from the flagellates (Buskey et al., 1983).

11.1.2. Swimming Speed

Swimming speeds of nauplii are determined over short, straight distances although they swim naturally along convoluted pathways. Paffenhöfer et al. (1996) describe the pathways of *Eucalanus*, *Paracalanus* and *Temora* species, including the changing orientation of the body. Nauplii normally swim at speeds of 0.5 to 2 mm s^{-1} (Table 62). In *Eurytemora affinis* swimming speeds are sensitive indicators of sublethal levels of pollution (Sullivan et al., 1983). In the presence of low dosages of cadmium or copper, swimming was influenced before any change of the development rates was evident.

Estimates of swimming speeds of copepodids and adults are known for a variety of species (Table 63). In general, swimming speed increases as the copepodids grow through the stages (Buskey, 1994). The swimming paths of different species and individuals differ and vertical and horizontal components of swimming can be determined to estimate a net to gross displacement ratio (NGDR). This ratio is the distance between the starting-point and the end-point of the path of swimming (net displacement) related to the total distance travelled from the starting-point to reach the end-point (gross displacement) (Buskey et al., 1986). Thus an NGDR approaching unity indicates a straight path while one much less than unity results from convoluted swimming. The majority of speeds recorded in Table 63 are for straight paths over short distances and range from about 1 to 20 mm s^{-1}, representing approximately 1 to 5 body lengths s^{-1}. Some higher speeds, up to 70 mm s^{-1} for *Euchirella rostrata*, approach those of escape reactions.

The swimming speeds listed in Table 63 were determined in the laboratory under experimental conditions. In contrast, Wiebe et al. (1992) estimated swimming speeds of adults of oceanic species in the genera *Euchirella*, *Pleuromamma*, *Scolecithrix* and *Undeuchaeta* from the amplitude and timings of their diel vertical migrations. Calculated speeds range from about 10 to 50 mm s^{-1}. Estimates made by Roe (1984) from the rates of diel vertical migration of seven oceanic species range from 9 to 47 mm s^{-1}, upwards and downwards rates of migration being virtually the same. Hattori (1989) calculated upwards and downwards speeds of 3.5 to 13.5 mm s^{-1} for the migrations of CV, females and males of *Metridia pacifica*, *M. okhotensis* and *Pleuromamma scutullata*; there was little difference between the upwards and downwards speeds or between those of the different stages, the average speeds being 5.5 to 8.5 mm s^{-1}. These speeds are of the same order as those in Table 63. Morris et al. (1985) have

Table 62 Swimming speeds of calanoid nauplii. Some rates are for swimming upwards, others are unspecified.

Species		Speed mm s^{-1}	Body lengths s^{-1}	Authority
Acartia clausi				
Upwards	NIII	0.15		Landry and Fagerness, 1988
	NIV	0.3		Landry and Fagerness, 1988
	NV	0.39		Landry and Fagerness, 1988
	NVI	0.16		Landry and Fagerness, 1988
A. tonsa	NI–NVI	0.2–2.0		Buskey, 1994
Calanus finmarchicus		1.23	2	Hardy and Bainbridge, 1954
C. pacificus	NII	0.9	3.3	Greene and Landry, 1985
	NIII	1.1	3.3	Greene and Landry, 1985
	NV	1.1	2.2	Greene and Landry, 1985
	NVI	0.9	1.5	Greene and Landry, 1985
Upwards	NIII	0.62		Landry and Fagerness, 1988
	NV	0.77		Landry and Fagerness, 1988
	NVI	0.51		Landry and Fagerness, 1988
Centropages velificatus	NIII–NVI	2.7–3.0		Paffenhöfer et al., 1996
Eucalanus hyalinus	NIV–NVI	0.26–1.23		Paffenhöfer et al., 1996
E. pileatus	NIV–NVI	0.53–1.06		Paffenhöfer et al., 1996
Eurytemora affinis		1.0–2.4	7–16	Sullivan et al., 1983
Paracalanus aculeatus	NIV–NVI	0.44–1.01		Paffenhöfer et al., 1996
P. quasimodo	NV–NVI	0.93–1.69		Paffenhöfer et al., 1996
Pseudocalanus sp.				
Upwards	NIII	0.37		Landry and Fagerness, 1988
	NIV	0.4		Landry and Fagerness, 1988
	NV	0.53		Landry and Fagerness, 1988
	NVI	0.23		Landry and Fagerness, 1988
Temora longicornis	NII	1.5		van Duren and Videler, 1995
	NVI	2		van Duren and Videler, 1995
T. stylifera	NIV–NVI	0.35–0.41		Paffenhöfer et al., 1996

developed a model describing the swimming of a copepod from which they estimate the swimming speed of *Pleuromamma xiphias* to range from 9 to 32 mm s^{-1} (Table 63).

Swimming speeds are often higher in males than in females (Table 63). Van Duren and Videler (1996) suggest that this reflects the mate-seeking activity of the male while the lower speed in the female maximizes energy intake and conservation. The males may be at greater risk of predation because of their swimming speeds.

Rates of acceleration of copepods were calculated for a number of species by Minkina (1983). Neustonic copepods, living in association with the surface film in the ocean, increase their swimming speed under conditions of increased pressure (Champalbert, 1978). The effects of increased turbulence are described by Yamazaki and Squires (1996).

As with nauplii, sublethal concentrations of pollutants decrease swimming speeds of older stages. For example, Cowles (1983) found that exposure of female *Centropages hamatus* to low concentrations of crude oil altered their swimming and feeding activity although they recovered quickly on being transferred to clean sea water.

11.1.3. Body Density and Sinking Rate

Copepods are denser than the sea water in which they live, whether it be fully saline or brackish. The density of sea water at 25 °C and salinity 35‰ is 1.025 g cm^{-3}. Densities of copepods are shown in Table 64. The single exception is *Pareuchaeta biloba* whose sinking rate was zero meaning that it was neutrally buoyant.

Sinking rates are variable between stages and individuals of the same species, dependent upon the physiological state of the stage or individual. The antennules are usually held outstretched to increase drag during sinking. Svetlichny (1980) measured sinking rates of some species with the antennae against the body and with them outstretched; the range of sinking speed determined by Svetlichny (1980) in Table 64 reflects these two modes, the faster speed resulting from closing of the antennules.

Environmental temperature alters the sinking rate and, according to Rudyakov (1972), the sinking rates of copepods conform to Stokes Law, already discussed in the context of the sinking rates of faecal pellets. The data in Table 64 suggest that some deep sea species may have greater sinking rates than epipelagic and coastal species. The integuments of some species are very robust while other species, such as the jelly copepods referred to elsewhere, have light integuments that may be reflected in their sinking rates. *Labidocera acutifrons* has the fastest sinking speed in Table 64; Parker (1901) determined the density of *L. aestiva* as 1.109 in sea water of density 1.082.

Table 63 Swimming speeds of copepodids and adults of calanoid copepods. Rates of swimming upwards, downwards and for different durations are noted.

Species		Speed mm s^{-1}	Body lengths s^{-1}	Authority
Acartia clausi	(2 min)	9.3		Hardy and Bainbridge, 1954
	(30 min)	2.5		Hardy and Bainbridge, 1954
	(60 min)	2.4		Hardy and Bainbridge, 1954
Upwards	CI	0.23		Landry and Fagerness, 1988
A. hudsonica		1–6	1–6	Buskey et al., 1983
A. tonsa		2–8	2–8	Buskey et al., 1986
		3–5		Tiselius, 1992
	CV–CVI	1–10		Buskey, 1994
Calanus finmarchicus				
Upwards	(2 min)	18.3	5	Hardy and Bainbridge, 1954
	(30 min)	10.1	3	Hardy and Bainbridge, 1954
	(60 min)	4.2	1.3	Hardy and Bainbridge, 1954
Downwards	(2 min)	29.7	9	Hardy and Bainbridge, 1954
	(30 min)	16.6	5	Hardy and Bainbridge, 1954
	(60 min)	13	4	Hardy and Bainbridge, 1954
		2.0–5.0	1–2	Buskey and Swift, 1985
		10	3–4	Hirche, 1987
C. helgolandicus		5–40		Minkina, 1981
		20		Minkina and Pavlova, 1981
C. pacificus	CI	0.3	0.4	Greene and Landry, 1985
	CII	1.2	1	Greene and Landry, 1985
	CIII	2.2	1.4	Greene and Landry, 1985
	CIV	3.3	1.8	Greene and Landry, 1985
	CV	3.9	1.5	Greene and Landry, 1985
	CVI	6.7	2.4	Greene and Landry, 1985
Upwards	CI	0.16		Landry and Fagerness, 1988
C. hyperboreus		1.1–1.3	0.2	Buskey and Swift, 1985
Centropages hamatus		5.5		Cowles, 1983

Species	Condition			Reference
C. typicus		7.2	5.2	Tiselius and Jonsson, 1990
C. velificatus		1.9	1.1	Tiselius and Jonsson, 1990
	CI–CII	1.8–2.9		Bundy and Paffenhöfer, 1996
		0.63–1.12		Paffenhöfer et al., 1996
Centropages sp.				
Upwards	(2 min)	15.2		Hardy and Bainbridge, 1954
	(30 min)	8.9		Hardy and Bainbridge, 1954
	(60 min)	8.5		Hardy and Bainbridge, 1954
Epischura lacustris		2.8–27.4		Wong and Sprules, 1986
Eucalanus pileatus	CI–CII	0.37–1.01		Paffenhöfer et al., 1996
Euchaeta elongata		6.3		Greene and Landry, 1985
E. marina		19-25		Pavlova, 1981
E. norvegica				
Downwards	(2 min)	37.7	6	Hardy and Bainbridge, 1954
	(30 min)	22.9	4	Hardy and Bainbridge, 1954
	(60 min)	19.6	3	Hardy and Bainbridge, 1954
		1.1–1.3	0.2	Buskey and Swift, 1985
E. rimana	♀	7	2	Yen, 1988
	♂ Upwards	7.5	2.3	Yen, 1988
	Downwards	5	1.6	Yen, 1988
Euchirella curticauda		16–20		Pavlova, 1981
E. rostrata		7–70		Minkina, 1981
Eurytemora affinis	♀	2.5		Katona, 1970
	♂	5.0		Katona, 1970
Heterorhabdus spp.		2.9		Buskey and Swift, 1990
Labidocera jollae	♀	10	3–4	Lillelund and Lasker, 1971
	♂	4	2	Lillelund and Lasker, 1971

Table 63 Continued.

Table 63 Continued.

Species		Speed mm s^{-1}	Body lengths s^{-1}	Authority
L. trispinosa	♀	6.2	2.3	Lillelund and Lasker, 1971
	♂	5.5	2.3	Lillelund and Lasker, 1971
L. wollastoni		8.3	3.5	Hardy and Bainbridge, 1954
Limnocalanus macrurus		0.6–9.0	3	Wong and Sprules, 1986
Lucicutia spp.		2.9		Buskey and Swift, 1990
Metridia longa		5.7–7.0	3	Buskey and Swift, 1985
M. lucens		4.1–8.5	3	Buskey and Swift, 1985
M. pacifica		8–25	3.5–10	Enright, 1977a
		0.8–11.8	5	Wong, 1988a
Paracalanus aculeatus		0.6–1.0		Bundy and Paffenhöfer, 1996
		0.44–0.54		Paffenhöfer *et al.*, 1996
P. parvus	(2 min)	9.1	10	Hardy and Bainbridge, 1954
	(30 min)	1.2	1.3	Hardy and Bainbridge, 1954
	(60 min)	0.9	1	Hardy and Bainbridge, 1954
		0.6	0.7	Tiselius and Jonsson, 1990
P. quasimodo	CI–CII	0.56–1.26		Paffenhöfer *et al.*, 1996
Pleuromamma abdominalis		13.7		Buskey and Swift, 1990
P. gracilis		13.4		Buskey and Swift, 1990
P. xiphias		9–32	1.5–5.2	Morris *et al.*, 1985
		14.7		Buskey and Swift, 1990
Pseudocalanus elongatus		0.5	0.4	Tiselius and Jonsson, 1990
P. minutus		2–4		Buskey, 1984
		2.2–2.7	1.9–2.4	Wong, 1988b

Pseudocalanus sp.		3.3		Greene and Landry, 1985
	CI	0.16	1.8	Landry and Fagerness, 1988
Rhincalanus nasutus		5–29		Pavlova, 1981
Scolecithix danae		19		Pavlova, 1981
Senecella calanoides		1.5–20.8		Wong and Sprules, 1986
Temora longicornis	Upwards (2 min)	5.3	3.5	Hardy and Bainbridge, 1954
	(30 min)	2.5	1.7	Hardy and Bainbridge, 1954
	(60 min)	1.4	1	Hardy and Bainbridge, 1954
	CI	2.7–6.1	4	Buskey and Swift, 1985
	CI	2		van Duren and Videler, 1995
	♀	4.5		van Duren and Videler, 1995
	♂	9		van Duren and Videler, 1995
	♀ fed	3.2–4.2	3	van Duren and Videler, 1996
	starved	4.8	3	van Duren and Videler, 1996
	♂	4.6–4.8	3	van Duren and Videler, 1996
T. stylifera	CI–CII	0.45–0.87		Paffenhöfer *et al.*, 1996

Lipid stores are often considered to affect or even control buoyancy (Sargent and Henderson, 1986) but no quantitative estimates are available.

Sinking rates of *Acartia hudsonica* were decreased by infestations of epizooic peritrich ciliates (Weissman *et al.*, 1993). The sinking rates of uninfested copepods increased with their increasing prosome length (PL):

$$\text{Sinking speed (mm s}^{-1}) = 1.801 \text{ PL (mm)} - 0.695$$

No such relationship was present for infested copepods.

11.1.4. Escape Reaction

The escape of nauplii of *Acartia hudsonica* from the flow field of predatory *Temora longicornis* was investigated by Yen and Fields (1992). The nauplii became entrained in the flow field that had velocities of up to 2 mm s^{-1}. These authors mapped the contours of velocity within the field in front of the antennules. The nauplii escape at points in the field where the change in velocity over distance, the shear, is high. The authors suggest that the setae on the nauplius are bent by the shear, so eliciting an escape response. Escape speeds greater than 10 mm s^{-1} have been measured in nauplii (Table 65).

The ability of copepodids and adults to escape "predators" was first examined by substituting a pipette or siphoning tube in which the rate of flow of the water could be adjusted. This provided a rough estimate of escape velocities of the experimental animals. Singarajah (1969, 1975) found that copepods were extremely successful in escaping the siphon at speeds of 2.6 mm s^{-1}. More recently, video recording has been used and accurate speeds over distance measured (Table 65). Escape velocities are much greater than normal swimming velocities, approaching 100 body lengths s^{-1}.

Normal swimming velocities of copepods, coupled with their body size, means that they have a Reynolds number of 1 and live in a viscous environment. Their potential predators have Reynolds numbers greater than 1 and live in an inertial world. The escape reaction of the copepods transposes them from the viscous into the inertial habitat (Naganuma, 1996). The Reynolds number (Re) defines these habitats:

$$\text{Re} = (\rho/\eta)\text{LU}$$

where L, in m, is the body length and U, in m s^{-1}, the velocity of the copepod; ρ, the fluid density, is $1.025 \times 10^3 \text{ kg m}^{-3}$ and η, the fluid viscosity, is

Table 64 Density and sinking speeds of calanoid copepods.

Species		Density g cm^{-3}	Sinking mm s^{-1}	Authority
Acartia clausi		1.071	0.8–1.3	Svetlichny, 1980
	♂		0.8–1.8	Svetlichny, 1980
	calm		0.3	Tiselius and Jonsson, 1990
	turbulence, low		1.2	Saiz and Alcaraz, 1992a
	turbulence, low		5	Saiz and Alcaraz, 1992a
	turbulence, high		10	Saiz and Alcaraz, 1992a
A. hudsonica			0.4–2.0	Weissman *et al.*, 1993
A. negligens		1.042	1.04	Svetlichny, 1980
A. tonsa			0.99	Jacobs, 1961
			0.6–0.8	Jonsson and Tiselius, 1990
Calanoides carinatus		1.049–1.054	3.0–3.6	Svetlichny, 1980
	♂		5.2	Svetlichny, 1980
Calanus finmarchicus			5	Apstein, 1910
	♀	1.043–1.047	1.9–2.2	Gross and Raymont, 1942
	♂		2.3–3.1	Gross and Raymont, 1942
	CV		0.8–3.5	Gross and Raymont, 1942
C. pacificus			2.6	Landry and Fagerness, 1988
C. propinquus	♀		7.8	Rudyakov, 1972
	♂		4.4	Rudyakov, 1972
Candacia maxima			4.7	Rudyakov, 1972
C. pachydactyla		1.084	6.1	Svetlichny, 1980
	♂		8.1	Svetlichny, 1980
Centropages hamatus			1.4	Tiselius and Jonsson, 1990
C. typicus			1	Tiselius and Jonsson, 1990
C. violaceus	♂		3.0	Svetlichny, 1980
Clausocalanus arcuicornis			2.1	Rudyakov, 1972

Table 64 Continued.

		Density g cm³	Sinking mm s⁻¹	Authority
Eucalanus attenuatus			4.7–8.6	Svetlichny, 1980
Euchaeta marina		1.071	4.4	Svetlichny, 1980
E. rimana	♂		5	Yen, 1988
Euchirella curticauda			9.6	Svetlichny, 1980
E. venusta		1.056		Svetlichny, 1980
Euchirella sp.		1.078	10.6	Svetlichny, 1980
	CV		3.2	Rudyakov, 1972
Gaetanus sp.	♂	1.061	7.2	Svetlichny, 1980
Gaidius tenuispinus			5.2	Rudyakov, 1972
Haloptilus longicornis			0.6	Svetlichny, 1980
Labidocera acutifrons	♂		14.8	Svetlichny, 1980
			11.0	Svetlichny, 1980
Lucicutia sp.			1.1	Rudyakov, 1972
Metridia gerlachei			5.2	Rudyakov, 1972
M. longa	♀		5.1	Apstein, 1910
M. lucens	CV		2.7	Rudyakov, 1972
			1.3	Rudyakov, 1972
Neocalanus cristatus			0.6	Landry and Fagerness, 1988
N. gracilis		1.076	5.3	Rudyakov, 1972
			9.3	Svetlichny, 1980
N. robustior	CV		2.6	Rudyakov, 1972
	♀	1.048–1.061	5.9–9.0	Svetlichny, 1980

N. tonsus	♀		6.1	Rudyakov, 1972
	CV		2.9	Rudyakov, 1972
	CIV		2.7	Rudyakov, 1972
Paracalanus parvus			0.6	Tiselius and Jonsson, 1990
Pareuchaeta antarctica	♀		0.6	Rudyakov, 1972
	♂		3.6	Rudyakov, 1972
	CV		0.5	Rudyakov, 1972
P. biloba			0	Apstein, 1910
P. norvegica			6.6	Svetlichny, 1980
Pleuromamma abdominalis	♂	1.075	8.2	Rudyakov, 1972
P. robusta			3.8	Svetlichny, 1980
Pontellina plumata			3.8	Apstein, 1910
Pseudocalanus elongatus			1.7	Tiselius and Jonsson, 1990
			1.1	
Pseudodiaptomus coronatus	♀		2.32	Jacobs, 1961
	♀ with egg mass		2.78	Jacobs, 1961
Rhincalanus nasutus		1.025	6.0	Svetlichny, 1980
Scolecithrix sp.		1.130	2.9	Svetlichny, 1980
Temora longicornis			2.5	Apstein, 1910
				Tiselius and Jonsson, 1990
Undeuchaeta plumosa		1.084	9.8	Svetlichny, 1980
Undinula vulgaris		1.072–1.089	5.8–9.8	Svetlichny, 1980
	♂		6.3–8.6	Svetlichny, 1980

Table 65 Escape speeds, lunge speeds or jump (dart) speeds of nauplii, copepodids and adult calanoid copepods. Most are measured in mm s^{-1}, some in body lengths (bls) s^{-1}.

Species		Speed		Authority
		mm s^{-1}	bls s^{-1}	
Nauplii, mixed species		13–28	87–187	Tiselius and Jonsson, 1990
Nauplii, Acartia hudsonica		0.48–1.02	2–4	Yen and Fields, 1992
Acartia clausi	calm	15		Saiz and Alcaraz, 1992a
	turbulence, low	20		Saiz and Alcaraz, 1992a
	turbulence, high	35		Saiz and Alcaraz, 1992a
A. fossae	jump	87		Yen, 1988
A. tonsa	escape	100	93	Buskey et al., 1986
Calanus finmarchicus	CV,	160 mean speed		Haury et al., 1980
	maximum speed	370		Haury et al., 1980
C. helgolandicus	escape	80–700		Minkina, 1981
Euchaeta marina	escape	449		Pavlova, 1981
E. rimana	♀ lunge	142	60	Yen, 1988
	♂ escape	360	150	Yen, 1988
Euchirella curticauda	escape	319		Pavlova, 1981
E. rostrata	escape	800		Minkina, 1981
Eurytemora affinis		10–14	67–93	Sullivan et al., 1983
Labidocera trispinosa	jump	23–90		Vlymen, 1970
Rhincalanus nasutus	escape	462		Pavlova, 1981
	maximum	2085		Pavlova, 1981
Temora longicornis	escape		>100	van Duren and Videler, 1996

1.1×10^{-3} kg m^{-1} s^{-1}. The majority of calanoids are in the body length range of 0.5 to 5 mm (Figure 3) and swim normally at speeds of 1 to 20 mm s^{-1}; this results in Re values of 0.5 to 150, at the lower values of which viscosity dominates speed. The escape reactions of these copepods have speeds of 20 to about 400 mm s^{-1} (Table 65), representing an Re range of 10 to about 1850. These Re values are comparable with those of predatory chaetognaths and fish.

Turbulence stimulates escape responses. Hwang and Strickler (1994) found that a threshold speed of turbulent particles of 0.84 mm s^{-1} triggers the responses in *Centropages hamatus*. Persistent conditions of turbulence cause habituation to the stimuli and a decrease in the frequency of elicited escape responses (Hwang and Strickler, 1994; Hwang *et al.*, 1994). Habituation occurs after the first 50 or so escape responses (Hwang and Strickler, 1994). Turbulence increases food-intake but repeated escape responses demand so much energy that the gross growth efficiency tends to decrease during turbulent conditions. Habituation of the escape responses may partially alleviate this energy drain.

According to Buskey and Swift (1985), simulated bioluminescent flashes of dinoflagellates cause increases in the swimming speeds of *Acartia hudsonica*, *Calanus finmarchicus*, *Metridia longa*, *M. lucens* and *Temora longicornis* but not of *Calanus hyperboreus* or *Pareuchaeta norvegica*. The swimming speed of *Calanus finmarchicus* increases to some 20 mm s^{-1}, well below the escape speed of 160 mm s^{-1} estimated for CV of *C. helgolandicus* (Table 65). Consequently, stimulatory flashes of light increase swimming speed but do not apparently induce escape reactions.

11.1.5. Response to Predators

Ohman (1988b) reviews the factors, including mobility and escape reactions, that make copepods less vulnerable to predators. The escape reaction removes the copepod from the path of the oncoming predator, detected by the copepod through the fluid disturbance made by the predator. Copepods may react to shadows thrown by an approaching gelatinous predator. Buskey *et al.* (1986) found that *Acartia tonsa* responds to a sharp decrease in light intensity with a bout of high-speed swimming, at speeds as great as 100 mm s^{-1}, which resulted in lateral displacement of the copepod. Haury *et al.* (1980) suggest that CV of *Calanus finmarchicus* alter the distance of escape according to the size of the predator, a larger predator eliciting a stronger escape reaction. Escape distances are of the order of 10 to 20 mm. The escape reaction, however, is used as a last resort when the predator is about to capture the copepod.

A copepod that is relatively still will broadcast fewer signals of its presence in the fluid environment. Van Duren and Videler (1996) found that *Temora longicornis* slowed its swimming speed when placed in water previously inhabited by fish, both females and males having an average speed of ~2.5 mm s^{-1}. This infers chemosensing of predators. The decrease in swimming activity may make them less detectable by the predatory fish. Nauplii, being relatively slow swimmers, are often cannibalized in particle feeding species. Pollutants can also depress swimming speeds presumably through their toxic action on the physiology of the copepods. In the presence of sublethal dosages of copper and cadmium, nauplii of *Eurytemora affinis* are more liable to predation (Sullivan *et al.*, 1983).

The liability of copepods to predation varies between species. For example, Kimmerer and McKinnon (1989) studied the distributions of the two common species *Acartia tranteri* and *Paracalanus indicus* in an Australian Bay. They found that the latter is more vulnerable to visual predation by fish, possibly having a weaker escape response, and that this means that it is virtually absent from the bay but occurred offshore. *Acartia tranteri*, with a probably stronger escape response, is dominant within the bay. Avoidance of visual predators is considered to be one of the benefits obtained when a copepod performs a diel vertical migration, avoiding the well-lit surface waters during the day. Bollens and Frost (1989a) showed that *Acartia hudsonica* in an enclosure reacted to freely swimming fish by escaping downwards but showed no such reaction to caged fish. They concluded that the escape response was not elicited chemically but rather through visual or mechanosensory means. Later, Bollens *et al.* (1994) showed that *A. hudsonica* would respond to mimics of fish by migrating downwards, presumably through visual and/or mechanical cues.

11.1.6. Rhythmic Activity

The evidence for diel rhythms of swimming activity in calanoids derives mainly from observations of feeding activity. Svetlichny and Yarkina (1989), however, found that a circadian rhythm of locomotion, observed predominantly in the feeding appendages, of *Calanus helgolandicus* persists over a period of 37 d in animals maintained in constant darkness and temperature. The neustonic species *Anomalocera patersoni*, *Labidocera wollastoni* and *Pontella mediterranea* have diel rhythms of swimming activity persisting for at least 2 or 3 d under conditions of constant darkness in the laboratory (Champalbert, 1978, 1979).

An endogenous, circatidal rhythm of swimming has been demonstrated in *Eurytemora affinis* by Hough and Naylor (1992). Copepods collected from the Conwy Estuary, North Wales, on a tide when tidal amplitudes were increasing, showed peak activity at about 1.5 h before expected high tide.

Copepods collected, on the other hand, on a tide when tidal amplitudes were decreasing, showed peak activity after the time of expected high tide. These rhythms persisted over at least five tidal cycles. This pattern of swimming would aid the animals in maintaining their position within the estuary.

These few observations suggest that endogenous rhythms may be quite common, especially in species that live in tidal and/or hyperbenthic environments. Diel rhythms of swimming are difficult to distinguish from those of feeding because the same appendages are involved, and most species need to swim to feed. Any widespread occurrence of endogenous rhythms of swimming and feeding will affect the interpretation of laboratory observations on feeding.

11.1.7. Energetic Cost

The energetic cost of swimming to the copepod has been of interest for some time but has as yet not been directly measured. Vlymen (1970) concludes that the swimming upwards by *Labidocera trispinosa* during its diel vertical migration expends slightly less than 0.3% of its basal metabolic rate, a negligible quantity. Vlymen's methods of observation are criticized by Enright (1977b) and answered by Vlymen (1977). The difficulties encountered by Vlymen and Enright are discussed by Strickler (1977) who points out that copepods swim along a three-dimensional path at variable velocity, have powers of acceleration as high as 12 000 mm s^{-2}, are decelerated in the viscous medium at rates of the order of 2500 to 5000 mm s^{-2}, and that conditions for filming are critical. Deceleration takes place in about 30 ms and is further discussed by Lehman (1977). The overall conclusion is that Vlymen's original estimate of the cost of swimming is probably too low at 0.3% of the basic metabolic rate. Svetlichny and Kurbatov (1987) estimate that between 0.02 and 60% of the basal metabolic rate is required by copepods of 0.4 to 10 mm body length to maintain themselves at a constant depth; energy cost of migrating 100 m is estimated as 13 to 120% of basal metabolism.

Other indirect methods of estimating energy expenditure have been tried. The mechanics and power strokes of legs have been studied using models of the legs (Svetlichny and Kurbatov, 1983) and later compared with the direct results obtained by measurements of living copepods (Svetlichny, 1992b). Weight-specific respiration rate is directly correlated with the amount of energy expended on swimming (Svetlichny, 1992a). Petipa and Ovstrovskaya (1989) examine the loss of body dry weight or carbon during vertical migration as an estimate of energy expenditure on swimming.

Morris *et al.* (1985) present a model of swimming in copepods derived from data on *Pleuromamma xiphias*. Hirche (1987) criticizes their model because it takes no account of the contribution of the antennules to locomotion. An adaptive behaviour network model is developed by Keiyu *et al.* (1994) with three types of links (informative, excitatory and inhibitory) and three types of nodes (database, behaviour and conditional).

The ratio of active to basal metabolism in *P. xiphias* was 3 at swimming speeds of 32 mm s^{-1}. This is much lower than the ratio of 7.8 to 15.6 found by Minkina (1981) for *Calanus helgolandicus* but much higher than Vlymen's (1970) estimate of 0.3. Copepods swim in a variety of modes and the patterns of paths followed vary between individuals and species and also with time. The energy expenditure of the escape reaction is obviously large relative to that of cruising. Much further work is still required before satisfactory values for the energy expended in different modes of swimming will be known.

No information appears to be available on the variable hydrodynamic shape of copepods. The body of *Rhincalanus nasutus* looks more streamlined than those of the other species illustrated in Figure 2 (p. 4). The head of *Gaetanus miles*, *G. pileatus* and some *Haloptilus* species have a forwardly directed spine (Figure 84) that may reduce turbulence generated by its forward momentum. A similar sharp, anterior projection of the head is present in *Arietellus armatus*. *Gaetanus latifrons* has a more dorsally

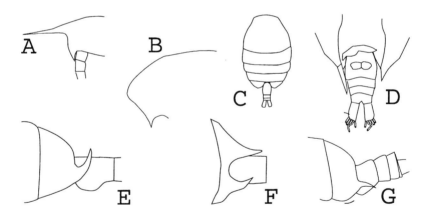

Figure 84 Possible hydrodynamic features of copepods.
A, the anterior spine found on the head of species such as *Gaetanus miles* and *G. pileatus*. B, the bulbous anterior region of the head of *Centraugaptilus horridus*. C, the nearly spherical body of *Phaenna spinifera*. The posterior end of the last prosome segment of D, *Scolecithrix danae* with asymmetrical points, E, *Paracomantenna* sp., F, *Neoscolecithrix* sp., and G, *Puchinia obtusa*.

directed spine (Figure 2). A more bulbous head is present in *Centraugaptilus horridus* (Figure 84) that is reminiscent of the bulbous bows on ships. Some species have almost spherical bodies and others have very blunt heads, especially among benthopelagic species such as some *Pseudocyclops* species. Some *Lucicutia* and *Temora* species have broad heads but the body then tapers posteriorly and terminates in a long urosome. There is a variety of architecture of the posterior end of the prosome (Figure 84). Many species have no spinal extensions while others have very pronounced ones. It is not known whether these act as fairings to the body or have some function during mating.

These are just some morphological features that may contribute to either increased streamlining of the body or to producing extra drag on it. Some are very prominent and must provide a benefit to the copepod.

11.2. SPATIAL DISTRIBUTION

Copepods are not randomly distributed in the sea but their distributions are affected by the structure of the water masses on large and small scales. An excellent introduction to this topic is given by Cassie (1963), Steele (1977), Mackas *et al.* (1985) and Piontkovski and Williams (1995). Knowledge of social aggregations of pelagic invertebrates is reviewed by Ritz (1995) who mentions those of copepods briefly. Spatial heterogeneity within a copepod population devolves from two broad sources. The first is the physical and chemical properties of the sea area, or environment, in which they are living and the effects of these properties are discussed in the next chapter. The second is the physiological and behavioural properties of the copepods themselves, resulting in interactions between individuals. This can be construed as social behaviour. It also encompasses the reactions of the copepods to their biological environment, including responses to patches of potential food organisms and to predators.

Attempts have been made to classify assemblages of planktonic organisms that are above the background concentrations in the surrounding sea (Omori and Hamner, 1982). Terms such as patch, aggregation, shoal, school, swarm, etc. have been used but are of restricted general value although they are useful in describing the different behaviour of a species or its congeners. Omori and Hamner use aggregation, school and swarm in their discussion of assemblages of plankton. The term school has only been found to apply relatively rarely to copepods, since it involves active directional swimming of an assemblage. The other terms used here are aggregation and swarm.

11.2.1. Aggregations

The terms aggregation, association and patch have been used in discussing the non-random distributions of planktonic organisms in general and are used synonymously here. The term aggregation simply refers to the horizontal or vertical regions of the distributions of a population of copepods that contain densities of copepods significantly, usually more than 3 to 5 times, above background concentrations outside the aggregation. It also infers that the physical hydrographic features of the environment are probably dominant in its formation, maintenance and dispersion. The size of aggregations range from a few metres to many kilometres. They are common in the offshore and inshore environments and reference to some examples are given in Table 66.

Patchiness is present on several simultaneous scales. Haury (1976a,b) examines patchiness, on the vertical scale, in the California Current and gives a comparison of the associations in the Current with those in the North Pacific central gyre. *Metridia pacifica* and *Rhincalanus nasutus* were patchily distributed vertically between the surface and 250 m depth in the water column of the Current at a scale of 10 m. The structure of this patchiness may be modified by diel vertical migration of the copepods. Aggregations in the Current were, on average, probably less than 100 m in size compared with a size probably larger than a few hundred metres in the central gyre. This situation must reflect the greater variability and rates of change within the water mass of the California Current relative to those in the gyre. Star and Mullin (1981) did not find major differences in patchiness between coastal and oceanic regions but did find some correspondence with the distribution of chlorophyll. Kawamura (1974) describes the occurrence of surface aggregations of *Neocalanus tonsus* in oceanic regions to the south and west of Australia. Aggregations often discolour the sea surface. They tend to be elongated in shape with an irregular outline. Their length was usually 100 to several 100 m. Tsuda *et al.* (1993) examine the patchy distribution of *Neocalanus cristatus* over a 2519 km track in the North Pacific. They identified 150 aggregations ranging in length from a few metres to 6.6 km and with a maximum recorded density of 1230 individuals m^{-3}. The average distance between aggregations at night on the track was 5.8 km, the majority being less than this, but some separated by tens of kilometres. On the microscale, the most common length of aggregation was 10 to 20 m but the range of density within these was the same as on the meso- to megascale. Maintenance of aggregations across all these scales cannot be simply by behavioural mechanisms alone nor by the physical structure of the water masses alone but, as the authors suggest, through interaction of the behaviour of the copepods with the physics of the environment. Steele and Henderson (1992) present a model of patchiness for herbivorous plankton emphasizing non-linear interactions between the animals and their food but suggest that this model, although useful, is probably too simplistic.

The persistence in time of aggregations is undoubtedly very variable. Cushing and Tungate (1963) followed the southeasterly drift of an aggregation of *Calanus finmarchicus* in the western North Sea for a period of 66 d. The aggregation had already formed at the beginning of the observations and was still recognizable on the 66th day when observations were concluded. The persistence of large aggregations in offshore areas can probably be frequently measured in weeks and months.

The density of copepods in aggregations is only a few times greater than the background density in the surrounding sea. Consequently, nearest neighbour distances (NND) between individuals will be large. It, therefore, seems probable that the reaction of one individual to another is not so important in maintaining the aggregation as the combined, and independent, reactions of the individuals to the environment. There may be threshold NNDs that initiate responses of one individual to another such that an aggregation becomes more cohesive and where individuals are no longer independent of one another.

The complexity of factors probably involved in the maintenance of an aggregation is emphasized by the study of Wishner *et al.* (1988) on a very large multispecies aggregation of copepods, with a possible area of about 2500 km^2, in the Great South Channel off New England. Maximal densities of over 4×10^4 copepods m^{-3} were recorded. The aggregation was dominated by *Calanus finmarchicus* which occurred at an average density of 6.3×10^3 m^{-3}. Sharply defined borders of the aggregation did not coincide with any marked hydrographic features of the region except one border that was adjacent to feeding right whales, whose predation may have contributed to its definition. One feature of the Great South Channel, not referred to by the authors, is its greater depth relative to the surrounding areas. It is possible that it is the availability of a deeper water column within the Channel that has initiated the formation, and contributed to the maintenance, of the aggregation. To the north, on the Scotian Shelf, Herman *et al.* (1991) found aggregations of *Calanus* species in the deep basins where weak internal circulations allow them to be resident throughout the year. Canyons at the edges of continental shelves may also act as concentrators of diel-migrating copepods and other zooplankton (Koslow and Ota, 1981). The densities of copepods in such depressions (Table 66, depth-related assemblages) may be enough to allow behavioural interaction among individuals although there is no physiological evidence at present to support this suggestion, because of the extent of the NNDs. Abrupt changes in bottom topography, as present on continental slopes, seamounts, the edges of banks and ridges, or around islands surrounded by deep waters can result in apparent aggregation of impinging pelagic species whose daytime depth

Table 66 Assemblages of calanoid copepods. The form of the assemblages, whether a ball or carpet over the sediment surface, or a swarm of different or undefined shape, is given. The diameter of the assemblage, the density of copepods within it, and the nearest neighbour distance (NND) are also indicated. The size of assemblages is usually given as the diameter or length (m); the units are stated when it is expressed as an area.

Species	Form of assemblage	Size of assemblage in m	Density No. m^{-3}	NND in mm	Authority
Aggregations					
Acartia clausi		25–100	<3.5 × 10^4		Anraku, 1975
Calanus finmarchicus		2500 km^2	6.2 × 10^3		Wishner *et al.*, 1988
C. finmarchicus	depth-related		2 × 10^4		Sameoto and Herman, 1990
	depth-related		4.2 × 10^4	33	Wishner *et al.*, 1988
C. glacialis	depth-related		2.5 × 10^3		Sameoto and Herman, 1990
C. hyperboreus	depth-related		1.5 × 10^3		Sameoto and Herman, 1990
Calanus spp.	depth-related		1–4 × 10^3		Herman *et al.*, 1991
C. pacificus			10^4 × 10^5		Omori and Hamner, 1982
Centropages orsinii		0.5			Hamner and Carleton, 1979
C. typicus			2 × 10^{2**}		Saiz *et al.*, 1992b
Heterocope septentrionalis			3.5 × 10^4		Hebert *et al.*, 1980
Neocalanus cristatus		10–20	3 × 10^2	34	Tsuda *et al.*, 1993
N. plumchrus			10^{4*}		Mackas and Louttit, 1988
N. plumchrus CIII–CV		1–2 km	5.5 × 10^3		Kawamura and Hirano, 1985
N. tonsus	layer	100–500	2.4 × 10^4	36	Kawamura, 1974
Paracalanus parvus		25–150	<4 × 10^3		Anraku, 1975
Pleuromamma abdominalis		70			Wiebe, 1970
P. gracilis		110			Wiebe, 1970
Mean of many species		100–300			Fasham *et al.*, 1974

Swarms				
Acartia australis	ball	$0.6–1.0$	$10^5–10^6$	10	Hamner and Carleton, 1979
	carpet				Omori and Hamner, 1982
A. bispinosa	carpet	<0.3			Hamner and Carleton, 1979
A. clausi	carpet	$0.1–0.3$	3×10^5		Ueda et al., 1983
A. erythraea	ball	0.2			Ueda et al., 1983
A. hamata	ovoid				Ueda et al., 1983
A. japonica	ball	$0.1–0.5$			Ueda et al., 1983
A. omorii	carpet		3×10^5		Tanaka et al., 1987
	carpet		$10^4–10^5$		Kimoto et al., 1988
	ball	$0.1–0.3$	$2–5 \times 10^5$		Kimoto et al., 1988
	ball	$0.1–0.2$	$2 \times 10^4–10^5$		Kimoto et al., 1988
			$1–6 \times 10^3$		Nomura et al., 1993
			7.5×10^6		Nomura et al., 1993
A. sinjiensis	ellipsoid				
	ball	<1.0	2×10^6	9	Ueda et al., 1983
	misty cloud		1.1×10^5	23	Emery, 1968
A. spinata	ball	$0.1–0.6$	$10^4–10^5$		Ueda et al., 1983
A. steueri	ball	$0.3–0.5$	2.1×10^5		Tanaka et al., 1987
	ball	$0.1–0.5$	$4 \times 10^4–10^5$		Kimoto et al., 1988
	ball	$0.1–1.0$	$5 \times 10^4–10^5$		Kimoto et al., 1988
A. tonsa			10^8	3	Haury and Yamazaki, 1995
Acartia sp.	carpet		4×10^4		Ohtsuka and Kimoto, 1989
Calanus finmarchicus	layer		10^7	5	Wiborg, 1976
	layer		$10^5–10^6$	6–12	Haury and Yamazaki, 1995
C. pacificus diapausing CV	layer at 450 m		2.6×10^7	4	Alldredge et al., 1984
Centropages abdominalis	ellipsoid		$1–4 \times 10^3$		Nomura et al., 1993
			8.5×10^6		Nomura et al., 1993
Labidocera pavo	disc	$0.1–0.2$			Ueda et al., 1983
School					
Labidocera pavo	disc		10^4	50	Omori and Hamner, 1982

* Aggregation caused by reaction to velocity gradients at river plume margin.
** At a density front.

coincides with depths on these features. Such mesopelagic species may become members of the benthopelagic fauna during the day and occur at higher densities than they do in the oceanic regions.

Wiebe (1970) found that *Pleuromamma* species were in almost circular aggregations of 27 to 31 m diameter at 90 m depth while at night at 20 m depth they occurred in aggregations of 110 to 140 m diameter. The average densities within these aggregations were 3 to 4 times the background densities. The diel vertical migration alters the morphology of the aggregation, frequently involving a degree of dispersion at night.

11.2.2. Swarms

Marshall and Orr (1955) have reviewed the early literature on observations of *Calanus* species swarming at the surface in the summer months. They form red patches that can be seen from some distance. The reasons for such swarming remain unknown as do the factors initiating it. It has been observed in *Calanus finmarchicus* but is thought to be performed also by *C. helgolandicus* in the northeast Atlantic, including the Norwegian and North Seas. Surface swarming of other planktonic organisms, such as euphausiids (Mauchline, 1980), are equally enigmatic.

Calanus species form subsurface aggregations that have very high densities and are classed as swarms in Table 66. An extremely dense layer of diapausing CV *Calanus pacificus* was observed, from a submersible, at 450 m depth, and 100 m above the bottom, in the Santa Barbara Basin, California. The layer had a mean thickness of 20 ± 3 m; densities of copepods reduced to less than 5×10^2 m^{-3} within 10 m above and below the layer. Minimum NND was measured as 4 mm so justifying this assemblage being considered as a swarm.

Most information on swarms has been obtained with the aid of SCUBA gear in shallow water, especially in tropical and subtropical regions (Kimoto et al., 1988). Species of *Acartia*, *Centropages* and *Labidocera* were found to swarm (Table 66) near the sea bed over sandy and rocky substrata and in association with coral reefs. The swarms seen were usually small in size, less than a metre in diameter, often associated with physical features of the topography such as the faces of coral reefs (Alldredge and King, 1977), in the lee of obstructions to flowing water currents, or over depressions in the sea bed (Omori and Hamner, 1982). The swarms are often near-globular balls but can be disc-shaped, ovoid or irregular; Emery (1968) describes a swarm of *Acartia spinata* as a misty cloud. Ball-like swarms may coalesce and spill to the sea bed as carpet-like layers, sometimes with a thickness approaching 2 m, a few centimetres above the surface of the sediment (Omori and Hamner, 1982). Disturbed swarms can re-form (Emery, 1968).

Swarms are usually monospecific but can be composed of different developmental stages, including adult males and females. Sometimes swarms form during the day and disperse at night, as found in *Acartia australis* by Hamner and Carleton (1979).

11.2.3. Schools

The occurrence of schools presupposes the ability to swim against any current present. Schools are moving formations of organisms, more or less tightly packed, that behave as a single organism. Copepods are relatively weak swimmers and, consequently, there are few observations of schooling of copepods. Omori and Hamner (1982) observed schools of *Labidocera pavo* shaped like a lens and 1 to 5 m in length. Disturbance of the school caused it to tighten, a feature of the schooling of other taxa. Kimoto *et al.* (1988) observed swarms of *Acartia omorii* and *A. steueri* swimming against a current to maintain position relative to a topographic feature of the sea bed but this may not be an example of true schooling behaviour.

11.2.4. Multispecies Aggregations and Swarms

Most aggregations and swarms consist of a single species but multispecies aggregations and swarms are not uncommon. The basins on the Scotian Shelf have such aggregations (Wishner *et al.*, 1988; Sameoto and Herman, 1990) comprising *Calanus* and other species such as *Pseudocalanus minutus*. Goswami and Rao (1981) observed a mixed aggregation of pontellid copepods in a bay in the Andaman Islands (Table 67). Haury and Wiebe (1982) concluded, from a series of Longhurst-Hardy Plankton Recorder samples, that oceanic zooplankton, including copepods, occurred in multispecies aggregations. A similar conclusion was drawn by Mauchline and Gordon (1986) about aggregations of benthopelagic plankton on the slope of the Rockall Trough, northeastern Atlantic, as deduced from the analyses of the stomach contents of macrourid fish.

Most observations, however, are on swarms of mixed composition in shallow coastal regions. The most detailed study is that of Kimoto *et al.* (1988) involving *Acartia* species, *Eurytemora pacifica* and *Pseudodiaptomus nihonkaiensis*. These species occurred in various combinations in single swarms (Table 67). Cyclopoid copepods of the genus *Oithona* often form a component of mixed swarms of *Acartia* species (Omori and Hamner, 1982). One of the most interesting is the swarm of *Centropages orsinii* containing mysids and fish eggs (Table 67). The immobile eggs must have accumulated within an eddy and been joined by the copepods and mysids but all were of

Table 67 Evidence for multispecies assemblages of copepods along with other planktonic organisms.

Species	Type of assemblage	Volume or length of assemblage	Density No. m^{-3}	Total density No. m^{-3}	Authority
Centropages orsinii	swarm	0.5 m^3	–		Hamner and Carleton, 1979
Anisomysis pelewensis			–		
Fish eggs			–		
Pontella spinipes	aggregation		9.1×10^4	1.4×10^5	Goswami and Rao, 1981
P. securifer			1.4×10^4		
P. princeps			1.3×10^4		
Pontellopsis regalis			1.4×10^4		
Pontellina plumata			5.1×10^3		
Labidocera acuta			1.0×10^3		
Calanopia elliptica			5.3×10^2		
Other copepods			5.6×10^3		
Acartia australis	swarm	0.6–1.0 m	–	10^5–10^6	Omori and Hamner, 1982
Oithona oculata					
Acartia sinjiensis	swarm	0.1–0.3 m			Ueda et al., 1983

A. erythraea	swarm	0.1–0.3 m		Ueda et al., 1983
A. omorii	swarm		1.4×10^5	Kimoto et al., 1988
A. steueri			1.7×10^5	
Eurytemora pacifica			7.0×10^4	
Other copepods			1.9×10^4	
Acartia steueri	swarm		3.3×10^5	Kimoto et al., 1988
Pseudodiaptomus nihonkaiensis			5.5×10^4	
Other copepods			2.1×10^5	
Tortanus longipes	carpet		2.7×10^2	Ohtsuka and Kimoto, 1989
T. rubidus				
Acartia omorii	aggregation		7.5×10^6	Nomoru et al., 1993
Centropages abdominalis			8.5×10^6	
Other copepods			1.0×10^4	

approximately the same body size. Auster *et al.* (1992) describes aggregations of mixed organisms that included myctophid fish, ctenophores, amphipods and euphausiids in which body size was similar regardless of the type of organism. They suggest that these multispecies aggregations enable more effective exploitation of common prey such as copepods as well as conferring defence against size-selective predators.

11.2.5. Mechano- and Chemoreception Distances

The aggregation of copepods raises the question of minimum distances between individuals that are required before one animal becomes aware of the presence of another. These will bear some relationship to the distances at which the copepods can detect the presence of food and predators. Haury and Yamazaki (1995) review the reaction distances of copepods to various stimuli and find that they range from <1 to 8 body lengths (bl). Supplementary data are given in Table 68 that tend to decrease the average detection distance, the 8 bl of *Centropages typicus* beginning to appear exceptionally large. The detection by males of pheromones exuded by females at some 20 mm (Table 68) really introduces another dimension. All other observations are presumed to measure the capabilities of the mechanoreception system of the copepods. The property or properties of the prey that are detected by the predatory copepod are not clearly understood. Yen (1988) suggests that *Euchaeta rimana* causes so much turbulence around itself that its mechanoreceptors on its antennules would be unable to detect turbulence of a prey organism. Instead, the potential prey organism may react to *E. rimana* by escaping and it is the high Reynolds number characteristic of the escape response that is detected by the *E. rimana*. Such escape could remove the prey some 5 to 8 mm from the *E. rimana* but may leave a trail of turbulence that the *E. rimana* can follow.

Since the mechanoreceptive distances appear to be short, is it possible that chemoreception is more important in location of organisms in the surrounding medium? This will be difficult to demonstrate satisfactorily because of the diffusive nature of the pheromones or exudates. The potential effects of environmental turbulence on the scent trails are briefly discussed by Strickler and Costello (1996). Kittredge *et al.* (1974) review early evidence of chemical recognition by males of female copepods but do not refer to Katona's (1973) observations on *Eurytemora affinis* (Table 68). They do, however, wonder if the circular pathways swum by male *Labidocera jollae*, similar to the loopings in Figure 83A, allow it to follow a chemical gradient to the female over a distance of a few centimetres. Dunham (1978) contends that this cycloid swimming is not a chemotaxis but

Table 68 Reaction distances of copepods to stimuli, the distances being given in mm and approximate body lengths. (After Haury and Yamazaki, 1995.)

Species	Distance mm	bl	Stimulus	Authority
Acartia hudsonica nauplii	1	4	Flow field *Temora longicornis*	Yen and Fields, 1992
A. tonsa	0.7	<1	Ciliate detection by antennules	Jonsson and Tiselius, 1990
Calanus finmarchicus CV	7	3	Flow around obstacle	Haury et al., 1980
Centropages typicus	13	8	Vibrating sphere	Haury and Yamazaki, 1995
Eurytemora affinis ♂	2–20	<20	Female pheromones	Katona, 1973
E. herdmani ♂	2–20	<20	Female pheromones	Katona, 1973
Diaptomus minutus	2	3	Attack by *Limnocalanus*	Wong et al., 1986
Euchaeta rimana	2	1	Attack distance	Yen, 1987b, 1988
Metridia gerlachei	1	<1	Stationary cyclopoid prey	Metz and Schmack-Schiel, 1995
Mixodiaptomus laciniatus	1	<1	Aquarium walls	Haury and Yamazaki, 1995
Pseudodiaptomus coronatus ♂	2–20	<20	Female pheromones	Katona, 1973
Tortanus forcipatus and *T. gracilis*	0.5	<1	Detection of prey	Uye and Kayano, 1994a

is a chemoklinokinetic reaction in which the male responds to a pheromone by increased frequency of turning at low concentrations and a decreased frequency at high concentrations resulting in its location of the source.

There are frequently changes in the swimming speed of copepods caused by changing concentrations of food that are probably through sensing exudates from the food organisms (Van Duren and Videler, 1995). Further, Van Duren and Videler (1996) have shown that female *Temora longicornis* react to exudates of male *T. longicornis*. They have also demonstrated that *T. longicornis* reacts to exudates of predatory fish by reducing their swimming speed. Consequently, there is increasing evidence of the prevalence of chemical communication, and response to it, among copepods. Nearest neighbour distances (NNDs) within aggregations have been examined by Haury and Yamazaki (1995) and are shown in Table 66 (p. 426). They are obviously larger than the average detection distances in Table 68. Yet copepods in disturbed swarms can re-form the swarms and aggregations with NNDs larger than any presently known detection limits are also able to maintain themselves. Admittedly, the individuals in some aggregations that are associated with bathymetric features, such as small or larger depressions in the sea bed, may be reacting individually to the physical environment rather than to each other. Secondary mechanisms, however, could be chemosensory reaction to exudates from the other individuals.

11.3. BIOLUMINESCENCE

Certain calanoid copepods bioluminesce but not in the sense that euphausiids do. There are no luminescent organs or photophores that emit light through the integument. Instead, subcuticular glands secrete luminous material that is extruded into the surrounding water, although Herring (1988) reports intracellular luminescence in *Hemirhabdus latus*. Table 69 lists the genera and species that Herring (1988) confirms as luminescent; more recent records of luminescence in these species are also cited. There are, in addition, scattered species in the genera Calanidae, Eucalanidae, Paracalanidae, Pseudocalanidae, Aetideidae, Euchaetidae, Scolecitrichidae, Centropagidae, Temoridae, Pontellidae and Candaciidae that have been recorded as luminescent but which Herring (1988) questions.

11.3.1. Luminescent Glands

The luminescence glands first appear in the later naupliar stages. Evstigneev (1982a,b) found that nauplii V and VI of *Pleuromamma* species luminesced

Table 69 Species and genera of copepods known to be bioluminescent. References are given in parentheses. (After Herring, 1988.)

Megacalanidae
 Megacalanus princeps (14)
Lucicutiidae
 Lucicutia aurita (14)
 L. clausi (14)
 L. flavicornis (2,10,11,12,14,20)
 L. gemina (14)
 L. grandis (14)
 L. magna (14)
 L. ovalis (14)
 L. sarsi (14)
 L. wolfendeni (14)
Heterorhabdidae
 Heterorhabdus norvegicus (14)
 H. papilliger (10,11,14)
 H. robustus (14)
 H. spinifrons (14)
 Hemirhabdus grimaldii (9,14)
 H. latus (14)
 Heterostylites longicornis (14)
 Disseta palumboi (9,14)
Augaptilidae
 Euaugaptilus bullifer (14)
 E. farrani (14)
 E. filiger (14)
 E. laticeps (1,9,14)
 E. magnus (1,9,14,21,22)
 E. nodifrons (14)
 E. periodosus (1,9,14)
 E. rectus (14)
 E. squamatus (14)
 E. truncatus/vicinus (14)

Augaptilidae
 Centraugaptilus horridus (14)
 C. cucullatus (14)
 C. rattrayi (14)
 Haloptilus longicirrus (14)
 Heteroptilus acutilobus (14)
 Pachyptilus eurygnathus (14)
Metridinidae
 Metridia gerlachei (14)
 M. longa (8,14,15,17,18)
 M. lucens (2,14,17,18)
 M. macrura (14)
 M. pacifica (14)
 M. princeps (1,14)
 Pleuromamma abdominalis
   (2,11,14,16,19)
 P. borealis (14,18)
 P. gracilis (2,3,4,11,12,14,16)
 P. indica (14)
 P. piseki (14)
 P. quadrungulata (2,14,16)
 P. robusta (14)
 P. xiphias (1,2,9,14,16,20)
 Gaussia princeps (5,6,14,19,20,21)

References: (1) Bannister & Herring, 1989; (2) Batchelder & Swift, 1989; (3) Batchelder *et al.*, 1990; (4) Batchelder *et al.*, 1992; (5) Bowlby & Case, 1991a; (6) Bowlby & Case, 1991b; (7) Buskey, 1992; (8) Buskey & Stearns, 1991; (9) Campbell & Herring, 1990; (10) Evstigneev, 1989; (11) Evstigneev, 1990a; (12) Evstigneev, 1990b; (13) Evstigneev, 1992; (14) Herring, 1988; (15) Lapota *et al.*, 1988a; (16) Lapota *et al.*, 1988b; (17) Lapota *et al.*, 1989; (18) Lapota *et al.*, 1992; (19) Latz *et al.*, 1988; (20) Latz *et al.*, 1990; (21) Widder, 1992; (22) Bannister, 1993b.

while Lapota *et al.* (1988a) recorded luminescence in nauplius IV of *Metridia longa*. The latter authors found that the intensity of the luminescence produced by the nauplii was about one-fiftieth of that of the adult stage. This difference is accounted for by the progressive addition of luminescent glands throughout the successive copepodid stages.

The histology of the glands has been examined by Clarke et al. (1962), Bannister and Herring (1989), Bowlby and Case (1991a) and Bannister (1993b). They consist of secretory vesicles enclosed in a sheath that open to the environment through a valve in the integument. The glands are usually single but are paired in the Augaptilidae, each pair discharging through a common opening to the outside (Bannister and Herring, 1989). Direct innervation of the luminous glands in the P3 to P5 of *Euaugaptilus magnus* has been described by Bannister (1993b) although how the nervous impulse produces release of the secretion is not understood. Luminous glands fluoresce under ultraviolet excitation but the response is irregular. Clarke *et al.* (1962) found that the luminous glands of Heterorhabdidae do not fluoresce, and Barnes and Case (1972) could not stimulate luminescence from antennal glands of *Gaussia princeps* which fluoresced. The luminescent glands of the Metridinidae and Augaptilidae fluoresce blue-green while the fluorescence of those of the Lucicutiidae is much yellower (Herring, 1988).

The glands occur at a variety of locations. Those in *Megacalanus princeps* and augaptilids are restricted to the exopods of some of the swimming legs. In contrast, species in the Metridinidae and Heterorhabdidae have them located in the legs and also subcuticularly over the body and even in the antennules. Those of the Lucicutiidae have a similar occurrence except that they are absent from the urosome. Maps of sites are given for a variety of species by David and Conover (1961), Clarke *et al.* (1962), Barnes and Case (1972), Evstigneev (1982a) and Bannister and Herring (1989).

11.3.2. The Luminous Secretions

Campbell and Herring (1990) found that imidazolopyrazine, coelenterazine, and its luciferase or photoprotein are responsible for the luminescence of copepods.

The luminous glandular secretions are discharged into the water. Herring (1988) discusses, in some detail, the effects of Reynolds numbers on the ejection of this material and also examines mechanisms, muscular and otherwise, of its expulsion from the subcuticular glands. It may remain coagulated or it may dissipate in the water. It may separate from the copepod or it may remain attached to the surface of the integument. All situations have been described as well as the possibility of it remaining within some of the glands – intracellular luminescence. Bowlby and Case (1991a) describe the secretions of species of the Metridinidae being discharged as droplets, suggesting that the glandular vesicles are hydrophobic or cohesive.

The wavelength of the luminescence peaks at about 470 to 490 nm and is usually unimodal, according to Herring (1988) who also notes bimodal emission spectra in *Pleuromamma* species. Latz *et al.* (1987, 1988) describe bimodal spectra, at 479 and 489 nm, in *Gaussia princeps*. The intensity of light produced has been measured in different ways. Herring (1988) lists maximum intensities in μW cm^{-2} and in terms of equivalent total photons per second. Intensities have been measured in some 20 species and are tabulated by Herring; intensities range from 0.05 to 4900 μW cm^{-2} × 10^{-5}, equivalent to a range of 0.03 to 377 photons s^{-1} × 10^{11}. There is considerable variation in the data owing to the variety of experimental techniques employed that make direct comparison of the results from the different species difficult. Comparable information on *Gaussia princeps* is provided by Bowlby and Case (1991b). There are a number of recent papers that quantify the intensity of the light in photons per flash (Lapota *et al.*, 1988a, b, 1989, 1992; Latz *et al.*, 1990) and others as photons per individual (Batchelder and Swift, 1989; Batchelder *et al.*, 1992). Some standardization of experimental procedures is necessary to provide comparative data.

11.3.3. Luminescent Behaviour

Copepods rarely luminesce spontaneously in the laboratory, although Nealson *et al.* (1986) found that *Pleuromamma borealis* did. Consequently, some form of stimulation is usually used in experimental work. This can be mechanical, such as an electrical stirrer (Bowlby and Case, 1991b; Buskey, 1992; Latz *et al.*, 1990), or withdrawal of water, by vacuum or simply draining it away, so that the animal is stranded on a filter (Buskey and Stearns, 1991; Clarke *et al.*, 1962; Lapota *et al.*, 1989). Light, from a flashlight (Lapota *et al.*, 1986) or photoflash (Buskey and Swift, 1985), can be used as a stimulus. Electrical stimulation is also employed by passing current through the container or aquarium in which the animal is held (David and Conover, 1961; Clarke *et al.*, 1962; Latz *et al.*, 1990; Bowlby and Case, 1991b; Widder, 1992). Sonic stimulation was used by Shevijrnogov (1972). Chemical stimulation by norepinephrine (Widder *et al.*, 1983) and by hydrogen peroxide and 5-hydroxytryptamine (serotonin) (Latz *et al.*, 1988) is also successful.

Repeated stimulation of the luminescence exhausts the potential of the copepods to luminesce further. Recovery after depletion requires about 24 h (Latz *et al.*, 1990; Bowlby and Case, 1991b). The kinetics of the luminescence is discussed by Herring (1988) and additional information is available from Bowlby and Case (1991b). Flash kinetics are highly dependent on a variety of factors; the form and type of stimulation and especially its frequency and duration, developmental stage of the copepod, its history, the extent of pre-stimulation and so depletion of the luminescent reserves, the number and degree of synchrony of responding glands and probably also factors such as experimental temperature. Evstigneev (1992) examined the effect of salinity on the luminescence of *Pleuromamma borealis*. In addition, Evstigneev and Bityukov (1986) showed that the intensity of the luminescence of *P. gracilis* was greater at night than during the day. There was no evidence of intrinsic rhythms. Herring (1988), discussing the lack of evidence, concludes that the synergistic effects of possible ambient light inhibition of and temperature effect on luminescence militates for maximal light emission by diel migrating species at the surface during the night.

Giesbrecht's (1895) original observation of seasonal changes in the ability of copepods to luminesce has not been confirmed although it is known that euphausiids exhibit such changes in their responses to stimulation (Mauchline, 1960; Tett, 1972).

The pattern of distribution of the luminescent glands over the body is often specific, the gland openings being components of the pore signatures of the species (see Chapter 2). Within species, males and females exhibit the same pattern. The luminescence emitted by females of *Pleuromamma* species is about twice the intensity of that emitted by the males (Bityukov and Evstigneev, 1982; Evstigneev, 1982b). The ecological or behavioural significance, if any, of the differences noted in the spectral emissions of species are unknown.

Measurement of naturally occurring bioluminescence in the sea is very difficult. The instruments hung in the water column cause disturbance, often stimulating organisms to bioluminesce. The identification of the source or sources of bioluminescence recorded in the water column is difficult because a variety of bioluminescent animals is usually present. The strength of the recording will depend on what species is luminescing and how far it is from the sensor. Buskey and Swift (1990) maintain that interactions between planktonic animals lead to spontaneous luminescence and describe a model predicting the frequency of such interactions.

The function of the luminescence is a further area of speculation. The cohesive secretions released by some species could act as decoys for predators. Widder (1992) has published pictures of video frames showing *Euaugaptilus magnus* and *Gaussia princeps* discharging luminous "boluses" during escape reactions. Other evidence suggests that luminosity can deter predators or, in the dark meso- and bathypelagic environment, temporarily blind them. The only evidence, so far, of its use for communication between individuals of a species derives from Buskey and Swift (1985) who suggest that it may act as a warning signal between individuals of *Metridia longa*. No evidence of its use in sex recognition is available. Buskey and Swift show that *M. longa* responds much better to simulated copepod bioluminescence, with flashes of 600 ms duration, than to dinoflagellate bioluminescence, with

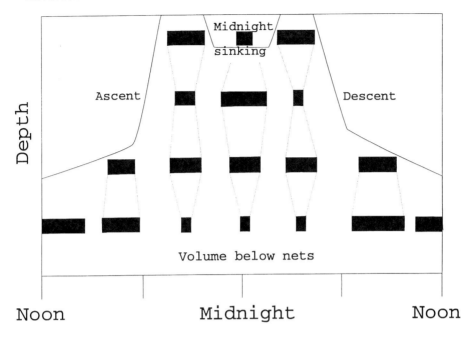

Figure 85 A schematic representation of the diel vertical migration of a copepod showing the periods of ascent and descent and a mid-darkness period of sinking from the immediate surface layer. The length of the bars represent the comparative numbers of copepods caught at each depth and time.

flashes of only 60 ms duration. Non-bioluminescent copepods, such as *Acartia hudsonica* and *Calanus finmarchicus*, show no such preference. Both *C. finmarchicus* and *Metridia longa*, however, are sensitive to the same wavelengths of light over a relatively broad range of approximately 460 to 560 nm (Buskey and Swift, 1985).

11.4. VERTICAL MIGRATION

Diel vertical migration has been known to occur in planktonic organisms for nearly 200 years (Cushing, 1951). The population lives at some distance from the surface of the sea during the day but as sunset approaches, usually about 2 h before sunset, the animals swim upwards (evening ascent) towards the surface (Figure 85). They are present there during the hours of darkness although a portion may sink passively during the night (midnight sinking). At dawn, the animals swim downwards to their daytime depths. There are many descriptions of this diel behaviour, encompassing a wide variety of species of copepods in different environmental circumstances (Cushing, 1951; Bainbridge, 1961; Harris, 1963). It was generally assumed that epipelagic and coastal copepods performed this diel behaviour and when species or populations were found to remain at the same depth by day and night, or even when they were shown to perform a reverse migration, at the surface during the day and at depth at night, they were treated as anomalous (Bainbridge, 1961).

One difficulty in the investigations of the migrations of continental shelf, coastal, and even estuarine populations is determining the numbers of individuals that remain in close association with the sea bed. The samples are usually collected by pelagic nets and inevitably there is an unsampled layer below the deepest net haul (Figure 85). Many pelagic, as opposed to hyperbenthic or benthopelagic, species have downward extensions of their pelagic populations into the hyperbenthic and benthopelagic environments within a few metres of the surface of the sediment (e.g. Stubblefield *et al.*, 1984; Imabayashi and Endo, 1986). There may be significant accumulations there during the day or seasonally that are not normally accounted for in studies of diel vertical migration. Estimating numbers at the bottom, such that they can be compared with those in the water column above, is very difficult because the two environments have to be sampled with different nets or equipment.

The phenomenon of diel vertical migration is now known to be much more complex than at first supposed. Some species, or some populations of a species, adhere more or less to the classical concept of the migration while others do not. Physical factors of the environment modify the behaviour and, in fact, can move the animals in such a way that a migration appears to be taking place. Banse (1964) notes the earlier records, especially in neritic environments, where, on re-examination of the data, it can be concluded that no active diel migration was present. He also discusses the sampling of stratified water columns, the effects of patchy distributions and gradients of temperature and salinity, and the interpretation of the results. The inherent difficulties in interpreting sample data like those in Figures 85 and 86 are discussed by Pearre (1979b). The principal problem is distinguishing information about the behaviour of the individual from that of the population. Is there, for example, synchronous upward and downward movement of all individuals within the population or is it only a small proportion that migrates? Do those that reach the surface first come from the upper leading edge of the daytime population or are they drawn from the bathymetric spectrum of the daytime distribution? Species living deeper in the oceanic water column during the day arrive later in the surface layers than those living at shallower depths (Wiebe *et al.*, 1992). This suggests that use of the leading edge concept may be justifiable in estimating the speeds of upward or downward swimming during the migration.

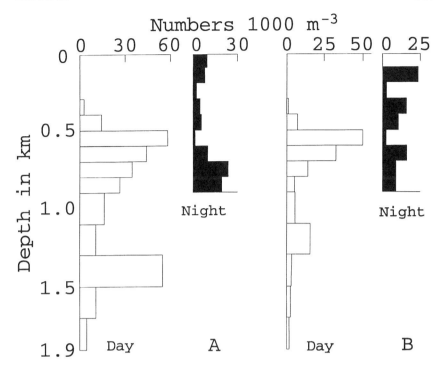

Figure 86 The day (open) and night (shaded) vertical distributions of A, *Pareuchaeta norvegica*, B, *Scaphocalanus magnus* in the Rockall Trough, northeast Atlantic. (Mauchline, unpublished.)

The problem of distinguishing the behaviour of individuals from that of the population is addressed in *Calanus finmarchicus* by Simard et al. (1985) in the lower St Lawrence estuary. The midnight sinking of copepods (Figure 85) is often a feature of the diel migration of this species. Simard et al. sampled two layers in the water column, the 0 to 30 m and the 30 to 100 m. Phytoplankton concentrations only occurred in the upper layer so that any copepods with full stomachs in the deeper layer were considered to have visited and fed in the upper one. The incidence of copepods in the upper layer increased sharply near dusk (Figure 87A) but decreased thereafter. Copepods with full stomachs appeared in the deeper horizon shortly after but the percentage showed a period of decrease during the night (Figure 87B). Simard et al. interpret these results as showing a midnight sinking to digest the food and a continuous exchange of individuals between the two horizons, peaking at dusk and again at dawn.

A bimodal distribution, such as that shown for *Pareuchaeta norvegica* in

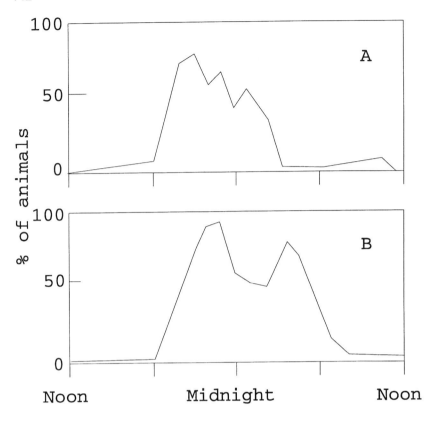

Figure 87 The feeding migration of *Calanus finmarchicus*. A, the percentage of animals in the total water column that occurred in the upper horizon (0–30 m). B, the percentage of copepods in the lower stratum (30–100 m) that had full stomachs. (After Simard *et al.*, 1985.)

the Rockall Trough (Figure 86A), is also quite common in many species (e.g. Marlowe and Miller, 1975; Hattori, 1989). There are often seasonal patterns; the summer warming of surface waters may be avoided by deep-living species that will enter the surface horizons only in colder seasons (e.g. Moraitou-Apostolopoulou, 1971).

11.4.1. Behaviour of Different Species

Studies of the vertical migration of very many species of copepods have been made. Much variation within and between species exists in the results and no attempt is made here to review the information on individual species.

Cushing (1951) reviews early work on species and Bainbridge (1961) lists more than 50 species of marine copepods on which field data exist. Further studies have been made since then on single species but they are only referred to later if they contribute some facet amplifying, or not conforming to, the general pattern. A few studies, from a variety of environments, have encompassed several to many species, including different developmental stages, and these are frequently the source of references to other less comprehensive investigations. Such multispecies studies are: Hure and Scotto di Carlo, 1969a,b, 1974; Moraitou-Apostolopoulou, 1971; Roe, 1972a–d, 1974, 1984; Binet, 1977; T.L. Hopkins, 1982; Daro, 1985; Harding *et al.*, 1986; Wishner and Allison, 1986; Ambler and Miller, 1987; Hopkins and Torres, 1988; Bennett and Hopkins, 1989; Hattori, 1989; Ward, 1989; Almeida Prado-Por, 1990; Peterson *et al.*, 1990b; Atkinson *et al.*, 1992a,b; Wiebe *et al.*, 1992; Mackas *et al.*, 1993; Pagano *et al.*, 1993; Żmijewska and Yen, 1993; Hays *et al.*, 1994; Tsuda and Sugisaki, 1994; Williams *et al.*, 1994.

11.4.2. Thermoclines, Haloclines and Oxyclines

The response to a thermocline varies between species and developmental stage and Cushing (1951) concluded that the migration was really only modified when the temperatures were near those defining the distributional limits of the species concerned. More modern results tend to confirm this, many species or developmental stages migrating through the thermocline while others tend to accumulate below it (Marlowe and Miller, 1975; Madhupratap *et al.*, 1981a; Sameoto, 1984; Williams and Conway, 1984; Fragopoulu and Lykakis, 1990; Saiz and Alcaraz, 1990; Morioka *et al.*, 1991).

The thermocline in middle to higher latitudes is a seasonal phenomenon and Williams (1985) shows that *Calanus finmarchicus* and *C. helgolandicus*, in the Celtic Sea, react differently to the thermocline and halocline. The former inhabited the colder more saline water below the thermocline while the latter lived above the thermocline in warmer and less saline water. The more northern-living *C. finmarchicus* is at its southern limit of geographical distribution in this region and consequently the temperatures within and above the thermocline present a barrier to its upward migration. The more southern *C. helgolandicus* may be restricted above the thermocline because of the lower temperatures below it. A similar difference was found between *Clausocalanus pergens* and *C. furcatus* in Patraikos Gulf, Ionian Sea. Fragopoulu and Lykakis (1990) state that *C. furcatus* is thermophilic whereas *C. pergens* is restricted to the colder water in the lower part of the thermocline. A seasonal thermocline occurs in Lindåspollene, a shallow land-locked fjord in Norway, and Magnesen et al. (1989) found that *Acartia longiremis*, *Centropages hamatus* and *Paracalanus parvus* had their modal distributions above the pycnocline during summer while *Pseudocalanus elongatus* and *Temora longicornis* occurred below it.

Oxygen minimum layers occur in many regions of the oceans and can present a barrier to the downward migration of some but not all copepods. Herman (1984), working on the Peru shelf, briefly reviews his own and earlier results.

11.4.3. Abrupt Bottom Topography

Abrupt changes in bottom topography can affect the vertical migrations of copepods. In areas with depths greater than 300 to 500 m and having banks, ridges or islands, or on the slopes of the continental shelves, there are vertical water columns that have insufficient depths for the natural range of diel vertical migration of many of the copepods in the adjacent deeper regions. Genin et al. (1994, 1995) found that the presence of banks resulted in a patchy distribution of diel migrating species. Gaps in the horizontal distributions of such species were present over the banks at night. The reason for this is that vertically migrating copepods are advected over the bank at night, some being eaten by predators originating from the bank. The bulk of the survivors, not having sufficient depth for their downward migration at daylight, are those that are advected off the bank to deeper waters. This may also result in an apparent aggregation of copepods around the edges of banks. It is analogous to the situations in depressions on the continental shelves, where the daytime depth ranges of species are not available on the shelf as a whole, and the copepods gather within the depressions.

The continental slopes, the slopes of islands with deep water around them, and seamounts are areas where there is restriction on the daytime depths of species. The result is impingement of deep water species on the slopes. Species that normally live in the oceanic, pelagic water column can become members of the benthopelagic fauna during the day, living in close association with the sediment surface. Such impingement of oceanic copepods takes place on the slopes of the Rockall Trough, northeastern Atlantic Ocean (Mauchline and Gordon, 1991). Species such as *Arietellus plumifer*, *Cephalophanes refulgens*, *Euaugaptilus magnus*, *Euchirella curticauda*, *Gaetanus kruppi*, *Lophothrix frontalis*, *Megacalanus princeps*, *Pareuchaeta barbata*, *P. norvegica*, *P. scotti*, *Pleuromamma robusta*, *Scottocalanus securifrons* and *Undeuchaeta plumosa* enter the benthopelagic environment and are subject to predation by the populations of benthopelagic and demersal fish.

11.4.4. Ontogenetic Migrations

Ontogenetic migrations are, in terms of diel periods, static because the scale of time involved is the generation time of the species. Examples are overwintering species where copepodids migrate to deeper water in the autumn, reside there throughout the winter, and migrate to shallower depths in the spring to reproduce. Ontogenetic diel vertical migrations are on diel time scales and, in the simplest terms, occur when the developmental stages of a species within the same water column have distinct bathymetric ranges over the diel cycle. The most common pattern of distribution is for the younger copepodids to live higher in the water column and older ones progressively deeper. *Euchaeta paraconcinna* was found to mirror this pattern, younger stages living progressively deeper (Binet, 1977). Some stages may perform a diel migration while others may reside at constant depth by day and night.

Ontogeny is bound to have an effect on the diel migratory behaviour because of several factors. Eggs are normally denser than sea water and sink while developing to hatching. One known exception is the egg of *Calanus hyperboreus*, which floats to the surface. There will then be a tendency for the early nauplii to occur at depth. Speed of swimming increases as development proceeds, especially after the transition from naupliar to copepodid stages and the acquirement of the swimming legs. Nutritional demands also change during development, and, in many cases, the diet changes. In subtropical oceanic species, some species had no ontogenetic or diel vertical migrations, others had an ontogenetic but no diel migration, while yet others had both types of migration (Ambler and Miller, 1987). Copepodid and adult *Euchaeta media* and *Pleuromamma* species share the same modal depths at night but progressively older copepodids occur at progressively greater depths during the day. The variety of patterns of diel migration between and within species may have been selected for to allow partitioning of the environmental resources, as discussed in the next chapter.

Recent detailed accounts of the ontogenetic diel vertical migrations of species in a variety of environments are listed in Table 70. It is most easily demonstrated in deeper water with water columns of 100 or more metres but it also occurs in shallow-living species as instanced by *Acartia* species and *Parvocalanus crassirostris* (Table 70) that occurred in water columns of 10 to 20 m.

Table 70 Recent analyses of ontogenetic diel vertical migrations of calanoid copepods.

Species	Region	Authority
Acartia clausi	San Juan Is., Washington	Landry, 1978b
A. hudsonica	Maizuru Bay, Japan	Ueda, 1987b
A. omorii	Maizuru Bay, Japan	Ueda, 1987b
	Shijiki Bay, Japan	Kimoto, 1988
Calanoides carinatus	Ivory Coast	Binet, 1977
	Benguela upwelling	Verheye and Field, 1992
Calanus finmarchicus	Barents Sea	Unstad and Tande, 1991
	Norwegian fjords	Tande, 1988b
C. glacialis	Barents Sea	Unstad and Tande, 1991
C. pacificus	Experimental tank	Huntley and Brooks, 1982
	Dabob Bay, Washington	Osgood and Frost, 1994a
C. sinicus	Japan Inland Sea	Uye *et al.*, 1990b; Huang *et al.*, 1992
Centropages chierchiae	Ivory Coast	Binet, 1977
Eucalanus crassus	Ivory Coast	Binet, 1977
E. monachus	Ivory Coast	Binet, 1977
E. pileatus	Ivory Coast	Binet, 1977
Euchaeta marina	Gulf of Mexico	Shuert and Hopkins, 1987
E. paraconcinna	Ivory Coast	Binet, 1977
Metridia longa	N Atlantic, North Sea	Hays, 1995
M. lucens	Dabob Bay, Washington	Osgood and Frost, 1994a
	N Atlantic, North Sea	Hays, 1995
Paracalanus crassirostris	Maizuru Bay, Japan	Ueda, 1987b
Pleuromamma spp.	Gulf of Mexico	Bennett and Hopkins, 1989
Temora stylifera	Ivory Coast	Binet, 1977
T. turbinata	Ivory Coast	Binet, 1977
Undinula vulgaris	Ivory Coast	Binet, 1977

11.4.5. Cues for Diel Migration

11.4.5.1. *Light*

Light is still considered to be the prime environmental factor controlling the diel vertical migration of copepods and other planktonic organisms. The following of isolumes, layers of constant light intensity, by animals such as euphausiids, appears to control the migrations of some species (Mauchline, 1980). Roe (1984), however, suggests that in oceanic populations none of the copepods can maintain themselves in a constant isolume because they do not swim fast enough and/or their vertical distributions are too diffuse.

Tranter et al. (1981) successfully caught a variety of shallow water copepods in a light-trap, especially at dusk and dawn or when the moon set. The species were *Acartia tranteri, Gladioferens pectinatus, Isias uncipes, Temora turbinata* and *Tortanus barbatus*. They ascribed this behaviour to a moving towards the light when light intensities decreased and a moving away from the light when light intensities increased. Demersal copepods emerged upwards from a subtidal sand flat into artificially darkened traps during the day and on moonlit nights (Alldredge and King, 1980), inferring that absence of light is a cue for emergence. Moon light influences the depth distribution of copepods at night as Jerling and Wooldridge (1992) show.

Many oceanic populations of copepods have a bimodal bathymetric distribution such as that for *Pareuchaeta norvegica* in Figure 86A (p. 441). This means that the attenuation of downwelling daylight in the deep oceanic water column results in the daytime range of a species dividing into two components, an upper migrating one and a lower non-migrating one. The depth of the division depends upon the photosensitivity of the individuals, the threshold for phototaxis (Figure 88), and the depth to which that intensity of downwelling light penetrates (Figure 88). This is also true in turbid coastal regions where light attenuation is very severe, as in Jervis Inlet (Figure 88). Stearns and Forward (1984b) show that the stimulus for the upward migration of *Acartia tonsa* is a relative decrease in the amount of light, that is a change in quantal intensity; a shift in wavelength alone evoked no response. The relative changes in intensity of the light, upwards or downwards, that initiates the upward and downward phases of the migration depends upon the intensity to which the copepods are adapted at that time. The relationship of the percentage change in light intensity required to stimulate the copepod is not linearly related to the intensity of light to which the copepod is adapted (Forward, 1988). The smallest percentage change, about 10%, stimulates *Acartia tonsa* adapted to a light intensity of 4.3 photons $m^{-2} s^{-1}$. Stimulation requires an increasing percentage change, up to nearly 100%, for *A. tonsa* adapted to light intensities lower or higher than the 4.3 photon level. The 4.3 photons $m^{-2} s^{-1}$ level corresponds to the intensity of light at the position in the water column where *A. tonsa* begins it evening ascent.

Detailed information, comparable to that on *Acartia tonsa*, is not available for other species. Comparable studies are required on offshore and oceanic species to examine the hypothesis that the relative change in light intensity is a cue initiating vertical migration. Diel-migrating species such as *Metridia longa* and *M. lucens* show seasonal variations in the amount of time that they spend in the surface layers that co-vary with the seasonal changes in night length (Hays, 1995). Copepods, like other marine crustaceans, are most sensitive to the blue-green region of the light spectrum, that is to lighting conditions at sunset and sunrise. Forward (1988) comments on the

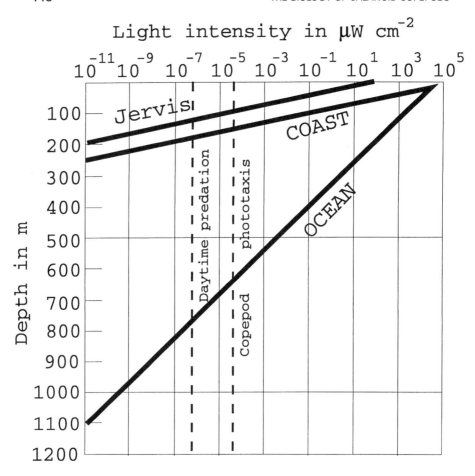

Figure 88 The attenuation of downwelling daylight in clearest ocean water (OCEAN), coastal water (COAST) and Jervis Inlet, British Columbia (Jervis). The thresholds for daytime predation by fish and for phototaxis of copepods and other Crustacea are indicated. (After Mackie, 1985.)

design of laboratory experiments examining the photobiology of copepods and other zooplankton. Reference to his review should be made for further information on all aspects of the photobiology of copepods.

There is still the problem of what cues the downwardly migrating copepods to stop swimming downwards. How do they identify their daytime depth? Hattori (1989) suggests that the migrating component of a bimodally distributed population recognizes the presence of the deeper non-migrating mode. It acts as a "stopper" to the downward migration. This is a difficult concept since the nearest neighbour distances are liable to be large in terms of reaction between the components of two populations.

11.4.5.2. *Predators*

Downward migration from the surface layers begins as light intensities increase at sunrise and most investigations assume that the reappearance of the appropriate isolume triggers the migration. There are, however, copepods that are subjected continuously to these light cues yet they do not respond or they respond seasonally or at apparently irregular intervals (Bollens and Frost, 1989b; Frost and Bollens, 1992; Bollens *et al.*, 1995). The presence of visually predating fish, or even fish mimics, has also been shown to induce a downward swimming response in some copepods, possibly by the copepods detecting the shadow of the predator – a shadow response (Bollens and Frost, 1989a,b, 1991a,b; Bollens *et al.*, 1992b, 1994). Bollens *et al.* (1994) state that "fish mimics provide the first direct experimental evidence in support of a mechanical or visual cue for the induction of downward daytime vertical migration in *Acartia hudsonica*". They suggest that the importance of this cue may be peculiar to this species of copepod. Ringelberg (1995), according to Bollens *et al.* (1995), misinterpreted his results on *A. hudsonica*, considering that he was stating that the daily cycle of changing light intensities was not important in the cuing of the vertical migration. It seems that, for species of copepods that do not, for one reason or another, have a stereotyped diel vertical migration, the presence of predators can cue the morning descent. The response to predators at the individual level is immediate, and Neill (1990) shows that this response can potentially develop as a population response within a period of 1 h. Forward (1988) discusses this shadow response to predators in some detail. The presence of predators may very well reinforce light cues for the downward migration in many species of regularly migrating copepods. Frost (1988), however, examining populations of *Calanus pacificus* in Dabob Bay, Washington, concludes that predator avoidance is the major selective force for the presence of a diel migration of this species in this region.

There is no valid reason why only a single cue should operate although one may override another. Bollens and Frost (1989b), for example, show that the irregularly migrating *Calanus pacificus* in Dabob Bay is strongly influenced by the presence of planktivorous fish in the surface 50 m of the water column. They obtain a significant relationship between increasing strength of the vertical migration and increasing abundance of planktivorous fish. Koslow (1979) raises the question of possible influence of the bioluminescence of dinoflagellates in modifying the timing or extent of the upward migration of copepods. He suggests that bioluminescence of dinoflagellates peaks after sunset and allows visual predators to locate copepods. Luminescing dinoflagellates have been shown to decrease feeding activity of copepods (Esaias and Curl, 1972; Buskey and Swift, 1983). Does this constitute cryptic behaviour, as suggested when discussing swimming speeds? Koslow suggests that copepods might best avoid the luminous dinoflagellates by changing their timing of arrival at the surface. In discussing the variance of the timing of migrations, Pearre (1979a) raises the question of visual predators other than fish. Predatory zooplankton, such as chaetognaths, euphausiids and other copepods are reacted to. They are avoided by the copepods, sometimes through adoption of a pattern of reverse migration (Ohman, 1990).

Hays et al. (1996) searched for effects of annual and longer-term fluctuations in the stocks of Atlantic herring on the diel vertical migration patterns of copepods in the North Sea, using the Plankton Recorder data for the years 1958 to 1994. Changes in the day:night ratio of numbers of *Calanus finmarchicus* at the surface were correlated with changes in the biomass of herring.

Copepods living in shallow water of a few metres depth are potentially prey of fish during the day. As described above, several of these species form aggregations and swarms and so obtain some defence. Fancett and Kimmerer (1985) found that the demersal *Pseudodiaptomus cornutus* and *P. colefaxi* remain close to the sediment surface during the day and do not feed. At night, they migrate upwards off the bottom and feed. Compared with the co-occurring species *Acartia tranteri*, which does not associate with the sediment and feeds continuously, the *Pseudodiaptomus* species have higher lipid contents and so are adapted to discontinuous food supplies; their rates of egg production are not impaired by the adoption of a cryptic life style during the day. Discontinuous feeding of *Acartia tranteri*, by contrast, decreased its rate of egg production.

11.4.5.3. *Food*

Copepods are often considered to migrate vertically to feed. The evidence is strong in some species but not in others (e.g. Yen, 1985; Harris, 1988). Diel feeding rhythms and/or rhythms of swimming activity could cue vertical migration. Changing light intensities would reinforce the rhythms, especially in herbivorous species. Undoubtedly, the situation is complex, and differs according to species and environment, and is also often modified by the physiology of individuals.

Mackas and Bohrer (1976) and Atkinson et al. (1992a) found evidence that the downward movement of herbivorous copepods is triggered by satiation. Well-fed *Neocalanus plumchrus* perform a diel migration but food-limited ones do not and Dagg (1985) suggests that in food-limiting situations, a diel migration is more liable to cost the copepod energetically than to locate a source of food. Dagg reviews the evidence of food-limitation modifying the diel migratory pattern. Huntley and Brooks (1982), studying *Calanus pacificus* experimentally in a deep tank, found that the amplitudes of the migrations decreased as the availability of food decreased. Results from a miniature water column employed by Bird and Kitting (1982), to study the tracking of vertical migrations of dinoflagellates by *Temora turbinata*, suggested that different types, distributions and concentrations of food organisms modified the behaviour of the copepod. Some supplementary observations on the copepods of the North Sea support this view, since Daro (1988) found that vertical migration is suspended when the phytoplankton concentrations are low.

Fiksen and Giske (1995), in modelling the vertical distribution and population dynamics of copepods, conclude that food supplies may influence population production more through decreased predation, by suppressing the upward diel migration, than through increased growth. They suggest that the intensity of diel migration changes with food density, from no migration during food scarcity, to maximum migration at intermediate food levels, and reduced migration again at high food densities.

The spatial distribution of patches of phytoplankton may modify the diel pattern of migration. As mentioned when discussing rates of egg production, correlations between the spatial distributions, vertically and horizontally, of phytoplankton and populations of herbivorous species of copepods may or may not be present (see also Napp *et al.*, 1988). A strong association with the phytoplankton concentrations infers a cessation of a regular diel vertical migration. The environment at the ice edge in high latitudes is an interphase often rich in ice algae. In the Fram Strait, Smith (1988), for example, found that *Calanus* species and other copepods aggregate in the upper 25 to 50 m of the water column and suspend their diel vertical migrations. Ice-melt caused release of the algae into the water column and formation of a brackish layer under the ice; female *C. glacialis* and *Pseudocalanus minutus* then ceased diel migration (Runge and Ingram, 1991).

11.4.5.4. *Tides*

Acartia longipatella, *A. natalensis* and *Pseudodiaptomus hessei* respond to tidal flows in the Sundays Estuary, South Africa. Wooldridge and Erasmus (1980) do not describe the responses of the *Acartia* species in terms of their vertical migration but rather their horizontal distributions.

Pseudodiaptomus hessei, however, remained associated with the sediment surface during flood and ebb tides but migrated upwards at slack water. Kimmerer and McKinnon (1987b) have shown that the direction of tidal flow cues the vertical migration of *Acartia tranteri* in a shallow bay, Westernport Bay, Australia. In a water column of 30 m, the mean depth of the population of adult *A. tranteri* was, on average, 3 m less than on the ebb than the flood tide. *Eurytemora affinis*, in the shallow Conwy Estuary, North Wales, has a circatidal rhythm with a period of peak activity about 1.5 h before the time of high tide (Hough and Naylor, 1992). Tidal responses are construed as mechanisms for maintenance of the population within a topographic location rather than as cues for a diel vertical migration.

11.5. RHYTHMS

The occurrence of endogenous rhythms within copepods has been reviewed by Forward (1988) in the context of zooplankton in general. Three studies, cited by Forward, found no evidence of an endogenous rhythm of vertical migration in *A. tonsa*. This is a shallow-living coastal species, unlike *Calanus helgolandicus* in which Harris (1963) observed an endogenous rhythm of vertical migration and Svetlichny and Yarkina (1989) found an endogenous rhythm in the activity of the feeding appendages, equated to locomotion. The observations are so few that conclusions about the prevalence or otherwise of the existence of endogenous rhythms in calanoid copepods cannot be drawn. Forward (1988), in reviewing their occurrence in a wide variety of zooplanktonic organisms, considers that they are important for the vertical migration of some species but that other species respond to exogenous factors of the environment.

Endogenous rhythms of swimming activity have been demonstrated in neustonic species of copepods by Champalbert (1978, 1979). This is a restricted environment, associated with the surface film, in which selective pressures will be different from those operable in the deeper water column. Likewise, circatidal rhythms found in *Acartia tranteri* by Kimmerer and McKinnon (1987b) and in *Eurytemora affinis* by Hough and Naylor (1992) have also been developed in restricted environments.

There are many observations of nocturnal peaks in feeding activity in a wide variety of species of copepods at night, reviewed when discussing periodicity of feeding in Chapter 5. They are derived primarily from environmental observations and not under experimental conditions of constant darkness, required to distinguish endogenous from exogenous rhythms. It is also important to consider other factors: Nott *et al.* (1985) suggest that regeneration of the gut epithelium, which is discarded to form the peritrophic membrane of the faecal pellets, may impose a diel periodicity in the feeding regime. Observations on individuals, rather than on groups or populations, are important. The physiological state of one individual varies from that of another at any one time (Båmstedt, 1988b). Kleppel *et al.* (1988b) point out that a high degree of asynchrony in feeding behaviour exists between individuals but, still, patterns exist at the population level. Does this imply that the individual responds to the population or is it simply an expression of the statistical mean of the activity of the individuals?

Consequently, there is a need for controlled observations on a variety of species from a variety of environments to assess the relative importance of endogenous and exogenous rhythms in the behaviour of the copepods.

11.6. CONCLUDING REMARKS

Considerable advances have been made in the understanding of the behaviour of calanoid copepods in recent years but much remains to be done. Energy expenditure on swimming depends to a considerable extent on the density or buoyancy of the copepods. Do they have any control over their buoyancy, within the diel cycle, through mechanisms such as alteration in their lipid stores? This is sometimes inferred (Sargent and Henderson, 1986) but remains to be demonstrated.

Further, definition of nearest neighbour distances at which one copepod responds to another, and responds to other co-occurring inhabitants of their environments, is required. This is especially pertinent where evidence of chemosensory response distances can be determined. Many populations of copepods have large nearest neighbour distances and, consequently, communication between individuals may be dependent upon chemosensory mechanisms. This may not be strictly between individuals but between an individual and the rest of the population, a population "smell".

Much work has been done on deep sound scattering layers (Farquhar, 1970; Andersen and Zahuranec, 1977; Hopkins and Evans, 1979). Sampling of these layers by conventional nets has shown the presence of copepods usually in association with other organisms such as mesopelagic fish and euphausiids. It is only relatively recently that high frequency echosounders have been employed. Richter (1985a) determined target strengths at 1.2 MHz using a mixed population, nauplii and copepodids, of *Calanus pacificus*. Target strengths of individual animals were proportional to their body volume rather than surface area. Volume scattering of populations, on the other hand, appeared to be a complicated function of individual body volumes or body carbon content. Richter (1985b) demonstrates the capability of a 1.2 MHz dual beam transducer to detect small zooplankton in a natural water column, although the actual species registering were not specifically identified. A deep scattering layer, detected with a 150 kHz sounder, occurred at approximately 2000 m depth in association with a hydrothermal plume on the Juan de Fuca Ridge in the northeast Pacific Ocean. Net sampling showed that copepods dominated the biomass, especially *Metridia assymetrica* and *Neocalanus plumchrus*. More recently, Guerin-Ancey and David (1993) adopted a multibeam-multifrequency echosounder with 7 discrete frequencies (75, 80, 90, 100, 110, 120 and 130 kHz). A comprehensive biological sampling programme was carried out simultaneously with the sonic profiles. This technique promises well for observation of populations with low diversity of species and developmental stages. Further development of echosounders may allow observation of a single copepod over time in a natural environment.

Pearre (1979b) presents an excellent review of the current problems facing studies of diel vertical migration of copepods. The sampling strategy is necessarily a compromise between resolution of space and that of time. A third factor is also involved, the resulting numbers of samples to be analysed. He suggests that new methods and approaches are required and some of the experiments done recently on behaviour of individuals have provided new insights. The behaviour of the individual in the sea is extremely difficult to study because of the range of depth at which the species live. Vertically oriented traps that will provide data on the individuals within the population that are actually moving upwards or downwards have been suggested as one approach. Such traps were successfully used by Harding *et al.* (1986), although they suspected that a proportion of the copepods migrating downwards might be avoiding the traps. Observations from submersibles are difficult because of the size of the individuals and impossible if lights have to be used. Examining components of the population, such as egg-bearing females, males, individual copepodid stages, the proportion of individuals feeding, have provided new information, not least of which is the recognition of the resting copepodids of overwintering populations.

Bohrer (1980) examined the vertical migrations of a number of species under experimental conditions in a tower tank. *Temora longicornis* and *Pseudocalanus minutus* show different bathymetric centres of abundance interpreted by Bohrer as a possibility that the species are reacting to each others' presence in the tower. The question arises whether such responses would be at the individual level, where much variation exists, or at the population level which Bohrer suggests. It may be, however, that these two species respond differently to the same environment and not primarily to each others' presence. The whole subject of one species' response to the presence of another, not just in predator-prey situations, but also those of co-occurring species including congeners, requires investigation.

The reasons why copepods perform diel vertical migrations have been discussed many times (Rudjakov, 1970; Lampert, 1989; Ohman, 1990), sometimes simplistically. An interesting study by Hays *et al.* (1994) examines correlations between body size, shape and colour and the presence of a diel migration. They found that larger species, >1 mm body width, were more likely to be migratory than species with a width <1 mm. Among the larger species, elongated copepods, such as *Eucalanus elongatus* and *Rhincalanus nasutus* were weak migrators. Carotenoid content was only important in the small copepods, those that were pigmented being more likely to migrate. Modern studies show that the reasons are often complex because the copepod is reacting to its physical, chemical and biological environment as well as frequently having specific physiological and metabolic requirements that have to be satisfied. As Ohman (1990) concludes, diel vertical migration is dynamic rather than a fixed, invariant behavioural trait within a population of a species. Consequently, the copepod will migrate to feed, breed, avoid visual predators or obtain a net gain in energy. It will suspend migration to exploit food patches, avoid predators, conserve energy or overwinter at depth. The priorities change between species, individuals, age groups, seasonally, and between different environments and populations. Frost (1988) concluded that in the population of *Calanus pacificus* in Dabob Bay, Washington the diel migration was not a foraging strategy optimizing individual growth rate but was unambiguously oriented towards avoidance of predators. Other populations will show the same and other priorities.

12. Distributional Ecology

12.1. Biomass of the Copepod Fauna 457
    12.1.1. Coastal and Shelf Regions 457
    12.1.2. The Oceanic Water Column 458
12.2. Associations of Copepods 460
    12.2.1. Diversity 461
    12.2.2. Dominance of Species 464
    12.2.3. Niches 466
    12.2.4. Concluding Remarks 470
12.3. Copepods of Pelagic Environments 470
    12.3.1. Fronts and Eddies 471
    12.3.2. Brackish Water 472
    12.3.3. Coastal and Shelf 479
    12.3.4. Oceanic 483
12.4. Restricted Environments 487
    12.4.1. Caves 488
    12.4.2. Neustonic 488
    12.4.3. Under Ice 490
    12.4.4. Hyperbenthic and Benthopelagic 496
    12.4.5. Commensal Species 503

As mentioned in the introduction, copepods are probably the most numerous multicellular organisms on earth. They are distributed throughout the oceans and their margins and extend into fresh waters. This account is confined to species that occur in the marine and brackish environments. Copepods have evolved, not in isolation, but as components of evolving ecosystems. Consequently, their life history strategies and distributions have been shaped, not only by the physical and chemical parameters of the environment, but also by their reactions and adaptations to the ecosystems in which they were and are evolving. Longhurst (1985) concludes that the structure of ancient ecosystems was probably no different from modern ones and that our current understanding can be extrapolated when considering selection pressures on evolving copepods.

12.1. BIOMASS OF THE COPEPOD FAUNA

Measurement of the biomass of copepods per unit area or volume of the water column in terms of wet or dry body weight or body carbon weight is of interest in studies of biological production. Biomass measurements are a product of the number and weight of copepods present but are sometimes measured as displacement volume m^{-3} (e.g. Allison and Wishner, 1986). This is an easier measurement to make, the copepods being allowed to settle in a container and the settled volume determined. As Wickstead (1961) points out, biomass measurements, equivalent to measurements of standing stock, can not be directly compared between tropical and high-latitude waters. Allowance has to be made for the higher rates of development and production in the warmer regions where the copepods are smaller in body size, with shorter generation times, than those in high latitudes. Wickstead reviews estimates of biomass of copepods in different geographical regions and finds that they can range from a few $mg\,m^{-3}$ in low latitudes to a seasonal maximum of about 5 g fresh weight m^{-3} in high latitudes. Estimates of biomass are frequently made from counts of individuals through transforming them, using body length to body weight regressions. Such regression equations are given in Tables 30 to 33 (pp. 223–232). Biomass, here, will be discussed in terms of numbers of copepods.

Densities of individuals within temperate and high-latitude populations, and those within populations in tropical and subtropical upwelling regions, fluctuate seasonally with consequent seasonal changes in their biomass. Degrees of aggregation within populations change seasonally and contribute to the variation. The largest source of error in estimating biomass arises from the general patchiness of distribution of individuals in both the vertical and horizontal planes. This patchiness is compounded when considering the biomass of a species in a region because the populations themselves are patchily distributed. Consequently, estimates of the biomass of the copepod fauna, as a whole, in a region can be only approximate. Examinations of the seasonal fluctuations in a copepod fauna, such as that of Regner (1984) in the central Adriatic Sea, are of limited value unless the dominant species are named.

12.1.1. Coastal and Shelf Regions

The densities of copepods vary greatly in coastal and shelf regions of the world. In the North Sea, for example, average annual densities ranged from a minimum of about 150 to a maximum of about 2000 m^{-3} of a 53 m water column (Roff *et al.*, 1988). Densities as high as 7000 individuals m^{-3} occur seasonally in some years. Coyle *et al.* (1990) found seasonal densities of up to 10 000 copepods m^{-3} in Auke Bay, Alaska. The densities measured in aggregations (Table 66, p. 426) range from 1000 to 10,000 m^{-3} and are common in restricted regions of, or at restricted times, in the columns in coastal regions.

12.1.2. The Oceanic Water Column

Zooplankton is irregularly distributed vertically in the surface 1000 to 1500 m of the oceanic water column, there being layers of increased concentration. The density of zooplankton then decreases exponentially below this surface region to depths of up to 8000 m (Koppelmann and Weikert, 1992). The polymodal distribution of the zooplankton in the upper 1000 to 1500 m is reflected in that of the copepod fauna which normally dominates the offshore plankton. The copepods frequently show several maxima of numerical abundance, between the surface and about 250 m depth, and again deeper at about 400 to 700 m depth, and these maxima are followed by a more regular decrease in abundance to bathyal and abyssal depths (Davis and Wiebe, 1985; Beckmann, 1988). Such fluctuations are shown for two species of copepods in Figure 86 (p. 441).

The numbers of copepods m^{-3} in the upper 200 to 300 m layer are variable, but a range of 20 to 60 individuals is average (Figure 89). The density decreases irregularly to a few individuals at 1000 m depth. The decrease then becomes exponential to 8000 m where only 5 to 13 individuals have been found in 100 m^3 of water (Beckmann, 1988). The numbers of genera and species also vary irregularly with depth to 1000 m (Figure 90). The influence of the diel migration of species is clear in the upper 300 m of the water column where more genera and species occur at night and this is reflected in greater numbers of individuals also being present.

Seasonality in temperate and high-latitude environments has a marked effect on the numbers of copepods occurring in the epipelagic and mesopelagic environments, that is between the surface and about 300 to 500 m depth. This part of the water column contains the majority of the diel, seasonal changes in vertically-migrating species strongly influenced by the surface phytoplankton production. Marked seasonal changes in water temperature occur in the upper 250 m of the water column but their amplitude decreases with increasing depth and this is reflected in decreases in the amplitudes of the seasonal range of population parameters of the copepods (Table 71, 460). The availability of food also changes. Surface copepods exploit, directly or indirectly, the phytoplankton production which is irregular in time and space.

The deeper living species also exploit the phytoplankton production as it is transferred to depth either through sedimentation, faecal pellets or

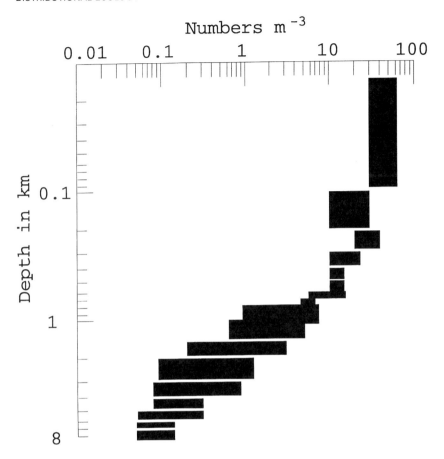

Figure 89   The approximate ranges in numbers of calanoid copepods m^{-3} at different depths in the oceanic water column. Collated from: Wheeler, 1970; Deevey and Brooks, 1977; Wishner, 1980a,b; Weikert, 1982b; Scotto di Carlo *et al.*, 1984; Davis and Wiebe, 1985; Beckmann, 1988; Roe, 1988; Weikert and Trinkaus, 1990.

downward movement of living or dead plankton that had direct access to the phytoplankton. These processes result in a more regular supply of food to the deeper-living species, although at markedly lower concentrations. The positive correlations in Table 71 are probably influenced by the decreasing environmental temperature with depth while the source of the negative correlations is probably the increasing environmental stability present with increasing depth. Thus in the deeper layers there are relatively constant numbers of copepods, although fewer and so at lower population densities.

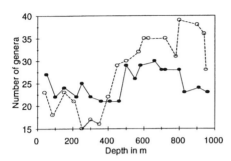 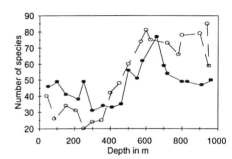

Figure 90   The numbers of genera, left, and species, right, recorded by day (open symbols) and night (black symbols) at different depths by Roe (1972a) off the Canary Islands.

Table 71   Positive and negative correlations between various life history parameters and depth of 12 *Euchaeta* species in the 2000 m water column of the Rockall Trough, northeastern Atlantic Ocean. (Mauchline, 1995.)

Positive correlation	
Prosome length within a species	
Spermatophore length	Decreasing
Egg diameter	temperature
Increasing generation time	
Negative correlation	
Variation in prosome length	Increasing
Variation in spermatophore length	environmental
Spermatophore production	stability
Amplitude of seasonal change	

12.2. ASSOCIATIONS OF COPEPODS

The description of marine environments as species-rich or species-poor in copepods does not take account of the other plankton co-occurring with the copepods and often influencing the diversity of copepods found there. Copepods are potential competitors among themselves for living space and resources and also compete directly with other members of the zooplankton. Major potential competitors of calanoid copepods in coastal situations are often cyclopoid copepods that can occur in vast numbers. The calanoids are also subject to predation that can modify their potential population sizes. Consequently, they do not live in isolation from the other planktonic organisms.

There are three interwoven topics to consider when examining the diversity of the faunas of calanoid copepods occurring in different parts of the marine environment. The first is the species richness of the copepod fauna, the second is the persistence or otherwise of the order of dominance of the species in the fauna and the third is the idea that individual species live within distinct and identifiable niches.

12.2.1. Diversity

Different numbers of species occur in different environments. For example, in a brackish pond the copepod fauna may be dominated by a single species and a second or third species may occur only occasionally whereas a sample of oceanic copepods usually consists of 100 or more species. The fauna may be dominated by a single species or by a single species plus several subdominant species, all the rest being rare. Consequently, there have been several attempts to produce an index that describes, quantitatively, the numbers of species and of individuals of each species present in an environment (Frontier, 1985). Such indices have been compared and reviewed by Heip and Engels (1974). In the following discussion, diversity and species richness, the number of species occurring in a region or water column, are used synonymously.

Identification of all calanoid copepods in a sample is often difficult and rare species may, on occasion, be overlooked and other species sometimes confused. Consequently, numbers of species determined in the sample or geographical region are often minimum numbers.

Diversity generally decreases from neritic coastal waters to inner regions such as enclosed bays, fjords and estuaries while the corresponding biomass tends to increase (Sautour and Castel, 1993). Diversity generally increases along the transect from neritic coastal waters to the open sea while the corresponding biomass decreases (Binet and Dessier, 1972; Regner, 1976; Sander and Moore, 1978). Diversity further increases with depth between the surface of the ocean and about 1500 m (Binet and Dessier, 1972), but decreases again below about 1500 m (Table 72, Deevey and Brooks, 1977). The species richness of the fauna in the pelagic water column at depths greater than 2000 m is considered to be large but decreasing with depth. No definitive data are available as very large samples are required to sample this sparse fauna adequately. Wishner (1980b), however, has counted the number of genera and species occurring in the benthopelagic environment within a few metres of the sea bed at three depths (Table 72). This fauna has a number of endemic species as well as downward extensions of the populations of pelagic species from the water column above. Examples of the numbers of species and genera occurring in a variety of environments

Table 72 The numbers of species and genera occurring in different environments. The depths sampled in oceanic environments are given. D, day hauls; N, night hauls.

Environment		Depth (m)	No. of species	No. of genera	Authority
Estuaries					
E North America	Igloolik		6	4	Turner, 1981
	Passamaquoddy Bay		15	11	Turner, 1981
	Woods Hole		17	13	Turner, 1981
	Block Island Sound		23	15	Turner, 1981
	Beaufort, NC		10	9	Turner, 1981
	North Inlet, SC		7	6	Turner, 1981
	Biscayne Bay		8	6	Turner, 1981
Coastal inlets					
Maizuru Bay, Japan			7	3	Ueda, 1991
Kumihama Bay, Japan			7	3	Ueda, 1991
Shijiki Bay, Japan			3	2	Ueda, 1991
Port Phillip, Westernport Bays, Australia			18	14	Kimmerer and McKinnon, 1985
Coastal bays					
Auke Bay, Alaska			16	11	Coyle et al., 1990
Discovery Bay, Jamaica			43	26	Webber and Roff, 1995a
Kaštela Bay, Adriatic			32		Regner, 1976
Arabian Gulf			26	15	Michel and Herring, 1984
Eastern India			32	23	Nair et al., 1981

Table 72 Continued.

Environment	Depth (m)	No. of species		No. of genera		Authority
		D	N	D	N	
Oceanic						
Off Canary Islands	40					
Northeast Atlantic	100	22	41	19	28	Roe, 1984
(44°N, 13°W)	250	36	46	28	29	Roe, 1984
	450	72	68	39	35	Roe, 1984
	600	76	67	38	36	Roe, 1984
	100–600	104		46		Roe, 1984
Sargasso Sea	0–500	88		49		Deevey and Brooks, 1977
	500–1000	178		57		Deevey and Brooks, 1977
	1000–1500	163		54		Deevey and Brooks, 1977
	1500–2000	111		38		Deevey and Brooks, 1977
Central Adriatic		57				Regner, 1976
Andaman Is., Indian Ocean		43		26		Madhupratap et al., 1981a
Benthopelagic environment						
San Diego Trough	1100–1200	68		27		Wishner, 1980b
E Tropical Pacific	2400–3000	39		17		Wishner, 1980b
Northeast Atlantic	2600–3200	30		16		Wishner, 1980b

is given in Table 72. The numbers of species present near the surface in middle to higher latitudes show diel changes because deeper living species migrate to the surface at night (e.g. Figure 90; Table 72, NE Atlantic data of Roe).

Diversity of copepods generally increases in the transect from high to low latitudes (Hattori and Motoda, 1983). Turner (1981) examined the diversity of the copepod faunas of 49 estuaries of eastern North America between 81°N and 18°N. There was a trend of lower numbers of calanoid species with decreasing latitude but also a tendency for highest diversity to occur in middle latitudes, between 30°N and 50°N. Species richness is highest in physically unstable estuaries and is reflected in the examples given in Table 72. Kimmerer (1991) discusses the potential importance of predation, depth and degree of enclosure in estuaries and embayments as factors determining the diversity of their copepod fauna.

Diversity changes seasonally. Tranter (1973) found that diversity is generally higher in winter than in summer in tropical and subtropical regions of the eastern Indian Ocean and Regner and Regner (1981) find the same in the central Adriatic Sea.

12.2.2. Dominance of Species

The copepod biomass in a region is usually dominated by a few species (Figure 91). One species is usually more common than any of the others but there may be up to 10 to 15 other species that are subdominant, all the rest being rarer. Hayward and McGowan (1979), McGowan and Walker (1979, 1985) and McGowan (1990) found surprising stability, in the North Pacific central gyre, in the order of dominance of the species. They examined the percentage similarity indices of samples relative to the intervals of time between their collection, including comparisons of day and night samples. Similarity indices were relatively constant at about 60 to 80% over the total time span of 24 and 42 months. None of the rare species ever became dominant or subdominant in the samples. Potential changes in dominance were examined relative to horizontal distance between the samples. A northern transect along 155°W, with sample intervals of 0.6 to 10 km and 18 to 160 km showed no substantial changes in dominance between 26°N and about 40° N. The sampling transect then passed through the transition zone at about 40°N to 44°N and entered the subarctic water masses where a substantial change in the copepod community was evident in samples north of 45°N. The semitropical central gyre community had changed to the subarctic community.

The relative stability of the copepod community in the North Pacific central gyre is not accompanied by comparable stability in shelf or coastal regions. McGowan and Walker (1985) and McGowan (1990) found that samples collected at intervals of up to 24 h and at spatial distances of 10 km in the California Current showed large changes in the order of dominance of species. Percentage similarity indices ranged between 12 and 80%. Samples taken at time intervals of about 72 to 120 h apart had indices ranging from almost zero to about 43%, indicating the replacement of one fauna with another. Patchiness in the two regions was on different scales. Haury (1976b) found that patches in the central gyre are larger than a few hundred metres while those in the California Current were probably less than 100 m.

The deep water column, below 1000 to 1500 m, where there is little seasonality in physical and biological properties, including breeding of the copepods, can be expected to have relatively stable populations (Mauchline, 1991, 1995). This will be reflected in the stability of the order of dominance of the species. The shape of the curves will be similar to those in Figure 91 but the total number of species will be smaller. One of the current problems with this habitat is the identification and enumeration of the large numbers of species likely to occur within the samples. Further, the samples have to be quite large, filtering 10^4 to 10^5 m^3 of water, to sample the fauna effectively. Wishner (1980a) examined dominance in the benthopelagic environment (Figure 92) at three different depths. The numbers of species decreased with depth. The northeast Atlantic station is interesting because of the abbreviated dominance present; the commonest species is only 12 times commoner than the rarest.

The curves shown in Figure 91 are for oceanic water where the number of species is large relative to those in shelf and coastal areas where the fauna usually consists of 20 to 30 species. Seasonal changes in dominance in coastal waters will be quite pronounced, there frequently being cold- and warm-water species present. Each dominant species maintains its dominance for periods of between 3 weeks and two or three months, depending upon the species involved. Many shelf and coastal faunas are subject to advective and dispersive processes changing the copepod fauna on a seasonal, shorter or longer time-scale. Mackas and Sefton (1982) examined the species composition and distribution in an open area off southern Vancouver Island during two successive summers. Large between-cruise variation in the geographical location of individual species was present. There were consistently, however, strong, alongshore and across-shelf changes in the composition of the copepod fauna. They concluded that advection was the important factor controlling the associations. This variability also applies to open bays with marked exchange with adjacent shelf waters. Koslow (1983) concluded that, in an area such as the North Sea, zooplankton abundance may be regulated by physical forcing rather than by predation by planktivorous fish as sometimes occurs within estuaries or bays.

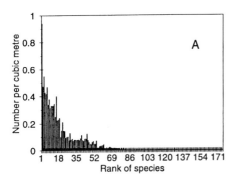 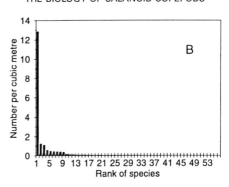

Figure 91 The numerical dominance of copepod species in the water column. A, the abundances of 174 species caught in seven cruises at depths of about 300 m in the North Pacific central gyre (after McGowan and Walker, 1985). B, the abundances of 56 species caught in the 100 m and 250 m depth layers at 44°N, 13°W in the northeast Atlantic (after Roe, 1984).

There are, however, some relatively stable habitats in coastal regions where advective and dispersive processes are minimized. Fjords and enclosed embayments often have restricted faunal compositions, often of the order of 10 species of copepods. The seasonal order of dominance of species is often consistent from year to year (e.g. Coyle *et al.*, 1990). Fulton (1984) found such consistent seasonal succession in estuaries of North Carolina, although control of community structure was seasonally determined by predation by planktivorous anchovies.

Estuaries, by their nature, have gradients of temperature and salinity and the number of species occurring in them is restricted. The fauna changes along the salinity gradient, species having different degrees of euryhalinity.

12.2.3. Niches

The large number of species of copepods occurring in the oceanic water column, coupled with the occurrence of as many as 10 to 15 congeneric species, raises the question as to how so many species exist together. Bainbridge (1972) suggested that two attributes permitted the co-existence of congeneric species, viz. different patterns of vertical migration between highly stratified water masses and size differences between species. It is likely, however, that the situation is more complex than this and additional factors must be involved.

Do copepod species partition the trophic resources of the water column? The evidence for niche separation of species of pelagic copepods in the open

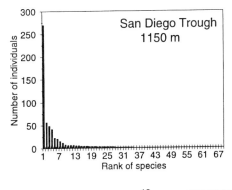
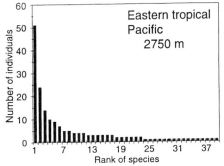
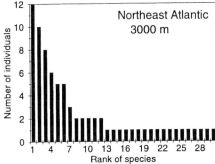

Figure 92 The numerical dominance of species in the benthopelagic environment at different depths and geographical locations. Samples were taken at 10 to 100 m above the bottom. (After Wishner, 1980b.)

ocean is very weak (McGowan and Walker, 1979, 1985; Hayward, 1980). Ambler and Miller (1987) examined the species in the upper 500 m of the water column of the North Pacific central gyre and concluded that congeneric species consistently had different depths of modal occurrence during day and night. A study, however, of five species of *Pleuromamma* in a transect between Honolulu and San Diego by Haury (1988) found no vertical separation of the species at night and only a division between the depth of occurrence of the smaller species and that of the larger during the day. Hayward (1980) found that copepods were not necessarily more abundant in parts of the North Pacific central gyre where more food appeared to be available. Twelve euchaetid species had different modal depths of occurrence in the 2000 m water column in the Rockall Trough (Mauchline, 1995), but there was much overlap in their vertical ranges. The vertical spread of the *Pleuromamma* species in the Gulf of Mexico is interpreted by Bennett and Hopkins (1989) as ensuring sanctuary for species maintenance in a predatory environment.

The above are all examinations in the vertical plane. The study of the Indian Ocean Candaciidae by Lawson (1977) is in the horizontal plane, or geographically, because all 18 species are epipelagic. Enough data were present to group 15 of the species on the basis of their distribution and morphology of their feeding appendages. Species within groups showed preferences for higher or lower environmental temperature, concentrations of food and zooplankton densities. In addition, the distributions also related differently to the oxygen minimum layer and salinity. The diel vertical migration behaviour of species also showed differences. Lawson (1977) suggested that these copepods may be specialist feeders on chaetognaths and each species may have a predilection for certain prey species. *Centropages typicus* dominated the other four *Centropages* species occurring off the northeastern United States (Grant, 1988) but the latter four species were separated in space and time.

It is not just congeneric species that may compete for resources. There are relatively few species in high latitudes where there may be several dominant-subdominant species. Williams (1988) found that the four species dominating the biomass in the surface 500 m of the northern North Atlantic, *Calanus finmarchicus*, *Pareuchaeta norvegica*, *Metridia lucens* and *Pleuromamma robusta*, had different modal depths of occurrence. In addition, their seasonally different reproductive periods, different diets, and different migratory behaviour further minimized competition. A similar vertical separation of species in the 3000 m water column of the Greenland Sea gyre is indicated by Richter (1995) and by Atkinson *et al.* (1992a) for Antarctic species near South Georgia. The large subarctic copepods, *Neocalanus cristatus*, *N. flemingeri*, *N. plumchrus* and *Eucalanus bungii* appear to partition the water column. Mackas *et al.* (1993) suggest that they are responding to the local intensities of turbulent mixing in the water column but that their feeding strategies may also be involved. Mackas (1984) had already shown that variability in a continental shelf ecosystem is greater across the offshore axis than along the axis parallel to the shore.

The young stages of a calanoid species are also potential competitors among themselves. The ontogenetic migrations performed by some, or the different modal depths of occurrence of the copepodid stages, may function to partition resources (Krause and Trahms, 1982; Tande, 1988b; Verheye and Field, 1992; Lopez *et al.*, 1993). Different body size alone is not a good indicator of lack of potential competition between species because size-related functions of feeding, development and growth rates may vary between species (Frost, 1980).

Potential competition between species also exists in shallow coastal regions, including estuaries. Ueda (1991), studying three coastal bays in Japan, sampled from the mouth to the head of each bay. The modal distance of each species from the most seaward station was determined; this is analogous to the modal depth of occurrence of a species in a water column. The overlap of the distributions of the species is great although their modal distances of occurrence are different. Brackish water species are restricted to the less saline parts of the bays but the factors determining the distributions of the other species are not clear. Concentration of food and/or predation pressure, rather than salinity distributions, are suspected as the controlling factors. Salinity distributions are undoubtedly important in determining the gross distributions in estuaries and embayments but do not seem to separate congeners into separate niches (Madhupratap, 1980). Estuaries are often subject to strong seasonal changes in their ranges of salinity, this especially being true of those experiencing seasonal monsoons. There is a seasonal succession in the order of dominance of the species of copepods (Wooldridge and Melville-Smith, 1979; Greenwood, 1981; Goswami, 1983) so decreasing competition. Hodgkin and Rippingale (1971), however, found that the extremely euryhaline *Gladioferens imparipes*, whose nauplii are severely predated by two less euryhaline species, *Acartia clausi* and *Sulcanus conflictus*, is restricted to the low salinity regions of the Swan River Estuary, Western Australia. Lakkis (1994) suggests that various factors are active in the coexistence of six *Acartia* species in Lebanese coastal waters as do Rodriguez and Jiménez (1990) for three *Acartia* species in Malaga Harbour, Spain. The genus *Acartia* is taxonomically complex and may consist of subgenera (Steuer, 1923). Ueda (1987a) suggests that coexisting *Acartia* species belonging to the same subgenus are segregated only in space while those belonging to different subgenera are segregated seasonally. Seasonal succession of species has been found in many areas (e.g. Ambler *et al.*, 1985). An early study by Tranter and Abraham (1971) on the occurrence of nine species of *Acartia* in the Cochin Backwaters takes account of the structure of the mandibles of each species. Some species are more sensitive to changing levels of food concentration than others and Skiver (1980) demonstrated interaction between *Acartia hudsonica*, *Temora longicornis* and *Pseudocalanus* sp. and less specialization in their diets as food became scarcer.

Demersal or benthopelagic species live in close association with the sea bed where niches are much more apparent. Swarms of such species, discussed earlier, often position themselves in different regions of coral reefs. Jacoby and Greenwood (1991) describe the distributions of four species of *Pseudodiaptomus* and nine species of *Stephos*, that are potential competitors on the Great Barrier Reef and in Moreton Bay, Queensland. The numbers of each species, whose mean body sizes are different, caught in traps show substratum-related, seasonal, lunar and diel variations. Ohtsuka *et al.* (1996b) point out that congeneric species exist together in both shallow and deep environments and that they belong primarily to the Aetideidae, Arietellidae, Diaixidae, Phaennidae, Pseudocyclopidae, Pseudocyclopiidae, Ridgewayiidae and Stephidae.

12.2.4. Concluding Remarks

The size of a population of copepods can be controlled by predators (e.g. Davis, 1984a), especially in shelf, coastal and relatively enclosed environments where species richness is reduced. The predators often select specific size categories of prey and consequently some species of copepods in an environment are more liable to predation than others. This will then modify the potential order of dominance within the copepod fauna. The modification may be over a short seasonal period and may not be discerned in changes in the average annual order of dominance. Colebrook (1981) showed persistence in the abundance of zooplankton of the northeast Atlantic over a period of 30 years but there was also a relationship between the size of the overwintering population and the subsequent population in the following year (Colebrook, 1985b). Obviously, the size of a subsequent generation or population of a copepod will depend to some extent, but not totally, on the numbers of adults that survive to breed. Other factors such as viability of the eggs, survival of the nauplii and adequate food have a role. The persistence of an order of dominance within a copepod fauna on a year-to-year basis suggests a degree of 'buffering', resilience, or recovery within the populations of individual species on time scales shorter than annual. *Pareuchaeta norvegica* feeds on overwintering stocks of *Calanus finmarchicus* in Norwegian fjords (Bathmann *et al.*, 1990). A copepod species that is subject to regular seasonal predation must have a portion of its population production that can be consumed by the predator without affecting the sizes of subsequent generations of the copepod in the long term. The reverse must also take place in years when the predation pressure, for one reason or another, is much less than normal. The population of the copepod does not produce subsequent generations of excess size and so the order of dominance remains the same. This thesis is very simplistic and year-to-year fluctuations do take place in stocks of individual species. It would be interesting, however, to understand the source of resilience or buffering within the populations.

12.3. COPEPODS OF PELAGIC ENVIRONMENTS

Calanoid copepods occur throughout the marine and brackish water environments of the oceans. Widely distributed species in coastal waters are eurythermic and euryhaline while those in the deep sea are eurybathic. Many species can not tolerate wide variations in one or more of these variables. Consequently, endemic species adapted to particular variables occur. Physical properties of the environment causing dispersion of populations are more active on pelagic populations in higher than lower latitudes. This arises from two principal sources. The scales of baroclinic eddies are much shorter, of the order of 5 km, in higher latitudes compared with a scale of about 150 km in a subtropical gyre, while generation times of copepods range from about 10 d in low latitudes to about 300 d in high latitudes (Huntley and Niiler, 1995). Consequently, a cohort of copepods in higher latitudes is liable to be dispersed and the species to occur over broad biogeographic regions. These dispersive mechanisms encourage the occurrence of species in isolated populations in such restricted environments as fjords or coastal basins where suitable characteristics exist for their survival.

12.3.1. Fronts and Eddies

The oceans, coastal waters, including estuaries and fjords, have many boundaries or fronts that affect the distributions and rates of production of copepods. Le Fèvre (1986) has reviewed the biology of frontal systems and reference should be made there for detailed information. Defined boundaries can exist between oceanic water masses, coastal water masses, fresh water-marine water interfaces, at pycnoclines and thermoclines, between the edges of eddies or rings, and at sediment-water or air-water interfaces. Aggregations of copepods, and other organisms, tend to form at boundaries or fronts in the vertical or horizontal planes (Petipa, 1985). Kiørboe (1991) and Piontkovski et al. (1995) argue that population production of copepods is greater at spatial-temporal discontinuities in the water column and in regions where such discontinuities are common. An examination of a western Irish Sea frontal system over a period of more than a year (Scrope-Howe and Jones, 1985) concluded that concentrations of copepods within the front were associated with phytoplankton patches. The front rarely influenced the overall production levels of copepods in this region but could modify their diel migration (Fogg, 1985). Boundaries exist on different time scales, some being transient and broken down by mixing, others being more permanent (Petipa, 1985).

The fast-flowing western boundary currents of the oceans transport large amounts of heat, salt, phytoplankton and zooplankton from equatorial into temperate regions. The currents are the Gulf Stream in the North Atlantic, the Kuroshio in the North Pacific and the East Australian Current in the Tasman Sea. Large warm-core eddies or rings break off from these currents on their poleward sides, with copepods and other fauna entrained in them. (Ashjian, 1993; Ashjian and Wishner, 1993). Such eddies, which often measure several hundred kilometres in diameter, can entrain water and organisms at their fronts with adjacent water masses. A warm-core ring of the Kuroshio laterally entrained neritic copepods (Yamamoto and Nishizawa, 1986). On the other hand, because of the persistence in time of some eddies, the copepod fauna within can develop under the conditions of the eddy and be further modified by, for instance, predation within the eddy or intrusions of water plus copepods from outside the eddy into its core (Tranter *et al.*, 1983; Haury, 1984; Davis and Wiebe, 1985; Bradford and Chapman, 1988; Wiebe *et al.*, 1992).

On a much smaller scale, Brylinski *et al.* (1988) examined the discontinuity between open sea waters, flowing eastward, and inshore waters, adjacent to the French coast in the English Channel. Adult *Temora longicornis* were larger in the coastal than in the open sea waters suggesting that the front had been established for some weeks. The copepod populations at fronts are usually different either side of the front while the population associated with the front can be more dense than outside it and modified in composition through reproduction and behaviour of the species. These aspects are discussed by Boucher (1984), Kahru *et al.* (1984), Moal *et al.* (1985), Richardson (1985), Smith and Vidal (1986), Boucher *et al.* (1987), Ibanez and Boucher (1987), Atkinson *et al.* (1990), Saiz *et al.* (1992b), Fernández *et al.* (1993) and Seguin *et al.* (1994). Studies of the faunal structure at fronts have to take account of distributions in space and time, the changing physical parameters of the environment, and the behaviour of the individual species (Fromentin *et al.*, 1993).

All sectors of the marine environment, both onshore and offshore, are influenced to a greater or lesser extent by eddies and turbulence. There are energetic and relatively quiet regions in the oceans and enclosed seas (Piontkovski *et al.*, 1995).

Cyclonic eddies, relatively stable in time, can develop behind a headland in coastal regions (Alldredge and Hamner, 1980; Verheye *et al.*, 1992) but still allowing entrainment of recruits from outside the eddy. Uneven bottom topography, whether in coastal and shelf regions or in the form of seamounts or islands in oceanic regions, can encourage turbulence and the formation of eddies and frontal systems. The effect of the Cobb Seamount, west of Vancouver Island, on the distributions of copepods was examined by Dower and Mackas (1996). Wind can break down pycnoclines and thermoclines, cause upwelling and on- and off-shore currents. All these factors affect the distribution and development of populations of copepods within a region.

12.3.2. Brackish Water

Brackish environments occur in estuaries and fjords, in enclosed bays or lagoons and more or less open bays in the coastline, in an enclosed sea such as the Baltic Sea, and even in offshore regions associated with river outflows such as that of the Amazon or those of some fjordic coastlines.

Brackish environments often exist behind beaches and were once part of the ocean. Similarly, the innermost parts of fjords may become cut off from the outer, fully saline regions. Tuborg Lake in Ellesmere Island, Canadian Arctic, is such a fjordic situation and is a brackish environment with salinities of about 10‰ at the surface and 28‰ below depths of about 50 m extending to 140 m. Two species, *Drepanopus bungei* and *Limnocalanus grimaldii*, have isolated populations present.

Brackish water species can invade fresh waters. Species in the families Temoridae, Centropagidae and Diaptomidae dominate the fresh water calanoid copepods. The marine glacial relict centropagid species *Limnocalanus macrurus* inhabits brackish environments of North Europe and North America but also occurs in fresh water, Wilson (1972) recording it in Cedar Lake at some 474 m altitude in the western foothills of the Cascade Mountains in Washington State, USA. *Eurytemora affinis* is a dominant species in brackish and estuarine waters of northwestern Europe and North America. Around 1958, it has spread, probably in ballast water of ships passing through the St Lawrence River and Erie Canal, to the Laurentian Great Lakes (Saunders III, 1993). It has also spread to reservoirs of the southern Great Plains, but how it has done so is not obvious. Saunders III discusses possible mechanisms in some detail.

One feature of estuaries is the decrease that takes place in the body size of the copepods between the seaward and landward ends. The most important estuarine species are those of the genus *Acartia* that dominate the biomass of most shallow, confined bays and lagoons. Kimmerer (1991) suggests that the spatial pattern of abundance of copepods in enclosed bays and estuaries is consistent (Figure 93). The neritic zone has a species of *Paracalanus* along with other moderate- to large-sized species and numerical dominance within this fauna is not clear. Further into the estuary or bay, a species of *Acartia* is dominant but this in turn is replaced by an *Acartia* species of smaller body size in the innermost region of the bay. In most cases, it co-occurs with a cyclopoid species of *Oithona* and, at higher temperatures, or lower latitudes, with *Parvocalanus crassirostris*. He suggests that this pattern may result from different patterns of recruitment and mortality. The neritic species at the seaward end have a deficit of birth over mortality while the species in the innermost part of the bay have either an excess of birth over mortality or a behavioural trait maintaining them in that region of the bay. Birth minus mortality in the populations of neritic and bay species changes with distance into the bay but in opposite directions (Figure 93). Intermediate species may have intermediate mortality rates. Soetaert and Herman (1994) examined the fate of neritic plankton drifting into the Westerschelde Estuary on the ingoing tides and concluded that they died,

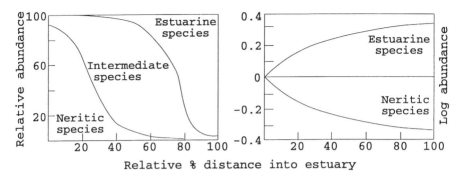

Figure 93 Patterns of distribution of copepods in estuaries and bays. Left, the relative percentage abundance of neritic, intermediate and estuarine species related to the distance into the estuary. Right, the abundance at different distances into the estuary or bay relative to the abundance at the mouth of the estuary or bay of neritic and estuarine species. (After Kimmerer, 1991.)

the apparent standing stock of such species being a result of continuous supply.

Mortality might be expected in an estuary when a tropical storm deposits heavy rain and wind breaks down any stratification and induces greater rates of flushing. Stubblefield and Vecchione (1985) examined such an event in a very shallow estuary discharging into the Gulf of Mexico. The pre-storm abundance of the total population of copepods was greater than that post-storm but there were no significant effects on the *Acartia tonsa* population.

There is a potential sampling problem in estuaries, as in other parts of the marine environment, when trying to quantify the distributions of the species of copepods. Pelagic nets can sample the water column adequately except in the near-bottom few metres. Stubblefield *et al.* (1984) found that a major portion of the population of *Acartia tonsa* was associated with the sediment surface and sampled only by an epibenthic sledge. This benthopelagic population shows diel changes in size that contribute to difficulties of obtaining quantitative horizontal samples in a tidal environment (Lee and McAlice, 1979; Gagnon and Lacroix, 1981).

12.3.2.1. *River Plumes*

Nutrients in river flows enhance productivity of inshore and offshore coastal regions (Chervin *et al.*, 1981). The plume of the Rhône River flows in the surface 1 m offshore from its mouth in the northwestern Mediterranean. This low salinity surface layer is rich in nutrients and overlies oligotrophic deeper waters. Concentrations of phytoplankton tend to gather at the pycnocline to which the copepods from deeper levels migrate to feed (Pagano et al., 1993). The plume is often recognizable for some distance offshore from its outflow from the estuary, as far as 50 to 100 km in the case of the Mississippi River in the northern Gulf of Mexico (Dagg and Whitledge, 1991; Tester and Turner, 1991). The copepod faunas of the Hudson River estuary and its subsequent plume in Long Island Sound have different dominant species (Stepien et al., 1981). Within the estuary, *Eurytemora affinis* and *Acartia tonsa* dominate but, more seaward, tend to be replaced by *Eurytemora americana* and *Acartia hudsonica*. The plume outside the estuary is dominated by *Pseudocalanus minutus, Centropages typicus, Temora longicornis, Acartia tonsa* and *Paracalanus parvus*. The rank order of dominance of the species changes seasonally in each region.

In the Mississippi plume in the Gulf of Mexico, *Acartia tonsa* is dominant up to 100 km offshore (Tester and Turner, 1991). A detailed study of the copepod faunas of this plume and regions adjacent to it in the Gulf of Mexico has been made by Ortner et al. (1989).

12.3.2.2. Estuaries

Copepods, both calanoid and cyclopoid, dominate the fauna of estuaries numerically and by weight. The copepod fauna is variable in composition being, first of all, dependent upon the geographical location of the estuary and, secondly, upon the size, topography and degree of mixing with adjacent coastal waters. Reviews of the general characteristics of estuaries are given in Ketchum (1983) and Heip et al. (1995). An estuary has a region where salinity is reduced and it is here that the endemic species of such genera as *Acartia, Eurytemora* and *Gladioferens* occur. The more saline seaward regions have a wider variety of species related to the adjacent coastal waters.

Average biomass of copepods in estuaries is usually lower than 200 mg C m^{-3} but can be as high as 1 g C m^{-3} (Heip et al., 1995). Densities can be over 10^5 individuals m^{-3}.

The distribution of the species within an estuary is related to salinity but other factors are also important (Bradley, 1991; Wellershaus and Soltanpou-Gargari, 1991). There is considerable variation, in time and space, of the distributions of individual species within the same estuary and between estuaries that are not explained by variations in salinity and temperature alone. The salinity/temperature tolerances of individual species are not well known although those of *Eurytemora affinis* have been investigated by Roddie et al. (1984) and Gonzalez and Bradley (1994), of *Paracalanus aculeatus* by Bhattacharya (1986), and of *Acartia tonsa* by Tester and Turner

(1991). The latter authors conclude that *A. tonsa* may be restricted to estuarine habitats because its nauplii survive best at salinities less than 25‰ and temperatures greater than 15°C.

The species can be classified into four types in a transect from the landward to seaward regions (Collins and Williams, 1981):

a. truly estuarine species restricted to regions of very low salinity, close to 1‰;
b. estuarine and marine, the intermediate species in Figure 93;
c. euryhaline marine species;
d. stenohaline marine species.

Seasonal succession of dominants in estuaries takes place and is controlled by the interaction of temperature and salinity but other factors, as yet undetermined, are also involved (Wooldridge and Melville-Smith, 1979). Studies showing seasonal succession of species and changes in the rankings of the dominant species in estuaries are those of Wooldridge and Melville-Smith (1979), Stepien *et al.* (1981), Collins and Williams (1982), Turner *et al.* (1983), Ambler *et al.* (1985), Bradley (1991), Soetaert and Rijswijk (1993).

Predators, invertebrates such as mysids as well as fish, can select one species relative to another (Fulton III, 1983) and potentially alter the proportions of their occurrence. Predation by *Acartia tonsa* on nauplii of other species aided changes in its dominance in an estuary near Beaufort, North Carolina (Fulton III, 1984).

Estuarine species may have behaviour patterns that maintain them within estuaries. They can associate with the sediment surface during the ebb tide but move upwards into the water column on the flood tide (Kimmerer and McKinnon, 1987b; Hough and Naylor, 1991, 1992). Current velocity increases with distance off the bottom so that, by selectively migrating upwards and downwards during different stages of the tidal cycle, a species can alter its position within the estuary (Wooldridge and Erasmus, 1980). An anadromous migration of *Eurytemora affinis* is described in the Elbe estuary by Heckman (1986). Adults and late copepodids inhabit the brackish water region of the estuary in autumn and winter. The increased fresh water run-off and rising water temperatures in late winter or early spring cause *E. affinis* to migrate to the shallower, fresh water regions of the estuary to breed. The resulting nauplii drift back to the brackish water section where they develop into copepodids. Heckman maintains that the adults of each subsequent generation migrate to the fresh water region to breed. Soltanpour-Gargari and Wellershaus (1987), however, concluded that, in the Elbe and Weser Estuaries, the distribution of this species was probably controlled entirely by the occurrence of salinities of 0.5 to 1.0‰. A study of the distribution of *E. affinis* in the Gironde Estuary, southwest

France, suggests that behavioural traits are not involved but that the hydrodynamic processes alone within the estuary explain its distribution and retention (Castel and Veiga, 1990). The effects on development times and breeding of *E. herdmani* of different ranges of environmental temperature and salinity were investigated by George (1985). Higher temperature reduced age at sexual maturity while there was a window of salinity between about 20 and 30‰ that favoured maximum fecundity.

Various aspects of the biology, persistence and impact of the copepods within the ecosystems of estuaries are discussed by Gagnon and Lacroix (1981, 1983), C.B. Miller (1983), Cummings and Ruber (1987), Stearns *et al.* (1987), Ough and Bayly (1989), Tackx *et al.* (1990), Jouffre *et al.* (1991), Castel (1993), Heip *et al.* (1995).

12.3.2.3. *Fjords*

Fjords are deep, often greater than 200 m depth, estuaries carved out by glacier action in middle to high latitudes. They occur in the coastlines of western Scandinavia, Scotland and Iceland at latitudes of 55°N to 72°N, the coastlines of Greenland and eastern Canada between 50°N and 75°N, the western coast of North America, from 50°N to 60°N, the western coast of South America, from 40°S to 55°S, and those of western New Zealand between 45°S to 47°S. Fjords penetrate for some distance inland, a few as much as 150 km, and can have one or several sills that may restrict their communication with adjacent coastal water. The topography and biology of fjords is reviewed in Freeland *et al.* (1980).

Fjords with sills often have deep water basins that have depths greater than the daytime depths of species of copepods in not only adjacent coastal regions but also adjacent oceanic regions. This is illustrated in Figure 94 where an oceanic population, during its diel vertical migration, can be advected on to the shelf and into a fjord where it can survive in the deep basins. These basins can form refugia for species that cannot survive the average depths of the coastal shelves. Their penetration into, or carriage out of, the fjords is determined by the exchange characteristics at the time of advection (Lewis and Thomas, 1986; Aksnes *et al.*, 1989; Krause and Kattner, 1989; Kaartvedt and Svendsen, 1995). Most fjords receive the discharges of one or more rivers so that there is a net outflow, usually over the surface of the saline deeper layers. These saline deeper layers may be stable for periods of weeks or months so that the populations of copepods in them are isolated. The copepods may not enter the out-going surface layers because of the low salinity present there. Immigration during this period is often possible as the tidal, saline inflow enters the fjord, over the sill and under the fresh water surface layer, mixing down into the trapped

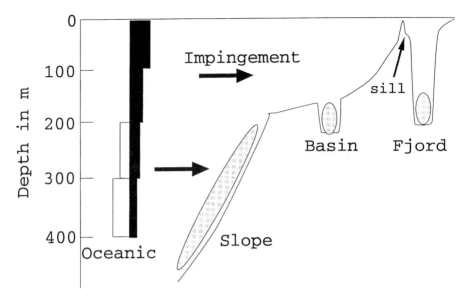

Figure 94 Schematic diagram of the impingement and survival of an oceanic species on the continental shelf and in a fjord. Lateral impingement can take place from the mesopelagic into the benthopelagic environment of the slope. Lateral impingement at night during the diel migration (black histogram) carries it over the shelf but, because the day-depth of the population (open histogram) is deeper than depths at the edge of the shelf, it finds refuge in the basins of the shelf and fjords. There, the deepest parts have environmental conditions equivalent to those in its upper daytime depths in the ocean.

saline water of the inner basin or basins (Ladurantaye *et al.*, 1984; Lindahl and Hernroth, 1988).

The conditions inside the fjord determine which immigrants survive and which do not. Stone (1980) lists all species occurring in the Queen Charlotte Strait, northern Vancouver Island and those that survive inside the fjord, Knight Inlet in British Columbia. Cold-water species in the Strait showed higher survival within the fjord than warm-water species. Thus, the copepod fauna of fjords reflects the community outside it but may be modified (Matthews and Heimdal, 1980; Hirche, 1984). The fauna of adjacent fjords, similar topographically, can vary (Fosshagen, 1980).

One difference between the fauna of a fjord and its adjacent coastal waters may be a result of its function as a refuge for deep-water species not normally occurring in coastal regions. Fjords can have depths of 500 m or more in their basins and these are suitable as habitats for mesopelagic and bathypelagic, offshore species. The slope species, *Pareuchaeta norvegica*, has established populations in the deeper fjords and sea lochs of Norway and Scotland. Oceanic species such as *Eucalanus bungii, Heterorhabdus tanneri, Neocalanus plumchrus, Scaphocalanus brevicornis* and *Spinocalanus brevicaudatus* have been recorded frequently in fjords. Such species occur in Norwegian fjords (Giske *et al.*, 1990), in fjords in British Columbia (Koeller, 1977; Krause and Lewis, 1979; Stone, 1980; Mackas and Anderson, 1986; Shih and Marhue, 1991), Chile (Hirakawa, 1989), and South Georgia (Ward, 1989). Most fjords investigated in Scotland, Iceland, Greenland, Chile and New Zealand are always more or less oxygenated but some in western and southern Scandinavia and in British Columbia are permanently anoxic or are flushed at intervals of several years (Jensen *et al.*, 1985). These latter fjords can often have relatively few species of copepods present even though the outside community is relatively rich.

Fjords can have relatively simple food webs because of the restricted numbers of species present. Consequently, they are attractive as natural enclosures for the study of ecological processes. Such studies are those of Giske *et al.* (1990) and Bollens *et al.* (1992a) on the trophic interactions between zooplankton and fish and many of the studies on the Balsfjorden and Korsfjorden in Norway (see papers by Bakke, 1977; Båmstedt and Holt, 1978; Matthews and Heimdal, 1980; Grønvik and Hopkins, 1984; Båmstedt, 1988a; Hopkins *et al.*, 1989; Båmstedt *et al.*, 1990).

12.3.3. Coastal and Shelf

Coastal species of copepods belong to a wide variety of genera. There are neritic species close to the shore that form populations within bays and the outer parts of estuaries. This fauna gradates into a shelf fauna that is a mixture of coastal neritic species and more open water, epipelagic species. The controlling factors that determine which species occur are temperature regimes, the seaward extent of the shelf and the depths of water available in the water column. Some shelf regions are broad while others are narrow. Broad regions such as those in the northeast and northwest Atlantic and the western Pacific have diverse faunas that become less diverse in higher latitudes. The copepods of narrow shelves tend to be dominated by species advected from the adjacent oceanic areas. There was no clear separation between neritic and oceanic communities off northeast Tenerife (Corral Estrada and Pereiro Muñoz, 1974).

The shelves are turbulent environments with along-shore currents. Diel vertical migration coupled with the variations in speed and depth of the surface under-currents determine whether a species remains on or is moved off the shelf (Binet, 1977; Boucher, 1988). The bottom topography can cause formation of eddies while fresh water run-off and incursion of water masses from adjacent regions result in frontal systems. Offshore banks can be refugia for neritic species and the edges of the banks have concentrations of deeper-living species. Drifting of pelagic copepods over banks or coral reefs at night may subject them to increased predation. The turbulence of the environment encourages patchy distributions of the copepods.

Coastal upwelling of nutrients results in blooms of phytoplankton with consequent effects on the populations of copepods. There are frequently across-shelf gradients of temperature, salinity, depth and tidal forces that result in changes in the fauna of copepods. Some of these gradients may be enhanced seasonally by upwelling events and monsoons as well as by the more widespread normal seasonal changes of light and temperature in middle and higher latitudes. Shelves, therefore, are variable environments encouraging opportunistic species of copepods able to exploit resources that are variable in time and space. Seasonal succession is a constant feature of coastal and shelf environments. It occurs, for example, off Oregon (Peterson and Miller, 1977) and at Syowa Station, Antarctica (Tanimura et al., 1986) where there are winter and summer dominants. In Saronikos Gulf, Greece, the winter fauna is dominated by *Clausocalanus pergens* and *Ctenocalanus vanus*, the summer by *Clausocalanus furcatus* and *Temora stylifera*, and the autumn by *Paracalanus parvus* (Siokou-Frangou, 1996). Seasonal succession of species takes place in upwelling regions, mentioned below.

Shelf species commonly belong to such genera as *Acartia*, *Aetideus*, *Calanus*, *Calanoides*, *Calocalanus*, *Candacia*, *Centropages*, *Clausocalanus*, *Ctenocalanus*, *Eucalanus*, *Euchaeta*, *Labidocera*, *Metridia*, *Microcalanus*, *Neocalanus*, *Paracalanus*, *Pseudocalanus*, *Temora*, and *Tortanus* among others. The species occurring in any one region depends to a considerable extent upon its geographical location.

12.3.3.1. *Across-shelf Gradients*

Many species of copepods living on the shelf extend their distributions seaward into the adjacent oceanic epipelagic regime, sometimes with a biomass maximum associated with the water column above the continental slope at the shelf break (e.g. Herman et al., 1981; Timonin et al., 1992). More commonly, oceanic epipelagic species intrude on the shelf. Hopkins et al. (1981) examined the landward distribution of oceanic species on to the western Florida continental shelf in the Gulf of Mexico. Such penetration was correlated with the bathymetry of the sea bed on the shelf and the minimum diel depth of occurrence of the species in the oceanic environment offshore, as illustrated in Figure 94. Epipelagic species penetrate farthest and only four out of 53 species occurred on the shelf at depths shallower than their minimum diel oceanic depths. This means that the deeper-living species will occur in depressions or basins on the shelf in the way that *Calanus finmarchicus* inhabits the deep basins on the Nova Scotian shelf (Sameoto and Herman, 1990). Deeper-living oceanic species can be moved on to the shelf during upwelling events but most are of seasonal occurrence. Castro *et al.* (1993), however, describe a population of *Rhincalanus nasutus* that is recruited to the shelf of the Auroco Gulf in Chile that survives and breeds there at depths less than 60 m. A similar inshore breeding population of another upwelling species, *Calanoides carinatus*, occurs in the southern Benguela Current (Verheye *et al.*, 1991). Species use the deep onshore current and the surface offshore current, coupled with a diel vertical migration between them, to migrate on to or off the shelf or to maintain themselves on the shelf (Castro *et al.*, 1993).

Neritic copepods are small in body size whereas larger species occur offshore. This gradient of size was found even within a species, *Temora longicornis*, in the eastern English Channel by Pessotti *et al.* (1986). Inshore waters tend to be tidally mixed and inhabited by small copepods such as *Acartia*, *Centropages* and *Temora* species while offshore waters tend to be stratified seasonally and inhabited by larger copepods (Williams *et al.*, 1994). There can, therefore, be community gradients across a shelf, such as the Scotian Shelf (Tremblay and Roff, 1983a). This shelf is complex hydrographically, receiving cold inshore water from the Gulf of St Lawrence and injections of offshore slope water. There is an across-shelf gradient of increasing temperature and salinity that is reflected in the changing fauna of copepods. Similar gradients occur in the broader adjacent shelf area to the southwest of Nova Scotia. Koslow *et al.* (1989) emphasize the importance of the wind on nutrient availability in this region and the subsequent abundance of copepods. Cooney and Coyle (1982) found that regions of the southeastern Bering Sea less than 80 m in depth are dominated by small species such as *Acartia longiremis* and *Pseudocalanus* spp. whereas larger oceanic species are rare. There is an accumulation, in a narrow band, of these large species at the shelf break in the spring.

Seasonality of occurrence of species on the shelf is common. It is most extreme in regions where active upwelling takes place, as discussed below. On-shelf transport of *Neocalanus cristatus*, *N. plumchrus* and *Eucalanus bungii* takes place seasonally in the northern Gulf of Alaska (Cooney, 1986). These species occur in the surface waters of the Gulf of Alaska in the period from late autumn to the summer and are subject to Ekman transport over the shelf. Seasonal patchiness in production of phytoplankton can affect the distributions of copepods. A gradient of phytoplankton production occurs across the shelf of the southeastern Bering Sea (Smith and Vidal, 1984). Spring blooms of phytoplankton produce a greater increase in the sizes of populations of copepods over the middle of the shelf than on the outer shelf.

12.3.3.2. *Upwelling*

Upwelling of deep, colder, and nutrient-rich water onto the continental shelves occurs on various scales. There can be short-period intrusions onto the shelf in very localized regions such as around the Izu Islands, Japan (Toda, 1989) or there can be seasonal monsoonal events such as in the Banda Sea (Arinardi, 1991). The short-period intrusions are exploited by neritic shelf species such as *Paracalanus* and *Pseudocalanus* species whereas the seasonal upwellings are usually dominated by *Calanoides* species that have developed a pattern of life history that exploits these events. In the northern part of the Benguela Current, upwelling is steadier and less strongly pulsed than in its southern part. Upwelling in the northern part is seasonal, being most active between May and November, but that off Lüderitz, 27°S, is continuous (Verheye *et al.*, 1992). Upwelling in the southern Benguela also extends for 6 to 8 months but short-period intrusions can occasionally take place during the winter (Verheye *et al.*, 1991). Upwelling is more vigorous than in the northern Benguela, especially south of 32°S. There are major upwelling centres in both the northern and southern regions so that the degree of upwelling is not uniform between 18°S and 34°S (Verheye *et al.*, 1992).

Species of the genus *Calanoides* are associated with upwelling (Table 73). Two, *C. carinatus* and *C. philippinensis*, perform ontogenetic migrations. The CV of *C. carinatus* sinks to colder water at 500 to 1000 m depth when the surface water temperatures rise after the upwelling season. It has lipid reserves and remains at depth until the next upwelling season. *Calanoides philippinensis* appears to have a similar life history in the Banda Sea (Arinardi, 1991). Verheye *et al.* (1991) suggest that *C. carinatus* has developed a complex, opportunistic life history to exploit both the long seasonal period of upwelling in the southern Benguela and the short-period, irregular upwellings that also occur. In addition to the deep-resting CV strategy, it has also developed a short-term resting CV and adult female on the outer shelf at depths of 100 to 300 m that can exploit short-term events. A third component of the population lives close inshore, is permanently active, and has no resting stages. These three components of the populations are mixed together during the main seasonal upwelling when the species breeds. The upwelling events off Oregon, USA, are detailed by Peterson *et al.* (1979) who describe mechanisms for the maintenance of the different species of copepods on the shelf.

The copepod faunas of some upwelling regions, such as the Benguela system, are well known while information on those of others is less comprehensive (Table 73).

12.3.4. Oceanic

The oceanic water column extends to a maximum depth of around 8000 m and calanoid copepods occur at all depths (Figure 89, p. 459). The numbers of individuals, however, decreases markedly below depths of about 1000 m so that at depths greater than 2000 m there is only one individual present in every 10 m^3 of water. A useful concept is dividing the oceanic water column into epipelagic, mesopelagic and bathypelagic zones. The vertical extent of these zones is not constant geographically and cannot be defined accurately even at a single location. This is because light penetrates to different depths at different locations and times depending upon the turbidity of the water column. In addition, the response of individuals to light and food varies within a species so that a population occupies a range of depth within the water column. *Pareuchaeta norvegica* is eurybathic (Figure 86, p. 441), extending from epipelagic depths at the surface during the night to bathypelagic depths below 1000 m by both day and night. It also occurs in fjords at depths of 100 to 200 m but these waters are turbid.

Mackie (1985), examining the fauna and conditions in the Georgia Strait, British Columbia, divides the water column into:

a. the epipelagic zone, the surface 50 m with variable temperature and salinity resulting from seasonal factors, winds, tides and fresh water run-off;
b. mesopelagic zone, 50 to 175 m depth, the twilight zone with not enough light for photosynthesis but enough to evoke responses from the copepods;
c. bathypelagic zone, below 175 m depth, the zone where no downwelling daylight penetrates.

He points out that, even in this abbreviated water column, the different species are not distributed neatly in the three regimes but most frequently occur in both the meso- and bathypelagic zones. *Pareuchaeta norvegica* and other oceanic meso- and bathypelagic species are advected to fjords and live in Mackie's bathypelagic zone, often developing dense and persistent populations.

The concept of epi-, meso- and bathypelagic regimes in the oceanic water column is useful although it is difficult to ascribe many species to one or other of them. The most interchange takes place between the epi- and mesopelagic regimes. Both receive downwelling daylight but there is not enough in the mesopelagic to allow photosynthesis. There are distinctive epipelagic species belonging to genera such as *Acartia, Acrocalanus, Bestiolina, Calanus, Calocalanus, Candacia, Clausocalanus, Cosmocalanus, Eucalanus, Euchaeta, Haloptilus, Ischnocalanus, Labidocera, Mecynocera,*

Table 73 The species associated with regions of upwelling.

Atlantic Ocean
Northwest Africa, 17°N (Weikert, 1982a, 1984; Brenning, 1985): *Calanoides carinatus, Centropages chierchiae, Eucalanus* spp., *Metridia lucens, Paracalanus parvus, Temora stylifera*

Off Guinea, 10°N (Bainbridge, 1972):
Paracalanus parvus

Ivory Coast shelf, 4 to 5°N (Binet, 1978, 1979):
Calanoides carinatus, Eucalanus spp., *Undinula vulgaris, Euchaeta paraconcinna*
Congolese continental shelf, 5°S (Verheye, 1991):
Calanoides carinatus

Northern Benguela, 18° to 28°S (Brenning, 1985; Timonin et al., 1992):
Calanoides carinatus, Paracalanus parvus, P. scotti, Metridia lucens, Rhincalanus nasutus
Southern Benguela, 28° to 34°S (Attwood and Peterson, 1989; Verheye, 1991; Verheye et al., 1991, 1992; Walker and Peterson, 1991; Hutchings et al., 1995):
Calanoides carinatus, Calanus agulhensis, Centropages brachiatus, Rhincalanus nasutus, Clausocalanus arcuicornis, Ctenocalanus vanus, Paracalanus parvus

Agulhas Bank, 35° to 36°S (Verheye et al., 1992):
Calanus agulhensis

Off Brazil, southwest Atlantic at 23°S (Valentin et al., 1987):
Calanoides carinatus, Ctenocalanus vanus

Indian Ocean
Off Somalia, northwest Indian Ocean near 5°N and 10°N (Smith, 1982, 1984, 1995): *Calanoides carinatus, Clausocalanus spp., Eucalanus monachus, E. crassus, Rhincalanus nasutus*

Banda Sea, Indonesia 4° to 7°S (Fleminger, 1986; Arinardi, 1991): *Calanoides philippinensis, Rhincalanus nasutus, Eucalanus dentatus, E. mucronatus, Euchaeta marina, Scolecithrix danae, Pleuromamma abdominalis*

Pacific Ocean
Izu Islands, Japan, 34°N (Toda, 1989): *Paracalanus parvus*

Off Washington, 47°N (Landry et al., 1991): *Pseudocalanus mimus*

Off Oregon, 44° to 45°N (Peterson et al., 1979; Wroblewski, 1982): *Pseudocalanus mimus, Calanus marshallae*

Off California, 39°N (Smith and Lane, 1991): *Eucalanus californicus*

Peruvian upwelling, 15°S (Smith, 1978; Boyd et al., 1980; Dagg et al., 1980; Judkins, 1980; Paffenhöfer, 1982; Boyd and Smith, 1983): *Eucalanus inermis, E. subtenuis, Centropages brachiatus, Calanus chilensis, Euaetideus acutus, Lucicutia flavicornis, Paracalanus denudatus, Scolecithrix bradyi*

Neocalanus, Paracalanus, Phaenna, Pontella, Pontellopsis, Temora and *Undinula*. They live in the surface 250 m, or shallower, and undergo diel migration to the surface at night. There are also migrating species that enter the epipelagic regime from the mesopelagic at night. Many do not reach the immediate surface layer, within 50 to 100 m of the surface. They belong to a variety of genera including *Arietellus, Chiridius, Chirundina, Euaetideus, Eucalanus, Euchaeta, Euchirella, Gaetanus, Heterorhabdus, Lophothrix, Lucicutia, Neocalanus, Pareuchaeta, Pleuromamma, Rhincalanus, Scaphocalanus, Scottocalanus* and *Undeuchaeta*.

The bathypelagic environment, below about 500 to 700 m, receives no downwelling daylight to which the copepods can respond. There are downward extensions of the populations of many shallower-living species but there are also many species that are more or less confined to this environment (Weikert and Koppelmann, 1996). They belong to such genera as *Amallothrix, Batheuchaeta, Bathycalanus, Bathypontia, Bradycalanus, Cephalophanes, Centraugaptilus, Chiridiella, Cornucalanus, Disco, Disseta, Euaugaptilus, Euchirella, Gaussia, Hemirhabdus, Heterorhabdus, Lophothrix, Lucicutia, Megacalanus, Mesorhabdus, Metridia, Onchocalanus, Pareuchaeta, Pseudeuchaeta, Pseudochirella, Phyllopus, Scaphocalanus, Scolecithricella, Scoecithrix, Spinocalanus, Valdiviella, Xanthocalanus*.

12.3.4.1. *Vertical Distribution*

The density of copepods decreases exponentially with depth throughout the bathypelagic environment (Figure 89, p. 459) and, consequently, most species are very rare in deep samples. Weikert and Koppelmann (1996) state that the decrease in density of copepods below 2500 m is better fitted by a power regression. Vertical distributions of individual species within the bathypelagic are not well known. Much more information is available on the more common meso- and epipelagic species. The following papers, and references therein, describe the bathymetric distributions of a wide variety of species: Vinogradov, 1968; Roe, 1972a-d, 1984; Angel and Fasham, 1973, 1974; Minoda, 1972; Deevey and Brooks, 1977; Hure, 1980; Judkins *et al.*, 1980; Pipe and Coombs, 1980; T.L. Hopkins, 1982; Schulz, 1982; Vives, 1982; Weikert, 1982b; Almeida Prado-Por, 1983; Bird, 1983; Cummings, 1983; Herman, 1983; Southward and Barrett, 1983; Fransz *et al.*, 1984; Longhurst *et al.*, 1984; Scotto di Carlo *et al.*, 1984; Hutchings, 1985; Vinogradov *et al.*, 1985; Y.-Q. Chen, 1986; Sameoto, 1986; Ambler and Miller, 1987; Bennett and Hopkins, 1989; Kosobokova, 1989; Fragopoulou and Lykakis, 1990; Hirakawa *et al.*, 1990; Weikert and Trinkaus, 1990; Arinardi, 1991; Atkinson *et al.*, 1992a; Markhaseva and Raszhivin, 1992; Hopkins *et al.*, 1993; Miller,

1993a; Weikert and Koppelmann, 1993; Żmijewski, 1993; Mauchline, 1995; Richter, 1995.

The depth distributions of epi- and mesopelagic species vary in time and space and there is overlapping, even between congenerics. Haury *et al.* (1990) suggest that the turbulent shear flow within water columns can result in the mixing together of species normally separated vertically in less energetic regimes. The vertical and horizontal distributions of species in the Gulf Stream is related to the vertical shifts in isopycnals and different current velocities and directions (Wishner and Allison, 1986).

According to Richter (1995), there is no epipelagic copepod fauna in high northern latitudes of the Greenland Sea. He defines three distributional strategies:

a. herbivorous species undergoing winter diapause in the bathypelagic and surfacing in the summer;
b. permanent mesopelagic residents that partition the water column;
c. permanent bathypelagic residents that partition the water column.

The seasonal migrants in the North Atlantic are *Calanus* and *Pseudocalanus* species and are paralleled in the North Pacific by *Neocalanus*, *Eucalanus* and *Pseudocalanus* species. The Antarctic also has its seasonal migrants, *Calanoides*, *Calanus*, and *Rhincalanus* species, and lacks endemic epipelagic fauna. This must be a response to the abbreviated period of seasonal phytoplankton production. An endemic epipelagic fauna is present at latitudes lower than 60° to 50°.

Bainbridge (1972), compared the vertical distribution of epipelagic species in the eastern and western tropical and subtropical Atlantic, and found that many have a much greater depth range in the western than in the eastern Atlantic. He considered that this may be associated with the greater depth of the layer of warm surface water in the west where it often extends down to 200 m or more.

Synoptic studies of the behaviour of species across their distributional range are lacking at present.

12.4. RESTRICTED ENVIRONMENTS

There are several environments within the marine habitat that have been colonized by calanoid copepods. Marine caves scattered around the margins of the oceans appear to have an endemic fauna. Such ecosystems are geographically isolated from each other in many cases and have only recently received attention. The immediate sub-surface layer of the oceans and the coasts also has an endemic fauna of copepods, the neuston. Some have developed special morphological features adapting them for this life. Another specialized, but extensive, environment is that immediately under the sea ice at high latitudes. There is an algal fauna associated with the underside of the ice that several calanoid species exploit. The last region discussed in this section is the interface between the sea bed and the overlying water column. This environment extends from the shallow brackish water lagoons, estuaries and bays offshore across the shelf and down into the abyssal ocean. Calanoids live in association with the sediments and substrata along this transect.

12.4.1. Caves

Anchihaline caves are inhabited by a variety of platycopioid and calanoid copepods (Table 74). Salinity in the caves generally ranges between about 18‰ and fully saline, there frequently being a fresh water layer at the surface. The copepods that have invaded caves all belong to hyperbenthic families, living in close association with the sediments. Investigations of the biology and physiology of the cave copepods, especially their tolerance of low oxygen concentrations, have not been made. They all appear to be euryhaline.

12.4.2. Neustonic

The neuston is the association of zoo- and phytoplankton living under the surface film of the ocean, copepods representing some 50 to 90% of the diverse fauna (Holdway and Maddock, 1983). This environment has been sub-divided on the basis of the organisms occurring there (Hattori et al., 1983). Neuston was originally described as a surface association of organisms in fresh water environments and components such as the epineuston, organisms on the aerial side of the surface film, are not so prominent in the marine environment. The neustonic species of copepods occur in the hyponeuston, the layer between the surface film and about 5 cm depth. It is not a permanent association but changes daily and seasonally. Hempel and Weikert (1972) recognized this and defined three components of the hyponeuston:

a. Euneuston are organisms whose maximum abundance occurs in the hyponeuston by both day and night.
b. Facultative neuston are organisms whose maximum abundance occurs in the hyponeuston only during certain diel periods, usually at night.

Table 74 Cave-dwelling copepods.

Species	Region	Authority
Platycopiidae		
Antrisocopia prehensilis	Bermuda	Fosshagen and Iliffe, 1985
Nanocopia minuta	Bermuda	Fosshagen and Iliffe, 1988
Epacteriscidae		
Enantiosus cavernicola	Bahamas	Barr, 1984
Epacteriscus rapax	Florida, Colombia	Fosshagen, 1973
Erobonectes macrochaetus	Caicos Is.	Fosshagen and Iliffe, 1994
E. nesioticus	Bermuda	Fosshagen and Iliffe, 1985
Boholiniidae		
Boholina spp.	Philippines	Fosshagen and Iliffe, 1989
Pseudocyclopidae		
Pseudocyclops sp.	Bermuda	Sket and Iliffe, 1980
Ridgewayiidae		
Brattstromia longicaudata	Belize	Fosshagen and Iliffe, 1991
Exumella mediterranea	Balearic Is., Sardinia	Jaume and Boxshall, 1995a
Ridgewayia marki	Bermuda	Da Rocha and Iliffe, 1993
Arietellidae		
Metacalanus sp.	Lanzarote, Canary Is.	Ohtsuka *et al.*, 1993a
Paramisophria galapagensis	Galapagos Is.	Ohtsuka *et al.*, 1993a
P. reducta	Lanzarote, Canary Is.	Ohtsuka *et al.*, 1993a
Fosshageniidae		
Fosshagenia ferrari	Bahamas	Suárez-Morales and Iliffe, 1996
Pontellidae		
Calanopia americana	Bermuda	Sket and Iliffe, 1980
Pseudocyclopiidae		
Stygocyclopia balearica	Balearic Is.	Jaume and Boxshall, 1995b
Paracyclopia gitana	Balearic Is.	Carola and Razouls, 1996
P. naessi	Bermuda	Da Rocha and Iliffe, 1993
Stephidae		
Stephos balearensis	Balearic Is.	Carola and Razouls, 1996
S. margalefi	Balearic Is.	Riera *et al.*, 1991
Meiostephos leamingtonensis	Bermuda	Da Rocha and Iliffe, 1993

c. Pseudoneuston are organisms whose maximum concentrations do not occur in the hyponeuston but are deeper and a portion of their populations enter the hyponeuston at night during diel vertical migration.

The commonest neustonic copepods belong to the family Pontellidae (Table 75). Some species belong to the euneuston but others are facultative neuston or pseudoneuston. All genera of pontellids have neustonic representatives. Other families of calanoids have genera, representatives of which can occur in the hyponeuston but usually in the pseudoneuston. Species in such genera as *Acartia*, *Calanus*, *Candacia*, *Centropages*, *Clausocalanus* and *Temora* occur in the hyponeuston at night (e.g. Champalbert, 1971b; Trela, 1989) and a few may persist irregularly there during the day.

A morphological adaptation to a neustonic existence has been developed in some Pontellidae. A structure, consisting of two semicircles of closely spaced setules, is present on a flattened area of the anterior dorsal surface of the cephalosome (Ianora *et al.*, 1992b). The semicircles of setae consist of dense rows of setae with branched ends, the rows thinning or swelling into clumps (Figure 95). It functions to attach the copepod to the surface film, so conserving energy. The form of the structure is peculiar to the species and occurs in all copepodids but not in the nauplii. Not all pontellids seem to possess this structure because Ianora *et al.* (1992b) did not find it present in either *Labidocera wollastoni* or *Pontellina plumata* and quote W.M. Pennell who did not find it in *Labidocera aestiva* or *L. nerii*. Physiological adaptations of pontellids to a neustonic existence are reviewed by Champalbert (1985). They include the production of diapause eggs when environmental conditions become adverse.

The occurrence of pontellids in the hyponeuston is detailed by Geinrikh (1969, 1974), Turner *et al.* (1979), Turner and Collard (1980).

12.4.3. Under Ice

The ice-water interface of polar regions forms a substratum for the development of a rich flora of microalgae. A variety of copepods exploit this source of food. The distribution of sea ice is different in the two polar regions, although both regions expand and contract coverage seasonally (Conover and Huntley, 1991). Most of the ice in the Antarctic is drifting pack ice whereas that in the Arctic is more or less permanently frozen. The species of copepods in the two regions are different (Table 76). Both polar environments have deep water under the ice but shallow coastal environments exist around the Antarctic continent and in the Arctic in such regions as the Bering Sea, north Greenland and the Canadian Arctic. Most of the

Table 75 Neustonic species of copepods and their densities of occurrence.

Genus or species	Area	Density Nos. m^{-1}	Authority
Pontellidae			
Anomalocera patersoni	Mediterranean	0.92	Champalbert, 1971a
Calanopia minor	Kuroshio Current	0.1–1	Matsuo and Marumo, 1982
Labidocera spp.	Central North Pacific	0.1–1	Sherman, 1963
	Central South Pacific	0.02–8.25	Sherman, 1964
	Kuroshio Current	0.02–0.50	Matsuo and Marumo, 1982
	Suruga Bay, Japan		Hattori *et al.*, 1983
Pontella spp.	Central North Pacific	<0.1	Sherman, 1963
	Central South Pacific	<2.2	Sherman, 1964
	Kuroshio Current	0.02–4.20	Matsuo and Marumo, 1982
	Suruga Bay, Japan	0.08–0.14	Hattori *et al.*, 1983
Parapontella brevicornis	Mediterranean	1.86	Champalbert, 1971a
Pontellina morii	Kuroshio Current	<0.2	Matsuo and Marumo, 1982
P. plumata	Central North Pacific	<0.2	Sherman, 1963
	Central South Pacific		Sherman, 1964
	Suruga Bay, Japan	0.11–0.14	Hattori *et al.*, 1983
Pontellopsis spp.	Central North Pacific	<0.05	Sherman, 1963
	Central South Pacific	<0.16	Sherman, 1964
	Kuroshio Current	0.03–1.58	Matsuo and Marumo, 1982
	Suruga Bay, Japan	0.02–0.42	Hattori *et al.*, 1983
Candaciidae			
Candacia aethiopica	Suruga Bay, Japan	0.03	Hattori *et al.*, 1983
C. bipinnata	Suruga Bay, Japan	0.03	Hattori *et al.*, 1983
C. curta	Suruga Bay, Japan	0.02	Hattori *et al.*, 1983
C. pachydactyla	Suruga Bay, Japan	0.02	Hattori *et al.*, 1983
Paracandacia simplex	Suruga Bay, Japan	0.03	Hattori *et al.*, 1983
Centropagidae			
Isias clavipes	Mediterranean		Champalbert, 1971a

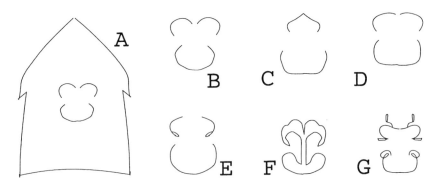

Figure 95 The surface structure attachment of pontellid copepods. A, *Anomalocera patersoni* showing position of the structure on the head. The comparative shapes of the rows of setae in the structures of B, *Anomalocera patersoni*; C, *Pontella mediterranea*; D, *Pontella atlantica*; E, *Pontella lobiancoi*; F, *Pontellopsis villosa* and G, *Pontellopsis regalis*. (After Ianora et al., 1992b.)

earlier work consisted of sampling over deep water in both hemispheres and it is only relatively recently that an ice flora and fauna have been recognized. Emphasis has also been on the spring blooming of this flora but Hoshiai *et al.* (1996) also examined an autumnal bloom.

Sampling of copepods at the ice-water interface is difficult but is often done by scuba diving. Macaulay and Daly (1987), however, describe a folding net, on the principal of an umbrella, that can be fished vertically through a hole drilled in the ice. A net of mouth diameter of about 0.5 m can be deployed through an ice core hole of 20 cm diameter in ice of 6 to 10 m thickness. Three nets, an umbrella net, a collapsible free-fall net and a diver-operated push net are described by Kirkwood and Burton (1987).

Conover and Huntley (1991) have recently reviewed knowledge of copepods in both hemispheres and list the species associated with the ice. Their list for the Arctic is the same as that in Table 76 except that they include the carnivorous *Pareuchaeta* species; these are primarily *P. glacialis* and *P. norvegica* that will undoubtedly exploit concentrations of copepods and copepodids at the ice-water interface. Their list of species, however, for the Antarctic differs markedly from that in Table 76. The only species common to the two lists are *Drepanopus* spp. and *Paralabidocera antarctica*. The other species in their list are the dominant Antarctic species *Calanoides acutus*, *Calanus propinquus*, *Metridia gerlachei* and *Rhincalanus gigas* and the carnivorous *Pareuchaeta antarctica*. These species do occur under the ice but are not so intimately associated with it as the other species in Table 76. They exploit it opportunistically, probably more so at the edges of the pack ice than nearer the continent.

Table 76 Calanoid copepods associated with polar ice. Usually caught in diver-operated nets 0 to 5 m below the ice.

Antarctic species		
Ctenocalanus citer	McMurdo Sound	Foster, 1987
	W Weddell Sea	Menshenina and Melnikov, 1995
C. vanus	Syowa Station	Hoshiai and Tanimura, 1986
Drepanopus bispinosus	Ellis Fjord	Kirkwood and Burton, 1987
Metridia gerlachei	McMurdo Sound	Foster, 1987
Microcalanus pygmaeus	W Weddell Sea	Menshenina and Melnikov, 1995
Paralabidocera		
antarctica	W Weddell Sea	Menshenina and Melnikov, 1995
	Ellis Fjord	Kirkwood and Burton, 1987
	McMurdo Sound	Forster, 1987
	Syowa Station	Hoshiai and Tanimura, 1986; Tanimura *et al.*, 1996
P. grandispinosa	McMurdo Sound	Waghorn and Knox, 1988
Stephos longipes	McMurdo Sound	Foster, 1987
	W Weddell Sea	Menshenina and Melnikov, 1995
	SE Weddell Sea	Kurbjeweit *et al.*, 1993
	E Weddell Sea	Schnack-Schiel *et al.*, 1995
Arctic species		
Acartia longiremis	Frobisher Bay	Grainger and Mohammed, 1986
	E Bering Sea	Coyle and Cooney, 1988
Calanus glacialis	E Greenland Sea	Hirche *et al.*, 1991
	Northeast Greenland	Hirche *et al.*, 1994
	Frobisher Bay	Grainger and Mohammed, 1986
	Resolute Passage	Fortier *et al.*, 1995
	Hudson Bay	Tremblay *et al.*, 1989
	Hudson Bay	Tourangeau and Runge, 1991
C. finmarchicus	E Greenland Sea	Hirche *et al.*, 1991
	Northeast Greenland	Hirche *et al.*, 1994
C. hyperboreus	E Greenland Sea	Hirche *et al.*, 1991
	Northeast Greenland	Hirche *et al.*, 1994
	Resolute Passage	Fortier *et al.*, 1995
	Hudson Bay	Tremblay *et al.*, 1989
C. marshallae	E Bering Sea	Coyle and Cooney, 1988
Eurytemora herdmani	E Bering Sea	Coyle and Cooney, 1988
Metridia longa	E Greenland Sea	Hirche *et al.*, 1991
	Resolute Passage	Fortier *et al.*, 1995
	Hudson Bay	Runge and Ingram, 1988
Pseudocalanus acuspes	Resolute Passage	Fortier *et al.*, 1995
P. minutus	Hudson Bay	Runge and Ingram, 1988, 1991
P. newmani	Hudson Bay	Runge and Ingram, 1988
Pseudocalanus spp.	Hudson Bay	Tremblay *et al.*, 1989

12.4.3.1. *Antarctic Ocean*

Paralabidocera antarctica and the dominant *Stephos longipes* occur in greater numbers immediately under the ice than at 5 m depth, *Ctenocalanus citer* is more evenly distributed within this layer, while *Microcalanus pygmaeus* is more common at 5 m depth than shallower in the offshore regions of the western Weddell Sea (Menshenina and Melnikov, 1995). *Stephos longipes* inhabits the under-ice environment throughout its life cycle. In the offshore regions of the southeastern Weddell Sea, according to Kurbjeweit *et al.* (1993), *S. longipes* is also the dominant under-ice species. It is replaced by *Paralabidocera antarctica* in inshore regions of the Indian and Pacific sectors of the Antarctic and by *P. grandispinosa* in McMurdo Sound, both these species being endemic to the coasts. Small copepods, *Ctenocalanus citer*, *Microcalanus pygmaeus* and *Stephos longipes* dominate the region of the ice edge while larger species, *Calanoides acutus*, *Calanus propinquus*, *Metridia gerlachei* and *Rhincalanus gigas* are dominant offshore of the Antarctic Peninsula (Schnack *et al.*, 1985).

The copepod fauna under the sea ice at Syowa Station (69° 00′S, 39°35′E) was studied throughout the year by Tanimura *et al.* (1986, 1996). The total depth of water was 10 m and the copepods were sampled through a hole drilled in the ice. The eight species of calanoids caught, included the seasonally dominant *Ctenocalanus vanus*, *Microcalanus pygmaeus*, *Paralabidocera antarctica* and *Stephos longipes* and the rarer *Calanus propinquus*, *Pareuchaeta* sp., *Metridia gerlachei* and *Scolecithicella glacialis*.

Copepods live in the interstices between the ice platelets of the sea ice (Hoshiai and Tanimura, 1986). These interstices are filled with sea water and are the regions in which the ice algae grow. By coring the ice at Syowa Station, they showed that *Paralabidocera antarctica* is a more-or-less permanent occupant of the sea ice while *Ctenocalanus vanus* lives there irregularly. The former is dependent upon the ice algae for its nutrition (Hoshiai *et al.*, 1987).

The biology of *Stephos longipes* has been examined in detail by Kurbjeweit *et al.* (1993) and Schnack-Schiel *et al.* (1995) in the southeastern and eastern Weddell Sea respectively. It appears to have a one-year life cycle closely coupled to the melt-and-growth cycle of the sea ice. Large numbers of adults and nauplii live in the sea ice in late winter and early spring. The ice melts and the summer population of copepods is found in the water column. Re-formation of the ice in the autumn results in its occupancy by the nauplii which overwinter there. A second population, dominated by the CIV, migrates to deep water in the autumn where it overwinters. Similarly, *Paralabidocera antarctica* has a one-year life cycle, overwinters in the NIV and NV and is intimately associated with the ice-algae (Tanimura *et al.*, 1996).

12.4.3.2. Arctic Ocean

Resolute Passage is one of the northernmost regions in the northern hemisphere that has sea ice that melts in the summer. The under-ice fauna is dominated by *Pseudocalanus acuspes* whose breeding commences well before the release of ice algae and is probably supported by lipid stores (Fortier *et al.*, 1995). The diel rhythms of feeding and vertical migration do not appear to be coupled to exploit the ice algae maximally (Hattori and Saito, 1995). The species occurring in Resolute Passage (Table 76) use the ice algae in different ways (Conover *et al.*, 1991). *Pseudocalanus acuspes* develops from the CIII to mature adult by feeding on algae, *Calanus glacialis* feeds on it by diel migration during the period of maturation of their gonads. The eggs of *C. hyperboreus* are buoyant and float up to the ice-water interphase where the nauplii also occur; there was little evidence that the later copepodids and adult *C. hyperboreus* feed on the algae.

The inshore region of southeastern Hudson Bay has a depth of about 75 m, and a seasonal ice cover, about 100 cm thick, with associated ice algae. The dominant copepods present are three species of *Pseudocalanus* and *Calanus glacialis*. Runge *et al.* (1991) conclude that the principal source of nutrition for egg production of the copepods is the sedimenting ice algae during and immediately after the algal bloom at the ice-water interface.

There appear to be few comparable studies to those in the Antarctic in which the penetration of copepods into crevices in the ice has been examined. Kern and Carey (1983) cored the under side of the seasonal ice in the coastal region of the Beaufort Sea. Calanoid copepods occurred infrequently in the cores but these authors could not determine whether the copepods entered the coring device prior to it being inserted into the ice or actually occurred within the ice.

12.4.3.3. Ice-edge Zones

Marginal ice zones in polar regions are areas of increased productivity (Hirche *et al.*, 1991). In the eastern Bering Sea, Coyle and Cooney (1988) found the dominant copepods associated with the ice-edge to be small-sized species such as *Acartia longiremis*, *Calanus marshallae*, *Eurytemora herdmani* and *Pseudocalanus* spp. Comparable ice-edge species in the Greenland Sea are *Calanus finmarchicus*, *C. glacialis*, *C. hyperboreus*, and *Metridia longa*. The concentrations of ice algae at the ice-edge did not appear to support egg production in *Calanus* spp. and *Metridia longa* in the Greenland Sea (Smith, 1990); egg production was greater away from the ice-edge.

The ice-edge in the western Weddell Sea overlays depths of more than

1000 m. Hopkins and Torres (1988) found that the temperature in the surface 50 m layer was more than 1°C cooler under the ice than in the adjacent open sea and suggest that this may be linked to the lower densities of nauplii and copepodids found in the upper 50 m under the ice.

12.4.3.4. *Ice Cover at Lower Latitudes*

According to Hattori and Saito (1994), Saroma Ko Lagoon in northern Hokkaido, Japan, is the southernmost marine environment in the northern hemisphere that is covered by sea ice each winter. As elsewhere, microalgae flourish at the ice-water interphase, chlorophyll *a* concentrations being approximately 200 times those in the water column at 1 to 9 m depth. Under-ice copepods were found to be dominated by *Pseudocalanus* spp. by Nishiyama *et al.* (1987) and by *Acartia longiremis* and *Pseudocalanus* spp. by Hattori and Saito (1994). Fortier *et al.* (1995) found that the ice algae did not appear to stimulate egg production of *Pseudocalanus* spp.

12.4.4. Hyperbenthic and Benthopelagic

The environment close to the sediment surface in both shallow and deep waters is inhabited by copepods, which usually dominate the fauna numerically. Some species are endemic to this environment. They are variously referred to as hyperbenthic, demersal, planktobenthic or, when they occur in deep water of 200 m or more depth, benthopelagic. Many of the species normally living in the pelagic water column have downward extensions of their populations into this environment in coastal, shelf and deep oceanic environments. The surface of the sea bed in shallow waters is accessible to observation by divers and so the relationship of the fauna to the topography of the region can be determined. This is not true at greater depths where samples are mostly obtained remotely from the vicinity of the sea bed. Pelagic samplers are often subject to damage if they come in contact with the bottom and so they are towed at measured heights above the sea bed. Species caught in them, but not higher in the water column, are assumed to be those associated with the sea bed, or at least living in the nepheloid layer immediately above the sea bed. Hence, the term benthopelagic is used as it does not assume that such species actually live in contact with the substratum as do many of the shallow-water hyperbenthic species.

A list of genera in which some or all species are hyperbenthic or benthopelagic in habit is given in Table 77. The list is not exhaustive because of the difficulties in determining whether the deeper-living species are endemic to this environment. The benthopelagic and hyperbenthic species tend to be small in body size and are often robustly built. Studies of their biology are few and much more investigation is required.

12.4.4.1. *Shallow Waters*

Coastal waters and estuaries with water columns of a few metres have few endemic demersal species. The normally pelagic species, holoplanktonic, of genera such as *Acartia, Calanus, Centropages, Eurytemora, Paracalanus* and *Temora* often have their vertical distributions foreshortened and occur in the benthic boundary layer (Imabayashi and Endo, 1986). They usually outnumber the true demersal species. Trapping of demersal species on areas of sand flats or coral reefs can, through the design of the trap, exclude holoplanktonic species (Robichaux *et al.*, 1981). Separating the two faunas is not as yet possible in deeper environments.

The copepods associated with coral reefs have been examined in some detail. The Great Barrier Reef is inhabited by species in the genera *Pseudocyclops, Pseudodiaptomus, Ridgewayia* and *Stephos* which occur in traps but are absent from pelagic net tows. Consequently, they are deemed hyperbenthic in habit unlike co-occurring species in the traps, of the genera *Acartia, Canthocalanus, Paracalanus, Parvocalanus, Temora*, etc. that also occurred in the pelagic samples (McWilliam *et al.*, 1981; Jacoby and Greenwood, 1988, 1989). These demersal copepods have substratum preferences and there are differences in their diel and seasonal emergence times from the substrata. Jacoby and Greenwood (1988) conclude that the *Stephos* spp. probably swarm near the bottom or around coral reef formations without settling on the substratum as do the *Pseudocyclops* and *Pseudodiaptomus* species.

Little is known about the biology of hyperbenthic copepods. They dominate the demersal fauna and show seasonal cycles of abundance (e.g. Lewis and Boers, 1991). Smith *et al.* (1979), in experimental incubations of demersal plankton, concluded that they did not feed to any extent on the phytoplankton in the water column during their periods of emergence, but this requires further investigation. Fancett and Kimmerer (1985) state that *Pseudodiaptomus cornutus* and *P. colefaxi* do not feed during the day when they are on the bottom and that this discontinuous feeding is possible through storage of lipids. Those that settle on the substratum during the day rise into the water column at night, even when the column is only a few metres in depth. Those that swarm around topographical features, such as coral heads, rise and disperse at night. Their populations are often subject to predation by visually hunting fish and other predators. Fancett and Kimmerer (1985) describe *Pseudodiaptomus* species attaching themselves

Table 77 Hyperbenthic and benthopelagic calanoid copepods. The habitats of or depths at which the former and the depths at which the latter have been caught are indicated.

Species	Notes	Authority
Platycopiidae		
Platycopia spp.	Littoral-ca 120	Fosshagen and Iliffe, 1988
Pseudocyclopidae		
Pseudocyclops spp.	Littoral	Jacoby and Greenwood, 1988, 1989
	Coral lagoons	Madhupratap et al., 1991
Ridgewayiidae		
Exumella spp.	7–15	Jaume and Boxshall, 1995a
Placocalanus spp.	3–53	Ohtsuka et al., 1996b
Ridgewayia spp.	Coral reefs	Jacoby and Greenwood, 1989
Augaptilidae		
Pachyptilus sp.	1300	Gowing and Wishner, 1986
Arietellidae		
Campaneria latipes	1234–1260	Ohtsuka et al., 1994
Crassarietellus spp.	3974–4060	Ohtsuka et al., 1994
Metacalanus spp.	5–8	Ohtsuka, 1984, 1985b
Paramisophria spp.	Tidal to 200	Ohtsuka et al., 1993a
Paraugaptiloides magnus	1060–1697	Ohtsuka et al., 1994
Pilarella longicornis	135	Ohtsuka et al., 1994
Rhapidophorus wilsoni	?	Ohtsuka et al., 1994
Sarsarietellus abysallis	1090	Heinrich, 1993
Scutogerulus pelophilus	1383–1397	Bradford, 1969b
Hyperbionychidae		
Hyperbionyx pluto	3870–4036	Ohtsuka et al., 1993b
Centropagidae		
Gladioferens spp.	Estuarine	Rippingale, 1994
Pseudodiaptomidae		
Pseudodiaptomus spp.	Coral reefs	Jacoby and Greenwood, 1988, 1989
Bathypontiidae		
Alloiopodus pinguis	1238–1697	Bradford, 1969b
Bathypontia spp.	1200	Wishner, 1980b
Temorites sp.	1259–1332	Gowing and Wishner, 1992
Zenkevitchiella spp.	1300	Gowing and Wishner, 1986

Table 77   Continued.

Species	Notes	Authority
Ryocalanidae		
Ryocalanus spp.	1300–3000	Markhaseva and Ferrari, 1995
Spinocalanidae		
Damkaeria falcifera	1380	Fosshagen, 1983
Isaacsicalanus paucisetus	Hydrothermal vent	Fleminger, 1983
Spinocalanus sp.	1300	Gowing and Wishner, 1986
	2945–3100	Gowing and Wishner, 1992
Teneriforma sp.	1300	Gowing and Wishner, 1986
Aetideidae		
Aetideopsis magna	1750–1822 m	Grice and Hülsemann, 1970
Bradyidius bradyi	240	Matthews, 1964
B. brevispinus	1194–1278	Bradford, 1969b
B. dentatus	1357–1697	Bradford, 1969b
B. plinioi	100–150	Campaner, 1978a, 1986
	775–880	Gowing and Wishner, 1992
B. robustus	1184–1193	Bradford, 1969b
B. similis	35–150	Fosshagen, 1978
B. spinibasis	1690	Bradford, 1969b
Chiridius armatus	240	Matthews, 1964
Chiridius spp.	1300	Gowing and Wishner, 1986
Comantenna brevicornis	240	Matthews, 1964
C. crassa	1234–1260	Bradford, 1969b
C. recurvata	1750–1822	Grice and Hülsemann, 1970
Crassantenna comosa	1383–1397	Bradford, 1969b
C. mimorostrata	1234–1260	Bradford, 1969b
Euchirella spp.	775–1332	Gowing and Wishner, 1992
Gaidius spp.	1200	Wishner, 1980b
G. pungens	775–1332	Gowing and Wishner, 1992
Lutamator elegans	130	Alvarez, 1984
L. hurleyi	1357	Bradford, 1969b
Paracomantenna spp.	75–200	Campaner, 1978a, 1986
Pseudeuchaeta flexuosa	1690	Bradford, 1969b
P. magna	1690	Bradford, 1969b
Pterochirella tuerkayi	1318	Schulz, 1990
Diaixidae		
Diaixis asymmetrica	1750–1822	Grice and Hülsemann, 1970
D. hibernica	35–150	Fosshagen, 1978
Diaixis sp.	1300	Gowing and Wishner, 1986

Table 77 Continued.

Species	Notes	Authority
Euchaetidae		
Pareuchaeta sp.	1200	Wishner, 1980b
	1300	Gowing and Wishner, 1986
Mesaiokeratidae		
Mesaiokeras kaufmanni	20–27	Fosshagen, 1978
M. nanseni	10–680	Fosshagen, 1978
Parkiidae		
Parkius karenwishnerae	2945–3010	Ferrari and Markhaseva, 1996
Phaennidae		
Brachycalanus rothlisbergi	20	Othman and Greenwood, 1988a
Brachycalanus spp.	72–150	Campaner, 1978b, 1986
Xanthocalanus alvinae	1750–1822	Grice and Hülsemann, 1970
X. distinctus	1750–1822	Grice and Hülsemann, 1970
X. elongatus	1750–1822	Grice and Hülsemann, 1970
X. fallax	240	Matthews, 1964
X. macrocephalon	1750–1822	Grice and Hülsemann, 1970
X. marlyae	100–150	Campaner, 1978b, 1986
X. minor	240	Matthews, 1964
Xanthocalanus spp.	1300	Gowing and Wishner, 1986
	775–3100	Gowing and Wishner, 1992
Pseudocyclopiidae		
Pseudocyclopia spp.	50–120	Jaume and Boxshall, 1995b
Scolecitrichidae		
Amallothrix spp.	1200	Wishner, 1980b
Amallophora macilenta	1750–1822	Grice and Hülsemann, 1970
A. rotunda	1750–1822	Grice and Hülsemann, 1970
Scaphocalanus spp.	1300	Gowing and Wishner, 1986
	775–1332	Gowing and Wishner, 1992
Scolecithricella pseudoculata	100	Campaner, 1986
Scolecithricella sp.	1300	Gowing and Wishner, 1986
	1259–1332	Gowing and Wishner, 1992
Scopalatum sp.	1300	Gowing and Wishner, 1986
Stephidae		
Stephos spp.	Coral reefs	Jacoby and Greenwood, 1988
S. canariensis	Anchihaline pool	Boxshall *et al.*, 1990

Table 77 Continued.

Species	Notes	Authority
Tharybidae		
Parundinella		
emarginata	1750–1822	Grice and Hülsemann, 1970
Rythabis atlantica	2860	Schulz and Beckmann, 1995
Tharybis angularis	2860	Schulz and Beckmann, 1995
T. crenata	2860	Schulz and Beckmann, 1995
Neoscolecithrix sp.	1300	Gowing and Wishner, 1986
Undinella altera	1750–1822	Grice and Hülsemann, 1970
U. compacta	1750–1822	Grice and Hülsemann, 1970
U. hampsoni	1750–1822	Grice and Hülsemann, 1970
Undinella spp.	1300	Gowing and Wishner, 1986
	775–3100	Gowing and Wishner, 1992

to the walls of the container and to detritus by using the long setae on the antennules. This immobility will protect them from detection by the predator. *Gladioferens imparipes*, an estuarine species, have setae on the dorsal surface of the prosome with which they attach themselves to underwater surfaces (Rippingale, 1994). *Stephos* and *Paramisophria* species swim with their left-hand sides ventrally opposed to the sediment. The left-hand side of the prosome of *P. platysoma* is flattened so that, when it lies on its side, it makes a good contact with the surface of the substratum and remains immobile (Ohtsuka and Mitsuzumi, 1990). Finally, *Placocalanus* species, with their short modified antennules and flattened bodies, may burrow into the sediment for short periods (Ohtsuka *et al.*, 1996b).

12.4.4.2. *Deeper Waters*

Deeper waters of 100 m or more are not observable by scuba diving. Benthopelagic copepods occur in fjords or on the continental shelf at depths of 100 to over 500 m (Table 77). Other species have been identified in the deep-sea environment.

The number of copepods decreases with depth in the ocean (Figure 89) but their numbers increase in the nepheloid, or benthic boundary layer, by about a factor 2 to 10 over the numbers in the layer immediately above (Angel, 1990).

Beckmann (1988) describes the copepod fauna of the upper nepheloid layer between 200 and 1500 m above the sea bed in the northeast Atlantic in the BIOTRANS area around 47°N, 20°W. Representatives of the families

Metridinidae, Lucicutiidae and Heterorhabdidae and the species *Foxtonia barbatula* comprised about 70% of all calanoid copepods caught. At depths exceeding 1000 m, the benthopelagic copepods are dominated by species in the superfamilies Arietelloidea, Bathypontioidea and Clausocalanoidea, the latter containing the Tharybidae (Schulz and Beckmann, 1995).

As with the shallow water species, there is little information available about the biology of the deeper-living benthopelagic species. Some produce eggs that have adhesive outer membranes that attach to surfaces on the sea bed (Matthews, 1964). It is not known, however, how widespread this behaviour is. Gowing and Wishner (1986), examining the stomach contents of copepods caught 1 m above the sea bottom at a depth of about 1300 m in the Santa Catalina Basin, California, concluded that they were predominantly detritivores. The stomach contents of the different species reflected what occurred in the water at that depth, indicating that no selective feeding was taking place. Some selection by the benthopelagic copepods was observed on an eastern tropical Pacific seamount by Gowing and Wishner (1992). The species appeared to select newly sedimented particles from the ocean surface and also bacteria-like bodies, indicating a degree of opportunistic feeding.

12.4.4.3. *Seep Sites and Hydrothermal Vents*

Organisms living above cold seeps in Sagami Bay, Japan, at a depth of 1160 m, were collected by a submersible (Toda *et al.*, 1994). Copepods were dominant, 14 species of calanoids occurring in the samples. The most common were of the families Aetideidae and Phaennidae but the species are not named.

Hydrothermal vents in the floor of the oceans have an associated fauna (Grassle, 1986). Fleminger (1983) described a spinocalanid, *Isaacsicalanus paucisetus*, associated with a vent in the eastern Pacific. It swarms in the vicinity of the vents, 920 individuals m^{-3} being counted (Smith, 1985). A hydrothermal vent field on the Endeavour Ridge around 48°N, 129°W, in the northeastern Pacific at depths of 2000 m, generates a plume that provides an additional source of nutrition in the deep water column. Burd *et al.* (1992) and Burd and Thomson (1995) describe two sonic scattering layers, one in mid-depth at 400 to 900 m, a second at 1200 to 1900 m overlying the hydrothermal plume. The dominant copepods present at depths greater than 1700 m, in areas without deep scattering layers, were the typically deep-living species *Metridia asymmetrica*, *Spinocalanus brevicaudatus*, *S. brevicornis*, and *S. longicornis*. The copepods in the scattering layers over the vent field were dominated by the CVs of shallow-living species *Neocalanus plumchrus* and *N. cristatus*, although the typical deep-living species were also present and at greater abundance than outside the influence of the plume. The *Neocalanus* species appeared to be opportunistically exploiting the detritus and bacteria within the plume.

A diverse fauna of copepods is associated with the hydrothermal vent field in the Guaymas Basin, Gulf of California (Wiebe *et al.*, 1988). The vents are at a depth of about 2000 m and more than 65 species of calanoids occurred in samples taken about 100 m above them. Many could not be assigned to known species and so may represent endemic vent species. Two epi- or mesopelagic species, *Calanus minor* and *Rhincalanus nasutus* occurred in the samples. Mauchline (unpublished) found that *R. nasutus* occurred at the greatest depths sampled by closing nets, 1900 m, in the northeastern Atlantic, and Wheeler (1970) caught it between 2200 and 4100 m in the western Atlantic. Consequently, the occurrence of these species at this depth in the Gulf of California may only represent normal downward extensions of their populations rather than any special linkage with the presence of the vents.

12.4.5. Commensal Species

Some copepods are known to associate with houses of the giant larvacean, *Bathochordaeus charon*. These houses can be tens of centimetres in diameter offering surfaces for colonization (Ferrari and Steinberg, 1993; Steinberg *et al.*, 1994). The larvaceans occur at depths of 100 to 500 m in Monterey Bay and other locations. Two scolecitrichid species, *Scopalatum vorax* and *Scolecithicella lobophora*, along with *Metridia pacifica* were present on or within the houses. *Scopalatum vorax* was mainly associated with the inner food-concentration filters and was occasionally thought to be feeding on material from their surface. Alldredge (1976), in the Gulf of California, and Ohtsuka and Kubo (1991), in the Inland Sea of Japan and the adjacent Pacific Ocean, found *Scolecithrix danae* feeding on the filters. The latter also found remains of larvaceans, including their faecal pellets in the guts of *Candacia bipinnata*, *C. catula* and *Paracandacia truncata*.

The only other apparent association between a calanoid and another organism is that of the hyperbenthic *Ridgewayia fosshageni* with the sea anemone *Bartholomea annulata*. Humes and Smith (1974) found that *R. fosshageni* formed pelagic aggregations in the close proximity to this anemone in preference to rocks or another anemone, *Stoichactis helianthus*.

13. Geographical Distribution

13.1. Introductions of Calanoids 505
13.2. Faunal Provinces and Large Marine Ecosystems 509
    13.2.1. Copepods of the Faunal Provinces 509
    13.2.2. Copepods of the LMEs 510
13.3. Identification of Species 516
13.4. Concluding Remarks 516

According to Sewell (1948), the pelagic copepods divide into two groups that can be separated by depth. These are, first, those that inhabit the upper 400 to 500 m of the oceans, including shelf and coastal regions and, second, those usually found at greater depths. This broad distinction is frequently contradicted by species such as *Calanus finmarchicus*, a recognized shelf species, also occurring in the meso- to bathypelagic, and *Pareuchaeta norvegica*, present in Norwegian and Scottish fjords, but also present as a resident population at 1200 to 1600 m in the Rockall Trough (see Vinogradov, 1997). Nevertheless, the concept divides the majority of species into shallow-living ones, about which something is known of their geographical distributions, and deep-living ones, about which much is inferred.

    The major ocean basins are connected by routes through the Arctic and around South America, South Africa and through the East Indies. Some copepod species, especially those of the bathypelagic zone, usually occur in two or more oceans, although some are endemic to one ocean. Coastal species, as would be expected, are frequently endemic to restricted geographical regions. Dunbar (1979) has reviewed the relationships between the oceans and discusses the distribution of plankton. He relates marine biogeographic zones, or faunal provinces, with terrestrial zones. Sewell (1956) places the distributions of copepods in the context of continental drift. A comparison of the copepod faunas of the north Atlantic and north Pacific regions is made by Parsons and Lalli (1988) while Halim (1990) assesses migration of copepods from the Red Sea to the eastern Mediterranean.

    Sewell (1948) discusses the vertical distribution of species in some detail, noting the surface occurrence of normally deep-living species and the occurrence at considerable depths, 3000 m in the case of the normally neustonic *Anomalocera patersoni*, of surface-living species. Sewell's paper is still important today as a general introduction to the study of the zoogeography of calanoid copepods because it examines the potential importance of biological and behavioural parameters against the background of the physical and chemical parameters of the oceans. More recently, Reid *et al.* (1978) describe the distributions of species of plankton relative to the hydrography of the oceans. Some copepods are used as examples. They compare the circulation patterns within the oceans. The Atlantic is the warmest ocean and this is reflected by more northward extensions of northern temperate species than are found in the Pacific.

The geographical distributions of most species of copepods are not well known. Many species, especially in the oceanic meso- and bathypelagic, have been recorded rarely and the geographical coverage of the sampling has not been uniform. Samples of plankton have been collected from the surface few hundred metres in many regions but sampling to depths greater than 1000 m has been relatively uncommon. It was first done, to any extent, in the late 1800s during the international oceanic expeditions. This produced a large number of samples that required many years to sort and identify the species contained in them. The first period of description of new species, 1890 to about 1930 (Figure 17, p. 50) derives principally from these expeditions and much of the work was purely descriptive of distributions along the track of ships. The second period of description of new species, 1950 to the present time (Figure 17, p. 50), is a second wave of oceanic exploration that has identified further deep-sea species and species in hyperbenthic and benthopelagic environments. The geographic coverage is still far from complete and, in some cases, reflects the geographic locations of investigators with the necessary expertise or interest.

The distributions of even the most commonly investigated species (Table 2, p. 7) are frequently incompletely known. Those of oceanic species can often be inferred and assumptions made about restricted regions where no samples are available. Coastal species, in several cases, are a different problem.

13.1. INTRODUCTIONS OF CALANOIDS

The chance dispersal of planktonic copepods to regions distant from their previously known distribution appears to have become more common since 1950 (Fleminger and Hendrix Kramer, 1988; Hedgpeth, 1993). Most, but not all, of these introductions are suspected as being through transport in ballast water. Carlton (1985) tabulates information on the organisms found, including calanoid copepods. Williams *et al.* (1988) sampled the ballast water from 31 bulk cargo carriers arriving in Australian ports from Japan and found *Calanus sinicus, Centropages abdominalis, C. yamadai, Labidocera bipinnata* and *Pontellopsis tenuicauda*, all endemic to Japanese coastal waters. The number of species of organisms identified in the ballast water decreased as voyage time increased from 9 to 17 d. These species are not recorded as established in Australia. Carlton and Geller (1993) examined ballast water on board 159 cargo ships in Coos Bay, Oregon coming from 25 Japanese ports and recorded a diversity of fauna after journeys of 11 to 21 d. The most common organisms were copepods, calanoid and cyclopoid. Not all species in ballast water will survive when the water is discharged in a foreign port. It will depend, to a large degree, on the temperature and salinity of the environment receiving the discharge. Further, the size of the population being discharged and the dispersion characteristics of the environment will control potential colonization. Several subsequent injections may be required before the species becomes established as a resident.

Several foreign copepods have been identified in various regions of the world (Table 78). They are, as would be expected, all estuarine pelagic or hyperbenthic species except for the records of invasion of eastern Mediterranean species into the Black Sea. The mechanism of transport of *Clausocalanus arcuicornis* and *Calocalanus pavo* is unknown but that of *Acartia tonsa* may have been within ballast water. *Labidocera* species have colonized the Eastern Mediterranean from the Red Sea. Lakkis (1984) identified *L. agilis* and *L. orsinii* on the basis of copepodids IV and V but these identifications require confirmation (Fleminger and Hendrix Kramer, 1988). Movement of species from the Red Sea to the eastern Mediterranean may be aided by ballast water although *Acartia centrura* could have migrated through the Suez Canal (Berdugo, 1974).

Pseudodiaptomus species are circumglobal and hyperbenthic in tropical and temperate, shallow coastal waters (Walter, 1989). The transport of *Pseudodiaptomus marinus* to San Francisco Bay may have been in ballast water but Fleminger and Hendrix Kramer (1988) believe it was transferred to Mission Bay, San Diego, through aquaculture programmes using shipments of live oysters and mussels from Japanese coastal waters. *Acartia tonsa* is largely distributed in the western Atlantic and the Indo-Pacific and was recorded in Europe about 1927 (Gaudy and Viñas, 1985). The earliest European record, by examination of old stored samples, is in the Zuiderzee (Redeke, 1934) where it appeared between 1912 and 1916. Since then, it has appeared in the Baltic Sea, being first observed in 1925 (Elmgren, 1984) and throughout the Mediterranean to the Black Sea. It is tempting to assume

Table 78 Species of copepods that have colonized new regions. The dates of first records in some regions are approximate.

Species	New location	Year	Native location	Authority
Acartia centrura	E Mediterranean		Red Sea	Berdugo, 1974
Acartia fossae	E Mediterranean		Red Sea	Lakkis, 1984
A. margalefi	Southampton, UK	1995		Castro-Longoria and Williams, 1996
A. omorii	S Chile	1983	Japan	Hirakawa, 1988
A. tonsa	Zuiderzee	1912–1916	Indo-Pacific	Redeke, 1934
	Baltic	1925		Elmgren, 1984
	Dunkirk, France	1980		Brylinski, 1981
	Marseilles	1985		Gaudy and Viñas, 1985
	N Adriatic	1987		Belmonte et al., 1994
	Black Sea	1990		Belmonte et al., 1994
Calanopia elliptica	E Mediterranean	1964	Red Sea	Berdugo, 1968
C. media	E Mediterranean	1965	Red Sea	Berdugo, 1968
Calocalanus pavo	Black Sea		Mediterranean	Belmonte et al., 1994
Centropages abdominalis	S Chile	1983	Japan	Hirakawa, 1986
C. typicus	Gulf of Mexico	1985	N Atlantic	McAden et al., 1987
Clausocalanus arcuicornis	Black Sea		Mediterranean	Pavlova, 1964
Labidocera detruncata	E Mediterranean		Red Sea	Lakkis, 1984
L. madurae	E Mediterranean		Red Sea	Lakkis, 1984
L. pavo	E Mediterranean		Red Sea	Lakkis, 1984
Pseudocalanus elongatus	N Aegean Sea	1982	Black Sea?	Siokou-Frangou, 1985
Pseudodiaptomus forbesi	San Francisco Bay	1988	China	Orsi and Walter, 1991
P. inopinus	Columbia River	1990	Asia	Cordell et al., 1992
P. marinus	Hawaii	1964	Japan	Jones, 1966a
	San Diego	1986	Japan	Fleminger and Hendrix Kramer, 1988
Sinocalanus doerrii	San Francisco Bay	1986	Japan	Orsi and Walter, 1991
	San Francisco Bay	1978	China	Orsi et al., 1983
S. sinensis	Japan	1981	China	Hiromi and Ueda, 1987

that its penetration through the Mediterranean started in the west and progressively reached the Black Sea (Table 78).

Acartia margalefi was recorded originally (Alcaraz, 1976) from a ria of Vigo, Spain and then at Southampton, in southern England (Table 78). It may have been confused with *A. clausi*, to which it is similar but smaller; it is also suggested as synonymous with *A. lefevreae* Bradford, 1976 (Castro-Longoria and Williams, 1996). This last species has been recorded from Killary harbour, western Ireland (Ryan *et al.*, 1986), Brest in northwestern France (Bradford, 1976), Vigo in northwestern Spain (Alcaraz, 1983), and the Mediterranean (Bradford, 1976). This is a disjunctive distribution, even if the two species are synonymous, and highlights two problems in deciding whether or not a species is a foreign invader: its native distribution is just not known or there is confusion in its taxonomic identification.

Disjunctive distributions are recorded and that of *Centropages typicus* in Table 78 is probably real because of the high incidence of sampling in regions between the Gulf of Mexico and the coast of South Carolina where the next nearest population occurs (Turner, 1981). *Pseudodiaptomus acutus* occurs in Jamaica, its next nearest southern population being in Brazil (Bowman, 1978b). On a larger scale again, *Candacia pachydactyla* occurs in the tropical Atlantic and Indian Oceans but not around South Africa or in the Benguela Current (Satyanarayana Rao, 1979).

Disjunctive distributions across shorter geographical distances can arise through transport of the copepods in water masses. The distributional range of a species can expand and contract, often on a regular seasonal basis. This can be enhanced by an irregularity in movements of water masses (e.g. Coyle *et al.*, 1990). Mesopelagic species from the northeast Atlantic at about 55°N can enter the northeastwards-flowing slope current along the eastern boundary of the Rockall Trough and be transported to the North Sea, the Norwegian Sea, and to the Norwegian coast where deep-water refugia exist within some of the fjords. Temperate species are exported along this route towards the Barents Sea. Down-current recruitment from one region to another must be fairly common. The southwestward flowing currents over the Nova Scotian continental shelf may aid recruitment of *Calanus* species to its deep basins from the Gulf of St Lawrence (Sameoto and Herman, 1990). This idea is further amplified by Plourde and Runge (1993) who suggest that the large population of *C. finmarchicus* within the lower St Lawrence Estuary may be exported to the Nova Scotian shelf. Exportation and recruitment of this species in the North Sea is discussed by Fransz and Diel (1985). The North Pacific species *Neocalanus cristatus* is exported southwards but can neither survive in Sagami Bay, eastern Japan (Oh *et al.*, 1991) nor in the southwestern warm region of the Japan Sea (Ikeda *et al.*, 1990).

13.2. FAUNAL PROVINCES AND LARGE MARINE ECOSYSTEMS

Biogeographic boundaries of oceanic faunal provinces are shown in Figure 96. These are primarily derived from the distributions of species of mesopelagic fish in the Atlantic Ocean and species of euphausiids in the Pacific Ocean, relative to the physical structure of the oceans. Van der Spoel and Heyman (1983) have examined the geographic distributions of planktonic organisms, including calanoid copepods, relative to such provinces and could draw no firm conclusions. The distributions of some 150 species of calanoid copepods have been partially or wholly collated (Table 79) but do not necessarily conform with the boundaries of the provinces or large marine ecosystems (LMEs) in Figure 96. The potential significance of diel, seasonal and ontogenetic vertical migrations is emphasized by Van der Spoel and Heyman (1983) as a mechanism for the stabilization of oceanic and shelf distributional patterns.

Haury (1986) discusses the concept of core regions within ecosystems and highlights the factors determining the limits to the distributions of species and how different species do not necessarily react in the same way to changes that take place across transitional zones or fronts. Boundaries of several faunal provinces are not sharply defined in terms of the distributions of many species of copepods that apparently cross them.

The faunal provinces in Figure 96 are pertinent not only to the oceanic distributions of copepods but also influence the faunas of enclosed seas and continental shelf regions. These latter regions have been divided into identifiable LMEs (Figure 96). There is a broad distinction between the composition of the copepod faunas of the oceanic faunal provinces and the generally shallower LMEs.

13.2.1. Copepods of the Faunal Provinces

The copepod faunas of the Antarctic and Arctic Oceans are much less diverse than those of the Atlantic, Pacific and Indian Oceans. Diversity is not much greater in the subarctic and subantarctic regions but increases in the temperate and is greatest in the subtropical and tropical regions. In these latter regions, there is more similarity between the epipelagic copepods of the Indian and Pacific Oceans than between the Indian and Atlantic Oceans (Madhupratap and Haridas, 1986). Other general features of the distributions of oceanic species, both horizontally and vertically are discussed in Chapter 11.

13.2.2. Copepods of the LMEs

Sherman (1994) states: "The LMEs are regions of ocean space encompassing coastal areas from river basins and estuaries on out to the seaward boundary of continental shelves and the seaward margins of coastal current systems. They are relatively large regions on the order of 200 000 km^2 or larger, characterized by distinct bathymetry, hydrography, productivity, and trophically dependent populations." The currently recognized LMEs have been primarily identified on the basis of oceanographical criteria and the distribution of commercial fisheries. It remains to define them in terms of ecological units (Ray and Hayden, 1993).

Knowledge of the copepod fauna of individual LMEs (Figure 96) is very variable. Perhaps the best known are the Gulf of Alaska (LME 2), California Current (LME 3), the Gulf of Mexico (LME 5), the Northeast US

Figure 96 The faunal provinces and the Large Marine Ecosystems (LMEs) of the oceans.
The faunal provinces, after Backus (1986), are: I, Arctic; II, subarctic; III, northern temperate; IV, northern subtropical; V, tropical; VI, southern subtropical; VII, southern temperate; VIII, subantarctic; IX, Antarctic.
The 49 Large Marine Ecosystems (LMEs) (Sherman, 1994) are:

1. Eastern Bering Sea
2. Gulf of Alaska
3. California Current
4. Gulf of California
5. Gulf of Mexico
6. Southeast US Continental Shelf
7. Northeast US Continental Shelf
8. Scotian Shelf
9. Newfoundland Shelf
10. West Greenland Shelf
11. Insular Pacific-Hawaiian
12. Caribbean Sea
13. Humboldt Current
14. Patagonian Shelf
15. Brazil Current
16. Northeast Brazil Shelf
17. East Greenland Shelf
18. Iceland Shelf
19. Barents Sea
20. Norwegian Shelf
21. North Sea
22. Baltic Sea
23. Celtic-Biscay Shelf
24. Iberian Coastal
25. Mediterranean Sea
26. Black Sea
27. Canary Current
28. Guinea Current
29. Benguela Current
30. Agulhas Current
31. Somali Coastal Current
32. Arabian Sea
33. Red Sea
34. Bay of Bengal
35. South China Sea
36. Sulu-Celebes Seas
37. Indonesian Seas
38. N. Australian Shelf
39. Great Barrier Reef
40. New Zealand Shelf
41. East China Sea
42. Yellow Sea
43. Kuroshio Current
44. Sea of Japan
45. Oyashio Current
46. Sea of Okhotsk
47. West Bering Sea
48. Faroe Plateau
49. Antarctic

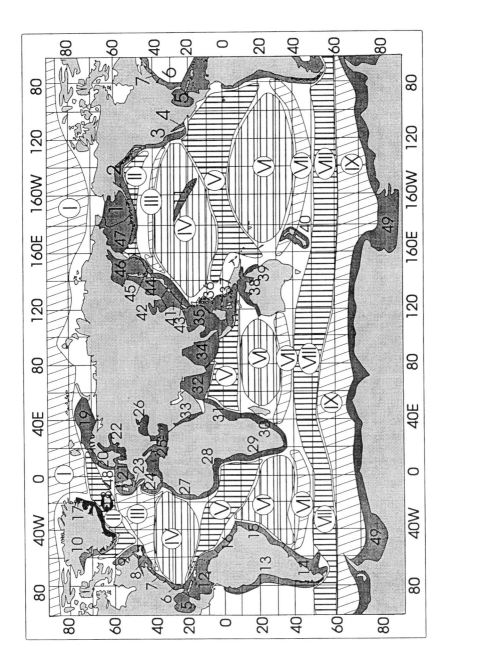

Table 79 Sources of distribution maps of species. Total distributions (World) are described as well as partial distributions confined to one ocean or region.

Species	Region	Reference
Acartia clausi	World	1, 4
Aetideus armatus	N Atlantic	4
Anomalocera patersoni	N Atlantic	4
Calanoides carinatus	N Atlantic	4
Calanus australis	World	15
C. finmarchicus	World	4, 15
C. glacialis	World	4, 15
C. helgolandicus	World	4, 15
C. hyperboreus	N. Atlantic	4
C. marshallae	World	15
C. (Nannocalanus) minor	N. Atlantic	4
C. pacificus	World	15
C. propinquus	World	11
C. simillimus	World	11
C. sinicus	World	15
Candacia armata	N Atlantic	4
C. bipinnata	Atlantic, Indian	4, 10
C. bradyi	Indian	10
C. catula	Indian	10, 15
C. cheirura	Indian	10
C. curta	Atlantic, Indian	4, 10
C. discaudata	Indian	10
C. ethiopica	Atlantic, Indian	4, 10, 15
C. guggenheimi	Indian	10
C. longimana	Atlantic, Indian	4, 10, 15
C. norvegica	N Atlantic	4
C. pachydactyla	Atlantic, Indian	10, 12
C. samassae	Indian	10
C. tenuimana	Atlantic, Indian	4, 10
C. tuberculata	Indian	10
C. varicans	Indian	10
Candaciidae	Indian	2
Centropages abdominalis	E Pacific	8, 13
C. aucklandicus	New Zealand	14
C. australiensis	Australasia	14
C. bradyi	N Atlantic	4
C. brachiatus	World	8, 13, 15
C. chierchiae	World	4, 14
C. furcatus	E Pacific	8, 13
C. hamatus	Atlantic, Pacific	4, 8, 13
C. typicus	Atlantic, Pacific	4, 8, 13
C. velificatus	Atlantic, Pacific	8, 13
C. violaceus	N Atlantic	4
Clausocalanus arcuicornis	World	15
C. farrani	World	15, 17

Table 79 Continued.

Species	Region	Reference
C. jobei	World	15, 17
C. laticeps	World	11
C. lividus	World	11
C. minor	World	15
C. paululus	World	15
Clausocalanus	7 spp. in Atlantic	9
Ctenocalanus vanus	N Atlantic	4
Epilabidocera longipedata	Pacific	3
Eucalanus attenuatus	World	4, 15
E. bungii	World	11, 13, 15
E. californicus	World	11, 13, 15
E. crassus	N Atlantic	4
E. elongatus	World	4, 11, 13, 15
E. hyalinus	World	11, 13, 15
E. inermis	World	11, 13, 15
E. langae	World	15
E. monachus	N Atlantic	4
E. mucronatus	N Atlantic	4
E. parki	World	15
E. pileatus	N. Atlantic	4
E. sewelli	World	15
E. subtenuis	World	15
Euchaeta acuta	Atlantic, Indian	4, 6
E. concinna	Indian	6
E. longicornis	Indian	6
E. marina	Atlantic, Indian	4, 6
E. media	Indian	6
E. paraacuta	Indian	6
E. plana	Indian	6
E. pubera	Atlantic, Indian	4, 6
E. spinosa	N Atlantic	4
E. tenuis	Indian	6
E. indica	Indian	6
Euchirella rostrata	N Atlantic	4
Haloptilus acutifrons	Indian	5
H. longicornis	N Atlantic	4
Heterorhabdus abyssalis	N Atlantic	4
H. norvegicus	N Atlantic	4
H. papilliger	N Atlantic	4
Isias clavipes	N Atlantic	4
Labidocera species and species groups		
L. acuta	Pacific	3
L. acutifrons	Pacific	3
L. aestiva spp. group	America	8

Table 79 Continued.

Species	Region	Reference
L. detruncata	Pacific	3
L. lubbocki	World	15
L. diandra	World	15
L. jollae	World	15
L. jollae spp. group	West N America	8
L. trispinosa spp. group	America	8
L. wollastoni	N Atlantic	4
Mecynocera clausi	N Atlantic	4
Mesocalanus tenuicornis	N Atlantic	4
Metridia longa	N Atlantic	4
M. lucens	N Atlantic	4
Neocalanus gracilis	N Atlantic	4
N. plumchrus	World	15
Paracandacia bispinosa	Atlantic, Indian	4, 10
P. simplex	Atlantic, Indian	4, 10
P. truncata	Indian	10
Pareuchaeta aequatorialis	World	19
P. barbata	World	18
P. comosa	World	19
P. confusa	World	19
P. gracilicauda	World	19
P. hebes	N Atlantic	4
P. norvegica	N Atlantic	4
P. pseudotonsa	World	4, 19
P. sarsi	World	19
P. scotti	World	19
P. tonsa	World	19
P. tuberculata	World	19
P. vorax	World	19
Parapontella brevicornis	N Atlantic	4
Pleuromamma abdominalis	Atlantic, Indian	4, 16
P. borealis	Atlantic, Indian	4, 16
P. gracilis	Atlantic, Indian	4, 16
P. indica	Indian	16
P. piseki	Atlantic, Indian	4, 16
P. quadrungulata	Indian	16
P. robusta	N Atlantic	4
P. xiphias	Atlantic, Indian	4, 16
Pontella agassizi	Pacific	3
P. danae	Pacific	3
P. denticauda	Pacific	3
P. fera	Pacific	3
P. princeps	Pacific	3
P. tenuiremis	Pacific	3
P. valida	Pacific	3
Pontellina morii	World	7, 11, 12, 15

Table 79 Continued.

Species	Region	Reference*
P. platychela	World	7, 11, 15, 17
P. plumata	World	7, 12, 15
P. sobrina	World	7, 11, 12, 15
Pontellopsis villosa	Pacific	3
Pseudocalanus elongatus	N Atlantic	4
Rhincalanus cornutus	Atlantic, Indian	12
R. nasutus	Atlantic	1, 4
R. rostrifrons	Indian	17
Temora discaudata	World	8, 13, 15
T. longicornis	Atlantic, Pacific	4, 8, 13
T. stylifera	Atlantic, Pacific	4, 8, 13
T. turbinata	Atlantic, Pacific	8, 13
Tortanus discaudatus	N Atlantic	4
Undeuchaeta major	N Atlantic	4
U. plumosa	N Atlantic	4
Undinula vulgaris	N Atlantic	4

*(1) Sewell, 1948; (2) Jones, 1966b; (3) Geinrikh, 1969; (4) Oceanographic Laboratory, Edinburgh, 1973; (5) Stephen and Sarala Devi, 1973; (6) Tanaka, 1973; (7) Fleminger and Hülsemann, 1974; (8) Fleminger, 1975; (9) Williams and Wallace, 1975; (10) Lawson, 1977; (11) Reid *et al.*, 1978; (12) Shih, 1979, 1986a; (13) Haedrich and Judkins, 1979; (14) Bradford, 1980; (15) Van der Spoel and Heyman, 1983; (16) Saraswathy, 1986; (17) Shih, 1986a; (18) Mauchline, 1992b; (19) Park, 1994b.

Continental Shelf (LME 7), the Scotian Shelf (LME 8), the Patagonian Shelf (LME 14), the North Sea (LME 21), the Baltic Sea (LME 22), parts of the Mediterranean (LME 25), the Benguela Current (LME 29), the Great Barrier Reef (LME 39), the Kuroshio (LME 43) and the Antarctic (LME 49). Information on most of the remaining 75% of LMEs is much less comprehensive, certainly as regards the copepods.

A broad study is required like that of Turner (1981) who examined the distributions of the estuarine copepods of eastern North America. Such work allows examination of the changes in species composition and diversity with latitude and other factors and represents a broad geographical range encompassing several contiguous LMEs. Conversely, equally valuable studies can be based on taxonomic groups, for example that of Park (1994b) on the genus *Pareuchaeta* on a world-wide basis or that of Lawson (1977) on the family Candaciidae on an ocean-wide basis. More attention should be paid to changing densities within the distribution of a species, relative to Haury's concept of core regions within ecosystems and, by extrapolation, core regions within the distribution of a species. This is difficult because of the variety of investigators and sampling gear. The attempt has to be made, especially against the background of current studies on biodiversity and climate change.

13.3. IDENTIFICATION OF SPECIES

Describing the geographical distribution of a species depends on its correct identification in each of the localities. The recognition of closely related species is often very difficult and sometimes involves considerable experience of their morphology. Many recognized species of calanoid copepods exhibit morphological variation but whether this is within-species variation or arises because more than one species is present is often unknown. Many examples of confusion of species during studies of their geographical distributions exist.

Shih (1979) uses the example of *Calanus finmarchicus* whose geographical distribution before 1948 was frequently assumed to be cosmopolitan. At that time, the genus also included the three large, high latitude species *C. hyperboreus*, *C. propinquus* and *C. simillimus* but *C. helgolandicus* and *C. ponticus* had also been described. The distinctive *Nannocalanus minor*, now usually included in the genus *Calanus* but still frequently ascribed to its own genus, had also been described.

The following species were separated from *C. finmarchicus* between 1948 and 1977: *C. australis*, *C. chilensis*, *C. glacialis*, *C. marshallae*, *C. pacificus* and *C. sinicus*. Then, nearly 20 years later, De Decker *et al.* (1991) distinguished *C. agulhensis* around the Agulhas Bank off South Africa and Hülsemann (1994) *C. jaschnovi* in the western Pacific as valid species. Hülsemann (1991b) confirmed the status of *C. ponticus* in the Black Sea as a valid species but changed its name to *C. euxinus*. A recent morphometric study of *C. chilensis* by Marin *et al.* (1994) provides further evidence of the status of that species.

Similar taxonomic problems exist within other genera such as *Pseudocalanus* examined by Frost (1989), and the family Euchaetidae studied by Park (1994a,b). Species in these genera are also common and widespread. The general consequence of these problems, discussed in more detail in Chapter 3, is that there is considerable doubt involved in collating records of the occurrence of many species from the literature to produce distributional maps.

13.4. CONCLUDING REMARKS

Collation of the distributions of copepod species with the boundaries of faunal provinces and LMEs is currently meeting with much difficulty. Many species are distributed across several provinces, occurring, for example, in the northern temperate, subarctic and Arctic North Atlantic or within the subtropical and tropical provinces of the oceans. Such broad distributions were noted by Van der Spoel and Heyman (1983). Consequently, in many cases there is a somewhat tenuous relationship between the faunal provinces and the distributions of individual species. Van der Spoel (1986) maps the distributions of a variety of planktonic organisms, including that of *Calanus helgolandicus*, in the North Atlantic. The northern and southern limits of its distribution do not coincide with the boundaries of the North Atlantic Cold Water. The most significant boundaries are those between the temperate and subtropical provinces in both hemispheres. Here, relatively aseasonal environmental characteristics are replaced by seasonal ones. This is reflected by the decreased richness of species on the higher latitude sides of these boundaries.

Calanus helgolandicus belongs to the northern temperate Atlantic faunal province. Records of it occurring in the North Sea and off western Norway are common, there being an input of North Atlantic water to the North Sea around the north of Scotland. *Calanus finmarchicus* inhabits the northern regions of the northern temperate Atlantic faunal province, and the whole of the subarctic Atlantic province, and is relatively common within the Arctic. Distributions of copepods that cross faunal provinces are common. Patterns of distribution of *Pareuchaeta* species are discussed in considerable detail by Park (1994b) relative to their phylogenetics and to faunal provinces. Some species are endemic to either eutrophic or oligotrophic regions. The eutrophic species have more restricted geographical distributions than the oligotrophic ones. When a pair of *Pareuchaeta* species dominate a eutrophic region, they are likely to belong to separate species groups (Park, 1994b).

Thus, a species has a core region of geographical distribution. The ocean currents transfer a portion of the population downstream and may even, through mixing, transfer it into an adjacent water mass. Conditions there may be or become, as the species is further transported within that region, unfavourable and so the species does not establish a breeding population. Conversely, such expatriated species can, on occasion, establish breeding populations that give rise to disjunctive distributional patterns. Thus, some fjordic environments around the world have isolated populations of normally offshore species. Marcus (1986), Shih (1986a,b) and Van der Spoel (1986) extrapolate such expatriation to situations where distinct forms of a species, and new species, evolve.

14. Copepods in Ecosystems

14.1. Ecosystems 519
    14.1.1. Food webs 519
    14.1.2. Predation 521
    14.1.3. Exploitation by man 522
    14.1.4. Parasites 523
14.2. Fate of Faecal Pellets 525
14.3. Perturbations within Ecosystems 526
14.4. The Biology of Copepods 527

The calanoid copepods are predominantly members of the pelagic plankton although they also inhabit the hyperbenthic and benthopelagic environments in association with the sea bed in shallow and deep waters. There are probably many new species of calanoid copepods yet to be described, especially in environments close to the sea bed from shallow, coastal waters and also from the deep sea. For instance, Madhupratap et al. (1991), examining the zooplankton of the lagoons of the Laccadives in the eastern Bay of Bengal, refer to two undescribed species of the genus *Pseudocyclops* and a new *Calanopia* species. Park (1994a) suggests that many more species of the *Pareuchaeta malayensis* group of species will be discovered through further deep-sea sampling. Finally, some new species will be recognized and separated from currently existing species. Yet another undescribed species of the genus *Calanus*, that has already received much attention, is present in the eastern equatorial Pacific (Hülsemann, 1994).

The majority of the different compartments of the marine environment have probably now been sampled for calanoid copepods, with more attention paid to some than others. Consequently, it is unlikely that many new forms of life history or physiology will be discovered among the copepods. Their role and interactions within marine ecosystems have been investigated, primarily in a descriptive context, and much remains to be done.

14.1. ECOSYSTEMS

Copepods live in association with many other kinds of organisms whose living space and nutritional resources they share. Sherman's papers describing Large Marine Ecosystems (LMEs) are linked with the distributions of copepods in Chapter 13. The identity of an ecosystem is difficult to define because its detailed species composition changes in time and space. Contiguous ecosystems usually have varying proportions of species in common. An ecosystem is a holistic concept of an association of organisms, the distribution and density of one or more of the dominant species often prescribing the boundaries of the system. Alternatively, the system may be defined by the distribution of physical characteristics of the environment such as a gyre, current, coral reef or hydrothermal vent. Such examples encompass a wide range of scales of size, an oceanic gyre being on a 10^6 m while a coral reef may be on a metre scale. This can represent a difference of about 10^{12} m^3 on a volume basis. This vast difference in the sizes of associations that can be considered as identifiable ecosystems results in large differences in their relative complexity, often exemplified by the trophic relationships in time and space between the individual component species.

An ecosystem is recognizable on time scales longer than a year as instanced by the persistent orders of dominance among species from year to year, for example. those of communities off the eastern United States (Sherman et al., 1983). There can, however, be seasonal fluctuations in the orders of the dominant species on time scales of weeks or months. Trophic relationships, articulated by describing food webs, are variable on shorter time scales of days or weeks. Changes in dominance result primarily from different species having their seasonal maxima in numbers at different times of the year. Variation in the relationships within a food web derive from seasonal changes in the trophic resources available as well as by the imposition of seasonal predation by migratory predators. Finally, there are the parasites within the ecosystem that infect the copepods, often on a seasonal basis.

14.1.1. Food Webs

Copepods, as eggs, nauplii, copepodids and adults and as individual species and faunas, interact with their biological environment. Some calanoid species compete with some cyclopoid species for living space and resources. Herbivorous species can, in some areas and circumstances, control populations of phytoplankton (e.g. Frost, 1987; Springer et al., 1989). They must also influence the distributions and sizes of the populations of their predators (Springer and Roseneau, 1985).

Seasonal changes in resources derive primarily from seasonality in phytoplankton production, resulting in cascades of larval production by benthic, planktonic and fish populations within the ecosystem. The timings of maxima of phytoplankton and all groups of meso- and macroplankton within a region of the North Sea are examined over a fifteen-year period by Roff *et al.* (1988), who found that it switched seasonally between being food limited and predator controlled. According to Lonsdale *et al.* (1996), egg production of *Acartia tonsa* is food-limited in summer but not in winter or spring in Long Island bays. Eggs and larvae are nutritionally rich and represent a major increase to the concentration of food particles of a size spectrum most frequently exploited within the ecosystem. The larvae of a species, because of their smaller body size, have different connections within the web from those of the adults.

Food webs, such as those described in Balsfjord, Norway by C.C.E. Hopkins *et al.* (1989) and Falk-Petersen *et al.* (1990) or in McMurdo Sound, Ross Sea by T.L. Hopkins (1987), the Scotia Sea by T.L. Hopkins *et al.* (1993), and in the western Weddell Sea by Hopkins and Torres (1989) look complex but, in fact, are considerably simplified. They are primarily qualitative descriptions of the principal trophic connections between species integrated over the year or season. The period examined by T.L. Hopkins (1987) was limited, sampling taking place between 18 and 21 February only. Hopkins examined the size ranges of the species within the web as well as the diversity within the diets of the species, providing a picture, simplified to a degree, of an oceanic food web existent within a three-day period. The amount of information required to describe most webs in any greater detail is large and whether it would contribute to a greater understanding is questionable.

A later study by T.L. Hopkins *et al.* (1993), on the structure of the food web in the region of the Weddell-Scotia confluence, found three major trophic pathways to the apex predators which comprised birds, seals and minke whales.

a. Phytoplankton + protozoans → copepods + krill → *Electrona antarctica* → flying seabirds.
b. Phytoplankton + protozoans → krill → seals + minke whale.
c. Phytoplankton + protozoans → krill → cephalopods → penguins + fur seal.

The third pathway was not adequately examined because birds catch squid much more efficiently than do trawls. This illustrates the difficulties of quantifying connections within food webs. Quantifying the biomasses of species and their developmental stages in time and space within a web is the first major sampling problem encountered. The second is quantifying, in time and space, the dietary constituents of these components of the web. Identification of materials within the stomach contents of copepods is discussed in Chapter 5 and the same problems are present in most other types of organisms. The final parameter to be defined is the rate of consumption, over time, of the food of each component in the system.

Parts of food webs can be examined experimentally (e.g. Greve, 1995; Kivi *et al.*, 1996) and provide insights into the functioning of the web as a whole.

14.1.2. Predation

The effects of predation within an ecosystem can be varied in space and time. Calanoid copepods, because of their numerical dominance, can affect the densities of their prey. They can limit the duration of the phytoplankton bloom in some instances through grazing it at a higher rate than the rate of production of the phytoplankton. In other instances, they can affect their own or the production of another species of copepod by grazing available eggs and nauplii.

A list of predators of copepods would consist of a vast array of organisms within the plankton, micronekton and nekton but would also include benthic invertebrates, mammals and birds. The predation consists of two general components. The first originates from the normal background predation resulting from predators more or less continuously present within the ecosystem. The second is intermittent, usually seasonal and results from migratory predators. This latter component of the predation pressure is often difficult to quantify. Wishner *et al.* (1988), in studying the distribution of a large multispecies aggregation of copepods off New England, could not correlate its well-defined borders with physical features of the environment. They suggested that one border, adjacent to a feeding ground of whales, may have been defined by the predation of the whales.

Much further work is required to identify quantitatively the mortality within copepod populations originating from predation and separating it from mortality from other causes. According to Mehner (1996), there are few calculations of the impact of 0-group fish on populations of copepods. Such fish have high metabolic rates and are of such a size as potentially to predate all developmental stages of copepods. Mehner's study in the Baltic found less impact on the copepods than expected. The viability of the eggs of copepods after they have passed through the guts of fish also reduces the effects of predation. Further investigation of the impacts of fish on populations of copepods are required.

Some species of copepods, and other organisms, are more liable to predation than others. Mauchline and Gordon (1986) examined the diets of some 33 commonly caught species of demersal fish within the Rockall Trough, northeastern Atlantic across a bathymetric range of 400 to 2900 m. A total of more than 230 species of prey organisms were recorded in the stomachs and many more species existed in the environment but were not recorded as items of prey. Of the 230 recorded species, only some 30, planktonic, benthopelagic and demersal species, occurred regularly within the stomachs of a variety of species of fish. The fish examined represented a variety of families and genera inhabiting different compartments over the bathymetric range. So why were there only 30 common species of prey? One common feature of these prey species is their probable occurrence in aggregations but this is also a feature of many species that were not recorded within the stomachs. These 30 species may be either more eurybathic and/or dominant numerically within their respective taxon and bathymetric range in the Rockall Trough than most other species. The species of copepods among these 30 prey species are *Pareuchaeta norvegica*, *Pleuromamma robusta*, *Aetideopsis multiserrata*, *Calanus helgolandicus*, *Heterorhabdus norvegicus*, *Xanthocalanus profundus* and *Xanthocalanus* spp. These could be termed "professional prey species" in that they function as food for other organisms within the ecosystem. They must thus have life styles that are different from species that are rarely preyed on by other organisms. Their populations must have rates of production related to their degree of exploitation as well as to processes of physiological mortality. They must contribute an element of predation-immunity to co-occurring non-prey species. Multispecies patches, or aggregations, of copepods and other organisms (Mauchline and Gordon, 1986) may primarily function to protect non-prey species as well as conferring a degree of protection on those commonly preyed upon.

14.1.3. Exploitation by Humans

The body size of copepods precludes them being commercially fished on a large scale. Some species can be caught in sufficient numbers, for example fjordic concentrations of *Pareuchaeta norvegica*, to raise the question of their feasibility as a commercial crop. The integument is indigestible although the flavour of the copepods is reminiscent of shrimp or prawn. The catch usually has to be boiled first in water and the emergent oils decanted. They are then fried in vegetable oil and served on buttered toast. They are delicious but second helpings usually result in indigestion.

Wickstead (1967) reviews the feasibility of harvesting plankton, and copepods in particular, in Indo-Pacific waters. He argues that they are nutritionally satisfactory and a suitable source of protein. A similarly optimistic study of the potential exploitation of the North Atlantic species *Calanus finmarchicus* is that of Wiborg (1976). This species was used in Norway from about 1960, in a deep-frozen state, as a supplementary food in the culture of salmonids. The copepod was caught in a 12 m² beam trawl whose mouth had a coarse net covering it to exclude larger organisms, especially medusae, from the catch. The net was either towed or anchored in a coastal current. The fishery was seasonal between the end of April and the end of June, when the best catches occurred in the late evening or at night. Wiborg suggests that the patchy distribution and short fishing season should make overfishing of the species difficult but modern technology could probably overcome this.

Wiborg's study is referred to by Omori (1978b) in reviewing the zooplankton fisheries of the world. The only other copepod commercially exploited at that time was *Neocalanus plumchrus* in the North Pacific. A seasonal fishery occurred off Kinkazan, Japan where it was exploited for pet food and ground bait.

Thus copepods are a potential resource of protein but their exploitation on a commercial basis is difficult at an economic cost. Further, they are involved in so many food chains that overfishing, even within a restricted geographic region, could have consequences on other commercial fisheries.

14.1.4. Parasites

Parasitic organisms, or endosymbionts, are a component of all ecosystems. There are also epibionts, or ectosymbionts, that attach themselves to the surfaces of their hosts but do not penetrate through the integument to the underlying tissues. The most comprehensive account of such organisms associated with copepods is that of Sewell (1951) on species collected from the Arabian Sea. He illustrates many of them and reviews the earlier literature. More recent reviews are those of Hiromi *et al.* (1985), Ho and Perkins (1985) and Théodoridès (1989).

14.1.4.1. *Endosymbionts*

Parasites living within the guts and other tissues of copepods represent a variety of organisms such as bacteria, dinoflagellates, ellobiopsids, gregarine sporozoans, digenean trematode, and cestode and nematode worms (Ho and Perkins, 1985). There is no confirmed infection of a calanoid copepod by a virus (Théodoridès, 1989).

Endosymbionts can be responsible for significant rates of mortality in populations of copepods. Kimmerer and McKinnon (1990) found that a parasitic dinoflagellate, *Atelodinium* sp., on female *Paracalanus indicus* in Port Phillip Bay, Australia caused about one third of the daily mortality. Females were primarily parasitized, CVs less so, and no males were infected. The infection is internally located and requires an incubation time of 17.5 to 37.6 hours for the development and release of the dinospores from the resultant dead body of the copepod. They estimated from the data of Ianora *et al.* (1990) that a similar degree of mortality was caused by the parasitizing of *P. parvus* by the ciliate *Syndinium* sp. in the Gulf of Naples. The latter authors review the genera of copepods parasitized by syndinians. They also observed infestations of another dinoflagellate, *Blastodinium contortum*, and of bacteria and fungi in several species of copepods. *Blastodinium* sp. is also reported from the gut of the Antarctic *Pareuchaeta antarctica* by Øresland (1991). They did not occur in the non-feeding males, suggesting that the spores are ingested with the food. The effects of this parasite on other copepods are reviewed although no effects were evident in *P. antarctica*.

An ectoparasitic dinoflagellate, *Syltodinium listii*, described by Drebes (1988) attaches its stalk to the eggs of *Acartia*, *Pseudocalanus* and *Temora* species, and sucks out their contents in about 90 min. Drebes describes its life history and reviews other dinoflagellates parasitizing eggs. No estimates of projected effects on the populations of copepods are given.

Cryptoniscid isopods of the family Bopyridae infect copepods. The adult parasite lives on a prawn and the female releases lecithotrophic, epicaridium larvae from a marsupium. These swim in the surface layers of the sea until they encounter and attach to a copepod. The epicaridium larva metamorphoses on the copepod to a microniscid larva that in turn develops to a cryptoniscid. This latter larva detaches and finds its final host, the prawn. Owens and Rothlisberg (1995) list some 14 species of calanoid copepods that act as intermediate hosts in the Gulf of Carpentaria, Australia; infestation was commonest in two species, *Canthocalanus pauper* and *Euchaeta concinna*.

14.1.4.2. *Ectosymbionts*

Many kinds of organisms attach themselves to the surfaces of the bodies of copepods. They include bacteria, diatoms, fungi, ciliates and epicaridean crustaceans and Ho and Perkins (1985) review their occurrence. They list ellobiopsids in this category but, although their reproductive organs protrude externally from the integument, they are endosymbionts.

The bacteria attached to the deep-sea *Neocalanus cristatus* are described by Nagasawa and Terazaki (1987) and Nagasawa (1989) reviews those on other copepods. Effects on the copepods are unknown.

Hiromi *et al.* (1985) suggest that epizoic diatoms found on copepods may be specific to the copepods but the benefits conferred on the diatoms by this association are as yet conjectural as are possible effects on the copepods. Infestation by such organisms tends to be seasonal as Weissman *et al.* (1993) found in peritrich ciliates attached to *Acartia hudsonica* in Long Island Sound. These stalked ciliates did not penetrate the integument of the copepod but were solely attached to its surface. Peak infection occurred in late spring and was not related to age or developmental stage of the copepod. Infected nauplii had lower survival rates but normal growth rates while infected adults had lower than average sinking rates. Weissman *et al.* (1993) conclude that such infections affect the survival rates of the copepods and so influence rates of production within the population. An experimental study by Heerkloss *et al.* (1990) also suggests a marked mortality of nauplii of *Eurytemora affinis* originating from the attachment of ciliates.

14.1.4.3. Conclusions

Reference should be made to the review papers mentioned at the beginning of this section, especially that of Théodoridès (1989), for detailed information of parasites. The life histories of most parasites are unknown as are their effects in controlling population sizes of their hosts. Many parasites have probably been overlooked. Ho and Perkins (1985) and Théodoridès (1989) review the state of knowledge, or rather lack of it, pointing out that there are few studies on the effects of parasites on the health of the copepods and consequently little information on the influence of parasites within ecosystems.

14.2. FATE OF FAECAL PELLETS

There is much interest in the rates of sedimentation of organic matter from the surface layers to the deeper water column, including the contribution made to this by faecal pellets (Turner and Ferrante, 1979; Andersen and Nival, 1988; Voss, 1991; Longhurst and Williams, 1992). The dominance by numbers of copepods in most regions of the oceans results in a considerable production of faecal pellets by them, as discussed in Chapter 6. The rates of production and the volumes of pellets produced by different species of copepods are given in Tables 24 and 25 (pp. 198 and 200). The rate of sedimentation of the pellets depends on the populations of copepods and the food eaten by them. Diets of diatoms, dinoflagellates and coccolithophorids, which have mineral skeletons, result in faecal pellets with faster sinking rates (Voss, 1991; González *et al.*, 1994). Emerson and Roff

(1987) concluded that small faecal pellets, those produced primarily by nauplii and early copepodids, were grazed in the water column while the larger pellets reached depth. Bathmann et al. (1987) concluded that most pellets were broken down or reingested by coprophagy in the water column during May and June at a station in the Norwegian Sea. A similar situation was described on the southeastern continental shelf of the United States by Hofmann et al. (1981) and experimentally by Lampitt et al. (1990). The pellets are fed on by a wide variety of organisms including nauplii (Green et al., 1992) and copepodids of the copepod species themselves.

The data on fluxes of faecal pellets to the deep layers of the ocean cannot necessarily be ascribed to named species of copepods. Some species of copepods produce distinctive pellets but many pellets cannot be identified with the species producing them. Coprophagy is common and a pellet may be consumed more than once. It is probable that it is eaten by a larger animal than originally produced it; in this case it would be likely to be incorporated within a larger faecal pellet.

Thus, the downward flux of faecal pellets, along with carcasses of dead animals and post-bloom phytodetritus, transfers part of the production from shallower to deeper ecosystems, in both the deep-sea and coastal water column. It can have a significant function within the ecosystems concerned.

14.3. PERTURBATIONS WITHIN ECOSYSTEMS

As inferred in the previous section, individual ecosystems have many stable and recognizable features that appear to be a result of many relatively chaotic components and events. Contiguous ecosystems usually share a proportion of their species. The geographical distributions, and their constraints, of many species are unknown and studies are required of the breadth of that of Turner (1981) who examines the distributions of the estuarine copepods of eastern North America. This allows examination of the changes in species composition and diversity with latitude, and other factors, and represents a broad biogeographical range over several contiguous LMEs. Conversely, equally valuable studies can be based on taxonomic groups such as that of Park (1994b), on genus *Pareuchaeta* on a world-wide basis, or of Lawson (1977), on the family Candaciidae, on an ocean-wide basis. These are carnivorous copepods and so correlation of their distributions with those of their prey species are pertinent. How common or otherwise are coinciding predator and prey distributions?

Reid (1967) maps world distributions of anchovies. They occur in all of the eastern boundary currents (California, Peru, Canary and Benguela currents, and off western Australia), in three of the western boundary currents (Kuroshio, East Australia and Brazil currents) but not in the Gulf Stream or Agulhas Current. They occur also in the Sea of Japan, the area south of Australia, around north New Zealand, and in the Mediterranean and Black Sea. In summer, they extend into the North Sea, Baltic and the Sea of Azov but retreat from these in cold winters, though they remain in the Black Sea throughout the year. It would be interesting to know the trophic resources exploited by the different populations of anchovies in the different regions. It is tempting to suppose that predators such as these, along with others such as sardines, herring, species of mesopelagic fish and possibly squid cause seasonal perturbations within the ecosystems through their exploitation of selected components.

Such perturbations, caused by the medusa *Aurelia aurita*, were observed in the Kiel Bight by Behrends and Schneider (1995). They monitored the populations over a five-year period, between 1990 and 1994. The numbers of total zooplankton organisms and copepods were inversely related to the numbers of medusae present. Not all species of copepods were affected. Significant changes in the numbers of *Paracalanus* and *Pseudocalanus* were present but not of *Centropages hamatus* and *Acartia* species. The trophic structure of the zooplankton web altered, fine-filter feeders and raptorial feeders dominating when numbers of medusae were low while coarse-filter feeders dominated when the medusae were numerous. Aaser *et al*. (1995), experimentally examining a brackish environment, suggest that a mysid, *Neomysis integer*, grazes *Eurytemora affinis* and can do so to such an extent that it reduces the grazing pressure of the copepod on the phytoplankton community. Perturbation at one locus in a system will have effects at other locations, and not necessarily instantaneously.

14.4. THE BIOLOGY OF COPEPODS

The biology of the copepods, primarily because they are dominant components of the ecosystems, must have a major influence in the design and functioning of these systems. As Verity and Smetacek (1996) point out, traditional, and modern, studies of pelagic marine ecology are biased towards factors regulating growth and production of organisms at the species, population and community level. Resource acquisition or bottom-up forcing is accentuated much more than predation or top-down forcing in considering the structuring of ecosystems. They further state that "The bulk of large-scale biogenic cycling and production of fish food is dependent on relatively few taxa and morphotypes". As pointed out above, when discussing predation within ecosystems, there are only a few species of copepods in the Rockall Trough that contribute significantly to the food resources of the benthopelagic and pelagic fish. Verity and Smetacek (1996) term these key species and briefly review the significance of the following in ecosystems: the pennate diatom, *Fragilariopsis* (*Nitzschia*) *kerguelensis*; the coccolithophorid, *Emiliana huxleyi*; the polymorphic taxon of phytoplankton, Phaeocystis; and the euphausiid crustacean, *Euphausia superba*.

Some species within certain genera of copepods are key species in the ecosystems that they dominate. In the genus *Calanus*, *C. finmarchicus* and *C. glacialis* are such in the North Atlantic and Arctic. Farther south, *Calanus helgolandicus* is important. In the Antarctic, *C. simillimus* and *C. propinquus* are prominent. The North Pacific has another genus of the family Calanidae represented by *Neocalanus plumchrus*, *N. cristatus* and *N. flemingeri*. Yet another genus of the Calanidae, *Calanoides*, has key species within regions of upwelling. The other species within the genus *Calanus* are important within the geographical regions in which they occur but they may not all be important enough to be key species within their respective ecosystems. There are other genera that are known to have key species in large or smaller ecosystems. Coastal ecosystems have key species representing the genera *Acartia*, *Centropages*, *Eurytemora*, *Paracalanus*, *Parvocalanus*, *Pseudocalanus* and *Pseudodiaptomus*. Offshore systems have species of *Ctenocalanus*, *Eucalanus*, *Euchaeta*, *Rhincalanus*. Deep-sea mesopelagic ecosystems have representatives of the genera *Heterorhabdus*, *Metridia*, *Pleuromamma* and *Pareuchaeta* while bathypelagic systems have species of *Euaugaptilus*, *Heterorhabdus*, *Lucicutia*, *Metridia* and *Pareuchaeta*. The ecosystems of the bathypelagic environment are least known but those studied in the North Atlantic, Levantine Basin of the Mediterranean Sea, and the Red Sea have different copepod faunas (Weikert and Koppelmann, 1993), and so different key species.

There are more than 200 genera of calanoid copepods but only about 20 of these subscribe key species. This may be an under-estimation because some others such as *Aetideus*, *Aetideopsis*, *Candacia*, *Clausocalanus*, *Pseudocyclopia*, *Spinocalanus*, *Stephos*, *Temora*, *Undeuchaeta* and *Xanthocalanus* may have key species in restricted environments or small ecosystems. Nevertheless, it is clear that only a few genera or morphotypes provide the key species. This is also true of other phyla, and Verity and Smetacek (1996) suggest that the reason for this is that "they may hold space because they sequester resources (bottom-up structuring) better than their rivals, or because they avoid or inhibit predators (top-down structuring)". Key species of copepods are near the bottom end of complex oceanic food webs but in the middle of less complex coastal webs. In both situations, however, their primary role, described above when discussing predation in ecosystems, is as a professional prey species. Consequently, they must be sequestering resources efficiently because their populations provide a crop for the predators. At the other extreme, a species such as the deep-sea *Valdiviella insignis* (Figure 8B, p. 22), that only produces two young per brood, must have a very different life strategy. There seems little room for a contribution to the resources of predators and it follows that it must have very efficient mechanisms for avoiding them. What these might be are unknown.

The conservative shape of the body throughout the pelagic genera of calanoid copepods has developed, according to Verity and Smetacek (1996), to enable predator-avoidance rather than food-gathering. Nothing is known about development of chemical defence by copepods. Likewise, although a few copepods have prominent spines (Figure 2, p. 4), their function is not known. Behavioural defence, such as the escape reaction, diel vertical migration and aggregations, whether single or multispecies, have been studied as have aspects of life history strategies such as diapausing eggs, nauplii and copepodids. The production of diapause stages is usually construed as a mechanism of avoidance of severe chemical and physical environmental conditions rather than of predators. Predator or grazer control of population sizes of prey organisms and the phenomenon of trophic cascading are reviewed by Verity and Smetacek (1996). They conclude that "biogeochemical fluxes are spearheaded by key taxa, perhaps 'keystone species' (*sensu* Bond, 1993) whose environmental adaptations, whether through behavior, morphology, physiology, or life history, are so strong that they direct trophic relations".

There is no clear evidence at present that the sizes of populations of key species of marine pelagic copepods are controlled by their predators. What is clear is that these species, within their life histories, provide a crop that is exploited by their predators. They are successful species as shown by their dominance within their communities. This could result from bottom-up control and their dominance makes them significant within food webs. But why these species and not other species? How have they been selected? What features have they got in common with key species from other taxa and phyla and at other positions in the food webs? These questions increase the importance of studies of the behaviour of copepods as described by Price *et al.* (1988). Morphological and histological studies are still important as emphasized by that of Nishida and Ohtsuka (1996) who discovered that *Heterorhabdus* species have poison fangs for disabling their prey. A colloquium (Marine Zooplankton Colloquium 1, 1989) identified several areas of investigation for the future, among which the following are pertinent to this discussion and their relevance to copepods amplified here:

a. Characterization of individual small-scale behaviours leading to a better understanding of the dynamics and function of aggregation and dispersal. The determination of nearest neighbour distance (NND) coupled with investigations of the functioning of chemo- and mechanosensory sensilla, production of exocrenes, determination of detection and interaction distances between individuals, between individuals and aggregations, and between adjacent aggregations should be done concurrently.

b. Determination of how the variance of the mean values of environmental variables affects physiology and behaviour. This approach emphasizes rates of change relative to the amplitude of the variation about the mean rather than to the mean itself.

c. Interrelationships between birth, death and growth rates of populations of key species, at all developmental stages, to environmental conditions, both present and past. This will provide data on the interaction between key copepods themselves and between key species in other phyla. Rates of mortality ascribed to predators should be examined in detail to test the top-down control of population sizes of key species.

d. Trophic studies of the developmental stages of key species should emphasize changing roles and linkages within the system.

e. Long-term observations of population and community dynamics that would allow analyses of interannual variability and its causes should be coupled with simultaneous data on key species in other phyla, to amplify the interdependence between elements within the system.

These investigations should not be confined to key species of copepods but should apply in parallel to key species of other phyla. The modern call is for multidisciplinary investigations and they certainly have relevance to the study of the biology of copepods. They cannot, however, be at the expense of the individualist for whom there will always be a role, whether in the discovery of poison fangs, adhesive organs, new species and environments of copepods or as contributors of new concepts to multidisciplinary studies. Much work remains to be done on the basic biology of copepods before we will fully understand their place within the ecosystems of the world.

REFERENCES

Aaser, H.F., Jeppesen, E. and Søndergaard, M. (1995). Seasonal dynamics of the mysid *Neomysis integer* and its predation on the copepod *Eurytemora affinis* in a shallow hypertrophic brackish lake. *Marine Ecology Progress Series* **127**, 47–56.

Abou Debs, C. (1984). Carbon and nitrogen budget of the calanoid copepod *Temora stylifera*: effect of concentration and composition of food. *Marine Ecology Progress Series* **15**, 213–223.

Abou Debs, C. and Nival, P. (1983). Étude de la ponte et du développement embryonnaire en relation avec la température et la nourriture chez *Temora stylifera* Dana (Copepoda: Calanoida). *Journal of Experimental Marine Biology and Ecology* **72**, 125–145.

Aksnes, D.L. (1996). Natural mortality, fecundity and development time in marine planktonic copepods – implications of behaviour. *Marine Ecology Progress Series* **131**, 315–316.

Aksnes, D.L. and Høisæter, T.J. (1987). Obtaining life table data from stage-frequency distributional statistics. *Limnology and Oceanography* **32**, 514–517.

Aksnes, D.L. and Magnesen, T. (1983). Distribution, development, and production of *Calanus finmarchicus* (Gunnerus) in Lindåspollene, western Norway, 1979. *Sarsia* **68**, 195–208.

Aksnes, D.L. and Magnesen, T. (1988). A population dynamics approach to the estimation of production of four calanoid copepods in Lindåspollene, western Norway. *Marine Ecology Progress Series* **45**, 57–68.

Aksnes, D.L., Aure, J., Kaartvedt, S., Magnesen, T. and Richard, J. (1989). Significance of advection for the carrying capacities of fjord populations. *Marine Ecology Progress Series* **50**, 263–274.

Alcaraz, M. (1976). Description of *Acartia margalefi*, a new species of pelagic copepod, and its relationship with *A. clausi*. *Investigacion Pesquera* **40**, 59–74.

Alcaraz, M. (1983). Coexistence and segregation of congeneric pelagic copepods: spatial distribution of the *Acartia* complex in the ria of Vigo (NW of Spain). *Journal of Plankton Research* **5**, 891–900.

Alcaraz, M. and Wagensberg, M. (1978). Análisis de series temporales: proporción sexual y densidad de poblaciones en copépodos. *Investigacion Pesquera* **42**, 155–165.

Alcaraz, M., Paffenhöfer, G.-A. and Strickler, J.R. (1980). Catching the algae: a first account of visual observations on filter-feeding calanoids. *In* "Evolution and Ecology of Zooplankton Communities" (W.C. Kerfoot, ed.), pp. 241–248. University Press of New England, Hanover, NH.

Alcaraz, M., Saiz, E., Marrasé, C. and Vaqué, D. (1988). Effects of turbulence on the

development of phytoplankton biomass and copepod populations in marine microcosms. *Marine Ecology Progress Series* **49**, 117–125.
Alekseev, V.R. and Fryer, G., eds (1996). Diapause in the Crustacea. *Hydrobiologia* **320**, 1–241.
Ali-Khan, S. and Ali-Khan, J. (1982). Seven new records of the family Lucicutiidae from Pakistan (Copepoda, Calanoida). *Crustaceana* **43**, 265–270.
Ali-Khan, S. and Ali-Khan, J. (1984). Nine new records of the family Augaptilidae from Pakistan (Copepoda, Calanoida). *Crustaceana* **47**, 303–313.
Allan, J.D. and Daniels, R.E. (1982). Life table evaluation of chronic exposure of *Eurytemora affinis* (Copepoda) to kepone. *Marine Biology* **66**, 179–184.
Allan, J.D., Kinsey, T.G. and James, M.C. (1976). Abundances and production of copepods in the Rhode River subestuary of Chesapeake Bay. *Chesapeake Science* **17**, 86–92.
Allan, J.D., Richman, S., Heinle, D.R. and Huff, R. (1977). Grazing in juvenile stages of some estuarine calanoid copepods. *Marine Biology* **43**, 317–331.
Alldredge, A.L. (1976). Discarded appendicularian houses as sources of food, surface habitats, and particulate organic matter in planktonic environments. *Limnology and Oceanography* **21**, 14–23.
Alldredge, A.L. and King, J.M. (1977). Distribution, abundance, and substrate preferences of demersal reef zooplankton at Lizard Island Lagoon, Great Barrier Reef. *Marine Biology* **41**, 317–333.
Alldredge, A.L. and King, J.M. (1980). Effects of moonlight on the vertical migration patterns of demersal zooplankton. *Journal of Experimental Marine Biology and Ecology* **44**, 133–156.
Alldredge, A.L., Robison, B.H., Fleminger, A., Torres, J.J., King, J.M. and Hamner, W.M. (1984). Direct sampling and *in situ* observation of a persistent copepod aggregation in the mesopelagic zone of the Santa Barbara Basin. *Marine Biology* **80**, 75–81.
Allison, S.K. and Wishner, K.F. (1986). Spatial and temporal patterns of zooplankton biomass across the Gulf Stream. *Marine Ecology Progress Series* **31**, 233–244.
Almeida Prado-Por, M.S. (1983). The diversity and dynamics of Calanoida (Copepoda) in the northern Gulf of Elat (Aqaba), Red Sea. *Oceanologica Acta* **6**, 139–145.
Almeida Prado-Por, M.S. (1990). A diel cycle of vertical distribution of the Calanoidea (Crustacea: Copepoda) in the northern Gulf of Aqaba (Elat). *Bulletin de l'Institut Océanographique, Monaco*, Spécial No. 7, 109–116.
Alstyne, K.L. van (1986). Effects of phytoplankton taste and smell on feeding behavior of the copepod *Centropages hamatus*. *Marine Ecology Progress Series* **34**, 187–190.
Alvarez, M.J.P. (1984). Two bottom living Copepoda Calanoida Aetideidae – *Bradyidius plinioi* and *Lutamator elegans* n.sp. collected in Brazilian waters. *Boletim Zoologia Universidade de São Paulo* **8**, 93–106.
Alvarez, M.P.J. (1985). Revision of the genus *Neoscolecithrix* (Copepoda, Calanoida) and description of *N. caetanoi*, n.sp., collected off Brazil. *Revista Brasileira de Zoologia* **3**, 197–207.
Alvarez, M.P.J. (1986). New calanoid copepods (Aetideidae) of the genera *Comantenna, Mesocomantenna*, new genus, and *Paracomantenna* off the Brazilian coast. *Journal of Crustacean Biology* **6**, 858–877.
Alvarez, V. and Kewalramani, H.G. (1970). Naupliar development of *Pseudodiaptomus ardjuna* Brehm (Copepoda). *Crustaceana* **18**, 269–276.
Alvarez, V. and Matthews, J.B.L. (1975). Experimental studies on the deep-water

pelagic community of Korsfjorden, western Norway. Feeding and assimilation by *Chiridius armatus* (Crustacea, Copepoda). *Sarsia* **58**, 67–78.
Ambler, J.W. (1985). Seasonal factors affecting egg production and viability of eggs of *Acartia tonsa* Dana from East Lagoon, Galveston, Texas. *Estuarine, Coastal and Shelf Science* **20**, 743–760.
Ambler, J.W. (1986a). Effect of food quantity and quality on egg production of *Acartia tonsa* Dana from East Lagoon, Galveston, Texas. *Estuarine, Coastal and Shelf Science* **23**, 183–196.
Ambler, J.W. (1986b). Formulation of an ingestion function for a population of *Paracalanus* feeding on mixtures of phytoplankton. *Journal of Plankton Research* **8**, 957–972.
Ambler, J.W. and Frost, B.W. (1974). The feeding behavior of a predatory planktonic copepod, *Tortanus discaudatus*. *Limnology and Oceanography* **19**, 446–451.
Ambler, J.W. and Miller, C.B. (1987). Vertical habitat-partitioning by copepodites and adults of subtropical oceanic copepods. *Marine Biology* **94**, 561–577.
Ambler, J.W., Cloern, J.E. and Hutchinson, A. (1985). Seasonal cycles of zooplankton from San Francisco Bay. *Hydrobiologia* **129**, 177–197.
Andersen, N.R. and Zahuranec, B.J., eds (1977). Oceanic Sound Scattering Prediction. Plenum Press, New York and London. 859 pp.
Andersen, V. and Nival, P. (1988). A pelagic ecosystem model simulating production and sedimentation of biogenic particles: role of salps and copepods. *Marine Ecology Progress Series* **44**, 37–50.
Anderson, J.T. (1990). Seasonal development of invertebrate zooplankton on Flemish Cap. *Marine Ecology Progress Series* **67**, 127–140.
Anderson, J.T. and Warren, W.G. (1991). Comparison of catch rates among small and large Bongo samplers for *Calanus finmarchicus* copepodite stages. *Canadian Journal of Fisheries and Aquatic Sciences* **48**, 303–308.
Anderson, T.R. (1992). Modelling the influence of food C:N ratio, and respiration on growth and nitrogen excretion in marine zooplankton and bacteria. *Journal of Plankton Research* **14**, 1645–1671.
Anderson, T.R. and Hessen, D.O. (1995). Carbon or nitrogen limitation in marine copepods? *Journal of Plankton Research* **17**, 317–331.
Andrews, K.J.H. (1966). The distribution and life-history of *Calanoides acutus* (Giesbrecht). *Discovery Reports* **34**, 117–162.
Andrews, J.C. (1983). Deformation of the active space in the low Reynolds number feeding current of calanoid copepods. *Canadian Journal of Fisheries and Aquatic Sciences* **40**, 1293–1302.
Andronov, V.N. (1974). Phylogenetic relation of large taxa within the suborder Calanoida (Crustacea, Copepoda). *Zoologichesky Zhurnal* **53**, 1002–1012.
Andronov, V.N. (1978). Diaixidae (Copepoda, Calanoida) of the western coast of Africa. *Byulletene Moskovskogo Obshchestva Ispȳtatelei Pirodȳ, Otdel Biologii* **84**, (4), 90–102.
Andronov, V.N. (1986a). Bottom Copepoda in the area of Cape Blanc (Islamic Republic of Mauritania). 2. The family Pseudocyclopiidae. *Zoologichesky Zhurnal* **65**, 295–298.
Andronov, V.N. (1986b). Bottom Copepoda in the area of Cape Blanc (Islamic Republic of Mauritania). 3. The family Pseudocyclopidae. *Zoologichesky Zhurnal* **65**, 456–462.
Andronov, V.N. (1991). On renaming of some taxa in Calanoida (Crustacea). *Zoologichesky Zhurnal* **70**, (6), 133–134.
Andronov, V.N. (1992). *Ryocalanus admirabilis* sp.n. (Copepoda, Calanoida,

Ryocalanidae). *Zoologichesky Zhurnal* **71**, (7), 140–144.
Angel, M.V. (1990). Life in the benthic boundary layer: connections to the mid-water and sea floor. *Philosophical Transactions of the Royal Society of London* **331A**, 15–28.
Angel, M.V. and Fasham, M.J.R. (1973). SOND cruise 1965: factor and cluster analyses of the plankton results, a general survey. *Journal of the Marine Biological Association of the United Kingdom* **53**, 185–231.
Angel, M.V. and Fasham, M.J.R. (1974). SOND cruise 1965: further factor analyses of the plankton data. *Journal of the Marine Biological Association of the United Kingdom* **54**, 879–894.
Anonymous (1990). Opinion 1613. *Lucicutia* Giesbrecht & Schmeil, 1898: conserved, and *Pseudaugaptilus longiremis* Sars, 1907: specific name conserved (both Crustacea, Copepoda). *Bulletin of Zoological Nomenclature* **47**, 226–227.
Anraku, M. (1975). Microdistribution of marine copepods in a small inlet. *Marine Biology* **30**, 79–87.
Anraku, M. and Omori, M. (1963). Preliminary study of the relationship between the feeding habit and the structure of the mouth-parts of marine copepods. *Limnology and Oceanography* **8**, 116–126.
Apstein, C. (1910). Hat ein Organismus in der tiefe gelebt, in der er gefischt ist? *Internationale Revue der gesamten Hydrobiologie und Hydrographie* **3**, 17–33.
Arashkevich, E.G. (1977). Duration of food digestion in marine copepods. *Polskie Archiwum Hydrobiologii* **24**, Supplement, 431–438.
Arashkevich, E.G. and Cahoon, L. (1980). Feeding patterns of dominant species of Copepoda in Peruvian upwelling region. *Polskie Archiwum Hydrobiologii* **27**, 457–469.
Arashkevich, E.G., Drits, A.V. and Timonin, A.G. (1996). Diapause in the life cycle of *Calanoides carinatus* (Kroyer) (Copepoda, Calanoida). *Hydrobiologia* **320**, 197–208.
Arcos, F. and Fleminger, A. (1986). Distribution of filter-feeding calanoid copepods in the eastern equatorial Pacific. *Reports of the California Cooperative Oceanic Fisheries Investigations* **27**, 170–187.
Arinardi, O.H. (1991). Vertical distribution of calanoid copepods in the Banda Sea, Indonesia, during and after upwelling period. *Bulletin of Plankton Society of Japan*, Special Volume, 291–298.
Arinardi, O.H., Baars, M.A. and Oosterhuis, S.S. (1990). Grazing in tropical copepods, measured by gut fluorescence, in relation to seasonal upwelling in the Banda Sea (Indonesia). *Netherlands Journal of Sea Research* **25**, 545–560.
Armstrong, D.A., Verheye, H.M. and Kemp, A.D. (1991). Short-term variability during an anchor station study in the southern Benguela upwelling system: fecundity estimates of the dominant copepod, *Calanoides carinatus*. *Progress in Oceanography* **28**, 167–188.
Arnaud, J., Brunet, M. and Mazza, J. (1980). Structure et ultrastructure comparées de l'intestin chez plusieurs espèces de Copépodes Calanoides (Crustacea). *Zoomorphologie* **95**, 213–233.
Arnaud, J., Brunet, M. and Mazza, J. (1988a). Labral glands in *Centropages typicus* (Copepoda, Calanoida). I. Sites of synthesis. *Journal of Morphology* **197**, 21–32.
Arnaud, J., Brunet, M. and Mazza, J. (1988b). Labral glands in *Centropages typicus* (Copepoda, Calanoida). II. Sites of secretory release. *Journal of Morphology* **197**, 209–219.
Arndt, H. (1989). Zooplankton production and its consumption by planktivores in

a Baltic inlet. *In* "Proceedings of the Twenty First European Marine Biology Symposium" (R.Z. Klekowski, E. Styczyńska and L. Falkowski, eds), pp. 205–214. Polish Academy of Sciences.

Arndt, H. and Schnese, W. (1986). Population dynamics and production of *Acartia tonsa* (Copepoda: Calanoida) in the Darss-Zingst estuary, southern Baltic. *Ophelia*, Supplement **4**, 329–334.

Arndt, H., Kramer, H.-J., Heerkloss, R. and Schröder, C. (1985). Zwei Methoden zur bestimmung populationsdynamischer parameter von zooplanktern unter laborbedingungen mit ersten ergebnissen an *Eurytemora affinis* (Copepoda, Calanoida) und *Synchaeta cecilia* (Rotatoria, Monogononta). *Wissenschaftliche Zeitschrift der Universität Rostock*, N-Reihe **34**, (6), 17–21.

Arnott, G.H., Brand, G.W. and Kos, L.C. (1986). Effects of food quality and quantity on the survival, development, and egg production of *Gladioferens pectinatus* (Brady) (Copepoda: Calanoida). *Australian Journal of Marine and Freshwater Research* **37**, 467–473.

Ashjian, C.J. (1993). Trends in copepod species abundances across and along a Gulf Stream meander: evidence for entrainment and detrainment of fluid parcels from the Gulf Stream. *Deep-Sea Research* **40A**, 461–482.

Ashjian, C.J. and Wishner, K.F. (1993a). Temporal and spatial changes in body size and reproductive state of *Nannocalanus minor* (Copepoda) females across and along the Gulf Stream. *Journal of Plankton Research* **15**, 67–98.

Ashjian, C.J. and Wishner, K.F. (1993b). Temporal persistence of copepod species groups in the Gulf Stream. *Deep-Sea Research* **40A**, 483–516.

Atema, J. (1985). Chemoreception in the sea: adaptations of chemoreceptors and behaviour to aquatic stimulus conditions. *Symposia of the Society for Experimental Biology* **39**, 387–423.

Atkinson, A. (1989a). Distribution of six major copepod species around South Georgia in early summer. *Polar Biology* **9**, 353–363.

Atkinson, A. (1989b). Distribution of six major copepod species around South Georgia during an austral winter. *Polar Biology* **10**, 81–88.

Atkinson, A. (1991). Life cycles of *Calanoides acutus*, *Calanus simillimus* and *Rhincalanus gigas* (Copepoda: Calanoida) within the Scotia Sea. *Marine Biology* **109**, 79–91.

Atkinson, A. (1995). Omnivory and feeding selectivity in five copepod species during spring in the Bellingshausen Sea, Antarctica. *ICES Journal of Marine Science* **52**, 385–396.

Atkinson, A. (1996). Subantarctic copepods in an oceanic, low chlorophyll environment: ciliate predation, food selectivity and impact on prey populations. *Marine Ecology Progress Series* **130**, 85–96.

Atkinson, A., Ward, P., Peck, J.M. and Murray, A.W.A. (1990). Mesoscale distribution of zooplankton around South Georgia. *Deep-Sea Research* **37A**, 1213–1227.

Atkinson, A., Ward, P. Williams, R. and Poulet, S.A. (1992a). Diel vertical migration and feeding of copepods at an oceanic site near South Georgia. *Marine Biology* **113**, 583–593.

Atkinson, A., Ward, P. Williams, R. and Poulet, S.A. (1992b). Feeding rates and diel vertical migration of copepods near South Georgia: comparison of shelf and oceanic sites. *Marine Biology* **114**, 49–56.

Atkinson, A., Shreeve, R.S., Pakhomov, E.A., Priddle, J., Blight, S.P. and Ward, P. (1996a). Zooplankton response to a phytoplankton bloom near South Georgia, Antarctica. *Marine Ecology Progress Series* **144**, 195–210.

Atkinson, A., Ward, P. and Murphy, E.J. (1996b). Diel periodicity of subantarctic copepods: relationships between vertical migration, gut fullness and gut evacuation rate. *Journal of Plankton Research* **18**, 1387–1405.

Attwood, C.G. and Peterson, W.T. (1989). Reduction in fecundity and lipids of the copepod *Calanus australis* (Brodskii) by strongly pulsed upwelling. *Journal of Experimental Marine Biology and Ecology* **129**, 121–131.

Auster, P.J., Griswold, C.A., Youngbluth, M.J. and Bailey, T.G. (1992). Aggregations of myctophid fishes with other pelagic fauna. *Environmental Biology of Fishes* **35**, 133–139.

Ayukai, T. (1987). Feeding by the planktonic calanoid copepod *Acartia clausi* Giesbrecht on natural suspended particulate matter of varying quantity and quality. *Journal of Experimental Marine Biology and Ecology* **106**, 137–149.

Ayukai, T. (1990). Fecal pellet production by two species of planktonic calanoid copepods fed on naturally occurring particles. *Bulletin of Plankton Society of Japan* **37**, 167–169.

Ayukai, T. and Nishizawa, S. (1986). Defecation rate as a possible measure of ingestion rate of *Calanus pacificus* (Copepoda: Calanoida). *Bulletin of Plankton Society of Japan* **33**, 3–10.

Backus, R.H. (1986). Biogeographic boundaries in the open ocean. *In* "Pelagic Biogeography", Unesco Technical papers in Marine Science **49**, 9–13.

Bailey, K.M. and Yen, J. (1983). Predation by a carnivorous marine copepod, *Euchaeta elongata* (Esterly), on eggs and larvae of the Pacific hake, *Merluccius productus*. *Journal of Plankton Research* **5**, 71–82.

Bainbridge, R. (1952). Underwater observations on the swimming of marine zooplankton. *Journal of the Marine Biological Association of the United Kingdom* **31**, 107–112.

Bainbridge, R. (1961). Migrations. *In* "The Physiology of Crustacea" (T.H. Waterman, ed.), Volume II, 431–463. Academic Press, New York and London.

Bainbridge, V. (1972). The zooplankton of the Gulf of Guinea. *Bulletin of Marine Ecology* **8**, 61–97.

Bakke, J.L.W. (1977). Ecological studies on the deep-water pelagic community of Korsfjorden, Western Norway. Population dynamics of *Euchaeta norvegica* (Crustacea, Copepoda) from 1971 to 1974. *Sarsia* **63**, 49–55.

Bakke, J.L.W. and Valderhaug, V.A. (1978). Ecological studies on the deep-water pelagic community of Korsfjorden, Western Norway. Population biology, biomass, and calorie content of *Chiridius armatus* (Crustacea, Copepoda). *Sarsia* **63**, 247–254.

Baldacci, F., Arnaud, J., Brunet, M. and Mazza, J. (1985a). Differenciation des gonades et ovogenese chez *Calanus helgolandicus* (Copepoda, Calanoida). *Rapports et Procès-Verbaux des Réunions Commission Internationale pour l'Exploration Scientifique de la Mer Méditerranée* **29**, (9), 213–214.

Baldacci, F., Arnaud, J., Brunet, M. and Mazza, J. (1985b). Etude structurale et ultrastructurale du tube digestif chez les copepodites et les adultes de *Calanus helgolandicus* (Copepoda, Calanoida). *Rapports et Procès-Verbaux des Réunions Commission Internationale pour l'Exploration Scientifique de la Mer Méditerranée* **29**, (9), 215–217.

Båmstedt, U. (1984). Diel variations in the nutritional physiology of *Calanus glacialis* from Lat. 78°N in the summer. *Marine Biology* **79**, 257–267.

Båmstedt, U. (1986). Chemical composition and energy content. *In* "The Biological Chemistry of Marine Copepods" (E.D.S. Corner and S.C.M. O'Hara, eds), pp. 1–58. Clarendon Press, Oxford.

Båmstedt, U. (1988a). The macrozooplankton community of Kosterfjorden, western Sweden. Abundance, biomass, and preliminary data on the life cycles of dominant species. *Sarsia* **73**, 107–124.
Båmstedt, U. (1988b). Ecological significance of individual variablity in copepod bioenergetics. *Hydrobiologia* **167/168**, 43–59.
Båmstedt, U. and Ervik, A. (1984). Local variations in size and activity among *Calanus finmarchicus* and *Metridia longa* (Copepoda, Calanoida) overwintering on the west coast of Norway. *Journal of Plankton Research* **6**, 843–857.
Båmstedt, U. and Holt, M.R. (1978). Experimental studies on the deep-water pelagic community of Korsfjorden, Western Norway. Prey-size preference and feeding of *Euchaeta norvegica* (Copepoda). *Sarsia* **63**, 225–236.
Båmstedt, U. and Matthews, J.B.L. (1975). Studies on the deep-water pelagic community of Korsfjorden, western Norway. The weight and biochemical composition of *Euchaeta norvegica* Boeck in relation to its life cycle. *In* "Proceedings of the 9th European Marine Biology Symposium 1975", pp. 311–327. Aberdeen University Press.
Båmstedt, U. and Skjoldal, H.R. (1980). RNA concentration of zooplankton: relationship with size and growth. *Limnology and Oceanography* **25**, 304–316.
Båmstedt, U. and Tande, K.S. (1985). Respiration and excretion rates of *Calanus glacialis* in arctic waters of the Barents Sea. *Marine Biology* **87**, 259–266.
Båmstedt, U. and Tande, K. (1988). Physiological responses of *Calanus finmarchicus* and *Metridia longa* (Copepod: Calanoida) during the winter-spring transition. *Marine Biology* **99**, 31–38.
Båmstedt, U., Tande, K.S. and Nicolajsen, H. (1985). Ecological investigations on the zooplankton community of Balsfjorden, northern Norway: physiological adaptations in *Metridia longa* (Copepoda) to the overwintering period. *In* "Marine Biology of Polar Regions and Effects of Stress on Marine Organisms" (J.S. Gray and M.E. Christiansen, eds), pp. 313–327. John Wiley & Sons Ltd, Chichester.
Båmstedt, U., Håkanson, J.L., Brenner-Larsen, J., Björnsen, P.K., Geertz-Hansen, O. and Tiselius, P. (1990). Copepod nutritional condition and pelagic production during autumn in Kosterfjorden, western Sweden. *Marine Biology* **104**, 197–208.
Båmstedt, U., Eilertsen, H.C., Tande, K.S., Slagstad, D. and Skjoldal, H.R. (1991). Copepod grazing and its potential impact on the phytoplankton development in the Barents Sea. *Polar Research* **10**, 339–353.
Ban, S. (1992). Effects of photoperiod, temperature, and population density on induction of diapause egg production in *Eurytemora affinis* (Copepoda: Calanoida) in Lake Ohnuma, Hokkaido, Japan. *Journal of Crustacean Biology* **12**, 361–367.
Ban, S. (1994). Effect of temperature and food concentration on post-embryonic development, egg production and adult body size of calanoid copepod *Eurytemora affinis*. *Journal of Plankton Research* **16**, 721–735.
Ban, S. and Minoda, T. (1991). The effect of temperature on the development and hatching of diapause and subitaneous eggs in *Eurytemora affinis* (Copepoda: Calanoida) in Lake Ohnuma, Hokkaido, Japan. *Bulletin of Plankton Society of Japan*, Special Volume, 299–308.
Ban, S. and Minoda, T. (1992). Hatching of diapause eggs of *Eurytemora affinis* (Copepoda: Calanoida) collected from lake-bottom sediments. *Journal of Crustacean Biology* **12**, 51–56.
Ban, S. and Minoda, T. (1994). Induction of diapause egg production in *Eurytemora affinis* by their own metabolites. *Hydrobiologia* **292/293**, 185–189.

Bandaranayake, W.M. and Gentien, P. (1982). Carotenoids of *Temora turbinata*, *Centropages furcatus*, *Undinula vulgaris* and *Euchaeta russelli*. *Comparative Biochemistry and Physiology* **72B**, 409–414.

Bannister, N.J. (1993a). Distribution and structure of sub-cuticular glands in the copepod *Temora longicornis*. *Journal of the Marine Biological Association of the United Kingdom* **73**, 97–107.

Bannister, N.J. (1993b). Innervation of luminous glands in the calanoid copepod *Euaugaptilus magnus*. *Journal of the Marine Biological Association of the United Kingdom* **73**, 417–423.

Bannister, N.J. and Herring, P.J. (1989). Distribution and structure of luminous cells in four marine copepods. *Journal of the Marine Biological Association of the United Kingdom* **69**, 523–533.

Banse, K. (1964). On the vertical distribution of zooplankton in the sea. *Progress in Oceanography* **1**, 55–125.

Banse, K. (1984). Reply. *Canadian Journal of Fisheries and Aquatic Sciences* **41**, 833–834.

Banse, K. and Mosher, S. (1980). Adult body mass and annual production/biomass relationships of field populations. *Ecological Monographs* **50**, 355–379.

Barnes, A.T. and Case, J.F. (1972). Bioluminescence in the mesopelagic copepod, *Gaussia princeps* (T. Scott). *Journal of Experimental Marine Biology and Ecology* **8**, 53–71.

Barr, D.J. (1984). *Enantiosus cavernicola*, a new genus and species of demersal copepod (Calanoida: Epacteriscidae) from San Salvador Island, Bahamas. *Proceedings of the Biological Society of Washington* **97**, 160–166.

Barr, D.J. and Ohtsuka, S. (1989). *Pseudocyclops lepidotus*, a new species of demersal copepod (Calanoida: Pseudocyclopidae) from the northwestern Pacific. *Proceedings of the Biological Society of Washington* **102**, 331–338.

Barthel, K.-G. (1983). Food uptake and growth efficiency of *Eurytemora affinis* (Copepoda: Calanoida). *Marine Biology* **74**, 269–274.

Barthel, K.-G. (1988). Feeding of three *Calanus* species on different phytoplankton assemblages in the Greenland Sea. *Meeresforschung* **32**, 92–106.

Bartram, W.C. (1980). Experimental development of a model for the feeding of neritic copepods on phytoplankton. *Journal of Plankton Research* **3**, 25–51.

Batchelder, H.P. (1985). Seasonal abundance, vertical distribution, and life history of *Metridia pacifica* (Copepoda: Calanoida) in the oceanic subarctic Pacific. *Deep-Sea Research* **32A**, 949–964.

Batchelder, H.P. (1986). A staining technique for determining copepod gonad maturation: application to *Metridia pacifica* from the northeast Pacific Ocean. *Journal of Crustacean Biology* **6**, 227–231.

Batchelder, H.P. and Miller, C.B. (1989). Life history and population dynamics of *Metridia pacifica*: results from simulation modelling. *Ecological Modelling* **48**, 113–136.

Batchelder, H.P. and Swift, E. (1989). Estimated near-surface mesoplanktonic bioluminescence in the western North Atlantic during July 1986. *Limnology and Oceanography* **34**, 113–128.

Batchelder, H.P. and Williams, R. (1995). Individual-based modelling of the population dynamics of *Metridia lucens* in the North Atlantic. *ICES Journal of Marine Science* **52**, 469–482.

Batchelder, H.P., Swift, E. and Keuren, J.R. Van (1990). Pattern of planktonic bioluminescence in the northern Sargasso Sea: seasonal and vertical distribution. *Marine Biology* **104**, 153–164.

Batchelder, H.P., Swift, E. and Keuren, J.R. Van (1992). Diel patterns of planktonic bioluminescence in the northern Sargasso Sea. *Marine Biology* **113**, 329–339.
Bathmann, U.V., Noji, T.T., Voss, M. and Peinert, R. (1987). Copepod fecal pellets: abundance, sedimentation and content at a permanent station in the Norwegian Sea in May/June 1986. *Marine Ecology Progress Series* **38**, 45–51.
Bathmann, U.V., Noji, T.T. and von Bodungen, B. (1990). Copepod grazing potential in late winter in the Norwegian Sea – a factor in the control of spring phytoplankton growth? *Marine Ecology Progress Series* **60**, 225–233.
Bathman, U.V., Makarov, R.R., Spirodonov, V.A. and Rohardt, G. (1993). Winter distribution and overwintering strategies of the Antarctic copepod species *Calanoides acutus*, *Rhincalanus gigas* and *Calanus propinquus* (Crustacea: Calanoida) in the Weddell Sea. *Polar Biology* **13**, 333–346.
Bautista, B. and Harris, R.P. (1992). Copepod gut contents, ingestion rates and grazing impact on phytoplankton in relation to size structure of zooplankton and phytoplankton during a spring bloom. *Marine Ecology Progress Series* **82**, 41–50.
Bautista, B., Rodriguez, V. and Jimenez, F. (1988). Short-term feeding rates of *Acartia grani* in natural conditions: diurnal variation. *Journal of Plankton Research* **10**, 907–920.
Bautista, B., Harris, R.P., Tranter, P.R.G. and Harbour, D. (1992). *In situ* copepod feeding and grazing rates during a spring bloom dominated by *Phaeocystis* sp. in the English Channel. *Journal of Plankton Research* **14**, 691–703.
Bautista, B., Harris, R.P., Rodriguez, V. and Guerrero, F. (1994). Temporal variability in copepod fecundity during two different spring bloom periods in coastal waters off Plymouth (SW England). *Journal of Plankton Research* **16**, 1367–1377.
Bayly, I.A.E. (1969). The body fluids of some centropagid copepods: total concentration and amounts of sodium and magnesium. *Comparative Biochemistry and Physiology* **28**, 1403–1409.
Bayly, I.A.E. (1979). Further contributions to a knowledge of the centropagid genera *Boeckella*, *Hemiboeckella* and *Calamoecia* (athalassic calanoid copepods). *Australian Journal of Marine and Freshwater Research* **30**, 103–127.
Bayly, I.A.E. (1982). The genus *Drepanopus* (Copepoda: Calanoida): a review of species in Antarctic and Sub-Antarctic waters, with a description of *D. bispinosus*, sp.nov. *Australian Journal of Marine and Freshwater Research* **33**, 161–172.
Bayly, I.A.E. (1992). Fusion of the genera *Boeckella* and *Pseudoboeckella* (Copepoda) and revision of their species from South America and sub-Antarctic islands. *Revista Chilena de Historia Natural* **65**, 17–63.
Beckman, B.R. and Peterson, W.T. (1986). Egg production by *Acartia tonsa* in Long Island Sound. *Journal of Plankton Research* **8**, 917–925.
Beckmann, W. (1984). Mesozooplankton distribution on a transect from the Gulf of Aden to the central Red Sea during the winter monsoon. *Oceanologica Acta* **7**, 87–102.
Beckmann, W. (1988). The zooplankton community in the deep bathyal and abyssal zones of the eastern North Atlantic. Preliminary results and data lists from MOCNESS hauls during cruise 08 of the RV "Polarstern". *Bericht zur Polarforschung*, Reports on Polar Research **42**, 1–57.
Behrends, G. and Schneider, G. (1995). Impact of *Aurelia aurita* medusae (Cnidaria, Scyphozoa) on the standing stock and community composition of mesozooplankton in the Kiel Bight (western Baltic Sea). *Marine Ecology Progress Series* **127**, 39–45.
Bělehrádek, J. (1935). Temperature and living matter. *Protoplasma*, Monograph **8**, 1–277.

Bellantoni, D.C. and Peterson, W.T. (1987). Temporal variability in egg production rates of *Acartia tonsa* Dana in Long Island Sound. *Journal of Experimental Marine Biology and Ecology* **107**, 199–208.
Belmonte, G. (1992). Diapause egg production in *Acartia* (*Paracartia*) *latisetosa* (Crustacea, Copepoda, Calanoida). *Bolletino di Zoologia* **59**, 363–366.
Belmonte, G. and Puce, M. (1994). Morphological aspects of subitaneous and resting eggs from *Acartia josephinae* (Calanoida). *Hydrobiologia* **292/293**, 131–135.
Belmonte, G., Mazzocchi, M.G., Prusova, I.Yu. and Shadrin, N.V. (1994). *Acartia tonsa*: a species new for the Black Sea fauna. *Hydrobiologia* **292/293**, 9–15.
Belmonte, G., Castello, P., Piccini, M.R., Quarta, S., Rubino, F., Geraci, S. and Boero, F. (1995). Resting stages in marine sediments off the Italian coast. *In* "Biology and Ecology of Shallow Coastal Waters" (A. Eleftheriou, A.D. Ansell and C.J. Smith, eds), pp. 59–65. Olsen & Olsen, Fredensborg.
Bennett, J.L. and Hopkins, T.L. (1989). Aspects of the ecology of the calanoid copepod genus *Pleuromamma* in the eastern Gulf of Mexico. *Contributions in Marine Science* **31**, 119–136.
Benon, P. (1977). Influence des rejets d'eau chaude de la centrale E.D.F. Martigues-Ponteau, sur les populations zooplanctoniques. *Téthys* **8**, 63–82.
Berardi, G. (1953). Apparecchio per una precisa valutazione volumetrica di campioni di plancton. *Memorie dell' Instituto Italiano di Idrobiologia* **7**, 221–228.
Berdugo, V. (1968). Sur la présence dans la Méditerranée orientale de deux espèces du genre *Calanopia* (Copepoda, Calanoida). *Rapports et Procès-Verbaux des Réunions Commission Internationale pour l'Exploration Scientifique de la Mer Méditerranée* **19** (3), 445–446.
Berdugo, V. (1974). On the occurrence of *Acartia centrura* (Copepoda: Calanoida) in the neritic waters of the eastern Mediterranean. *Rapports et Procès-Verbaux des Réunions Commission Internationale pour l'Exploration Scientifique de la Mer Méditerranée* **22** (9), 85–86.
Berggreen, U., Hansen, B. and Kiørboe, T. (1988). Food size spectra, ingestion and growth of the copepod *Acartia tonsa* during development: implications for determination of copepod production. *Marine Biology* **99**, 341–352.
Bernard, M. (1958). Revision des *"Calocalanus"* (Copepodes Calanoida), avec description d'un genre nouveaux et deux espèces nouvelles. *Bulletin de la Société Zoologique de France* **83**, 1–15.
Bernard, M. (1964). Le développement nauplien de deux copépodes carnivores: *Euchaeta marina* (Prestandr.) et *Candacia armata* (Boeck) et observations sur le cycle de l'astaxanthine au cours de l'ontogénèse. *Pelagos* **2**, 51–71.
Bernard, M. (1965). Observations sur la ponte et le développement larvaire en aquarium d'un copépode pélagique prédateur: *Candacia armata* Boeck. *Rapports et Procès-Verbaux des Réunions Commission Internationale pour l'Exploration Scientifique de la Mer Méditerranée* **18**, (2), 345–348.
Bhattacharya, S.S. (1986). Individual and combined effects of salinity and temperature on the calanoid copepod *Paracalanus aculeatus* Giesbrecht. *Syllogeus* **58**, 220–228.
Binet, D. (1977). Cycles biologiques et migrations ontogenétiques chez quelques copépodes pélagiques des eaux Ivoiriennes. *Cahiers O.R.S.T.O.M., Series Oceanographie* **15**, 111–138.
Binet, D. (1978). Analyse globale des populations de copépodes pélagiques du plateau continental Ivoirien. *Cahiers O.R.S.T.O.M., Series Oceanographie* **16**, 19–61.

Binet, D. (1979). Le zooplancton du plateau continental Ivoirien. Essai de synthèse écologique. *Oceanologica Acta* **2**, 397–410.
Binet, D. and Dessier, A. (1972). Premières données sur les copépodes pélagiques de la région Congolaise. *Cahiers O.R.S.T.O.M., Séries Océanographie* **10**, (3), 243–250.
Binet, D. and Suisse de Sainte Claire, E. (1975). Le copépode planctonique *Calanoides carinatus* répartition et cycle biologique au large de la Côte d'Ivoire. *Cahiers O.R.S.T.O.M., Séries Océanographie* **13**, 15–30.
Bird, D.F. and Prairie, Y.T. (1985). Practical guidelines for the use of zooplankton length-weight regression equations. *Journal of Plankton Research* **7**, 955–960.
Bird, J.L. (1983). Relationships between particle-grazing zooplankton and vertical phytoplankton distributions on the Texas continental shelf. *Estuarine, Coastal and Shelf Science* **16**, 131–144.
Bird, J.L. and Kitting, C.L. (1982). Laboratory studies of a marine copepod (*Temora turbinata* Dana) tracking dinoflagellate migrations in a miniature water column. *Contributions in Marine Science* **25**, 27–44.
Bishop, J.W. and Greenwood, J.G. (1994). Nitrogen excretion by some demersal macrozooplankton in Heron and One Tree Reefs, Great Barrier Reef, Australia. *Marine Biology* **120**, 447–453.
Bityukov, E.P. and Evstigneev, P.V. (1982). Main characteristics of luminescence and its species specificity in copepods of the genus *Pleuromamma*. *Ekologiya Morya* **11**, 53–62.
Björnberg, T.K.S. (1966). The developmental stages of *Undinula vulgaris* (Dana) (Copepoda). *Crustaceana* **11**, 65–76.
Björnberg, T.K.S. (1967a). Four new species of Megacalanidae (Crustacea: Copepoda). *Biology of Antarctic Seas III, Antarctic Research Series* **11**, 73–90.
Björnberg, T.K.S. (1967b). The larvae and young forms of *Eucalanus* Dana (Copepoda) from tropical Atlantic waters. *Crustaceana* **12**, 59–73.
Björnberg, T.K.S. (1972). Developmental stages of some tropical and subtropical planktonic marine copepods. *Studies on the Fauna of Curaçao and Other Caribbean Islands* **40**, 1–185.
Björnberg, T.K.S. (1975). New species of marine bathypelagic copepods collected off South America. *Ciencia e Cultura* **27**, 175–188.
Björnberg, T.K.S. (1981). Copepoda. *In* "Atlas del zooplancton del Atlantico sudoccidental y métodos de trabajo con zooplancton marino" (D. Boltovskoy, ed.), pp. 587–679. Instituto Nacional Investigación y Desarrollo Pesquero, Mar del Plata, Argentina.
Björnberg, T.K.S. (1986a). Aspects of the appendages in development. *Syllogeus* **58**, 51–66.
Björnberg, T.K.S. (1986b). The rejected nauplius: a commentary. *Syllogeus* **58**, 232–236.
Björnberg, T.K.S. and Campaner, A.F. (1988). On *Gaussia* Wolfenden (Copepoda, Calanoida, Metridinidae). *Hydrobiologia* **167/168**, 351–356.
Björnberg, T.K.S. and Campaner, A.F. (1990). On the genus *Gaussia* and the species *G. asymmetrica* (Copepoda, Calanoida). *Crustaceana* **58**, 106–112.
Blades, P.I. (1977). Mating behavior of *Centropages typicus* (Copepoda: Calanoida). *Marine Biology* **40**, 57–64.
Blades, P.I. and Youngbluth, M.J. (1979). Mating behavior of *Labidocera aestiva* (Copepoda: Calanoida). *Marine Biology* **51**, 339–355.
Blades, P.I. and Youngbluth, M.J. (1980). Morphological, physiological, and behavioral aspects of mating in calanoid copepods. *In* "Evolution and Ecology of

Zooplankton Communities". *Special Symposium No. 3, American Society for Limnology and Oceanography* 39–51.

Blades, P.I. and Youngbluth, M.J. (1981). Ultrastructure of the male reproductive system and spermatophore formation in *Labidocera aestiva* (Crust.: Copepoda). *Zoomorphology* **99**, 1–21.

Blades-Eckelbarger, P.I. (1986). Aspects of internal anatomy and reproduction in the Copepoda. *Syllogeus* **58**, 26–50.

Blades-Eckelbarger, P.I. (1991a). Functional morphology of spermatophores and sperm transfer in calanoid copepods. *In* "Crustacean Sexual Biology" (R.T. Bauer and J.W. Martin, eds), pp. 246–270. Columbia University Press, New York.

Blades-Eckelbarger, P.I. (1991b). Comparative ultrastructure of lipid storage sites in female *Euchaeta marina* and *Pleuromamma xiphias* (Copepoda: Calanoida). *Marine Biology* **108**, 49–58.

Blades-Eckelbarger, P.I. and Youngbluth, M.J. (1982). The ultrastructure of spermatogenesis in *Labidocera aestiva* (Copepoda: Calanoida). *Journal of Morphology* **174**, 1–15.

Blades-Eckelbarger, P.I. and Youngbluth, M.J. (1984). The ultrastructure of oogenesis and yolk formation in *Labidocera aestiva* (Copepoda: Calanoida). *Journal of Morphology* **179**, 33–46.

Blades-Eckelbarger, P.I. and Youngbluth, M.J. (1988). Ultrastructure of the "pigment knob" of *Pleuromamma* spp. (Copepoda: Calanoida). *Journal of Morphology* **197**, 315–326.

Blades-Eckelbarger, P.I. and Youngbluth, M.J. (1991). Comparative ultrastructure of lipid storage sites in female *Euchaeta marina* and *Pleuromamma xiphias* (Copepoda: Calanoida). *Marine Biology* **108**, 49–58.

Blanco, J.M[a], Guerrero, F. and Rodríguez, V. (1995). The fate of comparisons of models in temperature-dependent growth of copepods: a reply to the comment by McLaren. *Journal of Plankton Research* **17**, 1391–1394.

Boak, A.C. and Goulder, R. (1983). Bacterioplankton in the diet of the calanoid copepod *Eurytemora* sp. in the Humber Estuary. *Marine Biology* **73**, 139–149.

Bohrer, R.N. (1980). Experimental studies on diel vertical migration. *In* "Evolution and Ecology of Zooplankton Communities" (W.C. Kerfoot, ed.), pp. 111–121. University Press of New England, Hanover, N.H.

Bollens, S.M. and Frost, B.W. (1989a). Predator-induced diel vertical migration in a planktonic copepod. *Journal of Plankton Research* **11**, 1047–1065.

Bollens, S.M. and Frost, B.W. (1989b). Zooplanktivorous fish and variable diel vertical migration in the marine planktonic copepod *Calanus pacificus*. *Limnology and Oceanography* **34**, 1072–1083.

Bollens, S.M. and Frost, B.W. (1990). UV light and vertical distribution of the marine planktonic copepod *Acartia hudsonica* Pinhey. *Journal of Experimental Marine Biology and Ecology* **137**, 89–93.

Bollens, S.M. and Frost, B.W. (1991a). Diel vertical migration in zooplankton: rapid individual response to predators. *Journal of Plankton Research* **13**, 1359–1365.

Bollens, S.M. and Frost, B.W. (1991b). Ovigerity, selective predation and variable diel vertical migration in *Euchaeta elongata* (Copepoda: Calanoida). *Oecologia (Berlin)* **87**, 155–161.

Bollens, S.M. and Stearns, D.E. (1992). Predator-induced changes in the diel feeding cycle of a planktonic copepod. *Journal of Experimental Marine Biology and Ecology* **156**, 179–186.

Bollens, S.M., Frost, B.W., Schwaninger, H.R., Davis, C.S., Way, K.L. and Landsteiner, M.C. (1992a). Seasonal plankton cycles in a temperate fjord and comments on the match-mismatch hypothesis. *Journal of Plankton Research* **14**, 1279–1305.

Bollens, S.M., Frost, B.W., Thoreson, D.S. and Watts, S.J. (1992b). Diel vertical migration in zooplankton: field evidence in support of the predator avoidance hypothesis. *Hydrobiologia* **234**, 33–39.

Bollens, S.M., Frost, B.W. and Cordell, J.R. (1994). Chemical, mechanical and visual cues in the vertical migration behavior of the marine planktonic copepod *Acartia hudsonica*. *Journal of Plankton Research* **16**, 555–564.

Bollens, S.M., Frost, B.W. and Cordell, J.R. (1995). Diversity of cues in zooplankton vertical migration – a reply to Ringelberg's comment. *Journal of Plankton Research* **17**, 656–658.

Bond, W.J. (1993). Keystone species. *In* "Biodiversity and Ecosystem Function" (E.D. Schulze and H.A. Mooney, eds), pp. 237–253. Springer-Verlag, New York.

Borchers, P. and Hutchings, L. (1986). Starvation tolerance, development time and egg production of *Calanoides carinatus* in the Southern Benguela Current. *Journal of Plankton Research* **8**, 855–874.

Bosch, F. van den and Gabriel, W. (1994). A model of growth and development in copepods. *Limnology and Oceanography* **39**, 1528–1542.

Böttger, R. and Schnack, D. (1986). On the effect of formaldehyde fixation on the dry weight of copepods. *Meeresforschung* **31**, 141–152.

Bottrell, H.H. and Robins, D.B. (1984). Seasonal variations in length, dry weight, carbon and nitrogen of *Calanus helgolandicus* from the Celtic Sea. *Marine Ecology Progress Series* **14**, 259–268.

Boucher, J. (1984). Localization of zooplankton populations in the Ligurian marine front: role of ontogenetic migration. *Deep-Sea Research* **31A**, 469–484.

Boucher, J. (1988). Space-time aspects in the dynamics of planktonic stages. *In* "Toward a Theory on Biological–Physical Interactions in the World Ocean" (B.J. Rothschild, ed.), pp. 203–214. Kluwer Academic Publishers.

Boucher, J. and Bovée, F.de (1970). *Mimocalanus distinctocephalus* Brodsky, 1950 (Copepoda Calanoida) genre nouveau pour la Méditerranée, nouvelle description. *Vie et Milieu* **21B**, 527–534.

Boucher, J., Ibanez, F. and Prieur, L. (1987). Daily and seasonal variations in the spatial distribution of zooplankton populations in relation to the physical structure in the Ligurian Sea Front. *Journal of Marine Research* **45**, 133–173.

Bowlby, M.R. and Case, J.F. (1991a). Ultrastructure and neuronal control of luminous cells in the copepod *Gaussia princeps*. *Biological Bulletin, Marine Biological Laboratory, Wood's Hole, Mass.*, **180**, 440–446.

Bowlby, M.R. and Case, J.F. (1991b). Flash kinetics and spatial patterns of bioluminescence in the copepod *Gaussia princeps*. *Marine Biology* **110**, 329–336.

Bowman, T.E. (1971). The distribution of calanoid copepods off the southeastern United States between Cape Hatteras and southern Florida. *Smithsonian Contributions to Zoology* **96**, 1–58.

Bowman, T.E. (1978a). The modified suture between segments 8 and 9 on the first antenna of some calanoid copepods. *Crustaceana* **35**, 113–118.

Bowman, T.E. (1978b). From Brazil to Jamaica: a range extension of the neritic calanoid copepod, *Pseudodiaptomus acutus*. *Crustaceana* **35**, 249–252.

Bowman, T.E. and Abele, L.G. (1982). Classification of the recent Crustacea. *In* "The Biology of Crustacea" (D.E. Bliss, ed.), Vol. 1, pp. 1–27. Academic Press, New York.

Boxshall, G.A. (1985). The comparative anatomy of two copepods, a predatory calanoid and a particle-feeding mormonilloid. *Philosophical Transactions of the Royal Society of London* **311B**, 303–377.

Boxshall, G.A. (1992). Copepoda. *In* "Microscopic Anatomy of Invertebrates", Vol. 7, pp. 347–384. Wiley-Liss Inc.

Boxshall, G.A., Stock, J.M. and Sánchez, E. (1990). A new species of *Stephos* Scott, 1892 (Copepoda: Calanoida) from an anchihaline lava pool on Lanzarote, Canary Islands. *Stygologia* **5**, 33–41.

Boyd, C.M. and Smith, S.L. (1983). Plankton, upwelling, and coastally trapped waves off Peru. *Deep-Sea Research* **30A**, 723–742.

Boyd, C.M., Smith, S.L. and Cowles, T.J. (1980). Grazing patterns of copepods in the upwelling system off Peru. *Limnology and Oceanography* **25**, 583–596.

Bradford, J.M. (1969a). Notes on anomalous, British Antarctic (Terra Nova) Expedition, copepod records in the Three Kings Islands (New Zealand region). *Transactions of the Royal Society of New Zealand, Biological Sciences* **11**, (8), 93–99.

Bradford, J.M. (1969b). New genera and species of benthic calanoid copepods from the New Zealand slope. *New Zealand Journal of Marine and Freshwater Research* **3**, 473–505.

Bradford, J.M. (1970a). Diurnal variation in vertical distribution of pelagic Copepoda of Kaikoura, New Zealand. *New Zealand Journal of Marine and Freshwater Research* **4**, 337–350.

Bradford, J.M. (1970b). Records of pelagic copepods of Kaikoura, New Zealand. *New Zealand Journal of Marine and Freshwater Research* **4**, 351–363.

Bradford, J.M. (1971a). *Aetidius* and *Euaetidius* (Copepoda: Calanoida) from the Atlantic and Pacific Oceans. *New Zealand Journal of Marine and Freshwater Research* **5**, 12–40.

Bradford, J.M. (1971b). New and little-known species of Heterorhabdidae (Copepoda: Calanoida) from the southwest Pacific. *New Zealand Journal of Marine and Freshwater Research* **5**, 120–140.

Bradford, J.M. (1971c). Fauna of the Ross Sea. Part 8. Pelagic Copepoda. *New Zealand Department of Scientific and Industrial Research*, Bulletin, **206**, 7–31.

Bradford, J.M. (1972). Systematics and ecology of New Zealand central east coast plankton sampled at Kaikoura. *New Zealand Department of Scientific and Industrial Research*, Bulletin **207**, 1–87.

Bradford, J.M. (1973). Revision of family and some generic definitions in the Phaennidae and Scolecithricidae (Copepoda: Calanoida). *New Zealand Journal of Marine and Freshwater Research* **7**, 133–152.

Bradford, J.M. (1974a). *Euchaeta marina* (Prestandrea) (Copepoda, Calanoida) and two closely related new species from the Pacific Ocean. *Pacific Science* **28**, 159–169.

Bradford, J.M. (1974b). New and little-known Arietellidae (Copepoda: Calanoida) mainly from the south-west Pacific. *New Zealand Journal of Marine and Freshwater Research* **8**, 523–533.

Bradford, J.M. (1976). Partial revision of the *Acartia* subgenus *Acartiura* (Copepoda: Calanoida: Acartiidae). *New Zealand Journal of Marine and Freshwater Research* **10**, 159–202.

Bradford, J.M. (1980). Zoogeography of some New Zealand neritic pelagic Crustacea and their close relatives. *Information Series of the Department of Scientific and Industrial Research, New Zealand* **137**, (2), 593–612.

Bradford, J.M. (1981). Records of *Pareuchaeta* (Copepoda: Calanoida) from

McMurdo Sound, Antarctica, with a description of three hitherto unknown males. *New Zealand Journal of Marine and Freshwater Research* **15**, 391–402.
Bradford, J.M. (1988). Review of the taxonomy of the Calanidae (Copepoda) and the limits to the genus *Calanus*. *Hydrobiologia* **167/168**, 73–81.
Bradford, J.M. and Chapman, B.E. (1988). Epipelagic zooplankton assemblages and a warm-core eddy off East Cape, New Zealand. *Journal of Plankton Research* **10**, 601–619.
Bradford, J.M. and Jillett, J.B. (1974). A revision of generic definitions in the Calanidae (Copepoda, Calanoida). *Crustaceana* **27**, 5–16.
Bradford, J.M. and Jillett, J.B. (1980). The marine fauna of New Zealand: pelagic calanoid copepods: family Aetideidae. *Memoirs of the New Zealand Oceanographic Institute* **86**, 1– 102.
Bradford, J.M., Haakonssen, L. and Jillett, J.B. (1983). The marine fauna of New Zealand: pelagic calanoid copepods: families Euchaetidae, Phaennidae, Scolecithricidae, Diaixidae, and Tharybidae. *Memoirs of the New Zealand Oceanographic Institute* **90**, 1–150.
Bradford-Grieve, J.M. (1994). The marine fauna of New Zealand: pelagic calanoid Copepoda: Megacalanidae, Calanidae, Paracalanidae, Mecynoceridae, Eucalanidae, Spinocalanidae, Clausocalanidae. *New Zealand Oceanographic Institute Memoir* **102**, 1–160.
Bradley, B.P. (1976). The measurement of temperature tolerance: verification of an index. *Limnology and Oceanography* **21**, 596–599.
Bradley, B.P. (1986a). Genetic expression of temperature tolerance in the copepod *Eurytemora affinis* in different salinity and temperature environments. *Marine Biology* **91**, 561–565.
Bradley, B.P. (1986b). Traits, problems and methods in copepod life history studies. *Syllogeus* **58**, 247–253.
Bradley, B.P. (1991). Seasonal succession in Chesapeake Bay. *Bulletin of Plankton Society of Japan*, Special Volume, 129–131.
Bradley, B.P. and Davis, R. (1991). Time to coma as a predictor of survival in male and female *Eurytemora affinis* (Poppe). *Bulletin of Plankton Society of Japan*, Special Volume, 321–328.
Bradley, B.P. and Ketzner, P.A. (1982). Genetic and non-genetic variability in temperature tolerance of the copepod *Eurytemora affinis* in five temperature regimes. *Biological Bulletin, Marine Biological Laboratory, Woods Hole, Mass.* **162**, 233–245.
Bradley, B.P., Hakimzadeh, R. and Vincent, J.S. (1988). Rapid responses to stress in *Eurytemora affinis*. *Hydrobiologia* **167/168**, 197–200.
Bradley, B.P., Lane, M.A. and Gonzalez, C.M. (1992). A molecular mechanism of adaptation in an estuarine copepod. *Netherlands Journal of Sea Research* **30**, 3–10.
Brady, G.S. (1883). Report on the Copepoda collected by H.M.S. Challenger during the years 1873–1876. *The Voyage of H.M.S. "Challenger"*, **8**, 1–142.
Brady, G.S. (1918).Copepoda. *Australasian Antarctic Expedition, 1911–14, Scientific Reports, Series C, Zoology and Botany* **5**, part 3, 1–48.
Brand, G.W. and Bayly, I.A.E. (1971). A comparative study of osmotic regulation in four species of calanoid copepod. *Comparative Biochemistry and Physiology* **38B**, 361–371.
Brander, K. (1992). A re-examination of the relationship between cod recruitment and *Calanus finmarchicus* in the North Sea. *ICES Marine Science Symposia* **195**, 393–401.

Brenning, U. (1985). Structure and development of calanoid populations (Crustacea, Copepoda) in the upwelling regions off North West and South West Africa. *Beiträge zur Meereskunde* **52**, 3–33.
Bresciani, J. (1986). The fine structure of the integument of free-living and parasitic copepods. A review. *Acta Zoologica (Stockholm)* **67**, (3), 125–145.
Brodsky, K.A. (1950). Calanoida of the far eastern seas and polar basin of the USSR. *Israel Program for Scientific Translations, Jerusalem* **1967**, 1–440.
Brodsky, K.A. (1976). A description of male *Metridia okhotensis* Brodsky (Calanoida, Metridiidae). *Explorations of the Fauna of the Seas XX(XXVIII), Marine Plankton (Systematics and Faunistics), Academy of Sciences USSR*, 4–5.
Brodsky, K.A., Vyshkvartseva, N.V., Kos, M.S. and Markhaseva, E.L. (1983). Copepod crustaceans (Copepoda: Calanoida) of the USSR seas and adjacent waters. *Oprediteli po Fauna SSSR* **135**, 1–357. (In Russian).
Broekhuizen, N. and McKenzie, E. (1995). Patterns of abundance of *Calanus* and smaller copepods in the North Sea: time series decomposition of two CPR data sets. *Marine Ecology Progress Series* **118**, 103–120.
Brunet, M., Cuoc, C., Arnaud, J. and Mazza, J. (1991). Tegumental glands in a copepod *Hemidiaptomus ingens*: structural, ultrastructural and cytochemical aspects. *Tissue and Cell* **23**, 733–743.
Brunet, M., Arnaud, J. and Mazza, J. (1994). Gut structure and digestive cellular processes in marine Crustacea. *Oceanography and Marine Biology: an Annual Review* **32**, 335–367.
Brylinski, J.M. (1981). Report on the presence of *Acartia tonsa* Dana (Copepoda) in the harbour of Dunkirk (France) and its geographical distribution in Europe. *Journal of Plankton Research* **3**, 255–260.
Brylinski, J.M. (1984a). Anomalies morphologiques chez le genre *Acartia* (Crustacea, Copepoda): description et essai de quantification. *Journal of Plankton Research* **6**, 961–966.
Brylinski, J.M. (1984b). Réponses spécifiques de la croissance de quelques copépodes aux variations de la température et de la quantité de nourriture. *Crustaceana*, Supplement **7**, 91–101.
Brylinski, J.M., Bentley, D. and Quisthoudt, C. (1988). Discontinuité écologique et zooplancton (copépodes) en Manche Orientale. *Journal of Plankton Research* **10**, 503–513.
Bucklin, A. and LaJeunesse, T.C. (1994). Molecular genetic variation of *Calanus pacificus* (Copepoda; Calanoida): preliminary evaluation of genetic structure and sub-specific differentiation based on mtDNA sequences. *Reports of the California Cooperative Oceanic Fisheries Investigations* **35**, 45–51.
Bucklin, A. and Marcus, N.H. (1985). Genetic differentiation of populations of the planktonic copepod *Labidocera aestiva*. *Marine Biology* **84**, 219–224.
Bucklin, A., Frost, B.W. and Kocher, T.D. (1992). DNA sequence variation of the mitochondrial 16S rRNA in *Calanus* (Copepoda; Calanoida): intraspecific and interspecific patterns. *Molecular Marine Biology and Biotechnology* **1**, 397–407.
Bucklin, A., Frost, B.W. and Kocher, T.D. (1995). Molecular systematics of six *Calanus* and three *Metridia* species (Calanoida: Copepoda). *Marine Biology* **121**, 655–664.
Bucklin, A., LaJeunesse, T.C., Curry, E., Wallinga, J. and Garrison, K. (1996). Molecular diversity of the copepod, *Nannocalanus minor*: genetic evidence of species and population structure in the North Atlantic Ocean. *Journal of Marine Research* **54**, 285–310.

Bundy, M.H. and Paffenhöfer, G.-A. (1993). Innervation of copepod antennules investigated using laser scanning confocal microscopy. *Marine Ecology Progress Series* **102**, 1–14.

Bundy, M.H. and Paffenhöfer, G.-A. (1996). Analysis of flow fields associated with freely swimming calanoid copepods. *Marine Ecology Progress Series* **133**, 99–113.

Burckhardt, R. and Arndt, H. (1987). Untersuchungen zur Konsumtion von Ciliaten durch Metazooplankter des Barther Boddens (südliche Ostsee). *Wissenschaftliche Zeitschrift der Wilhelm-Pieck-Universität Rostock* **36**, 22–26.

Burd, B.J. and Thomson, R.E. (1995). Distribution of zooplankton associated with the Endeavour Ridge hydrothermal plume. *Journal of Plankton Research* **17**, 965–997.

Burd, B.J., Thomson, R.E. and Jamieson, G.S. (1992). Composition of a deep scattering layer overlying a mid- ocean ridge hydrothermal plume. *Marine Biology* **113**, 517–526.

Burkill, P.H. and Kendall, T.F. (1982). Production of the copepod *Eurytemora affinis* in the Bristol Channel. *Marine Ecology Progress Series* **7**, 21–31.

Busch, A. and Brenning, U. (1992). Studies on the status of *Eurytemora affinis* (Poppe, 1880) (Copepoda, Calanoida). *Crustaceana* **62**, 13–38.

Buskey, E.J. (1984). Swimming pattern as an indicator of the roles of copepod sensory systems in the recognition of food. *Marine Biology* **79**, 165–175.

Buskey, E.J. (1992). Epipelagic planktonic bioluminescence in the marginal ice zone of the Greenland Sea. *Marine Biology* **113**, 689–698.

Buskey, E.J. (1994). Factors affecting feeding selectivity of visual predators on the copepod *Acartia tonsa*: locomotion, visibility and escape responses. *Hydrobiologia* **292/293**, 447–453.

Buskey, E.J. and Stearns, D.E. (1991). The effects of starvation on bioluminescence potential and egg release of the copepod *Metridia longa*. *Journal of Plankton Research* **13**, 885–893.

Buskey, E.J. and Swift, E. (1983). Behavioral responses of the coastal copepod *Acartia hudsonica* (Pinhey) to simulated dinoflagellate bioluminescence. *Journal of Experimental Marine Biology and Ecology* **72**, 43–58.

Buskey, E.J. and Swift, E. (1985). Behavioral responses of oceanic zooplankton to simulated bioluminescence. *Biological Bulletin, Marine Biological Laboratory, Wood's Hole, Mass.*, **168**, 263–275.

Buskey, E.J. and Swift, E. (1990). An encounter model to predict natural planktonic bioluminescence. *Limnology and Oceanography* **35**, 1469–1485.

Buskey, E.J., Mills, L. and Swift, E. (1983). The effects of dinoflagellate bioluminescence on the swimming behavior of a marine copepod. *Limnology and Oceanography* **28**, 575–579.

Buskey, E.J., Mann, C.G. and Swift, E. (1986). The shadow response of the estuarine copepod *Acartia tonsa* (Dana). *Journal of Experimental Marine Biology and Ecology* **103**, 65–75.

Buskey, E.J., Mann, C.G. and Swift, E. (1987). Photophobic responses of calanoid copepods: possible adaptive value. *Journal of Plankton Research* **9**, 857–870.

Buskey, E.J., Baker, K.S., Smith, R.C. and Swift, E. (1989). Photosensitivity of the oceanic copepods *Pleuromamma gracilis* and *Pleuromamma xiphias* and its relationship to light penetration and daytime depth distribution. *Marine Ecology Progress Series* **55**, 207–216.

Butler, E.I., Corner, E.D.S. and Marshall, S.M. (1969). On the nutrition and metabolism of zooplankton. VI. Feeding efficiency of *Calanus* in terms of nitrogen and phosphorus. *Journal of the Marine Biological Association of the United Kingdom* **49**, 977–1001.

Butler, M. and Dam, H.G. (1994). Production rates and characteristics of fecal pellets of the copepod *Acartia tonsa* under simulated phytoplankton bloom conditions: implications for vertical fluxes. *Marine Ecology Progress Series* **114**, 81–91.

Cahoon, L.B. (1981). Reproductive response of *Acartia tonsa* to variations in food ration and quality. *Deep-Sea Research* **28A**, 1215–1221.

Cahoon, L.B. (1982). The use of mucus in feeding by the copepod *Euchirella venusta* Giesbrecht. *Crustaceana* **43**, 202–204.

Cairns, A.A. (1967). The zooplankton of Tanquary Fjord, Ellesmere Island, with special reference to the calanoid copepods. *Journal of the Fisheries Research Board of Canada* **24**, 555–568.

Calbet, A. and Alcaraz, M. (1996). Effects of constant and fluctuating food supply on egg production rates of *Acartia grani* (Copepoda: Calanoida). *Marine Ecology Progress Series* **140**, 33–39.

Campaner, A.F. (1977). New definition of the Arietellidae (Copepoda, Calanoida), with the description of a new genus and species, and separation of the Phyllopidae fam.n. *Ciència Cultura* **29**, 811–818.

Campaner, A.F. (1978a). On some new planktobenthic Aetideidae and Phaennidae (Copepoda, Calanoida) from the Brazilian continental shelf. I. Aetideidae. *Ciència e Cultura* **30**, 863–876.

Campaner, A.F. (1978b). On some new planktobenthic Aetideidae and Phaennidae (Copepoda, Calanoida) from the Brazilian continental shelf. II. Phaennidae. *Ciència e Cultura* **30**, 966–982.

Campaner, A.F. (1984). Some taxonomic problems within the Arietellidae (Calanoida). *Crustaceana*, Supplement **7**, 102–109.

Campaner, A.F. (1986). Planktobenthic copepods from the southern Brazilian continental shelf. *Syllogeus* **58**, 259–266.

Campaner, A.F. (1989). Supplementary description of *Macandrewella chelipes* (Giesbrecht, 1896) from the Gulf of Elat (Copepoda: Calanoida: Scolecithricidae), and comments on its relationships with *Scottocalanus* Sars and *Scolecocalanus* Farran. *Israel Journal of Zoology* **35**, 229–235.

Campbell, A.K. and Herring, P.J. (1990). Imidazolopyrazine bioluminescence in copepods and other marine organisms. *Marine Biology* **104**, 219–225.

Campbell, M.H. (1934). The life history and postembryonic development of the copepods *Calanus tonsus* Brady and *Euchaeta japonica* Marukawa. *Journal of the Biological Research Board of Canada* **1**, 1–65.

Caparroy, P. and Carlotti, F. (1996). A model for *Acartia tonsa*: effect of turbulence and consequences for the related physiological processes. *Journal of Plankton Research* **18**, 2139–2177.

Carillo, B.-G.E., Miller, C.B. and Wiebe, P.H. (1974). Failure of interbreeding between Atlantic and Pacific populations of the marine calanoid copepod *Acartia clausi* Giesbrecht. *Limnology and Oceanography* **19**, 452–458.

Carlotti, F. and Nival, S. (1991). Individual variability of development in laboratory-reared *Temora stylifera* copepodites: consequences for the population dynamics and interpretation in the scope of growth and development rules. *Journal of Plankton Research* **13**, 801–813.

Carlotti, F. and Nival, P. (1992a). Model of copepod growth and development: moulting and mortality in relation to physiological processes during an individual moult cycle. *Marine Ecology Progress Series* **84**, 219–233.

Carlotti, F. and Nival, P. (1992b). Moulting and mortality rates of copepods related to age within stage: experimental results. *Marine Ecology Progress Series* **84**, 235–243.
Carlotti, F. and Radach, G. (1996). Seasonal dynamics of phytoplankton and *Calanus finmarchicus* in the North Sea as revealed by a coupled one-dimensional model. *Limnology and Oceanography* **41**, 522–539.
Carlotti, F., Krause, M. and Radach, G. (1993). Growth and development of *Calanus finmarchicus* related to the influence of temperature: experimental results and conceptual model. *Limnology and Oceanography* **38**, 1125– 1134.
Carlsson, P., Granéli, E., Finenko, G. and Maestrini, S.Y. (1995). Copepod grazing on a phytoplankton community containing the toxic dinoflagellate *Dinophysis acuminata*. *Journal of Plankton Research* **17**, 1925–1938.
Carlton, J.T. (1985). Transoceanic and interoceanic dispersal of coastal marine organisms: the biology of ballast water. *Oceanography and Marine Biology: an Annual Review* **23**, 313–371.
Carlton, J.T. and Geller, J.B. (1993). Ecological roulette: the global transport of nonindigenous marine organisms. *Science* **261**, 78–82.
Carman, K.R., Thistle, D., Ertman, S.C. and Foy, M. (1991). Nile red as a probe for lipid-storage products in benthic copepods. *Marine Ecology Progress Series* **74**, 307–311.
Carola, M. (1994). Checklist of the marine planktonic Copepoda of southern Africa and their worldwide geographic distribution. *South African Journal of Marine Science* **14**, 225–253.
Carola, M. and Razouls, C. (1996). Two new species of Calanoida from a marine cave on Minorca Island, Mediterranean Sea: *Stephos balearensis* new species (Stephidae) and *Paracyclopia gitana* new species (Pseudocyclopiidae). *Bulletin of Marine Science* **58**, 344–352.
Cassie, R.M. (1963). Microdistribution of plankton. *Oceanography and Marine Biology: an Annual Review* **1**, 223–252.
Castel, J. (1993). Long-term distribution of zooplankton in the Gironde estuary and its relation with river flow and suspended matter. *Cahiers de Biologie marine* **34**, 145–163.
Castel, J. and Feurtet, A. (1989). Dynamics of the copepod *Eurytemora affinis hirundoides* in the Gironde Estuary: origin and fate of its production. *Scientia Marina* **53**, 577–584.
Castel, J. and Feurtet, A. (1992). Fecundity and mortality rates of the copepod *Eurytemora affinis* in the Gironde Estuary. *In* "Marine Eutrophication and Population Dynamics" (G. Colombo, I. Ferrari, V.U. Ceccherelli and R. Rossi, eds), pp. 143–149. Olsen & Olsen, Fredensborg.
Castel, J. and Veiga, J. (1990). Distribution and retention of the copepod *Eurytemora affinis hirundoides* in a turbid estuary. *Marine Biology* **107**, 119–128.
Castel, J., Courties, C. and Poli, J.-M. (1983). Dynamique du copépode *Eurytemora hirundoides* dans l'estuaire de la Gironde: effet de la température. *Oceanologica Acta*, No.Sp 57–61.
Castro-Longoria, E. and Williams, J.A. (1996). First report of the presence of *Acartia margalefi* (Copepoda: Calanoida) in Southampton Water and Horsea Lake, U.K. *Journal of Plankton Research* **18**, 567–575.
Castro, L.R., Bernal, P.A. and Gonzalez, H.E. (1991). Vertical distribution of copepods and the utilization of the chlorophyll a-rich layer within Concepcion Bay, Chile. *Estuarine, Coastal and Shelf Science* **32**, 243–256.
Castro, L.R., Bernal, P.A. and Trocoso, V.A. (1993). Coastal intrusion of copepods:

mechanisms and consequences on the population biology of *Rhincalanus nasutus*. *Journal of Plankton Research* **15**, 501–515.

Cataletto, B. and Fonda Umani, S. (1994). Seasonal variations in carbon and nitrogen content of *Acartia clausi* (Copepoda, Calanoida) in the Gulf of Trieste (Northern Adriatic Sea). *Hydrobiologia* **292/293**, 283–288.

Cataletto, B., Feoli, E., Fonda Umani, S. and Cheng-Yong, S. (1995). Eleven years of time-series analysis on the net-zooplankton community in the Gulf of Trieste. *ICES Journal of Marine Science* **52**, 669–678.

Cervelli, M., Battaglia, B., Bisol, P.M., Comaschi Scaramuzza, A. and Meneghetti, F. (1995). Genetic differentiation in the genus *Acartia* from the lagoon of Venice. *Vie et Milieu* **45**, 117–122.

Cervetto, G., Gaudy, R., Pagano, M., Saint-Jean, L., Verriopoulos, G., Arfi, R. and Leveau, M. (1993). Diel variations in *Acartia tonsa* feeding, respiration and egg production in a Mediterranean coastal lagoon. *Journal of Plankton Research* **15**, 1207–1228.

Champalbert, G. (1971a). Variations nycthémérales du plancton superficiel. I. Holohyponeuston et hétérohyponeuston. *Journal of Experimental Marine Biology and Ecology* **6**, 23–33.

Champalbert, G. (1971b). Variations nycthémérales du plancton superficiel. II. Espèces non caractéristique de l'hyponeuston et hyponeuston nocturne. *Journal of Experimental Marine Biology and Ecology* **6**, 55–70.

Champalbert, G. (1978). Rythmes d'activité natatoire de quelques copépodes hyponeustoniques (*Anomalocera patersoni* Templeton, *Pontella mediterranea* Claus, *Labidocera wollastoni* Lubbock) en fonction des conditions d'éclairement et de pression. *Journal of Experimental Marine Biology and Ecology* **35**, 233–249.

Champalbert, G. (1979). Rythme d'activité natatoire de deux espèces de copépodes pontellidés (*Anomalocera patersoni* et *Pontella mediterranea*). *Téthys* **9**, 83–85.

Champalbert, G. (1985). L'hyponeuston permanent. Approche ecophysiologique de la répartition des Pontellidae (Crustacea, Copepoda). *Téthys* **11**, 264–274.

Chapman, P.M. (1981). Evidence for dissolved glucose uptake from seawater by *Neocalanus plumchrus* (Arthropoda, Copepoda). *Canadian Journal of Zoology* **59**, 1618–1621.

Chaudron, Y., Poulet, S.A., Laabir, M., Ianora, A. and Miralto, A. (1996). Is hatching success of copepod eggs diatom density-dependent? *Marine Ecology Progress Series* **144**, 185–193.

Checkley, D.M. (1980a). The egg production of a marine planktonic copepod in relation to its food supply: laboratory studies. *Limnology and Oceanography* **25**, 430–446.

Checkley, D.M. (1980b). Food limitation of egg production by a marine, planktonic copepod in the sea off southern California. *Limnology and Oceanography* **25**, 991–998.

Checkley, D.M. and Miller, C.A. (1989). Nitrogen isotope fractionation by oceanic zooplankton. *Deep-Sea Research* **36A**, 1449–1456.

Checkley, D.M.(Jr), Dagg, M.J. and Uye, S. (1992). Feeding, excretion and egg production by individuals and populations of the marine, planktonic copepods, *Acartia* spp. and *Centropages furcatus*. *Journal of Plankton Research* **14**, 71–96.

Cheer, A.Y.L. and Koehl, M.A.R. (1987). Paddles and rakes: fluid flow through bristled appendages of small organisms. *Journal of Theoretical Biology* **129**, 17–40.

Chen, F. and Li, S.-J. (1991). On seasonal distribution and diapause eggs in *Tortanus* from Xiamen waters. *Acta Oceanologica Sinica* **13**, 721–727.
Chen, Q.-C. and Zhang, S.-Z. (1965). The planktonic copepods of the Yellow Sea and the East China Sea. I. Calanoida. *Studia Marina Sinica* **7**, 20–131.
Chen, Y.-Q. (1986). The vertical distribution of some pelagic copepods in the eastern tropical Pacific. *Reports of the California Cooperative Oceanic Fisheries Investigations* **27**, 205–227.
Chervin, M.B., Malone, T.C. and Neale, P.J. (1981). Interactions between suspended organic matter and copepod grazing in the plume of the Hudson River. *Estuarine, Coastal and Shelf Science* **13**, 169–183.
Chiba, T. (1953). Some observations on the morphological characteristics especially on the reproductive organ of Copepoda, *Calanus darwinii*. *Journal of the Shimonoseki College of Fisheries* **3**, 235–238.
Chiba, T. (1956). Studies on the development and the systematics of Copepoda. *Journal of the Shimonoseki College of Fisheries* **6**, 1–90.
Childress, J.J. (1977). Effects of pressure, temperature and oxygen on the oxygen-consumption rate of the midwater copepod *Gaussia princeps*. *Marine Biology* **39**, 19–24.
Chisholm, L.A. and Roff, J.C. (1990a). Size-weight relationships and biomass of tropical neritic copepods off Kingston, Jamaica. *Marine Biology* **106**, 71–77.
Chisholm, L.A. and Roff, J.C. (1990b). Abundances, growth rates, and production of tropical neritic copepods off Kingston, Jamaica. *Marine Biology* **106**, 79–89.
Chojnacki, J. (1983). Standard weights of the Pomeranian Bay copepods. *Internationale Revue der gesamten Hydrobiologie* **68**, 435–441.
Chojnacki, J. (1986). Biomass estimation of *Temora longicornis* on the basis of geometric method. *Syllogeus* **58**, 534–538.
Christou, E.D. and Verriopoulos, G.C. (1993a). Length, weight and condition factor of *Acartia clausi* (Copepoda) in the eastern Mediterranean. *Journal of the Marine Biological Association of the United Kingdom* **73**, 343–353.
Christou, E.D. and Verriopoulos, G.C. (1993b). Analysis of the biological cycle of *Acartia clausi* (Copepoda) in a meso-oligotrophic coastal area of the eastern Mediterranean Sea using time-series analysis. *Marine Biology* **115**, 643–651.
Ciszewski, P. and Witek, Z. (1977). Production of older stages of copepods *Acartia bifilosa* Giesb. and *Pseudocalanus elongatus* Boeck in Gdansk Bay. *Polskie Archiwum Hydrobiologie* **24**, 449–459.
Clarke, A. and Peck, L.S. (1991). The physiology of polar marine zooplankton. *Polar Research* **10**, 355–369.
Clarke, A., Holmes, L.J. and Hopkins, C.C.E. (1987). Lipid in an Arctic food chain: *Calanus, Bolinopsis, Beroe*. *Sarsia* **72**, 41–48.
Clarke, G.L., Conover, R.J., David, C.N. and Nicol, J.A.C. (1962). Comparative studies of luminescence in copepods and other pelagic marine animals. *Journal of the Marine Biological Association of the United Kingdom* **42**, 541–564.
Cohen, R.E. and Lough, R.G. (1981). Length-weight relationships for several copepods dominant in the Georges Bank-Gulf of Maine area. *Journal of Northwest Atlantic Fishery Science* **2**, 47–52.
Colebrook, J.M. (1978). Continuous plankton records: zooplankton and environment, North-east Atlantic and North Sea, 1948–1975. *Oceanologica Acta* **1**, 9–23.
Colebrook, J.M. (1981). Continuous plankton records: persistence in time-series of annual means of abundance of zooplankton. *Marine Biology* **61**, 143–149.
Colebrook, J.M. (1982). Continuous plankton records: persistence in time-series and

the population dynamics of *Pseudocalanus elongatus* and *Acartia clausi*. *Marine Biology* **66**, 289–294.

Colebrook, J.M. (1985a). Sea surface temperature and zooplankton, North Sea, 1948 to 1983. *Journal du Conseil International pour l'Exploration de la Mer* **42**, 179–185.

Colebrook, J.M. (1985b). Continuous plankton records: overwintering and annual fluctuations in abundance of zooplankton. *Marine Biology* **84**, 261–265.

Colebrook, J.M., Robinson, G.A., Hunt, H.G., Roskell, J., John, A.W.G., Bottrell, H.H., Lindley, J.A., Collins, N.R. and Halliday, N.C. (1984). Continuous plankton records: a possible reversal in the downward trend in the abundance of the plankton of the North Sea and the Northeast Atlantic. *Journal du Conseil International pour l'Exploration de la Mer* **41**, 304–306.

Collins, N.R. and Williams, R. (1981). Zooplankton of the Bristol Channel and Severn Estuary. The distribution of four copepods in relation to salinity. *Marine Biology* **64**, 273–283.

Collins, N.R. and Williams, R. (1982). Zooplankton communities in the Bristol Channel and Severn Estuary. *Marine Ecology Progress Series* **9**, 1–11.

Comita, G.W. and Tommerdahl, D.M. (1960). The postembryonic developmental instars of *Diaptomus siciloides* Lilljeborg. *Journal of Morphology* **107**, 297–355.

Conley, W.J. and Turner, J.T. (1985). Omnivory by the coastal marine copepods *Centropages hamatus* and *Labidocera aestiva*. *Marine Ecology Progress Series* **21**, 113–120.

Conover, R.J. (1956). Oceanography of Long Island Sound, 1952–1954. VI. Biology of *Acartia clausi* and *A. tonsa*. *Bulletin of the Bingham Oceanographic Collection* **15**, 156–233.

Conover, R.J. (1965). An intersex in *Calanus hyperboreus*. *Crustaceana* **8**, 153–158.

Conover, R.J. (1966a). Assimilation of organic matter by zooplankton. *Limnology and Oceanography* **11**, 338–345.

Conover, R.J. (1966b). Factors affecting the assimilation of organic matter by zooplankton and the question of superfluous feeding. *Limnology and Oceanography* **11**, 346–354.

Conover, R.J. (1967). Reproductive cycle, early development, and fecundity in laboratory populations of the copepod *Calanus hyperboreus*. *Crustaceana* **13**, 61–72.

Conover, R.J. (1988). Comparative life histories in the genera *Calanus* and *Neocalanus* in high latitudes of the northern hemisphere. *Hydrobiologia* **167/168**, 127–142.

Conover, R.J. and Huntley, M. (1991). Copepods in ice-covered seas – distribution, adaptations to seasonally limited food, metabolism, growth patterns and life cycle strategies in polar seas. *Journal of Marine Systems* **2**, 1–41.

Conover, R.J. and Poulet, S.A. (1986). Physiological methods for determining copepod production. *Syllogeus* **58**, 85–99.

Conover, R.J., Bedo, A.W. and Spry, J.A. (1988a). Arctic zooplankton prefer living ice algae: a caution for zooplankton excretion measurements. *Journal of Plankton Research* **10**, 267–282.

Conover, R.J., Bedo, A.W., Herman, A.W., Head, E.J.H., Harris, L.R. and Horne, E.P.W. (1988b). Never trust a copepod – some observations on their behavior in the Canadian Arctic. *Bulletin of Marine Science* **43**, 650–662.

Conover, R.J., Harris, L.R., Bedo, A.W. (1991). Copepods in cold oligotrophic waters – how do they cope? *Bulletin of Plankton Society of Japan*, Special Volume, 177–199.

Conway, D.V.P., McFadzen, I.R.B. and Tranter, P.R.G. (1994). Digestion of copepod eggs by larval turbot *Scophthalmus maximus* and egg viability following gut passage. *Marine Ecology Progress Series* **106**, 303–309.

Cooney, R.T. (1986). The seasonal occurrence of *Neocalanus cristatus*, *Neocalanus plumchrus*, and *Eucalanus bungii* over the shelf of the northern Gulf of Alaska. *Continental Shelf Research* **5**, 541–533.

Cooney, R.T. and Coyle, K.O. (1982). Trophic implications of cross-shelf copepod distributions in the southeastern Bering Sea. *Marine Biology* **70**, 187–196.

Cordell, J.R., Morgan, C.A. and Simenstad, C.A. (1992). Occurrence of the Asian calanoid copepod *Pseudodiaptomus inopinus* in the zooplankton of the Columbia River estuary. *Journal of Crustacean Biology* **12**, 260–269.

Corkett, C.J. (1967). The copepodid stages of *Temora longicornis* (O.F. Müller 1792) (Copepoda). *Crustaceana* **12**, 261–273.

Corkett, C.J. (1970). Techniques for breeding and rearing marine calanoid copepods. *Helgoländer wissenschaftliche Meeresuntersuchungen* **20**, 318–324.

Corkett, C.J. (1984). Observations on development in copepods. *Crustaceana*, Supplement **7**, 150–153.

Corkett, C.J. and McLaren, I.A. (1969). Egg production and oil storage by the copepod *Pseudocalanus* in the laboratory. *Journal of Experimental Marine Biology and Ecology* **3**, 90–105.

Corkett, C.J. and McLaren, I.A. (1970). Relationships between development rate of eggs and older stages of copepods. *Journal of the Marine Biological Association of the United Kingdom* **50**, 161–168.

Corkett, C.J. and McLaren, I.A. (1978). The biology of *Pseudocalanus*. *Advances in Marine Biology* **15**, 1–231.

Corkett, C.J. and Zillioux, E.J. (1975). Studies on the effect of temperature on the egg laying of three species of calanoid copepods in the laboratory (*Acartia tonsa*, *Temora longicornis* and *Pseudocalanus elongatus*). *Bulletin of Plankton Society of Japan* **21**, 77–85.

Corkett, C.J., McLaren, I.A. and Sevigny, J.-M. (1986). The rearing of the marine calanoid copepods *Calanus finmarchicus* (Gunnerus), *C. glacialis* Jaschnov and *C. hyperboreus* Kroyer with comment on the equiproportional rule. *Syllogeus* **58**, 539–546.

Corner, E.D.S. (1972). Laboratory studies related to zooplankton production in the sea. *Symposia of the Zoological Society of London* **29**, 185–201.

Corner, E.D.S. and O'Hara, S.C.M., eds (1986). The Biological Chemistry of Marine Copepods. Clarendon Press, Oxford. 349 pp.

Corner, E.D.S., Cowey, C.B. and Marshall, S.M. (1967). On the nutrition and metabolism of zooplankton. V. Feeding efficiency of *Calanus finmarchicus*. *Journal of the Marine Biological Association of the United Kingdom* **47**, 259–270.

Corner, E.D.S., Head, R.N. and Kilvington, C.C. (1972). On the nutrition and metabolism of zooplankton. VIII. The grazing of *Biddulphia* cells by *Calanus helgolandicus*. *Journal of the Marine Biological Association of the United Kingdom* **52**, 847–861.

Corner, E.D.S., O'Hara, S.C.M., Neal, A.C. and Eglinton, G. (1986). Copepod faecal pellets and the vertical flux of biolipids. *In* "The Biological Chemistry of Marine Copepods" (E.D.S. Corner and S.C.M. O'Hara, eds.), pp. 260–321. Clarendon Press, Oxford.

Corral Estrada, J. (1972a). La familia Calocalanidae (Copepoda, Calanoida) en aguas del archipélago Canario. *Boletin del Instituto Español de Oceanografia* **149**, 1–47.
Corral Estrada, J. (1972b). Copepoda. Sub-order: Calanoida. Family: Calocalanidae (Paracalanidae part.). *Conseil International pour l'Exploration de la Mer, Zooplankton Sheet* (138), 1–7.
Corral Estrada, J. and Pereiro Muñoz, J.A. (1974). Estudio de las asociaciones de copépodos planctónicos en una zona de las islas Canarias. *Boletin del Instituto Español de Oceanografia* **175**, 1–32.
Costello, J.H., Strickler, J.R., Marrasé, C., Trager, G., Zeller, R. and Freise, A.J. (1990). Grazing in a turbulent environment: behavioral response of a calanoid copepod, *Centropages hamatus*. *Proceedings of the National Academy of Sciences* **87**, 1648–1652.
Cowie, G.L. and Hedges, J.L. (1996). Digestion and alteration of the biochemical constituents of a diatom (*Thalassiosira weissflogii*) ingested by an herbivorous zooplankton (*Calanus pacificus*). *Limnology and Oceanography* **41**, 581–594.
Cowles, T.J. (1983). Effects of exposure to sublethal concentrations of crude oil on the copepod *Centropages hamatus*. II. Activity patterns. *Marine Biology* **78**, 53–57.
Cowles, T.J. and Strickler, J.R. (1983). Characterization of feeding activity patterns in the planktonic copepod *Centropages typicus* Kroyer under various food conditions. *Limnology and Oceanography* **28**, 106–115.
Cowles, T.J., Olson, R.J. and Chisholm, S.W. (1988). Food selection by copepods: discrimination on the basis of food quality. *Marine Biology* **100**, 41–49.
Coyle, K.O. and Cooney, R.T. (1988). Estimating carbon flux to pelagic grazers in the ice-edge zone of the eastern Bering Sea. *Marine Biology* **98**, 299–306.
Coyle, K.O., Paul, A.J. and Ziemann, D.A. (1990). Copepod populations during the spring bloom in an Alaskan subarctic embayment. *Journal of Plankton Research* **12**, 759–797.
Crawford, P. and Daborn, G.R. (1986). Seasonal variations in body size and fecundity in a copepod of turbid estuaries. *Estuaries* **9**, 133–141.
Crisafi, P. (1965). Les copépodes du detroit de Messine oeufs, stades naupliens et segmentation du corps du copépode pélagique *Pontella mediterranea* Claus. *Rapports et Procès-Verbaux des Réunions Commission Internationale pour l'Exploration Scientifique de la Mer Méditerranée* **18** (2), 411–416.
Cummings, E. and Ruber, E. (1987). Copepod colonization of natural and artificial substrates in a salt marsh pool. *Estuarine, Coastal and Shelf Science* **25**, 637–645.
Cummings, J.A. (1983). Habitat dimensions of calanoid copepods in the western Gulf of Mexico. *Journal of Marine Research* **41**, 163–188.
Cuoc, C., Arnaud, J., Brunet, M. and Mazza, J. (1989a). Structure, ultrastructure et essai d'interprétation fonctionnelle de l'appareil génital femelle d'*Hemidiaptomus ingens provinciae* et de *Mixodiaptomus kupelwieseri* (Copepoda, Calanoida). I. Les oviductes. *Canadian Journal of Zoology* **67**, 2569–2578.
Cuoc, C., Arnaud, J., Brunet, M. and Mazza, J. (1989b). Structure, ultrastructure et essai d'interprétation fonctionnelle de l'appareil génital femelle d'*Hemidiaptomus ingens provinciae* et de *Mixodiaptomus kupelwieseri* (Copepoda, Calanoida). II. L'aire génitale. *Canadian Journal of Zoology* **67**, 2579–2587.
Currie, M.E. (1918). Exuviation and variation of plankton copepods with special reference to *Calanus finmarchicus*. *Transactions of the Royal Society of Canada*, Section IV, 207–233.

Cushing, D.H. (1951). The vertical migration of planktonic Crustacea. *Biological Reviews* **26**, 158–192.
Cushing, D.H. (1959). On the nature of production in the sea. *Fishery Investigations*, London. Ser.II **22**, (6), 1–40.
Cushing, D.H. (1968). Grazing by herbivorous copepods in the sea. *Journal du Conseil International pour l'Exploration de la Mer* **32**, 70–82.
Cushing, D.H. and Tungate, D.S. (1963). Studies on a *Calanus* patch. I. The identification of a *Calanus* patch. *Journal of the Marine Biological Association of the United Kingdom* **43**, 327–337.
Daan, R., Gonzalez, S.R. and Klein Breteler, W.C.M. (1988). Cannibalism in omnivorous calanoid copepods. *Marine Ecology Progress Series* **47**, 45–54.
Dagg, M. (1977). Some effects of patchy food environments on copepods. *Limnology and Oceanography* **22**, 99–107.
Dagg, M. (1978). Estimated, *in situ*, rates of egg production for the copepod *Centropages typicus* (Krøyer) in the New York Bight. *Journal of Experimental Marine Biology and Ecology* **34**, 183–196.
Dagg, M.J. (1983). A method for the determination of copepod feeding rates during short time intervals. *Marine Biology* **75**, 63–67.
Dagg, M.J. (1985). The effects of food limitation on diel migratory behavior in marine zooplankton. *Archives fur Hydrobiologie Beiheft Ergebnisse Limnologie* **21**, 247–255.
Dagg, M.J. (1991). *Neocalanus plumchrus* (Marukawa): life in the nutritionally-dilute subarctic Pacific Ocean and the phytoplankton-rich Bering Sea. *Bulletin of Plankton Society of Japan*, Special Volume, 217–225.
Dagg, M.J. (1995). A comment on "Variability due to feeding activity of individual copepods" (Paffenhöfer, 1994). *Journal of Plankton Research* **17**, 903–905.
Dagg, M.J. and Grill, D.W. (1980). Natural feeding rates of *Centropages typicus* females in the New York Bight. *Limnology and Oceanography* **25**, 597–609.
Dagg, M.J. and Littlepage, J.L. (1972). Relationships between growth rate and RNA, DNA, protein and dry weight in *Artemia salina* and *Euchaeta elongata*. *Marine Biology* **17**, 162–170.
Dagg, M.J. and Turner, J.T. (1982). The impact of copepod grazing on the phytoplankton of Georges Bank and the New York Bight. *Canadian Journal of Fisheries and Aquatic Sciences* **39**, 979–990.
Dagg, M.J. and Walser, W.E. (1986). The effect of food concentration on fecal pellet size in marine copepods. *Limnology and Oceanography* **31**, 1066–1071.
Dagg, M.J. and Walser, W.E. (1987). Ingestion, gut passage, and egestion by the copepod *Neocalanus plumchrus* in the laboratory and in the subarctic Pacific Ocean. *Limnology and Oceanography* **32**, 178–188.
Dagg, M.J. and Whitledge, T.E. (1991). Concentrations of copepod nauplii associated with the nutrient-rich plume of the Mississippi River. *Continental Shelf Research* **11**, 1409–1423.
Dagg, M.J. and Wyman, K.D. (1983). Natural ingestion rates of the copepods *Neocalanus plumchrus* and *N. cristatus* calculated from gut contents. *Marine Ecology Progress Series* **13**, 37–46.
Dagg, M., Cowles, T., Whitledge, T., Smith, S., Howe, S. and Judkins, D. (1980). Grazing and excretion by zooplankton in the Peru upwelling system during April 1977. *Deep-Sea Research* **27A**, 43–59.
Dagg, M.J., Vidal, J., Whitledge, T.E., Iverson, R.L. and Goering, J.J. (1982). The feeding, respiration and excretion of zooplankton in the Bering Sea during a spring bloom. *Deep-Sea Research* **29A**, 45–63.

Dagg, M.J., Frost, B.W. and Walser, W.E. (1989). Copepod diel migration, feeding, and vertical flux of phaeopigments. *Limnology and Oceanography* **34**, 1062–1071.
Dahms, H.-U. (1995). Dormancy in the Copepoda – an overview. *Hydrobiologia* **306**, 199–211.
Dakin, W.J. and Colefax, A.N. (1940). The plankton of the Australian coastal waters of New South Wales. *Publications of the University of Sydney, Department of Zoology* **1**, 1– 215.
Dam, H.G. (1986). Short-term feeding of *Temora longicornis* Müller in the laboratory and the field. *Journal of Experimental Marine Biology and Ecology* **99**, 149–161.
Dam, H.G. and Peterson, W.T. (1988). The effect of temperature on the gut clearance rate constant of planktonic copepods. *Journal of Experimental Marine Biology and Ecology* **123**, 1–14.
Dam, H.G. and Peterson, W.T. (1991). *In situ* feeding behavior of the copepod *Temora longicornis*: effects of seasonal changes in chlorophyll size fractions and female size. *Marine Ecology Progress Series* **71**, 113–123.
Dam, H.G., Peterson, W.T. and Okubo, A. (1991). A simple mathematical analysis of the limitations to inferring feeding behavior of zooplankton from gut content. *Marine Ecology Progress Series* **69**, 41–45.
Dam, H.G., Peterson, W.T. and Bellantoni, D.C. (1994). Seasonal feeding and fecundity of the calanoid copepod *Acartia tonsa* in Long Island Sound: is omnivory important in egg production? *Hydrobiologia* **292/293**, 191–199.
Damkaer, D.M. (1975). Calanoid copepods of the genera *Spinocalanus* and *Mimocalanus* from the central Arctic Ocean, with a review of the Spinocalanidae. *National Oceanic and Atmospheric Administration, Technical Report*, National Marine Fisheries Service CIRC-**391**, 1–88.
Daro, M.H. (1980). Field study of the diel feeding of a population of *Calanus finmarchicus* at the end of a phytoplankton bloom. "*Meteor*" *Forschungsergebnisse* **22A**, 123–132.
Daro, M.H. (1985). Feeding rhythms and vertical distribution of marine copepods. *Bulletin of Marine Science* **37**, 487–497.
Daro, M.H. (1988). Migratory and grazing behavior of copepods and vertical distribution of phytoplankton. *Bulletin of Marine Science* **43**, 710–729.
Daro, M.H. and Gijsegem, B. van (1984). Ecological factors affecting weight, feeding, and production of five dominant copepods in the Southern Bight of the North Sea. *Rapports et Procès-Verbaux des Réunions Conseil International pour l'Exploration de la Mer* **183**, 226–233.
Da Rocha, C.E.F. and Iliffe, T.M. (1993). New cyclopoids (Copepoda) from anchialine caves in Bermuda. *Sarsia* **78**, 43–56.
David, C.N. and Conover, R.J. (1961). Preliminary investigation on the physiology and ecology of luminescence in the copepod *Metridia lucens*. *Biological Bulletin, Marine Biological Laboratory, Wood's Hole, Mass.*, **121**, 92–107.
Davis, C.C. (1949). The pelagic Copepoda of the northeastern Pacific Ocean. *University of Washington Publications in Biology* **14**, 1–118.
Davis, C.C. (1968). Mechanisms of hatching in aquatic invertebrate eggs. *Oceanography and Marine Biology: an Annual Review* **6**, 325–376.
Davis, C.C. (1977). *Sagitta* as food for *Acartia*. *Astarte* **10**, 1–3.
Davis, C.C. (1982). A preliminary quantitative study of the zooplankton from Conception Bay, Insular Newfoundland, Canada. *Internationale Revue der gesamten Hydrobiologie* **67**, 713–747.

Davis, C.S. (1984a). Predatory control of copepod seasonal cycles on Georges Bank. *Marine Biology* **82**, 31–40.
Davis, C.S. (1984b). Food concentrations on Georges Bank: non-limiting effect on development and survival of laboratory reared *Pseudocalanus* sp. and *Paracalanus parvus* (Copepoda: Calanoida). *Marine Biology* **82**, 41–46.
Davis, C.S. (1987). Components of the zooplankton production cycle in the temperate ocean. *Journal of Marine Research* **45**, 947–983.
Davis, C.S. and Alatalo, P. (1992). Effects of constant and intermittent food supply on life-history parameters in a marine copepod. *Limnology and Oceanography* **37**, 1618–1639.
Davis, C.S. and Wiebe, P.H. (1985). Macrozooplankton biomass in a warm-core Gulf Stream ring: time series changes in size structure, taxonomic composition, and vertical distribution. *Journal of Geophysical Research* **90**, 8871–8884.
Davis, C.S., Flierl, G.R., Wiebe, P.H. and Franks, P.J.S. (1991). Micropatchiness, turbulence and recruitment in plankton. *Journal of Marine Research* **49**, 109–151.
Dawson, J.K. and Knatz, G. (1980). Illustrated key to the planktonic copepods of San Pedro Bay, California. *Technical Report of the Allan Hancock Foundation* **2**, 1–106.
Deason, E.E. (1980). Potential effect of phytoplankton colony breakage on the calculation of zooplankton filtration rates. *Marine Biology* **57**, 279–286.
De Decker, A.H.B., Kaczmaruk, B.Z. and Marska, G. (1991). A new species of *Calanus* (Copepoda, Calanoida) from South African waters. *Annals of the South African Museum* **101**, 27–44.
Deevey, G.B. (1948). The zooplankton of Tisbury Great Pond. *Bulletin of the Bingham Oceanographic Collection* **12**, 1–44.
Deevey, G.B. (1960a). The zooplankton of the surface waters of the Delaware Bay region. *Bulletin of the Bingham Oceanographic Collection* **17** (2), 1–53.
Deevey, G.B. (1960b). Relative effects of temperature and food on seasonal variations in length of marine copepods in some eastern American and western European waters. *Bulletin of the Bingham Oceanographic Collection* **17** (2), 54–86.
Deevey, G.B. (1972). A new species of *Temoropia* (Copepoda: Calanoida) from the Sargasso Sea. *Proceedings of the Biological Society of Washington* **84**, 359–370.
Deevey, G.B. (1973). *Paraugaptilus* (Copepoda:Calanoida): two species, one new, from the Sargasso Sea. *Proceedings of the Biological Society of Washington* **86**, 247–260.
Deevey, G.B. and Brooks, A.L. (1977). Copepods of the Sargasso Sea off Bermuda: species composition, and vertical and seasonal distribution between the surface and 2000 m. *Bulletin of Marine Science* **27**, 256–291.
DeFrenza, J., Kirner, R.J., Maly, E.J. and Leeuwen, H.C. van (1986). The relationships of sex size ratio and season to mating intensity in some calanoid copepods. *Limnology and Oceanography* **31**, 491–496.
Delille, D. and Razouls, S. (1994). Community structures of heterotrophic bacteria of copepod fecal pellets. *Journal of Plankton Research* **16**, 603–615.
Demott, W.R. (1988). Discrimination between algae and detritus by freshwater and marine zooplankton. *Bulletin of Marine Science* **43**, 486–499.
Devi, C.Rama and Reddy, Y.Ranga (1989a). The complete postembryonic development of *Paradiaptomus greeni* (Gurney, 1906) (Copepoda, Calanoida) reared in the laboratory. *Crustaceana* **56**, 141–161.
Devi, C.Rama and Reddy, Y.Ranga (1989b). The complete postembryonic

development of *Allodiaptomus raoi* Kiefer, 1936 (Copepoda, Calanoida) reared in the laboratory. *Crustaceana* **56**, 246–266.

Dexter, B.L. (1981). Setogenesis and molting in planktonic crustaceans. *Journal of Plankton Research* **3**, 1–13.

Dexter, B.L. (1986). Developmental grazing capabilities of *Pseudocalanus* sp. and *Acartia clausi* (CI to adult): a comparative study of feeding. *Syllogeus* **58**, 547–551.

Diaz Zaballa, J. and Gaudy, R. (1996). Seasonal variations in the zooplankton and in the population structure of *Acartia tonsa* in a very eutrophic area: La Habana Bay (Cuba). *Journal of Plankton Research* **18**, 1123–1135.

Dickson, R.R., Kelly, P.M., Colebrook, J.M., Wooster, W.S. and Cushing, D.H. (1988). North winds and production in the eastern North Atlantic. *Journal of Plankton Research* **10**, 151–169.

Dickson, R.R., Colebrook, J.M. and Svendsen, E. (1992). Recent changes in the summer plankton of the North Sea. *ICES Marine Science Symposium* **195**, 232–242.

Diel, S. (1991). On the life history of dominant copepod species (*Calanus finmarchicus*, *C. glacialis*, *C. hyperboreus*, *Metridia longa*) in the Fram Strait. *Berichte zur Polarforschung* **88**, 1–113.

Diel, S. and Klein Breteler, W.C.M. (1986). Growth and development of *Calanus* spp. (Copepoda) during spring phytoplankton succession in the North Sea. *Marine Biology* **91**, 85–92.

Diel, S. and Tande, K. (1992). Does the spawning of *Calanus finmarchicus* in high latitudes follow a reproducible pattern? *Marine Biology* **113**, 21–31.

Digby, P.S.B. (1950). The biology of the small planktonic copepods of Plymouth. *Journal of the Marine Biological Association of the United Kingdom* **29**, 393–438.

Digby, P.S.B. (1954). The biology of the marine planktonic copepods of Scoresby Sound, East Greenland. *Journal of Animal Ecology* **23**, 298–338.

Doi, M., Singhagraiwan, S., Singhagraiwan, T., Kitade, M. and Ohno, A. (1994). Culture of the calanoid copepod, *Acartia sinjiensis*, in outdoor tanks in the tropics. *Thai Marine Fisheries Research Bulletin* **5**, 27–36.

Donaghay, P.L. (1985). An experimental test of the relative significance of food quality and past feeding history to limitation of egg production of the estuarine copepod *Acartia tonsa*. *Archives fur Hydrobiologie Beiheft Ergebnisse Limnologie* **21**, 235–245.

Donaghay, P.L. (1988). Role of temporal scales of acclimation, food quality and trophic dominance in controlling the evolution of copepod feeding behavior. *Bulletin of Marine Science* **43**, 469–485.

Dower, J.F. and Mackas, D.L. (1996). "Seamount effects" in the zooplankton community near Cobb Seamount. *Deep-Sea Research I* **43**, 837–858.

Dower, J.F., Miller, T.J. and Leggett, W.C. (1997). The role of microscale turbulence in the feeding ecology of larval fish. *Advances in Marine Biology* **31**, 170–220.

Downs, J.N. and Lorenzen, C.J. (1985). Carbon: phaeopigment ratios in zooplankton fecal pellets as an index of herbivorous feeding. *Limnology and Oceanography* **30**, 1024–1036.

Drach, P. (1939). Mue et cycle d'intermue chez les crustaces Decapodes. *Annales de l'Institut Océanographique* **19**, 103–391.

Drapun, I.E. (1982). Variability in *Metridia lucens* in the northwestern Atlantic Ocean. *Doklady Academy Sciences U.S.S.R., Biological Sciences Section (English Transl.)* **265**, 462–464.

Drebes, G. (1988). *Syltodinium listii* gen. et spec. nov., a marine ectoparasitic dinoflagellate on eggs of copepods and rotifers. *Helgoländer Meeresuntersuchungen* **42**, 583–591.

Drits, A.V., Pasternak, A.F. and Kosobokova, K.N. (1993). Feeding, metabolism and body composition of the Antarctic copepod *Calanus propinquus* Brady with special reference to its life cycle. *Polar Biology* **13**, 13–21.

Drits, A.V., Pasternak, A.F. and Kosobokova, K.N. (1994). Physiological characteristics of the antarctic copepod *Calanoides acutus* during the late summer in the Weddell Sea. *Hydrobiologia* **292/293**, 201–207.

Dudley, P.L. (1986). Aspects of general body shape and development in Copepoda. *Syllogeus* **58**, 7–25.

Dunbar, M.J. (1979). The relation between oceans. *In* "Zoogeography and Diversity in Plankton" (S. Van der Spoel and A.C. Pierrot-Bults, eds), pp. 112–125. Edward Arnold, London.

Dunham, P.J. (1978). Sex pheromones in Crustacea. *Biological Reviews* **53**, 555–583.

Durbin, A.G. and Durbin, E.G. (1981). Standing stock and estimated production rates of phytoplankton and zooplankton in Narragansett Bay, Rhode Island. *Estuaries* **4**, 24–41.

Durbin, A.G., Durbin, E.G. and Wlodarczyk, E. (1990). Diel feeding behavior in the marine copepod *Acartia tonsa* in relation to food availability. *Marine Ecology Progress Series* **68**, 23–45.

Durbin, E.G. and Durbin, A.G. (1978). Length and weight relationships of *Acartia clausi* from Narragansett Bay, R.I. *Limnology and Oceanography* **23**, 958–969.

Durbin, E.G., Durbin, A.G., Smayda, T.J. and Verity, P.G. (1983). Food limitation of production by adult *Acartia tonsa* in Narragansett Bay, Rhode Island. *Limnology and Oceanography* **28**, 1199–1213.

Dussart, B.H. and Defaye, D. (1995). Copepoda. Introduction to the Copepoda. *Guides to Identification of Microinvertebrates of Continental Waters of the World* **7**, 1–277.

Ederington, M.C., McManus, G.B. and Harvey, H.R. (1995). Trophic transfer of fatty acids, sterols, and a triterpenoid alcohol between bacteria, a ciliate, and the copepod *Acartia tonsa*. *Limnology and Oceanography* **40**, 860–867.

Ellis, S.G. and Small, L.F. (1989). Comparison of gut-evacuation rates of feeding and non-feeding *Calanus marshallae*. *Marine Biology* **103**, 175–181.

Elmgren, R. (1984). Trophic dynamics in the enclosed, brackish Baltic Sea. *Rapports et Procès-Verbaux des Réunions Conseil International pour l'Exploration de la Mer* **183**, 152–169.

Elofsson, R. (1966). The nauplius eye and frontal organs of the non-Malacostraca (Crustacea). *Sarsia* **25**, 1–128.

Elofsson, R. (1970). A presumed new photoreceptor in copepod crustaceans. *Zeitschrift für Zellforschung und Mikroskopische Anatomie* **109**, 316–326.

Elofsson, R. (1971). The ultrastructure of a chemoreceptor organ in the head of copepod crustaceans. *Acta Zoologica* **52**, 299–315.

Emerson, C.W. and Roff, J.C. (1987). Implications of fecal pellet size and zooplankton behaviour to estimates of pelagic-benthic carbon flux. *Marine Ecology Progress Series* **35**, 251–257.

Emery, A.R. (1968). Preliminary observations on coral reef plankton. *Limnology and Oceanography* **13**, 293–303.

Enright, J.T. (1977a). Copepods in a hurry: sustained high-speed upward migration. *Limnology and Oceanography* **22**, 118–125.

Enright, J.T. (1977b). Problems in estimating copepod velocity. *Limnology and Oceanography* **22**, 160–162.
Esaias, W.E. and Curl, H.C. (1972). Effect of dinoflagellate bioluminescence on copepod ingestion rates. *Limnology and Oceanography* **17**, 901–906.
Escaravage, V. and Soetaert, K. (1993). Estimating secondary production for the brackish Westerschelde copepod population *Eurytemora affinis* (Poppe) combining experimental data and field observations. *Cahiers de Biologie marine* **34**, 201–214.
Escribano, R. and Rodriguez, L. (1994). Life cycle of *Calanus chilensis* Brodsky in Bay of San Jorge, Antofagasta, Chile. *Hydrobiologia* **292/293**, 289–294.
Estep, K.W., MacIntyre, F., Hjörleifsson, E. and Sieburth, J.McN. (1986). MacImage: a user-friendly image-analysis system for the accurate mensuration of marine organisms. *Marine Ecology Progress Series* **33**, 243–253.
Esterly, C.O. (1905). The pelagic Copepoda of the San Diego Region. *University of California Publications, Zoology* **2** (4), 113–233.
Esterly, C.O. (1906). Additions to the copepod fauna of the San Diego Region. *University of California Publications, Zoology* **3**, 53–92.
Esterly, C.O. (1908). The light recipient organ of the copepod, *Eucalanus elongatus*. *Bulletin of the Museum of Comparative Zoölogy, Harvard* **53**, 55.
Evans, F. (1977). Seasonal density and production estimates of the commoner planktonic copepods of Northumberland coastal waters. *Estuarine and Coastal Marine Science* **5**, 223–241.
Evans, F. (1981). An investigation into the relationship of sea temperature and food supply to the size of the planktonic copepod *Temora longicornis* Müller in the North Sea. *Estuarine, Coastal and Shelf Science* **13**, 145–158.
Evstigneev, P.V. (1982a). On location of luminescent organs in *Pleuromamma* species. *Ecologiya Morya* **10**, 72–75.
Evstigneev, P.V. (1982b). Changes in the characteristics of bioluminescent signals during ontogenesis of copepods of the genus *Pleuromamma*. *Biologiya Morya* **5**, 55–59.
Evstigneev, P.V. (1989). Irritability of some plankton bioluminescents. *Ekologiya Morya* **32**, 72–77.
Evstigneev, P.V. (1990a). Studies on excitation of copepods by their motor and bioluminescent response. *Ekologiya Morya* **35**, 80–86.
Evstigneev, P.V. (1990b). Bioluminescence of copepods as a function of the stimulation frequency. *Ekologiya Morya* **36**, 92–96.
Evstigneev, P.V. (1992). Effect of medium salinity on bioluminescence of copepods. *Ekologiya Morya* **42**, 31–36.
Evstigneev, P.V. and Bityukov, E.P. (1986). On diurnal rhythm of luminescence of marine Copepoda and the effect of temperature on it. *Ekologiya Morya* **24**, 87–92.
Faber, D.J. (1966). Free-swimming copepod nauplii of Narragansett Bay with a key to their identification. *Journal of the Fisheries Research Board of Canada* **23**, 189–205.
Fahl, K. (1995). Biochemische Untersuchungen zum Lipidstoffwechsel antarktischer copepoden. *Berichte zur Polarforschung* **157**, 1–135.
Falk-Petersen, S., Hopkins, C.C.E. and Sargent, J.R. (1990). Trophic relationships in the pelagic, Arctic food web. *In* "Trophic Relationships in the Marine Environment" (M. Barnes and R.N. Gibson, eds.), pp. 315–333. Aberdeen University Press.
Fancett, M.S. and Kimmerer, W.J. (1985). Vertical migration of the demersal

copepod *Pseudodiaptomus* as a means of predator avoidance. *Journal of Experimental Marine Biology and Ecology* **88**, 31–43.
Farmer, L. (1980). Evidence for hyporegulation in the calanoid copepod, *Acartia tonsa*. *Comparative Biochemistry and Physiology* **65A**, 359–362.
Farquhar, G.B., ed. (1970). Proceedings of an International Symposium on Biological Sound Scattering in the Ocean. Maury Center for Ocean Science, Department of the Navy, Washington, D.C. 629 pp.
Farran, G.P. (1908). Second report on the Copepoda of the Irish Atlantic Slope. *Department of Agriculture and Technical Instruction for Ireland, Fisheries Branch, Scientific Investigations*, 1906, II, 1–104.
Farran, G.P. (1936). Copepoda. *Great Barrier Reef Expedition, 1928–1929* **5**, 73–142.
Farran, G.P. (1948). Copepoda. *Conseil International pour l'Exploration de la Mer, Zooplankton Identification Sheets*, Nos. 11–15.
Farran, G.P. (1951). Copepoda. *Conseil International pour l'Exploration de la Mer, Zooplankton Identification Sheets*, Nos. 11–17, 32–35, 37–40.
Fasham, M.J.R., Angel, M.V. and Roe, H.S.J. (1974). An investigation of the spatial pattern of zooplankton using the Longhurst-Hardy plankton recorder. *Journal of Experimental Marine Biology and Ecology* **16**, 93–112.
Fedotova, N.A. (1975). About seasonal character composition and zooplankton distribution in the southwestern Sakhalin. *Izvestia TINRO* **96**, 57–79.
Felgenhauer, B.E. (1987). Techniques for preparing crustaceans for scanning electron microscopy. *Journal of Crustacean Biology* **7**, 71–76.
Fernández, E., Cabal, J., Acuña, J.L., Bode, A., Botas, A. and García-Soto, C. (1993). Plankton distribution across a slope current-induced front in the southern Bay of Biscay. *Journal of Plankton Research* **15**, 619–641.
Fernández, F. (1979). Nutrition studies in the nauplius larva of *Calanus pacificus* (Copepoda: Calanoida). *Marine Biology* **53**, 131–147.
Ferrari, F. (1978). Spermatophore placement in the copepod *Euchaeta norvegica* Boeck 1872 from deepwater dumpsite 106. *Proceedings of the Biological Society of Washington* **91**, 509–521.
Ferrari, F. (1980). *Pseudochirella squalida* Grice & Hülsemann, 1967, from continental slope waters off Delaware (Copepoda: Calanoida). *Proceedings of the Biological Society of Washington* **93**, 536–550.
Ferrari, F.D. (1984). Pleiotropy and *Pleuromamma*, the looking-glass copepods (Calanoida). *Crustaceana*, Supplement **7**, 166–181.
Ferrari, F.D. (1985). Postnaupliar development of a looking-glass copepod, *Pleuromamma xiphias* (Giesbrecht, 1889), with analyses of distributions of sex and asymmetry. *Smithsonian Contributions to Zoology* **420**, 1–55.
Ferrari, F.D. (1988). Developmental patterns in numbers of ramal segments of copepod post-maxillipedal legs. *Crustaceana* **54**, 256–293.
Ferrari, F.D. (1991). Using patterns of appendage development to group taxa of *Labidocera*, Diaptomidae and Cyclopidae (Copepoda). *Bulletin of Plankton Society of Japan, Special Volume*, 115–128.
Ferrari, F.D. (1992). Development of copepod swimming legs with unequal numbers of ramal segments. *Journal of Crustacean Biology* **12**, 392–396.
Ferrari, F.D. (1993a). Exceptions to the rule of development that anterior is older among serially homologous segments of postmaxillipedal legs in copepods. *Journal of Crustacean Biology* **13**, 763–768.
Ferrari, F.D. (1993b). Ecology of *Metridia gerlachei* Giesbrecht in the western Bransfield Strait, Antarctica – comment. *Deep-Sea Research* **40A**, 1711–1712.

Ferrari, F.D. (1995). Six copepodid stages of *Ridgewayia klausruetzleri*, a new species of copepod crustacean (Ridgewayiidae: Calanoida) from the barrier reef in Belize, with comments on appendage development. *Proceedings of the Biological Society of Washington* **108**, 180–200.

Ferrari, F. and Dojiri, M. (1987). The calanoid copepod *Euchaeta antarctica* from Southern Ocean Atlantic Sector midwater trawls, with observations on spermatophore dimorphism. *Journal of Crustacean Biology* **7**, 458–480.

Ferrari, F.D. and Hayek, L.-A.C. (1990). Monthly differences in distribution of sex and asymmetry in a looking-glass copepod, *Pleuromamma xiphias*, off Hawaii. *Journal of Crustacean Biology* **10**, 114–127.

Ferrari, F.D. and Markhaseva, E.L. (1996). *Parkius karenwishnerae*, a new genus and species of calanoid copepod (Parkiidae, new family) from the benthopelagic waters of the eastern tropical Pacific Ocean. *Proceedings of the Biological Society of Washington* **109**, 264–285.

Ferrari, F.D. and Steinberg, D.K. (1993). *Scopalatum vorax* (Esterly, 1911) and *Scolecithricella lobophora* Park, 1970, calanoid copepods (Scolecithricidae) associated with a pelagic tunicate in Monterey Bay. *Proceedings of the Biological Society of Washington* **106**, 467–489.

Fiedler, P.C. (1982). Zooplankton avoidance and reduced grazing responses to *Gymnodinium splendens* (Dinophyceae). *Limnology and Oceanography* **27**, 961–965.

Fiksen, Ø. and Giske, J. (1995). Vertical distribution and population dynamics of copepods by dynamic optimization. *ICES Journal of Marine Science* **52**, 483–503.

Finenko, G.A. and Romanova, Z.A. (1991). Nutrition and survivability of the Black Sea *Acartia clausi* Giesbr. on detritus. *Ekologiya Morya* **38**, 55–60.

Fish, C.J. (1936). The biology of *Calanus finmarchicus* in the Gulf of Maine and Bay of Fundy. *Biological Bulletin, Marine Biological Laboratory, Woods Hole, Mass.*, **70**, 118–141.

Fisher, L.R. (1960). Vitamins. In "The Physiology of Crustacea" (T.H. Waterman, ed.), Vol. I, pp. 259–289. Academic Press, New York.

Fisher, L.R. (1964). Vitamin A and carotenoids in certain invertebrates. VII. Crustacea: Copepoda. *Journal of the Marine Biological Association of the United Kingdom* **44**, 685–692.

Fisher, N.S. and Reinfelder, J.R. (1991). Assimilation of selenium in the marine copepod *Acartia tonsa* studied with a radiotracer ratio method. *Marine Ecology Progress Series* **70**, 157–164.

Fisher, N.S., Nolan, C.V. and Fowler, S.W. (1991). Assimilation of metals in marine copepods and its biogeochemical implications. *Marine Ecology Progress Series* **71**, 37–43.

Fleminger, A. (1957). New calanoid copepods of *Pontella* Dana and *Labidocera* Lubbock with notes on the distribution of the genera in the Gulf of Mexico. *Tulane Studies in Zoology* **5** (2), 19–34.

Fleminger, A. (1967). Taxonomy, distribution and polymorphism in the *Labidocera jollae* group with remarks on evolution within the group (Copepoda: Calanoida). *Proceeding of the United States National Museum* **120** (3567), 1–61.

Fleminger, A. (1973). Pattern, number, variability, and taxonomic significance of integumental organs (sensilla and glandular pores) in the genus *Eucalanus* (Copepoda, Calanoida). *Fishery Bulletin, National Oceanic and Atmospheric Administration of the United States* **71**, 965–1010.

Fleminger, A. (1975). Geographical distribution and morphological divergence in

American coastal zone planktonic copepods of the genus *Labidocera*. In "Estuarine Research, Vol. 1, Chemistry, Biology and the Estuarine System" (L.E. Cronin, ed.), pp. 392–419. Academic Press, New York.

Fleminger, A. (1979). *Labidocera* (Copepoda, Calanoida): new and poorly known Caribbean species with a key to species in the western Atlantic. *Bulletin of Marine Science* **29**, 170–190.

Fleminger, A. (1983). Description and phylogeny of *Isaacsicalanus paucisetus*, n. ge., n. sp. (Copepoda: Calanoida: Spinocalanidae) from an East Pacific hydrothermal vent site (21°N). *Proceedings of the Biological Society of Washington* **96**, 605–622.

Fleminger, A. (1985). Dimorphism and possible sex change in copepods of the family Calanidae. *Marine Biology* **88**, 273–294.

Fleminger, A. (1986). The pleistocene equatorial barrier between the Indian and Pacific Oceans and a likely cause for Wallace's line. *Unesco Technical Papers in Marine Science* **49**, 84–97.

Fleminger, A. (1988). *Parastephos esterlyi*, a new species of copepod (Stephidae: Calanoida: Crustacea) from San Diego Bay, California. *Proceedings of the Biological Society of Washington* **101**, 309–313.

Fleminger, A. and Hendrix Kramer, S. (1988). Recent introduction of an Asian estuarine copepod, *Pseudodiaptomus marinus* (Copepoda: Calanoida), into southern California embayments. *Marine Biology* **98**, 535–541.

Fleminger, A. and Hülsemann, K. (1974). Systematics and distribution of the four sibling species comprising the genus *Pontellina* Dana (Copepoda, Calanoida). *Fishery Bulletin, National Oceanic and Atmospheric Administration of the United States* **72**, 63–120.

Fleminger, A. and Hülsemann, K. (1977). Geographical range and taxonomic divergence in North Atlantic *Calanus* (*C. helgolandicus*, *C. finmarchicus* and *C. glacialis*). *Marine Biology* **40**, 233–248.

Fleminger, A. and Hülsemann, K. (1987). Geographical variation in *Calanus helgolandicus* s.l. (Copepoda, Calanoida) and evidence of recent speciation of the Black Sea population. *Biological Oceanography* **5**, 43–81.

Fleminger, A. and Moore, E. (1977). Two new species of *Labidocera* (Copepoda, Calanoida) from the western north Atlantic region. *Bulletin of Marine Science* **27**, 520–529.

Fleminger, A., Othman, B. and Greenwood, J. (1982). The *Labidocera pectinata* group; an Indo-West Pacific lineage of planktonic copepods with descriptions of two new species. *Journal of Plankton Research* **4**, 245–270.

Flinkman, J., Vuorinen, I. and Christiansen, M. (1994). Calanoid copepod eggs survive passage through fish digestive tracts. *ICES Journal of Marine Science* **51**, 127–129.

Flint, M.V., Drits, A.V. and Pasternak, A.F. (1991). Characteristic features of body composition and metabolism in some interzonal copepods. *Marine Biology* **111**, 199–205.

Fogg, G.E. (1985). Biological activities at a front in the western Irish Sea. In "Proceedings of the Nineteenth European Marine Biology Symposium" (P.E. Gibbs, ed.), pp. 87–95. Cambridge University Press, Cambridge.

Fortier, L., Fortier, M. and Demers, S. (1995). Zooplankton and larval fish community development: comparative study under first-year sea ice at low and high latitudes in the northern hemisphere. *Proceedings of the NIPR Symposium on Polar Biology* **8**, 11–19.

Forward, R.B. (1988). Diel vertical migration: zooplankton photobiology and

behaviour. *Oceanography and Marine Biology: an Annual Review* **26**, 361–393.
Fosshagen, A. (1967). Two new species of Calanoid copepods from Norwegian fjords. *Sarsia* **29**, 307–320.
Fosshagen, A. (1968a). Marine biological investigations in the Bahamas. 4. Pseudocyclopidae (Copepoda, Calanoida) from the Bahamas. *Sarsia* **32**, 39–62.
Fosshagen, A. (1968b). Marine biological investigations in the Bahamas. 8. Bottom-living Arietellidae (Copepoda, Calanoida) from the Bahamas with remarks on *Paramisophria cluthae* T. Scott. *Sarsia* **35**, 57–64.
Fosshagen, A. (1970). Marine biological investigations in the Bahamas. 15. *Ridgewayia* (Copepoda, Calanoida) and two new genera of calanoids from the Bahamas. *Sarsia* **44**, 25–58.
Fosshagen, A. (1973). A new genus and species of bottom-living calanoid (Copepoda) from Florida and Colombia. *Sarsia* **52**, 145–154.
Fosshagen, A. (1978). *Mesaiokeras* (Copepoda, Calanoida) from Colombia and Norway. *Sarsia* **63**, 177–183.
Fosshagen, A. (1980). How the zooplankton community may vary within a single fjord system. *In* "Fjord Oceanography" (H.J. Freeland, D.M. Farmer and C.D. Levings, eds), NATO Conference Series, IV Marine Sciences, Vol. 4, pp. 399–405. Plenum Press, New York and London.
Fosshagen, A. (1983). A new genus of calanoid copepod from the Norwegian Sea. *Sarsia* **68**, 257–262.
Fosshagen, A. and Iliffe, T.M. (1985). Two new genera of Calanoida and a new Order of Copepoda, Platycopioida, from marine caves on Bermuda. *Sarsia* **70**, 345–358.
Fosshagen, A. and Iliffe, T.M. (1988). A new genus of Platycopioida (Copepoda) from a marine cave on Bermuda. *Hydrobiologia* **167/168**, 357–361.
Fosshagen, A. and Iliffe, T.M. (1989). *Boholina*, a new genus (Copepoda: Calanoida) with two new species from an anchialine cave in the Philippines. *Sarsia* **74**, 201–208.
Fosshagen, A. and Iliffe, T.M. (1991). A new genus of calanoid copepod from an anchialine cave in Belize. *Bulletin of Plankton Society of Japan*, Special Volume, 339–346.
Fosshagen, A. and Iliffe, T.M. (1994). A new species of *Erobonectes* (Copepoda, Calanoida) from marine caves on the Caicos Islands, West Indies. *Hydrobiologia* **292/293**, 17–22.
Foster, B.A. (1987). Composition and abundance of zooplankton under the spring sea-ice of McMurdo Sound, Antarctica. *Polar Biology* **8**, 41–48.
Fragopoulu, N. and Lykakis, J.J. (1990). Vertical distribution and nocturnal migration of zooplankton in relation to the development of the seasonal thermocline in Patraikos Gulf. *Marine Biology* **104**, 381–387.
Fransz, H.G. (1988). Vernal abundance, structure and development of epipelagic copepod populations of the eastern Weddell Sea (Antarctica). *Polar Biology* **9**, 107–114.
Fransz, H.G. and Diel, S. (1985). Secondary production of *Calanus finmarchicus* (Copepoda: Calanoidea) in a transitional system of the Fladen Ground area (northern North Sea) during the spring of 1983. *In* "Proceedings of the Nineteenth European Marine Biology Symposium" (P.E. Gibbs, ed.), pp. 123–133. Cambridge University Press, Cambridge.
Fransz, H.G. and Van Arkel, W.G. (1980). Zooplankton activity during and after the phytoplankton spring bloom at the central station in the FLEX box, northern

North Sea, with special reference to the calanoid copepod *Calanus finmarchicus* (Gunn.). *"Meteor" Forschungsergebnisse* **22A**, 113–121.

Fransz, H.G., Miquel, J.C. and Gonzalez, S.R. (1984). Mesozooplankton composition, biomass and vertical distribution, and copepod production in the stratified central North Sea. *Netherlands Journal of Sea Research* **18**, 82–96.

Fransz, H.G., Gonzalez, S.R. and Klein Breteler, C.M. (1989). Fecundity as a factor controlling the seasonal population cycle in *Temora longicornis* (Copepoda, Calanoida). *In* "Reproduction, Genetics and Distributions of Marine Organisms" (J.S. Ryland and P.A. Tyler, eds), pp. 83–90. Olsen & Olsen, Fredensborg, Denmark.

Fransz, H.G., Colebrook, J.M., Gamble, J.C. and Krause, M. (1991). The zooplankton of the North Sea. *Netherlands Journal of Sea Research* **28**, 1–52.

Fransz, H.G., Gonzalez, S.R., Cadée, G.C. and Hansen, F.C. (1992). Long-term change of *Temora longicornis* (Copepoda, Calanoida) abundance in a Dutch tidal inlet (Marsdiep) in relation to eutrophication. *Netherlands Journal of Sea Research* **30**, 23–32.

Fraser, A.J., Sargent, J.R. and Gamble, J.C. (1989). Lipid class and fatty acid composition of *Calanus finmarchicus* (Gunnerus), *Pseudocalanus* sp. and *Temora longicornis* Muller from a nutrient-enriched seawater enclosure. *Journal of Experimental Marine Biology and Ecology* **130**, 81–92.

Frasson-Boulay, M.F. (1973). Les cellules à guanine de l'ocelle médian de *Pontella mediterranea* Claus (Crustacé; Copépode). *Compte rendu hebdomadaire des séances de l'Académie des Sciences, Paris* **276**, 3323–3325.

Freeland, H.J., Farmer, D.M. and Levings, C.D., eds (1980). "Fjord Oceanography." NATO Conference Series, IV Marine Sciences, Vol. 4, pp. 1–715. Plenum Press, New York and London.

Friedman, M.M. (1980). Comparative morphology and functional significance of copepod receptors and oral structures. *In* "Evolution and Ecology of Zooplankton Communities" (W.C. Kerfoot, ed.), pp. 185–197. University Press of New England, Hanover, N.H.

Frolander, H.F., Miller, C.B., Flynn, M.J., Myers, S.C. and Zimmerman, S.T. (1973). Seasonal cycles of abundance in zooplankton populations of Yaquina Bay, Oregon. *Marine Biology* **21**, 277–288.

Fromentin, J.-M. and Planque, B. (1996). *Calanus* and environment in the eastern North Atlantic. II. Influence of the North Atlantic oscillation on *C. finmarchicus* and *C. helgolandicus*. *Marine Ecology Progress Series* **134**, 111–118.

Fromentin, J.-M., Ibanez, F. and Legendre, P. (1993). A phytosociological method for interpreting plankton data. *Marine Ecology Progress Series* **93**, 285–306.

Froneman, P.W., Pakhomov, E.A., Perissinotto, R. and McQuaid, C.D. (1996). Role of microplankton in the diet and daily ration of Antarctic zooplankton species during the austral summer. *Marine Ecology Progress Series* **143**, 15–23.

Frontier, S. (1985). Diversity and structure in aquatic ecosystems. *Oceanography and Marine Biology: an Annual Review* **23**, 253–312.

Frost, B.W. (1972). Effects of size and concentration of food particles on the feeding behavior of the marine planktonic copepod *Calanus pacificus*. *Limnology and Oceanography* **17**, 805–815.

Frost, B.W. (1974). *Calanus marshallae*, a new species of calanoid copepod closely allied to the sibling species *C. finmarchicus* and *C. glacialis*. *Marine Biology* **26**, 77–99.

Frost, B.W. (1975). A threshold feeding behavior in *Calanus pacificus*. *Limnology and Oceanography* **20**, 263–266.

Frost, B.W. (1980). The inadequacy of body size as an indicator of niches in zooplankton. *In* "Evolution and Ecology of Zooplankton Communities" (W.C. Kerfoot, ed.), pp. 742–753. University Press of New England, Hanover, NH.

Frost, B.W. (1985). Food limitation of the planktonic marine copepods *Calanus pacificus* and *Pseudocalanus* sp. in a temperate fjord. *Ergebnisse Hydrobiologie* **21**, 1–13.

Frost, B.W. (1987). Grazing control of phytoplankton stock in the open subarctic Pacific Ocean: a model assessing the role of mesozooplankton, particularly the large calanoid copepods *Neocalanus* spp. *Marine Ecology Progress Series* **39**, 49–68.

Frost, B.W. (1988). Variability and possible adaptive significance of diel vertical migration in *Calanus pacificus*, a planktonic marine copepod. *Bulletin of Marine Science* **43**, 675–694.

Frost, B.W. (1989). A taxonomy of the marine calanoid copepod genus *Pseudocalanus*. *Canadian Journal of Zoology* **67**, 525–551.

Frost, B.W. and Bollens, S.M. (1992). Variability of diel vertical migration in the marine planktonic copepod *Pseudocalanus newmani* in relation to its predators. *Canadian Journal of Fisheries and Aquatic Sciences* **49**, 1137–1141.

Frost, B. and Fleminger, A. (1968). A revision of the genus *Clausocalanus* (Copepoda: Calanoida) with remarks on distributional patterns in diagnostic characters. *Bulletin of the Scripps Institution of Oceanography* **12**, 1–235.

Frost, B.W., Landry, M.R. and Hassett, R.P. (1983). Feeding behavior of large calanoid copepods *Neocalanus cristatus* and *N. plumchrus* from the subarctic Pacific Ocean. *Deep-Sea Research* **30A**, 1–13.

Fryd, M., Haslund, O.H. and Wohlgemuth, O. (1991). Development, growth and egg production of the two copepod species *Centropages hamatus* and *Centropages typicus* in the laboratory. *Journal of Plankton Research* **13**, 683–689.

Fulton, J. (1972). Keys and references to the marine Copepoda of British Columbia. *Fisheries Research Board of Canada, Technical Report* **313**, 1–63.

Fulton, J. (1973). Some aspects of the life history of *Calanus plumchrus* in the Strait of Georgia. *Journal of the Fisheries Research Board of Canada* **30**, 811–815.

Fulton III, R.S. (1983). Interactive effects of temperature and predation on an estuarine zooplankton community. *Journal of Experimental Marine Biology and Ecology* **72**, 67–81.

Fulton III, R.S. (1984). Predation, production and the organization of an estuarine copepod community. *Journal of Plankton Research* **6**, 399–415.

Gagnon, M. and Lacroix, G. (1981). Zooplankton sample variability in a tidal estuary: an interpretative model. *Limnology and Oceanography* **26**, 401–413.

Gagnon, M. and Lacroix, G. (1983). The transport and retention of zooplankton in relation to a sill in a tidal estuary. *Journal of Plankton Research* **5**, 289–303.

Gamble, J.C. (1978). Copepod grazing during a declining spring phytoplankton bloom in the northern North Sea. *Marine Biology* **49**, 303–315.

Garcia-Pamanes, J., Lara Lara, J.R. and Gaxiola Castro, G. (1991). Daily zooplankton filtration rates off Baja California. *Estuarine, Coastal and Shelf Science* **32**, 503–510.

Garcia-Rodriguez, M. (1985). Contribución al conocimiento de la biología de *Centropages ponticus* Karavaev (Copepoda, Calanoida). *Bolletin del Instituto Español de Oceanografia* **2**, (2), 47–52.

Gardner, G.A. and Szabo, I. (1982). British Columbia pelagic marine Copepoda: an identification manual and annotated bibliography. *Canadian Special Publication Fisheries and Aquatic Science* **62**, 1–535.

Garmew, T.G., Hammond, S., Mercantini, A., Morgan, J., Neunert, C. and Fornshell, J.A. (1994). Morphological variability of geographically distinct populations of the estuarine copepod *Acartia tonsa*. *Hydrobiologia* **292/293**, 149–156.
Gaskin, D.E. (1982). "The Ecology of Whales and Dolphins." Heinemann, London. 459 pp.
Gatten, R.R., Sargent, J.R., Forsberg, T.E.V., O'Hara, S.C.M. and Corner, E.D.S. (1980). On the nutrition and metabolism of zooplankton. XIV. Utilization of lipid by *Calanus helgolandicus* during maturation and reproduction. *Journal of the Marine Biological Association of the United Kingdom* **60**, 391–399.
Gaudy, R. (1961). Note sur les stades larvaires de *Temora stylifera* Dana. *Recueil des Travaux de la Station Marine d'Endoume* **36** (22), 115–122.
Gaudy, R. (1962). Biologie des copepodes pelagiques du Golfe de Marseille. *Recueil des Travaux de la Station Marine d'Endoume* **42** (27), 93–184.
Gaudy, R. (1965). Sur une nouvelle espece d'Arietellidae (Copepoda, Calanoida): *Paraugaptilus mozambicus*. *Recueil des Travaux de la Station Marine d'Endoume* **54** (38), 123–127.
Gaudy, R. (1971). Etude expérimentale de la ponte chez trois espèces de copépodes pélagiques (*Centropages typicus*, *Acartia clausi* et *Temora stylifera*). *Marine Biology* **9**, 65–70.
Gaudy, R. (1972). Contribution a la connaissance du cycle biologique des copépodes du Golfe de Marseille. 2. Étude du cycle biologique de quelques espèces caractéristiques. *Téthys* **4**, 175–242.
Gaudy, R. (1973). Sur une collection de Copépodes récoltés par le bathyscaphe "Archimède" dans la région des Açores. *Téthys* **4**, 947–952.
Gaudy, R. (1974). Feeding four species of pelagic copepods under experimental conditions. *Marine Biology* **25**, 125–141.
Gaudy, R. (1984). Biological cycle of *Centropages typicus* in the north-western Mediterranean neritic waters. *Crustaceana*, Supplement **7**, 200–213.
Gaudy, R. (1989). The role of zooplankton in the nitrogen cycle of a Mediterranean brackish lagoon. *Scientia Marina* **53**, 609–616.
Gaudy, R. (1992). Seasonal variations of some population parameters of the copepod *Acartia tonsa* in a Mediterranean lagoon. *In* "Marine Eutrophication and Population Dynamics" (G. Colombo, I. Ferrari, V.U. Ceccherelli and R. Rossi, eds), pp. 151–155. Olsen & Olsen, Fredensborg.
Gaudy, R. and Boucher, J. (1983). Relation between respiration, excretion (ammonia and inorganic phosphorus) and activity of amylase and trypsin in different species of pelagic copepods from the Indian Ocean equatorial area. *Marine Biology* **75**, 37–45.
Gaudy, R. and Boucher, J. (1989). Respiration, excretion and digestive enzyme activity of certain copepod species in the Indian Ocean as dependent on the strategy of their nutrition. *Ekologiya Morya* **33**, 16–26.
Gaudy, R. and Pagano, M. (1987). Biologie d'un copépode des mares temporaires du littoral méditerranéen français: *Eurytemora velox*. III. Reproduction. *Marine Biology* **94**, 335–345.
Gaudy, R. and Pagano, M. (1989). Nutrition of copepods under conditions of the Mediterranean lagoon as a function concentration of food particles and temperature. *Ekologiya Morya* **33**, 57–67.
Gaudy, R. and Viñas, M.D. (1985). Premiere signalisation en Méditerranée du copépode pélagique *Acartia tonsa*. *Rapports et Procès-Verbaux des Réunions Commission Internationale pour l'Exploration Scientifique de la Mer Méditerranée* **29** (9), 227–229.

Gaudy, R., Moraitou-Apostolopoulou, M., Pagano, M., Saint-Jean, L. and Verriopoulos, G. (1988). Salinity a decisive factor in the length of cephalothorax of *Acartia clausi* from three different areas (Greece and Ivory Coast). *Rapports et Procès-Verbaux des Réunions Commission Internationale pour l'Exploration Scientifique de la Mer Méditerranée* **31** (2), 233.

Gaudy, R., Pagano, M., Cervetto, G., Saint-Jean, L., Verriopoulos, G. and Beker, B. (1996). Short term variations in feeding and metabolism of *Acartia tonsa* (pelagic copepod) in the Berre lagoon (France). *Oceanologica Acta* **19**, 635–644.

Gauld, D.T. (1957a). A peritrophic membrane in calanoid copepods. *Nature, London* **179**, 325–326.

Gauld, D.T. (1957b). Copulation in calanoid copepods. *Nature, London* **180**, 510.

Gauld, D.T. (1959). Swimming and feeding in crustacean larvae: the nauplius larva. *Proceedings of the Zoological Society of London* **132**, 31–50.

Gauld, D.T. (1966). The swimming and feeding of planktonic copepods. *In* "Some Contemporary Studies in Marine Science" (H. Barnes, ed.), pp. 313–334. George Allen and Unwin Ltd, London.

Gehrs, C.W. and Robertson, A. (1975). Use of life tables in analyzing the dynamics of copepod populations. *Ecology* **56**, 665–672.

Geinrikh (=Heinrich), A.K. (1962a). On the production of copepods in the Bering Sea. *International Revue der gesamten Hydrobiologie* **47**, 465–469.

Geinrikh (=Heinrich), A.K. (1962b). The life histories of plankton animals and seasonal cycles of plankton communities in the oceans. *Journal du Conseil International pour l'Exploration de la Mer* **27**, 15–24.

Geinrikh (=Heinrich), A.K. (1969). The ranges of neuston copepods in the Pacific Ocean. *Zoologichesky Zhurnal* **48**, 1456–1466.

Geinrikh (=Heinrich), A.K. (1974). On neustonic pontellids (Pontellidae, Copepoda) of the southern Atlantic. *Transactions of the P.P. Shirshov Institute of Oceanology* **98**, 43–50.

Geinrikh (=Heinrich), A.K. (1982). On tempory reconstructions of planktonic populations in the Norwegian and Sargasso Seas. *Oceanology* **22**, 213–215.

Geinrikh (=Heinrich), A.K. (1987). New species and subspecies of *Pontella* (Copepoda, Pontellida) from the eastern part of the Indian Ocean. *Zoologichesky Zhurnal* **66**, 932–938.

Geinrikh (=Heinrich), A.K. (1989). A new and rare species of the genus *Pontella* (Copepoda, Pontellidae) from a region to the north-west of Madagascar. *Zoologichesky Zhurnal* **68**, 130–135.

Geinrikh (=Heinrich), A.K. (1990). Calanoida copepods of the plankton collections from the submarine rises of the Nazca Ridge. *Transactions of the P.P. Shirshov Institute of Oceanology* **124**, 15–26.

Geinrikh (=Heinrich), A.K. (1993). Two species of the family Arietellidae (Copepoda, Calanoida) from the southwestern Indian Ocean. *Zoologichesky Zhurnal* **72** (8), 5–10.

Geinrikh (=Heinrich), A.K. (1995). A new species of the genus *Paraugaptilus* (Copepoda, Arietellidae) from the south-western part of the Indian Ocean. *Zoologichesky Zhurnal* **74** (4), 149–153.

Geinrikh (=Heinrich), A.K. (1996). Two new species of the genus *Paramisophria* (Copepoda, Arietellidae) from the Mozambique Strait. *Zoologichesky Zhurnal* **75**, 507–515.

Geletin, Yu.V. (1976). The ontogenetic abdomen formation in copepods of genera *Eucalanus* and *Rhincalanus* (Calanoida, Eucalanidae) and new system of these copepods. *Explorations of the Fauna of the Seas XVIII (XXVI), Functional*

Morphology, Growth and Development of Invertebrates of Sea Plankton. Academy of Sciences of the U.S.S.R., 75–93.
Genin, A., Greene, C., Haury, L., Wiebe, P., Gal, G., Kaartvedt, S., Meir, E., Fey, C. and Dawson, J. (1994). Zooplankton patch dynamics; daily gap formation over abrupt topography. *Deep-Sea Research* **41A**, 941–951.
Genin, A., Gal, G. and Haury, L. (1995). Copepod carcasses in the ocean. II. Near coral reefs. *Marine Ecology Progress Series* **123**, 65–71.
George, V.S. (1985). Demographic evaluation of the influence of temperature and salinity on the copepod *Eurytemora herdmani*. *Marine Ecology Progress Series* **21**, 145–152.
Geptner, M.V. (=Heptner) (1969). Systematic studies of *Lucicutia polaris* Brodsky, 1950 (Copepoda, Lucicutiidae) and description of *L. pseudopolaris* sp.n. from the Polar Basin. *Zoologichesky Zhurnal* **48** (2), 197–206.
Geptner, M.V. (=Heptner) (1971). On the copepod fauna of the Kurile-Kamchatka Trench. The families Euchaetidae, Lucicutiidae, Heterorhabdidae. *Trudy Instituta Okeanologii, Akademiya nauk SSSR* **92**, 73–161.
Geptner, M.V. (=Heptner) (1972a). New species of deep-water genera *Disseta* and *Heterorhabdus* (Copepoda, Calanoida) from the Pacific. *Zoologichesky Zhurnal* **51** (11), 1645–1650.
Geptner, M.V. (=Heptner) (1972b). A review of the generic structure of the family Heterorhabdidae (Copepoda, Calanoida). *Byulleten' Moskovskogo Obshchestva ispytatelei prirody* **77** (6), 54–64.
Geptner, M.V. (=Heptner) (1986). On the Calanoida (Copepoda) fauna in the Kurile-Kamchatka Trough. 2. Vertical distribution and geographical distribution of species from the families Euchaetidae and Lucicutiidae. *Sbornik Trudy Zoologichesky Museum* **24**, 3–58.
Geptner, M.V. (=Heptner) (1987). On the fauna of copepods (Calanoida) of the Kurile-Kamchatka Trench. New species of the genus *Pareuchaeta*. *Zoologichesky Zhurnal* **66**, 1177–1188.
Gerber, R.P. and Gerber, M.B. (1979). Ingestion of natural particulate organic matter and subsequent assimilation, respiration and growth by tropical lagoon zooplankton. *Marine Biology* **52**, 33–43.
Gibbons, S.G. (1933). A study of the biology of *Calanus finmarchicus* in the north-western North Sea. *Fisheries Scotland, Scientific Investigations*, 1933 (1), 1–24.
Gibbons, S.G. (1936). *Calanus finmarchicus* and other copepods in Scottish waters in 1933. *Fisheries Scotland, Scientific Investigations*, 1936 (2), 1–37.
Gibson, V.R. and Grice, G.D. (1976). Developmental stages of *Pontella meadi* Wheeler (Copepoda: Calanoida). *Journal of the Fisheries Research Board of Canada* **33**, 847–854.
Gibson, V.R. and Grice, G.D. (1977). The developmental stages of *Labidocera aestiva* Wheeler 1900 (Copepoda, Calanoida). *Crustaceana* **32**, 7–20.
Giesbrecht, W. (1892). Systematik und Faunistik der pelagischen Copepoden des Golfes von Neapel und der angrenzenden Meeres-Abschnitte. *Fauna und Flora des Golfes von Neapel* **19**, 1–831.
Giesbrecht, W. (1895). Mittheilungen über Copepoden. 8. Über das Leuchten der pelagischen Copepoden und das thierische Leuchten im Allgemeinen. *Mittheilungen aus der Zoologischen Station zu Neapel* **11**, 648–689.
Gieskes, W.W. and Kraay, G.W. (1977). Continuous plankton records: changes in the plankton of the North Sea and its eutrophic Southern Bight from 1948 to 1975. *Netherlands Journal of Sea Research* **11**, 334–364.

Gifford, D.J. (1993). Protozoa in the diets of *Neocalanus* spp. in the oceanic subarctic Pacific Ocean. *Progress in Oceanography* **32**, 223–237.

Gifford, D.J. and Dagg, M.J. (1988). Feeding of the estuarine copepod *Acartia tonsa* Dana: carnivory vs. herbivory in natural microplankton assemblages. *Bulletin of Marine Science* **43**, 458–468.

Giguere, L.A., St-Piere, J.F., Bernier, B., Vézina, A. and Rondeau, J.-G. (1989). Can we estimate the true weight of zooplankton samples after chemical preservation? *Canadian Journal of Fisheries and Aquatic Sciences* **46**, 522–527.

Gilbert, J.J. and Williamson, C.E. (1983). Sexual dimorphism in zooplankton (Copepoda, Cladocera, and Rotifera). *Annual Review of Ecology and Systematics* **14**, 1–33.

Gill, C.W. (1985). The response of a restrained copepod to tactile stimulation. *Marine Ecology Progress Series* **21**, 121–125.

Gill, C.W. (1986). Suspected mechano- and chemosensory structures of *Temora longicornis* (Copepoda: Calanoida). *Marine Biology* **93**, 449–457.

Gill, C.W. (1987). Recording the beat patterns of the second antennae of calanoid copepods with a microimpedence technique. *Hydrobiologia* **148**, 73–78.

Gill, C.W. and Crisp, D.J. (1985a). Sensitivity of intact and antennule amputated copepods to water disturbance. *Marine Ecology Progress Series* **21**, 221–227.

Gill, C.W. and Crisp, D.J. (1985b). The effect of size and temperature on the frequency of limb beat of *Temora longicornis* Müller (Crustacea: Copepoda). *Journal of Experimental Marine Biology and Ecology* **86**, 185–196.

Gill, C.W. and Harris, R.P. (1987). Behavioural responses of the copepods *Calanus helgolandicus* and *Temora longicornis* to dinoflagellate diets. *Journal of the Marine Biological Association of the United Kingdom* **67**, 785–801.

Gill, C.W. and Poulet, S.A. (1986). Utilization of a computerized micro-impedance system for studying the activity of copepod appendages. *Journal of Experimental Marine Biology and Ecology* **101**, 193–198.

Gill, C.W. and Poulet, S.A. (1988a). Impedence traces of copepod appendage movements illustrating sensory feeding behaviour. *Hydrobiologia* **167/168**, 303–310.

Gill, C.W. and Poulet, S.A. (1988b). Responses of copepods to dissolved free amino acids. *Marine Ecology Progress Series* **43**, 269–276.

Giron, F. (1963). Copépodes de la Mer d'Alboran (Campagnes du "Président-Théodore-Tissier", juin 1957). *Revue des Travaux de l'Institut des Pêches maritimes* **27**, 355–402.

Giske, J., Aksnes, D.L., Baliño, B.M., Kaartvedt, S., Lie, U., Nordeide, T.J., Salvanes, A.G.V., Wakili, S.M. and Aadnesen, A. (1990). Vertical distribution and trophic interactions of zooplankton and fish in Masfjorden, Norway. *Sarsia* **75**, 65–81.

Glasser, J.W. (1984). Analysis of zooplankton feeding experiments: some methodological considerations. *Journal of Plankton Research* **6**, 553–569.

Godlewska, M. (1989). Energy flow through the main phytophagous species in the Bransfield Strait and Drake Passage during SIBEX 1983/1984. *In* "Proceedings of the Twenty First European Marine Biology Symposium" (R.Z. Klekowski, E. Styczyńska and L. Falkowski, eds), pp. 291–296. Polish Academy of Sciences.

Gonzalez, C.R.M. and Bradley, B.P. (1994). Salinity stress proteins in *Eurytemora affinis*. *Hydrobiologia* **292/293**, 461–468.

González, H.E., González, S.R. and Brummer, G.-J.A. (1994). Short-term sedimentation pattern of zooplankton, faeces and microplankton at a permanent station in the Bjørnafjorden (Norway) during April-May 1992. *Marine Ecology Progress Series* **105**, 31–45.

Gopalakrishnan, T.C. (1973). A new species of *Macandrewella* (Copepoda: Calanoida) from off Cochin, south west coast of India. *Handbook to the International Zooplankton Collections, Indian Ocean Biological Centre* **5**, 180–189.

Gordejeva, K.T. (1974a). New species of planktonic Copepoda Calanoida from the tropic Atlantic and Mediterranean Sea. *Zoologichesky Zhurnal* **53**, 841–847.

Gordejeva, K.T. (1974b). New species of the genus Disco (Copepoda) from the tropical zone of the Atlantic and South Seas. *Zoologichesky Zhurnal* **53**, 1148–1154.

Gordejeva, K.T. (1975a). A new family, new genera and species of Copepoda (Calanoida) from the Atlantic and South Seas. *Zoologichesky Zhurnal* **54**, 188–194.

Gordejeva, K.T. (1975b). New species of Calanoida from the Central-American seas. *Zoologichesky Zhurnal* **54**, 1887–1890.

Gordejeva, K.T. (1976). New species of tropic Copepoda from the Atlantic and South Seas. *Zoologichesky Zhurnal* **55**, 1398–1401.

Gorsky, G., Dallot, S., Sardou, J., Fenaux, R., Carré, C. and Palazzoli, I. (1988). C and N composition of some northwestern Mediterranean zooplankton and micronekton species. *Journal of Experimental Marine Biology and Ecology* **124**, 133–144.

Goswami, S.C. (1977). Development & predation in a calanoid copepod *Tortanus forcipatus* (Giesbrecht). *Indian Journal of Marine Science* **6**, 154–159.

Goswami, S.C. (1978a). Developmental stages, growth & sex ratio in *Pseudodiaptomus binghami* Sewell (Copepoda: Calanoida). *Indian Journal of Marine Science* **7**, 103–109.

Goswami, S.C. (1978b). Developmental stages of calanoid copepods, *Labidocera pavo* Giesbrecht & *L. minuta* Giesbrecht. *Indian Journal of Marine Science* **7**, 288–294.

Goswami, S.C. (1983). Coexistence and succession of copepod species in the Mandovi and Zuari Estuaries, Goa. *Mahasagar* **16**, 251–258.

Goswami, S.C. and Rao, T.S.S. (1981). Copepod swarm in the Campbell Bay (Andaman Sea). *Indian Journal of Marine Sciences* **10**, 274–275.

Goswami, U. and Goswami, S.C. (1973). Karyological studies on the genus *Acartia* (Copepoda). *Current Science* **42**, 242–243.

Goswami, U. and Goswami, S.C. (1974). Cytotaxonomical studies on some calanoid copepods. *The Nucleus* **17**, 109–113.

Goswami, U. and Goswami, S.C. (1978). A note on the karyotypes of some calanoid copepods. *Mahasagar* **11**, 111–113.

Goswami, U. and Goswami, S.C. (1979a). Cytotaxonomical studies in family Eucalanidae (Copepoda). *Mahasagar* **12**, 103–108.

Goswami, U. and Goswami, S.C. (1979b). Karyology of the genus *Labidocera* Lubbock from the Laccadive Sea (Lakshadweep). *Indian Journal of Marine Sciences* **8**, 259–261.

Goswami, U. and Goswami, S.C. (1982). Chromosomal aberrations found in *Paracalanus aculeatus* (Giesbrecht) at the time of solar eclipse. *Mahasagar* **15**, 59–62.

Goswami, U. and Goswami, S.C. (1984). A note on chromosomes of *Pontellopsis herdmani* and *Pontella princeps* (Copepoda) from the Laccadive Sea. *Mahasagar* **17**, 129–132.

Goswami, U. and Goswami, S.C. (1985). Karyology of genera *Calanopia* & *Pontellina* with a note on cytotaxonomical & evolutionary relationship in family

Pontellidae. *Mahasagar* **18**, 433–438.
Gottfried, M. and Roman, M.R. (1983). Ingestion and incorporation of coral-mucus detritus by reef zooplankton. *Marine Biology* **72**, 211–218.
Gowing, M.M. and Wishner, K.F. (1986). Trophic relationships of deep-sea calanoid copepods from the benthic boundary layer of the Santa Catalina Basin, California. *Deep-Sea Research* **33A**, 939–961.
Gowing, M.M. and Wishner, K.F. (1992). Feeding ecology of benthopelagic zooplankton on an eastern tropical Pacific seamount. *Marine Biology* **112**, 451–467.
Grad, G. and Maly, E.J. (1988). Sex size ratios and their influence on mating success in a calanoid copepod. *Limnology and Oceanography* **33**, 1629–1634.
Grad, G. and Maly, E.J. (1992). Further observations relating sex size ratios to mating success in calanoid copepods. *Journal of Plankton Research* **14**, 903–913.
Graeve, M., Hagen, W. and Kattner, G. (1994a). Herbivorous or omnivorous? On the significance of lipid compositions as trophic markers in Antarctic copepods. *Deep-Sea Research I* **41 A**, 915–924.
Graeve, M., Kattner, G. and Hagen, W. (1994b). Diet-induced changes in the fatty acid composition of Arctic herbivorous copepods: experimental evidence of trophic markers. *Journal of Experimental Marine Biology and Ecology* **182**, 97–110.
Grainger, E.H. (1959). The annual oceanographic cycle at Igloolik in the Canadian Arctic. I. The zooplankton and physical and chemical observations. *Journal of the Fisheries Research Board of Canada* **16**, 453–501.
Grainger, E.H. and Mohammed, A.A. (1986). Copepods in Arctic sea ice. *Syllogeus* **58**, 303–310.
Grant, G.C. (1988). Seasonal occurrence and dominance of *Centropages* congeners in the Middle Atlantic Bight, USA. *Hydrobiologia* **167/168**, 227–237.
Grassle, J.F. (1986). The ecology of deep-sea hydrothermal vent communities. *Advances in Marine Biology* **23**, 301–362.
Green, E.P., Harris, R.P. and Duncan, A. (1991). The naupliar development of marine calanoid copepods under high and low food conditions. *Bulletin of Plankton Society of Japan*, Special Volume, 347–362.
Green, E.P., Harris, R.P. and Duncan, A. (1992). The production and ingestion of faecal pellets by nauplii of marine calanoid copepods. *Journal of Plankton Research* **14**, 1631–1643.
Green, E.P., Harris, R.P. and Duncan, A. (1993). The seasonal abundance of the copepodite stages of *Calanus helgolandicus* and *Pseudocalanus elongatus* off Plymouth. *Journal of the Marine Biological Association of the United Kingdom* **73**, 109–122.
Greene, C.H. (1988). Foraging tactics and prey-selection patterns of omnivorous and carnivorous calanoid copepods. *Hydrobiologia* **167/168**, 295–302.
Greene, C.H. and Landry, M.R. (1985). Patterns of prey selection in the cruising calanoid predator *Euchaeta elongata*. *Ecology* **66**, 1408–1416.
Greenwood, J.G. (1981). Occurrences of congeneric pairs of *Acartia* and *Pseudodiaptomus* species (Copepoda, Calanoida) in Moreton Bay, Queensland. *Estuarine, Coastal and Shelf Science* **13**, 591–596.
Greve, W. (1995). Mutual predation causes bifurcations in pelagic ecosystems: the simulation model PLITCH (PLanktonic swITCH), experimental tests, and theory. *ICES Journal of Marine Science* **52**, 505–510.
Greze, V.N. and Baldina, E.P. (1964). Population dynamics and the annual production of *Acartia clausi* Giesbr. and *Centropages kröyeri* Giesbr. in the neritic

zone of the Black Sea. *Trudy Sevastopol'skoi Biologicheskoi Stantsii* **18**, 249–261.

Greze, V.N., Baldina, E.P. and Bileva, O.K. (1968). Production of planktonic copepods in neritic zone of the Black Sea. *Okeanologiya* **8**, 1066–1070.

Grice, G.D. (1959). A new species of *Haloptilus* (Copepoda: Calanoida) from equatorial waters of the east-central Pacific Ocean. *Journal of the Washington Academy of Sciences* **49**, 193–195.

Grice, G.D. (1962a). Calanoid copepods from equatorial waters of the Pacific Ocean. *Fishery Bulletin, United States Fish and Wild Life Service* **61**, 171–246.

Grice, G.D. (1962b). Copepods collected by the nuclear submarine Seadragon on a cruise to and from the North Pole, with remarks on their geographic distribution. *Journal of Marine Research* **20**, 97–109.

Grice, G.D. (1963). A revision of the genus *Candacia* (Copepoda: Calanoida) with an annotated list of the species and a key for their identification. *Zoölogische Mededeelingen, Leiden* **38** (10), 171–194.

Grice, G.D. (1969). The developmental stages of *Pseudodiaptomus coronatus* Williams (Copepoda, Calanoida). *Crustaceana* **16**, 291–301.

Grice, G.D. (1971). The developmental stages of *Eurytemora americana* Williams, 1906 and *Eurytemora herdmani* Thompson & Scott, 1897 (Copepoda, Calanoida). *Crustaceana* **20**, 145–158.

Grice, G.D. (1972). The existence of a bottom-living calanoid copepod fauna in deep water with descriptions of five new species. *Crustaceana* **23**, 219–242.

Grice, G.D. (1973). *Alrhabdus johrdeae*, a new genus and species of benthic calanoid copepods from the Bahamas. *Bulletin of Marine Science* **23**, 942–947.

Grice, G.D. and Gibson, V.R. (1977). Resting eggs in *Pontella meadi* (Copepoda: Calanoida). *Journal of the Fisheries Research Board of Canada* **34**, 410–412.

Grice, G.D. and Gibson, V.R. (1981). Hatching of eggs of *Pontella mediterranea* Claus (Copepoda: Calanoida). *Vie et Milieu* **31**, 49–51.

Grice, G.D. and Gibson, V.R. (1982). The developmental stages of the calanoid copepod *Labidocera wollastoni* (Lubbock) with observations on its eggs. *Cahiers de Biologie marine* **23**, 215–225.

Grice, G.D. and Hülsemann, K. (1965). Abundance, vertical distribution and taxonomy of calanoid copepods at selected stations in the northeast Atlantic. *Journal of Zoology London* **146**, 213–262.

Grice, G.D. and Hülsemann, K. (1967). Bathypelagic calanoid copepods of the western Indian Ocean. *Proceedings of the United States National Museum* **122** (3583), 1–67.

Grice, G.D. and Hülsemann, K. (1970). New species of bottom-living calanoid copepods collected in deepwater by the DSRV Alvin. *Bulletin of the Museum of Comparative Zoology, Harvard* **139**, 185–230.

Grice, G.D. and Lawson, T.J. (1976). Resting eggs in the marine calanoid copepod *Labidocera aestiva* Wheeler. *Crustaceana* **30**, 9–12.

Grice, G.D. and Marcus, N.H. (1981). Dormant eggs of marine copepods. *Oceanography and Marine Biology: an Annual Review* **19**, 125–140.

Griffiths, A.M. and Frost, B.W. (1976). Chemical communication in the marine planktonic copepods *Calanus pacificus* and *Pseudocalanus* sp. *Crustaceana* **30**, 1–8.

Grigg, H. and Bardwell, S.J. (1982). Seasonal observations on moulting and maturation in stage V copepodites of *Calanus finmarchicus* from the Firth of Clyde. *Journal of the Marine Biological Association of the United Kingdom* **62**, 315–327.

Grigg, H., Bardwell, S.J. and Tyzack, S. (1981). Patterns of variation in the prosome length of overwintering stage V copepodites of *Calanus finmarchicus* in the Firth of Clyde. *Journal of the Marine Biological Association of the United Kingdom* **61**, 885–899.

Grigg, H., Holmes, L.J. and Bardwell, S.J. (1985). Seasonal observations on the biometry and development in copepodite stage V of *Calanus finmarchicus* from the Firth of Clyde. *Marine Biology* **88**, 73–83.

Grigg, H., Bardwell, S.J. and Holmes, L.J. (1987). Comparative observations on the biometry and development of *Calanus finmarchicus* and *C. helgolandicus* in copepodite stage V, with comments on other Calanidae. *Marine Biology* **96**, 253–262.

Grigg, H., Holmes, L.J. and Bardwell, S.J. (1989). Patterns of variation in the dry body weight of *Calanus finmarchicus* in copepodite stage V during autumn and winter in the Firth of Clyde. *Journal of the Marine Biological Association of the United Kingdom* **69**, 101–122.

Grindley, J.R. (1963). The Pseudodiaptomidae (Copepoda: Calanoida) of southern African waters, including a new species, *Pseudodiaptomus charteri*. *Annals of the South African Museum* **46**, 373–391.

Grindley, J.R. (1984). The zoogeography of the Pseudodiaptomidae. *Crustaceana*, Supplement **7**, 217–228.

Grobben, C. (1881). Die Entwicklungsgeschichte von *Cetochilus septentrionalis* Goodsir. *Arbeiten aus dem Zoologischen Institut der Universität Wien* **3**, 242–282.

Grønvik, S. and Hopkins, C.C.E. (1984). Ecological investigations of the zooplankton community of Balsfjorden, Northern Norway: generation cycle, seasonal vertical distribution, and seasonal variations in body weight and carbon and nitrogen content of the copepod *Metridia longa* (Lubbock). *Journal of Experimental Marine Biology and Ecology* **80**, 93–107.

Gross, F. and Raymont, J.E.G. (1942). The specific gravity of *Calanus finmarchicus*. *Proceedings of the Royal Society of Edinburgh* **61B**, 288–296.

Gruzov, L.N. and Alekseyeva, L.G. (1970). Weight characteristics of copepods from the equatorial Atlantic. *Oceanology, Moscow* **10**, 871–879.

Guerin-Ancey, O. and David, P.M. (1993). Use of a multibeam-multifrequency sounder to study the distribution of small zooplankton. *Deep-Sea Research* **40A**, 119–128.

Guerrero, F., Blanco, J.M^a. and Rodríguez, V. (1994). Temperature-dependent development in marine copepods: a comparative analysis of models. *Journal of Plankton Research* **16**, 95–103.

Guglielmo, L. and Ianora, A., eds (1995). Atlas of Marine Zooplankton (Straits of Magellan). Copepods. Springer Verlag, Berlin, Heidelberg, New York. 279 pp.

Gurney, R. (1931). "British Fresh-water Copepoda", Vol. 1. Ray Society, London.

Hada, A. and Uye, S.-i. (1991). Cannibalistic feeding behaviour of the brackish-water copepod *Sinocalanus tenellus*. *Journal of Plankton Research* **13**, 155–166.

Hada, A., Uye, S.-i. and Onbé, T. (1986). The seasonal life cycle of *Sinocalanus tenellus* (Copepoda: Calanoida) in a brackish-water pond. *Bulletin of Plankton Society of Japan* **33**, 29–41.

Haedrich, R.L. and Judkins, D.C. (1979). Macrozooplankton and its environment. In "Zoogeography and Diversity in Plankton" (S. Van der Spoel and A.C. Pierrot-Bults, eds), pp. 4–28. Edward Arnold, London.

Hagen, W. (1988). Zur Bedeutung der Lipide im antarktischen Zooplankton. *Berichte zur Polarforschung* **49**, 1–129.

Hagen, W. and Schnack-Schiel, S.B. (1996). Seasonal lipid dynamics in dominant Antarctic copepods: energy for overwintering or reproduction? *Deep-Sea Research I* **43**, 139–158.

Hagen, W., Kattner, G. and Graeve, M. (1993). *Calanoides acutus* and *Calanus propinquus*, Antarctic copepods with different lipid storage modes via wax esters or triacylglycerols. *Marine Ecology Progress Series* **97**, 135–142.

Hagen, W., Kattner, G. and Graeve, M. (1995). On the lipid biochemistry of polar copepods: compositional differences in the Antarctic calanoids *Euchaeta antarctica* and *Euchirella rostromagna*. *Marine Biology* **123**, 451–457.

Hairston, N.G., Jr and Brunt, R.A. Van (1994). Diapause dynamics of two diaptomid copepod species in a large lake. *Hydrobiologia* **292/293**, 209–218.

Hairston, N.G., Jr and Kearns, C.M. (1995). The interaction of photoperiod and temperature in diapause timing: a copepod example. *Biological Bulletin, Marine Biological Laboratory, Woods Hole, Mass.* **189**, 42–48.

Hairston, N.G., Braner, M. and Twombly, S. (1987). Perspective on prospective methods for obtaining life table data. *Limnology and Oceanography* **32**, 517–520.

Håkanson, J.L. (1987). The feeding condition of *Calanus pacificus* and other zooplankton in relation to phytoplankton pigments in the California Current. *Limnology and Oceanography* **32**, 881–894.

Halim, Y. (1990). On the potential migration of Indo-Pacific plankton through the Suez Canal. *Bulletin de l'Institut Océanographique Monaco*, Spécial No. 7, 11–27.

Hallberg, E. and Hirche, H.-J. (1980). Differentiation of mid-gut in adults and over-wintering copepodids of *Calanus finmarchicus* (Gunnerus) and *C. helgolandicus* Claus. *Journal of Experimental Marine Biology and Ecology* **48**, 283–295.

Hammer, R.M. (1978). Scanning electron microscope study of the spermatophore of *Acartia tonsa* (Copepoda: Calanoida). *Transactions of the American Microscopical Society* **97**, 386–389.

Hamner, W.M. (1985). The importance of ethology for investigation of marine zooplankton. *Bulletin of Marine Science* **37**, 414–424.

Hamner, W.M. (1988). Behavior of plankton and patch formation in pelagic ecosystems. *Bulletin of Marine Science* **43**, 752–757.

Hamner, W.M. and Carleton, J.H. (1979). Copepod swarms: attributes and role in coral reef ecosystems. *Limnology and Oceanography* **24**, 1–14.

Hansen, B. and Bech, G. (1996). Bacteria associated with a marine planktonic copepod in culture. I. Bacterial genera in seawater, body surface, intestines and fecal pellets and succession during fecal pellet degradation. *Journal of Plankton Research* **18**, 257–273.

Hansen, B., Hansen, P.J. and Nielsen, T.G. (1991). Effects of large nongrazable particles on clearance and swimming behaviour of zooplankton. *Journal of Experimental Marine Biology and Ecology* **152**, 257–269.

Hansen, B., Bjørnsen, P.K. and Hansen, P.J. (1994). The size ratio between planktonic predators and their prey. *Limnology and Oceanography* **39**, 395–403.

Hansen, B., Fotel, F.L., Jensen, N.J. and Madsen, S.D. (1996). Bacteria associated with a marine planktonic copepod in culture. II. Degradation of faecal pellets produced on a diatom, a nanoflagellate or a dinoflagellate diet. *Journal of Plankton Research* **18**, 275–288.

Hansen, F.C. and Boekel, W.H.M.van (1991). Grazing pressure of the calanoid copepod *Temora longicornis* on a *Phaeocystis* dominated spring bloom in a Dutch

tidal inlet. *Marine Ecology Progress Series* **78**, 123–129.
Hapette, A.M. and Poulet, S.A. (1990). Variation of vitamin C in some common species of marine plankton. *Marine Ecology Progress Series* **64**, 69–79.
Harbison, G.R. and McAlister, V.L. (1980). Fact and artifact in copepod feeding experiments. *Limnology and Oceanography* **25**, 971–981.
Harding, G.C.H. (1973). Decomposition of marine copepods. *Limnology and Oceanography* **18**, 670–673.
Harding, G.C.H. (1974). The food of deep-sea copepods. *Journal of the Marine Biological Association of the United Kingdom* **54**, 141–155.
Harding, G.C.H., Vass, W.P., Hargrave, B.T. and Pearre, S. (1986). Diel vertical movements and feeding activity of zooplankton in St. Georges Bay, N.S., using net tows and a newly developed passive trap. *Canadian Journal of Fisheries and Aquatic Sciences* **43**, 952–967.
Harding, J.P. (1963). The chromosomes of *Calanus finmarchicus* and *C. helgolandicus*. *Crustaceana* **6**, 81–88.
Harding, J.P. and Marshall, S.M. (1955). Triploid nauplii of *Calanus finmarchicus*. *Nature, London* **175**, 175.
Hardy, A.C. and Bainbridge, R. (1954). Experimental observations on the vertical migrations of planktonic animals. *Journal of the Marine Biological Association of the United Kingdom* **33**, 409–448.
Hargis, J.R. (1977). Comparison of techniques for the measurement of zooplankton filtration rates. *Limnology and Oceanography* **22**, 942–945.
Haridas, P., Madhupratap, M. and Ohtsuka, S. (1994). *Pseudocyclops lakshmi*, a new species (Pseudocyclopidae: Calanoida: Copepoda) from the Laccadives, India. *Proceedings of the Biological Society of Washington* **107**, 151–163.
Harris, J.E. (1963). The role of endogenous rhythms in vertical migration. *Journal of the Marine Biological Association of the United Kingdom* **43**, 153–166.
Harris, J.R.W. (1983). The development and growth of *Calanus* copepodites. *Limnology and Oceanography* **28**, 142–147.
Harris, R.P. (1988). Interactions between diel vertical migratory behavior of marine zooplankton and the subsurface chlorophyll maximum. *Bulletin of Marine Science* **43**, 663–674.
Harris, R.P. (1994). Zooplankton grazing on the coccolithophore *Emiliania huxleyi* and its role in inorganic carbon flux. *Marine Biology* **119**, 431–439.
Harris, R.P. and Malej, A. (1986). Diel patterns of ammonium excretion and grazing rhythms in *Calanus helgolandicus* in surface stratified waters. *Marine Ecology Progress Series* **31**, 75–85.
Harris, R.P. and Paffenhöfer, G.-A. (1976a). Feeding, growth and reproduction of the marine planktonic copepod *Temora longicornis* Müller. *Journal of the Marine Biological Association of the United Kingdom* **56**, 675–690.
Harris, R.P. and Paffenhöfer, G.-A. (1976b). The effect of food concentration on cumulative ingestion and growth efficiency of two small marine planktonic copepods. *Journal of the Marine Biological Association of the United Kingdom* **56**, 875–888.
Hart, R.C. (1990). Copepod post-embryonic development: pattern, conformity, and predictability. The realities of isochronal equiproportional development, and trends in the copepodid-naupliar duration ratio. *Hydrobiologia* **206**, 175–206.
Hart, R.C. and McLaren, I.A. (1978). Temperature acclimation and other influences on embryonic duration in the copepod *Pseudocalanus* sp. *Marine Biology* **45**, 23–30.
Hassett, R.P. and Blades-Eckelbarger, P. (1995). Diel changes in gut-cell morphology

and digestive activity of the marine copepod *Acartia tonsa*. *Marine Biology* **124**, 59–69.
Hassett, R.P. and Landry, M.R. (1983). Effects of food-level acclimation on digestive enzyme activities and feeding behavior of *Calanus pacificus*. *Marine Biology* **75**, 47–55.
Hassett, R.P. and Landry, M.R. (1985). Time scales in behavioral, biochemical, and energetic adaptations to food-limiting conditions by a marine copepod. *Archives fur Hydrobiologie Beiheft Ergebnisse Limnologie* **21**, 209–221.
Hassett, R.P. and Landry, M.R. (1988). Short-term changes in feeding and digestion by the copepod *Calanus pacificus*. *Marine Biology* **99**, 63–74.
Hassett, R.P. and Landry, M.R. (1990a). Seasonal changes in feeding rate, digestive enzyme activity, and assimilation efficiency of *Calanus pacificus*. *Marine Ecology Progress Series* **62**, 203–210.
Hassett, R.P. and Landry, M.R. (1990b). Effects of diet and starvation on digestive enzyme activity and feeding behavior of the marine copepod *Calanus pacificus*. *Journal of Plankton Research* **12**, 991–1010.
Hattori, H. (1989). Bimodal vertical distribution and diel migration of the copepods *Metridia pacifica*, *M. okhotensis* and *Pleuromamma scutullata* in the western North Pacific Ocean. *Marine Biology* **103**, 39–50.
Hattori, H. and Motoda, S. (1983). Regional difference in zooplankton communities in the western North Pacific Ocean (CSK data). *Bulletin of Plankton Society of Japan* **30**, 53–63.
Hattori, H. and Saito, H. (1994). Vertical distribution and feeding behavior of copepods under sea-ice in Lake Saroma. *Proceedings of the NIPR Symposium on Polar Biology* **7**, 298–301.
Hattori, H. and Saito, H. (1995). Diel changes in the vertical distribution and feeding activity of copepods in ice-covered Resolute Passage, Canadian Arctic. *Proceedings of the NIPR Symposium on Polar Biology* **8**, 69–73.
Hattori, H., Yuki, K., Zaitsev, Yu.P. and Motoda, S. (1983). A preliminary observation on the neuston in Suruga Bay. *La Mer* **21**, 11–20.
Haury, L.R. (1976a). Small-scale pattern of a California Current zooplankton assemblage. *Marine Biology* **37**, 137–157.
Haury, L.R. (1976b). A comparison of zooplankton patterns in the California Current and North Pacific central gyre. *Marine Biology* **37**, 159–167.
Haury, L.R. (1984). An offshore eddy in the California Current system. Part IV: plankton distributions. *Progress in Oceanography* **13**, 95–111.
Haury, L.R. (1986). Patches, niches, and oceanic biogeography. *Unesco Technical Papers in Marine Science* **49**, 126–131.
Haury, L.R. (1988). Vertical distribution of *Pleuromamma* (Copepoda: Metridinidae) across the eastern North Pacific Ocean. *Hydrobiologia* **167/168**, 335–342.
Haury, L. and Weihs, D. (1976). Energetically efficient swimming behavior of negatively buoyant zooplankton. *Limnology and Oceanography* **21**, 797–803.
Haury, L.R. and Wiebe, P.H. (1982). Fine-scale multi-species aggregations of oceanic zooplankton. *Deep-Sea Research* **29A**, 915–921.
Haury, L.R. and Yamazaki, H. (1995). The dichotomy of scales in the perception and aggregation behavior of zooplankton. *Journal of Plankton Research* **17**, 191–197.
Haury, L.R., Kenyon, D.E. and Brooks, J.R. (1980). Experimental evaluation of the avoidance reaction of *Calanus finmarchicus*. *Journal of Plankton Research* **2**, 187–202.

Haury, L.R., Yamazaki, H. and Itsweire, E.C. (1990). Effects of turbulent shear flow on zooplankton distribution. *Deep-Sea Research* **37A**, 447–461.

Haury, L., Fey, C., Gal, G., Hobday, A. and Genin, A. (1995). Copepod carcasses in the ocean. I. Over seamounts. *Marine Ecology Progress Series* **123**, 57–63.

Hay, S. (1995). Egg production and secondary production of common North Sea copepods: field estimates with regional and seasonal comparisons. *ICES Journal of Marine Science* **52**, 315–327.

Hay, S.J., Evans, G.T. and Gamble, J.C. (1988). Birth, growth and death rates for enclosed populations of calanoid copepods. *Journal of Plankton Research* **10**, 431–454.

Hay, S.J., Kiørboe, T. and Matthews, A. (1991). Zooplankton biomass and production in the North Sea during the Autumn Circulation Experiment, October 1987 – March, 1988. *Continental Shelf Research* **11**, 1453–1476.

Hays, G.C. (1995). Ontogenetic and seasonal variation in the diel migration of the copepods *Metridia lucens* and *Metridia longa*. *Limnology and Oceanography* **40**, 1461–1465.

Hays, G.C., Carr, M.R. and Taylor, A.H. (1993). The relationship between Gulf Stream position and copepod abundance derived from the Continuous Plankton Recorder Survey: separating biological signal from sampling noise. *Journal of Plankton Research* **15**, 1359–1373.

Hays, G.C., Procter, C.A., John, A.W.G. and Warner, A.J. (1994). Interspecific differences in the diel vertical migration of marine copepods: the implications of size, color and morphology. *Limnology and Oceanography* **39**, 1621–1629.

Hays, G.C., Warner, A.J. and Lefevre, D. (1996). Long-term changes in the diel vertical migration behaviour of zooplankton. *Marine Ecology Progress Series* **141**, 149–159.

Hayward, T.L. (1980). Spatial and temporal feeding patterns of copepods from the North Pacific central gyre. *Marine Biology* **58**, 295–309.

Hayward, T.L. (1981). Mating and depth distribution of an oceanic copepod. *Limnology and Oceanography* **26**, 374–377.

Hayward, T.L. and McGowan, J.A. (1979). Pattern and structure in an oceanic zooplankton community. *American Zoologist* **19**, 1045–1055.

Head, E.J.H. (1986). Estimation of Arctic copepod grazing rates *in vivo* and comparison with *in vitro* methods. *Marine Biology* **92**, 371–379.

Head, E.J.H. (1988). Copepod feeding behavior and the measurement of grazing rates *in vivo* and *in vitro*. *Hydrobiologia* **167/168**, 31–41.

Head, E.J.H. (1992a). Gut pigment accumulation and destruction by Arctic copepods *in vitro* and *in situ*. *Marine Biology*, **112**, 583–592.

Head, E.J.H. (1992b). Comparison of the chemical composition of particulate material and copepod faecal pellets at stations off the coast of Labrador and in the Gulf of St. Lawrence. *Marine Biology* **112**, 593–600.

Head, E.J.H. and Harris, L.R. (1985). Physiological and biochemical changes in *Calanus hyperboreus* from Jones Sound, N.W.T. during the transition from summer feeding to over-wintering condition. *Polar Biology* **4**, 99–106.

Head, E.J.H. and Harris, L.R. (1992). Chlorophyll and carotenoid transformation and destruction by *Calanus* spp. grazing on diatoms. *Marine Ecology Progress Series* **86**, 229–238.

Head, E.J.H. and Harris, L.R. (1994). Feeding selectivity by copepods grazing on natural mixtures of phytoplankton by HPLC analysis of pigments. *Marine Ecology Progress Series* **110**, 75–83.

Head, E.J.H. and Harris, L.R. (1996). Chlorophyll destruction by *Calanus* spp.

grazing on phytoplankton: kinetics, effects of ingestion rate and feeding history, and a mechanistic interpretation. *Marine Ecology Progress Series* **135**, 223–235.
Head, E.J.H., Harris, L.R. and Debs, C.A. (1985). Effect of daylength and food concentration on *in situ* diurnal feeding rhythms in Arctic copepods. *Marine Ecology Progress Series* **24**, 281–288.
Head, E.J.H., Harris, L.R. and Debs, C.A. (1986). Long term grazing experiments with Arctic copepods. *Journal of Experimental Marine Biology and Ecology* **100**, 271–286.
Head, E.J.H., Bedo, A. and Harris, L.R. (1988). Grazing, defecation and excretion rates of copepods from inter-island channels of the Canadian Arctic archipelago. *Marine Biology* **99**, 333–340.
Heberer, G. (1930). Die Struktur der Oocyten von *Eucalanus elongatus* Dana mit bemerkungen über den bau des weiblichen genitalapparates. (Cytologische mitteilungen I.). *Zeitschrift für Wissenschaftliche Zoologie* **136**, 155–194.
Heberer, G. (1932). Untersuchungen über Bau und Funktion der Genitalorgane der copepoden. I. Der männliche genitalapparat der calanoiden copepoden. *Zeitschrift für Mikroskopische-Anatomische Forschung* **31**, 250–424.
Heberer, G. (1937). Weitere Ergebnisse uber Bildung und Bau der Spermatophoren und Spermatophorenkoppelapparate bei calanoiden Copepoden. *Verhandlungen der Deutschen zoologischen Gesellschaft* **39**, 86–93.
Hebert, P.D.A., Good, A.G. and Mort, M.A. (1980). Induced swarming in the predatory copepod *Heterocope septentrionalis*. *Limnology and Oceanography* **25**, 747–750.
Heckman, C.W. (1986). The anadromous migration of a calanoid copepod, *Eurytemora affinis* (Poppe, 1880) in the Elbe Estuary. *Crustaceana* **50**, 176–181.
Hedgpeth, J.W. (1993). Foreign invaders. *Science* **261**, 34–35.
Heerkloss, R. and Ring, M. (1989). Ingestion rate and amylase activity in the estuarine copepod *Eurytemora affinis*. *Wissenschaftliche Zeitschrift der Universität Rostock*, N-Reihe **38**, 13–17.
Heerkloss, R., Brenning, U., Ihlenfeld, R. and Franke, R. (1990). Influence of temperature and epizoic ciliates on the growth of *Eurytemora affinis* (Poppe) (Calanoida, Copepoda) under laboratory conditions. *Wissenschaftliche Zeitschrift der Universität Rostock*, N-Reihe **39**, S12–S15.
Heerkloss, R., Schiewer, U., Wasmund, N. and Kühner, E. (1993). A long-term study of zooplankton succession in enclosures with special reference to *Eurytemora affinis* (Poppe), Calanoida, Copepoda. *Rostocker Meeresbiologische Beiträge* **1**, 25–35.
Heinle, D.R. (1966). Production of a calanoid copepod, *Acartia tonsa*, in the Patuxent Estuary. *Chesapeake Science* **7**, 59–74.
Heinle, D.R. (1969). Temperature and zooplankton. *Chesapeake Science* **10**, 186–209.
Heinle, D.R. (1970). Population dynamics of exploited cultures of calanoid copepods. *Helgoländer wissenschaftliche Meeresuntersuchungen* **20**, 360–372.
Heinle, D.R. and Flemer, D.A. (1975). Carbon requirements of a population of the estuarine copepod *Eurytemora affinis*. *Marine Biology* **31**, 235–247.
Heinrich, A.K. = Geinrikh, A.K.
Heip, C. and Engels, P. (1974). Comparing species diversity and evenness indices. *Journal of the Marine Biological Association of the United Kingdom* **54**, 559–563.
Heip, C.H.R., Goosen, N.K., Herman, P.M.J., Kromkamp, J., Middelburg, J.J. and Soetaert, K. (1995). Production and consumption of biological particles in

temperate tidal estuaries. *Oceanography and Marine Biology: an Annual Review* **33**, 1–149.
Hempel, G. and Weikert, H. (1972). The neuston of the subtropical and boreal North-eastern Atlantic Ocean. A review. *Marine Biology* **13**, 70–88.
Heptner, M.V. = Geptner, M.V.
Herman, A.W. (1983). Vertical distribution patterns of copepods, chlorophyll, and production in northeastern Baffin Bay. *Limnology and Oceanography* **28**, 709–719.
Herman, A.W. (1984). Vertical copepod aggregations and interactions with chlorophyll and production on the Peru shelf. *Continental Shelf Research* **3**, 131–146.
Herman, A.W. (1989). Vertical relationships between chlorophyll, production and copepods in the eastern tropical Pacific. *Journal of Plankton Research* **11**, 243–261.
Herman, A.W. (1992). Design and calibration of a new optical plankton counter capable of sizing small zooplankton. *Journal of the Marine Biological Association of the United Kingdom* **39**, 395–415.
Herman, A.W., Mitchell, M.R. and Young, S.W. (1984). A continuous pump sampler for profiling copepods and chlorophyll in the upper oceanic layers. *Deep-Sea Research* **31A**, 439–450.
Herman, A.W., Sameoto, D.D., Shunnian, C., Mitchell, M.R., Petrie, B. and Cochrane, N. (1991). Sources of zooplankton on the Nova Scotia Shelf and their aggregations within deep-shelf basins. *Continental Shelf Research* **11**, 211–238.
Hernández-León, S. and Gómez, M. (1996). Factors affecting the respiration/ETS ratio in marine zooplankton. *Journal of Plankton Research* **18**, 239–255.
Hernández-León, S., Almeida, C. and Montero, I. (1995). The use of aspartate transcarbamylase activity to estimate growth rates in zooplankton. *ICES Journal of Marine Science* **52**, 377–383.
Heron, G.A. and Bowman, T.E. (1971). Postnaupliar developmental stages of the copepod crustaceans *Clausocalanus laticeps*, *C. brevipes* and *Ctenocalanus citer* (Calanoida: Pseudocalanidae). In "Biology of the Antarctic Seas, 4". *Antarctic Research Series* **17**, 141–165.
Herring, P.J. (1988). Copepod luminescence. *Hydrobiologia* **167/168**, 183–195.
Hilton, I.F. (1931). The oogenesis of *Calanus finmarchicus*. *Quarterly Journal of Microscopical Science* **74**, 193–222.
Hirakawa, K. (1979). Seasonal changes of population structure of a calanoid copepod, *Calanus pacificus*, in Funka Bay, Hokkaido. *Bulletin of Plankton Society of Japan* **26**, 49–58.
Hirakawa, K. (1986). A new record of the planktonic copepod *Centropages abdominalis* (Copepoda, Calanoida) from Patagonian waters, southern Chile. *Crustaceana* **51**, 296–299.
Hirakawa, K. (1988). New records of the North Pacific coastal planktonic copepods, *Acartia omorii* (Acartiidae) and *Oithona davisae* (Oithonidae) from southern Chile. *Bulletin of Marine Science* **42**, 337–339.
Hirakawa, K. (1989). Planktonic copepods from Aysén Fjord and adjacent waters, southern Chile. *Proceedings of the NIPR Symposium on Polar Biology* **2**, 46–50.
Hirakawa, K. (1991). Vertical distribution and reproduction of planktonic copepods in Toyama Bay, southern Japan Sea, with special reference to *Metridia pacifica*. *Bulletin of Plankton Society of Japan*, Special Volume, 373–382.
Hirakawa, K. and Imamura, A. (1993). Seasonal abundance and life history of

Metridia pacifica (Copepoda: Calanoida) in Toyama Bay, southern Japan Sea. *Bulletin of Plankton Society of Japan* **40**, 41–54.

Hirakawa, K., Ikeda, T. and Kajihara, N. (1990). Vertical distribution of zooplankton in Toyama Bay, southern Japan Sea, with special reference to Copepoda. *Bulletin of Plankton Society of Japan* **37**, 111–126.

Hirakawa, K., Sasagawa, I. and Hattori, Y. (1995). Mandibular masticatory edge of *Neocalanus cristatus* (Krøyer, 1848) (Copepoda: Calanoida) reared in the laboratory. *Bulletin of Plankton Society of Japan* **42**, 167–171.

Hirche, H.-J. (1974). Die Copepoden *Eurytemora affinis* Poppe und *Acartia tonsa* Dana und ihre Besiedlung durch *Myoschiston centropagidarum* Precht (Peritricha) in der Schlei. *Kieler Meeresforschungen* **30**, 43–64.

Hirche, H.-J. (1980). The cultivation of *Calanoides carinatus* Krøyer (Copepoda: Calanoida) under different temperature and food conditions – with a description of eggs and nauplii. *Journal of the Marine Biological Association of the United Kingdom* **60**, 115–125.

Hirche, H.-J. (1983). Overwintering of *Calanus finmarchicus* and *Calanus helgolandicus*. *Marine Ecology Progress Series* **11**, 281–290.

Hirche, H.-J. (1984). Seasonal distribution of *Calanus finmarchicus* (Gunnerus) and *C. helgolandicus* (Claus) in a Swedish fjord. *Crustaceana*, Supplement **7**, 233–241.

Hirche, H.-J. (1987). Temperature and plankton. II. Effect on respiration and swimming activity in copepods from the Greenland Sea. *Marine Biology* **94**, 347–356.

Hirche, H.-J. (1989a). Spatial distribution of digestive enzyme activities of *Calanus finmarchicus* and *Calanus hyperboreus* in Fram Strait/Greenland Sea. *Journal of Plankton Research* **11**, 431–443.

Hirche, H.-J. (1989b). Egg production of the Arctic copepod *Calanus glacialis*: laboratory experiments. *Marine Biology* **103**, 311–318.

Hirche, H.-J. (1990). Egg production of *Calanus finmarchicus* at low temperature. *Marine Biology* **106**, 53–58.

Hirche, H.-J. (1991). Distribution of dominant calanoid copepod species in the Greenland Sea during late fall. *Polar Biology* **11**, 351–362.

Hirche, H.-J. and Bohrer, R.N. (1987). Reproduction of the Arctic copepod *Calanus glacialis* in Fram Strait. *Marine Biology* **94**, 11–17.

Hirche, H.-J. and Kattner, G. (1993). Egg production and lipid content of *Calanus glacialis* in spring: indication of a food-dependent and food-independent reproductive mode. *Marine Biology* **117**, 615–622.

Hirche, H.-J. and Mumm, N. (1992). Distribution of dominant copepods in the Nansen Basin, Arctic Ocean, in summer. *Deep-Sea Research* **39A**, Supplement, S485–S505.

Hirche, H.-J. and Niehoff, B. (1996). Reproduction of the Arctic copepod *Calanus hyperboreus* in the Greenland Sea – field and laboratory observations. *Polar Biology* **16**, 209–219.

Hirche, H.-J., Baumann, M.E.M., Kattner, G. and Gradinger, R. (1991). Plankton distribution and the impact of copepod grazing on primary production in Fram Strait, Greenland Sea. *Journal of Marine Systems* **2**, 477–494.

Hirche, H.-J., Hagen, W., Mumm, N. and Richter, C. (1994). The northeast water polynya, Greenland Sea. III. Meso- and macrozooplankton distribution and production of dominant herbivorous copepods during spring. *Polar Biology* **14**, 491–503.

Hiromi, J. and Ueda, H. (1987). Planktonic calanoid copepod *Sinocalanus sinensis*

(Centropagidae) from estuaries of Ariake-kai, Japan, with a preliminary note on the mode of introduction from China. *Proceedings of the Japan Society of Systematic Zoology* **35**, 19–26.

Hiromi, J., Kadota, S. and Takano, H. (1985). Diatom infestation of marine copepods (review). *Bulletin of the Tokai Regional Fisheries Research Laboratory* **117**, 37–45.

Hirose, E., Toda, H., Saito, Y. and Watanabe, H. (1992). Formation of the multi-layered fertilization envelope in the embryo of *Calanus sinicus* Brodsky (Copepoda: Calanoida). *Journal of Crustacean Biology* **12**, 186–192.

Hirota, R. (1981). Dry weight and chemical composition of the important zooplankton in the Setonaikai (Inland Sea of Japan). *Bulletin of Plankton Society of Japan* **28**, 19–24.

Hitchcock, G.L. (1982). A comparative study of the size dependent organic composition of marine diatoms and dinoflagellates. *Journal of Plankton Research* **4**, 363–377.

Ho, J.-s. (1990). Phylogenetic analysis of copepod orders. *Journal of Crustacean Biology* **10**, 528–536.

Ho, J.-s. (1994). Copepod phylogeny: a reconsideration of Huys and Boxshall's "parsimony versus homology". *Hydrobiologia* **292/293**, 31–39.

Ho, J.-s. and Perkins, P.S. (1985). Symbionts of marine Copepoda: an overview. *Bulletin of Marine Science* **37**, 586–598.

Hobson, K.A. and Welch, H.E. (1992). Determination of trophic relationships within a high Arctic marine food web using $\delta^{13}C$ and $\delta^{15}N$ analysis. *Marine Ecology Progress Series* **84**, 9–18.

Hodgkin, E.P. and Rippingale, R.J. (1971). Interspecies conflict in estuarine copepods. *Limnology and Oceanography* **16**, 573–576.

Hofmann, E.E., Klinck, J.M. and Paffenhöfer, G.-A. (1981). Concentrations and vertical fluxes of zooplankton fecal pellets on a continental shelf. *Marine Biology* **61**, 327–335.

Holdway, P. and Maddock, L. (1983). A comparative survey of neuston: geographical and temporal distribution patterns. *Marine Biology* **76**, 263–270.

Honjo, S. and Roman, M.R. (1978). Marine copepod fecal pellets: production, preservation and sedimentation. *Journal of Marine Research* **36**, 45–57.

Hopkins, C.C.E. (1977). The relationship between maternal body size and clutch size, development time and egg mortality in *Euchaeta norvegica* (Copepoda: Calanoida) from Loch Etive, Scotland. *Journal of the Marine Biological Association of the United Kingdom* **57**, 723–733.

Hopkins, C.C.E. (1978). The male genital system, and spermatophore production and function in *Euchaeta norvegica* Boeck (Copepoda: Calanoida). *Journal of Experimental Marine Biology and Ecology* **35**, 197–231.

Hopkins, C.C.E. (1982). The breeding biology of *Euchaeta norvegica* (Boeck) (Copepoda: Calanoida) in Loch Etive, Scotland: assessment of breeding intensity in terms of seasonal cycles in the sex ratio, spermatophore attachment, and egg-sac production. *Journal of Experimental Marine Biology and Ecology* **60**, 91–102.

Hopkins, C.C.E. and Evans, R.A. (1979). Diurnal and horizontal variations in a zooplankton sound scattering layer. *In* "Cyclic Phenomena in Marine Plants and Animals" (E. Naylor and R.G. Hartnoll, eds), pp. 375–382. Pergamon Press, Oxford.

Hopkins, C.C.E. and Machin, D. (1977). Patterns of spermatophore distribution and placement in *Euchaeta norvegica* (Copepoda: Calanoida). *Journal of the Marine Biological Association of the United Kingdom* **57**, 113–131.

Hopkins, C.C.E., Mauchline, J. and McLusky, D.S. (1978). Structure and function of the fifth pair of pleopods of male *Euchaeta norvegica* (Copepoda: Calanoida). *Journal of the Marine Biological Association of the United Kingdom* **58**, 631–637.

Hopkins, C.C.E., Grotnes, P.E. and Eliassen, J.-E. (1989). Organization of a fjord community at 70°N: the pelagic food web in Balsfjord, northern Norway. *Rapports et Procès-verbaux des Réunions Conseil International pour l'Exploration de la Mer* **188**, 146–153.

Hopkins, T.L. (1982). The vertical distribution of zooplankton in the eastern Gulf of Mexico. *Deep-Sea Research* **29A**, 1069–1083.

Hopkins, T.L. (1985). Food web of an Antarctic midwater ecosystem. *Marine Biology* **89**, 197–212.

Hopkins, T.L. (1987). Midwater food web in McMurdo Sound, Ross Sea, Antarctica. *Marine Biology* **96**, 93–106.

Hopkins, T.L. and Torres, J.J. (1988). The zooplankton community in the vicinity of the ice edge, western Weddell Sea, March 1986. *Polar Biology* **9**, 79–87.

Hopkins, T.L. and Torres, J.J. (1989). Midwater food web in the vicinity of a marginal ice zone in the western Weddell Sea. *Deep-Sea Research* **36A**, 543–560.

Hopkins, T.L., Milliken, D.M., Bell, L.M., McMichael, E.J., Heffernan, J.J. and Cano, R.V. (1981). The landward distribution of oceanic plankton and micronekton over the west Florida continental shelf as related to their vertical distribution. *Journal of Plankton Research* **3**, 645–658.

Hopkins, T.L., Lancraft, T.M., Torres, J.J. and Donnelly, J. (1993). Community structure and trophic ecology of zooplankton in the Scotia Sea marginal ice zone in winter (1988). *Deep-Sea Research* **40A**, 81–105.

Hoshiai, T. and Tanimura, A. (1986). Sea ice meiofauna at Syowa Station, Antarctica. *Memoirs of National Institute of Polar Research*, Special Issue **44**, 118–124.

Hoshiai, T., Tanimura, A. and Watanabe, K. (1987). Ice algae as food of an Antarctic ice-associated copepod, *Paralabidocera antarctica* (I.C. Thompson). *Proceedings of the NIPR Symposium on Polar Biology* **1**, 105–111.

Hoshiai, T., Tanimura, A. and Kudoh, S. (1996). The significance of autumnal ice biota in the ecosystem of ice-covered polar seas. *Proceedings of the NIPR Symposium on Polar Biology* **9**, 27–34.

Houde, S.E.L. and Roman, M.R. (1987). Effects of food quality on the functional ingestion response of the copepod *Acartia tonsa*. *Marine Ecology Progress Series* **40**, 69–77.

Hough, A.R. and Naylor, E. (1991). Field studies on retention of the planktonic copepod *Eurytemora affinis* in a mixed estuary. *Marine Ecology Progress Series* **76**, 115–122.

Hough, A.R. and Naylor, E. (1992). Endogenous rhythms of circatidal swimming activity in the estuarine copepod *Eurytemora affinis* (Poppe). *Journal of Experimental Marine Biology and Ecology* **161**, 27–32.

Howse, H.D., Woodmansee, R.A., Hawkins, W.E. and Perry, H.M. (1975). Ultrastructure of the heart of the copepod *Anomalocera ornata* Sutcliffe. *Transactions of the American Microscopical Society* **94**, 1–23.

Howse, H.D., Hawkins, W.E. and Perry, H.M. (1992). A note on the fine structure of myoskeletal junctions in *Acartia tonsa* Dana (Copepoda, Calanoida). *Gulf Research Reports* **8**, 431–434.

Huang, C., Uye, S. and Onbé, T. (1992). Ontogenetic diel vertical migration of the planktonic copepod *Calanus sinicus* in the Inland Sea of Japan. II. Late fall and

early spring. *Marine Biology* **113**, 391–400.
Huang, C., Uye, S. and Onbé, T. (1993). Geographic distribution, seasonal life cycle, biomass and production of a planktonic copepod *Calanus sinicus* in the Inland Sea of Japan and its neighboring Pacific Ocean. *Journal of Plankton Research* **15**, 1229–1246.
Hülsemann, K. (1966). A revision of the genus *Lucicutia* (Copepoda: Calanoida) with a key to its species. *Bulletin of Marine Science, Gulf and Caribbean* **16**, 702–747.
Hülsemann, K. (1967). Redescription of *Euaugaptilus mixtus* (Sars) (Copepoda, Calanoida). *Crustaceana* **12**, 163–166.
Hülsemann, K. (1985a). Two species of *Drepanopus* Brady (Copepoda Calanoida) with discrete ranges in the Southern Hemisphere. *Journal of Plankton Research* **7**, 909–925.
Hülsemann, K. (1985b). A new species of *Neoscolecithrix* Canu (Copepoda Calanoida) in Antarctic waters with remarks on the genus. *Polar Biology* **5**, 55–62.
Hülsemann, K. (1988a). *Tortanus sheni*, new name, replacement name for *Tortanus denticulatus* Shen and Lee, 1963 (Copepoda: Calanoida). *Journal of Crustacean Biology* **8**, 656.
Hülsemann, K. (1988b). *Pleuromamma princeps* Scott, 1894 (currently *Gaussia princeps*; Crustacea, Copepoda): proposed conservation of the specific name. *Bulletin of Zoological Nomenclature* **45**, 188–190.
Hülsemann, K. (1989). Case 2666. *Lucicutia* Giesbrecht in Giesbrecht & Schmeil, 1898: proposed conservation, and *Pseudaugaptilus longiremis* Sars, 1907: proposed conservation of the specific name (both Crustacea, Copepoda). *Bulletin of Zoological Nomenclature* **46**, 97–100.
Hülsemann, K. (1991a). The copepodid stages of *Drepanopus forcipatus* Giesbrecht, with notes on the genus and a comparison with other members of the family Clausocalanidae (Copepoda Calanoida). *Helgoländer wissenschaftliche Meeresuntersuchungen* **45**, 199–224.
Hülsemann, K. (1991b). *Calanus euxinus*, new name, a replacement name for *Calanus ponticus* Karavaev, 1894 (Copepoda: Calanoida). *Proceedings of the Biological Society of Washington* **104**, 620–621.
Hülsemann, K. (1991c). Tracing homologies in appendages during ontogenetic development of calanoid copepods. *Bulletin of Plankton Society of Japan, Special Volume*, 105–114.
Hülsemann, K. (1994). *Calanus sinicus* Brodsky and *C. jaschnovi*, nom. nov. (Copepoda: Calanoida) of the North-west Pacific Ocean: a comparison, with notes on the integumental pore pattern in *Calanus* s. str. *Invertebrate Taxonomy* **8**, 1461–1482.
Hülsemann, K. and Fleminger, A. (1975). Some aspects of copepodid development in the genus *Pontellina* Dana (Copepoda: Calanoida). *Bulletin of Marine Science* **25**, 174–185.
Hülsemann, K. and Fleminger, A. (1990). Taxonomic value of minute structures on the genital segment of *Pontellina* females (Copepoda: Calanoida). *Marine Biology* **105**, 99–108.
Humes, A.G. (1994). How many copepods? *Hydrobiologia* **292/293**, 1–7.
Humes, A.G. and Smith, W.L. (1974). *Ridgewayia fosshageni* n.sp. (Copepoda; Calanoida) associated with an actiniarian in Panama, with observations on the nature of the association. *Caribbean Journal of Science* **14**, 125–139.
Huntley, M. (1982). Yellow water in La Jolla Bay, California, July 1980. Suppression

of zooplankton grazing. *Journal of Experimental Marine Biology and Ecology* **63**, 81–91.
Huntley, M. (1988). Feeding biology of *Calanus*: a new perspective. *Hydrobiologia* **167/168**, 83–99.
Huntley, M. and Boyd, C. (1984). Food-limited growth of marine zooplankton. *The American Naturalist* **124**, 45–478.
Huntley, M. and Brooks, E.R. (1982). Effects of age and food availability on diel vertical migration of *Calanus pacificus*. *Marine Biology* **71**, 23–31.
Huntley, M. and Escritor, F. (1991). Dynamics of *Calanoides acutus* (Copepoda: Calanoida) in Antarctic coastal waters. *Deep-Sea Research* **38A**, 1145–1167.
Huntley, M.E. and Escritor, F. (1992). Ecology of *Metridia gerlachei* Giesbrecht in the western Bransfield Strait, Antarctica. *Deep-Sea Research* **39A**, 1027–1055.
Huntley, M.E. and Lopez, M.D.G. (1992). Temperature dependent production of marine copepods: a global synthesis. *The American Naturalist* **140**, 201–242.
Huntley, M.E. and Niiler, P.P. (1995). Physical control of population dynamics in the Southern Ocean. *ICES Journal of Marine Science* **52**, 457–468.
Huntley, M.E. and Nordhausen, W. (1995). Ammonium cycling by Antarctic zooplankton in winter. *Marine Biology* **121**, 457–467.
Huntley, M.E., Barthel, K.-G. and Star, J.L. (1983). Particle rejection by *Calanus pacificus*: discrimination between similarly sized particles. *Marine Biology* **74**, 151–160.
Huntley, M., Sykes, P., Rohan, S. and Marin, V. (1986). Chemically-mediated rejection of dinoflagellate prey by the copepods *Calanus pacificus* and *Paracalanus parvus*: mechanism, occurrence and significance. *Marine Ecology Progress Series* **28**, 105–120.
Huntley, M.E., Marin, V. and Escritor, F. (1987a). Zooplankton grazers as transformers of ocean optics: a dynamic model. *Journal of Marine Research* **45**, 911–945.
Huntley, M., Tande, K. and Eilertsen, H.C. (1987b). On the trophic fate of *Phaeocystis pouchetii* (Hariot). II. Grazing rates of *Calanus hyperboreus* (Krøyer) on diatoms and different size categories of *Phaeocystis pouchetii*. *Journal of Experimental Marine Biology and Ecology* **110**, 197–212.
Huntley, M.E., Ciminiello, P. and Lopez, M.D.G. (1987c). Importance of food quality in determining development and survival of *Calanus pacificus* (Copepoda: Calanoida). *Marine Biology* **95**, 103–113.
Huntley, M.E., Zhou, M. and Lopez, M.D.G. (1994). *Calanoides acutus* in Gerlache Strait, Antarctica II. Solving an inverse problem in population dynamics. *Deep-Sea Research II* **41A**, 209–227.
Hure, J. (1980). Vertical and horizontal distribution of oceanic copepods in the Adriatic Sea. *Acta Adriatica* **21**, 387–400.
Hure, J. and Scotto di Carlo, B. (1969a). Ripartizione quantitativa e distribuzione verticale dei Copepodi pelagici di profondità su una stazione nel Mar Tirreno ed una nell'Adriatico Meridionale. *Pubblicazioni della Stazione Zoologica di Napoli* **37**, 51–83.
Hure, J. and Scotto di Carlo, B. (1969b). Diurnal vertical migration of some deep-water copepods in the southern Adriatic (east Mediterranean). *Pubblicazioni della Stazione Zoologica di Napoli* **37**, 581–598.
Hure, J. and Scotto di Carlo, B. (1974). New patterns of diurnal vertical migration of some deep-water copepods in the Tyrrhenian and Adriatic Seas. *Marine Biology* **28**, 179–184.
Hutchings, L. (1985). Vertical distribution of mesozooplankton at an active

upwelling site in the Southern Benguela Current, December 1969. *Republic of South Africa, Sea Fisheries Research Institute, Investigational Report* **129**, 1–67.

Hutchings, L., Pillar, S.C. and Verheye, H.M. (1991). Estimates of standing stock, production and consumption of meso- and macrozooplankton in the Benguela ecosystem. *South African Journal of Marine Science* **11**, 499–512.

Hutchings, L., Verheye, H.M., Mitchell-Innes, B.A., Peterson, W.T., Huggett, J.A. and Painting, S.J. (1995). Copepod production in the southern Benguela system. *ICES Journal of Marine Science* **52**, 439–455.

Hutchins, D.A. and Bruland, K.W. (1994). Grazer-mediated regeneration and assimilation of Fe, Zn and Mn from planktonic prey. *Marine Ecology Progress Series* **110**, 259–269.

Huys, R. and Boxshall, G.A. (1991). Copepod Evolution. The Ray Society, London. 468 pp.

Hwang, J.-S. and Strickler, J.R. (1994). Effects of periodic turbulent events upon escape responses of a calanoid copepod *Centropages hamatus*. *Bulletin of Plankton Society of Japan* **41**, 117–130.

Hwang, J.-S. and Turner, J.T. (1995). Behaviour of cyclopoid, harpacticoid, and calanoid copepods from coastal waters of Taiwan. *P.S.Z.N.I.: Marine Ecology* **16**, 207–216.

Hwang, J.-S., Turner, J.T., Costello, J.H., Coughlin, D.J. and Strickler, J.R. (1993). A cinematographic comparison of behavior by the calanoid copepod *Centropages hamatus* Lilljeborg: tethered versus free-swimming animals. *Journal of Experimental Marine Biology and Ecology* **167**, 277–288.

Hwang, J.-S., Costello, J.H. and Strickler, J.R. (1994). Copepod grazing in turbulent flow: elevated foraging behavior and habituation of escape responses. *Journal of Plankton Research* **16**, 421–431.

Ianora, A. (1990). The effect of reproductive condition on egg production rates in the planktonic copepod *Centropages typicus*. *Journal of Plankton Research* **12**, 885–890.

Ianora, A. and Buttino, I. (1990). Seasonal cycles in population abundances and egg production rates in the planktonic copepods *Centropages typicus* and *Acartia clausi*. *Journal of Plankton Research* **12**, 473–481.

Ianora, A. and Poulet, S.A. (1993). Egg viability in the copepod *Temora stylifera*. *Limnology and Oceanography* **38**, 1615–1626.

Ianora, A. and Santella, L. (1991). Diapause embryos in the neustonic copepod *Anomalocera patersoni*. *Marine Biology* **108**, 387–394.

Ianora, A. and Scotto di Carlo, B. (1988). Observations on egg production rates and seasonal changes in the internal morphology of Mediterranean populations of *Acartia clausi* and *Centropages typicus*. *Hydrobiologia* **167/168**, 247–253.

Ianora, A., Scotto di Carlo, B. and Mascellaro, P. (1989). Reproductive biology of the planktonic copepod *Temora stylifera*. *Marine Biology* **101**, 187–194.

Ianora, A., Scotto di Carlo, B., Mazzocchi, M.G. and Mascellaro, P. (1990). Histomorphological changes in the reproductive condition of parasitized marine planktonic copepods. *Journal of Plankton Research* **12**, 249–258.

Ianora, A., Mazzocchi, M.G. and Grottoli, R. (1992a). Seasonal fluctuations in fecundity and hatching success in the planktonic copepod *Centropages typicus*. *Journal of Plankton Research* **14**, 1483–1494.

Ianora, A., Miralto, A. and Vanucci, S. (1992b). The surface attachment structure: a unique type of integumental formation in neustonic copepods. *Marine Biology* **113**, 401–407.

Ianora, A., Poulet, S.A. and Miralto, A. (1995). A comparative study of the inhibitory effect of diatoms on the reproductive biology of the copepod *Temora stylifera*. *Marine Biology* **121**, 533–539.

Ianora, A., Poulet, S.A., Miralto, A. and Grottoli, R. (1996). The diatom *Thalassiosira rotula* affects reproductive success in the copepod *Acartia clausi*. *Marine Biology* **125**, 279–286.

Ibanez, F. and Boucher, J. (1987). Anisotropie des populations zooplanctoniques dans la zone frontale de Mer Ligure. *Oceanologica Acta* **10**, 205–216.

Ikeda, T. (1970). Relationship between respiration rate and body size in marine plankton animals as a function of the temperature of habitat. *Bulletin of the Faculty of Fisheries Hokkaido University* **21**, 91–112.

Ikeda, T. (1974). Nutritional ecology of marine zooplankton. *Memoirs of Faculty of Fisheries Hokkaido University* **22**, 1–97.

Ikeda, T. (1977). The effect of laboratory conditions on the extrapolation of experimental measurements to the ecology of marine zooplankton. IV. Changes in respiration and excretion rates of boreal zooplankton species maintained under fed and starved conditions. *Marine Biology* **41**, 241–252.

Ikeda, T. (1988). Metabolism and chemical composition of crustaceans from the Antarctic mesopelagic zone. *Deep-Sea Research* **35A**, 1991–2002.

Ikeda, T. and Hirakawa, K. (1996). Early development and estimated life cycle of the mesopelagic copepod *Pareuchaeta elongata* in the southern Japan Sea. *Marine Biology* **126**, 261–270.

Ikeda, T. and Mitchell, A.W. (1982). Oxygen uptake, ammonia excretion and phosphate excretion by krill and other Antarctic zooplankton in relation to their body size and chemical composition. *Marine Biology* **71**, 283–298.

Ikeda, T. and Skjoldal, H.R. (1989). Metabolism and elemental composition of zooplankton from the Barents Sea during early Arctic summer. *Marine Biology* **100**, 173–183.

Ikeda, T., Hirakawa, K. and Kajihara, N. (1990). Some characteristics of a coldwater copepod *Calanus cristatus* from regions of the Japan Sea covered by the Tsushima Warm Current. *Bulletin of the Japan Sea National Fisheries Research Institute* **40**, 51–65.

Imabayashi, H. and Endo, T. (1986). Distribution of near-bottom plankton in the Ohta River estuary, Hiroshima Bay, in relation to salinity. *Bulletin of Plankton Society of Japan* **33**, 113–123.

Irigoien, X., Castel, J. and Sautour, B. (1993). *In situ* grazing activity of planktonic copepods in the Gironde estuary. *Cahiers de Biologie marine* **34**, 225–237.

Irigoien, X., Castel, J. and Gasparini, S. (1996). Gut clearance rate a predictor of food limitation situations. Application to two estuarine copepods: *Acartia bifilosa* and *Eurytemora affinis*. *Marine Ecology Progress Series* **131**, 159–163.

Ishii, H. (1990). *In situ* feeding rhythms of herbivorous copepods and the effect of starvation. *Marine Biology* **105**, 91–98.

Ishimaru, T., Nishida, S. and Marumo, R. (1988). Food size selectivity of zooplankton evaluated from the occurrence of coccolithophorids in the guts. *Journal of Plankton Society of Japan* **35**, 101–114.

Itoh, K. (1970). A consideration of feeding habits of planktonic copepods in relation to the structure of their oral parts. *Bulletin of Plankton Society of Japan* **17**, 1–10.

Ivanova, M.V. (1973). Growth patterns of copepod crustaceans. *Hydrobiological Journal* **9**, 15–21.

Iwasaki, H. and Kamiya, S. (1977). Cultivation of marine copepod *Pseudodiaptomus marinus* Sato. *Bulletin of Plankton Society of Japan* **24**, 44–54.

Iwasaki, H., Katoh, H. and Fujiyama, T. (1977). Cultivation of marine copepod, *Acartia clausi*, Giesbrecht. I. Factors affecting the generation time and egg production. *Bulletin of Plankton Society of Japan* **24**, 55–61.
Jacobs, J. (1961). Laboratory cultivation of the marine copepod *Pseudodiaptomus coronatus*. *Limnology and Oceanography* **6**, 443–446.
Jacobsen, T.R. and Azam, F. (1984). Role of bacteria in copepod fecal pellet decomposition: colonization, growth rates and mineralization. *Bulletin of Marine Science* **35**, 495–502.
Jacoby, C.A. and Greenwood, J.G. (1988). Spatial, temporal, and behavioral patterns of emergence of zooplankton in the lagoon of Heron Reef, Great Barrier Reef, Australia. *Marine Biology* **97**, 309–328.
Jacoby, C.A. and Greenwood, J.G. (1989). Emergent zooplankton in Moreton Bay, Queensland, Australia: seasonal, lunar, and diel patterns in emergence and distribution with respect to substrata. *Marine Ecology Progress Series* **51**, 131–154.
Jacoby, C.A. and Greenwood, J.G. (1991). Species-specific variations in emergence of coexisting *Stephos* and *Pseudodiaptomus* (Copepoda: Calanoida). *Bulletin of Plankton Society of Japan*, Special Volume, 405–418.
Jacoby, C.A. and Youngbluth, M.J. (1983). Mating behavior in three species of *Pseudodiaptomus* (Copepoda: Calanoida). *Marine Biology* **76**, 77–86.
Jacques, F. (1989). The setal system of crustaceans: types of setae, groupings, and fundamental morphology. In "Functional Morphology of Feeding and Grooming" (B.E. Felgenhauer, L. Watling and A.B. Thistle, eds), pp. 1–13. A.A. Balkema, Rotterdam.
Jansá, J. and Vives, F. (1992). Espectro de dimensiones de la fracción mesoplanctónica de la comunidad de copépodos del mar Balear. *Boletin Instituto Español de Oceanografia* **8**, 263–270.
Jaume, D. and Boxshall, G.A. (1995a). A new species of *Exumella* (Copepoda: Calanoida: Ridgewayiidae) from anchialine caves in the Mediterranean. *Sarsia* **80**, 93–105.
Jaume, D. and Boxshall, G.A. (1995b). *Stygocyclopia balearica*, a new genus and species of calanoid copepod (Pseudocyclopiidae) from anchihaline caves in the Balearic Islands (Mediterranean). *Sarsia* **80**, 213–222.
Jeffries, H.P. (1976). Succession of two *Acartia* species in estuaries. *Limnology and Oceanography* **7**, 354–364.
Jenkinson, I.R. and Wyatt, T. (1992). Selection and control of Deborah numbers in plankton ecology. *Journal of Plankton Research* **14**, 1697–1721.
Jensen, J.W., Nöst, T. and Stokland, Ø. (1985). The invertebrate fauna of a small fjord subject to wide ranges of salinity and oxygen content. *Sarsia* **70**, 33–43.
Jeong, H.J. (1994). Predation effects of the calanoid copepod *Acartia tonsa* on a population of the heterotrophic dinoflagellate *Ptoroperidinium* cf. *divergens* in the presence of co-occurring red-tide dinoflagellate prey. *Marine Ecology Progress Series* **111**, 87–97.
Jerling, H.L. and Wooldridge, T.H. (1991). Population dynamics and estimates of production for the calanoid copepod *Pseudodiaptomus hessei* in a warm temperate estuary. *Estuarine, Coastal and Shelf Science* **33**, 121–135.
Jerling, H.L. and Wooldridge, T.H. (1992). Lunar influence on distribution of a calanoid copepod in the water column of a shallow, temperate estuary. *Marine Biology* **112**, 309–312.
Johannessen, P.J. (1976). A new species and a new record of calanoid copepods from western Norway. *Sarsia* **60**, 19–24.

Johnson, J.K. (1980). Effects of temperature and salinity on production and hatching of dormant eggs of *Acartia californiensis* (Copepoda) in an Oregon estuary. *Fishery Bulletin, National Oceanic and Atmospheric Administration of the United States* **77**, 567–584.
Johnson, M.W. (1934a). The life history of the copepod *Tortanus discaudatus* (Thompson and Scott). *Biological Bulletin, Marine Biological Laboratory, Wood's Hole, Mass.*, **67**, 182–200.
Johnson, M.W. (1934b). The developmental stages of the copepod *Epilabidocera amphitrites* McMurrich. *Biological Bulletin, Marine Biological Laboratory, Wood's Hole, Mass.*, **67**, 466–483.
Johnson, M.W. (1935). The developmental stages of *Labidocera*. *Biological Bulletin, Marine Biological Laboratory, Wood's Hole, Mass.*, **68**, 397–421.
Johnson, M.W. (1936). *Pachyptilus pacificus* and *Centraugaptilus porcellus*, two new copepods from the North Pacific. *Bulletin of the Scripps Institution of Oceanography, Technical Series* **4**, 65–70.
Johnson, M.W. (1937). The developmental stages of the copepod *Eucalanus elongatus* Dana var. *bungii* Giesbrecht. *Transactions of the American Microscopical Society* **56**, 79–98.
Johnson, M.W. (1948). The postembryonic development of the copepod *Pseudodiaptomus euryhalinus* Johnson, and its phylogenetic significance. *Transactions of the American Microscopical Society* **67**, 319–330.
Johnson, M.W. (1958). *Bathycalanus sverdrupi*, n.sp., a copepod crustacean from great depths in the Pacific Ocean. *Proceedings of the Californian Academy of Science* **29** (6), 257–265.
Johnson, M.W. (1961). On zooplankton of some Arctic coastal lagoons of northwestern Alaska, with description of a new species of *Eurytemora*. *Pacific Science* **15**, 311–323.
Johnson, M.W. (1965). The nauplius larva of *Pontellopsis occidentalis* Esterly (Copepoda, Calanoida). *Transactions of the American Microscopical Society* **84**, 43–48.
Johnson, M.W. (1967). Some observations on the hatching of *Tortanus discaudatus* eggs subjected to low temperatures. *Limnology and Oceanography* **12**, 405–410.
Jónasdóttir, S.H. (1989). Effects of food concentration on egg-production rates of two species of *Pseudocalanus*: laboratory observations. *Journal of Experimental Marine Biology and Ecology* **130**, 33–43.
Jónasdóttir, S.H. (1994). Effects of food quality on the reproductive success of *Acartia tonsa* and *Acartia hudsonica*: laboratory observations. *Marine Biology* **121**, 67–81.
Jónasdóttir, S.H. and Kiørboe, T. (1996). Copepod recruitment and food composition: do diatoms affect hatching success? *Marine Biology* **125**, 743–750.
Jónasdóttir, S.H., Fields, D. and Pantoja, S. (1995). Copepod egg production in Long Island Sound, USA, as a function of the chemical composition of seston. *Marine Ecology Progress Series* **119**, 87–98.
Jones, E.C. (1966a). A new record of *Pseudodiaptomus marinus* Sato (Copepoda, Calanoida) from brackish waters of Hawaii. *Crustaceana* **10**, 316–317.
Jones, E.C. (1966b). The general distribution of species of the calanoid copepod family Candaciidae in the Indian Ocean with new records of species. *In* "Symposium on Crustacea", Vol. 1, pp. 399–405. Marine Biological Association of India.
Jonsson, Per R. and Tiselius, P. (1990). Feeding behaviour, prey detection and

capture efficiency of the copepod *Acartia tonsa* feeding on planktonic ciliates. *Marine Ecology Progress Series* **60**, 35–44.

Jossi, J.W. and Goulet, J.R. (1993). Zooplankton trends: US north-east shelf ecosystem and adjacent regions differ from north-east Atlantic and North Sea. *ICES Journal of Marine Science* **50**, 303–313.

Jouffre, D., Lam-Hoai, T., Millet, B. and Amanieu, M. (1991). Structuration spatiale des peuplements zooplanctoniques et fonctionnement hydrodynamique en milieu lagunaire. *Oceanologica Acta* **14**, 489–504.

Judkins, D.C. (1980). Vertical distribution of zooplankton in relation to the oxygen minimum off Peru. *Deep-Sea Research* **27A**, 475–487.

Judkins, D.C., Wirick, C.D. and Esaias, W.E. (1980). Composition, abundance and distribution of zooplankton in the New York Bight, September 1974–September 1975. *Fishery Bulletin, National Oceanic and Atmospheric Administration of the United States* **77**, 669–683.

Kaartvedt, S. and Svendsen, H. (1995). Effect of freshwater discharge, intrusions of coastal water, and bathymetry on zooplankton distribution in a Norwegian fjord system. *Journal of Plankton Research* **17**, 493–511.

Kahru, M., Elken, J., Kotta, I., Simm, M. and Vilbaste, K. (1984). Plankton distributions and processes across a front in the open Baltic Sea. *Marine Ecology Progress Series* **20**, 101–111.

Kang, Y.-S. (1996). Redescription of *Paracalanus parvus* and *P. indicus* (Copepoda: Paracalanidae) recorded in the Korean waters. *Journal of the Korean Fisheries Society* **29**, 409–413.

Kankaala, P. and Johansson, S. (1986). The influence of individual variation on length-biomass regressions in three crustacean zooplankton species. *Journal of Plankton Research* **8**, 1027–1038.

Kann, L.M. and Wishner, K. (1996). Genetic population structure of the copepod *Calanus finmarchicus* in the Gulf of Maine: allozyme and amplified mitochondrial DNA variation. *Marine Biology* **125**, 65–75.

Karanas, J.J., Dyke, H.V. and Worrest, R.C. (1979). Midultraviolet (UV-B) sensitivity of *Acartia clausii* Giesbrecht (Copepoda). *Limnology and Oceanography* **24**, 1104–1116.

Karanas, J.J., Worrest, R.C. and Dyke, H.V. (1981). Impact of UV-B radiation on the fecundity of the copepod *Acartia clausii*. *Marine Biology* **65**, 125–133.

Karentz, D., McEuen, F.S., Land, M.C. and Dunlap, W.C. (1991). Survey of mycosporine-like amino acid compounds in Antarctic marine organisms: potential protection from ultraviolet exposure. *Marine Biology* **108**, 157–166.

Karlson, K. and Båmstedt, U. (1994). Planktivorous predation on copepods. Evaluation of mandible remains in predator guts as a quantitative estimate of predation. *Marine Ecology Progress Series* **108**, 79–89.

Kasahara, S. and Uye, S. (1979). Calanoid copepod eggs in sea-bottom muds. V. Seasonal changes in hatching of subitaneous and diapause eggs of *Tortanus forcipatus*. *Marine Biology* **55**, 63–68.

Kasahara, S., Uye, S. and Onbé, T. (1974). Calanoid copepod eggs in sea-bottom muds. *Marine Biology* **26**, 167–171.

Kasahara, S., Uye, S. and Onbé, T. (1975a). Calanoid copepod eggs in sea-bottom muds. II. Seasonal cycles of abundance in the populations of several species of copepods and their eggs in the Inland Sea of Japan. *Marine Biology* **31**, 25–29.

Kasahara, S., Onbé, T. and Kamigaki, M. (1975b). Calanoid copepod eggs in sea-bottom muds. III. Effects of temperature, salinity and other factors on the hatching of resting eggs of *Tortanus forcipatus*. *Marine Biology* **31**, 31–35.

Katajisto, T. (1996). Copepod eggs survive a decade in the sediments of the Baltic Sea. *Hydrobiologia* **320**, 153–159.
Katona, S.K. (1970). Growth characteristics of the copepods *Eurytemora affinis* and *E. herdmani* in laboratory cultures. *Helgoländer wissenschaftliche Meeresuntersuchungen* **20**, 373–384.
Katona, S.K. (1971). The developmental stages of *Eurytemora affinis* (Poppe, 1880) (Copepoda, Calanoida) raised in laboratory cultures, including a comparison with the larvae of *Eurytemora americana* Williams, 1906, and *Eurytemora herdmani* Thompson & Scott, 1897. *Crustaceana* **21**, 5–20.
Katona, S.K. (1973). Evidence for sex pheromones in planktonic copepods. *Limnology and Oceanography* **18**, 574–583.
Katona, S.K. (1975). Copulation in the copepod *Eurytemora affinis* (Poppe, 1880). *Crustaceana* **28**, 89–95.
Katona, S.K. and Moodie, C.F. (1969). Breeding of *Pseudocalanus elongatus* in the laboratory. *Journal of the Marine Biological Association of the United Kingdom* **49**, 743–747.
Kattner, G. and Graeve, M. (1991). Wax ester composition of the dominant calanoid copepods of the Greenland Sea/Fram Strait region. *Polar Research* **10**, 479–485.
Kattner, G. and Hagen, W. (1995). Polar herbivorous copepods – different pathways in lipid biosynthesis. *ICES Journal of Marine Science* **52**, 329–335.
Kattner, G. and Krause, M. (1987). Changes in lipids during the development of *Calanus finmarchicus* s.l. from copepodid I to adult. *Marine Biology* **96**, 511–518.
Kattner, G., Hirche, H.J. and Krause, M. (1989). Spatial variability in lipid composition of calanoid copepods from Fram Strait, the Arctic. *Marine Biology* **102**, 473–480.
Kattner, G., Graeve, M. and Hagen, W. (1994). Ontogenetic and seasonal changes in lipid and fatty acid/alcohol compositions of the dominant Antarctic copepods *Calanus propinquus*, *Calanoides acutus* and *Rhincalanus gigas*. *Marine Biology* **118**, 637–644.
Kawamura, A. (1974). Food and feeding ecology in the southern sei whale. *Scientific Reports of the Whales Research Institute* **26**, 25–144.
Kawamura, A. and Hirano, K. (1985). The spatial scale of surface swarms of *Calanus plumchrus* Marukawa observed from consecutive plankton net catches in the northwestern North Pacific. *Bulletin of Marine Science* **37**, 626–633.
Keiyu, A.Y., Yamazaki, H. and Strickler, J.R. (1994). A new modelling approach for zooplankton behaviour. *Deep-Sea Research*, II **41**, 171–184.
Kerambrun, P. (1987). Composition chimique élémentaire (C, H, N) et équivalent énergie d'*Acartia clausi* (Crustacea: Copepoda), espèce importante dans la bioénergétique des écosystèmes côtiers de Méditerranée nord-occidentale. *Marine Biology* **95**, 115–121.
Kern, J.C. and Carey, A.G. (1983). The faunal assemblage inhabiting seasonal sea ice in the nearshore Arctic Ocean with emphasis on copepods. *Marine Ecology Progress Series* **10**, 159–167.
Ketchum, B.H., ed. (1983). "Ecosystems of the World 26. Estuaries and Enclosed Seas." Elsevier Scientific Publishing Company, Amsterdam. 500 pp.
Kimmerer, W.J. (1983). Direct measurement of the production: biomass ratio of the subtropical calanoid copepod *Acrocalanus inermis*. *Journal of Plankton Research* **5**, 1–14.
Kimmerer, W.J. (1984). Spatial and temporal variability in egg production rates of the calanoid copepod *Acrocalanus inermis*. *Marine Biology* **78**, 165–169.

Kimmerer, W.J. (1991). Predatory influences on copepod distributions in coastal waters. *Bulletin of Plankton Society of Japan*, Special Volume, 161–174.
Kimmerer, W.J. and McKinnon, A.D. (1985). A comparative study of the zooplankton in two adjacent embayments, Port Phillip and Westernport Bays, Australia. *Estuarine, Coastal and Shelf Science* **21**, 145–159.
Kimmerer, W.J. and McKinnon, A.D. (1987a). Growth, mortality, and secondary production of the copepod *Acartia tranteri* in Westernport Bay, Australia. *Limnology and Oceanography* **32**, 14–28.
Kimmerer, W.J. and McKinnon, A.D. (1987b). Zooplankton in a marine bay. II. Vertical migration to maintain horizontal distributions. *Marine Ecology Progress Series* **41**, 53–60.
Kimmerer, W.J. and McKinnon, A.D. (1989). Zooplankton in a marine bay. III. Evidence for influence of vertebrate predation on distributions of two common copepods. *Marine Ecology Progress Series* **53**, 21–35.
Kimmerer, W.J. and McKinnon, A.D. (1990). High mortality in a copepod population caused by a parasitic dinoflagellate. *Marine Biology* **107**, 449–452.
Kimoto, K. (1988). Segregation of vertical distribution of calanoid copepod *Acartia omorii* depending on the developmental stages in Shijiki Bay, western Kyushu, Japan. *Bulletin of the Seikai Regional Fisheries Research Laboratory* **66**, 35–39.
Kimoto, K., Uye, S.-i. and Onbé, T. (1986a). Growth characteristics of a brackish-water calanoid copepod *Sinocalanus tenellus* in relation to temperature and salinity. *Bulletin of Plankton Society of Japan* **33**, 43–57.
Kimoto, K., Uye, S.-i. and Onbé, T. (1986b). Egg production of a brackish-water calanoid copepod *Sinocalanus tenellus* in relation to food abundance and temperature. *Bulletin of Plankton Society of Japan* **33**, 133–145.
Kimoto, K., Nakashima, J. and Morioka, Y. (1988). Direct observations of copepod swarm in a small inlet of Kyushu, Japan. *Bulletin of the Seikai Regional Fisheries Research Laboratory* **66**, 41–58.
Kiørboe, T. (1989). Phytoplankton growth rate and nitrogen content: implications for feeding and fecundity in a herbivorous copepod. *Marine Ecology Progress Series* **55**, 229–234.
Kiørboe, T. (1991). Pelagic fisheries and spatio-temporal variability in zooplankton productivity. *Bulletin of Plankton Society of Japan*, Special Volume, 229–249.
Kiørboe, T. and Nielsen, T.G. (1990). Effects of wind stress on vertical water column structure, phytoplankton growth, and productivity of planktonic copepods. In "Trophic Relationships in the Marine Environment" (M. Barnes and R.N. Gibson, eds), pp. 28–40. Aberdeen University Press.
Kiørboe, T. and Nielsen, T.G. (1994). Regulation of zooplankton biomass and production in a temperate, coastal ecosystem. 1. Copepods. *Limnology and Oceanography* **39**, 493–507.
Kiørboe, T. and Sabatini, M. (1994). Reproductive and life cycle strategies in egg-carrying cyclopoid and free-spawning calanoid copepods. *Journal of Plankton Research* **16**, 1353–1366.
Kiørboe, T. and Sabatini, M. (1995). Scaling of fecundity, growth and development in marine planktonic copepods. *Marine Ecology Progress Series* **120**, 285–298.
Kiørboe, T. and Saiz, E. (1995). Planktivorous feeding in calm and turbulent environments, with emphasis on copepods. *Marine Ecology Progress Series* **122**, 135–145.
Kiørboe, T., Møhlenberg, F. and Nicolajsen, H. (1982). Ingestion rate and gut clearance in the planktonic copepod *Centropages hamatus* (Lilljeborg) in relation to food concentration and temperature. *Ophelia* **21**, 181–194.

Kiørboe, T., Møhlenberg, F. and Riisgård, H.U. (1985a). *In situ* feeding rates of planktonic copepods: a comparison of four methods. *Journal of Experimental Marine Biology and Ecology* **88**, 67–81.

Kiørboe, T., Møhlenberg, F. and Hamburger, K. (1985b). Bioenergetics of the planktonic copepod *Acartia tonsa*: relation between feeding, egg production and respiration, and composition of specific dynamic action. *Marine Ecology Progress Series* **26**, 85–97.

Kiørboe, T., Møhlenberg, F. and Tiselius, P. (1988). Propagation of planktonic copepods: production and mortality of eggs. *Hydrobiologia* **167/168**, 219–225.

Kiørboe, T., Kaas, H., Kruse, B., Møhlenberg, F., Tiselius, P. and Ærtebjerg, G. (1990). The structure of the pelagic food web in relation to water column structure in the Skagerrak. *Marine Ecology Progress Series* **59**, 19–32.

Kiørboe, T., Saiz, E. and Viitasalo, M. (1996). Prey switching behaviour in the planktonic copepod *Acartia tonsa*. *Marine Ecology Progress Series* **143**, 65–75.

Kirkwood, J.M. and Burton, H.R. (1987). Three new zooplankton nets designed for under-ice sampling; with preliminary results of collections made from Ellis Fjord, Antarctica during 1985. *Proceedings of the NIPR Symposium on Polar Biology* **1**, 112–122.

Kittredge, J.S., Takahashi, F.T., Lindsey, J. and Lasker, R. (1974). Chemical signals in the sea: marine allelochemicals and evolution. *Fishery Bulletin, National Oceanic and Atmospheric Administration of the United States* **72**, 1–11.

Kivi, K., Kuosa, H. and Tanskanen, S. (1996). An experimental study on the role of crustacean and microprotozoan grazers in the planktonic food web. *Marine Ecology Progress Series* **136**, 59–68.

Klein Breteler, W.C.M. (1980). Continuous breeding of marine pelagic copepods in the presence of heterotrophic dinoflagellates. *Marine Ecology Progress Series* **2**, 229–233.

Klein Breteler, W.C.M. (1982). The life stages of four pelagic copepods (Copepoda: Calanoida), illustrated by a series of photographs. *Netherlands Institute for Sea Research Publication Series* **6**, 1–32.

Klein Breteler, W.C.M. and Gonzalez, S.R. (1982). Influence of cultivation and food concentration on body length of calanoid copepods. *Marine Biology* **71**, 157–161.

Klein Breteler, W.C.M. and Gonzalez, S.R. (1986). Culture and development of *Temora longicornis* (Copepoda, Calanoida) at different conditions of temperature and food. *Syllogeus* **58**, 71–84.

Klein Breteler, W.C.M. and Laan, M. (1993). An apparatus for automatic counting and controlling density of pelagic food particles in cultures of marine organisms. *Marine Biology* **116**, 169–174.

Klein Breteler, W.C.M. and Schogt, N. (1994). Development of *Acartia clausi* (Copepoda, Calanoida) cultured at different conditions of temperature and food. *Hydrobiologia* **292/293**, 469–479.

Klein Breteler, W.C.M., Fransz, H.G. and Gonzalez, S.R. (1982). Growth and development of four calanoid copepod species under experimental and natural conditions. *Netherlands Journal of Sea Research* **16**, 195–207.

Klein Breteler, W.C.M., Schogt, N. and Gonzalez, S.R. (1990). On the role of food quality in grazing and development of life stages, and genetic change of body size during cultivation of pelagic copepods. *Journal of Experimental Marine Biology and Ecology* **135**, 177–189.

Klein Breteler, W.C.M., Schogt, N. and Meer, J. van der (1994). The duration of copepod life stages estimated from stage-frequency data. *Journal of Plankton Research* **16**, 1039–1057.

Klein Breteler, W.C.M., Gonzalez, S.R. and Schogt, N. (1995). Development of *Pseudocalanus elongatus* (Copepoda, Calanoida) cultured at different temperature and food conditions. *Marine Ecology Progress Series* **119**, 99–110.

Kleppel, G.S. (1992). Environmental regulation of feeding and egg production by *Acartia tonsa* off southern California. *Marine Biology* **112**, 57–65.

Kleppel, G.S. (1993). On the diets of calanoid copepods. *Marine Ecology Progress Series* **99**, 183–195.

Kleppel, G.S. and Burkart, C.A. (1995). Egg production and the nutritional environment of *Acartia tonsa*: the role of food quality in copepod nutrition. *ICES Journal of Marine Science* **52**, 297–304.

Kleppel, G.S. and Pieper, R.E. (1984). Phytoplankton pigments in the gut contents of planktonic copepods from coastal waters off southern California. *Marine Biology* **78**, 193–198.

Kleppel, G.S., Willbanks, L. and Pieper, R.E. (1985). Diel variation in body carotenoid content and feeding activity in marine zooplankton assemblages. *Journal of Plankton Research* **7**, 569–580.

Kleppel, G.S., Frazel, D., Pieper, R.E. and Holliday, D.V. (1988a). Natural diets of zooplankton off southern California. *Marine Ecology Progress Series* **49**, 231–241.

Kleppel, G.S., Pieper, R.E. and Trager, G. (1988b). Variability in the gut contents of individual *Acartia tonsa* from waters off Southern California. *Marine Biology* **97**, 185–190.

Knight-Jones, E.W. and Qasim, S.Z. (1967). Responses of Crustacea to changes in hydrostatic pressure. *In* "Symposium on Crustacea", III, pp. 1132–1150. Marine Biological Association of India.

Koehl, M.A.R. and Strickler, J.R. (1981). Copepod feeding currents: food capture at low Reynolds number. *Limnology and Oceanography* **26**, 1062–1073.

Koeller, P.A. (1977). Observations on some bathypelagic copepods living in British Columbia mainland inlets. *Journal of the Oceanographical Society of Japan* **33**, 219–226.

Koeller, P.A. and Littlepage, J.L. (1976). *Azygokeras columbiae*, a new genus and species of marine epibenthic copepod (Calanoida: Aetideidae) from British Columbia. *Journal of the Fisheries Research Board of Canada* **33**, 1547–1552.

Koeller, P.A., Barwell-Clarke, J.E., Whitney, F. and Takahashi, M. (1979). Winter condition of marine plankton populations in Saanich Inlet, B.C., Canada. III. Meso-zooplankton. *Journal of Experimental Marine Biology and Ecology* **37**, 161–174.

Koga, F. (1960a). Developmental stages of nauplius larvae of *Pareuchaeta russelli* (Farran). *Bulletin of the Japanese Society of Scientific Fisheries* **26**, 792–796.

Koga, F. (1960b). The nauplius larvae of *Centropages abdominalis* Sato. *Bulletin of the Japanese Society of Scientific Fisheries* **26**, 877–881.

Koga, F. (1968). On the pelagic eggs of Copepoda. *Journal of the Oceanographical Society of Japan* **24**, 16–20.

Koga, F. (1970). On the nauplius of *Centropages yamadai* Mori, Copepoda. *Journal of the Oceanographical Society of Japan* **26**, 195–202.

Koga, F. (1973). Life history of copepods especially of nauplius larvae ascertained mainly with cultivation of animals. *Bulletin of Plankton Society of Japan* **20**, 30–40.

Koga, F. (1984). The developmental stages of *Temora turbinata* (Copepoda: Calanoida). *Bulletin of Plankton Society of Japan* **31**, 43–52.

Komar, P.D., Morse, A.P., Small, L.F. and Fowler, S.W. (1981). An analysis of sinking

rates of natural copepod and euphausiid fecal pellets. *Limnology and Oceanography* **26**, 172–180.
Koomen, P. (1991). Integumental organs in the oral cavity of *Euchirella messinensis* (Claus, 1863) (Copepoda: Calanoida). *Bulletin of Plankton Society of Japan*, Special Volume, 437–450.
Koomen, P. and Vaupel Klein, J.C. von (1995). The suitability of various mounting media for permanent mounts of small chitinous crustaceans, with special reference to the observation of integumental organs. *Crustaceana* **68**, 428–437.
Koppelmann, R. and Weikert, H. (1992). Full-depth zooplankton profiles over the deep bathyal of the NE Atlantic. *Marine Ecology Progress Series* **86**, 263–272.
Koslow, J.A. (1979). Vertical migrators see the light? *Limnology and Oceanography* **24**, 783–784.
Koslow, J.A. (1983). Zooplankton community structure in the North Sea and northeast Atlantic: development and test of a biological model. *Canadian Journal of Fisheries and Aquatic Sciences* **40**, 1912–1924.
Koslow, J.A. and Ota, A. (1981). The ecology of vertical migration in three common zooplankters in the La Jolla Bight, April–August 1967. *Biological Oceanography* **1**, 107–134.
Koslow, J.A., Perry, R.I., Hurley, P.C.F. and Fournier, R.O. (1989). Structure and interannual variability of the plankton and its environment off southwest Nova Scotia in late spring and early summer. *Canadian Journal of Fisheries and Aquatic Sciences* **46**, Supplement 1, 44–54.
Kosobokova, K.N. (1989). Vertical distribution of plankton animals in the eastern part of the central Arctic Basin. *In* "Explorations of the Fauna of the Seas", 41 (49), Academy of Sciences USSR, pp. 24–32.
Kosobokova, K.N. (1993). Reproduction and fecundity of the White Sea copepod *Calanus glacialis* in experimental conditions. *Okeanologiya* **33**, 392–396.
Kosobokova, K.N. (1994). Reproduction of the calanoid copepod *Calanus propinquus* in the southern Weddell Sea, Antarctica: observations in the laboratory. *Hydrobiologia* **292/293**, 219–227.
Kouwenberg, J.H.M. (1993). Sex ratio of calanoid copepods in relation to population composition in the northwestern Mediterranean. *Crustaceana* **64**, 281–299.
Kovaleva, T.M. (1989). Effect of algae sizes and concentration on the consumption rate in two species of marine copepods. *Ekologiya Morya* **31**, 20–25.
Kovaleva, T.M. and Shadrin, N.V. (1987). The influence of prolonged fasting on the motor activity and fat consumption in *Pseudocalanus elongatus*. *Ekologiya Morya* **27**, 56–60.
Krause, E.P. and Lewis, A.G. (1979). Ontogenetic migration and the distribution of *Eucalanus bungii* (Copepoda; Calanoida) in British Columbia inlets. *Canadian Journal of Zoology* **57**, 2211–2222.
Krause, M. and Kattner, G. (1989). The influence of water exchange on zooplankton dynamics and species development in a south Norwegian fjord. *Journal of Plankton Research* **11**, 85–103.
Krause, M. and Trahms, J. (1982). Vertical distribution of copepods (all developmental stages) and other zooplankton during spring bloom in the Fladen Ground area of the North Sea. *Netherlands Journal of Sea Research* **16**, 217–230.
Krishnaswamy, S. (1948). A preliminary note on the eye of *Centropages furcatus* Dana. *Current Science* **17**, 190–191.
Krishnaswamy, S. (1959). A new species of Copepoda from the Eddystone shell gravel. *Journal of the Marine Biological Association of the United Kingdom* **38**, 543–546.

Krishnaswamy, S., Raymont, J.E.G., Woodhouse, M.A. and Griffin, R.L (1967). Studies on the fine structure of Copepoda. Observations on the fine structure of buttons on the setae of the maxilla and maxilliped of *Centraugaptilus horridus* (Farran). *Deep-Sea Research* **14**, 331–335.
Kršinić, F. (1990). A new type of zooplankton sampler. *Journal of Plankton Research* **12**, 337–343.
Kurbjeweit, F. and Buchholz, C. (1991). Structure and suspected functions of antennular sensilla and pores of three Arctic copepods (*Calanus glacialis, Metridia longa, Paraeuchaeta norvegica*). *Meeresforschung* **33**, 168–182.
Kurbjeweit, F., Gradinger, R. and Weissenberger, J. (1993). The life cycle of *Stephos longipes* – an example for cryopelagic coupling in the Weddell Sea (Antarctica). *Marine Ecology Progress Series* **98**, 255–262.
Kuz'micheva, V.I. (1985). Determination of the individual weight of copepods from body proportions. *Oceanology, Moscow* **25**, 673–676.
Laabir, M., Poulet, S.A., Ianora, A., Miralto, A. and Cueff, A. (1995a). Reproductive response of *Calanus helgolandicus*. II. *In situ* inhibition of embryonic development. *Marine Ecology Progress Series* **129**, 97–105.
Laabir, M., Poulet, S.A. and Ianora, A. (1995b). Measuring production and viability of eggs in *Calanus helgolandicus*. *Journal of Plankton Research* **17**, 1125–1142.
Ladurantaye, R.De, Therriault, J.-C., Lacroix, G. and Côté, R. (1984). Processus advectifs et répartition du zooplancton dans un fjord. *Marine Biology* **82**, 21–29.
Lakkis, S. (1984). On the presence of some rare copepods in the Levantine Basin. *Crustaceana*, Supplement **7**, 286–304.
Lakkis, S. (1994). Coexistence and competition within *Acartia* (Copepoda, Calanoida) congeners from Lebanese coastal water: niche overlap measurements. *Hydrobiologia* **292/293**, 481–490.
Lampert, W. (1989). The adaptive significance of diel vertical migration of zooplankton. *Functional Ecology* **3**, 21–27.
Lampitt, R.S., Noji, T. and Bodungen, B.von (1990). What happens to zooplankton faecal pellets? Implications for material flux. *Marine Biology* **104**, 15–23.
Lance, J. (1964). The salinity tolerances of some estuarine planktonic crustaceans. *Biological Bulletin, Marine Biological Laboratory, Woods Hole, Mass.*, **127**, 108–118.
Landry, M.R. (1975a). Dark inhibition of egg hatching of the marine copepod *Acartia clausi* Giesbr. *Journal of Experimental Marine Biology and Ecology* **20**, 43–47.
Landry, M.R. (1975b). Seasonal temperature effects and predicting development rates of marine copepod eggs. *Limnology and Oceanography* **20**, 434–440.
Landry, M.R. (1975c). The relationship between temperature and the development of life stages of the marine copepod *Acartia clausi* Giesbr. *Limnology and Oceanography* **20**, 854–857.
Landry, M.R. (1978a). Predatory feeding behavior of a marine copepod, *Labidocera trispinosa*. *Limnology and Oceanography* **23**, 1103–1113.
Landry, M.R. (1978b). Population dynamics and production of a planktonic marine copepod, *Acartia clausi*, in a small temperate lagoon on San Juan Island, Washington. *Internationale Revue der gesamten Hydrobiologie* **63**, 77–119.
Landry, M.R. (1980). Detection of prey by *Calanus pacificus*: implications of the first antennae. *Limnology and Oceanography* **25**, 545–549.
Landry, M.R. (1983). The development of marine calanoid copepods with comment on the isochronal rule. *Limnology and Oceanography* **28**, 614–624.

Landry, M.R. and Fagerness, V.L. (1988). Behavioral and morphological influences on predatory interactions among marine copepods. *Bulletin of Marine Science* **43**, 509–529.

Landry, M.R. and Lehner-Fournier, J.M. (1988). Grazing rates and behaviors of *Neocalanus plumchrus*: implications for phytoplankton control in the subarctic Pacific. *Hydrobiologia* **167/168**, 9–19.

Landry, M.R., Fagerness, V.L. and Peterson, W.K. (1991). Ontogenetic patterns in the distribution of *Pseudocalanus* spp. during upwelling off the coast of Washington, USA. *Bulletin of Plankton Society of Japan*, Special Volume, 451–466.

Landry, M.R., Peterson, W.K. and Fagerness, V.L. (1994a). Mesozooplankton grazing in the Southern California Bight. I. Population abundances and gut pigment contents. *Marine Ecology Progress Series* **115**, 55–71.

Landry, M.R., Lorenzen, C.J. and Peterson, W.K. (1994b). Mesozooplankton grazing in the Southern California Bight. II. Grazing impact and particulate flux. *Marine Ecology Progress Series* **115**, 73–85.

Lang, K. (1948). Copepoda "Notodelphyoida" from the Swedish west coast with an outline on the systematics of the Copepoda. *Arkiv für Zoologi* **40 A** (14), 1–36.

Lapota, D., Losee, J.R. and Geiger, M.L. (1986). Bioluminescence displays induced by pulsed light. *Limnology and Oceanography* **31**, 887–889.

Lapota, D., Bowman, T.E. and Losee, J.R. (1988a). Observations on bioluminescence in the nauplius of *Metridia longa* (Copepoda, Calanoida) in the Norwegian Sea. *Crustaceana* **54**, 314–320.

Lapota, D., Galt, C., Losee, J.R., Huddell, H.D., Orzech, J.K. and Nealson, K.H. (1988b). Observations and measurements of planktonic bioluminescence in and around a milky sea. *Journal of Experimental Marine Biology and Ecology* **119**, 55–81.

Lapota, D., Geiger, M.L., Stiffey, A.V., Rosenberger, D.E. and Young, D.K. (1989). Correlations of planktonic bioluminescence with other oceanographic parameters from a Norwegian fjord. *Marine Ecology Progress Series* **55**, 217–227.

Lapota, D., Rosenberger, D.E. and Lieberman, S.H. (1992). Planktonic bioluminescence in the pack ice and the marginal ice zone of the Beaufort Sea. *Marine Biology* **112**, 665–675.

Latz, M.I., Frank, T.M., Bowlby, M.R., Widder, E.A. and Case, J.F. (1987). Variability in flash characteristics of a bioluminescent copepod. *Biological Bulletin, Marine Biological Laboratory, Woods Hole, Mass.*, **173**, 489–503.

Latz, M.I., Frank, T.M. and Case, J.F. (1988). Spectral composition of bioluminescence of epipelagic organisms from the Sargasso Sea. *Marine Biology* **98**, 441–446.

Latz, M.I., Bowlby, M.R. and Case, J.F. (1990). Recovery and stimulation of copepod bioluminescence. *Journal of Experimental Marine Biology and Ecology* **136**, 1–22.

Lawrence, S.A. and Sastry, A.N. (1985). The role of temperature in seasonal variations in egg production by the copepod, *Tortanus discaudatus* (Thompson and Scott), in Narragansett Bay. *Journal of Experimental Marine Biology and Ecology* **91**, 151–167.

Lawrence, S.G., Ahmad, A. and Azam, F. (1993). Fate of particle-bound bacteria ingested by *Calanus pacificus*. *Marine Ecology Progress Series* **97**, 299–307.

Lawson, T.J. (1977). Community interactions and zoogeography of the Indian Ocean Candaciidae (Copepoda: Calanoida). *Marine Biology* **43**, 71–92.

Lawson, T.J. and Grice, G.D. (1970). The developmental stages of *Centropages typicus* Kröyer (Copepoda, Calanoida). *Crustaceana* **18**, 187–208.
Lawson, T.J. and Grice, G.D. (1973). The developmental stages of *Paracalanus crassirostris* Dahl, 1894 (Copepoda, Calanoida). *Crustaceana* **24**, 43–56.
Le Borgne, R. (1982). Zooplankton production in the eastern tropical Atlantic Ocean: net growth efficiency and P:B in terms of carbon, nitrogen and phosphorus. *Limnology and Oceanography* **27**, 681–698.
Le Borgne, R. (1986). The release of soluble end products of metabolism. *In* "The Biological Chemistry of Marine Copepods" (E.D.S. Corner and S.C.M. O'Hara, eds), pp. 109–164. Clarendon Press, Oxford.
Le Borgne, R., Blanchot, J. and Charpy, L. (1989). Zooplankton of Tikehau atoll (Tuamoto archipelago) and its relationship to particulate matter. *Marine Biology* **102**, 341–353.
Lebour, M.V. (1922). The food of plankton organisms. *Journal of the Marine Biological Association of the United Kingdom* **12**, 644–677.
Lee, B.-G. and Fisher, N.S. (1992). Decomposition and release of elements from zooplankton debris. *Marine Ecology Progress Series* **88**, 117–128.
Lee, B.-G. and Fisher, N.S. (1994). Effects of sinking and zooplankton grazing on the release of elements from planktonic debris. *Marine Ecology Progress Series* **110**, 271–281.
Lee, C.M. (1972). Structure and function of the spermatophore and its coupling device in the Centropagidae (Copepoda: Calanoida). *Bulletin of Marine Ecology* **8**, 1–20.
Lee, R.F. and Nevenzel, J.C. (1979). Wax esters in the marine environment: origin, and composition of the wax from Bute Inlet, British Columbia. *Journal of the Fisheries Research Board of Canada* **36**, 1519–1523.
Lee, W.Y. and McAlice, B.J. (1979). Sampling variability of marine zooplankton in a tidal estuary. *Estuarine and Coastal Marine Science* **8**, 565–582.
Le Fèvre, J. (1986). Aspects of the biology of frontal systems. *Advances in Marine Biology* **23**, 163–299.
Légier-Visser, M.F., Mitchell, J.G., Okubo, A. and Fuhrman, J.A. (1986). Mechanoreception in calanoid copepods. A mechanism for prey detection. *Marine Biology* **90**, 529–535.
Lehman, J.T. (1977). On calculating drag characteristics for decelerating zooplankton. *Limnology and Oceanography* **2**, 170–172.
Lenz, P.H. and Yen, J. (1993). Distal setal mechanoreceptors of the first antennae of marine copepods. *Bulletin of Marine Science* **53**, 170–179.
Lewis, A.G. and Ramnarine, A. (1969). Some chemical factors affecting the early development stages of *Euchaeta japonica* (Crustacea: Copepoda: Calanoida). *Journal of the Fisheries Research Board of Canada* **26**, 1347–1362.
Lewis, A.G. and Thomas, A.C. (1986). Tidal transport of planktonic copepods across the sill of a British Columbia fjord. *Journal of Plankton Research* **8**, 1079–1089.
Lewis, J.B. and Boers, J.J. (1991). Patchiness and composition of coral reef demersal zooplankton. *Journal of Plankton Research* **13**, 1273–1289.
Liang, D. and Uye, S. (1996). Population dynamics and production of the planktonic copepods in a eutrophic inlet of the Inland Sea of Japan. II. *Acartia omorii*. *Marine Biology* **125**, 109–117.
Liang, D., Uye, S.-i. and Onbé, T. (1994). Production and loss of eggs in the calanoid copepod *Centropages abdominalis* Sato in Fukuyama Harbor, the Inland Sea of Japan. *Bulletin of Plankton Society of Japan* **41**, 131–142.
Liang, D., Uye, S. and Onbé, T. (1996). Population dynamics and production of the

planktonic copepods in a eutrophic inlet of the Inland Sea of Japan. I. *Centropages abdominalis*. *Marine Biology* **124**, 527–536.
Libourel Houde, S.E. and Roman, M.R. (1987). Effects of food quality on the functional ingestion response of the copepod *Acartia tonsa*. *Marine Ecology Progress Series* **40**, 69–77.
Lillelund, K. and Lasker, R. (1971). Laboratory studies of predation by marine copepods on fish larvae. *Fishery Bulletin, National Oceanic and Atmospheric Administration of the United States* **69**, 655–667.
Lincoln, R.J. (1971). Observations of the effects of changes in hydrostatic pressure and illumination on the behaviour of some planktonic crustaceans. *Journal of Experimental Biology* **54**, 677–688.
Lindahl, O. and Hernroth, L. (1988). Large-scale and long-term variations in the zooplankton community of the Gullmar fjord, Sweden, in relation to advective processes. *Marine Ecology Progress Series* **43**, 161–171.
Lindley, J.A. (1986). Dormant eggs of calanoid copepods in sea-bed sediments of the English Channel and southern North Sea. *Journal of Plankton Research* **8**, 399–400.
Lindley, J.A. (1990). Distribution of overwintering calanoid copepod eggs in sea-bed sediments around southern Britain. *Marine Biology* **104**, 209–217.
Lindley, J.A. (1992). Resistant eggs of the Centropagoidea (Copepoda: Calanoida): a possible preadaptation to colonization of inland waters. *Journal of Crustacean Biology* **12**, 368–371.
Lindley, J.A. and Hunt, H.G. (1989). The distributions of *Labidocera wollastoni* and *Centropages hamatus* in the North Atlantic Ocean and the North Sea in relation to the role of resting eggs in the sediment. *In* "Reproduction, Genetics and Distributions of Marine Organisms" (J.S. Ryland and P.A. Tyler, eds), pp. 407–413. Olsen & Olsen, Fredensborg, Denmark.
Lindley, J.A., Gamble, J.C. and Hunt, H.G. (1995). A change in the zooplankton of the central North Sea (55° to 58°N): a possible consequence of changes in the benthos. *Marine Ecology Progress Series* **119**, 299–303.
Lindquist, A. (1959). Studien über das Zooplankton der Bottensee. I. Nauplien und Copepoditen von *Limnocalanus grimaldii* (de Guerne) (Copepoda, Calanoida). *Report of the Institute of Marine Research, Lysekil, Series Biology* **10**, 1–19.
Lindquist, A. (1961). Untersuchungen an *Limnocalanus* (Copepoda, Calanoida). *Report of the Institute of Marine Research, Lysekil, Series Biology* **13**, 1–124.
Li Shaojing and Fang Jinchuan (1983). The developmental stages of *Labidocera euchaeta* Giesbrecht. *Journal Xiamen University of Natural Sciences* **22**, 375–381.
Li Shaojing and Fang Jinchuan (1984). The developmental stages of *Calanopia thompsoni* A.Scott. *Journal Xiamen University of Natural Sciences* **23**, 392–397.
Li Shaojing, Chen Feng and Wang Guizhong (1989). Studies on the feature of eggs and its hatching rates of some planktonic copepods in Xiamen waters. *Journal of Xiamen University of Natural Sciences* **28**, 542–543.
Lock, A.R. and McLaren, I.A. (1970). The effect of varying and constant temperatures on the size of a marine copepod. *Limnology and Oceanography* **15**, 638–640.
Longhurst, A. and Williams, R. (1992). Carbon flux by seasonal vertical migrant copepods is a small number. *Journal of Plankton Research* **14**, 1495–1509.
Longhurst, A., Sameoto, D. and Herman, A. (1984). Vertical distribution of Arctic zooplankton in summer: eastern Canadian archipelago. *Journal of Plankton Research* **6**, 137–168.

Longhurst, A.R. (1985). The structure and evolution of plankton communities. *Progress in Oceanography* **15**, 1–35.

Lonsdale, D.J., Heinle, D.R. and Siegfried, C. (1979). Carnivorous feeding behavior of the adult calanoid copepod *Acartia tonsa* Dana. *Journal of Experimental Marine Biology and Ecology* **36**, 235–248.

Lonsdale, D.J., Cosper, E.M., Kim, W.-S., Doall, M., Divadeenam, A. and Jónasdóttir, S.H. (1996). Food web interactions in the plankton of Long Island bays, with preliminary observations on brown tide effects. *Marine Ecology Progress Series* **134**, 247–263.

Lopez, M.D.G. (1991). Molting and mortality depend on age and stage in naupliar *Calanus pacificus*: implication for development time of field cohorts. *Marine Ecology Progress Series* **75**, 79–89.

Lopez, M.D.G., Huntley, M.E. and Sykes, P.F. (1988). Pigment destruction by *Calanus pacificus*: impact on the estimation of water column fluxes. *Journal of Plankton Research* **10**, 715–734.

Lopez, M.D.G., Huntley, M.E. and Lovette, J.T. (1993). *Calanoides acutus* in Gerlache Strait, Antarctica. I. Distribution of late copepodite stages and reproduction during spring. *Marine Ecology Progress Series* **100**, 153–165.

Lowe, E. (1935). On the anatomy of a marine copepod, *Calanus finmarchicus* (Gunnerus). *Transactions of the Royal Society of Edinburgh* **58** (3), 561–603.

Lucks, R. (1937). Die Crustaceen und Rotatorien des Messinasees. *Bericht des Westpreussischen botanisch-zoologischen Vereins* **59**, 59–101.

Lutz, R.V., Marcus, N.H. and Chanton, J.P. (1992). Effects of low oxygen concentrations on the hatching and viability of eggs of marine calanoid copepods. *Marine Biology* **114**, 241–247.

Lutz, R.V., Marcus, N.H. and Chanton, J.P. (1994). Hatching and viability of copepod eggs at two stages of embryological development: anoxic/hypoxic effect. *Marine Biology* **119**, 199–204.

McAden, D.C., Greene, G.N. and Baker, W.B. Jr (1987). First record of *Centropages typicus* Krøyer (Copepoda: Calanoida) in the Gulf of Mexico. *Texas Journal of Science* **39**, 290–291.

Macaulay, M.C. and Daly, K.L. (1987). A collapsible opening-closing net for zooplankton sampling through the ice. *Journal of Plankton Research* **9**, 1069–1073.

McClatchie, S. (1992). Time-series measurement of grazing rates of zooplankton and bivalves. *Journal of Plankton Research* **14**, 183–200.

Macdonald, A.G., Gilchrist, I. and Teal, J.M. (1972). Some observations on the tolerance of oceanic plankton to high hydrostatic pressure. *Journal of the Marine Biological Association of the United Kingdom* **52**, 213–223.

McGowan, J.A. (1990). Species dominance-diversity patterns in oceanic communities. In "The Earth in Transition. Patterns and Processes of Biotic Impoverishment" (G.M. Woodwell, ed.), pp. 395–421. Cambridge University Press, Cambridge, New York.

McGowan, J.A. and Walker, P.W. (1979). Structure in the copepod community of the North Pacific central gyre. *Ecological Monographs* **49**, 195–226.

McGowan, J.A. and Walker, P.W. (1985). Dominance and diversity maintenance in an oceanic ecosystem. *Ecological Monographs* **55**, 103–118.

Mackas, D.L. (1984). Spatial autocorrelation of plankton community composition in a continental shelf ecosystem. *Limnology and Oceanography* **29**, 451–471.

Mackas, D.L. and Anderson, E.P. (1986). Small-scale zooplankton community variability in a northern British Columbia fjord system. *Estuarine, Coastal and Shelf Science* **22**, 115–142.

Mackas, D. and Bohrer, R. (1976). Fluorescence analysis of zooplankton gut contents and an investigation of diel feeding patterns. *Journal of Experimental Marine Biology and Ecology* **25**, 77–85.

Mackas, D.L. and Burns, K.E. (1986). Poststarvation feeding and swimming activity in *Calanus pacificus* and *Metridia pacifica*. *Limnology and Oceanography* **31**, 383–392.

Mackas, D.L. and Louttit, G.C. (1988). Aggregation of the copepod *Neocalanus plumchrus* at the margin of the Fraser River plume in the Strait of Georgia. *Bulletin of Marine Science* **43**, 810–824.

Mackas, D.L. and Sefton, H.A. (1982). Plankton species assemblages off southern Vancouver Island: geographic pattern and temporal variability. *Journal of Marine Research* **40**, 1173–1200.

Mackas, D.L., Denman, K.L. and Abbott, M.R. (1985). Plankton patchiness: biology in the physical vernacular. *Bulletin of Marine Science* **37**, 652–674.

Mackas, D.L., Sefton, H., Miller, C.B. and Raich, A. (1993). Vertical habitat partitioning by large calanoid copepods in the oceanic subarctic Pacific during spring. *Progress in Oceanography* **32**, 259–294.

Mackie, G.O. (1985). Midwater macroplankton of British Columbia studied by submersible Pisces IV. *Journal of Plankton Research* **7**, 753–777.

McKinnon, A.D. and Arnott, G. (1985). The developmental stages of *Gladioferens pectinatus* (Brady, 1899) (Copepoda: Calanoida). *New Zealand Journal of Marine and Freshwater Research* **19**, 21–42.

McKinnon, A.D. and Dixon, P. (1994). *Centropages acutus*, a new calanoid copepod from the Fly River estuary, Papua New Guinea. *The Beagle. Records of the Museums and Art Galleries of the Northern Territory* **11**, 9–14.

McKinnon, A.D. and Kimmerer, W.J. (1985). *Paramisophria variabilis*, a new arietellid (Copepoda: Calanoida) from hypersaline waters of Shark Bay, western Australia. *Records of the Australian Museum* **37**, 85–89.

McKinnon, A.D. and Kimmerer, W.J. (1988). A new species of calanoid copepod from Shark Bay, Western Australia. *Records Western Australian Museum* **14**, 171–176.

McKinnon, A.D. and Thorrold, S.R. (1993). Zooplankton community structure and copepod egg production in coastal waters of the central Great Barrier Reef Lagoon. *Journal of Plankton Research* **15**, 1387–1411.

McKinnon, A.D., Kimmerer, W.J. and Benzie, J.A.H. (1992). Sympatric sibling species within the genus *Acartia* (Copepoda, Calanoida): a case study from Westernport and Port Phillip Bays, Australia. *Journal of Crustacean Biology* **12**, 239–259.

McLaren, I.A. (1966). Predicting development rate of copepod eggs. *Biological Bulletin, Marine Biological Laboratory, Woods Hole, Mass.* **131**, 457–469.

McLaren, I.A. (1969). Population and production ecology of zooplankton in Ogac Lake, a landlocked fjord on Baffin Island. *Journal Fisheries Research Board of Canada* **26**, 1485–1559.

McLaren, I.A. (1974). Demographic strategy of vertical migration by a marine copepod. *The American Naturalist* **108**, 91–102.

McLaren, I.A. (1976). Inheritance of demographic and production parameters in the marine copepod *Eurytemora herdmani*. *Biological Bulletin, Marine Biological Laboratory, Woods Hole, Mass.* **151**, 200–213.

McLaren, I.A. (1978). Generation lengths of some temperate marine copepods:

estimation, prediction, and implications. *Journal of the Fisheries Research Board of Canada* **35**, 1330–1342.

McLaren, I.A. (1986). Is "structural" growth of *Calanus* potentially exponential? *Limnology and Oceanography* **31**, 1342–1346.

McLaren, I.A. (1995). Temperature-dependent development in marine copepods: comments on choices of models. *Journal of Plankton Research* **17**, 1385–1390.

McLaren, I.A. and Corkett, C.J. (1978). Unusual genetic variation in body size, development times, oil storage, and survivorship in the marine copepod *Pseudocalanus*. *Biological Bulletin, Marine Biological Laboratory, Woods Hole, Mass.* **155**, 347–359.

McLaren, I.A. and Corkett, C.J. (1981). Temperature-dependent growth and production by a marine copepod. *Canadian Journal of Fisheries and Aquatic Sciences* **38**, 77–83.

McLaren, I.A. and Corkett, C.J. (1984). Singular, mass-specific P/B ratios cannot be used to estimate copepod production. *Canadian Journal of Fisheries and Aquatic Sciences* **41**, 828–830.

McLaren, I.A. and Corkett, C.J. (1986). Life cycles and production of two copepods on the Scotian Shelf, eastern Canada. *Syllogeus* **58**, 362–368.

McLaren, I.A. and Leonard, A. (1995). Assessing the equivalence of growth and egg production of copepods. *ICES Journal of Marine Science* **52**, 397–408.

McLaren, I.A. and Marcogliese, D.J. (1983). Similar nucleus numbers among copepods. *Canadian Journal of Zoology* **61**, 721–724.

McLaren, I.A., Walker, D.A. and Corkett, C.J. (1968). Effects of salinity on mortality and development rate of eggs of the copepod *Pseudocalanus minutus*. *Canadian Journal of Zoology* **46**, 1267–1269.

McLaren, I.A., Corkett, C.J. and Zillioux, E.J. (1969). Temperature adaptation of copepod eggs from the Arctic to the tropics. *Biological Bulletin, Marine Biological Laboratory, Woods Hole, Mass.* **137**, 486–493.

McLaren, I.A., Sevigny, J.-M. and Corkett, C.J. (1988). Body sizes, development rates, and genome sizes among *Calanus* species. *Hydrobiologia* **167/168**, 275–284.

McLaren, I.A., Tremblay, M.J., Corkett, C.J. and Roff, J.C. (1989a). Copepod production on the Scotian Shelf based on life-history analyses and laboratory rearings. *Canadian Journal of Fisheries and Aquatic Sciences* **46**, 560–583.

McLaren, I.A., Laberge, E., Corkett, C.J. and Sévigny, J.-M. (1989b). Life cycles of four species of *Pseudocalanus* in Nova Scotia. *Canadian Journal of Zoology* **67**, 552–558.

McLaren, I.A., Sévigny, J.-M. and Corkett, C.J. (1989c). Temperature-dependent development in *Pseudocalanus* species. *Canadian Journal of Zoology* **67**, 559–564.

McLaren, I.A., Sévigny, J.-M. and Frost, B.W. (1989d). Evolutionary and ecological significance of genome sizes in the copepod genus *Pseudocalanus*. *Canadian Journal of Zoology* **67**, 565–569.

MacLellan, D.C. (1967). The annual cycle of certain calanoid species in West Greenland. *Canadian Journal of Zoology* **45**, 101–105.

MacLellan, D.C. and Shih, C.-t. (1974). Descriptions of copepodite stages of *Chiridius gracilis* Farran, 1908 (Crustacea: Copepoda). *Journal of the Fisheries Research Board of Canada* **31**, 1337–1349.

McWilliam, P.S., Sale, P.F. and Anderson, D.T. (1981). Seasonal changes in resident zooplankton sampled by emergence traps in One Tree Lagoon, Great Barrier Reef. *Journal of Experimental Marine Biology and Ecology* **52**, 185–203.

Madhupratap, M. (1980). Ecology of the coexisting copepod species in Cochin Backwaters. *Mahasagar* **13**, 45–52.
Madhupratap, M. and Haridas, P. (1986). Epipelagic calanoid copepods of the northern Indian Ocean. *Oceanologica Acta* **9**, 105–117.
Madhupratap, M. and Haridas, P. (1992). New species of *Pseudodiaptomus* (Copepoda: Calanoida) from the salt pans of the Gulf of Kutch, India and a comment on its speciation. *Journal of Plankton Research* **14**, 555–562.
Madhupratap, M. and Haridas, P. (1994). Descriptions of *Acartia* (*Euacartia*) *southwelli* Sewell 1914 and *Acartia* (*Euacartia*) *sarojus* n.sp. from India and status of the subgenus *Euacartia* Steuer 1923. *Hydrobiologia* **292/293**, 67–74.
Madhupratap, M., Nair, V.R., Nair, S.R.S. and Achuthankutty, C.T. (1981). Thermocline and zooplankton distribution. *Indian Journal of Marine Sciences* **10**, 262–265.
Madhupratap, M., Achuthankutty, C.T. and Nair, S.R.S. (1991). Zooplankton of the lagoons of the Laccadives: diel patterns and emergence. *Journal of Plankton Research* **13**, 947–958.
Madhupratap, M., Nehring, S. and Lenz, J. (1996). Resting eggs of zooplankton (Copepoda and Cladocera) from the Kiel Bay and adjacent waters (southwestern Baltic). *Marine Biology* **125**, 77–87.
Magnesen, T., Aksnes, D.L. and Skjoldal, H.R. (1989). Fine-scale vertical structure of a summer zooplankton community in Lindåspollene, western Norway. *Sarsia* **74**, 115–126.
Makarova, N.P. (1974). Regularities of Copepoda crustaceans linear growth. *Gydrobiologiya Zhurnal* **10**, 84–89.
Malej, A. and Harris, R.P. (1993). Inhibition of copepod grazing by diatom exudates: a factor in the development of mucus aggregates. *Marine Ecology Progress Series* **96**, 33–42.
Mallin, M.A., Burkholder, J.M., Larsen, L.M. and Glasgow, H.B. Jr (1995). Response of two zooplankton grazers to an ichthyotoxic estuarine dinoflagellate. *Journal of Plankton Research* **17**, 351–363.
Maly, E.J. (1996). A review of relationships among centropagid copepod genera and some species found in Australasia. *Crustaceana* **69**, 727–733.
Manwell, C., Baker, C.M.A., Ashton, P.A. and Corner, E.D.S. (1967). Biochemical differences between *Calanus finmarchicus* and *C. helgolandicus*. Esterases, malate and triose-phosphate dehydrogenases, adolase, "peptidases", and other enzymes. *Journal of the Marine Biological Association of the United Kingdom* **47**, 145–169.
Marcus, N.H. (1979). On the population biology and nature of diapause of *Labidocera aestiva* (Copepoda: Calanoida). *Biological Bulletin, Marine Biological Laboratory, Woods Hole, Mass.* **157**, 297–305.
Marcus, N.H. (1982a). The reversibility of subitaneous and diapause egg production by individual females of *Labidocera aestiva* (Copepoda: Calanoida). *Biological Bulletin, Marine Biological Laboratory, Woods Hole, Mass.* **162**, 39–44.
Marcus, N.H. (1982b). Photoperiodic and temperature regulation of diapause in *Labidocera aestiva* (Copepoda: Calanoida). *Biological Bulletin, Marine Biological Laboratory, Woods Hole, Mass.* **162**, 45–52.
Marcus, N.H. (1984). Recruitment of copepod nauplii into the plankton: importance of diapause eggs and benthic processes. *Marine Ecology Progress Series* **15**, 47–54.
Marcus, N.H. (1985a). Endogenous control of spawning in a marine copepod. *Journal of Experimental Marine Biology and Ecology* **91**, 263–269.

Marcus, N.H. (1985b). Population dynamics of marine copepods: the importance of genetic variation. *Bulletin of Marine Science* **37**, 684–690.

Marcus, N.H. (1986). Genetics, life histories, and pelagic biogeography. *Unesco Technical Papers in Marine Science* **49**, 182–185.

Marcus, N.H. (1987). Differences in the duration of egg diapause of *Labidocera aestiva* (Copepoda: Calanoida) from the Woods Hole, Massachusetts, region. *Biological Bulletin, Marine Biological Laboratory, Woods Hole, Mass.* **173**, 169–177.

Marcus, N.H. (1988). Photoperiodic conditions, food patchiness and fecundity. *Bulletin of Marine Science* **43**, 641–649.

Marcus, N.H. (1989). Abundance in bottom sediments and hatching requirements of eggs of *Centropages hamatus* (Copepoda: Calanoida) from Alligator Harbor region, Florida. *Biological Bulletin, Marine Biological Laboratory, Woods Hole, Mass.* **176**, 142–146.

Marcus, N.H. (1990). Calanoid copepod, cladoceran, and rotifer eggs in sea-bottom sediments of northern Californian coastal waters: identification, occurrence and hatching. *Marine Biology* **105**, 413–418.

Marcus, N.H. (1991). Planktonic copepods in a sub-tropical estuary: seasonal patterns in the abundance of adults, copepodites, nauplii, and eggs in the sea bed. *Biological Bulletin, Marine Biological Laboratory, Woods Hole, Mass.* **181**, 269–274.

Marcus, N.H. (1995). Seasonal study of planktonic copepods and their benthic resting eggs in northern California coastal waters. *Marine Biology* **123**, 459–465.

Marcus, N.H. (1996). Ecological and evolutionary significance of resting eggs in marine copepods: past, present, and future studies. *Hydrobiologia* **320**, 141–152.

Marcus, N.H. and Alatalo, P. (1989). Conditions for rearing *Calanus finmarchicus* (Gunnerus, 1770) (Copepoda, Calanoida) through multiple generations in the laboratory. *Crustaceana* **57**, 101–103.

Marcus, N.H. and Fuller, C.M. (1986). Subitaneous and diapause eggs of *Labidocera aestiva* Wheeler (Copepoda: Calanoida): differences in fall velocity and density. *Journal of Experimental Marine Biology and Ecology* **99**, 247–256.

Marcus, N.H. and Fuller, C.M. (1989). Distribution and abundance of eggs of *Labidocera aestiva* (Copepoda: Calanoida) in the bottom sediments of Buzzards Bay, Massachusetts, USA. *Marine Biology* **100**, 319–326.

Marcus, N.H. and Lutz, R.V. (1994). Effects of anoxia on the viability of subitaneous eggs of planktonic copepods. *Marine Biology* **121**, 83–87.

Marcus, N.H. and Schmidt-Gegenbach, J. (1986). Recruitment of individuals into the plankton: the importance of bioturbation. *Limnology and Oceanography* **31**, 206–210.

Marcus, N.H. and Taulbee, K. (1992). Potential effects of a resuspension event on the vertical distribution of copepod eggs in the sea bed: a laboratory simulation. *Marine Biology* **114**, 249–251.

Marcus, N.H., Lutz, R., Burnett, W. and Cable, P. (1994). Age, viability, and vertical distribution of zooplankton resting eggs from an anoxic basin: evidence of an egg bank. *Limnology and Oceanography* **39**, 154–158.

Marin, V. (1988a). Qualitative models of the life cycles of *Calanoides acutus*, *Calanus propinquus*, and *Rhincalanus gigas*. *Polar Biology* **8**, 439–446.

Marin, V. (1988b). Independent life cycles: an alternative to the asynchronism hypothesis for Antarctic calanoid copepods. *Hydrobiologia* **167/168**, 161–168.

Marin, V., Huntley, M.E. and Frost, B. (1986). Measuring feeding rates of pelagic

herbivores: analysis of experimental design and methods. *Marine Biology* **93**, 49–58.
Marin, V., Espinoza, S. and Fleminger, A. (1994). Morphometric study of *Calanus chilensis* males along the Chilean coast. *Hydrobiologia* **292/293**, 75–80.
Marine Zooplankton Colloquium 1 (1989). Future marine zooplankton research – a perspective. *Marine Ecology Progress Series* **55**, 197–206.
Markhaseva, E.L. (1986a). A new species of *Pseudochirella* (Copepoda, Calanoida) from the south-eastern Pacific. *Zoologichesky Zhurnal* **65**, 462–465.
Markhaseva, E.L. (1986b). Revision of the genus *Batheuchaeta* (Calanoida, Aetideidae). *Zoologichesky Zhurnal* **65**, 837–850.
Markhaseva, E.L. (1986c). New species of the genus *Pseudeuchaeta* (Calanoida, Aetideidae) from the Arctic and Pacific Oceans with notes on geographical distribution of *P. brevicauda*. *Zoologichesky Zhurnal* **65**, 1892–1898.
Markhaseva, E.L. (1989). Survey of the genus *Pseudochirella* (Copepoda, Calanoida). *In* "Explorations of the Fauna of the Seas 41 (49). Marine plankton" (M.G. Petrushevska and S.D. Stepanjants, eds), pp. 33–60. USSR Academy of Sciences, Leningrad.
Markhaseva, E.L. (1993). Two new species of *Bradyidius* with notes on *B.armatus* Giesbrecht (Crustacea, Copepoda: Aetideidae). *Zoosystematica Rossica* **2**, 47–53.
Markhaseva, E.L. (1996). Calanoid copepods of the family Aetideidae of the world ocean. *Trudy Zoologicheskogo Instituta Academiya nauk SSSR* **268**, 1–331.
Markhaseva, E.L. and Ferrari, F.D. (1995). Three new species of *Ryocalanus* from the eastern tropical Pacific (Crustacea, Copepoda: Ryocalanidae). *Zoosystematica Rossica* **4**, 63–70.
Markhaseva, E.L. and Raszhivin, V.Yu. (1992). Vertical distribution of Aetideidae (Copepoda, Calanoida) in the area of Kurile-Kamchatka Trench. *Okeanologiya* **32**, 888–896.
Marlowe, C.J. and Miller, C.B. (1975). Patterns of vertical distribution and migration of zooplankton at Ocean Station "P". *Limnology and Oceanography* **20**, 824–844.
Marrasé, C., Costello, J.H., Granata, T. and Strickler, J.R. (1990). Grazing in a turbulent environment: energy dissipation, encounter rates, and efficacy of feeding currents in *Centropages hamatus*. *Proceedings of the National Academy of Science* **87**, 1653–1657.
Marshall, S.M. (1924). The food of *Calanus finmarchicus* during 1923. *Journal of the Marine Biological Association of the United Kingdom* **13**, 473–479.
Marshall, S.M. (1949). On the biology of the small copepods in Loch Striven. *Journal of the Marine Biological Association of the United Kingdom* **28**, 45–122.
Marshall, S.M. (1973). Respiration and feeding in copepods. *Advances in Marine Biology* **11**, 57–120.
Marshall, S.M. and Orr, A.P. (1954). Hatching in *Calanus finmarchicus* and some other copepods. *Journal of the Marine Biological Association of the United Kingdom* **33**, 393–401.
Marshall, S.M. and Orr, A.P. (1955). "The Biology of a Marine Copepod, *Calanus finmarchicus* (Gunnerus)." Oliver & Boyd, London. 188 pp.
Marshall, S.M. and Orr, A.P. (1956). On the biology of *Calanus finmarchicus*. IX. Feeding and digestion in the young stages. *Journal of the Marine Biological Association of the United Kingdom* **35**, 587–603.
Marshall, S.M., Nicholls, A.G. and Orr, A.P. (1934). On the biology of *Calanus finmarchicus*. Part V. Seasonal distribution, size, weight and chemical composition

in Loch Striven in 1933 and their relation to the phytoplankton. *Journal of the Marine Biological Association of the United Kingdom* **19**, 793–828.

Marshall, S.M., Nicholls, A.G. and Orr, A.P. (1935). On the biology of *Calanus finmarchicus*. Part VI. Oxygen consumption in relation to environmental conditions. *Journal of the Marine Biological Association of the United Kingdom* **20**, 1–27.

Martens, P. (1980). Beiträge zum mesozooplankton des Nordsylter Wattenmeers. *Helgoländer wissenschaftliche Meeresuntersuchungen* **34**, 41–53.

Martens, P. (1981). On the *Acartia* species of the northern Wadden Sea off Sylt. *Kieler Meeresforschungen* **5**, 153–163.

Masuzawa, T., Koyama, M. and Terazaki, M. (1988). A regularity in trace element contents of marine zooplankton species. *Marine Biology* **97**, 587–591.

Matsuo, Y. and Marumo, R. (1982). Diurnal vertical migration of pontellid copepods in the Kuroshio. *Bulletin of Plankton Society of Japan* **29**, 89–98.

Matthews, J.B.L. (1964). On the biology of some bottom-living copepods (Aetideidae and Phaennidae) from western Norway. *Sarsia* **16**, 1–46.

Matthews, J.B.L. (1966). Experimental investigations of the systematic status of *Calanus finmarchicus* and *C. glacialis* (Crustacea: Copepoda). *In* "Some Contemporary Studies in Marine Science" (H. Barnes, ed.), pp. 479–492. George Allen and Unwin Ltd, London.

Matthews, J.B.L. (1972). The genus *Euaugaptilus* (Crustacea, Copepoda). New descriptions and a review of the genus in relation to *Augaptilus*, *Haloptilus* and *Pseudhaloptilus*. *Bulletin of the British Museum (Natural History), Zoology* **24** (1), 1–71.

Matthews, J.B.L. and Heimdal, B.R. (1980). Pelagic productivity and food chains in fjord systems. *In* "Fjord Oceanography" (H.J. Freeland, D.M. Farmer and C.D. Levings, eds), NATO Conference Series IV, Marine Sciences, Vol. 4, pp. 377–398. Plenum Press, New York and London.

Matthews, J.B.L., Hestad, L. and Bakke, J.L.W. (1978). Ecological studies in Korsfjorden, Western Norway. The generations and stocks of *Calanus hyperboreus* and *C. finmarchicus* in 1971–1974. *Oceanologica Acta* **1**, 277–284.

Mauchline, J. (1960). The biology of the euphausiid crustacean, *Meganyctiphanes norvegica* (M.Sars). *Proceedings of the Royal Society of Edinburgh* **67B**, 141–179.

Mauchline, J. (1973). Intermoult growth of species of Mysidacea. *Journal of the Marine Biological Association of the United Kingdom* **53**, 569–572.

Mauchline, J. (1977a). The integumental sensilla and glands of pelagic Crustacea. *Journal of the Marine Biological Association of the United Kingdom* **57**, 973–994.

Mauchline, J. (1977b). Growth of shrimps, crabs and lobsters – an assessment. *Journal du Conseil International pour l'Exploration de la Mer* **37**, 162–169.

Mauchline, J. (1980). The biology of mysids and euphausiids. *Advances in Marine Biology* **18**, 1–677.

Mauchline, J. (1988a). Egg and brood sizes of oceanic pelagic crustaceans. *Marine Ecology Progress Series* **43**, 251–258.

Mauchline, J. (1988b). Taxonomic value of pore pattern in the integument of calanoid copepods (Crustacea). *Journal of Zoology, London* **214**, 697–749.

Mauchline, J. (1989). Functional morphology of feeding in euphausiids. *In* "Functional Morphology of Feeding and Grooming in Crustacea" (B.E. Felgenhauer, L. Watling and A.B. Thistle, eds), pp. 173–184. A.A. Balkema, Rotterdam.

Mauchline, J. (1991). Some modern concepts in deep-sea pelagic studies: patterns of growth in the different horizons. *In* "Marine Biology, its Accomplishment and Future Prospect" (J. Mauchline and T. Nemoto, eds), pp. 107–130. Hokusen-sha, Tokyo.
Mauchline, J. (1992a). Restriction of body size spectra within species of deep-sea plankton. *Marine Ecology Progress Series* **90**, 1–8.
Mauchline, J. (1992b). Taxonomy, distribution and biology of *Euchaeta barbata* (=*E. farrani*) (Copepoda: Calanoida). *Sarsia* **77**, 131–142.
Mauchline, J. (1994a). Seasonal variation in some population parameters of *Euchaeta* species (Copepoda: Calanoida). *Marine Biology* **120**, 561–570.
Mauchline, J. (1994b). Spermatophore transfer in *Euchaeta* species in a 2000 m water column. *Hydrobiologia* **292/293**, 309–316.
Mauchline, J. (1995). Bathymetric adaptations of life history patterns of congeneric species (*Euchaeta*: Calanoida) in a 2000 m water column. *ICES Journal of Marine Science* **52**, 511–516.
Mauchline, J. and Fisher, L.R. (1969). The biology of euphausiids. *Advances in Marine Biology* **7**, 1–454.
Mauchline, J. and Gordon, J.D.M. (1986). Foraging strategies of deep-sea fish. *Marine Ecology Progress Series* **27**, 227–238.
Mauchline, J. and Gordon, J.D.M. (1991). Oceanic pelagic prey of benthopelagic fish in the benthic boundary layer of a marginal oceanic region. *Marine Ecology Progress Series* **74**, 109–115.
Mauchline, J. and Nemoto, T. (1977). The occurrence of integumental organs in copepodid stages of calanoid copepods. *Bulletin of Plankton Society of Japan* **24**, 108–114.
Mayzaud, P. (1986a). Digestive enzymes and their relation to nutrition. *In* "The Biological Chemistry of Marine Copepods" (E.D.S. Corner and S.C.M. O'Hara, eds), pp. 165–225. Clarendon Press, Oxford.
Mayzaud, P. (1986b). Enzymatic measurements of metabolic processes concerned with respiration and ammonia excretion. *In* "The Biological Chemistry of Marine Copepods" (E.D.S. Corner and S.C.M. O'Hara, eds), pp. 226–259. Clarendon Press, Oxford.
Mayzaud, P. and Conover, R.J. (1988). O:N atomic ratio as a tool to describe zooplankton metabolism. *Marine Ecology Progress Series* **45**, 289–302.
Mayzaud, P., Roche-Mayzaud, O. and Razouls, S. (1992). Medium term time acclimation of feeding and digestive enzyme activity in marine copepods: influence of food concentration and copepod species. *Marine Ecology Progress Series* **89**, 197–212.
Mazza, J. (1964). Le développement de quelques copépodes en Méditerranée. I. Les stades jeunes d' *Euchaeta acuta* Giesbrecht et d' *Euchaeta spinosa* Giesbrecht. *Revue des Travaux de l'Institut des Pêches Maritimes* **28**, 271–292.
Mazza, J. (1965). Le développement de quelques copépodes en Méditerranée. II. Les stades jeunes de *Gaetanus kruppi* Giesb., *Euchirella messinensis* Cl., *Chiridius poppei* Giesb., *Pseudaetideus armatus* (Boeck) et *Heterorhabdus spinifrons* Cl. *Revue des Travaux de l'Institut des Pêches maritimes* **29**, 285–320.
Mazza, J. (1966). Évolution de l'appareil buccal au cours du développement post-larvaire des Aetideidae et des Euchaetidae (Copépodes pélagiques) ses incidences sur le sex-ratio des adultes. *Vie et Milieu* **17**, 1027–1044.
Mazzocchi, M.G. and Ribera d'Alcalà, M. (1995). Recurrent patterns in zooplankton structure and succession in a variable coastal environment. *ICES Journal of Marine Science* **52**, 679–691.

Mehner, T. (1996). Predation impact of age 0-group fish on a copepod population in a Baltic Sea inlet as estimated by two energetic models. *Journal of Plankton Research* **18**, 1323–1340.

Meise-Munns, C., Green, J., Ingham, M. and Mountain, D. (1990). Interannual variability in the copepod populations of Georges Bank and the western Gulf of Maine. *Marine Ecology Progress Series* **65**, 225–232.

Melle, W. and Skjoldal, H.R. (1989). Zooplankton reproduction in the Barents Sea: vertical distribution of eggs and nauplii of *Calanus finmarchicus* in relation to spring phytoplankton development. *In* "Reproduction, Genetics and Distributions of Marine Organisms" (J.S. Ryland and P.A. Tyler, eds), pp. 137–145. Olsen & Olsen, Fredensborg, Denmark.

Mensah, M.A. (1974). The reproduction and feeding of the marine copepod *Calanoides carinatus* (Krøyer) in Ghanaian waters. *Ghana Journal of Science* **14**, 167–191.

Menshenina, L.L. and Melnikov, I.A. (1995). Under-ice zooplankton of the western Weddell Sea. *Proceedings of the NIPR Symposium on Polar Biology* **8**, 126–138.

Menu-Marque, S.A. and Zúñiga, L.R. (1994). *Boeckella diamantina* n.sp. (Calanoida, Centropagidae), from a high Andean lake in Mendoza, Argentina. *Hydrobiologia* **292/293**, 81–87.

Metz, C. and Schnack-Schiel, S.B. (1995). Observations on carnivorous feeding in Antarctic copepods. *Marine Ecology Progress Series* **129**, 71–75.

Michel, H.B. (1994). Antarctic Megacalanidae (Copepoda: Calanoida) and the distribution of the family. *Journal of the Marine Biological Association of the United Kingdom* **74**, 175–192.

Michel, H.B. and Herring, D.C. (1984). Diversity and abundance of Copepoda in the northwestern Arabian Gulf. *Crustaceana*, Supplement **7**, 326–335.

Middlebrook, K. and Roff, J.C. (1986). Comparison of methods for estimating annual productivity of the copepods *Acartia hudsonica* and *Eurytemora herdmani* in Passamaquoddy Bay, New Brunswick. *Canadian Journal of Fisheries and Aquatic Sciences* **43**, 656–664.

Miller, C.B. (1983). The zooplankton of estuaries. *In* "Ecosystems of the World 26. Estuaries and Enclosed Seas" (B.H. Ketchum, ed.), pp. 103–149. Elsevier Scientific Publishing Company, Amsterdam.

Miller, C.B. (1988). *Neocalanus flemingeri*, a new species of Calanidae (Copepoda: Calanoida) from the subarctic Pacific Ocean, with a comparative redescription of *Neocalanus plumchrus* (Marukawa) 1921. *Progress in Oceanography* **20**, 223–273.

Miller, C.B., ed. (1993a). Pelagic ecodynamics in the Gulf of Alaska. Results from the SUPER program. *Progress in Oceanography* **32**, 1–358.

Miller, C.B. (1993b). Development of large copepods during spring in the Gulf of Alaska. *Progress in Oceanography* **32**, 295–317.

Miller, C.B. and Clemons, M.J. (1988). Revised life history analysis for large grazing copepods in the subarctic Pacific Ocean. *Progress in Oceanography* **20**, 293–313.

Miller, C.B. and Grigg, H. (1991). An experimental study of the resting phase in *Calanus finmarchicus* (Gunnerus). *Bulletin of Plankton Society of Japan*, Special Volume, 479–493.

Miller, C.B. and Nielsen, R.D. (1988). Development and growth of large, calanoid copepods in the ocean subarctic Pacific, May 1984. *Progress in Oceanography* **20**, 275–292.

Miller, C.B. and Tande, K.S. (1993). Stage duration estimation for *Calanus* populations, a modelling study. *Marine Ecology Progress Series* **102**, 15–34.

Miller, C.B. and Terazaki, M. (1989). The life histories of *Neocalanus flemingeri* and *Neocalanus plumchrus* in the Sea of Japan. *Bulletin of Plankton Society of Japan* **36**, 27–41.

Miller, C.B., Johnson, J.K. and Heinle, D.R. (1977). Growth rules in the marine copepod genus *Acartia*. *Limnology and Oceanography* **22**, 326–335.

Miller, C.B., Nelson, D.M., Guillard, R.R.L. and Woodward, B.L. (1980). Effects of media with low silicic acid concentrations on tooth formation in *Acartia tonsa* Dana (Copepoda: Calanoida). *Biological Bulletin, Marine Biological Laboratory, Wood's Hole, Mass.* **159**, 349–363.

Miller, C.B., Frost, B.W., Batchelder, H.P., Clemons, M.J. and Conway, R.E. (1984a). Life histories of large, grazing copepods in a subarctic ocean gyre: *Neocalanus plumchrus*, *Neocalanus cristatus*, and *Eucalanus bungii* in the northeast Pacific. *Progress in Oceanography* **13**, 201–243.

Miller, C.B., Huntley, M.E. and Brooks, E.R. (1984b). Post-collection molting rates of planktonic, marine copepods: measurement, applications, problems. *Limnology and Oceanography* **29**, 1274–1289.

Miller, C.B., Nelson, D.M., Weiss, C. and Soeldner, A.H. (1990). Morphogenesis of opal teeth in calanoid copepods. *Marine Biology* **106**, 91–101.

Miller, C.B., Cowles, T.J., Wiebe, P.H., Copley, N.J. and Grigg, H. (1991). Phenology in *Calanus finmarchicus*; hypothesis about control mechanisms. *Marine Ecology Progress Series* **72**, 79–91.

Miller, D.D. and Marcus, N.H. (1994). The effects of salinity and temperature on the density and sinking velocity of eggs of the calanoid copepod *Acartia tonsa* Dana. *Journal of Experimental Marine Biology and Ecology* **179**, 235–252.

Minkina, N.I. (1981). Estimation by hydrodynamic method of energy expenditure of copepods (Copepoda, Crustacea) on swimming. *Doklady Akademiia Nauk SSSR, Biological Sciences Section, (CTC Translation)* **257**, 141–144.

Minkina, N.I. (1983). Space-time characteristics of copepod swimming. *Ekologiya Morya* **14**, 38–44.

Minkina, N.I. and Pavlova, E.V. (1981). Hydrodynamic drag and power at variable swimming in *Calanus helgolandicus* (Claus). *Ekologiya Morya* **7**, 63–75.

Minkina, N.I. and Pavlova, E.V. (1992). Concerning the difference between natural and laboratory respiration levels in sea copepods. *Ekologiya Morya* **40**, 77–84.

Minoda, T. (1972). Characteristics of the vertical distribution of copepods in the Bering Sea and south of the Aleutian chain, May–June, 1962. *In* "Biological Oceanography of the Northern North Pacific Ocean" (A.Y. Takenouti *et al.*, eds), pp. 323–331. Idemitsu Shoten, Tokyo.

Miralto, A., Ianora, A. and Poulet, S.A. (1995). Food type induces different reproductive responses in the copepod *Centropages typicus*. *Journal of Plankton Research* **17**, 1521–1534.

Miralto, A., Ianora, A., Poulet, S.A., Romano, G. and Laabir, M. (1996). Is fecundity modified by crowding in the copepod *Centropages*? *Journal of Plankton Research* **18**, 1033–1040.

Mizdalski, E. (1988). Weight and length data of zooplankton in the Weddell Sea in austral spring 1986 (Ant V/3). *Berichte zur Polarforschung* **55**, 1–72.

Moal, J., Samain, J.F., Koutsikopoulos, C., Coz, J.R.Le and Daniel, J.Y. (1985). Ushant thermal front: digestive enzymes and zooplankton production. *In* "Proceedings of the Nineteenth European Marine Biology Symposium" (P.E.

Gibbs, ed.), pp. 145–156. Cambridge University Press, Cambridge.

Mobley, C.T. (1987). Time-series ingestion rate estimates on individual *Calanus pacificus* Brodsky: interactions with environmental and biological factors. *Journal of Experimental Marine Biology and Ecology* **114**, 199–216.

Møhlenberg, F. (1987). A submersible net-pump for quantitative zooplankton sampling; comparison with conventional net sampling. *Ophelia* **27**, 101–110.

Moloney, C.L. and Gibbons, M.J. (1996). Sampling and analysis of gut contents in relation to environmental variability and diel vertical migration by herbivorous zooplankton. *Journal of Plankton Research* **18**, 1535–1556.

Montagnes, D.J.S., Lynn, D.H., Roff, J.C. and Taylor, W.D. (1988). The annual cycle of heterotrophic planktonic ciliates in the waters surrounding the Isles of Shoals, Gulf of Maine: an assessment of their trophic role. *Marine Biology* **99**, 21–30.

Moore, E. and Sander, F. (1977). A study of the offshore zooplankton of the tropical western Atlantic near Barbados. *Ophelia* **16**, 77–96.

Moore, E. and Sander, F. (1983). Physioecology of tropical marine copepods. II. Sex ratios. *Crustaceana* **44**, 113–122.

Moraitou-Apostolopoulou, M. (1969). Variability of some morphoecological factors in six pelagic copepods from the Aegean Sea. *Marine Biology* **3**, 1–3.

Moraitou-Apostolopoulou, M. (1971). Vertical distribution, diurnal and seasonal migration of copepods in Saronic Bay, Greece. *Marine Biology* **9**, 92–98.

Morales, C.E. (1987). Carbon and nitrogen content of copepod faecal pellets: effect of food concentration and feeding behavior. *Marine Ecology Progress Series* **36**, 107–114.

Morales, C.E., Bautista, B. and Harris, R.P. (1990). Estimates of ingestion in copepod assemblages: gut fluorescence in relation to body size. *In* "Trophic Relationships in the Marine Environment" (M. Barnes and R.N. Gibson, eds), pp. 565–577. Aberdeen University Press.

Morales, C.E., Bedo, A., Harris, R.P. and Tranter, P.G. (1991). Grazing of copepod assemblages in the north-east Atlantic: the importance of the small size fraction. *Journal of Plankton Research* **13**, 455–472.

Morales, C.E., Harris, R.P., Head, R.N. and Tranter, P.R.G. (1993). Copepod grazing in the oceanic northeast Atlantic during a 6 week drifting station: the contribution of size classes and vertical migrants. *Journal of Plankton Research* **15**, 185–211.

Mori, T. (1937). The Pelagic Copepoda from the Neighbouring Waters of Japan. Second Edition, 1964. The Soyo Company Inc., Tokyo, 150 pp., 80 plates.

Morioka, Y. (1975). A preliminary report on the distribution and life history of a copepod, *Pareuchaeta elongata*, in the vicinity of Sado Island, the Japan Sea. *Bulletin of the Japan Sea Regional Fisheries Laboratory* **26**, 41–56.

Morioka, Y. (1981). Zooplankton production in the Toyama Bay in March–May, 1978. *Bulletin of the Japan Sea Regional Fisheries Research Laboratory* **32**, 57–64.

Morioka, Y., Shinohara, F., Nakashima, J. and Irie, T. (1991). A diel vertical migration of the copepod *Calanus sinicus* in relation to well-developed thermocline in the Yellow Sea, October 1987. *Bulletin of the Seikai National Fisheries Research Institute* **69**, 79–85.

Morris, M.J. and Hopkins, T.L. (1983). Biochemical composition of crustacean zooplankton from the eastern Gulf of Mexico. *Journal of Experimental Marine Biology and Ecology* **69**, 1–19.

Morris, M.J., Gust, G. and Torres, J.J. (1985). Propulsion efficiency and cost of transport for copepods: a hydromechanical model of crustacean swimming. *Marine Biology* **86**, 283–295.

Mullin, M.M. (1968). Egg-laying in the planktonic copepod *Calanus helgolandicus* (Claus). *Crustaceana*, Supplement **1**, 29–34.

Mullin, M.M. (1979). Differential predation by the carnivorous marine copepod, *Tortanus discaudatus. Limnology and Oceanography* **24**, 774–777.

Mullin, M.M. (1991a). Production of eggs by the copepod *Calanus pacificus* in the southern California sector of the California Current system. *California Cooperative Oceanic Fisheries Investigations, Reports* **32**, 65–90.

Mullin, M.M. (1991b). Relative variability of reproduction and mortality in two pelagic copepod populations. *Journal of Plankton Research* **13**, 1381–1387.

Mullin, M.M. (1993). Reproduction of the oceanic copepod *Rhincalanus nasutus* off southern California, compared to that of *Calanus pacificus. California Cooperative Oceanic Fisheries Investigations, Reports* **34**, 89–99.

Mullin, M.M. (1995). Nauplii of the copepod, *Calanus pacificus*, off southern California in the El Niño winter-spring of 1992, and implications for larval fish. *Journal of Plankton Research* **17**, 183–189.

Mullin, M.M. and Brooks, E.R. (1967). Laboratory culture, growth rate, and feeding behavior of a planktonic marine copepod. *Limnology and Oceanography* **12**, 657–666.

Mullin, M.M. and Brooks, E.R. (1970). Growth and metabolism of two planktonic, marine copepods as influenced by temperature and type of food. *In* "Marine Food Chains" (J.H. Steele, ed.), pp. 74–95. Oliver and Boyd, Edinburgh.

Myers, R.A. and Runge, J.A. (1983). Predictions of seasonal natural mortality rates in a copepod population using life-history theory. *Marine Ecology Progress Series* **11**, 189–194.

Myers, R.A. and Runge, J. (1986). Temperature-dependent changes in copepod adult size: an evolutionary theory. *Syllogeus* **58**, 374–378.

Myklebust, R., Saetersdal, T. and Tjønneland, A. (1977). The membrane systems of the cardiac muscle cell of *Euchaeta norvegica* Boeck (Crustacea, Copepoda). *Cell Tissue Research* **180**, 283–292.

Næss, T. (1991a). Marine calanoid resting eggs in Norway: abundance and distribution of two copepod species in the sediment of an enclosed marine basin. *Marine Biology* **110**, 261–266.

Næss, T. (1991b). Tolerance of marine calanoid resting eggs: effects of freezing, desiccation and Rotenone exposure – a field and laboratory study. *Marine Biology* **111**, 455–459.

Næss, T. (1996). Benthic resting eggs of calanoid copepods in Norwegian enclosures used in mariculture: abundance, species composition and hatching. *Hydrobiologia* **320**, 161–168.

Naganuma, T. (1996). Calanoid copepods: linking lower-higher trophic levels by linking lower-higher Reynolds numbers. *Marine Ecology Progress Series* **136**, 311–313.

Nagaraj, M. (1988). Combined effects of temperature and salinity on the complete development of *Eurytemora velox* (Crustacea: Calanoida). *Marine Biology* **99**, 353–358.

Nagasawa, S. (1989). Supplementary records of copepods with attached bacteria. *Bulletin of Plankton Society of Japan* **36**, 63–64.

Nagasawa, S. (1992). Concurrent observations on gut interior and fecal pellets of marine crustaceans. *Journal of Plankton Research* **14**, 1625–1630.

Nagasawa, S. and Terazaki, M. (1987). Bacterial epibionts of the deep-sea copepod *Calanus cristatus* Krøyer. *Oceanologica Acta* **10**, 475–479.

Nair, S.R.S., Nair, V.R., Achuthankutty, C.T. and Madhupratap, M. (1981).

Zooplankton composition and diversity in western Bay of Bengal. *Journal of Plankton Research* **3**, 493–508.
Nakata, K., Nakano, H. and Kikuchi, H. (1994). Relationship between egg productivity and RNA/DNA ratio in *Paracalanus* sp. in the frontal waters of the Kuroshio. *Marine Biology* **119**, 591–596.
Napp, J.M., Brooks, E.R., Matrai, P. and Mullin, M.M. (1988). Vertical distribution of marine particles and grazers. II. Relation of grazer distribution to food quality and quantity. *Marine Ecology Progress Series* **50**, 59–72.
Nassogne, A. (1970). Influence of food organisms on the development and culture of pelagic copepods. *Helgoländer wissenschaftliche Meeresuntersuchungen* **20**, 333–345.
Nealson, K.H., Arneson, A.C. and Huber, M.E. (1986). Identification of marine organisms using kinetic and spectral properties of their bioluminescence. *Marine Biology* **91**, 77–83.
Neill, W.E. (1990). Induced vertical migration in copepods as a defence against invertebrate predation. *Nature, London* **345**, 524–526.
Nejstgaard, J.C., Båmstedt, U., Bagøien, E. and Solberg, P.T. (1995). Algal constraints on copepod grazing. Growth state, toxicity, cell size, and season as regulating factors. *ICES Journal of Marine Science* **52**, 347–357.
Nemoto, T. (1968). Chlorophyll pigments in the stomach of euphausiids. *Journal of the Oceanographical Society of Japan* **24**, 253–260.
Nemoto, T. and Saijo, Y. (1968). Trace of chlorophyll pigments in stomachs of deep sea zoo-plankton. *Journal of the Oceanographical Society of Japan* **24**, 46–48.
Nemoto, T., Mauchline, J. and Kamada, K. (1976). Brood size and chemical composition of *Pareuchaeta norvegica* (Crustacea: Copepoda) in Loch Etive, Scotland. *Marine Biology* **36**, 151–157.
Neunes, H.W. (1965). A simple key for common pelagic Mediterranean copepods. A tool for the identification of species in production and radioaccumulation studies. *Pubblicazioni della Stazione Zoologica di Napoli* **34**, 462–475.
Nicholls, A.G. (1934). The developmental stages of *Euchaeta norvegica*, Boeck. *Proceedings of the Royal Society of Edinburgh* **54**, 31–50.
Nichols, J.H. and Thompson, A.B. (1991). Mesh selection of copepodite and nauplius stages of four calanoid copepod species. *Journal of Plankton Research* **13**, 661–671.
Nielsen, T.G. and Hansen, B. (1995). Plankton community structure and carbon cycling on the western coast of Greenland during and after the sedimentation of a diatom bloom. *Marine Ecology Progress Series* **125**, 239–257.
Nielsen, T.G., Løkkegaard, B., Richardson, K., Pedersen, F.B. and Hansen, L. (1993). Structure of plankton communities in the Dogger Bank area (North Sea) during a stratified situation. *Marine Ecology Progress Series* **95**, 115–131.
Nishida, S. (1989). Distribution, structure and importance of the cephalic dorsal hump, a new sensory organ in calanoid copepods. *Marine Biology* **101**, 173–185.
Nishida, S. and Ohtsuka, S. (1996). Specialized feeding mechanism in the pelagic copepod genus *Heterorhabdus* (Calanoida: Heterorhabdidae), with special reference to the mandibular tooth and labral glands. *Marine Biology* **126**, 619–632.
Nishida, S., Oh, B.-C. and Nemoto, T. (1991). Midgut structure and food habits of the mesopelagic copepods *Lophothrix frontalis* and *Scottocalanus securifrons*. *Bulletin of Plankton Society of Japan*, Special Volume, 527–534.
Nishiyama, T., Tanimura, A., Watanabe, K., Fukuchi, M. and Aota, M. (1987).

Comparison of zooplankton abundance under sea ice between NIPR net and NORPAC net samplings, in Lagoon Saroma Ko, Hokkaido. *Proceedings of the NIPR Symposium on Polar Biology* **1**, 123–137.

Nival, S., Pagano, M. and Nival, P. (1990). Laboratory study of the spawning rate of the calanoid copepod *Centropages typicus*: effect of fluctuating food concentration. *Journal of Plankton Research* **12**, 535–547.

Nöges, T. (1992). Comparison of two methods of zooplankton grazing measurements. *Internationale Revue der gesamten Hydrobiologie* **77**, 665–672.

Noji, T.T., Estep, K.W., MacIntyre, F. and Norrbin, F. (1991). Image analysis of faecal material grazed upon by three species of copepods: evidence for coprorhexy, coprophagy and coprochaly. *Journal of the Marine Biological Association of the United Kingdom* **71**, 465–480.

Nomura, H., Ishimaru, T. and Murano, M. (1993). Dense swarms of calanoid copepods in Tokyo Bay, Japan. *Bulletin of Plankton Society of Japan* **39**, 147–149.

Norrbin, M.F. (1991). Gonad maturation as an indication of seasonal cycles for several species of small copepods in the Barents Sea. *Polar Research* **10**, 421–432.

Norrbin, M.F. (1994). Seasonal patterns in gonad maturation, sex ratio and size in some small, high-latitude copepods: implications for overwintering tactics. *Journal of Plankton Research* **16**, 115–131.

Norrbin, M.F. (1996). Timing of diapause in relation to the onset of winter in the high-latitude copepods *Pseudocalanus acuspes* and *Acartia longiremis*. *Marine Ecology Progress Series* **142**, 99–109.

Norrbin, M.F., Olsen, R.-E. and Tande, K.S. (1990). Seasonal variation in lipid class and fatty acid composition of two small copepods in Balsfjorden, northern Norway. *Marine Biology* **105**, 205–211.

Nott, J.A., Corner, E.D.S., Mavin, L.J. and O'Hara, S.C.M. (1985). Cyclical contributions of the digestive epithelium to faecal pellet formation by the copepod *Calanus helgolandicus*. *Marine Biology* **89**, 271–279.

Oberg, M. (1906). Die Metamorphose der Planktoncopepoden der Kieler Bucht. *Wissenschaftliche Meeresuntersuchungen, Abt. Kiel* **9**, 37–103.

Oceanographic Laboratory, Edinburgh (1973). Continuous plankton records: a plankton atlas of the North Atlantic and the North Sea. *Bulletin of Marine Ecology* **7**, 1–174.

O'Connors, H.B., Small, L.F. and Donaghay, P.L. (1976). Particle-size modification by two size classes of the estuarine copepod *Acartia clausi*. *Limnology and Oceanography* **21**, 300–308.

O'Connors, H.B., Biggs, D.C. and Ninivaggi, D.V. (1980). Particle-size-dependent maximum grazing rates for *Temora longicornis* fed natural particle assemblages. *Marine Biology* **56**, 65–70.

Ogilvie, H.S. (1953). Copepod nauplii (I). *Conseil International pour l'Exploration de la Mer, Zooplankton Sheet* **50**, 1–4.

Ogle, J. (1979). Adaptation of a brown water culture technique to the mass culture of the copepod *Acartia tonsa*. *Gulf Research Reports* **6**, 291–292.

Oh, B.-C., Terazaki, M. and Nemoto, T. (1991). Some aspects of the life history of the subarctic copepod *Neocalanus cristatus* (Calanoida) in Sagami Bay, central Japan. *Marine Biology* **111**, 207–212.

Ohman, M.D. (1985). Resource-satiated population growth of the copepod *Pseudocalanus* sp. *Archives fur Hydrobiologie Beiheft Ergebnisse Limnologie* **21**, 15–32.

Ohman, M.D. (1986). Predator-limited population growth of the copepod *Pseudocalanus* sp. *Journal of Plankton Research* **8**, 673–713.
Ohman, M.D. (1987). Energy sources for recruitment of the subantarctic copepod *Neocalanus tonsus*. *Limnology and Oceanography* **32**, 1317–1330.
Ohman, M.D. (1988a). Sources of variability in measurements of copepod lipids and gut fluorescence in the California coastal zone. *Marine Ecology Progress Series* **42**, 143–153.
Ohman, M.D. (1988b). Behavioral responses of zooplankton to predation. *Bulletin of Marine Science* **43**, 530–550.
Ohman, M.D. (1990). The demographic benefits of diel vertical migration by zooplankton. *Ecological Monographs* **60**, 257–281.
Ohman, M.D. (1996). Freezing and storage of copepod samples for analysis of lipids. *Marine Ecology Progress Series* **130**, 295–298.
Ohman, M.D. and Runge, J.A. (1994). Sustained fecundity when phytoplankton resources are in short supply: omnivory by *Calanus finmarchicus* in the Gulf of St. Lawrence. *Limnology and Oceanography* **39**, 21–36.
Ohman, M.D. and Wood, S.N. (1995). The inevitability of mortality. *ICES Journal of Marine Science* **52**, 517–522.
Ohman, M.D. and Wood, S.N. (1996). Mortality estimation for planktonic copepods: *Pseudocalanus newmani* in a temperate fjord. *Limnology and Oceanography* **41**, 126–135.
Ohman, M.D., Bradford, J.M. and Jillett, J.B. (1989). Seasonal growth and lipid storage of the circumglobal, subantarctic copepod, *Neocalanus tonsus* (Brady). *Deep-Sea Research* **36A**, 1309–1326.
Ohno, A. and Okamura, Y. (1988). Propagation of the calanoid copepod, *Acartia tsuensis*, in outdoor tanks. *Aquaculture* **70**, 39–51.
Ohno, A., Takahashi, T. and Taki, Y. (1990). Dynamics of exploited populations of the calanoid copepod, *Acartia tsuensis*. *Aquaculture* **84**, 27–39.
Ohtsuka, S. (1984). Calanoid copepods collected from the near-bottom in Tanabe Bay on the Pacific coast of the Middle Honshu, Japan. I. Arietellidae. *Publications of the Seto Marine Biological Laboratory* **29**, 359–365.
Ohtsuka, S. (1985a). A note on the feeding habit of a calanoid copepod, *Pontellopsis yamadai* Mori. *Publications of the Seto Marine Biological Laboratory* **30**, 145–149.
Ohtsuka, S. (1985b). Calanoid copepods collected from the near-bottom in Tanabe Bay on the Pacific coast of the Middle Honshu, Japan. II. Arietellidae (cont.). *Publications of the Seto Marine Biological Laboratory* **30**, 287–306.
Ohtsuka, S. (1992). Calanoid copepods collected from the near-bottom in Tanabe Bay on the Pacific coast of the Middle Honshu, Japan. IV. Pseudocyclopiidae. *Publications of the Seto Marine Biological Laboratory* **35**, 295–301.
Ohtsuka, S. and Boxshall, G.A. (1994). *Platycopia orientalis*, new species (Copepoda: Platycopioida), from the North Pacific, with descriptions of copepodid stages. *Journal of Crustacean Biology* **14**, 151–167.
Ohtsuka, S. and Hiromi, J. (1987). Calanoid copepods collected from the near-bottom in Tanabe Bay on the Pacific coast of the Middle Honshu, Japan. III. Stephidae. *Publications of the Seto Marine Biological Laboratory* **32**, 219–232.
Ohtsuka, S. and Kimoto, K. (1989). *Tortanus* (*Atortus*) (Copepoda: Calanoida) of southern Japanese waters, with description of two new species, *T.* (*A.*) *digitalis* and *T.* (*A.*) *ryukyuensis*, and discussion on distribution and swarming behavior of *Atortus*. *Journal of Crustacean Biology* **9**, 392–408.
Ohtsuka, S. and Kubo, N. (1991). Larvaceans and their houses as important food for

some pelagic copepods. *Bulletin of Plankton Society of Japan, Special Volume,* 535–551.
Ohtsuka, S. and Mitsuzumi, C. (1990). A new asymmetrical near-bottom calanoid copepod, *Paramisophria platysoma*, with observations of its integumental organs, behavior and *in-situ* feeding habit. *Bulletin of Plankton Society of Japan* **36**, 87–101.
Ohtsuka, S. and Onbé, T. (1989). Evidence of selective feeding on larvaceans by the pelagic copepod *Candacia bipinnata* (Calanoida: Candaciidae). *Journal of Plankton Research* **11**, 869–872.
Ohtsuka, S. and Onbé, T. (1991). Relationship between mouthpart structures and in situ feeding habits of species of the family Pontellidae (Copepoda: Calanoida). *Marine Biology* **111**, 213–225.
Ohtsuka, S., Fukuura, Y. and Go, A. (1987a). Description of a new species of *Tortanus* (Copepoda: Calanoida) from Kuchinoerabu Island, Kyushu, with notes on its possible feeding mechanism and in-situ feeding habits. *Bulletin of Plankton Society of Japan* **34**, 53–63.
Ohtsuka, S., Fleminger, A. and Onbé, T. (1987b). A new species of *Pontella* (Copepoda: Calanoida) from the Inland Sea of Japan with notes on its feeding habits and related species. *Journal of Crustacean Biology* **7**, 554–571.
Ohtsuka, S., Fosshagen, A. and Go, A. (1991). The hyperbenthic calanoid copepod *Paramisophria* from Okinawa, south Japan. *Zoological Science* **8**, 793–804.
Ohtsuka, S., Fosshagen, A. and Iliffe, T.M. (1993a). Two new species of *Paramisophria* (Copepoda, Calanoida, Arietellidae) from anchialine caves on the Canary and Galápagos Islands. *Sarsia* **78**, 57–67.
Ohtsuka, S., Roe, H.S.J. and Boxshall, G.A. (1993b). A new family of calanoid copepods, the Hyperbionycidae, collected from the deep-sea hyperbenthic community in the northeastern Atlantic. *Sarsia* **78**, 69–82.
Ohtsuka, S., Ohaye, S., Tanimura, A., Fukuchi, M., Hattori, H., Sasaki, H. and Matsuda, O. (1993c). Feeding ecology of copepodid stages of *Eucalanus bungii* in the Chukchi and northern Bering Seas in October 1988. *Proceedings of the NIPR Symposium on Polar Biology* **6**, 27–37.
Ohtsuka, S., Boxshall, G.A. and Roe, H.S.J. (1994). Phylogenetic relationships between arietellid genera (Copepoda: Calanoida), with the establishment of three new genera. *Bulletin of the Natural History Museum, London (Zoology)* **60**, 105–172.
Ohtsuka, S., Ueda, H. and Lian, G.-S. (1995). *Tortanus derjugini* Smirnov (Copepoda: Calanoida) from the Ariake Sea, western Japan, with notes on the zoogeography of brackish-water calanoid copepods in East Asia. *Bulletin of Plankton Society of Japan* **42**, 147–162.
Ohtsuka, S., Shimozu, M., Tanimura, A., Fukuchi, M., Hattori, H., Sasaki, H. and Matsuda, O. (1996a). Relationships between mouthpart structures and *in situ* feeding habits of five neritic calanoid copepods in the Chukchi and northern Bering Seas in October 1988. *Proceedings of the NIPR Symposium on Polar Biology* **9**, 153–168.
Ohtsuka, S., Fosshagen, A. and Soh, H.S. (1996b). Three new species of the demersal calanoid copepod *Placocalanus* (Ridgewayiidae) from Okinawa, southern Japan. *Sarsia* **81**, 247–263.
Ommanney, F.D. (1936). *Rhincalanus gigas* (Brady) a copepod of the southern macroplankton. *Discovery Report* **13**, 277–384.
Omori, M. (1978a). Some factors affecting on dry weight, organic weight and concentrations of carbon and nitrogen in freshly prepared and in preserved

zooplankton. *Internationale Revue der gesamten Hydrobiologie* **63**, 261–269.
Omori, M. (1978b). Zooplankton fisheries of the world: a review. *Marine Biology* **48**, 199–205.
Omori, M. and Hamner, W.M. (1982). Patchy distribution of zooplankton: behavior, population assessment and sampling problems. *Marine Biology* **72**, 193–200.
Omori, M. and Ikeda, T. (1984). "Methods in Marine Zooplankton Ecology." John Wiley & Sons, New York 332 pp.
Onbé, T., Hotta, T. and Ohtsuka, S. (1988). The developmental stages of the marine calanoid copepod *Labidocera rotunda* Mori. *Journal of the Faculty of Applied Biological Science, Hiroshima University* **27**, 79–91.
Oosterhuis, S.S. and Baars, M.A. (1985). On the usefulness of digestive enzyme activity as index for feeding activity in copepods. *Hydrobiological Bulletin* **19**, 89–100.
Øresland, V. (1991). Feeding of the carnivorous copepod *Euchaeta antarctica* in Antarctic waters. *Marine Ecology Progress Series* **78**, 41–47.
Øresland, V. (1995). Winter population structure and feeding of the chaetognath *Eukrohnia hamata* and the copepod *Euchaeta antarctica* in Gerlache Strait, Antarctic Peninsula. *Marine Ecology Progress Series* **119**, 77–86.
Øresland, V. and Ward, P. (1993). Summer and winter diet of four carnivorous copepod species around South Georgia. *Marine Ecology Progress Series* **98**, 73–78.
Orsi, J.J. and Walter, T.C. (1991). *Pseudodiaptomus forbesi* and *P. marinus* (Copepoda: Calanoida), the latest copepod immigrants to California's Sacramento-San Joaquin Estuary. *Bulletin of Plankton Society of Japan*, Special Volume, 553–562.
Orsi, J.J., Bowman, T.E., Marelli, D.C. and Hutchinson, A. (1983). Recent introduction of the planktonic calanoid copepod *Sinocalanus doerrii* (Centropagidae) from mainland China to the Sacramento-San Joaquin estuary of California. *Journal of Plankton Research* **5**, 357–375.
Ortner, P.B., Hill, L.C. and Cummings, S.R. (1989). Zooplankton community structure and copepod species composition in the northern Gulf of Mexico. *Continental Shelf Research* **9**, 387–402.
Osborn, T. (1996). The role of turbulent diffusion for copepods with feeding currents. *Journal of Plankton Research* **18**, 185–195.
Osgood, K.E. and Frost, B.W. (1994a). Ontogenetic diel vertical migration behaviors of the marine planktonic copepods *Calanus pacificus* and *Metridia lucens*. *Marine Ecology Progress Series* **104**, 13–25.
Osgood, K.E. and Frost, B.W. (1994b). Comparative life histories of three species of planktonic calanoid copepods in Dabob Bay, Washington. *Marine Biology* **118**, 627–636.
Ostrovskaya, N.A., Pavlova, E.V. and Sazhina, L.I. (1982). Quantitative characteristics of linear and weight growth of *Scolecithrix danae* (Lubbock) from the Indian Ocean tropical region. *Ekologiya Morya* **9**, 94–100.
Østvedt, O.-J. (1955). Zooplankton investigations from Weather Ship M in the Norwegian Sea, 1948–49. *Hvalrådets Skrifter* **40**, 1–93.
Ota, A.Y. and Landry, M.R. (1984). Nucleic acids as growth rate indicators for early developmental stages of *Calanus pacificus* Brodsky. *Journal of Experimental Marine Biology and Ecology* **80**, 147–160.
Othman, B.H.R. (1987). Two new species of *Tortanus* (Crustacea, Copepoda) from Sabah, Malaysia. *Malayan Nature Journal* **41**, 61–73.
Othman, B.H.R. and Greenwood, J.G. (1987). A new species of *Bradyidius*

(Copepoda, Calanoida) from the Gulf of Carpentaria, Australia. *Journal of Plankton Research* **9**, 1133–1141.
Othman, B.H.R. and Greenwood, J.G. (1988a). *Brachycalanus rothlisbergi*, a new species of planktobenthic copepod (Calanoida, Phaennidae) from the Gulf of Carpentaria, Australia. *Records Australian Museum* **40**, 353–358.
Othman, B.H.R. and Greenwood, J.G. (1988b). A new species of *Ridgewayia* (Copepoda, Calanoida) from the Gulf of Carpentaria. *Memoirs Australian Museum* **25**, 465–469.
Othman, B.H.R. and Greenwood, J.G. (1989). Two new species of copepods from the family Pseudocyclopidae (Copepoda, Calanoida). *Crustaceana* **56**, 63–77.
Othman, B.H.R. and Greenwood, J.G. (1992). *Paramisophria fosshageni*, a new species of copepod (Calanoida, Arietellidae) from the Gulf of Carpentaria. *Proceedings of the Royal Society of Queensland* **102**, 49–56.
Othman, B.H.R. and Greenwood, J.G. (1994). A new genus with three new species of copepods from the family Diaixidae (Crustacea: Calanoida), and a redefinition of the family. *Journal of Natural History* **28**, 987–1005.
Ough, K. and Bayly, I.A.E. (1989). Salinity tolerance, development rates and predation capabilities of *Sulcanus conflictus* Nicholls (Copepoda: Calanoida). *Estuarine, Coastal and Shelf Science* **28**, 195–209.
Owens, L. and Rothlisberg, P.C. (1995). Epidemiology of cryptonisci (Bopyridae: Isopoda) in the Gulf of Carpentaria, Australia. *Marine Ecology Progress Series* **122**, 159–164.
Owre, H.B. and Foyo, M. (1967). Copepods of the Florida Current. Fauna Caribaea 1, Crustacea, Part 1: Copepoda, pp. 1–137.
Paffenhöfer, G.-A. (1970). Cultivation of *Calanus helgolandicus* under controlled conditions. *Helgoländer wissenschaftliche Meeresuntersuchungen* **20**, 346–359.
Paffenhöfer, G.-A. (1971). Grazing and ingestion rates of nauplii, copepodids and adults of the marine planktonic copepod *Calanus helgolandicus*. *Marine Biology* **11**, 286–298.
Paffenhöfer, G.-A. (1976). Feeding, growth, and food conversion of the marine planktonic copepod *Calanus helgolandicus*. *Limnology and Oceanography* **21**, 39–50.
Paffenhöfer, G.-A. (1982). Grazing by copepods in the Peru upwelling. *Deep-Sea Research* **29A**, 145–147.
Paffenhöfer, G.-A. (1984a). Does *Paracalanus* feed with a leaky sieve? *Limnology and Oceanography* **29**, 155–160.
Paffenhöfer, G.-A. (1984b). Food ingestion by the marine planktonic copepod *Paracalanus* in relation to abundance and size distribution of food. *Marine Biology* **80**, 323–333.
Paffenhöfer, G.-A. (1988). Feeding rates and behavior of zooplankton. *Bulletin of Marine Science* **43**, 430–445.
Paffenhöfer, G.-A. (1991). Some characteristics of abundant subtropical copepods in estuarine, shelf and oceanic waters. *Bulletin of Plankton Society of Japan*, Special Volume, 201–216.
Paffenhöfer, G.-A. (1994). Variability due to feeding activity of individual copepods. *Journal of Plankton Research* **16**, 617–626.
Paffenhöfer, G.-A. and Gardner, W.S. (1984). Ammonium release by juveniles and adult females of the subtropical marine copepod *Eucalanus pileatus*. *Journal of Plankton Research* **6**, 505–513.
Paffenhöfer, G.-A. and Harris, R.P. (1976). Feeding, growth and reproduction of the marine planktonic copepod *Pseudocalanus elongatus*. *Journal of the Marine*

Biological Association of the United Kingdom **56**, 327–344.
Paffenhöfer, G.-A. and Harris, R.P. (1979). Laboratory culture of marine holozooplankton and its contribution to studies of marine planktonic food webs. Advances in Marine Biology **16**, 211–308.
Paffenhöfer, G.-A. and Knowles, S.C. (1978). Feeding of marine planktonic copepods on mixed phytoplankton. Marine Biology **48**, 143–152.
Paffenhöfer, G.-A. and Knowles, S.C. (1979). Ecological implications of fecal pellet size, production and consumption by copepods. Journal of Marine Research **37**, 35–49.
Paffenhöfer, G.-A. and Knowles, S.C. (1980). Omnivorousness in marine planktonic copepods. Journal of Plankton Research **2**, 355–365.
Paffenhöfer, G.-A. and Lewis, K.D. (1989). Feeding behavior of nauplii of the genus Eucalanus (Copepoda, Calanoida). Marine Ecology Progress Series **57**, 129–136.
Paffenhöfer, G.-A. and Lewis, K.D. (1990). Perceptive performance and feeding behavior of calanoid copepods. Journal of Experimental Marine Biology and Ecology **12**, 933–946.
Paffenhöfer, G.-A. and Stearns, D.E. (1988). Why is Acartia tonsa (Copepoda: Calanoida) restricted to nearshore environments. Marine Ecology Progress Series **42**, 33–38.
Paffenhöfer, G.-A., Strickler, J.R. and Alcaraz, M. (1982). Suspension-feeding by herbivorous calanoid copepods: a cinematographic study. Marine Biology **67**, 193–199.
Paffenhöfer, G.-A., Bundy, M.H., Lewis, K.D. and Metz, C. (1995). Rates of ingestion and their variability between individual calanoid copepods: direct observations. Journal of Plankton Research **17**, 1573–1585.
Paffenhöfer, G.-A., Strickler, J.R., Lewis, K.D. and Richman, S. (1996). Motion behavior of nauplii and early copepodid stages of marine planktonic copepods. Journal of Plankton Research **18**, 1699–1715.
Pagano, M. (1981a). Observations sur le cycle annuel d'Eurytemora velox (Lilljeborg, 1853) copépode calanoide des mares saumatres de Camargue. Rapports et Procès-Verbaux des Réunions Commission Internationale pour l'Exploration Scientifique de la Mer Méditerranée **27** (4), 145–146.
Pagano, M. (1981b). Etude biometrique et mesures de la biomasse a differents stades du developpement chez le copépode calanoide Eurytemora velox. Rapports et Procès-Verbaux des Réunions Commission Internationale pour l'Exploration Scientifique de la Mer Méditerranée **27** (4), 147–150.
Pagano, M. and Gaudy, R. (1986). Biologie d'un copépode des mares temporaires du littoral méditerranéen français: Eurytemora velox. II. Respiration et excrétion. Marine Biology **93**, 127–136.
Pagano, M. and Saint-Jean, L. (1989). Biomass and production of the calanoid copepod Acartia clausi in a tropical lagoon: Lagune Ebrié, Ivory coast. Scientia Marina **53**, 617–624.
Pagano, M. and Saint-Jean, L. (1993). Organic matter, carbon, nitrogen and phosphorus contents of the mesozooplankton, mainly Acartia clausi, in a tropical brackish lagoon (Ebrié Lagoon, Ivory Coast). Internationale Revue der gesamten Hydrobiologie **78**, 139–149.
Pagano, M. and Saint-Jean, L. (1994). In-situ metabolic budget for the calanoid copepod Acartia clausi in a tropical brackish water lagoon (Ebrié Lagoon, Ivory Coast). Hydrobiologia **272**, 147–161.
Pagano, M., Gaudy, R., Thibault, D. and Lochet, F. (1993). Vertical migrations and

feeding rhythms of mesozooplanktonic organisms in the Rhône River plume area (north-west Mediterranean Sea). *Estuarine, Coastal and Shelf Science* **37**, 251–269.
Park, C. and Landry, M.R. (1993). Egg production by the subtropical copepod *Undinula vulgaris*. *Marine Biology* **117**, 415–421.
Park, J.S. (1995a). The development of integumental pore signatures in the genus *Pleuromamma* (Copepoda: Calanoida). *Journal of the Marine Biological Association of the United Kingdom* **75**, 211–218.
Park, J.S. (1995b). Biology of Deep-sea Calanoid Copepod Genus *Pleuromamma* with Particular References to Phylogeny, Pore Signatures, Moulting and Life History. Ph.D Thesis, University of Stirling, Scotland. 208 pp.
Park, J.S. (1996). Intraspecific variation in integumental pore signatures in the genus *Pleuromamma* (Copepoda: Calanoida). *Journal of Natural History* **30**, 1007–1020.
Park, J.S. and Mauchline, J. (1994). Evaluation of integumental pore signatures of species of calanoid copepods (Crustacea) for interpreting inter-species relationships. *Marine Biology* **120**, 107–114.
Park, T. (1966). The biology of a calanoid copepod *Epilabidocera amphitrites* McMurrich. *La Cellule* **66**, 129–251.
Park, T. (1968). Calanoid copepods from the central North Pacific Ocean. *Fishery Bulletin, United States Fish and Wildlife Service* **66**, 527–572.
Park, T. (1970). Calanoid copepods from the Caribbean Sea and Gulf of Mexico. 2. New species and new records from plankton samples. *Bulletin of Marine Science, Gulf and Caribbean* **20**, 472–546.
Park, T. (1973). Calanoid copepods of the genus *Aetideus* from the Gulf of Mexico. *Fishery Bulletin, National Oceanic and Atmospheric Administration of the United States* **72**, 215–221.
Park, T. (1975a). Calanoid copepods of the genera *Gaetanus* and *Gaidius* from the Gulf of Mexico. *Bulletin of Marine Science* **25**, 9–34.
Park, T. (1975b). Calanoid copepods of the family Euchaetidae from the Gulf of Mexico and western Caribbean Sea. *Smithsonian Contributions to Zoology* **196**, 1–26.
Park, T. (1976). Calanoid copepods of the genus *Euchirella* from the Gulf of Mexico. *Contributions in Marine Science* **20**, 101–122.
Park, T. (1978). Calanoid copepods (Aetideidae and Euchaetidae) from Antarctic and Subantarctic waters. *Biology of Antarctic Seas VII, Antarctic Research Series* **27**, 91–290.
Park, T. (1980). Calanoid copepods of the genus *Scolecithricella* from Antarctic and Subantarctic waters. *Biology of Antarctic Seas IX, Antarctic Research Series* **31**, 25–79.
Park, T. (1982). Calanoid copepods of the genus *Scaphocalanus* from Antarctic and Subantarctic waters. *Biology of Antarctic Seas XI, Antarctic Research Series* **34**, 75–127.
Park, T. (1983a). Calanoid copepods of some scolecithricid genera from the Antarctic and Subantarctic waters. *Biology of Antarctic Seas XIII, Antarctic Research Series* **38**, 165–213.
Park, T. (1983b). Calanoid copepods of the family Phaennidae from Antarctic and Subantarctic waters. *Biology of Antarctic Seas XIV, Antarctic Research Series* **39**, 317–368.
Park, T. (1986). Phylogeny of calanoid copepods. *Syllogeus* **58**, 191–196.
Park, T. (1988). Calanoid copepods of the genus *Haloptilus* from Antarctic and

Subantarctic Waters. *Biology of Antarctic Seas XIX, Antarctic Research Series* **47**, 1–25.
Park, T. (1993). Calanoid copepods of the genus *Euaugaptilus* from Antarctic and Subantarctic waters. *Biology of Antarctic Seas XXII, Antarctic Research Series* **58**, 1–48.
Park, T. (1994a). Taxonomy and distribution of the marine calanoid copepod family Euchaetidae. *Bulletin of the Scripps Institution of Oceanography* **29**, 1–203.
Park, T. (1994b). Geographic distribution of the bathypelagic genus *Paraeuchaeta* (Copepoda, Calanoida). *Hydrobiologia* **292/293**, 317–332.
Parker, G.H. (1901). The reactions of copepods to various stimuli and the bearing of this on daily depth-migrations. *Bulletin of the United States Fish Commission* **21**, 103–123.
Parrish, K.K. and Wilson, D.F. (1978). Fecundity studies on *Acartia tonsa* (Copepoda: Calanoida) in standardized culture. *Marine Biology* **46**, 65–81.
Parsons, T.R. and Lalli, C.M. (1988). Comparative oceanic ecology of the plankton communities of the subarctic Atlantic and Pacific Oceans. *Oceanography and Marine Biology: an Annual Review* **26**, 317–359.
Parsons, T.R., LeBrasseur, R.J., Fulton, J.D. and Kennedy, O.D. (1969). Production studies in the Strait of Georgia. Part II. Secondary production under the Fraser River Plume, February to May, 1967. *Journal of Experimental Marine Biology and Ecology* **3**, 39–50.
Parsons, T.R., Takahashi, M. and Hargrave, B. (1984). Biological Oceanographic Processes. Third Edition. Pergamon Press, Oxford. 330 pp.
Pasternak, A.F. (1994). Gut fluorescence in herbivorous copepods: an attempt to justify the method. *Hydrobiologia* **292/293**, 241–248.
Pasternak, A.F. (1995). Gut contents and diel feeding rhythm in dominant copepods in the ice-covered Weddell Sea, March 1992. *Polar Biology* **15**, 583–586.
Paul, A.J., Coyle, K.O. and Ziemann, D.A. (1990). Variations in egg production rates by *Pseudocalanus* spp. in a subarctic Alaskan bay during the onset of feeding by larval fish. *Journal of Crustacean Biology* **10**, 648–658.
Pavlova, E.V. (1964). On the occurrence of Mediterranean species in the plankton of the Black Sea. *Zoologichesky Zhurnal* **43**, 1710–1713.
Pavlova, E.V. (1981). Motion rate of copepods from the Indian Ocean plankton. *Ekologiya Morya* **5**, 61–65.
Pavlova, E.V. (1994). Diel changes in copepod respiration rates. *Hydrobiologia* **292/293**, 333–339.
Pavlova, E.V., Parchevsky, V.P. and Prazukin, A.V. (1982). Studies of locomotion character in marine copepods by the basic components method under laboratory conditions. *Ekologiya Morya* **11**, 42–53.
Pavlova, E.V., Morozova, A.L., Gaudy, R. and Shchepkin, V.Ya. (1989). Application of biochemical substrates and respiration during short-term starvation of small planktonic Crustacea. *Ekologiya Morya* **33**, 41–46.
Pearre, S. (1979a). On the adaptive significance of vertical migration. *Limnology and Oceanography* **24**, 781–782.
Pearre, S. (1979b). Problems of detection and interpretation of vertical migration. *Journal of Plankton Research* **1**, 29–44.
Pearre, S. Jr (1980). The copepod width-weight relation and its utility in food chain research. *Canadian Journal of Zoology* **58**, 1884–1892.
Pedersen, G. and Tande, K.S. (1992). Physiological plasticity to temperature in *Calanus finmarchicus*. Reality or artefact? *Journal of Experimental Marine Biology and Ecology* **155**, 183–197.

Pedersen, K., Tande, K. and Ottesen, G.O. (1995). Why does a component of *Calanus finmarchicus* stay in the surface waters during the overwintering period in high latitudes? *ICES Journal of Marine Science* **52**, 523–531.
Peitsch, A. (1993). Difficulties in estimating mortality rates of *Eurytemora affinis* in the brackish water region of the Elbe estuary. *Cahiers de Biologie marine* **34**, 215–224.
Peitsch, A. (1995). Production rates of *Eurytemora affinis* in the Elbe estuary, comparison of field and enclosure estimates. *Hydrobiologia* **311**, 127–137.
Penry, D.L. and Frost, B.W. (1990). Re-evaluation of the gut-fullness (gut fluorescence) method for inferring ingestion rates of suspension-feeding copepods. *Limnology and Oceanography* **35**, 1207–1214.
Perissinotto, R. (1992). Mesozooplankton size-selectivity and grazing impact on the phytoplankton community of the Prince Edward Archipelago (Southern Ocean). *Marine Ecology Progress Series* **79**, 243–258.
Person-Le Ruyet, J. (1975). Élevage de copépodes calanoides. Biologie et dynamique des populations: premiers résultats. *Annals de l'Institut Océanographique* **51**, 203–221.
Person-Le Ruyet, J., Razouls, C. and Razouls, S. (1975). Biologie comparée entre espèces vicariantes et communes de copépodes dans un écosystème néritique en Méditerranée et en Manche. *Vie et Milieu* **25 B**, 283–312.
Pertzova, N.M. (1974). Life cycle and ecology of a thermophilous copepod *Centropages hamatus* in the White Sea. *Zoologichesky Zhurnal* **53**, 1013–1022.
Pertzova, N.M. (1981). Number of generations and their span in *Pseudocalanus elongatus* (Copepoda, Calanoida) in the White Sea. *Zoologichesky Zhurnal* **60**, 673–684.
Pessotti, E., Razouls, C. and Razouls, S. (1986). Distribution de taille d'une espèce de copépode en relation avec sa distribution spatiale. *Syllogeus* **58**, 409–419.
Peters, R.H. and Downing, J.A. (1984). Empirical analysis of zooplankton filtering and feeding rates. *Limnology and Oceanography* **29**, 763–784.
Peterson, W.T. (1985). Abundance, age structure and *in situ* egg production rates of the copepod *Temora longicornis* in Long Island Sound, New York. *Bulletin of Marine Science* **37**, 726–738.
Peterson, W.T. (1986). Development, growth, and survivorship of the copepod *Calanus marshallae* in the laboratory. *Marine Ecology Progress Series* **29**, 61–72.
Peterson, W.T. (1988). Rates of egg production by the copepod *Calanus marshallae* in the laboratory and in the sea off Oregon, USA. *Marine Ecology Progress Series* **47**, 229–237.
Peterson, W.T. and Bellantoni, D.C. (1987). Relationships between water-column stratification, phytoplankton cell size and copepod fecundity in Long Island Sound and off central Chile. *South African Journal of Marine Science* **5**, 411–421.
Peterson, W.T. and Dam, H.G. (1996). Pigment ingestion and egg production rates of the calanoid copepod *Temora longicornis*: implications for gut pigment loss and omnivorous feeding. *Journal of Plankton Research* **18**, 855–861.
Peterson, W.T. and Hutchings, L. (1995). Distribution, abundance and production of the copepod *Calanus agulhensis* on the Agulhas Bank in relation to spatial variations in hydrography and chlorophyll concentration. *Journal of Plankton Research* **17**, 2275–2294.
Peterson, W.T. and Kimmerer, W.J. (1994). Processes controlling recruitment of the marine calanoid copepod *Temora longicornis* in Long Island Sound: egg production, egg mortality, and cohort survival rates. *Limnology and Oceanog-*

raphy **39**, 1594–1605.

Peterson, W.T. and Miller, C.B. (1977). Seasonal cycle of zooplankton abundance and species composition along the central Oregon coast. *Fishery Bulletin, National Oceanic and Atmospheric Administration of the United States* **75**, 717–724.

Peterson, W.T. and Painting, S.J. (1990). Developmental rates of the copepods *Calanus australis* and *Calanoides carinatus* in the laboratory, with discussion of methods used for calculation of development time. *Journal of Plankton Research* **12**, 283–293.

Peterson, W.T., Miller, C.B. and Hutchinson, A. (1979). Zonation and maintenance of copepod populations in the Oregon upwelling zone. *Deep-Sea Research* **26A**, 467–494.

Peterson, W.T., Arcos, D.F., McManus, G.B., Dam, H., Bellantoni, D., Johnson, T. and Tiselius, P. (1988). The nearshore zone during coastal upwelling: daily variability and coupling between primary and secondary production off central Chile. *Progress in Oceanography* **20**, 1–40.

Peterson, W., Painting, S. and Barlow, R. (1990a). Feeding rates of *Calanoides carinatus*: a comparison of five methods including evaluation of the gut fluorescence method. *Marine Ecology Progress Series* **63**, 85–92.

Peterson, W.T., Painting, S.J. and Hutchings, L. (1990b). Diel variations in gut pigment content, diel vertical migration and estimates of grazing impact for copepods in the southern Benguela upwelling region in October 1987. *Journal of Plankton Research* **12**, 259–281.

Peterson, W.T., Tiselius, P. and Kiørboe, T. (1991). Copepod egg production, moulting and growth rates, and secondary production, in the Skagerrak in August, 1988. *Journal of Plankton Research* **13**, 131–154.

Petipa, T.S. (1985). Production and concentration of plankton at boundary water masses: perspectives of investigations. *In* "Proceedings of the Nineteenth European Marine Biology Symposium" (P.E. Gibbs, ed.), pp. 61–71. Cambridge University Press, Cambridge.

Petipa, T.S. and Ovstrovskaya, N.A. (1989). A new way for estimating active exchange and efficiency of chemical energy use in migration of copepods. *Ekologiya Morya* **33**, 54–56.

Petit, D. and Courties, C. (1976). *Calanoides carinatus* (Copépode pélagique) sur le Plateau Continental Congolais. I. Aperçu sur la répartition bathymétrique, géographique et biométrique des stades; générations durant la saison froide 1974. *Cahiers O.R.S.T.O.M., Series Océanographie* **14**, 177–199.

Pillai, P.P. (1971). On the post-naupliar development of the calanoid copepod *Labidocera pectinata* Thompson and Scott (1903). *Journal of the Marine Biological Association of India* **13**, 66–77.

Pillai, P.P. (1975a). Post-naupliar development of the calanoid copepod *Temora turbinata* (Dana), with remarks on the distribution of the species of the genus *Temora* in the Indian Ocean. *Journal of the Marine Biological Association of India* **17**, 87–95.

Pillai, P.P. (1975b). On the species of *Pontella* Dana and *Pontellopsis* Brady of the International Indian Ocean Expedition collections (1960–1965). *Journal of the Marine Biological Association of India* **17**, 129–146.

Pillai, P.P. (1980). A review of the calanoid copepod family Pseudodiaptomidae with remarks on the taxonomy and distribution of the species from the Indian Ocean. *Journal of the Marine Biological Association of India* **18**, 1976, 242–265.

Piontkovski, S.A. and Williams, R. (1995). Multiscale variability of tropical ocean zooplankton biomass. *ICES Journal of Marine Science* **52**, 643–656.

Piontkovski, S.A., Williams, R., Peterson, W. and Kosnirev, V.K. (1995). Relationship between oceanic mesozooplankton and energy of eddy fields. *Marine Ecology Progress Series* **128**, 35–41.
Pipe, R.K. and Coombs, S.H. (1980). Vertical distribution of zooplankton over the northern slope of the Wyville Thompson Ridge. *Journal of Plankton Research* **2**, 223–234.
Planque, B. and Fromentin, J.-M. (1996). *Calanus* and environment in the eastern North Atlantic. I. Spatial and temporal patterns of *C. finmarchicus* and *C. helgolandicus*. *Marine Ecology Progress Series* **134**, 101–109.
Plourde, S. and Runge, J.A. (1993). Reproduction of the planktonic copepod *Calanus finmarchicus* in the Lower St. Lawrence Estuary: relation to the cycle of phytoplankton production and evidence for a *Calanus* pump. *Marine Ecology Progress Series* **102**, 217–227.
Poli, J.M. and Castel, J. (1983). Cycle biologique en laboratoire d'un copépode planctonique de l'estuaire de la Gironde: *Eurytemora hirundoides* (Nordquist, 1888). *Vie et Milieu* **33**, 79–86.
Pond, D., Harris, R., Head, R. and Harbour, D. (1996). Environmental and nutritional factors determining seasonal variability in the fecundity and egg viability of *Calanus helgolandicus* in coastal waters off Plymouth, UK. *Marine Ecology Progress Series* **143**, 45–63.
Pond, D.W., Harris, R.P. and Brownlee, C. (1995). A microinjection technique using a pH-sensitive dye to determine the gut pH of *Calanus helgolandicus*. *Marine Biology* **123**, 75–79.
Porumb, F.I. (1974). Production des copépodes pélagiques dans les eaux roumaines de la mer Noire. *Rapports et Procès-Verbaux des Réunions Commission Internationale pour l'Exploration Scientifique de la Mer Méditerranée* **22** (9), 91–92.
Potts, G.W. (1976). A diver-controlled plankton net. *Journal of the Marine Biological Association of the United Kingdom* **56**, 959–962.
Poulet, S.A. (1974). Seasonal grazing of *Pseudocalanus minutus* on particles. *Marine Biology* **25**, 109–123.
Poulet, S.A. (1978). Comparison between five coexisting species of marine copepods feeding on naturally occurring particulate matter. *Limnology and Oceanography* **23**, 1126–1143.
Poulet, S.A. (1983). Factors controlling utilization of non-algal diets by particle-grazing copepods. A review. *Oceanologica Acta* **6**, 221–234.
Poulet, S.A. and Gill, C.W. (1988). Spectral analyses of movements made by the cephalic appendages of copepods. *Marine Ecology Progress Series* **43**, 259–267.
Poulet, S.A. and Marsot, P. (1978). Chemosensory grazing by marine calanoid copepods (Arthropoda: Crustacea). *Science* **200**, 1403–1405.
Poulet, S.A. and Marsot, P. (1980). Chemosensory feeding and food-gathering by omnivorous marine copepods. In "Evolution and Ecology of Zooplankton Communities" (W.C. Kerfoot, ed.), pp. 198–218. University Press of New England, Hanover, N.H.
Poulet, S.A. and Ouellet, G. (1982). The role of amino acids in the chemosensory swarming and feeding of marine copepods. *Journal of Plankton Research* **4**, 341–361.
Poulet, S.A., Samain, J.F. and Moal, J. (1986a). Chemoreception, nutrition and food requirements among copepods. *Syllogeus* **58**, 426–442.
Poulet, S.A., Harris, R.P., Martin-Jezequel, V., Moal, J. and Samain, J.-F. (1986b).

Free amino acids in copepod faecal pellets. *Oceanologica Acta* **9**, 191–197.
Poulet, S.A., Hapette, A.M., Cole, R.B. and Tabet, J.C. (1989). Vitamin C in marine copepods. *Limnology and Oceanography* **34**, 1331–1335.
Poulet, S.A., Williams, R., Conway, D.V.P. and Videau, C. (1991). Co-occurrence of copepods and dissolved free amino acids in shelf waters. *Marine Biology* **108**, 373–385.
Poulet, S.A., Ianora, A., Miralto, A. and Meijer, L. (1994). Do diatoms arrest embryonic development in copepods. *Marine Ecology Progress Series* **111**, 79–86.
Poulet, S.A., Ianora, A., Laabir, M. and Klein Breteler, W.C.M. (1995a). Towards the measurement of secondary production and recruitment in copepods. *ICES Journal of Marine Science* **52**, 359–368.
Poulet, S.A., Laabir, M., Ianora, A. and Miralto, A. (1995b). Reproductive response of *Calanus helgolandicus*. I. Abnormal embryonic and naupliar development. *Marine Ecology Progress Series* **129**, 85–95.
Powell, M.D. and Berry, A.J. (1990). Ingestion and regurgitation of living and inert materials by the estuarine copepod *Eurytemora affinis* (Poppe) and the influence of salinity. *Estuarine, Coastal and Shelf Science* **31**, 763–773.
Power, J.H. (1996). Simulations of the effect of advective-diffusive processes on observations of plankton abundance and population rates. *Journal of Plankton Research* **18**, 1881–1896.
Prahl, F.G., Eglinton, G., Corner, E.D.S., O'Hara, S.C.M. and Forsberg, T.E.V. (1984). Changes in plant lipids during passage through the gut of *Calanus*. *Journal of the Marine Biological Association of the United Kingdom* **64**, 317–334.
Prestidge, M.C., Harris, R.P. and Taylor, A.H. (1995). A modelling investigation of copepod egg production in the Irish Sea. *ICES Journal of Marine Science* **52**, 693–703.
Price, H.J. (1988). Feeding mechanisms in marine and freshwater zooplankton. *Bulletin of Marine Science* **43**, 327–343.
Price, H.J. and Paffenhöfer, G.-A. (1984). Effects of feeding experience in the copepod *Eucalanus pileatus*: a cinematographic study. *Marine Biology* **84**, 35–40.
Price, H.J. and Paffenhöfer, G.-A. (1985). Perception of food availability by calanoid copepods. *Archives fur Hydrobiologie Beiheft Ergebnisse Limnologie* **21**, 115–124.
Price, H.J. and Paffenhöfer, G.-A. (1986a). Capture of small cells by the copepod *Eucalanus elongatus*. *Limnology and Oceanography* **31**, 189–194.
Price, H.J. and Paffenhöfer, G.-A. (1986b). Effects of concentration on the feeding of a marine copepod in algal monocultures and mixtures. *Journal of Plankton Research* **8**, 119–128.
Price, H.J., Paffenhöfer, G.-A. and Strickler, J.R. (1983). Modes of cell capture in calanoid copepods. *Limnology and Oceanography* **28**, 116–123.
Price, H.J., Paffenhöfer, G.-A., Boyd, C.M., Cowles, T.J., Donaghay, P.L., Hamner, W.M., Lampert, W., Quetin, L.B., Ross, R.M., Strickler, J.R. and Youngbluth, M.J. (1988). Future studies of zooplankton behavior: questions and technological developments. *Bulletin of Marine Science* **43**, 853–872.
Purcell, J.E., White, J.R. and Roman, M.R. (1994). Predation by gelatinous zooplankton and resource limitation as potential controls of *Acartia tonsa* copepod populations in Chesapeake Bay. *Limnology and Oceanography* **39**, 263–278.
Rau, G.H., Hopkins, T.L. and Torres, J.J. (1991). $^{15}N/^{14}N$ and $^{13}C/^{12}C$ in Weddell

Sea invertebrates: implications for feeding diversity. *Marine Ecology Progress Series* **77**, 1–6.

Ray, G.C. and Hayden, B.P. (1993). Marine biogeographic provinces of the Bering, Chukchi, and Beaufort Seas. *In* "Large Marine Ecosystems: Stress, Mitigation, and Sustainability" (K. Sherman, L.M. Alexander and B.D. Gold, eds), pp. 175–184. American Association for the Advancement of Science Press, Washington, DC.

Raymont, J.E.G., Krishnaswamy, S., Woodhouse, M.A. and Griffin, R.L. (1974). Studies on the fine structure of Copepoda. Observations on *Calanus finmarchicus* (Gunnerus). *Proceedings of the Royal Society of London* **185 B**, 409–424.

Rayner, N.A. (1994). *Tropodiaptomus zambeziensis*, *T. bhangazii* and *T. capriviensis*, three new species of *Tropodiaptomus* (Copepoda, Calanoida) from southern Africa. *Hydrobiologia* **292/293**, 97–104.

Razouls, C. (1974). Variations annuelles quantitatives de deux espèces dominantes de copépodes planctoniques *Centropages typicus* et *Temora stylifera* de la région de Banyuls: cycles biologiques et estimations de la production. III. Dynamique des populations et calcul de leur production. *Cahiers de Biologie marine* **15**, 51–88.

Razouls, C. (1982). Répertoire mondial taxonomique et bibliographique provisoire des copépodes planctoniques marins et des eaux saumâtres. Divers systèmes de classification. Vol. 1, pp. 1–394 (SN 82 400 340, microfiche 83 093 39); Vol. 2, pp. 395–875 (SN 82 400 340, microfiche 83 03 40). Banyuls-sur-Mer and Institute d'Ethnologie, Paris.

Razouls, C. (1991). Bilan actuel des copépodes planctoniques marins et des eaux saumâtres. Corrections et complements. 240 pp. (Sn 91 400 537, microfiche 91 05 37). Banyuls-sur-Mer and Institute d'Ethnologie, Paris.

Razouls, C. (1992). Inventaire des copépodes planctoniques marins Antarctiques et Sub-Antarctiques. *Vie et Milieu* **42**, 337–343.

Razouls, C. (1993). Bilan taxonomique actuel des copépodes planctoniques marins et des eaux saumâtres. *Crustaceana* **64**, 300–313.

Razouls, C. (1994). Manuel d'identification des principales espèces de copépodes pélagiques antarctiques et subantarctiques. *Annales de l'Institut Océanographique* **70**, 1–204.

Razouls, C. (1995). Diversité et répartition géographique chez les copépodes pelágiques 1. Calanoida. *Annales de l'Institut Océanographique* **71**, 81–401.

Razouls, C. (1996). Diversité et répartition géographique chez les copépodes pelágiques 2. Platycopioida, Misophrioida, Mormonilloida, Cyclopoida, Poecilostomatoida, Siphonostomatoida, Harpacticoida, Monstrilloida. *Annales de l'Institut Océanographique* **72**, 1–149.

Razouls, C. and Carola, M. (1996). The presence of *Ridgewayia marki minorcaensis* n. spp. in the western Mediterranean. *Crustaceana* **69**, 47–55.

Razouls, C. and Durand, J. (1991). Inventaire des copépodes planctoniques Méditerranéens. *Vie et Milieu* **41**, 73–77.

Razouls, C. and Razouls, S. (1976). Dimensions, poids sec, valeur calorifique et courbes de croissance de deux populations naturelles de copépodes planktoniques en Méditerranée. *Vie et Milieu* **26 B**, 281–297.

Razouls, S. (1975). Fécondité, maturité sexuelle et différenciation de l'appareil génital des femelles de deux copépodes planctoniques: *Centropages typicus* et *Temora stylifera*. *Pubblicazioni della Stazione Zoologica di Napoli* **39**, Supplement **1**, 297–306.

Razouls, S. (1981). Étude expérimentale de la ponte des copépodes planctoniques

Temora stylifera et Centropages typicus. I. Influence des conditions expérimentales. Vie et Milieu **31**, 195–204.
Razouls, S. (1982). Ètude expérimentale de la ponte de deux copépodes pélagiques Temora stylifera et Centropages typicus. II. Dynamique des pontes. Vie et Milieu **32**, 11–20.
Razouls, S. (1985). Observations on the ecophysiology of a planktonic crustacean, Drepanopus pectinatus (Copepoda, Calanoida, Pseudocalanidae), from southern islands. In "Marine Biology of Polar Regions and Effects of Stress on Marine Organisms" (J.S. Gray and M.E. Christiansen, eds), pp. 123–139. J. Wiley & Sons, New York.
Razouls, S. and Apostolopoulou, M. (1977). Bilan énergétique de deux populations de copépodes pélagiques Temora stylifera et Centropages typicus, en relation avec la présence d'une thermocline. Vie et Milieu **27**, 13–25.
Razouls, S. and Razouls, C. (1988). Seasonal size distribution of developmental stages of sub-antarctic copepod. Hydrobiologia **167/168**, 239–246.
Razouls, S., Nival, S. and Nival, P. (1986). La reproduction de Temora stylifera: ses implications anatomiques en relation avec le facteur "nutrition". Journal of Plankton Research **8**, 875–889.
Razouls, S., Nival, P. and Nival, S. (1987). Development of the genital system in the copepodid stages of the calanoid copepod Temora stylifera Dana. Journal of the Marine Biological Association of the United Kingdom **67**, 653–661.
Razouls, S., Razouls, C. and Huntley, M. (1991). Development and expression of sexual maturity in female Calanus pacificus (Copepoda: Calanoida) in relation to food quality. Marine Biology **110**, 65–74.
Redden, A.M. and Daborn, G.R. (1991). Viability of subitaneous copepod eggs following fish predation on egg-carrying calanoids. Marine Ecology Progress Series **77**, 307–310.
Reddy, Y.Ranga and Devi, C.Rama (1985). The complete postembryonic development of Megadiaptomus hebes Kiefer, 1936 (Copepoda, Calanoida) reared in the laboratory. Crustaceana **48**, 40–63.
Reddy, Y.Ranga and Devi, C.Rama (1989). The complete postembryonic development of Heliodiaptomus contortus (Gurney, 1907) (Copepoda, Calanoida) reared in the laboratory. Crustaceana **57**, 113–133.
Reddy, Y.Ranga and Devi, C.Rama (1990). The complete postembryonic development of Heliodiaptomus cinctus (Gurney, 1907) (Copepoda, Calanoida), reared in the laboratory. Crustaceana **58**, 45–66.
Redeke, H.C. (1934). On the occurrence of two pelagic copepods, Acartia bifilosa and Acartia tonsa, in the brackish waters of the Netherlands. Journal du Conseil International pour l'Exploration de la Mer **9**, 39–45.
Reeve, M.R. and Walter, M.A. (1977). Observations on the existence of lower threshold and upper critical food concentrations for the copepod Acartia tonsa Dana. Journal of Experimental Marine Biology and Ecology **29**, 211–221.
Regner, D. (1976). On the copepods diversity in the central Adriatic in 1971. Rapports et Procès-Verbaux des Réunions Commission Internationale pour l'Exploration Scientifique de la Mer Méditerranée **23** (9), 95–96.
Regner, D. (1984). Seasonal and multiannual oscillations of copepod density in the central Adriatic. Crustaceana, Supplement **7**, 352–359.
Regner, D. (1991). Long-term investigations of copepods (zooplankton) in the coastal waters of the eastern middle Adriatic. Acta Adriatica **32**, 731–740.
Regner, D. and Regner, S. (1981). Diversity of some plankton taxocenosis in the central Adriatic. Rapports et Procès-Verbaux des Réunions Commission

Internationale pour l'Exploration Scientifique de la Mer Méditerranée **27** (7), 181–183.
Regner, D. and Vučetić, T. (1980). Seasonal and multiannual fluctuations of copepods in the Kaštela Bay (1960–1969). *Acta Adriatica* **21**, 101–122.
Reid, J.L. (1967). Oceanic environments of the genus *Engraulis* around the world. *Reports of the California Cooperative Oceanic Fisheries Investigations* **11**, 29–33.
Reid, J.L., Brinton, E., Fleminger, A., Venrick, E.L. and McGowan, J.A. (1978). Ocean circulation and marine life. *In* "Advances in Oceanography" (H. Charnock and Sir George Deacon, eds), pp. 65–130. Plenum Publishing Corporation, New York.
Reid, J.W. (1987). *Scolodiaptomus*, a new genus proposed for *Diaptomus* (*sensu lato*) *corderoi* Wright, and description of *Notodiaptomus brandorffi*, new species (Copepoda: Calanoida), from Brazil. *Journal of Crustacean Biology* **7**, 364–379.
Reinfelder, J.R., Fisher, N.S., Fowler, S.W. and Teyssié, J.-L. (1993). Release rates of trace elements and protein from decomposing planktonic debris. 2. Copepod carcasses and sediment trap particulate matter. *Journal of Marine Research* **51**, 423–442.
Richardson, K. (1985). Plankton distribution and activity in the North Sea/Skagerrak-Kattegat frontal area in April 1984. *Marine Ecology Progress Series* **26**, 233–244.
Richardson, K. (1997). Harmful or exceptional phytoplankton blooms in the marine ecosystem. *Advances in Marine Biology* **31**, 301–385.
Richman, S. and Rogers, J.N. (1969). The feeding of *Calanus helgolandicus* on synchronously growing populations of the marine diatom *Ditylum brightwellii*. *Limnology and Oceanography* **14**, 701–709.
Richman, S., Heinle, D.R. and Huff, R. (1977). Grazing by adult estuarine calanoid copepods of the Chesapeake Bay. *Marine Biology* **42**, 69–84.
Richter, C. (1995). Seasonal changes in the vertical distribution of mesozooplankton in the Greenland Sea Gyre (75°N): distribution strategies of calanoid copepods. *ICES Journal of Marine Science* **52**, 533–539.
Richter, K.E. (1985a). Acoustic scattering at 1.2 MHz from individual zooplankters and copepod populations. *Deep-Sea Research* **32A**, 149–161.
Richter, K.E. (1985b). Acoustic determination of small-scale distributions of individual zooplankters and zooplankton aggregations. *Deep-Sea Research* **32A**, 163–182.
Riera, T. (1983). Variabilidad morfométrica en *Temora stylifera* Dana, 1848. *Investigacion Pesquera* **47**, 363–396.
Riera, T., Vives, F. and Gili, J.-M. (1991). *Stephos margalefi* sp. nov. (Copepoda: Calanoida) from a submarine cave of Majorca Island (Western Mediterranean). *Oecologia Aquatica* **10**, 317–323.
Rijswijk, P. van, Bakker, C. and Vink, M. (1989). Daily fecundity of *Temora longicornis* (Copepoda Calanoida) in the Oosterschelde Estuary (SW Netherlands). *Netherlands Journal of Sea Research* **23**, 293–303.
Ringelberg, J. (1995). Is diel vertical migration possible without a rhythmic signal? Comments on a paper by Bollens *et al.* (1994). *Journal of Plankton Research* **17**, 653–655.
Rippingale, R.J. (1994). A calanoid copepod *Gladioferens imparipes*, holding to surfaces. *Hydrobiologia* **292/293**, 351–360.
Rippingale, R.J. and Hodgkin, E.P. (1974). Population growth of a copepod

Gladioferens imparipes Thomson. *Australian Journal of Marine and Freshwater Research* **25**, 351–360.

Rippingale, R.J. and Hodgkin, E.P. (1977). Food availability and salinity tolerance in a brackish water copepod. *Australian Journal of Marine and Freshwater Research* **28**, 1–7.

Ritz, D.A. (1995). Social aggregation in pelagic invertebrates. *Advances in Marine Biology* **30**, 155–216.

Robertson, A. (1968). The continuous plankton recorder: a method for studying the biomass of calanoid copepods. *Bulletin of Marine Ecology* **6**, 185–223.

Robertson, J.R. (1983). Predation of estuarine zooplankton on tintinnid ciliates. *Estuarine, Coastal and Shelf Science* **16**, 27–36.

Robertson, S.B. and Frost, B.W. (1977). Feeding by an omnivorous planktonic copepod *Aetideus divergens* Bradford. *Journal of Experimental Marine Biology and Ecology* **29**, 231–244.

Robichaux, D.M., Cohen, A.C., Reaka, M.L. and Allen, D. (1981). Experiments with zooplankton, or, will the real demersal plankton please come up? *P.S.Z.N.I: Marine Ecology* **2**, 77–94.

Roche-Mayzaud, O., Mayzaud, P. and Biggs, D.C. (1991). Medium-term acclimation of feeding and of digestive and metabolic enzyme activity in the neritic copepod *Acartia clausi*. I. Evidence from laboratory experiments. *Marine Ecology Progress Series* **69**, 25–40.

Roddie, B.D., Leakey, R.J.G. and Berry, A.J. (1984). Salinity-temperature tolerance and osmoregulation in *Eurytemora affinis* (Poppe) (Copepoda: Calanoida) in relation to its distribution in the zooplankton of the upper reaches of the Forth Estuary. *Journal of Experimental Marine Biology and Ecology* **79**, 191–211.

Rodriguez, V. and Durbin, E.G. (1992). Evaluation of synchrony of feeding behaviour in individual *Acartia hudsonica* (Copepoda, Calanoida). *Marine Ecology Progress Series* **87**, 7–13.

Rodriguez, V. and Jiménez, F. (1990). Co-existence within a group of congeneric species of *Acartia* (Copepoda: Calanoida): sexual dimorphism and ecological niche in *Acartia grani*. *Journal of Plankton Research* **12**, 497–511.

Rodríguez, V., Guerrero, F. and Bautista, B. (1995). Egg production of individual copepods of *Acartia grani* Sars from coastal waters: seasonal and diel variability. *Journal of Plankton Research* **17**, 2233–2250.

Roe, H.S.J. (1972a). The vertical distributions and diurnal migrations of calanoid copepods collected on the SOND cruise, 1965. I. The total population and general discussion. *Journal of the Marine Biological Association of the United Kingdom* **52**, 277–314.

Roe, H.S.J. (1972b). The vertical distributions and diurnal migrations of calanoid copepods collected on the SOND cruise, 1965. II. Systematic account: families Calanidae up to and including the Aetideidae. *Journal of the Marine Biological Association of the United Kingdom* **52**, 315–343.

Roe, H.S.J. (1972c). The vertical distributions and diurnal migrations of calanoid copepods collected on the SOND cruise, 1965. III. Systematic account: families Euchaetidae up to and including the Metridiidae. *Journal of the Marine Biological Association of the United Kingdom* **52**, 525–552.

Roe, H.S.J. (1972d). The vertical distributions and diurnal migrations of calanoid copepods collected on the SOND cruise, 1965. IV. Systematic account of families Lucicutiidae to Candaciidae. The relative abundance of the numerically most important genera. *Journal of the Marine Biological Association of the United Kingdom* **52**, 1021–1044.

Roe, H.S.J. (1974). Observations on the diurnal vertical migrations of an oceanic animal community. *Marine Biology* **28**, 99–113.
Roe, H.S.J. (1975). Some new and rare species of calanoid copepods from the northeastern Atlantic. *Bulletin of the British Museum (Natural History), Zoology* **28**, 295–372.
Roe, H.S.J. (1984). The diel migrations and distributions within a mesopelagic community in the north east Atlantic. 4. The copepods. *Progress in Oceanography* **13**, 353–388.
Roe, H.S.J. (1988). Midwater biomass profiles over the Madeira Abyssal Plain and the contribution of copepods. *Hydrobiologia* **167/168**, 169–181.
Roff, J.C. (1972). Aspects of the reproductive biology of the planktonic copepod *Limnocalanus macrurus* Sars, 1863. *Crustaceana* **22**, 155–160.
Roff, J.C. and Carter, J.C.H. (1972). Life cycle and seasonal abundance of the copepod *Limnocalanus macrurus* Sars in a high Arctic Lake. *Limnology and Oceanography* **17**, 363–370.
Roff, J.C. and Tremblay, M.J. (1984). Reply with additional notes on P/B ratios. *Canadian Journal of Fisheries and Aquatic Sciences* **41**, 830–833.
Roff, J.C., Middlebrook, K. and Evans, F. (1988). Long-term variability in North Sea zooplankton off the Northumberland coast: productivity of small copepods and analysis of trophic interactions. *Journal of the Marine Biological Association of the United Kingdom* **68**, 143–164.
Roff, J.C., Kroetsch, J.T. and Clarke, A.J. (1994). A radiochemical method for secondary production in planktonic Crustacea based on rate of chitin synthesis. *Journal of Plankton Research* **16**, 961–976.
Rolke, M. and Lenz, J. (1984). Size structure analysis of zooplankton samples by means of an automated image analyzing system. *Journal of Plankton Research* **6**, 637–645.
Roman, M.R. (1977). Feeding of the copepod *Acartia tonsa* on the diatom *Nitzschia closterium* and brown algae (*Fucus vesiculosus*) detritus. *Marine Biology* **42**, 149–155.
Roman, M.R. (1984). Utilization of detritus by the copepod, *Acartia tonsa*. *Limnology and Oceanography* **29**, 949–959.
Roman, M.R. and Rublee, P.A. (1980). Containment effects in copepod grazing experiments: a plea to end the black box approach. *Limnology and Oceanography* **25**, 982–990.
Roman, M.R. and Rublee, P.A. (1981). A method to determine *in situ* zooplankton grazing rates on natural particle assemblages. *Marine Biology* **65**, 303–309.
Roman, M.R., Ashton, K.A. and Gauzens, A.L. (1988). Day/night differences in the grazing impact of marine copepods. *Hydrobiologia* **167/168**, 21–30.
Romano, J.E. (1993). Estadios de madurez gonadal de hembras de tres especies de copépodos planctónicos: *Calanoides carinatus* (Kroyer, 1848), *Paracalanus parvus* (Claus, 1863) y *Acartia tonsa* (Dana, 1849). *Boletin Instituto Español de Oceanografia* **9**, 313–322.
Rose, M. (1933). Copépodes pélagiques. *Faune de France* **26**, 1–374.
Rosenberg, G.G. (1980). Filmed observations of filter feeding in the marine planktonic copepod *Acartia clausii*. *Limnology and Oceanography* **25**, 738–742.
Rothlisberg, P.C. (1985). Life history strategies. *Bulletin of Marine Science* **37**, 761–762.
Roy, S. and Poulet, S.A. (1990). Laboratory study of the chemical composition of aging copepod fecal material. *Journal of Experimental Marine Biology and Ecology* **135**, 3–18.

Roy, S., Harris, R.P. and Poulet, S.A. (1989). Inefficient feeding by *Calanus helgolandicus* and *Temora longicornis* on *Coscinodiscus wailesi*: quantitative estimation using chlorophyll-type pigments and effects on dissolved free amino acids. *Marine Ecology Progress Series* **52**, 145–153.

Rubenstein, D.I. and Koehl, M.A.R. (1977). The mechanisms of filter feeding: some theoretical considerations. *The American Naturalist* **111**, 981–994.

Rudjakov, J.A. (1970). The possible causes of diel vertical migrations of planktonic organisms. *Marine Biology* **6**, 98–105.

Rudyakov, Yu.A. (1972). Rate of passive vertical movement of planktonic organisms. *Oceanology, Moscow* **12**, 886–890.

Runge, J.A. (1980). Effects of hunger and season on the feeding behavior of *Calanus pacificus*. *Limnology and Oceanography* **25**, 134–145.

Runge, J.A. (1984). Egg production of the marine, planktonic copepod, *Calanus pacificus* Brodsky: laboratory observations. *Journal of Experimental Marine Biology and Ecology* **74**, 125–145.

Runge, J.A. (1985). Relationship of egg production of *Calanus pacificus* to seasonal changes in phytoplankton availability in Puget Sound, Washington. *Limnology and Oceanography* **30**, 382–396.

Runge, J.A. (1987). Measurement of egg production rate of *Calanus finmarchicus* from preserved samples. *Canadian Journal of Fisheries and Aquatic Science* **44**, 2009–2012.

Runge, J.A. and Ingram, R.G. (1988). Under-ice grazing by planktonic, calanoid copepods in relation to a bloom of ice microalgae in southeastern Hudson Bay. *Limnology and Oceanography* **33**, 280–286.

Runge, J.A. and Ingram, R.G. (1991). Under-ice feeding and diel migration by the planktonic copepods *Calanus glacialis* and *Pseudocalanus minutus* in relation to the ice algal production cycle in southeastern Hudson Bay, Canada. *Marine Biology* **108**, 217–225.

Runge, J.A. and Myers, R.A. (1986). Constraints on the evolution of copepod body size. *Syllogeus* **58**, 443–447.

Runge, J.A., McLaren, I.A., Corkett, C.J., Bohrer, R.N. and Koslow, J.A. (1985). Molting rates and cohort development of *Calanus finmarchicus* and *C. glacialis* in the sea off southwest Nova Scotia. *Marine Biology* **86**, 241–246.

Runge, J.A., Therriault, J.-C., Legendre, L., Ingram, R.G. and Demers, S. (1991). Coupling between ice microalgal productivity and the pelagic, metazoan food web in southeastern Hudson Bay: a synthesis of results. *Polar Research* **10**, 325–338.

Ryan, T.H., Rodhouse, P.G., Roden, C.M. and Hensey, M.P. (1986). Zooplankton fauna of Killary Harbour: the seasonal cycle of abundance. *Journal of the Marine Biological Association of the United Kingdom* **66**, 731–748.

Sabatini, M.E. (1989). Ciclo anual del copépodo *Acartia tonsa* Dana 1849 en la zona interna de la Bahía Blanca (Provincia de Buenos Aires, Argentina). *Scientia Marina* **53**, 847–856.

Sabatini, M.E. (1990). The developmental stages (copepodids I to VI) of *Acartia tonsa* Dana, 1849 (Copepoda, Calanoida). *Crustaceana* **59**, 53–61.

Safronov, S.G. (1989). Fecundity and peculiarities of changes in female reproductive system of *Calanus glacialis* (Copepoda, Calanoida). *Zoologichesky Zhurnal* **58** (7), 138–141.

Safronov, S.G. (1991). Biological characteristics of *Epilabidocera amphitrites* McMurrich (Copepoda, Crustacea). *Ekologiya Morya* **38**, 60–66.

Saito, H. and Taguchi, S. (1996). Diel feeding behavior of neritic copepods during spring and fall blooms in Akkeshi Bay, eastern coast of Hokkaido, Japan. *Marine Biology* **125**, 97–107.

Saito, H., Ogishima, T. and Taguchi, S. (1991). Gut clearance rate of boreal copepods *Eurytemora herdmani* Thompson and Scott (1897) and *Pseudocalanus* spp. at different food concentrations. *Bulletin of Plankton Society of Japan, Special Volume*, 563–572.

Saiz, E. (1994). Observations of the free-swimming behavior of *Acartia tonsa*: effects of food concentration and turbulent water motion. *Limnology and Oceanography* **39**, 1566–1578.

Saiz, E. and Alcaraz, M. (1990). Pigment gut contents of copepods and deep phytoplankton maximum in the western Mediterranean. *Journal of Plankton Research* **12**, 665–672.

Saiz, E. and Alcaraz, M. (1992a). Free-swimming behaviour of *Acartia clausi* (Copepoda: Calanoida) under turbulent water movement. *Marine Ecology Progress Series* **80**, 229–236.

Saiz, E. and Alcaraz, M. (1992b). Enhanced excretion rates induced by small-scale turbulence in *Acartia* (Copepoda: Calanoida). *Journal of Plankton Research* **14**, 681–689.

Saiz, E. and Kiørboe, T. (1995). Predatory and suspension feeding of the copepod *Acartia tonsa* in turbulent environments. *Marine Ecology Progress Series* **122**, 147–158.

Saiz, E., Alcaraz, M. and Paffenhöfer, G.-A. (1992a). Effects of small-scale turbulence on feeding rate and gross-growth efficiency of three *Acartia* species (Copepoda: Calanoida). *Journal of Plankton Research* **14**, 1085–1097.

Saiz, E., Rodriguez, V. and Alcaraz, M. (1992b). Spatial distribution and feeding rates of *Centropages typicus* in relation to frontal hydrographic structures in the Catalan Sea (Western Mediterranean). *Marine Biology* **112**, 49–56.

Salonen, K., Sarvala, J., Hakala, I. and Viljanen, M.L. (1976). The relation of energy and organic carbon in aquatic invertebrates. *Limnology and Oceanography* **21**, 724–730.

Salzen, E.A. (1956). The density of the eggs of *Calanus finmarchicus*. *Journal of the Marine Biological Association of the United Kingdom* **35**, 549–554.

Samain, J.F., Moal, J., Daniel, J.Y. and Le Coz, J.R. (1989). A tentative approach to zooplanktonic growth rate when temperature is considered a limiting factor. *In* "Proceedings of the Twenty First European Marine Biology Symposium" (R.Z. Klekowski, E. Styczyńska and L. Falkowski, eds), pp. 423–433. Polish Academy of Sciences.

Sameoto, D.D. (1978). Zooplankton sample variation on the Scotian Shelf. *Journal of the Fisheries Research Board of Canada* **35**, 1207–1222.

Sameoto, D.D. (1984). Environmental factors influencing diurnal distribution of zooplankton and ichthyoplankton. *Journal of Plankton Research* **6**, 767–792.

Sameoto, D.D. (1986). Influence of the biological and physical environment on the vertical distribution of mesozooplankton and micronekton in the eastern tropical Pacific. *Marine Biology* **93**, 263–279.

Sameoto, D.D. and Herman, A.W. (1990). Life cycle and distribution of *Calanus finmarchicus* in deep basins on the Nova Scotia shelf and seasonal changes in *Calanus* spp. *Marine Ecology Progress Series* **66**, 225–237.

Sander, F. and Moore, E. (1978). A comparative study of inshore and offshore copepod populations at Barbados, West Indies. *Crustaceana* **35**, 225–240.

Sander, F. and Moore, E.A. (1983). Physioecology of tropical marine copepods. I. Size variations. *Crustaceana* **44**, 83–93.

Santella, L. and Ianora, A. (1990). Subitaneous and diapause eggs in Mediterranean populations of *Pontella mediterranea* (Copepoda: Calanoida): a morphological study. *Marine Biology* **105**, 83–90.
Santos, B.A. (1992). Pastoreo de *Paracalanus parvus* (Claus, 1863) sobre dinoflagelados productores de mareas rojas en la plataforma argentina. *Boletín Instituto Español de Oceanografía* **8**, 255–261.
Sarala Devi, K. (1977). Two new records of *Haloptilus* (Copepoda: Calanoida) from the Indian Ocean. *In* "Proceedings of the Symposium on Warm Water Zooplankton" pp. 41–47. Special Publication of the National Institute of Oceanography, Goa.
Saraswathy, M. (1973). The genus *Gaussia* (Copepoda-Calanoida) with a description of *G.sewelli* sp.nov. from the Indian Ocean. *International Indian Ocean Expedition, Handbook to the International Zooplankton Collections* **5**, 190–195.
Saraswathy, M. (1986). *Pleuromamma* (Copepoda-Calanoida) in the Indian Ocean. *Mahasagar* **19**, 185–201.
Saraswathy, M. and Bradford, J.M. (1980). Integumental structures on the antennule of the copepod *Gaussia*. *New Zealand Journal of Marine and Freshwater Research* **14**, 79–82.
Saraswathy, M. and Santhakumari, V. (1982). Sex ratio of five species of pelagic copepods from Indian Ocean. *Mahasagar* **15**, 37–42.
Sargent, J.R. and Henderson, R.J. (1986). Lipids. *In* "The Biological Chemistry of Marine Copepods" (E.D.S. Corner and S.C.M. O'Hara, eds), pp. 58–108. Clarendon Press, Oxford.
Sargent, J.R. and Falk-Petersen, S. (1988). The lipid biochemistry of calanoid copepods. *Hydrobiologia* **167/168**, 101–114.
Sars, G.O. (1903). An account of the Crustacea of Norway with short descriptions and figures of all the species. Volume IV. Copepoda, Calanoida. Published by the Bergen Museum 171 pp.
Sars, G.O. (1925). Copépodes particulièrement bathypélagiques provenant des campagnes scientifiques du Prince Albert Ier de Monaco. *Résultats des Campagnes scientifiques Prince Albert I* **69** (1924), plates 1–127; (1925), text 1–408.
Satyanarayana Rao, T.S. (1979). Zoogeography of the Indian Ocean. *In* "Zoogeography and Diversity in Plankton" (S. Van der Spoel and A.C. Pierrot-Bults, eds), pp. 254–292. Edward Arnold, London.
Saunders III, J.F. (1993). Distribution of *Eurytemora affinis* (Copepoda: Calanoida) in the southern Great Plains, with notes on zoogeography. *Journal of Crustacean Biology* **13**, 564–570.
Saunders III, J.F. and Lewis, W.M. (1987). A perspective on the use of cohort analysis to obtain demographic data for copepods. *Limnology and Oceanography* **32**, 511–513.
Saupe, S.M., Schell, D.M. and Griffiths, W.B. (1989). Carbon-isotope ratio gradients in western Arctic zooplankton. *Marine Biology* **103**, 427–432.
Sautour, B. (1994). Broutage des copépodes planctoniques en laboratoire: effets dus aux incubations. *Cahiers de Biologie marine* **35**, 113–129.
Sautour, B. and Castel, J. (1993). Distribution of zooplankton populations in Marennes-Oléron Bay (France), structure and grazing impact of copepod communities. *Oceanologica Acta* **16**, 279–290.
Sautour, B., Artigas, F., Herbland, A. and Laborde, P. (1996). Zooplankton grazing impact in the plume of dilution of the Gironde estuary (France) prior to the spring bloom. *Journal of Plankton Research* **18**, 835–853.
Sazhin, A.F. (1985). Role of detritus and migrating animals in providing food for

deep-water plankton fauna of boreal and tropical Pacific Ocean. *Oceanology, Moscow* **25**, 530–535.
Sazhina, L.I. (1960). Razvitie Chernomorskikh Copepoda. I. Nauplial'nye stadii *Acartia clausi* Giesbr., *Centropages kröyeri* Giesbr., *Oithona minuta* Krotcz. *Trudy Sevastopol'skoi Biologicheskoi Stantsii* **13**, 49–67.
Sazhina, L.I. (1968). On hibernating eggs of marine Calanoida. *Zoologichesky Zhurnal* **47**, 1554–1556.
Sazhina, L.I. (1971). Fecundity of mass pelagic Copepoda in the Black Sea. *Zoologichesky Zhurnal* **50**, 586–588.
Sazhina, L.I. (1982). Nauplii of mass species of Atlantic Copepoda. *Zoologichesky Zhurnal* **61**, 1154–1164.
Sazhina, L.I. (1985). Fecundity and growth rate of copepods in different dynamic zones of equatorial countercurrent of the Indian Ocean. *Polskie Archiwum Hydrobiologii* **32**, 491–505.
Sazhina, L.I. (1986). Tropic copepods production in various pelagic accumulations of the Indian Ocean. *Ekologiya Morya* **24**, 52–59.
Schnack, S.B. (1978). Feeding habits of *Calanoides carinatus* (Krøyer) in the NW-African upwelling region. *In* "Symposium of the Canary Current, Upwelling and Living Resources". *Fishery Report of the Food and Agricultural Organization* **75**, 1–13.
Schnack, S.B. (1979). Feeding of *Calanus helgolandicus* on phytoplankton mixtures. *Marine Ecology Progress Series* **1**, 41–47.
Schnack, S.B. (1982). The structure of the mouth parts of copepods in Kiel Bay. *Meeresforschung* **29**, 89–101.
Schnack, S.B. (1983). On the feeding of copepods on *Thalassiosira partheneia* from the northwest African upwelling area. *Marine Ecology Progress Series* **11**, 49–53.
Schnack, S.B. (1985). Feeding by *Euphausia superba* and copepod species in response to varying concentrations of phytoplankton. *In* "Antarctic Nutrient Cycles and Food Webs" (W.R. Siegfried, P.R. Condy and R.M. Laws, eds), pp. 311–323. Springer-Verlag, Berlin.
Schnack, S.B. (1989). Functional morphology of feeding appendages in calanoid copepods. *In* "Functional Morphology of Feeding and Grooming in Crustacea" (B.E. Felgenhauer, L. Watling and A.B. Thistle, eds), pp. 137–151. A.A. Balkema, Rotterdam.
Schnack, S.B., Smetacek, V., von Bodungen, B. and Stegmann, P. (1985). Utilization of phytoplankton by copepods in Antarctic waters during spring. *In* "Marine Biology of Polar Regions and Effects of Stress on Marine Organisms" (J.S. Gray and M.E. Christiansen, eds), pp. 65–81. John Wiley & Sons Ltd, Chichester.
Schnack-Schiel, S.B. and Hagen, W. (1994). Life cycle strategies and seasonal variations in distribution and population structure of four dominant calanoid copepod species in the eastern Weddell Sea, Antarctica. *Journal of Plankton Research* **16**, 1543–1566.
Schnack-Schiel, S.B. and Hagen, W. (1995). Life-cycle strategies of *Calanoides acutus*, *Calanus propinquus*, and *Metridia gerlachei* (Copepoda: Calanoida) in the eastern Weddell Sea, Antarctica. *ICES Journal of Marine Science* **52**, 541–548.
Schnack-Schiel, S.B. and Mizdalski, E. (1994). Seasonal variations in distribution and population structure of *Microcalanus pygmaeus* and *Ctenocalanus citer* (Copepoda: Calanoida) in the eastern Weddell Sea, Antarctica. *Marine Biology* **119**, 357–366.
Schnack-Schiel, S.B., Hagen, W. and Mizdalski, E. (1991). Seasonal comparison of

Calanoides acutus and *Calanus propinquus* (Copepoda: Calanoida) in the southeastern Weddell Sea, Antarctica. *Marine Ecology Progress Series* **70**, 17–27.

Schnack-Schiel, S.B., Thomas, D., Dieckmann, G.S., Eicken, H., Gradinger, R., Spindler, M., Weissenberger, J., Mizdalski, E. and Beyer, K. (1995). Life cycle strategy of the Antarctic calanoid copepod *Stephos longipes*. *Progress in Oceanography* **36**, 45–75.

Schram, T.A., Svelle, M. and Opsahl, M. (1981). A new divided neuston sampler in two modifications: description, tests, and biological results. *Sarsia* **66**, 273–282.

Schulz, K. (1981). *Tharybis minor* sp.n. (Copepoda: Calanoida: Tharybidae) aus dem nordwestafrikanischen Auftriebsgebiet mit Anmerkungen zur Gattung *Tharybis* Sars. *Mitteilungen aus dem Hamburgischen zoologischen Museum und Institut* **78**, 169–177.

Schulz, K. (1982). The vertical distribution of calanoid copepods north of Cape Blanc. *Rapports et Procès-Verbaux des Réunions Conseil International pour l'Exploration de la Mer* **180**, 297–302.

Schulz, K. (1986). *Temoropia setosa* sp.n. (Copepoda: Calanoida: Temoridae) aus dem Kanarenstromgebiet (Nordost-Atlantik) mit Anmarkungen zur Gattung *Temoropia* T. Scott. *Mitteilungen aus dem Hamburgischen zoologischen Museum und Institut* **83**, 139–146.

Schulz, K. (1987). Zwei neue Arten der Gattung *Scaphocalanus* (Copepoda: Calanoida: Scolecitrichidae) aus dem subtropischen Nordostatlantik. *Mitteilungen aus dem Hamburgischen zoologischen Museum und Institut* **84**, 105–113.

Schulz, K. (1989). Notes on rare spinocalanid copepods from the eastern North Atlantic, with descriptions of new species of the genera *Spinocalanus* and *Teneriforma* (Copepoda: Calanoida). *Mitteilungen aus dem Hamburgischen zoologischen Museum und Institut* **86**, 185–208.

Schulz, K. (1990). *Pterochirella tuerkayi*, new genus, new species, an unusual copepod from the deep Gulf of Aden (Indian Ocean). *Mitteilungen aus dem Hamburgischen zoologischen Museum und Institut* **87**, 181–189.

Schulz, K. (1991). New species of the family Scolecithricidae (Copepoda: Calanoida) from the Arabian Sea. *Mitteilungen aus dem Hamburgischen zoologischen Museum und Institut* **88**, 197–209.

Schulz, K. (1992). *Kunihulsea arabica*, a new genus and species of calanoid copepod from the Arabian Sea. *Mitteilungen aus dem Hamburgischen zoologischen Museum und Institut* **89**, 175–180.

Schulz, K. (1993). New species of Discoidae from the eastern North Atlantic (Copepoda: Calanoida). *Mitteilungen aus dem Hamburgischen zoologischen Museum und Institut* **90**, 197–207.

Schulz, K. and Beckmann, W. (1995). New benthopelagic tharybids (Copepoda: Calanoida) from the deep North Atlantic. *Sarsia* **80**, 199–211.

Schweder, T. (1979). Patterns of spermatophore distributions in *Euchaeta norvegica*: a further analysis. *Journal of the Marine Biological Association of the United Kingdom* **59**, 369–371.

Sciandra, A., Gouze, J.-L. and Nival, P. (1990). Modelling the reproduction of *Centropages typicus* (Copepoda: Calanoida) in a fluctuating food supply: effect of adaptation. *Journal of Plankton Research* **12**, 549–572.

Scott, A. (1909). The Copepoda of the Siboga Expedition. Part I. Free-swimming, littoral and semi-parasitic Copepoda. *Siboga Expeditie* **29a**, 1–323.

Scotto di Carlo, B., Ianora, A., Fresi, E. and Hure, J. (1984). Vertical zonation

patterns for Mediterranean copepods from the surface to 3000 m at a fixed station in the Tyrrhenian Sea. *Journal of Plankton Research* **6**, 1031–1056.
Scrope-Howe, S. and Jones, D.A. (1985). Biological studies in the vicinity of a shallow-sea tidal mixing front V. Composition, abundance and distribution of zooplankton in the western Irish Sea, April 1980 to November, 1981. *Philosophical Transactions of the Royal Society of London* **310B**, 501–519.
Seguin, G., Errhif, A. and Dallot, S. (1994). Diversity and structure of pelagic copepod populations in the frontal zone of the eastern Alboran Sea. *Hydrobiologia* **292/293**, 369–377.
Sekiguchi, H. (1974). Relation between the ontogenetic vertical migration and mandibular gnathobase in pelagic copepods. *Bulletin of the Faculty of Fisheries, Mie University* **1**, 1–10.
Sekiguchi, H., McLaren, I.A. and Corkett, C.J. (1980). Relationship between growth rate and egg production in the copepod *Acartia clausi* hudsonica. *Marine Biology* **58**, 133–138.
Sellner, K.G., Olson, M.M. and Kononen, K. (1994). Copepod grazing in a summer cyanobacteria bloom in the Gulf of Finland. *Hydrobiologia* **292/293**, 249–254.
Sévigny, J.-M. and McLaren, I.A. (1988). Protein polymorphisms in six species of the genus *Calanus*. *Hydrobiologia,* **167/168**, 267–274.
Sévigny, J.-M. and Odense, P. (1985). Comparison of isoenzyme systems of calanoid copepods by use of ultrathin agarose gel isoelectric focusing techniques. *Comparative Biochemistry and Physiology* **80 B**, 455–461.
Sévigny, J.-M., McLaren, I.A. and Frost, B.W. (1989). Discrimination among and variation within species of *Pseudocalanus* based on the GPI locus. *Marine Biology* **102**, 321–327.
Sewell, R.B.S. (1929). The Copepoda of Indian Seas. Calanoida. *Memoirs of the Indian Museum* **10**, 1–221.
Sewell, R.B.S. (1932). The Copepoda of Indian Seas. Calanoida. *Memoirs of the Indian Museum* **10**, 223–407.
Sewell, R.B.S. (1947). The free-swimming planktonic Copepoda. *Scientific Reports of the John Murray Expedition, 1933–34, Zoology* **8**, 1–303.
Sewell, R.B.S. (1948). The free-swimming planktonic Copepoda. Geographical distribution. *Scientific Reports of the John Murray Expedition, 1933–34, Zoology* **8**, 317–592.
Sewell, R.B.S. (1949). The littoral and semi-parasitic Cyclopoida, the Monstrilloida and Notodelphoida. *Scientific Reports of the John Murray Expedition 1933–34, Zoology and Botany* **9** (2), 17–199.
Sewell, R.B.S. (1951). The epibionts and parasites of the planktonic Copepoda of the Arabian Sea. *Scientific Reports of the John Murray Expedition 1933–34, Zoology and Botany* **9** (4), 255–394.
Sewell, R.B.S. (1956). The continental drift and the distribution of Copepoda. *Proceedings of the Linnaean Society of London* **166**, 149–177.
Shadrin, N.V. and Popova, E.V. (1994). Variability of *Acartia clausi* in the Black Sea. *Hydrobiologia* **292/293**, 179–184.
Shadrin, N.V., Melnik, T.A. and Piontkovsky, S.A. (1983). Temperature effect on the motor and feeding activities of *Acartia clausi* Giesbr. (Copepoda) females. *Ekologiya Morya* **12**, 62–67.
Shaw, B.A., Harrison, P.J. and Andersen, R.J. (1994). Evaluation of the copepod *Tigriopus californicus* as a bioassay organism for the detection of chemical feeding deterrents produced by marine phytoplankton. *Marine Biology* **121**, 89–95.
Shchepkina, A.M., Trusevich, V.V. and Pavlovskaya, T.Ya. (1991). Peculiarities of

lipid composition in some representatives of the mass species of tropical zooplankton from the Atlantic and Indian Ocean. *Ekologiya Morya* **38**, 84–88.

Sherman, K. (1963). Pontellid copepod distribution in relation to surface water types in the central North Pacific. *Limnology and Oceanography* **8**, 214–227.

Sherman, K. (1964a). Pontellid copepod occurrence in the central South Pacific. *Limnology and Oceanography* **9**, 476–484.

Sherman, K. (1994b). Sustainability, biomass yields, and health of coastal ecosystems: an ecological perspective. *Marine Ecology Progress Series* **112**, 277–301.

Sherman, K., Green, J.R., Goulet, J.R. and Ejsymont, L. (1983). Coherence in zooplankton of a large northwest Atlantic ecosystem. *Fishery Bulletin, National Oceanic and Atmospheric Administration of the United States*, 81, 855–862.

Shevijrnogov, A.I. (1972). Characteristics of the bioluminescence impulses of *Metridia pacifica* under supersonic stimulation. *Trudỹ Moskovskogo Obshchestva Ispỹtateleĩ Prirodỹ* **39**, 124–126.

Shih, C.-T. (1979). East–west diversity. *In* "Zoogeography and Diversity in Plankton" (S. Van der Spoel and A.C. Pierrot-Bults, eds), pp. 87–102. Edward Arnold, London.

Shih, C.-T. (1986a). Longitudinal distribution of oceanic calanoids (Crustacea: Copepoda): an example of marine biogeography. *Syllogeus* **58**, 105–114.

Shih, C.-T. (1986b). Biogeography of oceanic zooplankton. *Unesco Technical Papers in Marine Science* **49**, 250–253.

Shih, C.-T. and Marhue, L. (1991). Calanoid copepods in deepwater fjords of northern British Columbia, Canada. *Bulletin of the Plankton Society of Japan*, Special volume, 585–591.

Shmeleva, A.A. (1965). Weight characteristics of the zooplankton of the Adriatic Sea. *Bulletin de l'Institut Océanographique* **65** (1351), 1–24.

Shmeleva, A.A. (1987a). New species of the genus *Calocalanus* (Copepoda, Calanoida) from the Bank Saia-de-Malia (the Indian Ocean). *Zoologichesky Zhurnal* **66**, 1576–1579.

Shmeleva, A.A. (1987b). New species of the genus *Calocalanus* (Copepoda, Calanoida) from Indian Ocean. *Vestnik Zoologichesky* (6), 8–13.

Shmeleva, A.A. and Kovalev, A.V. (1974). Cycles biologique des copepodes (Crustacea) de la mer Adriatique. *Bolletino di Pesca Piscicoltura e Idrobiologia* **29**, 49–70.

Shuert, P.G. and Hopkins, T.L. (1987). The vertical distribution and feeding ecology of *Euchaeta marina* in the eastern Gulf of Mexico. *Contributions in Marine Science* **30**, 49–61.

Shushkina, E.A., Kislyakov, Yu. Ya. and Pasternak, A.F. (1974). Estimating the productivity of marine zooplankton by combining the radiocarbon method with mathematical modeling. *Oceanology, Moscow* **14**, 259–265.

Siefert, D.L.W. (1994). The importance of sampler mesh size when estimating total daily egg production by *Pseudocalanus* spp. in Shelikov Strait, Alaska. *Journal of Plankton Research* **16**, 1489–1498.

Simard, Y., Lacroix, G. and Legendre, L. (1985). *In situ* twilight grazing rhythm during diel vertical migrations of a scattering layer of *Calanus finmarchicus*. *Limnology and Oceanography* **30**, 598–606.

Singarajah, K.V. (1969). Escape reactions of zooplankton: the avoidance of a pursuing siphon tube. *Journal of Experimental Marine Biology and Ecology* **3**, 171–178.

Singarajah, K.V. (1975). Escape reactions of zooplankton: effects of light and turbulence. *Journal of the Marine Biological Association of the United Kingdom* **55**, 627–639.
Siokou-Frangou, I. (1985). Sur la presence du copepode calanoide *Pseudocalanus elongatus* (Boeck) en Mer Egee du Nord. *Rapports et Procès-Verbaux des Réunions Commission Internationale pour l'Exploration Scientifique de la Mer Méditerranée* **29** (9), 231–233.
Siokou-Frangou, I. (1996). Zooplankton annual cycle in a Mediterranean coastal area. *Journal of Plankton Research* **18**, 203–223.
Sket, B. and Iliffe, T.M. (1980). Cave fauna of Bermuda. *International Revue der gesamten Hydrobiologie* **65**, 871–882.
Skiver, J. (1980). Seasonal resource partitioning patterns of marine calanoid copepods: species interactions. *Journal of Experimental Marine Biology and Ecology* **44**, 229–245.
Slagstad, D. and Tande, K.S. (1990). Growth and production dynamics of *Calanus glacialis* in an Arctic pelagic food web. *Marine Ecology Progress Series* **63**, 189–199.
Small, L.F., Fowler, S.W. and Ünlü, M.Y. (1979). Sinking rates of natural copepod fecal pellets. *Marine Biology* **51**, 233–241.
Smetacek, V.S. (1980). Zooplankton standing stock, copepod faecal pellets and particulate detritus in Kiel Bight. *Estuarine and Coastal Marine Science* **11**, 477–490.
Smith, D.F., Bulleid, N.C., Campbell, R., Higgins, H.W., Rowe, F., Tranter, D.J. and Tranter, H. (1979). Marine food-web analysis: an experimental study of demersal zooplankton using isotopically labelled prey species. *Marine Biology* **54**, 49–59.
Smith, S.L. (1978). Nutrient regeneration by zooplankton during a red tide off Peru, with notes on biomass and species composition of zooplankton. *Marine Biology* **49**, 125–132.
Smith, S.L. (1982). The northwestern Indian Ocean during the monsoons of 1979: distribution, abundance, and feeding of zooplankton. *Deep-Sea Research* **29A**, 1331–1353.
Smith, S.L. (1984). Biological indications of active upwelling in the northwestern Indian Ocean in 1964 and 1979, and a comparison with Peru and northwest Africa. *Deep-Sea Research* **31A**, 951–967.
Smith, S.L. (1985). Macrozooplankton of a deep-sea hydrothermal vent: *In situ* rates of oxygen consumption. *Limnology and Oceanography* **30**, 102–110.
Smith, S.L. (1988). Copepods in Fram Strait in summer: distribution, feeding and metabolism. *Journal of Marine Research* **46**, 145–181.
Smith, S.L. (1990). Egg production and feeding by copepods prior to the spring bloom of phytoplankton in Fram Strait, Greenland Sea. *Marine Biology* **106**, 59–69.
Smith, S.L. (1995). The Arabian Sea: mesozooplankton response to seasonal climate in a tropical ocean. *ICES Journal of Marine Science* **52**, 427–438.
Smith, S.L. and Hall, B.K. (1980). Transfer of radioactive carbon within the copepod *Temora longicornis*. *Marine Biology* **55**, 277–286.
Smith, S.L. and Lane, P.V.Z. (1985). Laboratory studies of the marine copepod *Centropages typicus*: egg production and development rates. *Marine Biology* **85**, 153–162.
Smith, S.L. and Lane, P.V.Z. (1987). On the life history of *Centropages typicus*: responses to a fall diatom bloom in the New York Bight. *Marine Biology* **95**, 305–313.

Smith, S. and Lane, P. (1991). The coastal jet off Pt. Arena, California: its role in secondary production of the copepod *Eucalanus californicus* Johnson. *Journal of Geophysical Research* **96**, 14849–14858.

Smith, S.L. and Vidal, J. (1984). Spatial and temporal effects of salinity, temperature and chlorophyll on the communities of zooplankton in the southeastern Bering Sea. *Journal of Marine Research* **42**, 221–257.

Smith, S.L. and Vidal, J. (1986). Variations in the distribution, abundance, and development of copepods in the southeastern Bering Sea in 1980 and 1981. *Continental Shelf Research* **5**, 215–239.

Snell, T.W. and Carmona, M.J. (1994). Surface glycoproteins in copepods: potential signals for mate recognition. *Hydrobiologia* **292/293**, 255–264.

Soetaert, K. and Herman, P.M.J. (1994). One foot in the grave: zooplankton drift into the Westerschelde estuary (The Netherlands). *Marine Ecology Progress Series* **105**, 19–29.

Soetaert, K. and Rijswijk, P.V. (1993). Spatial and temporal patterns of the zooplankton in the Westerschelde estuary. *Marine Ecology Progress Series* **97**, 47–59.

Soler, E., Río, J.G. del and Vives, F. (1988). Morphological and taxonomical revision of *Centropages ponticus* Karavaev, 1895 (Copepoda, Calanoida). *Crustaceana* **55**, 129–146.

Solokhina, E.V. (1992). Two forms of *Eurytemora pacifica* (Crustacea, Copepoda, Calanoida) from the Lagoon Gladkovskaya (the Commandor Islands). *Zoologichesky Zhurnal* **71** (8), 137–139.

Soltanpour-Gargari, A. and Wellershaus, S. (1987). Very low salinity stretches in estuaries – the main habitat of *Eurytemora affinis*, a plankton copepod. *Meeresforschung* **31**, 199–208.

Sömme, J. (1934). Animal plankton of the Norwegian coast waters and the open sea. I. Production of *Calanus finmarchicus* (Gunner.) and *Calanus hyperboreus* (Kröyer) in the Lofoten area. *Reports of the Norwegian Fisheries and Marine Investigations* **4**, (9), 1–163.

Southward, A.J. and Barrett, R.L. (1983). Observations on the vertical distribution of zooplankton, including post-larval teleosts, off Plymouth in the presence of a thermocline and a chlorophyll-dense layer. *Journal of Plankton Research* **5**, 599–618.

Springer, A.M. and Roseneau, D.G. (1985). Copepod-based food webs: auklets and oceanography in the Bering Sea. *Marine Ecology Progress Series* **21**, 229–237.

Springer, A.M., McRoy, C.P. and Turco, K.R. (1989). The paradox of pelagic food webs in the northern Bering Sea -II. Zooplankton communities. *Continental Shelf Research* **9**, 359–386.

Star, J.L. and Mullin, M.M. (1981). Zooplanktonic assemblages in three areas of the North Pacific as revealed by continuous horizontal transects. *Deep-Sea Research* **28A**, 1303–1322.

Stearns, D.E. (1986). Copepod grazing behavior in simulated natural light and its relation to nocturnal feeding. *Marine Ecology Progress Series* **30**, 65–76.

Stearns, D.E. and Forward, R.B., Jr (1984a). Photosensitivity of the calanoid copepod *Acartia tonsa*. *Marine Biology* **82**, 85–89.

Stearns, D.E. and Forward, R.B., Jr (1984b). Copepod photobehavior in a simulated light environment and its relation to nocturnal vertical migration. *Marine Biology* **82**, 91–100.

Stearns, D.E., Litaker, W. and Rosenberg, G. (1987). Impacts of zooplankton grazing and excretion on short-interval fluctuations in chlorophyll *a* and nitrogen concentrations in a well-mixed estuary. *Estuarine, Coastal and Shelf Science* **24**, 305–325.

Stearns, D.E., Tester, P.A. and Walker, R.L. (1989). Diel changes in the egg production rate of *Acartia tonsa* (Copepoda, Calanoida) and related environmental factors in two estuaries. *Marine Ecology Progress Series* **52**, 7–16.

Steedman, H.F., ed. (1976). Zooplankton fixation and preservation. *Monographs on Oceanographic Methodology* **4**, UNESCO, Paris, 350 pp.

Steele, D.H. and Steele, V.J. (1975). Egg size and duration of embryonic development in Crustacea. *Internationale Revue der gesamten Hydrobiologie* **60**, 711–715.

Steele, J.H., ed. (1977). Spatial Pattern in Plankton Communities. Plenum Press, New York and London. 470 pp.

Steele, J.H. and Henderson, E.W. (1992). A simple model for plankton patchiness. *Journal of Plankton Research* **14**, 1397–1403.

Steele, J.H. and Henderson, E.W. (1995). Predation control of plankton demography. *ICES Journal of Marine Science* **52**, 565–573.

Steinberg, D.K. (1995). Diet of copepods (*Scopalatum vorax*) associated with mesopelagic detritus (giant larvacean houses) in Monterey Bay, California. *Marine Biology* **122**, 571–584.

Steinberg, D.K., Silver, M.W., Pilskaln, C.H., Coale, S.L. and Paduan, J.B. (1994). Midwater zooplankton communities on pelagic detritus (giant larvacean houses) in Monterey Bay, California. *Limnology and Oceanography* **39**, 1606–1620.

Stephen, R. and Sarala Devi, K. (1973). Distribution of *Haloptilus acutifrons* (Copepoda, Calanoida) in the Indian Ocean with a description of the hitherto unknown male. *Handbook to the International Zooplankton Collections, Indian Ocean Biological Centre* **5**, 172–179.

Stephens, G.C. (1988). Epidermal amino acid transport in marine invertebrates. *Biochimica et biophysica acta* **947**, 113–138.

Stepien, J.C., Malone, T.C. and Chervin, M.B. (1981). Copepod communities in the estuary and coastal plume of the Hudson River. *Estuarine, Coastal and Shelf Science* **13**, 185–195.

Steuer, A. (1923). Bausteine zu einer Monographie der Copepodengattung *Acartia*. *Arbeiten aus dem Zoologischen Institut der Universität Innsbruck* **1** (5), 1–148.

Steuer, A. (1928). Ueber das sogenannte Leuchtorgan des Tiefsee-Copepoden *Cephalophanes* G.O. Sars. *Arbeiten aus dem Zoologischen Institut der Universität Innsbruck* **3**, 9–16.

Stock, J.H. and Vaupel Klein, J.C. von (1996). Mounting media revisited: the suitability of Reyne's fluid for small crustaceans. *Crustaceana* **69**, 794–798.

Stoecker, D.K. and Capuzzo, J.M. (1990). Predation on Protozoa: its importance to zooplankton. *Journal of Plankton Research* **12**, 891–908.

Stoecker, D.K. and Egloff, D.A. (1987). Predation by *Acartia tonsa* Dana on planktonic ciliates and rotifers. *Journal of Experimental Marine Biology and Ecology* **110**, 53–68.

Stone, D.P. (1980). The distribution of zooplankton communities in a glacial run-off fjord and exchanges with the open sea. *In* "Fjord Oceanography" (H.J. Freeland, D.M. Farmer and C.D. Levings, eds), NATO Conference Series IV, Marine Sciences, Vol. 4, pp. 291–297. Plenum Press, New York and London.

Støttrup, J.G. and Jensen, J. (1990). Influence of algal diet on feeding and egg-production of the calanoid copepod *Acartia tonsa* Dana. *Journal of Experimental Marine Biology and Ecology* **141**, 87–105.

Strickler, J.R. (1977). Observation of swimming performances of planktonic copepods. *Limnology and Oceanography* **22**, 165–170.
Strickler, J.R. (1982). Calanoid copepods, feeding currents, and the role of gravity. *Science, New York* **218** (4568), 158–160.
Strickler, J.R. (1985). Feeding currents in calanoid copepods: two new hypotheses. *Symposia of the Society of Experimental Biology* **39**, 459–485.
Strickler, J.R. and Costello, J.H. (1996). Calanoid copepod behavior in turbulent flows. *Marine Ecology Progress Series* **139**, 301–312.
Stubblefield, C.L. and Vecchione, M. (1985). Zooplankton distribution in a wind-driven estuary before and after a major storm. *Contributions in Marine Science* **28**, 55–67.
Stubblefield, C.L., Lascara, C.M. and Vecchione, M. (1984). Vertical distribution of zooplankton in a shallow turbid estuary. *Contributions in Marine Science* **27**, 93–104.
Suárez-Morales, E. and Iliffe, T.M. (1996). New superfamily of Calanoida (Copepoda) from an anchialine cave in the Bahamas. *Journal of Crustacean Biology* **16**, 754–762.
Sullivan, B.K. and Banzon, P.V. (1990). Food limitation and benthic regulation of populations of the copepod *Acartia hudsonica* Pinhey in nutrient-limited and nutrient-enriched systems. *Limnology and Oceanography* **35**, 1618–1631.
Sullivan, B.K. and McManus, L.T. (1986). Factors controlling seasonal succession of the copepods *Acartia hudsonica* and *A. tonsa* in Narragansett Bay, Rhode Island: temperature and resting egg production. *Marine Ecology Progress Series* **28**, 121–128.
Sullivan, B.K. and Ritacco, P.J. (1985). The response of dominant copepod species to food limitation in a coastal marine ecosystem. *Archives fur Hydrobiologie Beiheft Ergebnisse Limnologie* **21**, 407–418.
Sullivan, B.K., Miller, C.B., Peterson, W.T. and Soeldner, A.H. (1975). A scanning electron microscope study of the mandibular morphology of boreal copepods. *Marine Biology* **30**, 175–182.
Sullivan, B.K., Buskey, E., Miller, D.C. and Ritacco, P.J. (1983). Effects of copper and cadmium on growth, swimming and predator avoidance in *Eurytemora affinis* (Copepoda). *Marine Biology* **77**, 299–306.
Sunami, Y. and Hirakawa, K. (1996). Attempt for modeling of the variability in biomass of *Metridia pacifica* (Copepoda: Calanoida) in Toyama Bay, in the southern Japan Sea. *Journal of National Fisheries University* **44**, 143–149.
Suzuki, H., Sasaki, H., Shibata, K. and Tamate, H.B. (1995). PCR amplification of the 28S ribosomal RNA gene in calanoid copepods and its application to the detection of copepod DNA in marine detritus. *Bulletin of Plankton Society of Japan* **42**, 163–167.
Svensson, J.E. (1995). Predation risk increases with clutch size in a copepod. *Functional Ecology* **9**, 774–777.
Svetlichny, L.S. (1980). On certain dynamic parameters of tropical copepod passive submersion. *Ekologiya Morya* **2**, 28–33.
Svetlichny, L.S. (1983). Calculation of planktonic copepod biomass by means of coefficients of proportionality between volume and linear dimensions of the body. *Ekologiya Morya* **15**, 46–58.
Svetlichny, L.S. (1988). Morphology and functional parameters of body muscles of *Calanus helgolandicus* (Copepoda, Calanoida). *Zoologichesky Zhurnal* **67** (1), 23–30.
Svetlichny, L.S. (1991). Filming, tensometry and energy estimation of swimming

locomotion using mouth appendages in *Calanus helgolandicus* (Crustacea, Copepoda). *Zoologichesky Zhurnal* **70** (3), 23–29.

Svetlichny, L.S. (1992a). Dependence of the specific respiration rate of copepods on mechanical energy of locomotion. *Ekologiya Morya* **40**, 72–77.

Svetlichny, L.S. (1992b). Concerning biohydrodynamics of the steady and unsteady locomotion of copepods. *Ekologiya Morya* **40**, 84–89.

Svetlichny, L.S. and Kurbatov, B.V. (1983). Efficiency of *Calanus helgolandicus* locomotion in swimming and jumping. *Ekologiya Morya* **12**, 79–83.

Svetlichny, L.S. and Kurbatov, B.V. (1987). Effect of the body sizes on spatial-time and energy parameters of the vertical migration of copepods. *Ekologiya Morya* **27**, 78–85.

Svetlichny, L.S. and Yarkina, I.Ya (1989). Locomotion rhythms in *Calanus helgolandicus* (Crustacea, Copepoda). *Zoologichesky Zhurnal* **68** (6), 50–55.

Swadling, K.M. and Marcus, N.H. (1994). Selectivity in the natural diets of *Acartia tonsa* Dana (Copepoda: Calanoida): comparison of juveniles and adults. *Journal of Experimental Marine Biology and Ecology* **181**, 91–103.

Sykes, P.F. and Huntley, M.E. (1987). Acute physiological reactions of *Calanus pacificus* to selected dinoflagellates: direct observations. *Marine Biology* **94**, 19–24.

Tackx, M.L.M. and Daro, M.H. (1993). Influence of size dependent ^{14}C uptake rates by phytoplankton cells in zooplankton grazing measurements. *Cahiers de Biologie marine* **34**, 253–260.

Tackx, M. and Polk, P. (1986). Effect of incubation time and concentration of animals in grazing experiments using a narrow size range of particles. *Syllogeus* **58**, 604–609.

Tackx, M.L.M., Bakker, C., Francke, J.W. and Vink, M. (1989). Size and phytoplankton selection by Oosterschelde zooplankton. *Netherlands Journal of Sea Research* **23**, 35–43.

Tackx, M.L.M., Bakker, C. and Rijswijk, P.V. (1990). Zooplankton grazing pressure in the Oosterschelde (The Netherlands). *Netherlands Journal of Sea Research* **25**, 405–415.

Tackx, M.L.M., Zhu, L., De Coster, W., Billones, R. and Daro, M.H. (1995). Measuring selectivity of feeding by estuarine copepods using image analysis combined with microscopic and Coulter counting. *ICES Journal of Marine Science* **52**, 419–425.

Taguchi, S. and Ishii, H. (1972). Shipboard experiments on respiration, excretion, and grazing of *Calanus cristatus* and *C. plumchrus* (Copepoda) in the northern North Pacific. *In* "Biological Oceanography of the Northern North Pacific Ocean" (A.Y. Takenouti *et al.*, eds), pp. 419–431. Idemitsu Shoten, Tokyo.

Takeda, N. (1950). Experimental studies on the effect of external agencies on the sexuality of a marine copepod. *Physiological Zoölogy* **23**, 288–301.

Tanaka, M., Ueda, H. and Azeta, M. (1987). Near-bottom copepod aggregations around the nursery ground of the juvenile red sea bream in Shijiki Bay. *Nippon Suisan Gakkaishi* **53**, 1537–1544.

Tanaka, O. (1953). The pelagic copepods of the Izu region. *Records of Oceanographic Works in Japan* **1**, 126–137.

Tanaka, O. (1956a). The pelagic copepods of the Izu Region, middle Japan. Systematic account. I. Families Calanidae and Eucalanidae. *Publications of the Seto Marine Biological Laboratory* **5**, 251–272.

Tanaka, O. (1956b). The pelagic copepods of the Izu Region, middle Japan. Systematic account. II. Families Paracalanidae and Pseudocalanidae. *Publications*

of the Seto Marine Biological Laboratory **5**, 367–406.
Tanaka, O. (1957a). The pelagic copepods of the Izu Region, middle Japan. Systematic account. III. Family Aetideidae (Part 1). *Publications of the Seto Marine Biological Laboratory* **6**, 31–68.
Tanaka, O. (1957b). The pelagic copepods of the Izu Region, middle Japan. Systematic account. IV. Family Aetideidae (Part 2). *Publications of the Seto Marine Biological Laboratory* **6**, 169–207.
Tanaka, O. (1958). The pelagic copepods of the Izu Region, middle Japan. Systematic account. V. Family Euchaetidae. *Publications of the Seto Marine Biological Laboratory* **6**, 327–367.
Tanaka, O. (1960a). Pelagic Copepoda. Biological Results of the Japanese Antarctic Research Expedition. *Special Publications from the Seto Marine Biological Laboratory* **10**, 1–176.
Tanaka, O. (1960b). The pelagic copepods of the Izu Region, Middle Japan. Systematic account VI. Families Phaennidae and Tharybidae. *Publications of the Seto Marine Biological Laboratory* **8**, 85–135.
Tanaka, O. (1961). The pelagic copepods of the Izu Region, Middle Japan. Systematic account VII. Family Scolecithricidae (Part 1). *Publications of the Seto Marine Biological Laboratory* **9**, 139–190.
Tanaka, O. (1962). The pelagic copepods of the Izu Region, Middle Japan. Systematic account VIII. Family Scolecithricidae (Part 2). *Publications of the Seto Marine Biological Laboratory* **10**, 35–90.
Tanaka, O. (1963). The pelagic copepods of the Izu Region, Middle Japan. Systematic account IX. Families Centropagidae, Pseudodiaptomidae, Temoridae, Metridiidae and Lucicutiidae. *Publications of the Seto Marine Biological Laboratory* **11**, 7–55.
Tanaka, O. (1964a). The pelagic copepods of the Izu Region, Middle Japan. Systematic account X. Family Heterorhabdidae. *Publications of the Seto Marine Biological Laboratory* **12**, 1–37.
Tanaka, O. (1964b). The pelagic copepods of the Izu Region, middle Japan. Systematic account XI. Family Augaptilidae. *Publications of the Seto Marine Biological Laboratory* **12**, 39–91.
Tanaka, O. (1964c). The pelagic copepods of the Izu Region, Middle Japan. Systematic account XII. Families Arietellidae, Pseudocyclopidae, Candaciidae and Pontellidae. *Publications of the Seto Marine Biological Laboratory* **12**, 231–271.
Tanaka, O. (1965). The pelagic copepods of the Izu Region, middle Japan. Systematic account. XIII. Parapontellidae, Acartiidae and Tortanidae. *Publications of the Seto Marine Biological Laboratory* **12**, 379–407.
Tanaka, O. (1966). Neritic Copepoda Calanoida from the north-west coast of Kyushu. *Marine Biological Association of India, Symposium on Crustacea* **1**, 38–50.
Tanaka, O. (1973). On *Euchaeta* (Copepoda, Calanoida) of the Indian Ocean. *Handbook to the International Zooplankton Collections, Indian Ocean Biological Centre* **4**, 126–149.
Tanaka, O. and Omori, M. (1967). Large-sized pelagic copepods in the northwestern Pacific Ocean adjacent to Japan. *Information Bulletin of Planktology in Japan, Commemoration Number to Dr Y. Matsue*, 239–260.
Tanaka, O. and Omori, M. (1968). Additional report on calanoid copepods from the Izu Region. Part 1. *Euchaeta* and *Pareuchaeta*. *Publications of the Seto Marine Biological Laboratory* **16**, 219–261.
Tanaka, O. and Omori, M. (1969a). On *Euchirella* (Copepoda, Calanoida) collected

chiefly by the U.S. Steamer *Albatross* from the Pacific Ocean. *Publications of the Seto Marine Biological Laboratory* **17**, 33–65.
Tanaka, O. and Omori, M. (1969b). Additional report on calanoid copepods from the Izu Region. Part 2. *Euchirella* and *Pseudochirella*. *Publications of the Seto Marine Biological Laboratory* **17**, 155–169.
Tanaka, O. and Omori, M. (1970a). Additional report on calanoid copepods from the Izu Region. Part 3-A. *Euaetideus, Aetideopsis, Chiridius, Gaidius* and *Gaetanus*. *Publications of the Seto Marine Biological Laboratory* **18**, 109–141.
Tanaka, O. and Omori, M. (1970b). Additional report on calanoid copepods from the Izu Region. Part 3-B. *Chirundina, Undeuchaeta, Pseudeuchaeta, Valdiviella,* and *Chiridiella*. *Publications of the Seto Marine Biological Laboratory* **18**, 143–155.
Tanaka, O. and Omori, M. (1971). Additional report on calanoid copepods from the Izu Region. Part 4. *Haloptilus, Augaptilus, Centraugaptilus, Pseudaugaptilus,* and *Pachyptilus*. *Publications of the Seto Marine Biological Laboratory* **19**, 249–268.
Tanaka, O. and Omori, M. (1974). Additional report on calanoid copepods from the Izu Region. Part 5. *Euaugaptilus*. *Publications of the Seto Marine Biological Laboratory* **21**, 193–267.
Tanaka, O. and Omori, M. (1992). Additional report on calanoid copepods from the Izu Region Part 6. Phaennidae. *Publications of the Seto Marine Biological Laboratory* **35**, 253–271.
Tande, K.S. (1982). Ecological investigations on the zooplankton community of Balsfjorden, northern Norway: generation cycles, and variations in body weight and body content of carbon and nitrogen related to overwintering and reproduction in the copepod *Calanus finmarchicus* (Gunnerus). *Journal of Experimental Marine Biology and Ecology* **62**, 129–142.
Tande, K.S. (1988a). Aspects of developmental and mortality rates in *Calanus finmarchicus* related to equiproportional development. *Marine Ecology Progress Series* **44**, 51–58.
Tande, K.S. (1988b). An evaluation of factors affecting vertical distribution among recruits of *Calanus finmarchicus* in three adjacent high-latitude localities. *Hydrobiologia* **167/168**, 115–126.
Tande, K.S. (1991). *Calanus* in North Norwegian fjords and in the Barents Sea. *Polar Research* **10**, 389–407.
Tande, K.S. and Båmstedt, U. (1985). Grazing rates of the copepods *Calanus glacialis* and *C. finmarchicus* in Arctic waters of the Barents Sea. *Marine Biology* **87**, 251–258.
Tande, K.S. and Grønvik, S. (1983). Ecological investigations on the zooplankton community of Balsfjorden, northern Norway: sex ratio and gonad maturation cycle in the copepod *Metridia longa* (Lubbock). *Journal of Experimental Marine Biology and Ecology* **71**, 43–54.
Tande, K.S. and Hopkins, C.C.E. (1981). Ecological investigations on the zooplankton community of Balsfjorden, northern Norway: the genital system in *Calanus finmarchicus* and the role of gonad development in overwintering strategy. *Marine Biology* **63**, 159–164.
Tande, K.S., Hassel, A. and Slagstad, D. (1985). Gonad maturation and possible life cycle strategies in *Calanus finmarchicus* and *Calanus glacialis* in the northwestern part of the Barents Sea. *In* "Marine Biology of Polar Regions and Effects of Stress on Marine Organisms" (J.S. Gray and M.E. Christiansen, eds), pp. 141–155. J. Wiley & Sons, New York.
Tang, K.W., Chen, Q.C. and Wong, C.K. (1994). Diel vertical migration and gut

pigment rhythm of *Paracalanus parvus*, *P. crassirostris*, *Acartia erythraea* and *Eucalanus subcrassus* (Copepoda, Calanoida) in Tolo Harbour, Hong Kong. *Hydrobiologia* **292/293**, 389–396.

Tanimura, A., Fukuchi, M. and Ohtsuka, H. (1984). Occurrence and age composition of *Paralabidocera antarctica* (Calanoida, Copepoda) under the fast ice near Syowa Station, Antarctica. *Memoirs of National Institute of Polar Research*, Special Issue **32**, 81–86.

Tanimura, A., Fukuchi, M. and Hoshiai, T. (1986). Seasonal change in the abundance of zooplankton and species composition of copepods in the ice-covered sea near Syowa Station, Antarctica. *Memoirs of National Institute of Polar Research*, Special Issue **40**, 212–220.

Tanimura, A., Hoshiai, T. and Fukuchi, M. (1996). The life cycle strategy of the ice-associated copepod, *Paralabidocera antarctica* (Calanoida, Copepoda), at Syowa Station, Antarctica. *Antarctic Science* **8**, 257–266.

Tanskanen, S. (1994). Seasonal variability in the individual carbon content of the calanoid copepod *Acartia bifilosa* from the northern Baltic Sea. *Hydrobiologia* **292/293**, 397–403.

Tepper, B. and Bradley, B.P. (1989). Temporal changes in a natural population of copepods. *Biological Bulletin, Marine Biological Laboratory, Woods Hole, Mass.*, **176**, 32–40.

Terazaki, M. and Wada, M. (1988). Occurrence of large numbers of carcasses of the large, grazing copepod *Calanus cristatus* from the Sea of Japan. *Marine Biology* **97**, 177–183.

Tester, P.A. (1985). Effects of parental acclimation temperature and egg-incubation temperature on egg-hatching time in *Acartia tonsa* (Copepoda: Calanoida). *Marine Biology* **89**, 45–53.

Tester, P.A. (1986). Egg development time and acclimation temperature in *Acartia tonsa* (Dana). *Syllogeus* **58**, 475–480.

Tester, P.A. and Turner, J.T. (1988). Comparative carbon-specific ingestion rates of phytoplankton by *Acartia tonsa*, *Centropages velificatus* and *Eucalanus pileatus* grazing on natural phytoplankton assemblages in the plume of the Mississippi River (northern Gulf of Mexico continental shelf). *Hydrobiologia* **167/168**, 211–217.

Tester, P.A. and Turner, J.T. (1989). Zooplankton feeding ecology: feeding rates of the copepods *Acartia tonsa*, *Centropages velificatus* and *Eucalanus pileatus* in relation to the suspended sediments in the plume of the Mississippi River (Northern Gulf of Mexico continental shelf). *Scientia Marina* **53**, 231–237.

Tester, P.A. and Turner, J.T. (1990). How long does it take copepods to make eggs? *Journal of Experimental Marine Biology and Ecology* **141**, 169–182.

Tester, P.A. and Turner, J.T. (1991). Why is *Acartia tonsa* restricted to estuarine habitats? *Bulletin of Plankton Society of Japan*, Special Volume 603–611.

Tett, P.B. (1972). An annual cycle of flash induced luminescence in the euphausiid *Thysanoessa raschii*. *Marine Biology* **12**, 207–218.

Thébault, J.-M. (1985). Étude expérimentale de la nutrition d'un copépode commun (*Temora stylifera* Dana). Effets de la température et de la concentration de nourriture. *Journal of Experimental Marine Biology and Ecology* **93**, 223–234.

Théodoridès, J. (1989). Parasitology of marine zooplankton. *Advances in Marine Biology* **25**, 117–177.

Thompson, A.M., Durbin, E.G. and Durbin, A.G. (1994). Seasonal changes in maximum digestion rate of *Acartia tonsa* in Narragansett Bay, Rhode Island, USA. *Marine Ecology Progress Series* **108**, 91–105.

Thompson, B.M. (1982). Growth and development of *Pseudocalanus elongatus* and *Calanus* sp. in the laboratory. *Journal of the Marine Biological Association of the United Kingdom* **62**, 359-372.
Thompson, I.C. and Scott, A. (1903). Report on the Copepoda collected by Professor Herdman, at Ceylon, in 1902. *Ceylon Pearl Oyster Fisheries, Supplementary Reports* (7), 227-307.
Timonin, A.G., Arashkevich, E.G., Drits, A.V. and Semenova, T.N. (1992). Zooplankton dynamics in the northern Benguela ecosystem, with special reference to the copepod *Calanoides carinatus*. *South African Journal of Marine Research* **12**, 545-560.
Tiselius, P. (1988). Effects of diurnal feeding rhythms, species composition and vertical migration on the grazing impact of calanoid copepods in the Skagerrak and Kattegat. *Ophelia* **28**, 215-230.
Tiselius, P. (1989). Contribution of aloricate ciliates to the diet of *Acartia clausi* and *Centropages hamatus* in coastal waters. *Marine Ecology Progress Series* **56**, 49-56.
Tiselius, P. (1992). Behavior of *Acartia tonsa* in patchy food environments. *Limnology and Oceanography* **37**, 1640-1651.
Tiselius, P. and Jonsson, P.R. (1990). Foraging behaviour of six calanoid copepods: observations and hydrodynamic analysis. *Marine Ecology Progress Series* **66**, 23-33.
Tiselius, P., Nielsen, T.G., Breuel, G., Jaanus, A., Korshenko, A. and Witek, Z. (1991). Copepod egg production in the Skagerrak during SKAGEX, May-June 1990. *Marine Biology* **111**, 445-453.
Tiselius, P., Hansen, B., Jonsson, P., Kiørboe, T., Nielsen, T.G., Piontkovski, S. and Saiz, E. (1995). Can we use laboratory reared copepods for experiments? A comparison of feeding behaviour and reproduction between a field and a laboratory population of *Acartia tonsa*. *ICES Journal of Marine Science* **52**, 369-376.
Toda, H. (1989). Surface distributions of copepods in relation to regional upwellings around the Izu Islands in summer of 1988. *Journal of the Oceanographical Society of Japan* **45**, 251-257.
Toda, H. and Hirose, E. (1991). SEM and TEM observations on egg membranes of the two types of *Calanus sinicus* eggs. *Bulletin of Plankton Society of Japan*, Special Volume, 613-617.
Toda, T., Suh, H.-L. and Nemoto, T. (1989). Dry fracturing: a simple technique for scanning electron microscopy of small crustaceans and its application to internal observations of copepods. *Journal of Crustacean Biology* **9**, 409-413.
Toda, T., Kikuchi, T., Ohta, S. and Gamô, S. (1994). Benthopelagic zooplankton from a deep-sea cold-seep site in Sagami Bay. *Bulletin of Plankton Society of Japan* **41**, 173-176.
Tomas, C.R. and Deason, E.E. (1981). The influence of grazing by two *Acartia* species on *Olisthodiscus luteus* Carter. *P.S.Z.N.I. Marine Ecology* **2**, 215-223.
Tourangeau, S. and Runge, J.A. (1991). Reproduction of *Calanus glacialis* under ice in spring in southeastern Hudson Bay, Canada. *Marine Biology* **108**, 227-233.
Tranter, D.J. (1973). Seasonal studies of a pelagic ecosystem (meridian 110° E). *In* "The Biology of the Indian Ocean" (B. Zeitschel, ed.), pp. 487-520. Chapman & Hall Ltd, London.
Tranter, D.J. (1977). Further studies of plankton ecosystems in the eastern Indian Ocean. V. Ecology of the Copepoda. *Australian Journal of Marine and Freshwater Research* **28**, 593-625.

Tranter, D.J. and Abraham, S. (1971). Coexistence of species of Acartiidae (Copepoda) in the Cochin Backwater, a monsoonal estuarine lagoon. *Marine Biology* **11**, 222–241.

Tranter, D.J. and Fraser, J.H., eds (1968). Zooplankton sampling. *Monographs on Oceanographic Methodology* **2**, UNESCO, Paris, 174 pp.

Tranter, D.J., Bulleid, N.C., Campbell, R., Higgins, H.W., Rowe, F., Tranter, H.A. and Smith, D.F. (1981). Nocturnal movements of phototactic zooplankton in shallow water. *Marine Biology* **61**, 317–326.

Tranter, D.J., Tafe, D.J. and Sandland, R.L. (1983). Some zooplankton characteristics of warm-core eddies shed by the East Australian Current, with particular reference to copepods. *Australian Journal of Marine and Freshwater Research* **34**, 587–607.

Trela, P. (1989). Subsurface zooplankton (neuston) of Hornsund Fjord, Svalbard. *Rapports et Procès-verbaux des Réunions Conseil International pour l'Exploration de la Mer* **188**, 138–145.

Tremblay, C., Runge, J.A. and Legendre, L. (1989). Grazing and sedimentation of ice algae during and immediately after a bloom at the ice-water interface. *Marine Ecology Progress Series* **56**, 291–300.

Tremblay, M.J. and Roff, J.C. (1983a). Community gradients in the Scotian Shelf zooplankton. *Canadian Journal of Fisheries and Aquatic Sciences* **40**, 598–611.

Tremblay, M.J. and Roff, J.C. (1983b). Production estimates for Scotian Shelf copepods based on mass specific P/B ratios. *Canadian Journal of Fisheries and Aquatic Sciences* **40**, 749–753.

Trujillo-Ortiz, A. (1986). Life cycle of the marine calanoid copepod *Acartia californiensis* Trinast reared under laboratory conditions. *California Cooperative Oceanic Fisheries Investigations Reports* **27**, 188–204.

Trujillo-Ortiz, A. (1990). Porciento de eclosión, producción de huevos y tiempo de desarrollo de *Acartia californiensis*, Trinast (Copepoda, Calanoida) bajo condiciones de laboratorio. *Ciencia Marina* **16**, 1–22.

Trujillo-Ortiz, A. (1995). Alternative method for the calculation of mean time for the assessment of secondary production by true cohort analysis. *Journal of Plankton Research* **17**, 2175–2190.

Tschislenko, L.L. (1964). On the sex ratio in marine free living Copepoda. *Zoologichesky Zhurnal* **43**, 1400–1402.

Tsuda, A. (1994). Starvation tolerance of a planktonic marine copepod *Pseudocalanus newmani* Frost. *Journal of Experimental Marine Biology and Ecology* **181**, 81–89.

Tsuda, A. and Nemoto, T. (1984). Feeding of a marine copepod *Acartia clausi* on cultured red-tide phytoplankton. *Bulletin of Plankton Society of Japan* **31**, 79–80.

Tsuda, A. and Nemoto, T. (1987). The effect of food concentration on the gut clearance time of *Pseudocalanus minutus* Krøyer (Calanoida: Copepoda). *Journal of Experimental Marine Biology and Ecology* **107**, 121–130.

Tsuda, A. and Nemoto, T. (1988). Feeding of copepods on natural suspended particles in Tokyo Bay. *Journal of the Oceanographical Society of Japan* **44**, 217–227.

Tsuda, A. and Nemoto, T. (1990). The effect of food concentration on the fecal pellet size of the marine copepod *Pseudocalanus newmani* Frost. *Bulletin of Plankton Society of Japan* **37**, 83–90.

Tsuda, A. and Sugisaki, H. (1994). In situ grazing rate of the copepod population

in the western subarctic North Pacific during spring. *Marine Biology* **120**, 203–210.
Tsuda, A., Sugisaki, H., Ishimaru, T., Saino, T. and Sato, T. (1993). White-noise-like distribution of the oceanic copepod *Neocalanus cristatus* in the subarctic North Pacific. *Marine Ecology Progress Series* **97**, 39–46.
Tsytsugina, V.G. (1974). Karyological investigation of *Calanus* (*Neocalanus*) *robustior* Giesbrecht (Calanidae). *Zoologichesky Zhurnal* **53**, 1568–1569.
Turner, J.T. (1977). Sinking rates of fecal pellets from the marine copepod *Pontella meadii*. *Marine Biology* **40**, 249–259.
Turner, J.T. (1978). Scanning electron microscope investigations of feeding habits and mouthpart structures of three species of copepods of the family Pontellidae. *Bulletin of Marine Science* **28**, 487–500.
Turner, J.T. (1981). Latitudinal patterns of calanoid and cyclopoid diversity in estuarine waters of eastern North America. *Journal of Biogeography* **8**, 369–382.
Turner, J.T. (1984a). Zooplankton feeding ecology: contents of fecal pellets of the copepods *Temora turbinata* and *T. stylifera* from continental shelf and slope waters near the mouth of the Mississippi River. *Marine Biology* **82**, 73–83.
Turner, J.T. (1984b). Zooplankton feeding ecology: contents of fecal pellets of the copepods *Eucalanus pileatus* and *Paracalanus quasimodo* from continental shelf waters of the Gulf of Mexico. *Marine Ecology Progress Series* **15**, 27–46.
Turner, J.T. (1984c). Zooplankton feeding ecology: contents of fecal pellets of the copepods *Acartia tonsa* and *Labidocera aestiva* from continental shelf waters near the mouth of the Mississippi River. *P.S.Z.N.I. Marine Ecology* **5**, 265–282.
Turner, J.T. (1985). Zooplankton feeding ecology: contents of fecal pellets of the copepod *Anomalocera ornata* from continental shelf and slope waters of the Gulf of Mexico. *P.S.Z.N.I. Marine Ecology* **6**, 285–298.
Turner, J.T. (1987). Zooplankton feeding ecology: contents of fecal pellets of the copepod *Centropages velificatus* from waters near the mouth of the Mississippi River. *Biological Bulletin, Marine Biological Laboratory, Woods Hole, Mass.*, **173**, 377–386.
Turner, J.T. and Anderson, D.M. (1983). Zooplankton grazing during dinoflagellate blooms in a Cape Cod embayment, with observations of predation upon tintinnids by copepods. *P.S.Z.N.I. Marine Ecology* **4**, 359–374.
Turner, J.T. and Collard, S.B. (1980). Winter distribution of pontellid copepods in the neuston of the eastern Gulf of Mexico continental shelf. *Bulletin of Marine Science* **30**, 526–529.
Turner, J.T. and Ferrante, J.G. (1979). Zooplankton fecal pellets in aquatic ecosystems. *Bioscience* **29**, 670–677.
Turner, J.T. and Granéli, E. (1992). Zooplankton feeding ecology: grazing during enclosure studies of phytoplankton blooms from the west coast of Sweden. *Journal of Experimental Marine Biology and Ecology* **157**, 19–31.
Turner, J.T. and Roff, J.C. (1993). Trophic levels and trophospecies in marine plankton: lessons from the microbial food web. *Marine Microbial Food Webs* **7**, 225–248.
Turner, J.T. and Tester, P.A. (1989). Zooplankton feeding ecology: nonselective grazing by the copepods *Acartia tonsa* Dana, *Centropages velificatus* De Oliveira, and *Eucalanus pileatus* Giesbrecht in the plume of the Mississippi River. *Journal of Experimental Marine Biology and Ecology* **126**, 21–43.
Turner, J.T., Collard, S.B., Wright, J.C., Mitchell, D.V. and Steele, P. (1979). Summer distribution of pontellid copepods in the neuston of the eastern Gulf of Mexico

continental shelf. *Bulletin of Marine Science* **29**, 287–297.

Turner, J.T., Bruno, S.F., Larson, R.J., Staker, R.D. and Sharma, G.M. (1983). Seasonality of plankton assemblages in a temperate estuary. *P.S.Z.N.I. Marine Ecology* **4**, 81–99.

Turner, J.T., Tester, P.A. and Hettler, W.F. (1985). Zooplankton feeding ecology. A laboratory study of predation on fish eggs and larvae by the copepods *Anomalocera ornata* and *Centropages typicus*. *Marine Biology* **90**, 1–8.

Turriff, N., Runge, J.A. and Cembella, A.D. (1995). Toxin accumulation and feeding of the planktonic copepod *Calanus finmarchicus* exposed to the red-tide dinoflagellate *Alexandrium excavatum*. *Marine Biology* **123**, 55–64.

Uchima, M. (1988). Gut content analysis of neritic copepods *Acartia omorii* and *Oithona davisae* by a new method. *Marine Ecology Progress Series* **48**, 93–97.

Uchima, M. and Murano, M. (1988). Mating behavior of the marine copepod *Oithona davisae*. *Marine Biology* **99**, 39–45.

Ueda, H. (1978). Analysis of the generations of inlet copepods, with special reference to *Acartia clausi* in Maizuru Bay, Middle Japan. *Bulletin of Plankton Society of Japan* **25**, 55–66.

Ueda, H. (1986a). Redescriptions of the closely related calanoid copepods *Acartia japonica* and *A. australis* with remarks on their zoogeography. *Bulletin of Plankton Society of Japan* **33**, 11–20.

Ueda, H. (1986b). Reproductive isolation between the sympatric, closely related species *Acartia omorii* and *A. hudsonica* (Copepoda: Calanoida). *Bulletin of Plankton Society of Japan* **33**, 59–60.

Ueda, H. (1987a). Temporal and spatial distribution of the two closely related *Acartia* species *A. omorii* and *A. hudsonica* (Copepoda, Calanoida) in a small inlet water of Japan. *Estuarine, Coastal and Shelf Science* **24**, 691–700.

Ueda, H. (1987b). Small-scale ontogenetic and diel vertical distributions of neritic copepods in Maizuru Bay, Japan. *Marine Ecology Progress Series* **35**, 65–73.

Ueda, H. (1991). Horizontal distribution of planktonic copepods in inlet waters. *Bulletin of Plankton Society of Japan*, Special Volume, 143–160.

Ueda, H. and Hiromi, J. (1987). The *Acartia plumosa* T. Scott species group (Copepoda, Calanoida) with a description of *A. tropica* n. sp. *Crustaceana* **53**, 225–236.

Ueda, H., Kuwahara, A., Tanaka, M. and Azeta, M. (1983). Underwater observations on copepod swarms in temperate and subtropical waters. *Marine Ecology Progress Series* **11**, 165–171.

Ummerkutty, A.N.P. (1964). Studies on Indian copepods. 6. The post-embryonic development of two calanoid copepods, *Pseudodiaptomus aurivilli* Cleve and *Labidocera bengalensis* Krishnaswamy. *Journal of the Marine Biological Association of India* **6**, 48–60.

Umminger, B.L. (1968). Polarotaxis in copepods. II. The ultrastructural basis and ecological significance of polarized light sensitivity in copepods. *Biological Bulletin, Marine Biological Laboratory, Woods Hole, Mass.* **135**, 252–261.

Unstad, K.H. and Tande, K.S. (1991). Depth distribution of *Calanus finmarchicus* and *C. glacialis* in relation to environmental conditions in the Barents Sea. *Polar Research* **10**, 409–420.

Urban, J.L., McKenzie, C.H. and Deibel, D. (1992). Seasonal differences in the content of *Oikopleura vanhoeffeni* and *Calanus finmarchicus* faecal pellets: illustrations of zooplankton food web shifts in coastal Newfoundland waters. *Marine Ecology Progress Series* **84**, 255–264.

Urban, J.L., Deibel, D. and Schwinghamer, P. (1993). Seasonal variations in the densities of fecal pellets produced by *Oikopleura vanhoeffeni* (C. Larvacea) and *Calanus finmarchicus* (C. Copepoda). *Marine Biology* **117**, 607–613.

Uye, S.-I. (1980a). Development of neritic copepods *Acartia clausi* and *A. steueri*. I. Some environmental factors affecting egg development and the nature of resting eggs. *Bulletin of Plankton Society of Japan* **27**, 1–9.

Uye, S.-I. (1980b). Development of neritic copepods *Acartia clausi* and *A. steueri*. II. Isochronal larval development at various temperatures. *Bulletin of Plankton Society of Japan* **27**, 11–18.

Uye, S.-I. (1981). Fecundity studies of neritic calanoid copepods *Acartia clausi* Giesbrecht and *A. steueri* Smirnov: a simple empirical model of daily egg production. *Journal of Experimental Marine Biology and Ecology* **50**, 255–271.

Uye, S.-I. (1982a). Population dynamics and production of *Acartia clausi* Giesbrecht (Copepoda: Calanoida) in inlet waters. *Journal of Experimental Marine Biology and Ecology* **57**, 55–83.

Uye, S.-I. (1982b). Length-weight relationships of important zooplankton from the Inland Sea of Japan. *Journal of the Oceanographical Society of Japan* **38**, 149–158.

Uye, S.-I. (1983). Seasonal cycle in abundance of resting eggs of *Acartia steueri* Smirnov (Copepoda, Calanoida) in sea-bottom mud of Onagawa Bay, Japan. *Crustaceana* **44**, 103–105.

Uye, S.-I. (1984). Production ecology of marine planktonic copepods. *Bulletin of Plankton Society of Japan* **30**, 44–54.

Uye, S.-I. (1985). Resting egg production as a life history strategy of marine planktonic copepods. *Bulletin of Marine Research* **37**, 440–449.

Uye, S. (1986). Impact of copepod grazing on the red-tide flagellate *Chattonella antiqua*. *Marine Biology* **92**, 35–43.

Uye, S.-I. (1988). Temperature-dependent development and growth of *Calanus sinicus* (Copepoda: Calanoida) in the laboratory. *Hydrobiologia* **167/168**, 285–293.

Uye, S.-I. (1991). Temperature-dependent development and growth of the planktonic copepod *Paracalanus* sp. in the laboratory. *Bulletin of Plankton Society of Japan*, Special Volume, 627–636.

Uye, S.-I. (1996). Induction of reproductive failure in the planktonic copepod *Calanus pacificus* by diatoms. *Marine Ecology Progress Series* **133**, 89–97.

Uye, S. and Fleminger, A. (1976). Effects of environmental factors on egg development of several species of *Acartia* in southern California. *Marine Biology* **38**, 253–262.

Uye, S.-I. and Kaname, K. (1994). Relations between fecal pellet volume and body size for major zooplankters of the Inland Sea of Japan. *Journal of Oceanography* **50**, 43–49.

Uye, S.-I. and Kasahara, S. (1983). Grazing of various developmental stages of *Pseudodiaptomus marinus* (Copepoda: Calanoida) on naturally occurring particles. *Bulletin of Plankton Society of Japan* **30**, 147–158.

Uye, S.-I. and Kayano, Y. (1994a). Predatory feeding behavior of *Tortanus* (Copepoda, Calanoida): life-stage differences and the predation impact on small planktonic crustaceans. *Journal of Crustacean Biology* **14**, 473–483.

Uye, S.-I. and Kayano, Y. (1994b). Predatory feeding of the planktonic copepod *Tortanus forcipatus* on three different prey. *Bulletin of Plankton Society of Japan* **40**, 173–176.

Uye, S.-I. and Matsuda, O. (1988). Phosphorus content of zooplankton from the

Inland Sea of Japan. *Journal of the Oceanographical Society of Japan* **44**, 280–286.

Uye, S.-I. and Onbé, T. (1975). The developmental stages of *Pseudodiaptomus marinus* Sato (Copepoda, Calanoida) reared in the laboratory. *Bulletin of Plankton Society of Japan* **21**, 65–76.

Uye, S.-I. and Shibuno, N. (1992). Reproductive biology of the planktonic copepod *Paracalanus* sp. in the Inland Sea of Japan. *Journal of Plankton Research* **14**, 343–358.

Uye, S.-I. and Takamatsu, K. (1990). Feeding interactions between planktonic copepods and red-tide flagellates from Japanese coastal waters. *Marine Ecology Progress Series* **59**, 97–107.

Uye, S.-I. and Yamamoto, F. (1995). *In situ* feeding of the planktonic copepod *Calanus sinicus* in the Inland Sea of Japan, examined by the gut fluorescence method. *Bulletin of Plankton Society of Japan* **42**, 123–139.

Uye, S.-I. and Yashiro, M. (1988). Respiration rates of planktonic crustaceans from the Inland Sea of Japan with special reference to the effects of body weight and temperature. *Journal of the Oceanographical Society of Japan* **44**, 47–51.

Uye, S.-I., Kasahara, S. and Onbé, T. (1979). Calanoid copepod eggs in sea-bottom muds. IV. Effects of some environmental factors on the hatching of resting eggs. *Marine Biology* **51**, 151–156.

Uye, S.-i., Iwai, Y. and Kasahara, S. (1982). Reproductive biology of *Pseudodiaptomus marinus* (Copepoda: Calanoida) in the Inland Sea of Japan. *Bulletin of Plankton Society of Japan* **29**, 25–35.

Uye, S., Iwai, Y. and Kasahara, S. (1983). Growth and production of the inshore marine copepod *Pseudodiaptomus marinus* in the central part of the Inland Sea of Japan. *Marine Biology* **73**, 91–98.

Uye, S., Imaizumi, K. and Matsuda, O. (1990a). Size composition and estimated phosphorus regeneration rates of the copepod community along an estuarine-offshore transect in the Inland Sea of Japan. *Estuarine, Coastal and Shelf Science* **31**, 851–863.

Uye, S., Huang, C. and Onbé, T. (1990b). Ontogenetic diel vertical migration of the planktonic copepod *Calanus sinicus* in the Inland Sea of Japan. *Marine Biology* **104**, 389–396.

Vaas, P. and Pesch, G.G. (1984). A karyological study of the calanoid copepod *Eurytemora affinis*. *Journal of Crustacean Biology* **4**, 248–251.

Vaissière, R. (1961). Morphologie et histologie comparées des yeux des crustacés copépodes. *Archives de Zoologie Expérimentale et Générale* **100**, 1–126.

Valentin, J. (1972). La ponte et les œufs chez les copépodes du Golfe de Marseille: cycle annuel et étude expérimentale. *Téthys* **4**, 349–390.

Valentin, J.L., Monteiro-Ribas, W.M., Mureb, M.A. and Pessotti, E. (1987). Sur quelques zooplanctontes abondants dans l'upwelling de Cabo Frio (Brésil). *Journal of Plankton Research* **9**, 1195–1216.

Van der Spoel, S. (1986). What is unique about open-ocean biogeography; zooplankton? *Unesco Technical Papers in Marine Science* **49**, 254–260.

Van der Spoel, S. and Heyman, R.P. (1983). "A Comparative Atlas of Zooplankton: Biological Patterns in the Oceans." Wetenschappelijke uitgeverij Bunge, Utrecht. 186 pp.

Vanderploeg, H.A. and Ondricek-Fallscheer, R.L. (1982). Intersetule distances are a poor predictor of particle-retention efficiency in *Diaptomus sicilis*. *Journal of Plankton Research* **4**, 237–244.

Van Duren, L.A. and Videler, J.J. (1995). Swimming behaviour of developmental

stages of the calanoid copepod *Temora longicornis* at different food concentrations. *Marine Ecology Progress Series* **126**, 153-161.
Van Duren, L.A. and Videler, J.J. (1996). The trade-off between feeding, mate seeking and predator avoidance in copepods: behavioural responses to chemical cues. *Journal of Plankton Research* **18**, 805-818.
Vaupel Klein, J.C. von (1982a). Structure of integumental perforations in the *Euchirella messinensis* female (Crustacea, Copepoda, Calanoida). *Netherlands Journal of Zoology* **32**, 374-394.
Vaupel Klein, J.C. von (1982b). A taxonomic review of the genus *Euchirella* Giesbrecht, 1888 (Copepoda, Calanoida). II. The type species, *Euchirella messinensis* (Claus, 1863). A. The female of f.*typica*. *Zoologische Verhandelingen* **198**, 1- 131.
Vaupel Klein, J.C. von (1984). A primer of a phylogenetic approach to the taxonomy of the genus *Euchirella* (Copepoda, Calanoida). *Crustaceana*, Supplement **9**, 1-194.
Vaupel Klein, J.C. von (1986). On articulations in the furcal setae of calanoid copepods and their role in swimming movements. *Syllogeus* **58**, 494-501.
Vaupel Klein, J.C. von (1989). *Euchirella lisettae* spec.nov., a new calanoid copepod from the Pacific Ocean. *Crustaceana* **57**, 145-170.
Vaupel Klein, J.C. von (1995). *Pseudochirella major* (G.O. Sars, 1907) nov. comb., the valid name for *Pseudochirella fallax* G.O. Sars, 1920 (Copepoda, Calanoida, Aetideidae). *Crustaceana* **68**, 914-917.
Vaupel Klein, J.C. von (1996). Designation of a type-species for the genus *Pseudochirella* G.O. Sars, 1920 (Copepoda, Calanoida). *Crustaceana* **69**, 446-454.
Vaupel Klein, J.C. von and Koomen, L.B. (1994). The possible origin of mucus jets used for immobilizing prey in species of *Euchirella* (Copepoda, Calanoida, Aetideidae). I. Theoretical considerations in relation to swimming and feeding behaviour. *Crustaceana* **66**, 184-204.
Vaupel Klein, J.C. von and Rijerkerk, C.D.M. (1996). A detailed redescription of the *Pseudochirella obesa* female (Copepoda, Calanoida). I. General part; trunk and antennules. *Crustaceana* **69**, 567-593.
Verheye, H.M. (1991). Short-term variability during an anchor station study in the southern Benguela upwelling system; abundance, distribution and estimated production of mesozooplankton with special reference to *Calanoides carinatus* (Krøyer, 1849). *Progress in Oceanography* **28**, 91-119.
Verheye, H.M. and Field, J.G. (1992). Vertical distribution and diel migration of *Calanoides carinatus* (Krøyer, 1849) developmental stages in the southern Benguela upwelling region. *Journal of Experimental Marine Biology and Ecology* **158**, 123-140.
Verheye, H.M., Hutchings, L. and Peterson, W.T. (1991). Life history and population maintenance strategies of *Calanoides carinatus* (Copepoda: Calanoida) in the southern Benguela ecosystem. *South African Journal of Marine Science* **11**, 179-191.
Verheye, H.M., Hutchings, L., Huggett, J.A. and Painting, S.J. (1992). Mesozooplankton dynamics in the Benguela ecosystem, with emphasis on the herbivorous copepods. *South African Journal of Marine Research* **12**, 561-584.
Verity, P.G. and Paffenhöfer, G.-A. (1996). On assessment of prey ingestion by copepods. *Journal of Plankton Research* **18**, 1767-1779.
Verity, P.G. and Smetacek, V. (1996). Organism life cycles, predation, and the

structure of marine pelagic ecosystems. *Marine Ecology Progress Series* **130**, 277–293.
Vervoort, W. (1946). Biological results of the Snellius Expedition. XV. The bathypelagic Copepoda Calanoida of the Snellius Expedition. Families Calanidae, Eucalanidae, Paracalanidae, and Pseudocalanidae. *Temminckia* **8**, 1–181.
Vervoort, W. (1951). Plankton copepods from the Atlantic sector of the Antarctic. *Verhandelingen der Koninklijke Nederlandse Akademie van Wetenschappen, Afdeeling Natuurkunde,* Section 2 **47**, (4), 1–156.
Vervoort, W. (1957). Copepods from Antarctic and sub-Antarctic plankton samples. *Report of the British, Australian and New Zealand Antarctic Research Expedition, 1929–1931,* Series B **3**, 1–160.
Vervoort, W. (1963). Pelagic Copepoda. Part I. Copepoda Calanoida of the families Calanidae up to and including Euchaetidae. *Atlantide Report* **7**, 77–194.
Vervoort, W. (1964). Free-living Copepoda from Ifaluk Atoll in the Caroline Islands with notes on related species. *Bulletin of United States National Museum* **236**, 1–431.
Vervoort, W. (1965). Pelagic Copepoda. Part II. Copepoda Calanoida of the families Phaennidae up to and including Acartiidae, containing the description of a new species of Aetideidae. *Atlantide Report* **8**, 9–216.
Vervoort, W. (1986a). Bibliography of Copepoda, up to and including 1980. Part I (A-G). *Crustaceana,* Supplement **10**, 1–369.
Vervoort, W. (1986b). Bibliography of Copepoda, up to and including 1980. Part II (H-R). *Crustaceana,* Supplement **11**, 375–845.
Vervoort, W. (1988). Bibliography of Copepoda including: bibliography of Copepoda, up to and including 1980. Part III (S-Z), addenda et corrigenda, Copepoda bibliography supplement 1981–1985. *Crustaceana,* Supplement **12**, 847–1316.
Vidal, J. (1980a). Physioecology of zooplankton. I. Effects of phytoplankton concentration, temperature, and body size on the growth rate of *Calanus pacificus* and *Pseudocalanus* sp. *Marine Biology* **56**, 111–134.
Vidal, J. (1980b). Physioecology of zooplankton. II. Effects of phytoplankton concentration, temperature, and body size on the development and molting rates of *Calanus pacificus* and *Pseudocalanus* sp. *Marine Biology* **56**, 135–146.
Vidal, J. (1980c). Physioecology of zooplankton. III. Effects of phytoplankton concentration, temperature, and body size on the metabolic rate of *Calanus pacificus*. *Marine Biology* **56**, 195–202.
Vidal, J. (1980d). Physioecology of zooplankton. IV. Effects of phytoplankton concentration, temperature, and body size on the net production efficiency of *Calanus pacificus*. *Marine Biology* **56**, 203–211.
Vidal, J. and Smith, S.L. (1986). Biomass, growth, and development of populations of herbivorous zooplankton in the southeastern Bering Sea during spring. *Deep-Sea Research* **33A**, 523–556.
Vidal, J. and Whitledge, T.E. (1982). Rates of metabolism of planktonic crustaceans as related to body weight and temperature of habitat. *Journal of Plankton Research* **4**, 77–84.
Viitasalo, M. (1994). Seasonal succession and long-term changes of mesozooplankton in the northern Baltic Sea. *Finnish Marine Research* **263**, 3–39.
Viitasalo, M. and Katajisto, T. (1994). Mesozooplankton resting eggs in the Baltic Sea: identification and vertical distribution in laminated and mixed sediments. *Marine Biology* **120**, 455–465.

Viitasalo, M., Katajisto, T. and Vuorinen, I. (1994). Seasonal dynamics of *Acartia bifilosa* and *Eurytemora affinis* (Copepoda: Calanoida) in relation to abiotic factors in the northern Baltic Sea. *Hydrobiologia* **292/293**, 415–422.

Viitasalo, M., Vuorinen, I. and Saesmaa, S. (1995a). Mesozooplankton dynamics in the northern Baltic Sea: implications of variations in hydrography and climate. *Journal of Plankton Research* **17**, 1857–1878.

Viitasalo, M., Koski, M., Pellikka, K. and Johansson, S. (1995b). Seasonal and long-term variations in the body size of planktonic copepods in the northern Baltic Sea. *Marine Biology* **123**, 241–250.

Vijverberg, J. (1980). Effect of temperature in laboratory studies on development and growth of Cladocera and Copepoda from Tjeukemeer, The Netherlands. *Freshwater Biology* **10**, 317–340.

Vilela, M.H. (1972). The developmental stages of the marine calanoid copepod *Acartia grani* Sars bred in the laboratory. *Notas e Estudos do Instituto de Biologia Maritima* **40**, 1–20.

Villate, F., Ruiz, A. and Franco, J. (1993). Summer zonation and development of zooplankton within a shallow mesotidal system: the estuary of Mundaka. *Cahiers de Biologie marine* **34**, 131–143.

Vinogradov, M.E. (1968). "Vertical Distribution of the Oceanic Zooplankton." Israel Program for Scientific Translations, 1970, 339 pp.

Vinogradov, M.E. (1997). Some problems of vertical distribution of meso- and macroplankton in the ocean. *Advances in Marine Biology* **32**, 1–92.

Vinogradov, M.E., Flint, M.V. and Shushkina, E.A. (1985). Vertical distribution of mesoplankton in the open area of the Black Sea. *Marine Biology* **89**, 95–107.

Vinogradov, M.E., Arashkevich, E.G. and Ilchenko, S.V. (1992). The ecology of the *Calanus ponticus* population in the deeper layer of its concentration in the Black Sea. *Journal of Plankton Research* **14**, 447–458.

Vives, F. (1982). Sur les copépodes de la région CINECA (parties nord et centrale). *Rapports et Procès-Verbaux des Réunions Conseil International pour l'Exploration de la Mer* **180**, 289–296.

Vlymen, W.J. (1970). Energy expenditure of swimming copepods. *Limnology and Oceanography* **15**, 348–356.

Vlymen, W.J. (1977). Reply to comment by J.T. Enright. *Limnology and Oceanography* **22**, 162–165.

Volkman, J.K., Gatten, R.R. and Sargent, J.R. (1980). Composition and origin of milky water in the North Sea. *Journal of the Marine Biological Association of the United Kingdom* **60**, 759–768.

Voss, M. (1991). Content of copepod faecal pellets in relation to food supply in Kiel Bight and its effect on sedimentation rate. *Marine Ecology Progress Series* **75**, 217–225.

Vučetic, T. (1966). On the biology of *Calanus helgolandicus* (Claus) from the Veliko Jezero i. Mljet. *Acta Adriatica* **6** (11), 1–91.

Vuorinen, I. (1982). The effect of temperature on the rates of development of *Eurytemora hirundoides* (Nordquist) in laboratory culture. *Annales Zoologici Fennicae* **19**, 129–134.

Vuorinen, I. (1987). Vertical migration of *Eurytemora* (Crustacea, Copepoda): a compromise between the risks of predation and decreased fecundity. *Journal of Plankton Research* **9**, 1037–1046.

Vyshkvartseva, N.V. (1987). *Scaphocalanus acutocornis* sp.n. (Scolecithricidae, Copepoda) from the abyssopelagial Kuril-Kamchatka Trench. *Zoologichesky Zhurnal* **66**, 1573–1576.

Vyshkvartseva, N.V. (1989a). *Puchinia obtusa* gen.et sp.n. (Copepoda, Calanoida) from the ultra-abyssal of the Kuril-Kamchatsk Trench and the place of the genus in the family Scolecithricidae. *Zoologichesky Zhurnal* **68** (4), 29–38.

Vyshkvartseva, N.V. (1989b). On the systematic of the family Scolecithricidae (Copepoda, Calanoida). New genus *Archescolecithix* and redescription of the genus *Mixtocalanus* Brodsky, 1950. *In* "Explorations of the Fauna of the Seas, 41 (49). Marine Plankton" (M.G. Petrushevska and S.D. Stepanjants, eds), pp. 5–23. USSR Academy of Sciences, Leningrad.

Vyshkvartseva, N.V. (1994). *Senecella siberica* n. sp. and the position of the genus *Senecella* in Calanoida classification. *Hydrobiologia* **292/293**, 113–121.

Waghorn, E.J. and Knox, G.A. (1988). Summer tide-crack zooplankton at White Island, McMurdo Sound, Antarctica. *New Zealand Journal of Marine and Freshwater Research* **22**, 577–582.

Walker, D.R. and Peterson, W.T. (1991). Relationships between hydrography, phytoplankton production, biomass, cell size and species composition, and copepod production in the southern Benguela upwelling system in April 1988. *South African Journal of Marine Science* **11**, 289–305.

Walter, T.C. (1984). New species of *Pseudodiaptomus* from the Indo-Pacific with a classification of *P. aurivilli* and *P. mertoni* (Crustacea: Copepoda: Calanoida). *Proceedings of the Biological Society of Washington*, 97, 369–391.

Walter, T.C. (1986a). New and poorly known Indo-Pacific species of *Pseudodiaptomus* (Copepoda: Calanoida), with a key to the species groups. *Journal of Plankton Research* **8**, 129–168.

Walter, T.C. (1986b). The zoogeography of the genus *Pseudodiaptomus* (Calanoida: Pseudodiaptomidae). *Syllogeus* **58**, 502–508.

Walter, T.C. (1987). Review of the taxonomy and distribution of the demersal copepod genus *Pseudodiaptomus* (Calanoida: Pseudodiaptomidae) from southern Indo-West Pacific waters. *Australian Journal of Marine and Freshwater Research* **38**, 363–396.

Walter, T.C. (1989). Review of the New World species of *Pseudodiaptomus* (Copepoda: Calanoida), with a key to the species. *Bulletin of Marine Science* **45**, 590–628.

Walter, T.C. (1994). A clarification of two congeners, *Pseudodiaptomus lobipes* and *P. binghami* (Calanoida, Pseudodiaptomidae) from India, with description of *P. mixtus* sp.n. from Bangladesh. *Hydrobiologia* **292/293**, 123–130.

Wang, R. and Conover, R.J. (1986). Dynamics of gut pigment in the copepod *Temora longicornis* and the determination of in situ grazing rates. *Limnology and Oceanography* **31**, 867–877.

Wang, W.-X., Reinfelder, J.R., Lee, B.-G. and Fisher, N.S. (1996). Assimilation and regeneration of trace elements by marine copepods. *Limnology and Oceanography* **41**, 70–81.

Wang, Z. (1992). The effect of environmental factors on population dynamics of *Drepanopus bispinosus* (Calanoida: Copepoda) in Burton Lake, Antarctica. *Proceedings of the NIPR Symposium on Polar Biology* **5**, 151–162.

Ward, P. (1989). The distribution of zooplankton in an Antarctic fjord at South Georgia during summer and winter. *Antarctic Science* **1**, 141–150.

Ward, P. and Robins, D.B. (1987). The reproductive biology of *Euchaeta antarctica* Giesbrecht (Copepoda: Calanoida) at South Georgia. *Journal of Experimental Marine Biology and Ecology* **108**, 127–145.

Ward, P. and Shreeve, R.S. (1995). Egg production in three species of Antarctic calanoid copepods during an austral summer. *Deep-Sea Research I* **42**, 721–735.

Ward, P., Shreeve, R.S., Cripps, G.C. and Trathan, P.N. (1996a). Mesoscale distribution and population dynamics of *Rhincalanus gigas* and *Calanus simillimus* in the Antarctic polar open ocean and polar frontal zone during summer. *Marine Ecology Progress Series* **140**, 21–32.
Ward, P., Shreeve, R.S. and Cripps, G.C. (1996b). *Rhincalanus gigas* and *Calanus simillimus*: lipid storage patterns of two species of copepod in the seasonally ice-free zone of the Southern Ocean. *Journal of Plankton Research* **18**, 1439–1454.
Watling, L. (1989). A classification system for crustacean setae based on the homology concept. *In* "Functional Morphology of Feeding and Grooming in Crustacea" (B.E. Felgenhauer, L. Watling and A.B. Thistle, eds), pp. 15–26. A.A. Balkema, Rotterdam.
Watras, C.J. (1983). Male location by diaptomid copepods. *Journal of Plankton Research* **5**, 417–423.
Watson, N.H.F. (1986). Variability of diapause in copepods. *Syllogeus* **58**, 509–513.
Weatherby, T.M., Wong, K.K. and Lenz, P.H. (1994). Fine structure of the distal sensory setae on the first antennae of *Pleuromamma xiphias* Giesbrecht (Copepoda). *Journal of Crustacean Biology* **14**, 670–685.
Webb, D.G. and Weaver, A.J. (1988). Predation and the evolution of free spawning in marine calanoid copepods. *Oikos* **51**, 189–192.
Webber, M.K. and Roff, J.C. (1995a). Annual structure of the copepod community and its associated pelagic environment off Discovery Bay, Jamaica. *Marine Biology* **123**, 467–479.
Webber, M.K. and Roff, J.C. (1995b). Annual biomass and production of the oceanic copepod community off Discovery Bay, Jamaica. *Marine Biology* **123**, 481–495.
Weikert, H. (1977). Copepod carcasses in the upwelling region south of Cap Blanc, N.W. Africa. *Marine Biology* **42**, 351–355.
Weikert, H. (1982a). Some features of zooplankton distribution in the upper 200 m in the upwelling region off northwest Africa. *Rapports et Procès-Verbaux des Réunions Conseil International pour l'Exploration de la Mer* **180**, 280–288.
Weikert, H. (1982b). The vertical distribution of zooplankton in relation to habitat zones in the area of the Atlantis II Deep, central Red Sea. *Marine Ecology Progress Series* **8**, 129–143.
Weikert, H. (1984). Zooplankton distribution and hydrography in the Mauritanian upwelling region off northwestern Africa, with special reference to the calanoid copepods. *Meeresforschung* **30**, 155–171.
Weikert, H. and Koppelmann, R. (1993). Vertical structural patterns of deep-living zooplankton in the NE Atlantic, the Levantine Sea and the Red Sea: a comparison. *Oceanologica Acta* **16**, 163–177.
Weikert, H. and Koppelmann, R. (1996). Mid-water zooplankton profiles from the temperate ocean and partially landlocked seas. A re-evaluation of interoceanic differences. *Oceanologica Acta* **19**, 657–664.
Weikert, H. and Trinkaus, S. (1990). Vertical mesozooplankton abundance and distribution in the deep Eastern Mediterranean Sea SE of Crete. *Journal of Plankton Research* **12**, 601–628.
Weisse, T. (1983). Feeding of calanoid copepods in relation to *Phaeocystis pouchetii* blooms in the German Wadden Sea area of Sylt. *Marine Biology* **74**, 87–94.
Weissman, P., Lonsdale, D.L. and Yen, J. (1993). The effect of peritrich ciliates on the production of *Acartia hudsonica* in Long Island Sound. *Limnology and*

Oceanography **38**, 613–622.
Wellershaus, S. and Soltanpour-Gargari, A. (1991). Planktonic copepods in the very low salinity region in estuaries. *Bulletin of Plankton Society of Japan*, Special Volume, 133–142.
Wheeler, E.H. (1970). Atlantic deep-sea calanoid Copepoda. *Smithsonian Contributions to Zoology* **55**, 1–31.
White, J.R. and Dagg, M.J. (1989). Effects of suspended sediments on egg production of the calanoid copepod *Acartia tonsa*. *Marine Biology* **102**, 315–319.
White, J.R. and Roman, M.R. (1991). Measurement of zooplankton grazing using particles labelled in light and dark with [methyl-^{3}H] methylamine hydrochloride. *Marine Ecology Progress Series* **71**, 45–52.
White, J.R. and Roman, M.R. (1992a). Egg production by the calanoid copepod *Acartia tonsa* in the mesohaline Chesapeake Bay: the importance of food resources and temperature. *Marine Ecology Progress Series* **86**, 239–249.
White, J.R. and Roman, M.R. (1992b). Seasonal study of grazing by metazoan zooplankton in the mesohaline Chesapeake Bay. *Marine Ecology Progress Series* **86**, 251–261.
Wiadnyana, N.N. and Rassoulzadegan, F. (1989). Selective feeding of *Acartia clausi* and *Centropages typicus* on microzooplankton. *Marine Ecology Progress Series* **53**, 37–45.
Wiborg, K.F. (1976). Fishery and commercial exploitation of *Calanus finmarchicus* in Norway. *Journal du Conseil International pour l'Exploration de la Mer* **36**, 251–258.
Wickstead, J.H. (1959). A predatory copepod. *Journal of Animal Ecology* **28**, 69–72.
Wickstead, J.H. (1961). A quantitative and qualitative study of some Indo-West-Pacific plankton. *Fishery Publications, Colonial Office* **16**, 1–200.
Wickstead, J.H. (1962). Food and feeding in pelagic copepods. *Proceedings of the Zoological Society of London* **139**, 545–555.
Wickstead, J.H. (1967). Pelagic copepods as food organisms. *In* "Proceedings of the Symposium on Crustacea", Volume IV, pp. 1460–1465. Marine Biological Association of India. The Bangalore Press, Bangalore.
Wickstead, J. and Krishnaswamy, S. (1964). On *Ivellopsis elephas* (Brady), a rare calanoid copepod. *Crustaceana* **7**, 27–32.
Widder, E.A. (1992). Mixed light imaging system for recording bioluminescence behaviours. *Journal of the Marine Biological Association of the United Kingdom* **72**, 131–138.
Widder, E.A., Latz, M.I. and Case, J.F. (1983). Marine bioluminescence spectra measured with an optical multichannel detection system. *Biological Bulletin, Marine Biological Laboratory, Wood's Hole, Mass.*, **165**, 791–810.
Wiebe, P.H. (1970). Small-scale spatial distribution in oceanic zooplankton. *Limnology and Oceanography* **15**, 205–217.
Wiebe, P.H. (1971). A computer model study of zooplankton patchiness and its effects on sampling error. *Limnology and Oceanography* **16**, 29–38.
Wiebe, P.H. and Greene, C.H. (1994). The use of high frequency acoustics in the study of zooplankton spatial and temporal patterns. *Proceedings of the NIPR Symposium on Polar Biology* **7**, 133–157.
Wiebe, P.H. and Holland, W.R. (1968). Plankton patchiness: effects on repeated net tows. *Limnology and Oceanography* **13**, 315–321.
Wiebe, P.H., Morton, A.W., Bradley, A.M., Backus, R.H., Craddock, J.E., Barber, V., Cowles, T.J. and Flierl, G.R. (1985). New developments in the MOCNESS, an

apparatus for sampling zooplankton and micronekton. *Marine Biology* **87**, 313–323.
Wiebe, P.H., Copley, N., Van Dover, C., Tamse, A. and Manrique, F. (1988). Deep-water zooplankton of the Guaymas Basin hydrothermal vent field. *Deep-Sea Research* **35A**, 985–1013.
Wiebe, P.H., Copley, N.J. and Boyd, S.H. (1992). Coarse-scale horizontal patchiness and vertical migration of zooplankton in Gulf Stream warm-core ring 82-H. *Deep-Sea Research* **39A**, Supplement **1A**, S247–S278.
Willason, S.W., Favuzzi, J. and Cox, J.L. (1986). Patchiness and nutritional condition of zooplankton in the California Current. *Fishery Bulletin, National Oceanic and Atmospheric Administration of the United States* **84**, 157–176.
Williams, J.A. (1996). Blooms of *Mesodinium rubrum* in Southampton Water – do they shape mesozooplankton distribution? *Journal of Plankton Research* **18**, 1685–1697.
Williams, R. (1985). Vertical distribution of *Calanus finmarchicus* and *C. helgolandicus* in relation to the development of the seasonal thermocline in the Celtic Sea. *Marine Biology* **86**, 145–149.
Williams, R. (1988). Spatial heterogeneity and niche differentiation in oceanic zooplankton. *Hydrobiologia* **167/168**, 151–159.
Williams, R. and Conway, D.V.P. (1984). Vertical distribution, and seasonal and diurnal migration of *Calanus helgolandicus* in the Celtic Sea. *Marine Biology* **79**, 63–73.
Williams, R. and Conway, D.V.P. (1988). Vertical distribution and seasonal numerical abundances of the Calanidae in oceanic waters to the south-west of the British Isles. *Hydrobiologia* **167/168**, 259–266.
Williams, R. and Lindley, J.A. (1980a). Plankton of the Fladen Ground during FLEX 76. I. Spring development of the plankton community. *Marine Biology* **57**, 73–78.
Williams, R. and Lindley, J.A. (1980b). Plankton of the Fladen Ground during FLEX 76. III. Vertical distribution, population dynamics and production of *Calanus finmarchicus* (Crustacea: Copepoda). *Marine Biology* **60**, 47–56.
Williams, R. and Robins, D.B. (1982). Effects of preservation on wet weight, dry weight, nitrogen and carbon contents of *Calanus helgolandicus* (Crustacea: Copepoda). *Marine Biology* **71**, 271–281.
Williams, R. and Wallace, M.A. (1975). Continuous plankton records: a plankton atlas of the North Atlantic and North Sea: supplement I – the genus *Clausocalanus* (Crustacea: Copepoda, Calanoida) in 1965. *Bulletin of Marine Ecology* **8**, 167–184.
Williams, R., Collins, N.R. and Conway, D.V.P. (1983). The double LHPR system, a high speed micro- and macroplankton sampler. *Deep-Sea Research* **30A**, 331–342.
Williams, R., Conway, D.V.P. and Hunt, H.G. (1994). The role of copepods in the planktonic ecosystems of mixed and stratified waters of the European shelf seas. *Hydrobiologia* **292/293**, 521–530.
Williams, R.J., Griffiths, F.B., Van der Wal, E.J. and Kelley, J. (1988). Cargo vessel ballast water as a vector for the transport of non-indigenous marine species. *Estuarine, Coastal and Shelf Science* **26**, 409–420.
Wilson, C.B. (1932). The copepods of the Woods Hole region Massachusetts. *Bulletin of the Smithsonian Institution, United States National Museum* **158**, 1–635.
Wilson, C.B. (1950). Contributions to the biology of the Philippine Archipelago

and adjacent regions. Copepods collected by the United States Fisheries steamer "Albatross" from 1887 to 1909, chiefly in the Pacific Ocean. *Bulletin of the Smithsonian Institution, United States National Museum* **100** (14), (4), 141–441.

Wilson, D.F. and Parrish, K.K. (1971). Remating in a planktonic marine calanoid copepod. *Marine Biology* **9**, 202–204.

Wilson, M.S. (1972). Copepods of marine affinities from mountain lakes of western North America. *Limnology and Oceanography* **17**, 762–763.

Winberg, G.G., ed. (1971). Methods for the Estimation of Production of Aquatic Animals. Academic Press, London and New York. 175 pp.

Wirick, C.D. (1989a). Herbivores and the spatial distributions of the phytoplankton. I. The plankton market. *Internationale Revue der gesamten Hydrobiologie* **74**, 15–28.

Wirick, C.D. (1989b). Herbivores and the spatial distributions of the phytoplankton. II. Estimating grazing in planktonic environments. *Internationale Revue der gesamten Hydrobiologie* **74**, 249–259.

Wishner, K.F. (1980a). The biomass of the deep-sea benthopelagic plankton. *Deep-Sea Research* **27A**, 203–216.

Wishner, K.F. (1980b). Aspects of the community ecology of deep-sea, benthopelagic plankton with special attention to gymnopleid copepods. *Marine Biology* **60**, 179–187.

Wishner, K.F. and Allison, S.K. (1986). The distribution and abundance of copepods in relation to the physical structure of the Gulf Stream. *Deep-Sea Research* **33A**, 705–731.

Wishner, K., Durbin, E., Durbin, A., Macaulay, M., Winn, H. and Kenney, R. (1988). Copepod patches and right whales in the Great South Channel off New England. *Bulletin of Marine Research* **43**, 825–844.

Wlodarczyk, E., Durbin, A.G. and Durbin, E.G. (1992). Effect of temperature on lower feeding thresholds, gut evacuation rate, and diel feeding behavior in the copepod *Acartia hudsonica*. *Marine Ecology Progress Series* **85**, 93–106.

Wolfenden, R.N. (1911). Die marinen Copepoden: II. Die pelagischen Copepoden der Westwinddrift und des südlichen Eismeers. *Deutsche Südpolar-Expedition 1901–1903*, **12**, 181–380.

Wong, C.K. (1988a). The swimming behavior of the copepod *Metridia pacifica*. *Journal of Plankton Research* **10**, 1285–1290.

Wong, C.K. (1988b). Effects of competitors, predators, and prey on the grazing behavior of herbivorous calanoid copepods. *Bulletin of Marine Science* **43**, 573–582.

Wong, C.K. and Sprules, W.G. (1986). The swimming behavior of the freshwater calanoid copepods *Limnocalanus macrurus* Sars, *Senecella calanoides* Juday and *Epischura lacustris* Forbes. *Journal of Plankton Research* **8**, 79–90.

Woodmansee, R.A. (1958). The seasonal distribution of the zooplankton off Chicken Key in Biscayne Bay, Florida. *Ecology* **39**, 247–262.

Woods, S.M. (1969). Polyteny and size variation in the copepod *Pseudocalanus* from two semi-landlocked fjords on Baffin Island. *Journal of the Fisheries Research Board of Canada* **26**, 543–556.

Wooldridge, T. and Erasmus, T. (1980). Utilization of tidal currents by estuarine zooplankton. *Estuarine and Coastal Marine Science* **11**, 107–114.

Wooldridge, T. and Melville-Smith, R. (1979). Copepod succession in two South African estuaries. *Journal of Plankton Research* **1**, 329–341.

Wroblewski, J.S. (1982). Interaction of currents and vertical migration in maintain-

ing *Calanus marshallae* in the Oregon upwelling zone – a simulation. *Deep-Sea Research* **29A**, 665–686.
Yamamoto, T. and Nishizawa, S. (1986). Small-scale zooplankton aggregations at the front of a Kuroshio warm-core ring. *Deep-Sea Research* **33A**, 1729–1740.
Yamazaki, H. and Squires, K.D. (1996). Comparison of oceanic turbulence and copepod swimming speed. *Marine Ecology Progress Series* **144**, 299–301.
Yang, C.-M. (1977). The egg development of *Paracalanus crassirostris* Dahl, 1894 (Copepoda, Calanoida). *Crustaceana* **33**, 33–38.
Yassen, S.T. (1981). Méthode d'élevage de copépodes planctoniques au laboratoire (*Temora stylifera, Acartia clausi*) estimation du taux de mortalité. *Annales de l'Institut Océanographique* **57**, 125–132.
Yen, J. (1982). Sources of variability in attack rates of *Euchaeta elongata* Esterly, a carnivorous marine copepod. *Journal of Experimental Marine Biology and Ecology* **63**, 105–117.
Yen, J. (1983). Effects of prey concentration, prey size, predator life stage, predator starvation, and season on predation rates of the carnivorous copepod *Euchaeta elongata*. *Marine Biology* **75**, 69–77.
Yen, J. (1985). Selective predation by the carnivorous marine copepod *Euchaeta elongata*: laboratory measurements of predation rates verified by field observations of temporal and spatial feeding patterns. *Limnology and Oceanography* **30**, 577–597.
Yen, J. (1987). Predation by a carnivorous marine copepod, *Euchaeta norvegica* Boeck, on eggs and larvae of the North Atlantic cod *Gadus morhua* L. *Journal of Experimental Marine Biology and Ecology* **112**, 283–296.
Yen, J. (1988). Directionality and swimming speeds in predator-prey and male-female interactions of *Euchaeta rimana*, a subtropical marine copepod. *Bulletin of Marine Research* **43**, 395–403.
Yen, J. (1991). Predatory feeding behavior of an Antarctic marine copepod, *Euchaeta antarctica*. *Polar Research* **10**, 433–442.
Yen, J. and Fields, D.M. (1992). Escape responses of *Acartia hudsonica* (Copepoda) nauplii from the flow fields of *Temora longicornis* (Copepoda). *Archives fur Hydrobiologie Beiheft Ergebnisse Limnologie* **36**, 123–134.
Yen, J. and Nicholl, N.T. (1990). Setal array on the first antennae of a carnivorous marine copepod, *Euchaeta norvegica*. *Journal of Crustacean Biology* **10**, 218–224.
Yen, J., Lenz, P.H., Gassie, D.V. and Hartline, D.K. (1992). Mechanoreception in marine copepods: electrophysiological studies on the first antennae. *Journal of Plankton Research* **14**, 495–512.
Zagalsky, P.F., Clark, R.J.H. and Fairclough, D.P. (1983). Resonance raman and circular dichroism studies of the copepod (*Anomalocera patersoni*) and siphonophore (*Porpita* sp.) astaxanthin-proteins with an identical absorption maximum at 650 nm. *Comparative Biochemistry and Physiology* **75B**, 169–170.
Zaika, V.E. (1968). Age-structure dependence of the "specific production" in zooplankton populations. *Marine Biology* **1**, 311–315.
Zavala-Hamz, V.A., Alvarez-Borrego, J. and Trujillo-Ortiz, A. (1996). Diffraction patterns as a tool to recognize copepods. *Journal of Plankton Research* **18**, 1471–1484.
Zheng Zhong, Li Shaojing, Zhou Qiulin, Xu Zhenzu and Yang Qiwen (1989). Marine Planktology. China Ocean Press, Beijing and Springer-Verlag, Berlin, Heidelberg, New York, Tokyo. 438 pp.
Zheng Xiaoyan and Zheng Zhong (1989). Study on the relationship between

mandibular edge and feeding mechanism of Copepoda. *Oceanology Limnology Sinica* **20**, 308–313.

Zhong, X.-F. and Xiao, Y.-C. (1992). Resting eggs of *Acartia bifilosa* Giesbrecht and *A. pacifica* Steuer in Jiaozhou Bay. *Marine Sciences, Qingdao* **5**, 55–59.

Zillioux, E.J. and Gonzalez, J.G. (1972). Egg dormancy in a neritic calanoid copepod and its implications to overwintering in boreal waters. *In* Fifth European Marine Biology Symposium (B. Battaglia, ed.), pp. 217–230. Piccin Editore, Padova.

Zillioux, E.J. and Wilson, D.F. (1966). Culture of a planktonic calanoid copepod through multiple generations. *Science* **151**, 996–998.

Zinntae Zo (1982). The sequential taxonomic key: an application to some copepod genera. *Hydrobiologia* **96**, 9–13.

Żmijewska, M.I. (1993). Seasonal and spatial variations in the population structure and life histories of the Antarctic copepod species *Calanoides acutus, Calanus propinquus, Rhincalanus gigas, Metridia gerlachei* and *Euchaeta antarctica* (Calanoida) in Croker Passage (Antarctic Peninsula). *Oceanologia, Warsaw* **35**, 73–100.

Żmijewska, M.I. and Yen, J. (1993). Seasonal and diel changes in the abundance and vertical distribution of the Antarctic copepod species *Calanoides acutus, Calanus propinquus, Rhincalanus gigas, Metridia gerlachei* and *Euchaeta antarctica* (Calanoida) in Croker Passage (Antarctic Peninsula). *Oceanologia, Warsaw* **35**, 101–127.

Żurek, R. and Bucka, H. (1994). Algal size classes and phytoplankton-zooplankton interacting effects. *Journal of Plankton Research* **16**, 583–601.

Zvereva, J.A. (1975). *Valdiviella brodskyi* (Copepoda, Calanoida) from the Pacific Ocean and comparison of genital fields in some species of the genus. *Zoologichesky Zhurnal* **54**, 1890–1894.

Zvereva, J.A. (1976). About functional significance of morphological formations on genital segment of females *Pareuchaeta*. *In* "Explorations of the Fauna of the Seas, 18 (16). Functional Morphology, Growth and Development of Invertebrates of Sea Plankton", pp. 70–74. Academy of Sciences of the U.S.S.R., Zoological Institute, Leningrad.

Taxonomic Index

Note: Page references in *italics* refer to Figures; those in **bold** refer to Tables

Acartia 4, 37, 218, 236, 298
  behaviour 402, **427**, 428, 429, 445, 451
  ecology 469, 473, 475, 480, 481, 483, 490, 497, 524, 527, 528
  growth/development 298, 321, 329, **341**
  life history 347, 384, 389, 390, **395**
  nutrition **152**, 164, 166, 169, 177, 195, **208**
  reproduction 257, 265, **268**, 268, **269**, 272, **285**, 290
  size/weight 328, *332*, 333
  taxonomy **51**, 60, *61*, 61, 64, 95, 110–11
Acartia adriatica 110
  africana 110
  amboinensis 110
  australis 110, 111, **427**, 429, **430**
  bacorehuisensis 110
  baylyi 110
  bermudensis 110
  bifilosa 7, **26**, 110
    biochemistry 237, **238**
    ecology **375**, **395**, 398
    growth/development **301**
    nutrition 143, 155, **186**
    reproduction **269**
  bilobata 110
  bispinosa 110, **427**
  bowmani 110
  brevicornis 110
  californiensis **26**, 110
    ecology **395**
    growth **300**, **320**
    reproduction **269**, **285**
  centrura **47**, 110, 506, **507**
  chilkaensis 110
  clausi 7, **26**, 37, 60, 64, 110
    behaviour 405, 406, **408**, **410**, **415**, **418**, **426**, **427**, 446
    biochemistry 237, **238**, 240, **241**, 242, 250, 251
    chromosomes **47**
    distribution 508, **512**
    ecology 351, **355**, 363, 366–7, *367*, 369, **375**, 380, **395**, 398, 469
    growth/development 289, **300**, **310**, **311**, **312**, *316*, 316, **320**, 339, **341**
    life history 340, **345**, **386**
    nutrition 150, 156, 159, 161, 164, 167, **184**, **186**, **187**, **189**, 197, **198**
    physiology 215, 216, 217
    reproduction 260, 267, **269**, 281, 282, **285**, **288**, 290
    size/weight **228**, **331**, 332, 336, **337**
  danae 7, **26**, 110
  denticornis 110
  discaudata 7, 110, 216, **301**
  dubia 110
  dweepi 110
  ensifera 110
  enzoi 110
  erythraea 110, **269**, **427**, **431**
  fancetti 110
  floridana 110
  forcipata 110
  fossae 110, 149, **418**, **507**
  giesbrechti 110
  grani **26**, 110
    growth/development, **301**, **310**
    nutrition **186**
    reproduction 268, 282, **285**
  gravelyi **47**, 110
  hamata, 110, **427**
  hudsonica 110, 228
    behaviour 406, **410**, 414, **415**, **418**, 419, 420, **433**, 439, **446**, 449
    ecology 373, 374, **375**, 469, 475, 525
    growth/development 339
    nutrition 143, 157, **187**, **189**
    physiology, 217
    reproduction 258, **269**, 282, **285**, 291, 294

Acartia adriatica (contd.)
  iseana 110
  italica 110
  japonica 110, 111, **427**
  jilletti 110
  josephinae, 110, **269**
  kempi 110
  keralensis **47**, 110
  latisetosa 110, **269**
  laxa 110
  lefevreae 110, 508
  levequei 110
  lilljeborgi **26**, 110
  longipatella 110, 451
  longiremis **7**, **26**, 110
    behaviour 444
    biochemistry 247
    ecology **355**, 364, **375**, 481, **493**, 495, 496
    growth/development **310**
    life history **386**
    nutrition 143, 150, **189**
    reproduction 43, **255**, **269**
  longisetosa 110
  macropus 110
  major 110
  margalefi, **47**, 64, 110, 507, 508
  minor 110
  mossi 110
  natalensis 110, 451
  negligens **7**, **26**, **47**, 110, **355**, **415**
  nicolae 110
  omorii 110
    behaviour **427**, 429, **431**, **446**
    biochemistry **238**
    distribution **507**
    ecology **355**
    nutrition 157, 164, **200**, 203
    reproduction 258, **285**
    size 330
  pacifica 110, **269**
  pietschmanni, 110
  plumosa **47**, 110
  ransoni 110
  remivagantis 110
  sarojus 110
  sewelli 110
  simplex 110
  sinensis 110
  sinijensis 110, **269**, **301**, **427**, **430**
  southwelli **47**, 110
  spinata 110, **427**, 428
  spinicauda 22, **47**, 110, **269**
  steueri 110
    behaviour **427**, 429, **431**
    growth/development 298, **301**, **310**, **312**, **320**
    life history **386**
    reproduction **269**, **285**
  teclae 110, **269**
  tonsa **7**, 60, 110, 297
    behaviour 406, **408**, **410**, **415**, **418**, 419, **427**, **433**, 447, 452
    biochemistry 237, **238**, 240, **241**, 242, 243
    distribution 506, **507**
    ecology 351, **355**, 365, 369, 372, 374, **375**, 474, 475–6, 520
    growth/development **301**, **310**, **320**, **329**, **341**
    longevity 340, **345**
    morphology **26**, 36, 37, 38
    nutrition 143, 146, 151, 153, 154, 157, 159, 162, 164, 167, 171, 177, 178, 181, **183**, **184**, **186**, **187**, **189**, 192, **194**, 197, **198**, **200**, 203, 205, **206**, 209
    physiology 217
    reproduction **255**, 260, 262, 265, **269**, 272, 281, 282, **285**, **288**, 290, 291, 520
    size/weight 229, 328, 330, **331**, 333, *334*, *335*, 336, **337**
  tortaniformis 110
  tranteri 110, **229**
    behaviour 420, 447, 450, 452
    ecology 369, **375**
    reproduction **285**
    size/weight **229**, **337**, 340, **341**
  tropica 110
  tsuensis 110
    biochemistry **238**, **241**
    growth/development **301**, **320**
    reproduction **269**
    size/weight **229**, 332
  tumida 110
Acartiidae **51**, 110–11, 268, **269**
Acrocalanus **52**, 77, 120, 483
  andersoni 120
  gibber 120, **285**, **331**
  gracilis 120, **285**
  indicus 120
  inermis 37, **375**
  longicornis 120

TAXONOMIC INDEX

monachus, 120
Aetideidae 3, 434
  ecology 470, **499**, 502
  morphology 21, 27, 28
  taxonomy **53**, 124–9
Aetideopsis
  ecology 528
  nutrition **152**
  taxonomy **53**, 88, 90, 92, 124
Aetideopsis albatrossae 124
  antarctica 124
  armata 124
  carinata 124
  cristata 124
  divaricata 124
  inflata 124
  magna **499**
  minor 124
  modesta 124
  multiserrata 4, 124, 522
  retusa 124
  rostrata 124
  trichechus 124
  tumerosa, 124
Aetideus
  ecology 480, 528
  nutrition **152**, 166
  taxonomy **53**, 92, 124–5
Aetideus acutus 124
  arcuatus 124
  armatus 7, **26**, 124, **223**, 268, **512**
  australis 124
  bradyi 124
  divergens 124, 165, **188**, **189**
  giesbrechti 124
  mexicanus 124
  pseudarmatus 124
  truncatus 124
Alexandrium excavatum, 157
Alloiopodus **52**, 121
  pinguis 121, **498**
Alrhabdus **51**, 70, 106
  johrdeae 106
Amallophora **53**, 134
  elegans 124, 135
  impar 124, 136
  macilenta **500**
  obtusifrons 134
  rotunda **500**
Amallothrix
  ecology 486, **500**

  taxonomy **53**, 57, 81, 82, 134
Amallothrix arcuata 134
  dentipes 134
  emarginata 134
  falcifer 134
  farrani 136
  gracilis 134
  hadrosoma 134
  indica 134
  invenusta 134
  longispina 134
  parafalcifer 134
  profunda 136
  pseudoarcuata 134
  pseudopropinqua 134
  robusta 134
  sarsi 136
Amphascandria 17
Anawekia **53**, 78, 81, 130
  bilobata 130
  robusta 130
  spinosa 130
Anisomysis pelewensis **430**
Anomalocera **52**, 66, 95, 113
  opalus 113
  ornata 39, 113, **270**, 291
  patersoni 7, 49, 113, 218
    behaviour 406, 420
    biochemistry 243, 251
    distribution 505, **512**
    ecology 370, **491**
    morphology 30, **32**, *492*
    nutrition 150, 202
    reproduction 268, **270**, **288**
Antrisocopia **51**, 76, 99
  prehensilis 99, **489**
Archescolecithrix **53**, 81, 84, 134
  auropecten 134
Archidiaptomus **52**, 67, 115
  aroorus 115
Arietellidae, **51**, 101–3, 470, **489**, **498**
Arietelloidea
  ecology 502
  morphology 17
  taxonomy 50, **51**, *54*, *55*, 101–9
Arietellus **51**, 71, 74, 101, 486
  aculeatus 101
  armatus 101, 422
  giesbrechti 101
  minor 101
  mohri 101

Arietellus (contd.)
  pacificus 101
  pavoninus 101
  plumifer 101, 444
  setosus 101, 142
  simplex 101
  tripartitus 101
Artemia **188**, 297
Atelodinium 373, 524
Augaptilidae 29, **51**, 103–6, **435**, 436, **498**
Augaptilina **51**, 66, 103
  scopifera 103
Augaptilus
  morphology **32**
  taxonomy **51**, 55, 71, 74, 103
Augaptilus anceps 103
  cornutus 103
  glacialis 103
  lamellifer 103
  longicaudatus 103
  megalurus 103
  spinifrons 103
Aurelia aurita 527
Azygokeras **53**, 80, 81, 125
  columbiae 125

Bartholomea annulata 503
Batheuchaeta **53**, 88, 90, 125, 486
  anomala 125
  antarctica 125
  brodskyi 125
  enormis 125
  gurjanovae 125
  heptneri 125
  lamellata 125
  peculiaris 125
  pubescens 125
  tuberculata 125
Bathochordaeus 160
  charon 503
Bathycalanus **52**, 71, 119, 486
  bradyi 119
  eltaninae 119
  eximius 119
  inflatus 119
  princeps **4**, 119
  richardi 119
  sverdrupi 5, 119
  unicornis 119
Bathypontia **52**, 93, 121, 486, **498**
  elegans 121

  elongata 121
  inispina 121
  intermedia 121
  kanaevae 121
  longicornis 121
  longiseta 121
  major 121
  michelae 121
  minor 121
  regalis 121
  sarsi 121
  similis 121
  spinifera 121
Bathypontiidae **52**, 121–2, **152**, **498**
Bathypontioidea **52**, 54, 55, 121–2, 502
Bestiola 120
Bestiolina **52**, 77, 120, 483
  amoyensis 120
  inermis 120
  similis 120
  sinicus 120
  zeylonica 120
Blastodinium 524
  contortum 524
Boeckella **52**, 111, 113
Boholina **51**, 70, 73, 100, **489**
  crassicephala 100
  purgata 100
Boholiniidae **51**, 100, 489
Bopyridae 524
Brachycalanus **53**, 84, 132, **500**
  atlanticus 132
  bjornbergae 132
  minutus 132
  ordinarius 132
  rothlisbergi 132, **500**
Bradycalanus **52**, 71, 119, 486
  gigas 119
  pseudotypicus 119
  pseudotypicus enormis 5, 119
  sarsi 119
  typicus 119
Bradyetes **53**, 87, 90, 125
  brevis 125
  florens 125
  inermis 125
  matthei 125
Bradyidius
  nutrition **152**, 166
  taxonomy **53**, 78, 80, 88, 90, 92, 93, 125
Bradyidius angustus 125

armatus, 125
arnoldi, 125
bradyi 125, 268, **499**
brevispinus **499**
curtus 125
dentatus **499**
hirsutus 125
luluae 125
pacificus 125
plinioi 125, **499**
rakuma 125
robustus **499**
saanichi 125
similis 125, **499**
spinibasis **499**
spinifer 125
styliformis 125
subarmatus 125
tropicus 125
Brattstromia **51**, 70, 73, 100
 longicaudata 100, **489**

Calamoecia **52**, 111
Calanidae 222, 347, 365, 434, 528
 morphology 21, 30, 37
 taxonomy **52**, 118–19
Calanipeda **52**, 115
 aquae-dulcis 115, **375**
Calanoida 1
 phylogeny 2–3, *3*
 taxonomy **2**, 50, **51–3**, 65–97, 99–138
Calanoides
 ecology 480, 482, 487, 528
 growth 299
 nutrition **152**
 taxonomy **52**, 68, 118
Calanoides acutus **7**, 118
  biochemistry 245, 247, 248
  ecology **355**, 492, 494
  growth/development 344
  life history 385, **386**, 390
  nutrition 159, 170, **186, 189**, 196, **208**
  physiology 210
  reproduction 266, **285**, 292
  size/weight **223, 229, 234**
 carinatus **7**, **26**, 118, 298
  behaviour **415, 446**
  biochemistry 245
  distribution **512**
  ecology **356**, 365, **375**, 481, 482, **484, 485**

  growth/development **301, 320**
  life history 340, 383, 384, 385, **386**
  nutrition **183, 186, 189, 198**
  reproduction 255, **268**, 282, **285**
  size/weight **223, 229, 331**
 macrocarinatus 118
 natalis 118
 patagoniensis 118
 philippinensis 118, 482, **485**
Calanopia **52**, 93, 113, 518
 americana 113, **270, 489**
 aurivilli **47**, 113
 australica 113
 biloba 113
 elliptica **47**, 113, **430, 507**
 herdmani 113
 media 113, **507**
 minor **47**, 113, **491**
 parathompsoni 113
 sarsi 113
 sewelli 113
 seymouri 113
 thompsoni **26**, 113, **270**
Calanus 8, 41, 218, 236, 298, 346, 518
 behaviour 402, 405, 425, **426**, 428, 429, 451
 biochemistry 237, 240, 243, 245
 distribution 508
 ecology 366, 370, 374, 480, 483, 487, 490, 495, 497, 528
 growth/development 299, 330
 nutrition 140, 150, **152**, 168, 169, 172, **208**
 reproduction 265, 283
 taxonomy **52**, 60, 62, 63–4, 67, 118
Calanus agulhensis 60, 118, 516
 ecology **375, 484**
 size 338, *340*
 australis 118, 516
  distribution **512**
  growth **302, 320**
  reproduction **286**, 291
 chilensis 118, **356, 485**, 516
 euxinus 48, 60, 118, 516
 finmarchicus **7**, 10, **26**, 516
  behaviour 405, **408, 410, 415, 418**, 419, 425, **426, 427**, 428, **433**, 439, 441, **442**, 443, **446**, 450
  biochemistry 237, 240, 242, 247, 250
  chromosomes **47**, 48
  distribution 506, 508, **512**, 516, 517

Calanus agulhensis (contd.)
finmarchicus (contd.)
  ecology 351, **356**, 365, 369, 370, **376**, 380, 394, **395**, 398, 468, 470, 481, **493**, 495, 523, 528
  growth/development **302**, **310**, **311**, **312**, *318*, **320**, 323, 338, 339, **341**
  life history 385, **386**
  morphology *4, 22*, 28, 31, **32**, 32, 38, 39, 46, 47
  nutrition 150, **151**, 157, 163, 165, 172, **183**, **186**, **187**, **189**, **192**, 197, **200**, 203, 204, **206**, **208**
  physiology 215, 216, 217
  reproduction 43, 254, **255**, 260, 261, 265, 267, **268**, 273, 282, **286**, **288**, 289
  size/weight **229**, **234**
  taxonomy 49, 63, 64, 118
fonsecai 118
glacialis **7**, **32**, 47, 118, 516
  behaviour 405, **426**, **446**, 451
  biochemistry 247
  distribution **512**
  ecology **357**, 367, **376**, **493**, 495, 528
  growth/development **310**, **311**
  life history 385, **386**
  nutrition **186**, **190**
  physiology 216
  reproduction 281, 283, **286**, **288**, 292
  size/weight **234**, 334
helgolandicus **7**, 9, **26**, 63, 118, 297, 516
  behaviour 402, **410**, **418**, 419, 420, 422, 428, 443, 452
  biochemistry 237, **239**, 240, 251
  chromosomes **47**
  distribution **512**, 517
  ecology **357**, 365, **376**, **386**, **396**, 398, 522, 528
  growth/development **310**, **311**, 319, **320**, 321, 326, *327*, **341**
  morphology *40*, 41
  nutrition 145, 150, **151**, 166, 171, **187**, **192**, 197, **198**, **200**, 203, 205, **206**, **208**
  reproduction 43, **268**, 282, **286**, **288**, 290
  size/weight 221, **223**, **229**, 233
hyperboreus **7**, **26**, 49, 63, 118, 516
  behaviour 406, **410**, 419, **426**, 445
  biochemistry 247
  distribution **512**
  ecology **357**, 369, **376**, 399, **493**, 495
  growth/development **302**, **310**, **312**, 321

  life history 385, **386**
  morphology **32**, 37, 46
  nutrition 156, 166, **190**, **206**
  physiology 216
  reproduction 260, 267, 272, 277, 283, **286**, **288**, 292
  size/weight **229**, **234**, *332*
jaschnovi, 36, 60, 118, 516
magnellanicus 118
marshallae **32**, 118, 516
  distribution **512**
  ecology **357**, 365, **485**, **493**, 495
  growth/development **302**, **310**, **312**, 314, 316, **320**, 321, 323, *329*, **341**
  life history **386**
  nutrition 196, **208**
  reproduction 281, 282, **286**
  size/weight **230**, 328, 333, *334, 335*
minor **7**, **26**, **40**, 64, 118, 516
  distribution **512**
  ecology **357**, 503
  growth/development **302**
  nutrition **194**
  size/weight **223**, **331**
pacificus **7**, 118, 297, 516
  behaviour 406, **408**, **410**, **415**, **426**, **427**, 428, **446**, 449, 451, 453, 455
  biochemistry 245, 249
  distribution **512**
  ecology **357**, 365, 366, 371, **376**
  growth/development **302**, **310**, **312**, 315, *316*, 316, **320**, 326, 338, *340*, **341**
  life history **386**
  nutrition 140, 149, 154, 159, 163, 164, 171, 178, 182, **183**, **186**, **187**, 188, **190**, **192**, **198**, 203, 204
  reproduction **255**, 257, 259, **268**, 281, **286**, **288**, 289, 290
plumchrus 385
ponticus 516
propinquus **7**, 118, 516
  behaviour **415**
  biochemistry 226, 237, 240, 242, 243, 245, 246, 247, 248
  distribution **512**
  ecology **357**, 492, 494, 528
  life history **386**
  nutrition 163, 166, **190**, **208**
  physiology 218
  reproduction **268**, **286**
  size/weight **223**, **230**, **234**

TAXONOMIC INDEX

simillimus 118, 516
  biochemistry 247
  distribution **512**
  ecology **357**, 528
  life history **387**, 390
  nutrition **190**
  reproduction **286**
sinicus 22, 36, 118, 516
  behaviour **446**
  biochemistry **239**, **241**
  distribution 506, **512**
  ecology **357**, **376**
  growth/development **310**, **312**, **320**, 344
  nutrition **200**
  reproduction **268**
  size/weight **230**, **331**, 333
tenuicornis **223**
Calocalanus **52**, *61*, 77, 120, 480, 483
  aculeatus 120
  adriaticus 120
  africanus 120
  alboranus, 120
  antarcticus, 120
  atlanticus 120
  beklemishevi 120
  contractus, 120
  curtus 120
  dellacrocei 120
  elegans 120, **358**
  elongatus 120
  equilicauda 120
  fiolentus 120
  fusiformis 120
  gracilis 120
  gresei 120
  indicus 120
  kristalli 120
  latus 120
  lomonosovi 120
  longifurca 120
  longisetosus 120
  longispinus 120
  minor 120
  minutus 120
  monospinus 120
  namibiensis 120
  nanus 120
  neptunus 120
  omaniensis 120
  ovalis 120
  paracontractus 120

paralongatus 120
pavo **7**, **26**, 120, **358**, 506, **507**
pavoninus 120, **223**
plumatus 120
pseudocontractus 120
pubes 120
pyriformis 120
regini 120
sayademalja 120
spinosus 120
styliremis **7**, **26**, 120
tenuiculus 120
vinogradovi 120
vitjazi 120
vivesi 120
Campaneria **51**, 74, 101
  latipes 101, 103, **498**
Candacia
  ecology 480, 483, 490, 528
  morphology *16*, 36
  nutrition 160, 165, 166
  taxonomy **51**, *61*, 95, 111
Candacia aethiopica **7**, **26**, **32**, **491**
  armata **26**, 111, **302**, **376**, **512**
  bipinnata 111, **491**, 503, **512**
  bradyi 111, 160, **512**
  caribeannensis 111
  catula 111, 503, **512**
  cheirura 111, **512**
  columbiae 111
  curta 111, **491**, **512**
  curticauda **223**
  discaudata 111, **512**
  elongata 111
  ethiopica 111, **512**
  falcifera 111
  giesbrechti 111
  grandis 111
  guggenheimi 111, **512**
  guinensis 111
  ketchumi 111
  longimana 111, **512**
  magna 111
  maxima 111, **415**
  nigrocincta 111
  norvegica *111*, **512**
  pachydactyla 111, **223**, **415**, **491**, 508, **512**
  paenelongimana 111
  parafalcifera 111
  pofi 111
  rotunda 111

Candacia aethiopica (contd.)
  samassae 111, **512**
  tenuimana 111, **512**
  tuberculata 111, **512**
  varicans 111, **512**
Candaciidae
  behaviour 434
  distribution **512**, 515
  ecology **491**, 526
  nutrition **152**, 160
  taxonomy **51**, 111
Canthocalanus **52**, 68, 118, 497
  pauper 118, 524
Capitella 293
Centraugaptilus **51**, 71, 74, 103, 486
  cucullatus 103, **435**
  horridus 103, *422*, 423, **435**
  lucidus 103
  porcellus 103
  pyramidalis 68, 103
  rattrayi, 103, **435**
Centropages 218, 298
  behaviour 402, **411**, **427**, 428
  distribution 507
  ecology 468, 480, 481, 490, 497, 528
  life history 347, 389
  morphology 31
  nutrition **152**, **194**
  reproduction 46, 257, 261, **270**
  size *332*
  taxonomy **52**, *61*, 68, 73, 93, 112
Centropages abdominalis, 22, **26**, 112
  behaviour **431**
  biochemistry **239**, **241**
  distribution 506, **512**
  ecology 358
  growth/development **303**, **310**, **312**
  nutrition **200**
  reproduction **269**, 279, **286**
  size/weight **230**, 330
  acutus 112
  alcocki 112
  aucklandicus 112, **512**
  australiensis 112, **512**
  brachiatus 112, **376**, **484**, **485**, **512**
  bradyi 112, **512**
  brevifurcus 112
  calaninus 112
  caribbeanensis 112
  chierchiae **26**, 112, **303**, **446**, **484**, **512**
  dorsispinatus 112

elegans 112
elongatus 112
furcatus **7**, **26**, *32*, 112
  biochemistry 251
  chromosomes **47**
  distribution **512**
  growth/development **303**, **310**
  nutrition 159, **184**, **200**, 203, **208**
  reproduction **268**, **269**
gracilis 112
halinus 112
hamatus **7**, **26**, 112, 527
  behaviour 403, 409, **410**, **415**, 419, 444
  distribution **512**
  ecology **358**, 369, **376**, **396**
  growth/development **303**, 316, *317*, **320**, **342**
  life history 388
  nutrition 143, 161, 164, 177, **184**, **186**, **190**, 197, **206**
  physiology 216, 217
  reproduction **270**, 282, **286**
  size/weight, **230**, 330, 333
karachiensis 112
kröyeri **26**, 112, **376**, **396**
longicornis 112
mcmurrichi 112
natalensis 112
orsinii 112, **426**, 429, **430**
ponticus 112, **270**, **288**
sinensis 112
tenuiremis 112, **270**
trispinosus 112
typicus **7**, **26**, 112, 236
  behaviour 406, **411**, **415**, **426**, 432, **433**
  biochemistry 251
  distribution **507**, 508, **512**
  ecology **358**, 364, 366, **377**, **396**, 468, 475
  growth/development **303**, **310**, 315, *316*, **320**, **342**, 344
  longevity 345
  morphology **32**, *40*
  nutrition, 145, 150, 161, 169, **183**, **186**, **188**, **190**, 195, **206**, **208**
  reproduction 43, 259, 260, **270**, 281, 282, 283, **287**, **288**, 291, 293
  size/weight 223, **230**, **331**, 333, 335
velificatus 112
  behaviour 403, **408**, **411**
  distribution **512**

growth/development **304**, **342**
life history 383
nutrition 143–4, 164, **190**, **192**
reproduction 291
violaceus 112, **223**, **396**, **415**, **512**
yamadai **26**, 112, **270**, 506
Centropagidae 3, 6, 434
ecology 473, **491**, **498**
reproduction 44, 46, 257, 261, 268, **269**
taxonomy **52**, 111–13
Centropagoidea 50, **51–2**, *54*, *55*, 110–18, 389
Cephalophanes **53**, 132, 486
frigidus 132
refulgens *4*, 31, **32**, 132, 444
tectus 132
Chiridiella
ecology 486
taxonomy **53**, 66, 77, 88, 90, 93, 125–6
Chiridiella abyssalis 125
atlantica 125
bichela 125
bispinosa 125
brachydactyla 125
brooksi 125
chainae 125
gibba 125
kuniae 125
macrodactyla 125
megadactyla 125
ovata 125
pacifica 125
reducta 125
sarsi 125
smoki 125
subaequalis 125
trihamata 125
Chiridius
ecology 486, **499**
nutrition **152**, 166
taxonomy **53**, 88, 89, 92, 126
Chiridius armatus **32**, 32, 126
ecology **358**, **499**
nutrition **206**
reproduction 267
carnosus 126
gracilis 126, 267
longispinus 126
mexicanus 126
obtusifrons 126
pacificus 126

polaris 126
poppei 126
subantarcticus 126
subgracilis 126
Chirundina **53**, 92, 93, 126, 486
antarctica 126
indica 126
streetsi 126
Chirundinella **53**, 87, 90, 126
magna 126
Clausocalanidae **53**, 129–30
Clausocalanoidea 50, **53**, *55*, 124–38, 502
Clausocalanus
distribution **513**
ecology **377**, 480, 483, **485**, 490, 528
reproduction 265
taxonomy **53** 85, 129
Clausocalanus arcuicornis **7**, 129, **223**
behaviour **415**
distribution 506, **507**, **512**
ecology **358**, **484**
nutrition **198**
brevipes 129
dubius 129
farrani 129, **512**
furcatus *24*, **26**, 129, **223**
behaviour 443
ecology **358**, 480
ingens 129
jobei 129, **513**
laticeps 129, **513**
latipes 129
lividus 129, **513**
mastigophorus 129, **194**
minor 129, **513**
parapergens 129
paululus 129, **223**, 262, **358**, **513**
pergens, 129, 443, 480
Comantenna, **53**, 88, 90, 126
brevicornis, 126, **499**
crassa 126, **499**
curtisetosa 126
recurvata 126, **499**
Cornucalanus **53**, 81, 82, 132, 486
antarcticus 132
chelifer 132
indicus 132
notabilis 132
robustus 132
sewelli 132
simplex 132

Cosmocalanus **52**, 67, 119, 483
  caroli 119
  darwini **7**, 119, 262, **377**
Crassantenna **53**, 88, 126
  comosa 126, **499**
  mimorostrata 126, **499**
Crassarietellus **51**, 71, 101, **498**
Ctenocalanus **53**, 85, 129, 480, 528
  campaneri 129
  citer 6, 129, **304**, **358**, **493**, 494
  huysi 101
  tageae 129
  vanus **7**, **26**, 129, **223**, **513**
    ecology **358**, **377**, **396**, 480, **484**, **493**, 494
    growth **304**
Cyclopoida 1, **2**, *3*
Cyclops
  longicornis 49
  marina 49

Damkaeria **52**, 87, 89, 123
  falcifera 123, **499**
Daphnia magna 248, 344
Delius **52** 77, 121
  nudus 121
  sewelli 121
Derjuginia 127
Diaixidae **53**, 130, 470, **499**
Diaixis
  ecology **499**
  taxonomy **53**, 78, 81, 88, 89, 130
Diaixis asymmetrica 130, **499**
  centrura 130
  durani 130
  gambiensis 130
  helenae 130
  hibernica 130, **499**
  pygmaea 130
  tridentata 130
  trunovi 130
Diaptomidae 3, 6, **52**, 55, 113, 473
Diaptomus 257
  clavipes 367
  minutus **433**
Dinophysis acuminata 157
Disco **51**, 76, 106, 486
  atlanticus 106
  caribbeanensis 106
  creatus 106
  curtirostris 106

  elephantus 106
  erythraeus 106
  fiordicus 106
  hartmanni 106
  inflatus 106
  intermedius 106
  longus 66, 106
  marinus 106
  minutus 106
  oceanicus 106
  oviformis 106
  peltatus 106
  populosus 106
  robustipes 106
  tropicus 106
  vulgaris 106
Discoidae **51**, 106
Disseta **51**, 70, 73, 106–7, 486
  coelebs 106
  grandis 106
  magna 106
  palumboi 106, **435**
  scopularis 106
Drepanopsis orbus 125
Drepanopus
  ecology 492
  nutrition 160
  taxonomy **53**, 81, 82, 85, 89, 129
Drepanopus bispinosus 129, **358**, **387**, 390, **493**
  bungei 129, 473
  forcipatus 6, 129
  pectinatus 129, 174, **230**, 251, **331**, **358**

Emiliania huxleyi 159, 204, 528
Enantiosus **51**, 70, 73, 99
  cavernicola 99, **489**
Epacteriscidae 17, **51**, 99–100, **489**
Epacteriscioidea **51**, 54, 55, 99–100
Epacteriscus **51**, 68, 99
  rapax 99, **489**
Epilabidocera **52**, 66, 95, 113
  amphitrites **26**, 113
    ecology **358**
    morphology **32**, **38**, **39**, **46**
    reproduction **43**, **270**
  longipedata **270**, **513**
Epischura **52**, 116
  lacustris **411**
Erobonectes **51**, 67, 100
  macrochaetus 100, **489**

nesioticus 100, **489**
Euaetideus 486
  acutus **485**
Euaugaptilus 236
  ecology 486, 528
  nutrition **152**
  taxonomy **51**, 56, 61, 71, 73, 74, 76, 103–4
Euaugaptilus affinis 103
  aliquantus 103
  angustus 103
  antarcticus 103
  atlanticus 103
  austrinus 103
  brevirostratus 103
  brodskyi 103
  bullifer 103, **435**
  clavatus 103
  curtus 103
  digitatus 103
  diminutus 103
  distinctus 103, 104
  elongatus 103
  facilis 103
  fagettiae 103
  farrani 103, **435**
  fecundus 103
  filigerus *103*, **435**
  fundatus 103
  gibbus 103
  gracilis 103
  graciloides 103
  grandicornis 103
  hadrocephalus 104
  hecticus 68, 104
  hulsemannae 104
  humilis 104
  hyperboreus 104
  indicus 104
  laticeps 104, **435**
  latifrons 104
  longicirrhus 104
  longimanus 104
  longiseta 104
  luxus 104
  magnus 104, 267, **435**, 436, 438, 444
  malacus 104
  marginatus 104
  matsuei 104
  maxillaris 104
  mixtus 104
  modestus 104

niveus 104
nodifrons 104, **435**
nudus 104
oblongus 104
pachychaeta 104
pacificus 104
palumbii 104
parabullifer 104
paroblongus 104
pencillatus 104
perasetosus 104
periodosus **435**
placitus 38, 40, 104
propinquus 104
pseudaffinis 104
quaesitus 104
rectus 104, **435**
rigidus 104
roei 104
sarsi 104
similis 104
squamatus 104, **435**
sublongiseta 104
tenuicaudis 104
tenuispinus 104
truncatus 104, **435**
unisetosus 104
validus 104
vescus 104
vicinus 104, **435**
Eucalanidae **52**, 122, 340, 434
Eucalanoidea **52**, 54, 55, 122
Eucalanus 30, 215
  behaviour 407
  biochemistry 245
  ecology 480, 483, **484**, 486, 487, 528
  nutrition, **152**
  taxonomy, **52**, 62, 62–3, 76, 122
Eucalanus attenuatus 7, **26**, 122, **223**
    behaviour **416**
    chromosomes **47**
    distribution **513**
    nutrition **194**
  bungii 7, **26**, 47, 63, 122
    distribution **513**
    ecology 358, 468, 479, 481
    growth/development **341**, **342**
    life history **387**
    nutrition 167, 172, 203
    reproduction **287**
  californicus 63, 122, **485**, **513**

Eucalanus attenuatus (contd.)
  crassus **7**, **26**, 122, **223**
    behaviour **446**
    chromosomes **47**
    distribution **513**
    ecology **485**
    nutrition 148
  dentatus 122, **485**
  elongatus **7**, **26**, **32**, 32, 122
    behaviour 455
    chromosomes **47**
    distribution **513**
    nutrition 183
    reproduction 42–3
  hyalinus 122, 181
    behaviour **408**
    biochemistry 226, 243, 248, 250
    distribution **513**
    growth/development **304**
    longevity **345**
    nutrition 181, **183**
  inermis 122, 226, 243, 248, **485**, **513**
  langae 122, **513**
  longiceps 122
  monachus **47**, 122
    behaviour **446**
    distribution **513**
    ecology **387**, **485**
    physiology 217
  mucronatus **47**, 122, **485**, **513**
  muticus 122
  parki 122, **513**
  peruanus 122
  pileatus **26**, 122
    behaviour **408**, **411**, **446**
    distribution **513**
    growth/development **304**
    longevity **345**
    nutrition 162, 165, **190**, **198**, 199, **200**
  pseudoattenuatus **223**, **377**
  quadrisetosus 122
  sewelli 122, **513**
  subcrassus, **47**, 122
  subtenuis 62, 122, **194**, **223**, **485**, **513**
Euchaeta 236
  behaviour 405
  ecology **460**, 480, 483, 486, 528
  morphology 22
  nutrition 140, 148–9
  reproduction 266, 267, 274, **275**, 277, 279, 281

  taxonomy 53, *61*, 61, 63, 92, 130
Euchaeta acuta **7**, 130, **275**, **513**
  concinna 130, **230**, 524
    biochemistry **239**, **241**
    distribution **513**
    reproduction **275**
  elongata **411**
  indica 130, **275**, **513**
  longicornis 130, **275**, **513**
  magniloba 130
  marina **7**, 49, 130, 297
    behaviour **411**, **416**, **418**, **446**
    distribution **513**
    ecology **377**, **485**
    growth/development **304**, **342**
    morphology *24*, 24, **32**, 47–8
    reproduction **275**
    size/weight **223**
  marinella 130
  media 130, **275**, 445, **513**
  norvegica 411
  paraacuta 130, **513**
  paraconcinna 130, **223**
    behaviour 445, **446**
    ecology **484**
    growth **304**
    reproduction **275**
  plana 130, **230**, **239**, **241**, **513**
  pubera 130, **275**, **513**
  rimana 130
    behaviour 402, **411**, **416**, **418**, **432**, **433**
    nutrition 149, 163
  spinosa 130, **275**, **513**
  tenuis 130, **513**
  wrighti 130
Euchaetidae 222, 434
  ecology **500**
  growth/development 321
  morphology 21, 27
  nutrition 140, **152**
  reproduction 265, 266, 277, **278**, **280**
  taxonomy 53, 130–1
Euchirella 6, 236, 486, **499**
  behaviour 407, **416**
  ecology 486
  nutrition 140, 149, **152**
  reproduction 263, 267
  taxonomy 53, 92, 93, 126
Euchirella amoena 126
  bella 126
  bitumida 126

curticauda 126, **223**
  behaviour **411**, **416**, **418**, 444
  biochemistry 250
formosa 126
galeata 126
grandicornis 126
latirostris 126
lisettae 126
maxima 126
messinensis **7**, 29, 126, 149, 263
orientalis 126
paulinae 126
pseudopulchra 126
pseudotruncata 126
pulchra 126, **223**
rostrata **7**, 126, **223**
  behaviour 407, **411**, **418**
  distribution **513**
  nutrition 206
rostromagna 126, 153, 246, 247
similis 126
speciosa 126
splendens 126, **224**
tanseii 126
truncata 126
unispina 126
venusta 126, **416**
Euphausia
  pacifica 372
  superba 528
Eurytemora **52**, 93, 117, 298
  behaviour 405
  ecology 365, 475, 497, 528
  growth/development 299
  nutrition 154
  physiology 218
  reproduction 257, 265, 267, 279
  size/weight 333
Eurytemora *affinis* **7**, **26**, 60, 117, 346
  behaviour 407, **408**, **411**, **418**, 420, 432, **433**, 452
  biochemistry, **239**
  chromosomes **47**
  ecology **359**, 366, 367, 367, 369, 371, **377**, **396**, 398, 473, 475, 476, 525, 527
  growth/development 296, **304–5**, **310**, **320**
  life history **345**, 388
  nutrition 155, 165, 185, **186**, **190**, **206**
  physiology 215, 216
  reproduction 260, 262, **271**, 271, 272, 280, 281, **284**, **288**, 293
  size/weight 227, **230**, 335
  americana **26**, 117, **271**, **306**, **396**, 475
  anadyrensis 117
  arctica 117
  asymmetrica 117
  bilobata 117
  canadensis 117
  composita 117
  foveola 117
  gracilis 117
  graculicauda 117
  grimmi 117
  herdmani **7**, **26**, 63, 117
    behaviour **433**
    ecology 346, 374, **377**, 477, **493**, 495
    growth/development **306**, **310**, **312**, **320**, **342**
    reproduction 279, 282, **284**, 293
    size/weight **230**, 328, 333
  hirundo **26**, 60, 117
  hirundoides **26**, 60, 117, **288**
  inermis 117
  kieferi 117
  kurenkovi 117
  lacustris 117
  pacifica **26**, 117, **271**, 429, **431**
  raboti 117
  richingsi 117
  thompsoni 117
  transversalis 117
  velox **26**, 117
    ecology **359**
    growth/development **306**
    physiology 216
    reproduction 260, **271**, **284**, **288**
    size/weight **231**, 332
  wolteckeri 117
  yukonensis 117
Exumella **51**, 68, 73, 101, **498**
  mediterranea 101, **489**
  polyartha 101
  tuberculata 101

Farrania **53**, 81, 82, 129
  frigida 129
  lyra 129
  orbus 80, 125, 129
  pacifica 129
Fosshagenia **51**, 77, 109
  ferrarii 50, 109, **489**

Fosshageniidae **51**, 109, **489**
Fosshagenioidea 50, **51**, *54*, 109
Foxtonia **52**, 66, 123
  barbatula 123, 502
Fragilariopsis (Nitzschia) kerguelensis 528

Gaetanus
  behaviour **416**
  ecology 486
  morphology 28
  nutrition 140, **152**
  taxonomy **53**, 87, 90, 127
Gaetanus antarcticus 127
  armiger 127
  brachyurus 127
  brevicaudatus 127
  brevicornis 127
  campbellae 127
  curvicornis 127
  divergens 127
  intermedius 127
  kruppi 127, 250, 444
  latifrons *4*, 127, 328, 422
  microcanthus 127
  miles 127, 422, *422*
  minispinus 127
  minor 127
  paracurvicornis 127
  pileatus 127, 250, *422*, 422
  recticornis 127
  tenuispinus 127
  wolfendeni 127
Gaidiopsis **53**, 88, 127
  crassirostris 127
Gaidius
  ecology **499**
  nutrition **152**
  taxonomy **53**, 88, 90, 92, 93, 127
Gaidius affinis 127
  brevirostris 127
  brevispinus 127
  columbiae 127
  inermis 127
  intermedius 127
  minutus 127
  pungens 127, **499**
  robustus 127
  tenuispinus 224, **231**, **234**, **416**
  variabilis 127
Ganchosia **52**, 66, 117
  littoralis 117

Gaussia **51**, 67, 108, 486
  asymmetrica 108, 261
  princeps 108, 211, **435**, 436, 437, 438
  scotti 108
  sewelli 108
Gelyelloida 1, **2**, *3*
Gippslandia **52**, 67, 112
  estuarina 112
Gladioferens **52**, 67, 112, 475, **498**
  imparipes 30, 112, **331**
    ecology, 469, 501
    growth, **306**
    reproduction 280
  inermis 112
  pectinatus **26**, 112, 280, **306**, 365, 447
  spinosus 112
  symmetricus 112
Gonyaulax grindleyi 165
Gymnoplea **2**
Gyrodinium dorsum 164

Haloptilus
  behaviour 422
  ecology 483
  nutrition **152**
  taxonomy **51**, 56, 71, 73, 74, 104–5
Haloptilus aculeatus 104
  acutifrons 104, **513**
  angusticeps 104
  austini 104
  bulliceps 104
  caribbeanensis 104
  chierchiae 104, 105
  fertilis 104
  fons 104
  furcatus 104
  longicirrus 104, **435**
  longicornis **7**, 104, **359**, **416**, **513**
  major 104
  mucronatus 104
  ocellatus 104
  orientalis 105
  ornatus 105
  oxycephalus 105
  pacificus 105
  paralongicirrus 105
  plumosus 105
  pseudooxycephalus 105
  setuliger 105
  spiniceps 105
  tenuis 105

validus 105
Harpacticoida **2**, 2, *3*
Hemiboeckella **52**, 112
Hemidiaptomus
　ingens 29
　ingens provinciae 41
Hemirhabdus **51**, 70, 74, 107, 486
　falciformis 107
　grimaldii 107, **435**
　latus 107, 434, **435**
　truncatus 107
Heteramalla **53**, 82, 134
　dubia 134
　sarsi 134
Heterarthrandria 17
Heterocope **52**, 66, 117
　appendiculata 117
　borealis 117
　saliens 117
　septentrionalis 117, **426**
Heteroptilus **51**, 70, 74, 76, 105
　acutilobus 105, **435**
　attenuatus 105
Heterorhabdidae
　behaviour **435**, 436
　ecology 502
　morphology 20, 36
　nutrition **152**
　taxonomy **51**, 106–7
Heterorhabdus 6
　behaviour **411**
　ecology 486, 528, 529
　morphology 19
　nutrition 160, 164
　taxonomy **51**, 70, 74, 107
Heterorhabdus abyssalis, 107, **513**
　atlanticus 107
　austrinus 107, **231**
　brevicornis 107
　caribbeanensis 107
　clausi 107
　compactoides 107
　compactus 107
　egregius 107
　farrani 107, **224**, **231**, **234**
　fistulosus 107
　lobatus 107
　longispinus 107
　medianus 107
　nigrotinctus 107
　norvegicus 107, **435**, **513**, 522

pacificus 107
papilliger **7**, 107, **435**, **513**
proximus 107
pustulifera 107
robustoides 107
robustus 107, **435**
spinifer 107
spinifrons **7**, 107, **435**
spinosus 107
sub-spinifrons 107
tanneri 107, 479
tenuis 107
tropicus 107
vipera 107
Heterostylites **51**, 70, 73, 107
　longicornis 107, **435**
　major, 107
Hyperbionychidae **51**, 108, **498**
Hyperbionyx **51**, 71, 74, 108
　pluto, 108, **498**

Isaacsicalanus **52**, 87, 123
　paucisetus 123, **499**, 502
Ischnocalanus **52**, 77, 121, 483
　equalicauda 121
　gracilis 121
　plumulosus 121, **359**
　tenuis 121
Isias **52**, 70, 112, 405
　clavipes **7**, 112, **491**, **513**
　tropica 112
　uncipes 112, 447
Isokerandria 17
Ivellopsis **52**, 66, 114
　elephas 114

Jaschnovia **53**, 78, 127
　johnsoni 127
　tolli 127

Kunihulsea **52**, 87, 123
　arabica 123

Labidocera
　behaviour 428
　chromosomes **47**
　distribution 506, **513**
　ecology 480, 483, **491**
　morphology 23
　reproduction 46, 257, 266, **270**
　taxonomy **52**, 55, 61, 66, 93, 95, 114

Labidocera acuta **32**, **47**, 114, **430**, **513**
  acutifrons **26**, 114, **224**
    behaviour 409, **416**
    chromosomes **47**
    distribution **513**
  aestiva **26**, 64, 114
    distribution **513**
    ecology 490
    life history 389
    nutrition **190**, 203
    physiology 217
    reproduction 42, 43, 45, 259, 259, **268**, **268**, **270**, 272, 291, 293
  agilis 506
  antiguae 114
  barbadiensis 114
  barbudae 114
  bataviae **47**, 114
  bengalensis **26**, **47**, 114
  bipinnata 114, **270**, 506
  brunescens **26**, 114, **288**
  carpentariensis 114
  caudata 114
  cervi 114
  dakini 114
  detruncata 114, **507**, **514**
  diandra 114, 262, **514**
  discaudata **47**
  euchaeta **26**, **47**, 114
  farrani 114
  fluviatilis **26**, 114
  insolita 114
  jaafari 114
  japonica 114
  johnsoni 114
  jollae **26**, 114
    behaviour 405, **411**, 432
    distribution **514**
    nutrition **188**
  kolpos 114
  kröyeri **47**, 114
  laevidentata **47**, 114
  lubbockii 114, **514**
  madurae **47**, 114, **507**
  minuta **26**, **47**, 114
  mirabilis 114
  moretoni 114
  nerii 114, 490
  orsinii 114, 506
  panamae 114
  papuensis 114
  pavo **26**, **47**, 114, **427**, 429, **507**
  pectinata **47**, 114
  pseudacuta **47**, 114
  rotunda **26**, 114
  scotti 114, **270**
  tenuicauda 114
  trispinosa **26**, 114
    behaviour **412**, **418**, 421
    distribution **514**
    growth/development **306**
    nutrition 159, 163
    reproduction **270**
    size/weight **224**
  wilsoni 114
  wollastoni **7**, 114
    behaviour 406, **412**, 420
    distribution **514**
    ecology 490
    growth/development **306**
    life history 388
    morphology 30, **32**
    nutrition 150
    reproduction **270**
Lahmeyeria **52**, 93, 117
  turrisphari 117
Landrumius **53**, 85, 134
  antarcticus 134
  gigas 134
  insignis 134
  sarsi 134
  thorsoni 134
Leptodiaptomus minutus 272
Limacina helicina 160
Limnocalanus
  behaviour 405
  nutrition **152**
  reproduction 257, 266
  taxonomy **52**, 67, 112
Limnocalanus grimaldi **26**, 112, 473
  johanseni 112
  macrurus 112
    behaviour 406, **412**
    biochemistry **239**
    ecology 359, **396**, 473
    growth **306**
    life history 384
    reproduction 262
Lophothrix **53**, 82, 85, 134–5, 136, 486
  frontalis **40**, 40, 134, 444
  humilifrons 134
  latipes 134

quadrispinosa 135
similis 135
simplex 135
varicans 135
Lucicutia
  behaviour, **412**, **416**, 423
  ecology 486, 528
  taxonomy **51**, *61*, 70, 71, 73, 76, 108
Lucicutia anisofurcata 108
  anomala 108
  aurita 108, **435**
  bella 108
  bicornuta 108
  biuncata 108
  challengeri 108
  cinerea 108
  clausi 108, **435**
  curta 108
  curvifurcata 108
  flavicornis **7**, 108, **223**, **359**, **435**, **485**
  formosa 108
  gaussae 108
  gemina 108, **435**
  gigantissima 108
  grandis 108, **435**
  intermedia 108
  longicornis 108
  longifurca 108
  longiserrata 108
  longispina 108
  lucida 108
  macrocera 108
  magna 108, **435**
  major 108
  maxima 108
  oblonga 108
  orientalis 108
  pacifica 108
  pallida 108
  paraclausi 108
  parva 108
  pellucida 108
  pera 108
  polaris 108
  profunda 108
  pseudopolaris 108
  rara 108
  sarsi 108, **435**
  sewelli 108
  uschakovi 108
  wolfendeni 108, **435**

Lucicutiidae **51**, 108, **152**, **435**, 436, 502
Lutamator **53**, 88, 127
  elegans 127, **499**
  hurleyi 127, **499**

Macandrewella **53**, 78, 81, 84, 135
  agassizi 135
  asymmetrica 135
  chelipes 135
  cochinensis 135
  joanae 135
  mera 135
  scotti 135
  sewelli 135
Manaia **52**, 66, 117
  velificata 117
Mecynocera **52**, 77, 119, 120, 483
  clausi **7**, 119, **224**, **359**, **514**
Mecynoceridae 30, **52**, 119
Megacalanidae 30, **52**, 119, 277, 435
Megacalanoidea **52**, *54*, 55, 118–21
Megacalanus **52**, 71, 119, 486
  longicornis 119
  princeps 119, **435**, 436, 444
Menidia menidia 293
Mesaiokeras **53**, 87, 88, 89, 131
  heptneri 131
  kaufmanni 131, **500**
  nanseni 131, **500**
  semiplenus 131
  tantillus 131
Mesaiokeratidae **53**, 131, **500**
Mesocalanus **52**, 119, 140
  lighti 119
  tenuicornis **7**, 119, **359**, **396**, **514**
Mesocomantenna **53**, 88, 127
  spinosa 127
Mesodinium rubrum 157
Mesorhabdus **51**, 70, 73, 107, 486
  angustus 107
  brevicaudatus 107
  gracilis 107
Metacalanus **51**, 71, 74, 102, **489**, **498**
  acutioperculum 102
  aurivilli 102
  curvirostris 102
  inaequicornis 102
Metridia
  ecology 480, 486, 528
  taxonomy **51**, 64, 67, 109

Metridia alata 109
  andraeana 109
  asymmetrica 109, 454, 502
  bicornuta 109
  boecki 109
  brevicauda 109
  calypsoi 109
  curticauda 109
  discreta 109
  effusa 109
  gerlachei 6, 109
    behaviour **416**, **433**, **435**
    biochemistry 237, 240, 242, 243, 247
    ecology 359, 492, **493**, 494
    life history 384
    nutrition 159, 164, 166, **191**, **206**, **208**
    physiology 210
    size/weight **224**, **231**, **234**
  gurjanovae 109
  ignota 109
  longa **7**, 109
    behaviour 403, **412**, **416**, 419, **435**, 435, 438, 439, **446**, 447
    biochemistry 237, 240, 243
    distribution **514**
    ecology 359, 364, **493**, 495
    growth/development **310**
    life history 384, 385
    nutrition **187**, **191**
    physiology 216
    reproduction 254, **255**
    size/weight **231**, **234**
  lucens **7**, **26**, 37, 109, **231**
    behaviour **412**, **416**, 419, **435**, **446**, 447
    distribution **514**
    ecology 352, **359**, 377, 396, 468, **484**
    nutrition **188**
  macrura 109, **435**
  okhotensis 109, 407
  ornata 109
  pacifica 48, 109
    behaviour 403, 406, 407, **412**, 424, **435**
    ecology 352, **359**, 503
    growth/development **306**, **342**
    life history **387**
    nutrition 160, 188, **191**
    reproduction **268**
  princeps 109, **435**
  similis 109
  trispinosa 109
  venusta 109

Metridinidae 222
  behaviour **435**, 436
  ecology 347, 502
  nutrition **152**
  taxonomy **51**, 108–9
Microcalanus
  ecology 480
  nutrition 160, **188**
  size *332*
  taxonomy **53**, 88, 89, 129
Microcalanus pusillus **26**, 129, 254
  pygmaeus 129, 254, **306**, **359**, **493**, 494
Microdisseta **51**, 70, 73, 107
  minuta 107
Mimocalanus **52**, 87, 89, 123
  brodskii 123
  crassus 123
  cultrifer 123
  damkaeri 123
  heronae 123
  inflatus 123
  major 123
  nudus 123
  ovalis 123
  sulcifrons 123
Miostephos **53**, 85, 89, 137
  cubrobex 137
  leamingtonensis 137, **489**
Misophrioida 1, **2**, *3*
Mixodiaptomus
  kupelwieseri 41
  laciniatus **433**
Mixtocalanus **53**, 80, 84, 135
  alter 135
  robustus 135
  vervooti 135
Monacilla **52**, 78, 80, 123
  gracilis 123
  tenera 123
  typica 123
Monoculus finmarchicus 49
Monstrilloida **2**, 2, *3*
Mormonilloida 1, **2**, *3*

Nanocopia **51**, 93, 99
  minuta 5, 99, **489**
Neoaugaptilus distinctus 104
Neoboeckella **52**
Neocalanus
  behaviour 405
  ecology 374, 480, 486, 487, 503

morphology 30
nutrition 172
reproduction **255**, 283
taxonomy **52**, 68, 119
Neocalanus cristatus **7**, 119, 336
  behaviour **416**, 424, **426**
  distribution 508
  ecology **359**, 370, 468, 481, 502, 524, 528
  growth/development **342**
  life history **387**
  nutrition 140, 181, **191**, **206**
 flemingeri 35, 119
  biochemistry 237, 240, 242
  ecology **360**, 468, 528
  growth/development 316, **342**
  life history **387**
 gracilis **7**, **26**, **32**, 119, **224**
  behaviour **416**
  distribution **514**
  ecology **360**
  nutrition **194**
 minor 516
 plumchrus **7**, 35, 119
  behaviour **426**, 450, 454
  biochemistry 235, 237, 240, 243, **244**
  distribution **514**
  ecology **360**, 370, 372, **377**, 468, 479, 481, 502, 523, 528
  growth/development **306**, 317, **342**
  life history **387**
  nutrition 153, 156, 181, **186**, **191**, *193*, 196, 197, **200**, 203, **206**
  reproduction 267, 281, **287**, **288**
 robusitor 47, 119, **224**, **416**
 tonsus **26** 119
  behaviour **417**, 424, **426**
  biochemistry 237, 240, 242, 247, 251
  nutrition 181, **191**, **192**
  reproduction **288**, 292
Neocopepoda **2**
Neomysis integer 527
Neopontella **52**, 95, 113
 typica 113
Neorhabdus 107
Neoscolecithrix
 behaviour *422*
 ecology **501**
 taxonomy **53**, 78, 87, 90, 138
Neoscolecithrix antarctica 138
 catenoi 138

 farrani 138
 koehleri 138
 magna 138
 watersae 138
Nodularia spumigena 155

Oithona 160, **184**, 429, 473
 davisae 257
 oculata **430**
Oncaea 160
Onchocalanus **53**, 81, 82, 132, 486
 affinis 132
 cristatus 132
 hirtipes 132
 latus 132
 magnus 132
 paratrigoniceps 132
 scotti 132
 subcristatus 132
 trigoniceps 132
 wolfendeni 132

Pachyptilus
 ecology **498**
 nutrition **152**
 taxonomy **51**, 70, 76, 105, 106
Pachyptilus abbreviatus 105
 eurygnathus 105, **435**
 lobatus 105
 pacificus 105
Paivella **53**, 92, 127
 inaciae 127
 naporai 127
Paracalanidae 30, **52**, 120–1, 434
Paracalanus 298
 behaviour 405, 407
 biochemistry **239**, 249
 ecology 473, 480, 482, 486, 497, 527, 528
 growth/development 299, **307**, **311**, **313**, 314, **320**
 nutrition **152**, 181, **192**, 192, **200**, **206**
 reproduction 256, **268**, 280, **287**
 size/weight **331**, 333
 taxonomy **52**, 77, 121
Paracalanus aculeatus **7**, *24*, **26**, 37, 121
 behaviour **408**, **412**
 chromosomes **47**, 48
 ecology 475
 growth/development **306**, **320**, **343**
 life history 383

Paracalanus aculeatus (cont.)
  nutrition 143–4, **187**, 192
  size/weight **224**
 brevispinatus 121
 campaneri 121
 crassirostris **446**
 denudatus 121, **485**
 gracilis 121
 indicus 121, **343**, 373, 420, 524
 intermedius 121
 mariae 121
 nanus 121, **360**
 parvus 6, **7**, **26**, 37, 47, 121, **287**
  behaviour **412**, **417**, **426**, 444
  biochemistry **239**
  ecology **360**, 378, **397**, 475, 480, **484**, **485**, 524
  growth/development **306–7**, **311**, **313**, **343**
  longevity **345**
  nutrition 150, 161, **183**, **186**
  reproduction **255**, 282, **288**
  size/weight **224**, **231**
 ponticus 121
 quasimodo 121, **408**, **412**
 serratipes 121
 serrulus 121
 tropicus 121
Paracandacia **51**, 95, 111, 160, 165
 bispinosa 111, **514**
 simplex 111, **491**, **514**
 truncata 111, 503, **514**
 worthingtoni 111
Paracomantenna **53**, 88, 127, **422**, **499**
 gracilis 127
 magalyae 127
 minor 127
Paracyclopia **53**, 78, 133
 gitana 133, **489**
 naessi 133, **489**
Paradisco **51**, 71, 74, 106
 gracilis 106
 grandis 106
 mediterraneus 106
 nudus 106
Paralabidocera **51**, 95, 111
 antarctica 111, **361**, **387**, 492, **493**, 494
 grandispina 111, **493**, 494
 separabilis 111
Paramisophria **51**, 68, 102, 402, **498**, 501
 ammophila 102

 cluthae 102
 fosshageni 102
 galapagensis 102, **489**
 giselae 102
 itoi 102
 japonica 102
 ovata 102
 platysoma 102, 402, 501
 reducta 102, **489**
 rostrata 102
 spooneri 102
 variabilis 102
Parapontella **52**, 95, 113
 brevicornis 113, 406, **491**, **514**
Parapontellidae **52**, 113
Parapseudocyclops giselae 102
Parascaphocalanus **53**, 81, 85, 135
 zenkevitchi 135
Parastephos **53**, 87, 89, 137
 esterlyi 137
 occatum 137
 pallidus 137
Paraugaptiloides **51**, 74, 102
 magnus 102, **498**
Paraugaptilus **51**, 71, 74, 102, 267
 archimedi 102
 bermudensis 102
 buchani 102
 indicus 102
 magnus 102
 meridionalis 102
 mozambicus 102
 similis 102
Pareucalanus 122
Pareuchaeta 236, 515
 behaviour 405
 distribution 517
 ecology 486, 492, 494, **500**, 526, 528
 morphology 22, 44
 nutrition 140, 160, 166, 174, **206**
 reproduction 258, 262, *263*, 263, 266, 267, **275**, *277*, 279, *281*, 294
 size *332*
 taxonomy **53**, *61*, 61, 63, 92, 130, 131
Pareuchaeta abbreviata 131, **275**
 abrikosovi 131
 abyssalis 131
 abyssaloides 131
 aequatorialis 131, **514**
 affinis 131
 alaminae 131

altibulla 131
anfracta 131
antarctica 131
  behaviour **417**
  biochemistry 247
  ecology **360**, 492, 524
  nutrition 153, 162, **188**
  reproduction 263, **284**
  size/weight **224**, **231**, **234**
austrina 131
barbata 131, **275**, 444, **514**
biloba 131, 409, **417**
birostrata 131, **275**
bisinuata 131, **275**
bradyi 131
brevirostris 131
bulbirostris 131
californica 131
calva 131
comosa 131, **514**
confusa 131, **275**, **514**
copleyae 131
dactylifera 131
elongata **26**, 131
  ecology 372
  growth/development **307**, 343
  nutrition 163, 165, 166, **188**, 188
  reproduction **275**, 281
  size/weight **224**, **231**
eminens 131
erebi 131, 262, 263
euryhina 131
exigua 131
flava 131
glacialis 131, **275**, 492
gracilauda 131, **514**
gracilis 131, **224**, 263, **275**, **360**
grandiremis 131
guttata 131
hanseni 131, **275**
hastata 131
hebes 131, **224**, **275**, **360**, **514**
implicata 131
incisa 131
investigatoris 131
kurilensis 131, **275**
longisetosa 131
malayensis 131, 518
megaloba 131
mexicana 131
modesta 131

norvegica **7**, 131
  behaviour **417**, 419, *441*, 441, 444, 447
  distribution 504, **514**
  ecology 349, **350**, 351, **360**, 363, 364, *365*, 468, 470, 478, 483, 492, 522
  growth/development 315, 321
  life cycle 347
  morphology *16*, *22*, 30, 31, **32**, 36, 37, 39
  nutrition 140, 146, 163, **188**, **201**
  reproduction 44, 46, *258*, *259*, 259, 260, 263, 265, **275**, 279, 280, **284**, 294
oculata 131
orientalis 131
papilliger 131
parabreviata 131
paraprudens 131
parvula 131
pavlovskii 131
perplexa 131
plaxiphora 131
plicata 131
polaris 131
prima 131
propinqua 131
prudens 131
pseudotonsa 131, **360**, **514**
rasa 131, **275**
regalis 131
ribicunda 131
robusta 131
rotundirostris 131
rubra 131, **275**
russelli **26**, 131, 251, **275**, 321
sarsi 131, **275**, **514**
scaphula 131
scopaeorhina 131
scotti 131, **275**, 444, **514**
sesquipedalis 131
sibogae 131
similis 131, 262, 263
simplex 131, **275**
subtilirostris 131
tonsa 37, 131, **514**
tridentata 131
triloba 131
tuberculata 131, **514**
tumidula 131
tycodesma 131
vervoorti 131
vorax 131, **514**

Pareuchaeta abbreviata (cont.)
   weberi 131
Parkiidae **53**, 132, **500**
Parkius **53**, 132
   karenwishnerae 132, **500**
Parundinella **53**, 82, 84, 85, 87, 90, 138
   dakini 138
   emarginata 138, **501**
   manicula 138
   spinodenticula 138
Parvocalanus **52**, 77, 120, 121, 497, 528
   crassirostris **26**, 37, 121
      behaviour 445
      ecology 473
      reproduction 267, 294
   dubia 121
   elegans 121
   latus 121
   scotti 121
Pelagobia longicirrata 160
Peridinium trochoidum 164
Phaenna **53**, 78, 132, 486
   spinifera 132, *422*
   zetlandica 132
Phaennidae
   ecology 470, **500**, 502
   morphology 28, 29
   nutrition **152**
   taxonomy **53** 132–3
Phaeocystis 528
   puchetii 155–6
Phaeodactylum cornutum 290
Phyllopodidae **51**, 109
Phyllopus **51**, 71, 74, 109, 486
   aequalis 109
   bidentatus 109
   giesbrechti 109
   helgae 109
   impar 109
   integer 109
   mutatus 109
   muticus 109
Pilarella **51**, 73, 102
   longicornis 102, **498**
Placocalanus **51**, 66, 101, **498**, 501
   brevipes 101
   inermis 101
   insularis 101
   longicauda 101
   nannus 101
Platycopia **51**, 76, 77, 99, **498**

   inornata 99
   orientalis 99
   perplexa 99
   pygmaea 99
   robusta 99
   sarsi 99
   tumida 99
Platycopiidae 49, **51**, 99, **489**, **498**
Platycopioida 1, 2–3, *3*
   taxonomy **2**, **51**, *54*, *55*, 98–9
Platycopioidea 99
Pleuromamma 203
   behaviour 407, 428, 434, 437, 438, 445, **446**
   ecology 363, 467, 486, 528
   morphology 17, 30, 36
   physiology 217
   reproduction 257
   taxonomy **51**, *56*, 56, 61, *62*, 62, 67, 109
Pleuromamma abdominalis **7**, **26**, **32**, 56, 62, 109
   behaviour **412**, **417**, **426**, **435**
   distribution **514**
   ecology **485**
   size/weight **224**, **231**
   borealis *56*, 109, **435**, 437, 438, **514**
   gracilis **7**, *56*, 109
      behaviour **412**, **426**, **435**, 438
      distribution **514**
      ecology **361**
      physiology 217
      weight **224**, **231**
   indica *56*, 109, **435**, **514**
   piseki *56*, 109, **231**, **435**, **514**
   quadrungulata 56, 109, **435**, **514**
   robusta **7**, *56*, 56, 109
      behaviour **417**, **435**, 444
      distribution **514**
      ecology 468, 522
      growth/development *315*, 315
      morphology 32, *33*, **34**
      size/weight **224**
   scutullata *56*, 56, 109, 407
   wolfendeni 109
   xiphias **7**, 37, 48, **56**, 109
      behaviour 409, **412**, 421, 422, **435**
      distribution **14**
      ecology 363, 364
      nutrition **194**
      physiology 217
      size/weight **231**

Podoplea **2**
Poecilostomatoida **2**, 2, *3*
Pontella 16
  ecology 486, **491**
  taxonomy **52**, *61*, 66, 95, 114–15
Pontella agassizi 114, **514**
  alata 114
  andersoni 114
  asymmetrica 114
  atlantica **26**, 30, 114, *492*
  cerami 114
  chierchiae 114
  cristata 114
  danae 114, **514**
  denticauda 114, **514**
  diagonalis 114
  elegans 114
  fera 114, **514**
  forficula 114
  gaboonensis 114
  gracilis 114
  hanloni 114
  indica 114
  inermis 114
  investigatoris 114
  karachiensis 114
  kieferi 114
  latifurca 114
  lobiancoi 30, 114, *492*
  marplatensis 114
  meadii **26**, 114, 197, 217, **270**, **307**
  mediterranea **26**, 30, **32**, 114, *492*
    behaviour 420
    reproduction 268, **270**, **288**
  mimocerami 114
  natalis 114
  novae-zealandiae 114
  patagoniensis 114
  pennata 114
  polydactyla 114
  princeps **47**, 114, **430**, **514**
  pulvinata 114
  rostraticauda 114
  securifer 114, **430**
  sewelli 114
  sinica 115
  spinicauda 115
  spinipedata 115
  spinipes **32**, 115, **430**
  surrecta 115
  tenuiremis 115, **514**
  tridactyla 115
  valida 115, **514**
  whiteleggei 115
Pontellidae 434
  ecology **489**, 490, **491**
  nutrition 141, **152**
  reproduction 44, 46, 257, 261, 268, **270**
  taxonomy **52**, 113–15
Pontellina 27, **52**, 95, 115
  morii 115, **491**, **514**
  platychela 115, **515**
  plumata **7**, 30, 115
    behaviour **417**, **430**
    chromosomes **47**
    distribution **515**
    ecology 490, **491**
  sobrina 115, **515**
Pontellopsis 17, **52**, 95, 115, 486, **491**
  albatrossi 115
  armata 115
  bitumida 115
  brevis **26**, 115
  digitata 115
  elongatus 115
  globosa 115
  herdmani **47**, 115
  inflatodigitata 115
  krameri 115
  laminata 115
  lubbockii 115
  macronyx 115
  occidentalis **26**, 115
  pacifica 115
  perspicax 115
  pexa 115
  regalis 30, **32**, 115, *492*
    behaviour **430**
    nutrition 166, 203
  scotti 115
  sinuata 115
  strenua 115
  tasmaniensis 115
  tenuicauda 115, 506
  villosa 30, 115, *492*, **515**
  yamadai 115
Pontoptilus **51**, 70, 74, 105
  lacertosus 105
  mucronatus 105
  muticus 105
  ovalis 105
  pertenuis 105

Pontoptilus **51** (*cont.*)
　robustus 105
Prodisco **51**, 76, 106
　princeps 106
　secundus 106
Progymnoplea **2**, 99
Pseudaugaptilus **51**, 56, 71, 74, 105
　longiremis 105
　orientalis 105
　polaris 105
Pseudeuchaeta 20, **53**, 78, 127–8, 486
　arctica 127
　brevicauda **4**, 127
　flexuosa, 127, **499**
　magna 128, **499**
　major 128
　spinata 128
Pseudhaloptilus **51**, 106
　longimanus 70, 106
Pseudoboeckella **52**, 112–13
Pseudocalanidae 3, **152**, 434
Pseudocalanus 218, 236, 298, 346, 516
　behaviour 402, **408**, 413
　biochemistry **239**, 247
　chromosomes 48
　ecology **361**, 366, 372, **378**, **397**, 469, 480, 481, 482, 487, **493**, 495, 496, 524, 527, 528
　growth/development 298, 299, **307**, 338, **343**
　life history 347, 348, 389
　morphology 8, 25
　nutrition 140, 165, 167, **187**, **191**, 207
　physiology 215
　reproduction 43, 46, 257, 259, 261, 265, 267, 280, **284**, 289, 348
　size/weight **232**, *332*
　taxonomy **53**, 60, 64, 88, 89, 130
Pseudocalanus acuspes 43, 130
　　biochemistry 247
　　ecology **361**, 364, **493**, 495
　　growth/development **307**, **311**, **313**, 338
　　life history **387**
　　reproduction 254
　elongatus **7**, 130
　　behaviour **412**, **417**, 444
　　biochemistry 237
　　chromosomes **47**
　　distribution **507**, **515**
　　ecology **361**, 366, **378**, **397**, 398
　　growth/development **307**, **311**, **313**, 320, 321, **343**
　　life history **387**, 390
　　nutrition 145, 150, 161, 163, 166, 171, **187**, **191**, **198**, **201**, **206**, **284**
　　reproduction 260, **288**, 289, 293
　　size/weight **231**, 233
　major 130
　mimus 130, **485**
　minutus **7**, **26**, 130
　　behaviour **412**, 429, 451, 454
　　biochemistry **239**
　　chromosomes **47**
　　ecology **361**, **378**, **397**, 475, **493**
　　growth/development 298, **307**, **311**, **312**, 338
　　life history **387**
　　nutrition 145, 169, 172, **183**, **186**, **187**, 188
　　reproduction 260, 282, **284**, **288**
　　size/weight **231**, 328, 329, **331**
　moultoni 130, **284**, 292, **311**, **313**, **361**
　newmani 130
　　ecology **361**, 366, 368, **378**, **493**
　　growth/development **311**, **313**
　　nutrition 182, **183**, 197, **198**, **201**, 203
　　reproduction **284**, 292
Pseudochirella 236
　ecology 486
　nutrition **152**
　reproduction 263, 266, 267
　taxonomy **53**, *61*, 87, 90, 92, 93, 128
Pseudochirella accepta 128
　batillipa 128
　bilobata 128
　bowmani 128
　calcarata 128
　cryptospina 128
　dentata 128
　divaricata 128
　dubia 128
　elongata 128
　fallax 128
　formosa 128
　gibbera 128
　granulata 128
　gurjanovae 128
　hirsuta 128
　limata 128
　lobata 128
　major 128
　mariana 128

mawsoni 128
notacantha 128
obesa 128
obtusa 128
pacifica 128
palliata 128
polyspina 128
pustulifera 128
scopularis 128
semispina 128
spectabilis 128
spinosa 128
sqaulida 128, 262
tanakai 128
vervoorti 128
Pseudocyclopia **53**, 77, 133, **500**, 528
  caudata 133
  crassicornis 133
  giesbrechti 133
  insignis 133
  minor 133
  muranoi 133, 402
  stephoides 133
Pseudocyclopidae **51**, 100, 470, **489**, **498**
Pseudocyclopiidae **53**, 133, 470, **489**, **500**
Pseudocyclopoidea **51**, *54*, *55*, 100–1
Pseudocyclops 518
  behaviour 423
  ecology **489**, 497, **498**
  taxonomy **51**, 68, 100
Pseudocyclops arguinensis 100
  australis 100
  bahamensis 100
  bilobatus 100
  cokeri 100
  crassiremis 100
  gohari 100
  kulai 100
  lakshmi 100
  latens 100
  latisetosus 100
  lepidotus 100
  lerneri 100
  magnus 100
  mathewsoni 100
  minya 100
  mirus 100
  obtusatus 100
  oliveri 100
  pacificus 100
  paulus 100

pumilis 100
reductus 100
rostratus 100
rubrocinctus 100
simplex 100
spinulosus 100
steinitzi 100
umbraticus 100
xyphophorus 100
Pseudodiaptomidae **52**, 115–16, **498**
Pseudodiaptomus 23, 218, 298
  behaviour 450
  distribution 506
  ecology 469, 497, **498**, 528
  growth/development 296, 321
  reproduction 257, 258, 259, 260, 262, 266, 267, 279
  taxonomy **52**, 67, 115–16
Pseudodiaptomus acutus **26**, 115, **307**, **345**, 508
  americanus 115
  andamanensis 115
  annandalei 115
  ardjuna **26**, 115
  aurivilli **26**, **47**, 115
  australiensis 115
  batillipes 115
  bayli 115
  binghami **26**, 115
  bispinosus 115
  bowmani 115
  brehmi 115
  bulbiferus 115
  bulbosus 115
  burckhardti 115
  caritus 115
  charteri 115
  clevei 115
  cokeri 115, **308**, **345**
  colefaxi 115, 450, 497
  compactus 115
  cornutus 116, 497
  coronatus **26**, 116, 297
    behaviour **417**, **433**, 450
    growth/development **308**, 321
    longevity **345**
    nutrition 166
    physiology 217
    reproduction **288**
  cristobalensis 116
  culebrensis 116

Pseudodiaptomus acutus (cont.)
  dauglishi 116
  diadelus 116
  dubius 116
  euryhalinus **26**, 116, 260
  forbesi 116, **507**
  galapagensis 116
  galleti 116
  gracilis 116
  griggae 116
  hessei 116, **232**
    behaviour 451-2
    ecology **379**
    growth **308**, **311**, **320**
  heterothrix 116
  hickmani 116
  hypersalinus 116
  incisus 116
  inflatus 116
  inflexus 116
  inopinus 116, **507**
  ishigakiensis 116
  jonesi 116
  lobipes 116
  longispinosus 116
  malayalus 116
  marinus **26**, 116, 298
    biochemistry **239**
    distribution 506, **507**
    ecology **379**, 380
    growth/development **308**, **311**, **312**, **313**, **320**, **343**
    nutrition 157, **186**, **201**
    reproduction 280, 282, **284**
  marshi 116
  masoni 116
  mertoni 116
  mixtus 116
  nankauriensis 116
  nihonkaiensis 116, 429, **431**
  nostradamus 116
  occidentalis 116
  ornatus 116
  pacificus 116
  panamensis 116
  pankajus 116
  pauliani 116
  pelagicus 116
  penicillus 116
  philippinensis 116
  poplesia 116

  poppei 116
  richardi 116
  salinus 116
  serricaudatus **47**, 116
  sewelli 116
  smithi 116
  spatulus 116
  stuhlmanni 116
  tollingeri 116
  trihamatus 116
  trispinosus 116
  wrighti 116
Pseudolovenula **52**, 113
  magna 113
Pseudophaenna **53**, 85, 89, 135
  typica 135
Pseudotharybis **53**, 82, 128
  brevispinus 128
  dentatus 80, 128
  magnus 128
  robustus 80, 128
  spinibasis 128
  zetlandicus 128
Pterochirella **53**, 66, 128
  tuerkayi **128**, **499**
Puchinia **53**, 82, 135
  obtusa 135, 422

Racovitzanus **53**, 78, 82, 84, 135
  antarcticus 135
  levis 135
  pacificus 135
  porrectus 135
Rhapidophorus **51**, 102
  wilsoni 68, 102, **498**
Rhincalanus
  ecology 486, 487, 528
  morphology 24
  nutrition 140, **152**
  taxonomy **52**, 76, 122
Rhincalanus cornutus 24, **26**, 122, **224**
    chromosomes **47**
    distribution **515**
    nutrition **194**
  gigas **7**, **26**, 122
    biochemistry 247
    ecology **361**, 399, 492, 494
    life history **387**, 390
    nutrition **191**, **208**
    reproduction **255**, **287**
    size/weight **224**, **232**, **234**

nasutus **4**, **7**, **26**, 122, **232**, 297
  behaviour **413**, **417**, **418**, 422, 424, 455
  biochemistry **239**
  chromosomes **47**
  distribution **515**
  ecology 365, 366, 371, **379**, 481, **484**, **485**, 503
  growth/development **308**, 321, **343**
  nutrition 159, 166, **191**, **194**
  reproduction 265, **288**
  rostifrons 122, **515**
Rhinomaxillaris **53**, 123
  bathybia 123
Ridgewayia **51**, 70, 73, 101, 497, **498**
  canalis 101
  flemingeri 101
  fosshageni 101, 503
  gracilis 101
  klausruetzleri **26**, 101
  krishnaswamyi 101
  marki 101, **489**
  marki minorcaensis 101
  shoemakeri 101
  typica 101
  wilsoni 101
Ridgewayiidae 17, **51**, 100–1, 470, **489**, **498**
Ryocalanidae **52**, 122–3, **499**
Ryocalanoidea **52**, *54*, **55**, 122–3
Ryocalanus **52**, 78, 80, 122–3, **499**
  admirabilis 122
  asymmetricus 122
  bicornis 123
  bowmani 123
  infelix 123
Rythabis **53**, 84, 138
  atlantica 138, **501**

Sagitta
  elegans 372
  enflata 160
Sarsarietellus **51**, 68, 74, 103
  abyssalis 103, **498**
  natalis 103
Scaphocalanus
  ecology 486, **500**
  taxonomy **53**, 57, *61*, 80, 82, 84, 85, 134, 135–6
Scaphocalanus acuminatus 135
  acutocornis 135
  affinis 135
  amplius 135, 136
  angulifrons 136
  antarcticus 135
  bogorovi 136
  brevicornis 135, 479
  brevirostris 135
  californicus 136
  curtus 135, 136
  difficilis 135
  echinatus 135
  elongatus 136
  farrani 135
  insignis 136
  insolitus 136
  invalidus 135
  longifurca 135
  magnus 135, *441*
  major 135
  parantarcticus 135
  paraustralis 135
  pseudobrevirostris 135
  similis 135, 136
  subbrevicornis 135
  subcurtus 136
  subelongatus 136
Scolecithricella
  ecology 486, **500**
  taxonomy **53**, 57, 61, 80, 81, 84, 136
Scolecithricella abyssalis 136
  aspinosa 136
  avia 136
  canariensis 136
  cenotelis 136
  curticauda 134
  dentata 136
  denticulata 134
  glacialis 494
  globulosa 136
  incisa 134
  lanceolata 134
  lobata 134, 136
  lobophora 136, 503
  marquesae 134, 136
  minor **7**, 136
  modica 136
  neptuni 136
  obscura 136
  orientalis 136
  ovata 136
  pacifica 136
  paramarginata 136
  pearsoni 136

Scolecithricella abyssalis (cont.)
  polaris 136
  profunda 136
  propinqua 134
  pseudoculata 136, **500**
  schizosoma 136
  spinacantha 136
  spinata 134
  timida 134
  tropica 136
  unispinosa 136
  vespertina 136
  vittata 136
Scolecithrix
  behaviour 407, **417**
  ecology 486
  taxonomy **53**, 57, 78, 80, 84, 136
Scolecithrix aculeata 134
  birshteini 136
  bradyi 136, **485**
  ctenopus 136
  danae **7**, 136, *422*
    behaviour **413**
    ecology **485**, 503
    nutrition 154, **194**
    reproduction 41
    size/weight 222, **224**
  elaphas 134
  fowleri 136
  grata 136
  laminata 136
  longipes 136
  longispinosa 136
  magnus 134
  marginata 136
  maritima 136
  medius 134
  mollis 134
  nicobarica 136
  subdentata 136
  subvittata 136
  tenuipes 136
  tenuiserrata 136
  valens 134
  valida 134
Scolecitrichidae 29, 340, 434
  ecology **500**
  nutrition **152**
  taxonomy **53**, 134–7
Scolecocalanus **53**, 80, 81, 137
  galeatus 137
  lobatus 137
  spinifer 137
Scopalatum **53**, 81, 82, 137, **500**
  dubia 137
  farrani 137
  gibbera 137
  smithae 137
  vorax 137, **152**, 160, 503
Scottocalanus **53**, 82, 84, 137, 486
  corystes 137
  dauglishi 137
  farrani 137
  helenae 137
  infrequens 137
  investigatoris 137
  longispinus 137
  persecans 137
  rotundatus 137
  securifrons 137, 444
  sedatus 137
  setosus 137
  terranovae 137
  thomasi 137
  thorii 137
Scottolana **184**
Scottula
  abyssalis 103
  inaequicornis 102
Scutogerulus **51**, 68, 103
  pelophilus 101, 103, **498**
Senecella 3, **53**, 128
  calanoides 128, **413**
  siberica 128
Sinocalanus **52**, 67, 113
  doerrii 113, **507**
  sinensis 113, **507**
  solstitialis 113
  tenellus 113, **232**, 330
    biochemistry **239**, **241**
    ecology 365
    growth/development **308**, **311**, **343**
    nutrition 159, **201**
    reproduction **255**, 260, 266, **270**, **287**, **288**
Siphonostomatoida **2**, 2, *3*
Skeletonema costatum 290
Snelliaetideus 125
Sognocalanus **53**, 87, 90, 123
  confertus 123
Spicipes **53**, 66, 130
  nanseni 130

Spinocalanidae **52**, 123–4, **152**, **499**
Spinocalanoidea **52–3**, *54*, 55, 123–4
Spinocalanus
  ecology 486, **499**, 528
  taxonomy **53**, 78, 80, 124
Spinocalanus abruptus 124
  abyssalis 124
  angusticeps 124
  antarcticus 124
  aspinosus 124
  brevicaudatus 124, 479, 502
  brevicornis 502
  dispar 124
  hirtus 124
  hoplites 124
  horridus 124
  longicornis 124, 502
  macrocephalon 124
  magnus 124
  oligospinosus 124
  polaris 124
  profundalis 124
  similis 124
  spinosus 124
  terranovae 124
  usitatus 124
  validus 124
Stephidae **53**, 99, 137–8, 470, **489**, **500**
Stephos
  behaviour 402
  ecology 469, 497, **500**, 501, 528
  taxonomy **53**, 85, 88, 89, 137–8
Stephos antarcticus 137
  arcticus 137
  balearensis 137, **489**
  canariensis 137, **500**
  deichmannae 137
  exumensis 137
  fultoni 137
  gyrans 137
  kurilensis 137
  lamellatus 137
  longipes 137, **361**, **387**, **493**, 494
  lucayensis 137
  maculosus 137
  margalefi 137, **489**
  minor 137
  morii 137
  pacificus 137
  pentacanthos 137
  robustus 137
  rustadi 137
  scotti 137
  seclusum 138
  tropicus 138
  tsuyazakiensis 138
Stoichactis helianthus 503
Streblospio benedicti 293
Strombidium sulcatum 164
Stygocyclopia **53**, 78, 134
  balearica 134, **489**
Subeucalanus 122
Sulcanidae **52**, 116
Sulcanus **52**, 66, 116
  conflictus 116, **308**, **311**, 469
Sursamucro **53**, 80, 88, 128
  spinatus 128
Syltodinium listii 524
Syndinium 524

Talacalanus **53**, 81, 82, 132
  greeni 132
  maximus 132
Temora
  behaviour 402, 407, 423
  ecology **397**, 480, 481, 486, 490, 497, 524, 528
  life history 347, 389
  nutrition 152
  reproduction 257
  size *332*
  taxonomy **52**, 76, 93, 117
Temora discaudata **47**, 117, **515**
  kerguelensis 117
  longicornis 6, **7**, **26**, 298
    behaviour 403, **408**, **413**, 414, **417**, **418**, 419, 420, 434, 444, 454
    biochemistry 247, 251
    distribution **515**
    ecology **362**, 366, 369, 372, 373, **379**, 380, **397**, 469, 472, 475, 481
    growth/development **309**, **311**, **312**, **313**, **320**, **343**
    life history 340, 388
    nutrition 143, 145, 150, 156, 159, 161, 164, 166, 170, **184**, **186**, 187, **191**, 199, **201**, 204, 205, **206**, 207
    reproduction 257, **271**, **282**, **287**, 291
    size/weight **232**, **331**, 332, 333, 335, **337**
    taxonomy 49, 63, 117
  stylifera **7**, **26**, 117, 236
    behaviour **408**, **413**, 446

Temora discaudata (cont.)
  stylifera (cont.)
    chromosomes **47**
    distribution **515**
    ecology **362**, 364, 370, **379**, **397**, 480, **484**
    growth/development **309**, **311**, **312**, 315, 316, 338
    nutrition 181, **184**, **191**, **201**, **206**, **208**
    physiology 215
    reproduction 43, **255**, 260, 265, 283, **287**, 290
    size/weight **225**, 335
  turbinata **7**, **26**, 117
    behaviour 403, **446**, 447, 451
    biochemistry 251
    chromosomes **47**
    distribution **515**
    growth/development **309**, **320**, 343
    life history 383
    nutrition **198**, **201**
Temoridae 3, **52**, 116–17, 268, 434, 473
Temorites **52**, 76, 122, **498**
  brevis, 122
  discoveryae 66, 122
Temoropia **52**, 77, 117
  mayumbaensis 117
  minor 117
  setosa 117
Teneriforma **53**, 87, 89, 124, **499**
  meteorae 124
  naso 87, 124
  pentatrichodes 124
Thalassiosira 332
  partheneia 155
  wessflogii 164, 181
Tharybidae **53**, 99, 138, **152**, **500**, 502
Tharybis **53**, 80, 84, 87, 90, 138
  altera 138
  angularis 138, **501**
  asymmetrica 138
  compacta 138
  crenata 138, **501**
  fultoni 138
  macrophthalma 138
  magna 138
  megalodactyla 138
  minor 138
  neptuni 138
  sagamiensis 138
Tigriopus

  californicus 155
  fulvus 218
  japonicus 364
Tortanidae **52**, 117–18, 148, **152**, **271**
Tortanus **52**, 95, 117–18, 166, 480
  barbatus **47**, 117, 447
  bonjol 117
  bowmani 117
  brevipes 118
  capensis 118
  compernis 118
  denticulatus 118
  derjugini 118, **271**
  dextrilobatus 118
  digitalis 118
  discaudatus **7**, **26**, 118
    distribution **515**
    growth/development **311**
    nutrition 159, 163, **188**
    reproduction **271**, 279, 294
  erabuensis 118
  forcipatus **7**, 22, 118, **232**
    behaviour **433**
    biochemistry **239**, **241**
    chromosomes **47**
    nutrition 162, 163
    reproduction 268, **271**
  giesbrechti 118
  gracilis **26**, **47**, 118, 163, **433**
  longipes 118, **431**
  lophus 118
  murrayi 118
  recticauda 118
  rubidus 118, **431**
  ryukyuensis 118
  scaphus 118
  setacaudatus 118
  sheni 118
  sinensis 118
  spinicaudatus 118
  tropicus 118
  vermiculus 118

Undeuchaeta 236
  behaviour 407
  ecology 486, 528
  taxonomy **53**, 87, 93, 129
Undeuchaeta bispinosa 129
  incisa 129
  intermedia 129

TAXONOMIC INDEX

Undeuchaeta bispinosa (cont.)
  magna 129
  major 129, **515**
  plumosa 6, **7**, 129, **417**, 444, **515**
Undinella **53**, 87, 90, 93, 138, **501**
  acuta 138
  altera 138, **501**
  brevipes 138
  compacta 138, **501**
  frontalis 138
  gricei 138
  hampsoni 138, **501**
  oblonga 138
  spinifer 138
  stirni 138
Undinothrix **53**, 84, 137
  spinosa 137
Undinula **52**, 68, 119, 486
  vulgaris 7, **26**, 119
    behaviour **417**, **446**
    biochemistry 251
    distribution **515**
    ecology **379**, **484**
    growth/development **309**, **343**
    nutrition **186**, **201**, **206**, **208**
    reproduction 263, **287**
    size/weight **245**

Valdiviella 236
  ecology 486
  reproduction 267
  taxonomy **53**, 63, 93, 129
Valdiviella brevicornis 129
  brodskyi 129
  ignota 129
  imperfecta 129
  insignis 22, **22**, 129
    ecology 529
    life history 392
    reproduction 274, *277*, *279*, *281*, 283
  minor 129
  oligarthra 129

Wilsonidius **53**, 92, 129
  alaskaensis 129

Xantharus **53**, 81, 84, 132
  formosus 132

Xanthocalanus
  ecology 486, **500**, 522, 528
  taxonomy **53**, 61, 80, 81, 82, 84, 85, 133
Xanthocalanus agilis 133
  alvinae 133, **500**
  amabilis 133
  antarcticus 133
  borealis 133
  claviger 133
  cornifer 133
  crassirostris 133
  difficilis 133
  dilatus 133
  distinctus 133, **500**
  echinatus 133
  elongatus 133, **500**
  fallax **26**, 133, **500**
  giesbrechti 133
  gracilis 133
  greeni 132
  groenlandicus 133, 138
  harpagatus 133
  incertus 133
  irritans 133
  kurilensis 133
  legatus 133
  macilenta1 133
  macrocephalon 133, 138, **500**
  marlyae 133, **500**
  medius 133
  minor 133, **500**
  mixtus 133
  multispinus 133
  muticus 133
  obtusus 133
  oculata 133
  paraincertus 133, 138
  pavlovskii 133
  pectinatus 133
  penicillatus 133
  pinguis 133
  polaris 133
  profundus 133, 522
  propinquus 133
  pulcher 133
  rotunda 133
  serrata 133
  simplex 133
  soaresmoreirai 133
  squamatus 133
  subagilis, 133

tenuiremis 133
tenuiserratus 133
typicus 133

Zenkevitchiella **52**, 77, 122, **498**
   abyssalis 122
   atlantica 122
   crassa 122
   tridentae 76, 122

Subject Index

Note: Page references in *italics* refer to Figures; those in **bold** refer to Tables

Abnormalities 36
  chromosomal 48
Abundance *see* Population biology
Adenosine triphosphate (ATP) 249–50
Aesthetascs 18, 23
Aggregations 424–5, **426**, 428
  multispecies aggregations 429–32, **430–1**
Alimentary canal 39–41, *40*
Allometric growth 338
Amino acids 249
Anatomy 37–48
  circulatory system 39
  digestive system, 39–41 *40*
  endoskeleton 38
  excretory system 46
  muscular system 38
  nervous system 38–9
  oil sac, 47–8
  see also Morphology; Reproductive system
Antarctic Ocean
  ice-water interface **493**, 494
Antenna *18*, 19
  see also Swimming appendages
Antennules 17–19, *18*
  copepodids, 27
  feeding function, 140
  see also Swimming appendages
Apodemes 38
Apolysis *see* Moulting
Appendages *15*, *16*, 16, 18
  feeding appendages 140–3
  nauplius, 23–4
  see also individual appendages
Arctic Ocean
  ice-water interface **493**, 495
Ash content 233–5, *235*
Ash-free dry weight 233
Assemblages *see* Spatial distribution
Assimilation efficiency 205–6, **206**

  net growth efficiency 205–7, **208**
Associations 460–70
  diversity 461–4, **462–3**
  dominance of species 464–6, *466*, *467*
  niches 466–70
Astaxanthin 251
Asymmetry 36

Bacteria
  ectosymbionts 524
  in diet 154–5
  in faecal pellets 203
Bathypelagic zone 483
Behaviour *see* Learning; Feeding; Mating behaviour; Spatial distribution; Swimming activity; Vertical migration
Benthopelagic habit 496–7, **498–501**, 501–2
Biochemistry *see* Elemental composition; Organic components
Bioluminescence 434–9, **435**
  luminescent behaviour 437–9
  luminescent glands 434–6
  luminous secretions 436–7
Biomass 393, 398–9, 457–60
  coastal and shelf regions 457–8
  inter-annual fluctuations 394
  oceanic water column 458–9
  production/biomass (P/B) ratios 373–81, **375–9**
  see also Population biology
Body density, sinking rate and 409, 414, **415–17**
Body form *4*, 14–15
  see also Morphology
Body size *see* Size
Body weight 221–35, 333–7
  ash weight 233–5, *235*
  ash-free dry weight *227*, 233, **234**
  dry weight, 227–33, **228–2**, *334*, *335*, **337**

Body weight (*cont.*)
    wet weight 221–6, **223–5**, **226**
Brackish water 472–9
    estuaries *474*, 475–7
    fjords 349–51, **350**, 477–9, *478*
    river plumes 474–5
Brain 38
Breeding biology *see* Reproduction
Brood size 272–4, **275–6**, 276–81, *281*
    seasonal variation 281–2, 371
Buccal cavity 39

Caloric density 250
Cannibalism 159
Carbohydrate content 248
Carbon content 236–40, **238–9**
    carbon–nitrogen ratio 242
Carnivory *see* Predatory feeding
Carotenoids 251
    as dietary markers 151
Cave habitat 488, **489**
Cephalosome *15*, 16, 24–5
Chaetognath prey 160
Chemoreception
    feeding responses *144*, 144–6, 149
    reaction distances 432–4, **433**
    role in mate-seeking behaviour 257
Chitin 248
Chromosomes 48
    aberrations 48
    haploid number of **47**, 48
Ciliates
    ectosymbionts 525
    in diet 146, 156–7
Circulatory system 39
Citation frequency **7**
Classification 1–2, **2**, 50, **51–3**
Clutch size *see* Brood size
Coastal environments 470–80
    biomass 457–8
Coccolithophores, in diet 159
Colonizations 505–8, **507**
Commensalism 503
Communities *see* Associations
Competition 468–9
Condition factors (CF) 336
Continental shelves 444, *478*, 479–82
    across-shelf gradients 480–1
    biomass 457–8
    upwelling 482, **484–5**
Copepodid

feeding 167
morphology 24–8
segmentation development 24–7, *25*, **27**
swimming pattern 403–7
see also Development

Deep-sea environment 501–2
    life history strategies 390–2
Demographic analysis *see* Population biology
Density *see* Population biology
Depth distribution 458–9, *459*, *460*, 486–7, 504–5
    see also Vertical migration
Detritus feeding 153–4
Development 296
    development time **300–9**, *318*
    eggs, 298–9, **310–13**, *314*, 314
    median development time (MDS) 322–3
    food availability and *316*, 316
    models of, 317–21, *319*
        equiproportional development 318–19, 327
        isochronal development 319
        non-conformist development 321
        sigmoidal development 319–21
    segmentation 24–7, *25*, **27**
    sexual differentiation 43–4
    stage duration **317**, 317–23
    swimming legs **27**, *27*, 55
    see also Growth; Life history
Diapause eggs 268–72, **269–71**, **273**, 388–9
    see also Life history
Diatoms
    ectosymbionts 525
    in diet 145, 155, 290
Diel rhythms
    diet 169
    feeding periodicity 170–1, *442*, 450, 452–3
    swimming activity, 420–1
    see also Vertical migration
Diet 149–60
    bacteria 154–5
    ciliates 156–7
    coccolithophores 149
    detritus 153–4
    diatoms 155
    diel changes 169
    dietary requirements 168–9
    dinoflagellates 157, **158**

SUBJECT INDEX

dissolved organic matter (DOM), 153
  egg production and 290–2
  foraminiferans 159
  metazoans 159–60
  Phaeocystis pouchetii 155–6
  protozoans 156–9
  radiolarians 159
  selectivity 164–6
Digestion 41, 204–5, 209
  digestive enzymes 204
  gut clearance rate constant (K) 188, 192, *194*, 195–6
  gut evacuation rate 192, 195–6
  see also Assimilation efficiency
Digestive system, 39–41, *40*
Dinoflagellates
  in diet 157, **158–9**
  luminescence influence on vertical migration 171, 449–50
  parasitic dinoflagellates 373, 524
  swimming speed and 419
Dispersal 505–8, **507**
Dissolved organic matter (DOM) 153
Distribution *see* Geographical distribution; Spatial distribution
Diversity 461–4, **462–3**
DNA content 249
  RNA/DNA ratio 256
Dominance of species 464–6, *466*, *467*
Ductus ejaculatorius 45, 254, 260

Ecological niches 466–70
Ecosystems 3–5, 519–25
  food webs 519–21
  human exploitation 522–3
  key species 528–30
  large marine ecosystems (LMEs) 509, *510–11*, 510, 515–16, 519
  parasites 523–5
    ectosymbionts 524–5
    endosymbionts 523–4
  perturbations within 526–7
  predation 521–2
Ectosymbionts 524–5
Eddies 471–2
Egestion, rate of 195–6
  see also Faecal pellets
Egg laying 266–8
Egg number 256, 272–4, **275–6**, 276–81
  egg production rates 273, 289–90, 291–2
  food availability and 289–92

seasonal variation 281–2
stored lipids and 292
see also Fecundity
Eggs
  development time 298–9, **310–13**, *314*, 314
  egg mass 267, 279, 280, 281, 283, 299
  hatching 293–4
  identification 23
  morphology *22*, 22–3
  mortality 292–3, 366–7
  resting eggs in sediments 268–72, **269–71**, 388–9
  size 274–6, **275–6**, *277*, **278**, *279*, **280**, *280*
Electron transport system (ETS) 210
Elemental composition 236–45
  atomic ratios 243
  carbon 236–40, **238–9**
  carbon–nitrogen ratio 242
  hydrogen 242
  mineral composition 243–5, **244**
  nitrogen 240, **241**
  phosphorus **241**, 242–3
Embryonic development *see* Eggs
Endoskeleton 38
Endosymbionts 523–4
Energetic cost of swimming 421–3
Energy content 250
Environments *see* Pelagic environments; Restricted environments
Enzymes 252
  digestive 204
Epipelagic zone 483
Equiproportional development 318–19, *327*
Escape reaction 406, 414, **418**, 419
  response to predators 419–20
Estuaries 473–4, *474*, 475–7
Excretion 209–11, **213**, *214*
Excretory system 46
Eye *31*, 31, **32**

Faecal pellets 196–203
  fate of 525–6
  production of 196–9, *197*, **198**, **200–1**
  sinking rates, 199–203, *202*
Fatty acids 246–7
  as dietary markers 151, 153
Faunal provinces, 509, *510–11*, 516–17
Fecundity 272–3, 282–3, **284–7**, 288
  see also Egg number
Feeding
  copepodids 167

Feeding (cont.)
  detritus feeding 153–4
  feeding history influence 172–4, *173*
  herbivorous feeding, 142–3
  nauplii, 166–7
  see also Diet; Foraging; Omnivorous feeding; Particle feeding; Predatory feeding
Feeding appendages 140–3
Feeding current 143–4, 146–8, *147*
Feeding periodicity 167–72
  diel periodicities 170–1, 442, 450, 452–3
  dietary requirements 168–9
  gut fluorescence analysis 167–8
  seasonal periodicities 171–2
  short-term periodicities 170
  vertical migration and 170–1
Feeding rate 176–93
  criticism of laboratory methods 176–8, *179*
  filtering/clearance rates 180–4, *181*, *182*
  ingestion rate 185–93, **186**, **187**, **188**, **189–91**, **192**, **193**, 207
  predatory feeding, **184**, 184
Fertilization 43
Fertilization tubes 43, 262–3, 266
Filter feeding
  feeding appendages 141
  filtering/clearance rates 180–4, *181*, *182*
  see also Particle feeding
Fixation 10–11
Fjordic environments 349–51, **350**, 477–9, *478*
Food see Diet
Food availability
  as cue for diel vertical migration 450–1
  development and *316*, 316
  egg production and 289–92
  population regulation 372
Food capture 143–9
  food detection 145–6
  particle feeding 146–8
  predatory feeding 148–9, 162–4, *163*
Food webs 519–21
Foraging 160–6, *161*
  particulate feeding 161–2
  predatory feeding 162–4
  selectivity 164–6
Foraminiferans, in diet 159
Formalin 10–11
Free amino acids 249

Frontal organs *31*, 31, **32**
Fronts 471–2
Generation time 363, 374, *380*
Generations per year **355–62**
Generic key 57–60, *59*, **65**, 65–97
Genetic variation 63–4, 371
Genital antrum 41, 43
Genital somite 16, 25–7, *42*, 43, 63, 267
  spermatophore attachment *42*, 43, 262, *263*
Geographical distribution 504–17, **512–15**
  faunal provinces 509, *510–11*, 516–17
  introductions 505–8, **507**
  large marine ecosystems (LMEs) 509, *510–11*, 510, 515–16, 519
  see also Spatial distribution
Germinal site 42
Gnathobase 19–20, 141
Growth 296, 314–27, **315**, **320**, *336*
  allometric growth 338
  general concept of 326–7
  growth rate 316, 338–9, *339*, *340*, **341–3**, 344, *381*, *382*
  seasonal variation 371
  moulting 315–17
    intermoult duration 317–23
    moult increment 323–6, *324*, **325**
  net growth efficiency 205–7, **208**
  see also Development; Life history; Size
Gut clearance rate constant (K) 188, 192, *194*, 195–6
Gut contents analysis 149–51, **151**, **152**
Gut evacuation rate 192, 195–6
  see also Faecal pellets
Gut filling time 193
Gut fluorescence analysis 167–8
Gut passage time 188, *193*, **194**

Habitats see Pelagic environments; Restricted environments
Halocline responses 443
Hatching 293–4
Heart 39
Herbivorous feeding 142–3
  see also Diet; Phytoplankton feeding
High latitude life history strategies 384–90
  overwintering copepodids 385–8, **396–87**
  resting eggs in sediments 388–9
Hindgut 39, *40*
Human exploitation 522–3
Hydrodynamic features *422*, 422–3

Hydrogen content 242
Hydrothermal vents 502–3
Hyperbenthic habit 496–501, **498–501**

Ice-water interface 490–6, **493**
  Antarctic Ocean **493**, 494
  Arctic Ocean **493**, 495
  ice cover at lower latitudes 496
  ice-edge zones 495–6
Identification 57–64, 516
  eggs 23
  genera 65–97
  molecular genetics 63–4
  pore signatures 61–3, *62*
  species 60–1, 516
Ingestion rate *see* Feeding rate
Integument 28–35
  eye *31*, 31, **32**
  frontal organs *31*, 31, **32**
  moulting 32–5, *33*, **34**
  pore pattern signatures 30
  setae 28–30
  subintegumental glands 29–30
  see also Moulting
Intersexes 37
Introductions 505–8, **507**
Isochronal development 319

Key species 528–30
Key to genera 57–60, 59, **65**, 65–97

Labels 11–12
Labium 19, 39
Laboratory culture 218, 297–8
Labral glands 19, 39
Labrum 19, 39
Large marine ecosystems (LMEs) 509,
  510–11, 510, 515–16, 519
Larvacean prey 160
Learning 174
Length *see* Size
Life history 347–8, 352, 353
  deep-sea patterns 390–2
  high latitude patterns 384–90, *391*, *392*
    overwintering copepodids 385–8
    resting eggs 388–9
    resultant life histories 389–90
  strategies 381–92
  temperate patterns 383–4
  tropical and subtropical patterns 382–3
Light
  as cue for diel vertical migration 446–9,
    448
  response to 217–18
Lipid content 245–8
  stored lipid influence on egg production
    292
Loch Etive, Scotland 349–51, **350**
Longevity 339–44, **345**
Longhurst/Hardy Plankton Recorder
  (LHPR) 9
Luminescence *see* Bioluminescence
Luminescent glands 434–6

Mandible *18*, 19–20, 141
  edge index, *141*, 141–2
  moult cycle stages 35
  see also Swimming appendages
Mass mortalities 369–70
Mating behaviour 256–66
  attraction of the sexes 256–8
  interspecific mating 258
  mating position 258–61
  spermatophore attachment 261–3
    number attached 261–2
    placement position 262–3, *263*
  spermatophore production 264–6
  spermatophore transfer *258*, 258–60
  see also Reproduction
Maxilla 18, 20
  filtering setae 142–3, 148
Maxillary glands 46
Maxilliped *18*, 20
Maxillule *18*, 20
Mechanoreception
  feeding responses *144*, 144–6, 149
  reaction distances 432–4, **433**
  role in mate-seeking behaviour 257
Median development time (MDS) 322–3
Mesopelagic zone 483
Metabolism 211–15
Metasome *15*, 16, 24–5
Midgut 39–41, *40*
Migration, ontogenetic 391, 391–2, 392, 445,
  446
  see also Vertical migration
Mineral composition 243–5, **244**
Molecular genetics 63–4
Morphological variation 36–7
Morphology 14–28, *15*, *16*
  adults 16–21
  asymmetry 36

Morphology (cont.)
  body form 4, 14–15
  copepodid 24–8
  egg 22–3
  nauplius 23–4, 24
  see also Anatomy; Integument; Size
Mortality 366–70
  eggs 292–3, 366–7
  mass mortalities 369–70
  parasites as cause 523–4
  see also Population biology
Moulting 32–5, 33, 34, 296, 315–17
  intermoult duration 317–23
  moult increment 323–6, 324, 325
Mouth 19, 39
Mouthparts 140–3
Multiple Opening/Closing Net and Environmental Sensing System (MOCNESS) 9
Multispecies assemblages 429–32, 430–1
Muscular system 38

Nauplius
  feeding 166–7
  morphology 23–4, 24, 403
  sources of descriptions 26
  swimming pattern 402–3
  see also Development
Nearest neighbour distances (NND) 425, 428, 434
Nervous system 38
Net growth efficiency 205–7, 208
Net to gross displacement ratio (NGDR), 407
Neustonic environment 488–90, 491
Niches 466–70
Nitrogen content 240, 241
  carbon–nitrogen ratio 242
Nitrogen excretion 210, 213
Non-conformist development 321
Nucleic acids 249–50
  adenosine triphosphate (ATP) 249
Number of generations 355–62
Nutritional requirements 168–9

Oceanic water column 483–7
  vertical distribution 486–7
Ocelli 31, 31
Oesophagus 39, 40
Oil sac 47–8
Omnivorous feeding 142–3, 150

feeding appendages 141
  see also Diet
Ontogenetic migrations 391, 391–2, 392, 445, 446
Oocytes 42–3
Oogenesis 42–3, 273
Organic components 245–50
  adenosine triphosphate (ATP) 249–50
  carbohydrates 248
  chitin 248
  enzymes 252
  free amino acids 249
  lipids 245–8
  nucleic acids 249
  protein 245
Organs of Bellonci 31, 32
Organs of Gicklhorn 31, 31–2
Osmoregulation 215, 216
Ovary 41–3, 42
  ovarian development 254–6, 255
Overwintering stages 386–7
  copepodids 385–8
  diapause eggs 267–71, 268–72, 273, 388–9
Oviducts 41, 42, 43
Oxycline responses 443–4
Oxygen, response to 217
Oxygen consumption 210, 211
  see also Respiration

Parasites 373, 523–5
  ectosymbionts 524–5
  endosymbionts 523–4
Particle feeding 146–8
  feeding appendages 141
  foraging tactics 161–2
  selectivity 165–6
  see also Filter feeding
Patchy distribution 8
Pelagic environments 470–87
  brackish water 472–9
    estuaries 474, 475–7
    fjords 349–51, 350, 477–9, 478
    river plumes 474–5
  coastal and shelf habitats 478, 479–82
    across-shelf gradients 480–1
    biomass 457–8
    upwelling 482, 484–5
  eddies 471–2
  fronts 471–2
  oceanic environment 483–7
    vertical distribution 486–7

Perivitelline space 22–3
Pheromones 256–7
Phosphorus content **241**, 242–3
Phylogeny 2, *3*, 49–56, *54*, 55–6
Phytoplankton feeding 6–8
  feeding rate **187**
  feeding responses *144*, 144
Polyvinyl lactophenol 12
Population biology 348, 352–81
  breeding seasons 354, 363
  density 457–9
    coastal and shelf regions 457–8
    oceanic water column 458–9, *459*
  generation time, 363, 374, *380*
  mortality 366–70
  population maintenance 392–8
    annual fluctuations 393
    inter-annual fluctuations, 394–8, **395–7**
  population regulation 370–3
    brood size seasonal variation 371
    food availability 372
    genetic aspects 371
    growth rate variation 371
    parasites 373
    predation 372–3
  production 373–81
  seasonal changes in stage structure 353–4
  sex ratio 363–6, *365*
  size frequency 354
  see also Biomass; Life history
Pore pattern signatures 30, 50, 55, 56, *56*
  species identification 61–3, *62*
Predators
  as cue for diel vertical migration 449–50
  population regulation 372–3, 470
  predation effects 521–2
  responses to 419–20
Predatory feeding 142–3
  ecological effects of 521
  feeding appendages 140–1, *142*
  feeding rate **184**, 184, **188**
  food capture 148–9, 162–4, *163*
  foraging tactics 162–4
  selectivity 165
  see also Diet
Preservation of samples 10–11
Pressure, response to 218
Prey *see* Diet; Predatory feeding
Production 373–81
  production/biomass (P/B) ratios 373–81, **375–9**

see also Egg number
Prosome *15*, 16
  musculature 38
Protein content 245
Protozoa, in diet 156–9

Radiolarians, in diet 159
Reproduction 253–93
  fecundity 272–3, 282–3, **284–7**, **288**
  seasonality, 253–6, 354, 363
  see also Eggs; Mating behaviour; Spawning
Reproductive system
  female 41–4, *42*
  male 44–6, *45*
Respiration 209–11, **212**, *213*
Restricted environments 487–503
  caves 488, **489**
  hyperbenthic and benthopelagic 496–503, **498–501**
    deeper waters 501–2
    seep sites and hydrothermal vents 502–3
    shallow waters 497–501
  neustonic 488–90, **491**
  under ice 490–6, **493**
    Antarctic Ocean **493**, 494
    Arctic Ocean **493**, 495
    ice cover at lower latitudes 496
    ice-edge zones 495–6
Reynolds number 147–8
Rhythmic activity 420–1, 452–3
  see also Diel rhythms; Feeding periodicity; Vertical migration
River plumes 474–5
RNA content 249
RNA/DNA ratio 256

Salinity
  osmoregulation and 215
  responses to 216
Sampling 8–10, 349–52
  error 349, 350
  preservation of samples 10–11
Scanning Electron Microscope (SEM) 12
Schools **427**, 429
Scientific literature 5–8
  citation frequency **7**
Seep sites 502–3
Segmentation development 24–7, *25*, **27**
Semi-permanent mounts 12

Sensillae 29–30
Setae 28–30
  filtering setae 142–3, 148
  replacement at moulting 33–5, **35**
  spinal and setal formula *21*, 21
Sex ratio 363–6
  sampling error 349
  seasonal variation 364, *365*
Sexual differences
  copepodids 27–8
  development of 43–4
  integument 30
Sigmoidal development 319–21
Sinking rate
  body density and 409, 414, **415–17**
  faecal pellets 199–203, *202*
Size *5*, 5
  body length 327–33, *329*, **331**, *332*
  egg 274–6, **275–6**, *277*, **278**, *279*, **280**, *280*
  size frequency 354
  variation 36
  volume 235–6
  see also Body weight; Growth
Spatial distribution 423–34
  aggregations 424–5, **426**, 428
  multispecies assemblages 429–32, **430–1**
  schools **427**, 429
  swarms **427**, 428–9
  see also Geographical distribution
Spawning 266–88, **268**
  egg laying 266–8
  resting eggs in sediments 268–72, **269–71**
Species diversity 461–4, **462–3**
Species dominance 464–6, *466*, *467*
Species identification *see* Identification
Spermathecae 41, 43
Spermatogenesis 45
Spermatophores 43, 44–6, *45*, 254, 261, *264*
  attachment *42*, 43, 261–3
    number attached 261–2
    placement position 262–3, *263*
  production of 264–6
  transfer *258*, 258–60
Spermatozoa 44–6, 260
Spinal and setal formula *21*, 21
Stains 12
Starvation tolerance 182, **183**
Steedman's fluid 10
Stomach contents analysis 149–51, **151**, **152**
Subintegumental glands 29–30
Subtropical life history strategies 382–3

Survival curves 367, 367–8, *368*
  see also Mortality
Swarms **427**, 428–9
  multispecies swarms 429–32, **430–1**
Swimming activity 401–23
  energetic cost 421–3
  escape reaction 414, **418**, 419
  hydrodynamic features *422*, 422–3
  rhythmic activity 420–1
  sinking rate, body density and 409, 414, **415–17**
  swimming pattern 402–7, *404*
    copepodid, 403–7
    nauplius, 402–3
  swimming speed 407–9, **408**, **410–13**
Swimming appendages *18*, 21, 402–6, *403*
  development **27**, 27, 55

Temperate life history strategies 383–4
Temperature
  development time and 298–9, 344
  egg production and 289–90
  osmoregulation and 215
  responses to 216
  size relationship 329–30, **331**
  weight relationship 335, **337**
Testis 44, 46
Thermal tolerance 216
Thermocline responses 443–4
Tides, as cue for diel vertical migration 451–2
Trace element composition 243–5, **244**
Triacylglycerols 246–8
Tropical life history strategies 382–3

Upwelling 482, **484–5**
Urosome *15*, 16–17, 24–5
  musculature 38
UV irradiation 217–18

Variation
  genetic 63–4, 371
  growth rate 316
  morphological 36–7
Vas deferens 44
Ventral nerve cord 38–9
Vertical distribution 458–9, *459*, *460*, 486–7, 504–5
Vertical migration *439*, 439–52, *441*, 454–5
  cues for 446–52
    food availability 450–1

light 446–9, *448*
  predators 449–50
  tides 451–2
diel feeding periodicities and 170–1, *442*
effects of abrupt bottom topography 444
halocline responses 443
multispecies studies 442–3
ontogenetic migrations 445, **446**
oxycline responses 444
thermocline responses 443–4

Vitamin content 250–1
Vitellogenesis 43
Volume 235–6

Water content 226–7, *227*
Wax esters 246–8
Weight *see* Body weight

Zoogeography *see* Geographical distribution

Cumulative Index of Titles

Note: **Titles of papers** have been converted into subjects and a specific article may therefore appear more than once

Abyssal and hadal zones, zoogeography, **32**, 325
Abyssal macrobenthos, trophic structure, **32**, 427
Acetabularia, marine alga, recent advances in research, **14**, 123
Algal–invertebrate interactions, **3**, 1
Antarctic benthos, **10**, 1
Antarctic fishes, comparative physiology, **24**, 321
Ascidians
  biology, **9**, 1
  physiology, **12**, 2
Atlantic, Northeast, meiobenthos, **30**, 1

Baltic Sea, autrophic and heterotrophic picoplankton, **29**, 73
Barnacles, growth, **22**, 199
Bathyal zone, biogeography, **32**, 389
Benthic marine infaunal studies, development and application of analytical methods, **26**, 169
Benthos
  abyssal macrobenthos, trophic structure, **32**, 427
  Antarctic, **10**, 1
  Northeast Atlantic meiobenthos, **30**, 1
  sampling methods, **2**, 171
Blood groups, marine animal, **2**, 85
Blue whiting, North Atlantic, population biology, **19**, 257
Brachiopods, living, biology, **28**, 175
Bryozoans, marine, physiology and ecology, **14**, 285
Bullia digitalis, **25**, 179

Calanoid copepods, biology of, **33**
Cephalopods
  flotation mechanisms in modern and fossil, **11**, 197
  recent studies on spawning, embryonic development, and hatching, **25**, 85
Chaetognaths, biology, **6**, 271
Cladocerans, marine, reproductive biology, **31**, 80
Climatic changes, biological response in the sea, **14**, 1
Clupeid fish
  behaviour and physiology, **1**, 262
  biology, **20**, 1
  parasites, **24**, 263

Copepods
    association with marine invertebrates, **16**, 1
    calanoid, biology of, **33**
    respiration and feeding, **11**, 57
Coral reefs
    adaptations to physical environmental stress, **31**, 222
    assessing effects of stress, **22**, 1
    biology, **1**, 209
    communities, modification relative to past and present prospective Central American seaways, **19**, 91
    ecology and taxonomy of *Halimeda*: primary producer of coral reefs, **17**, 1
Crustaceans, spermatophores and sperm transfer, **29**, 129
Ctenophores, nutritional ecology, **15**, 249

Diel vertical migrations of marine fishes: an obligate or facultative process?, **26**, 115
Donax serra, **25**, 179

Echinoids, photosensitivity, **13**, 1
Eels, North Atlantic freshwater, breeding, **1**, 137
Effluents, effects on marine and estuarine organisms, **3**, 63
Environmental simulation experiments upon marine and estuarine animals, **19**, 133
Euphausiids, biology, **7**, 1, **18**, 373

Fish
    alimentary canal and digestion in teleosts, **13**, 109
    Antarctic fish, comparative physiology, **24**, 321
    artificial propagation of marine fish, **2**, 1
    clupeid behaviour and physiology, **1**, 262
    clupeid biology, **10**, 1
    clupeid parasites, **24**, 263
    diel vertical migrations, obligate or facultative process?, **26**, 115
    diseases, **4**, 1
    egg quality, **26**, 71
    gustatory system, **13**, 53
    migration, physiological mechanisms, **13**, 241
    North Atlantic freshwater eels, **1**, 137
    nutrition, **10**, 383
    parasites in deep-sea environment, **11**, 121
    photoreception and vision, **1**, 171
    predation on eggs and larvae of marine fishes and the recruitment problem, **25**, 1
    production and upwelling, **9**, 255
    year class strength, and plankton production, update of matchmismatch hypothesis, **26**, 249
Fish farming, estuarine, **8**, 119
Fish larvae
    appraisal of condition measures for marine fish larvae, **31**, 217
    field investigations of the early life stages of marine fish, **28**, 1
    turbulence and feeding ecology, role of microscale, **31**, 170

Fish migration, physiological mechanisms in the migration of marine and amphihaline fish, **13**, 248
Fisheries, management of resources, **6**, 1
Fisheries and seabird communities, competition, **20**, 225
Frontal systems, aspects of biology, **23**, 163

Gastropods
    comparison between *Donax serra* and *Bullia digitalis*, bivalve molluscs, **25**, 179
    intertidal, ecology, **16**, 111
    marine, burrowing habit, **28**, 389
Gonatid squids, subarctic North Pacific, ecology, biogeography, niche diversity and role in ecosystem, **32**, 243
Gonionemus, erratic distribution: the occurrence in relation to oyster distribution, **14**, 251

Habitat selection by aquatic invertebrates, **10**, 271
Halibut *Hippoglossus hippoglossus*, biology, **26**, 1
Halimeda, **17**, 1
Herring *Clupea harengus* L.
    and other clupeids
        behaviour and physiology, **1**, 262
        biology, **20**, 1
    relationships with its parasites, **24**, 263
Human affairs, marine biology, **15**, 233
Hybridization in the sea, **31**, 2
Hydrothermal vent communities, deep sea, ecology, **23**, 301
Hydrothermal vent fauna, mid-Atlantic ridge, ecology and biogeography, **32**, 93

Indo-West Pacific region, mangrove swamps and forests, fauna and flora, **6**, 74
Isopoda, oniscid, biology of the genus *Tylos*, **30**, 89

Japan, scallop industry, **20**, 309
Japanese oyster culture industry, recent developments, **21**, 1

Learning by marine invertebrates, **3**, 1

Mangrove swamps and forests of Indo-West Pacific region, general account of fauna and flora, **6**, 74
Marine animals
    blood groups, **2**, 85
    neoplasia, **12**, 151
Marine toxins
    venomous and poisonous animals, **3**, 256
    venomous and poisonous plants and animals, **21**, 59
Meiobenthos of the Deep Northeast Atlantic, **30**, 1
Mesoplankton
    distribution patterns, **32**, 9
    and macroplankton, some problems of vertical distribution in ocean, **32**, 1
Metabolic energy balance in marine invertebrates, influence of temperature on maintenance, **17**, 329
Microbiology, marine, present status of some aspects, **2**, 133

Molluscs
   hosts for symbioses, **5**, 1
   wood-boring teredinids, biology, **9**, 336
Molluscs, bivalve
   effects of environmental stress, **22**, 101
   and gastropods: comparison between *Donax serra* and *Bullia digitalis*, **25**, 179
   rearing, **1**, 1
   scatological studies, **8**, 307
Mysids, biology, **18**, 1

^{15}N, natural variations in the marine environment, **24**, 389
Nazca submarine ridge, composition and distribution of fauna, **32**, 145
Nitrogen cycle, and phosphorus cycle, plankton, **9**, 102

Oil pollution of the seas, problems, **8**, 215
Oniscid isopods, biology of the genus *Tylos*, **30**, 89
Oysters, living, speciation, **13**, 357

Particulate and organic matter in sea water, **8**, 1
Pelagic invertebrates, social aggregation, **30**, 155
Penaeida, biology, **27**, 1
Petroleum hydrocarbons and related compounds, **15**, 289
Phoronida, biology, **19**, 1
Phosphorus cycle, and nitrogen cycle, plankton, **9**, 102
Pigments of marine invertebrates, **16**, 309
Plankton
   distribution, vertical, in the ocean, **32**, 4
   laboratory culture of marine holozooplankton and its contribution to studies of planktonic food webs, **16**, 211
   and nitrogen and phosphorus cycles of sea, **9**, 102
   oceanic phytoplankton, outline of geographical distribution, **32**, 527
   parasitology of marine zooplankton, **25**, 112
   phytoplankton, circadian periodicities in natural populations, **12**, 326
   phytoplankton blooms, harmful or exceptional, **31**, 302
   picoplankton, Baltic Sea, **29**, 73
   production, and year class strength in fish populations: update of the matchmismatch hypothesis, **26**, 249
Pollution studies with marine plankton
   Part 1: Petroleum hydrocarbons and related compounds, **15**, 289
   Part 2: Heavy metals, **15**, 381
Pseudocalanus, biology, **15**, 1
Pycnogonida, biology, **24**, 1

Sala y Gómez submarine ridge, composition and distribution of fauna, **32**, 145
Salmon, acclimatization experiments, Southern Hemisphere, **17**, 398
Scallop industry in Japan, **20**, 309
Sea anemones, nutrition, **22**, 65
Seabird communities, and fisheries, competition, **20**, 225
Seamounts, biology, **30**, 305
Seaweeds
   of economic importance, aspects of biology, **3**, 105

population and community ecology, **23**, 1
Shrimps, pelagic, biology, **12**, 233
Siphonophore biology, **24**, 97
Sole *(Solea solea)*, Bristol Channel, **29**, 215
Spermatophores and sperm transfer in marine crustaceans, **29**, 129
Squids
    gonatid squids, subarctic North Pacific, ecology, biogeography, niche diversity and role in ecosystem, **32**, 243
    oceanic, review of systematics and ecology, **4**, 93

Taurine in marine invertebrates, **9**, 205
Teredinid molluscs, wood-boring, biology, **9**, 336
Tropical marine environment, aspects of stress, **10**, 217
Turbulence
    feeding ecology of larval fish, role of microscale, **31**, 170
    phytoplankton cell size, and structure of pelagic food webs, **29**, 1
Tylos biology, **30**, 89

Cumulative Index of Authors

Ahmed, J., **13**, 357
Akberali, H. B., **22**, 102
Allen, J. A., **9**, 205
Ansell, A. D., **28**, 175
Arakawa, K. Y., **8**, 307
Arnaud, F., **24**, 1
Bailey, K. M., **25**, 1
Bailey, R. S., **19**, 257
Balakrishnan Nair, M., **9**, 336
Bamber, R. N., **24**, 1
Bett, B. J., **30**, 1
Blaxter, J. H. S., **1**, 262, **20**, 1
Boletzky, S. V., **25**, 85
Boney, A. D., **3**, 105
Bonotto, S., **14**, 123
Bourget, E., **22**, 200
Branch, G. M., **17**, 329
Brinkhurst, R. O., **26**, 169
Brown, A. C., **25**, 179, **28**, 389, **30**, 89
Brown, B. E., **22**, 1, **31**, 222
Bruun, A. F., **1**, 137
Burd, B. J., **26**, 169
Campbell, J. I., **10**, 271
Carroz, J. E., **6**, 1
Chapman, A. R. O., **23**, 1
Cheng, T. C., **5**, 1
Clarke, M. R., **4**, 93
Collins, M. J., **28**, 175
Corkett, C. J., **15**, 1
Corner, E. D. S., **9**, 102, **15**, 289
Cowey, C. B., **10**, 383
Crisp, D. J., **22**, 200
Curry, G. B., **28**, 175
Cushing, D. H., **9**, 255, **14**, 1, **26**, 249
Cushing, J. E., **2**, 85
Dall, W., **27**, 1
Davenport, J., **19**, 133
Davis, A. G., **9**, 102, **15**, 381
Davies, H. C., **1**, 1

Dell, R. K., **10**, 1
Denton, E. J., **11**, 197
Dickson, R. R., **14**, 1
Dinet, A., **30**, 1
Dower, J. F., **31**, 170
Edwards, C., **14**, 251
Egloff, D. A., **31**, 80
Emig, C. C., **19**, 1
Evans, H. E., **13**, 53
Ferrero, T., **30**, 1
Ferron, A., **30**, 217
Fisher, L. R., **7**, 1
Fofonoff, P. W., **31**, 80
Fontaine, M., **13**, 241
Furness, R. W., **20**, 225
Galkin, S. V., **32**, 93
Gardner, J. P. A., **31**, 2
Garrett, M. R., **9**, 205
Gebruk, A. V., **32**, 93
Ghirardelli, E., **6**, 271
Gilpin-Brown, J. B., **11**, 197
Glynn, P. W., **19**, 91
Gooday, A. J., **30**, 1
Goodbody, I., **12**, 2
Gotto, R. V., **16**, 1
Grassle, J. F., **23**, 301
Gulland, J. A., **6**, 1
Harris, R. P., **16**, 211
Haug, T., **26**, 1
Heath, M. R., **28**, 1
Hickling, C. F., **8**, 119
Hill, B. J., **27**, 1
Hillis-Colinvaux, L., **17**, 1
Holliday, F. G. T., **1**, 262
Holme, N. A., **2**, 171
Holmefjord, I., **26**, 71
Horwood, J., **29**, 215
Houde, E. D., **25**, 1
Howard, L. S., **22**, 1
Hunter, J. R., **20**, 1

James, M. A., **28**, 175
Kapoor, B. G., **13**, 53, **13**, 109
Kennedy, G. Y., **16**, 309
Kiørboe, T., **29**, 1
Kjørsvik, E., **26**, 71
Kuosa, H., **29**, 73
Kuparinen, J., **29**, 73
Lambshead, P. J. D., **30**, 1
Le Fèvre, J., **23**, 163
Leggett, W. C., **30**, 217, **31**, 170
Loosanoff, V. L., **1**, 1
Lurquin, P., **14**, 123
Macdonald, J. A., **24**, 321
Mackenzie, K., **24**, 263
Mackie, G. O., **24**, 97
McLaren, I. A., **15**, 1
Macnae, W., **6**, 74
Mangor-Jensen, A., **26**, 71
Marshall, S. M., **11**, 57
Mauchline, J., **7**, 1, **18**, 1, **33**, 1–660
Mawdesley-Thomas, L. E., **12**, 151
Mazza, A., **14**, 123
Meadows, P. S., **10**, 271
Millar, R. H., **9**, 1
Miller, T. J., **31**, 170
Millot, N., **13**, 1
Mironov, A. N., **32**, 144
Montgomery, J. C., **24**, 321
Moore, H. B., **10**, 217
Moskalev, L. I., **32**, 93
Naylor, E., **3**, 63
Neilson, J. D., **26**, 115
Nelson-Smith, A., **8**, 215
Nemec, A., **26**, 169
Nesis, K. N., **32**, 144, **32**, 243
Newell, R. C., **17**, 329
Nicol, J. A. C., **1**, 171
Noble, E. R., **11**, 121
Odendaal, F. J., **30**, 89
Omori, M., **12**, 233
Onbé, T., **31**, 80
Owens, N. J. P., **24**, 389
Paffenhöfer, G. A., **16**, 211
Parin, N. V., **32**, 144
Peck, L. S., **28**, 175
Perry, R. I., **26**, 115
Pevzner, R. A., **13**, 53

Pfannkuche, O., **30**, 1
Pugh, P. R., **24**, 97
Purcell, J. E., **24**, 97
Reeve, M. R., **15**, 249
Rhodes, M. C., **28**, 175
Richardson, K., **31**, 302
Riley, G. A., **8**, 1
Ritz, D. A., **30**, 155
Rogers, A. D., **30**, 305
Rothlisberg, P. C., **27**, 1
Russell, F. E., **3**, 256, **21**, 60
Russell, F. S., **15**, 233
Ryland, J. S., **14**, 285
Saraswathy, M., **9**, 336
Sargent, J. R., **10**, 383
Scholes, R. B., **2**, 133
Semina, H. J., **32**, 527
Shelbourne, J. E., **2**, 1
Shewan, J. M., **2**, 133
Sindermann, C. J., **4**, 1
Smit, H., **13**, 109
Sokolova, M. N., **32**, 427
Soltwedel, T., **30**, 1
Sournia, A., **12**, 326
Southward, A. J., **32**, 93
Staples, D. J., **27**, 1
Stenton-Dozey, J. M. E., **25**, 179
Stewart, L., **17**, 397
Subramoniam, T., **29**, 129
Taylor, D. L., **11**, 1
Théodoridès, J., **25**, 117
Trueman, E. R., **22**, 102, **25**, 179, **28**, 389
Underwood, A. J., **16**, 111
Van-Praët, M., **22**, 66
Vanreusel, A., **30**, 1
Ventilla, R. F., **20**, 309, **21**, 1
Vereshchaka, A. L., **32**, 93
Verighina, I. A., **13**, 109
Vinogradov, M. E., **32**, 1
Vinogradova, N. G., **32**, 325
Vincx, M., **30**, 1
Walters, M. A., **15**, 249
Wells, M. J., **3**, 1
Wells, R. M. G., **24**, 321
Yonge, C. M., **1**, 209
Zezina, O. N., **32**, 389